SAP PRESS e-books

Print or e-book, Kindle or iPad, workplace or airplane: Choose where and how to read your SAP PRESS books! You can now get all our titles as e-books, too:

- By download and online access
- For all popular devices
- And, of course, DRM-free

Convinced? Then go to www.sap-press.com and get your e-book today.

Payment and Bank Communication Management with SAP®

SAP PRESS

SAP PRESS is a joint initiative of SAP and Rheinwerk Publishing. The know-how offered by SAP specialists combined with the expertise of Rheinwerk Publishing offers the reader expert books in the field. SAP PRESS features first-hand information and expert advice, and provides useful skills for professional decision-making.

SAP PRESS offers a variety of books on technical and business-related topics for the SAP user. For further information, please visit our website: *www.sap-press.com*.

Stoil Jotev
Configuring SAP S/4HANA Finance (3rd Edition)
2025, 744 pages, hardcover and e-book
www.sap-press.com/5920

Carlson, Carlson, Lasecki
Treasury and Risk Management with SAP S/4HANA: The Comprehensive Guide
2025, 820 pages, hardcover and e-book
www.sap-press.com/5907

Fuhr, Heyne, Teichelmann, Tuncer, Walter
Tax with SAP S/4HANA: Configuration and Determination
2022, 506 pages, hardcover and e-book
www.sap-press.com/5495

Dirk Neumann, Lawrence Liang
Cash Management with SAP S/4HANA (2nd Edition)
2021, 561 pages, hardcover and e-book
www.sap-press.com/5169

Anand Seetharaju, Mayank Sharma
General Ledger Accounting with SAP S/4HANA
2023, 886 pages, hardcover and e-book
www.sap-press.com/5630

Adrian Matys, Jean-Michele Szczecina

Payment and Bank Communication Management with SAP®

Editor Meagan White
Acquisitions Editor Emily Nicholls
Copyeditors Melinda Rankin
Cover Design Graham Geary
Photo Credit iStockphoto: 1335295270/© piranka
Layout Design Vera Brauner
Production Hannah Lane
Typesetting SatzPro, Germany
Printed and bound in the United States of America, on paper from sustainable sources

ISBN 978-1-4932-2682-5
1st edition 2025

© 2025 by:
Rheinwerk Publishing, Inc.
2 Heritage Drive, Suite 305
Quincy, MA 02171
USA
info@rheinwerk-publishing.com
+1.781.228.5070

Represented in the E.U. by:
Rheinwerk Verlag GmbH
Rheinwerkallee 4
53227 Bonn
Germany
service@rheinwerk-verlag.de
+49 (0) 228 42150-0

Library of Congress Cataloging-in-Publication Control Number: 2025024814

All rights reserved. Neither this publication nor any part of it may be copied or reproduced in any form or by any means or translated into another language, without the prior consent of Rheinwerk Publishing.

Rheinwerk Publishing makes no warranties or representations with respect to the content hereof and specifically disclaims any implied warranties of merchantability or fitness for any particular purpose. Rheinwerk Publishing assumes no responsibility for any errors that may appear in this publication.

"Rheinwerk Publishing" and the Rheinwerk Publishing logo are registered trademarks of Rheinwerk Verlag GmbH, Bonn, Germany. SAP PRESS is an imprint of Rheinwerk Verlag GmbH and Rheinwerk Publishing, Inc.

All screenshots and graphics reproduced in this book are subject to copyright © SAP SE, Dietmar-Hopp-Allee 16, 69190 Walldorf, Germany.

SAP, ABAP, ASAP, Concur Hipmunk, Duet, Duet Enterprise, ExpenseIt, SAP ActiveAttention, SAP Adaptive Server Enterprise, SAP Advantage Database Server, SAP ArchiveLink, SAP Ariba, SAP Business ByDesign, SAP Business Explorer (SAP BEx), SAP BusinessObjects, SAP BusinessObjects Explorer, SAP BusinessObjects Web Intelligence, SAP Business One, SAP Business Workflow, SAP BW/4HANA, SAP C/4HANA, SAP Concur, SAP Crystal Reports, SAP EarlyWatch, SAP Fieldglass, SAP Fiori, SAP Global Trade Services (SAP GTS), SAP GoingLive, SAP HANA, SAP Jam, SAP Leonardo, SAP Lumira, SAP MaxDB, SAP NetWeaver, SAP PartnerEdge, SAPPHIRE NOW, SAP PowerBuilder, SAP PowerDesigner, SAP R/2, SAP R/3, SAP Replication Server, SAP Roambi, SAP S/4HANA, SAP S/4HANA Cloud, SAP SQL Anywhere, SAP Strategic Enterprise Management (SAP SEM), SAP SuccessFactors, SAP Vora, TripIt, and Qualtrics are registered or unregistered trademarks of SAP SE, Walldorf, Germany.

All other products mentioned in this book are registered or unregistered trademarks of their respective companies.

No part of this book may be used or reproduced in any manner for the purpose of training artificial intelligence technologies or systems. In accordance with Article 4(3) of the Digital Single Market Directive 2019/790, Rheinwerk Publishing, Inc. expressly reserves this work from text and data mining.

Contents at a Glance

1	Payments Within a Corporation	21
2	Bank Connectivity	93
3	Bank Account Management	151
4	Advanced Payment Management and In-House Banking	215
5	SAP Multi-Bank Connectivity	349
6	SAP S/4HANA Finance for Cash Management	419
7	SAP Bank Communication Management	481
8	Mapping Format Data	511
9	Bank Statements	539
10	The Payment Factory	583
11	Outlook on Payments and Bank Communication with SAP	607

Contents

Preface 13

1   Payments Within a Corporation   21

1.1   Payments as a Business Imperative 22
1.2   Payments with SAP 26
    1.2.1   Payments Management Within Financial Accounting 27
    1.2.2   Bank Account Management 29
    1.2.3   SAP Multi-Bank Connectivity and Bank Connectivity 34
    1.2.4   SAP S/4HANA Finance for Advanced Payments Management 35
    1.2.5   In-House Banking Versus SAP In-House Cash 42
    1.2.6   SAP Bank Communication Management 50
    1.2.7   SAP S/4HANA Finance for Cash Management 56
1.3   Payments Processing 59
    1.3.1   Payments in the Name Of 59
    1.3.2   Direct Debit 61
    1.3.3   Intercompany Payments 63
    1.3.4   Payments on Behalf Of 71
    1.3.5   Collections/Receivables on Behalf Of 76
    1.3.6   Cash Pooling 80
    1.3.7   Treasury Payments 83
    1.3.8   Preprocessing and Postprocessing 88
1.4   Summary 91

2   Bank Connectivity   93

2.1   Connectivity Standards and Security in Payments 93
    2.1.1   SWIFT 95
    2.1.2   Host to Host 108
    2.1.3   API Usage 108
    2.1.4   EBICS 109
    2.1.5   Bank Collection and Forwarding Service 120
2.2   Financial Standards 121
    2.2.1   Payment Instruments 121

		2.2.2	Payment Schemes	130
		2.2.3	Payment Systems	137
		2.2.4	Payment Formats	142
		2.2.5	Bank Fee Analysis	148
	2.3	Summary		149

3 Bank Account Management 151

	3.1	Functions and Processes		151
		3.1.1	Bank Keys	152
		3.1.2	Creating Banks	153
		3.1.3	Uploading, Migrating, and Replicating Bank Accounts	181
		3.1.4	Bank Account Approvals	183
		3.1.5	Manage Bank Account Hierarchies App	187
		3.1.6	Manage Bank Account Reviews App	188
		3.1.7	Closing Bank Accounts	189
		3.1.8	Manage Bank Chains App	190
		3.1.9	Foreign Bank Accounts App	191
		3.1.10	Bank Relationship Overview App	192
		3.1.11	Power of Attorney for Banking Transactions	193
		3.1.12	Bank Fees	197
	3.2	Configuration		201
		3.2.1	Define Basic Settings	201
		3.2.2	Enable Payment Approval	204
		3.2.3	Number Ranges for Technical Accounts ID	205
		3.2.4	Define Settings for Bank Account Master Data	206
		3.2.5	Define Settings for Bank Account Contract Types	209
		3.2.6	Define Field Status Groups	212
	3.3	Summary		214

4 Advanced Payment Management and In-House Banking 215

	4.1	Advanced Payment Management Functions and Processes		215
		4.1.1	Key Components	216
		4.1.2	Functions and Apps	230
		4.1.3	Master Data	245

4.2	In-House Banking Functions and Processes	260
	4.2.1 Prerequisites	261
	4.2.2 Account Management Lifecycle	262
	4.2.3 General Functions	264
	4.2.4 Payments: In-House Operations	282
4.3	Configuration	295
	4.3.1 Advanced Payment Management	295
	4.3.2 In-House Banking	331
4.4	Integration of Intelligent Services	339
	4.4.1 Function and Processes	340
	4.4.2 Intelligent Service Integration	341
	4.4.3 Configuration	343
4.5	Summary	347

5 SAP Multi-Bank Connectivity 349

5.1	Introduction to SAP Multi-Bank Connectivity	350
	5.1.1 System Landscape and Components	351
	5.1.2 Cloud Foundry Migration	354
5.2	Connectivity Options	356
	5.2.1 EBICS	357
	5.2.2 Host-to-Host	360
	5.2.3 SWIFT Connectivity	365
	5.2.4 Member Banks	366
5.3	Functions and Processes	366
	5.3.1 Connector Monitor Transaction	367
	5.3.2 Manage Bank Messages App	373
	5.3.3 Pull Messages	377
	5.3.4 Push Messages	378
	5.3.5 Pickup Files	379
	5.3.6 Handling of Sensitive Data	382
	5.3.7 SWIFTRef Integration	383
5.4	Onboarding	385
5.5	Configuration	387
	5.5.1 SAP S/4HANA Finance Configuration and Prerequisites	387
	5.5.2 Basic Configuration	392
	5.5.3 Inbound and Outbound Processing	402

		5.5.4	Business Add-Ins	413
		5.5.5	Authorizations	415
	5.6	Summary		417

6 SAP S/4HANA Finance for Cash Management   419

	6.1	Functions and Processes		419
		6.1.1	Bank Account Balances	420
		6.1.2	Cash Position	424
		6.1.3	Cash Flow Analyzer	425
		6.1.4	Short-Term Positioning	438
		6.1.5	One Exposure from Operations	446
		6.1.6	Manage Memo Records	452
		6.1.7	Import Memo Records	454
		6.1.8	Bank-to-Bank Transfers	456
		6.1.9	Free-Form Payments	459
		6.1.10	Cash Concentration	462
	6.2	Configuration		467
		6.2.1	Planning Levels	467
		6.2.2	Planning Groups	470
		6.2.3	Liquidity Items	471
		6.2.4	Memo Records	472
		6.2.5	Cash Pools	476
		6.2.6	Define Source Application Accounting	477
		6.2.7	Activate Source Application	478
	6.3	Summary		479

7 SAP Bank Communication Management   481

	7.1	Functions and Processes		482
		7.1.1	Payment Approvals	484
		7.1.2	Payment Run	487
		7.1.3	Batching and Processing Payments	492
	7.2	Configuration		498
		7.2.1	Basic Configuration	498

		7.2.2	Batching Rules Configuration	500
		7.2.3	Workflow Activation and Configuration	502
		7.2.4	Alerts	506
		7.2.5	Useful BAdIs	509
7.3	Summary			509

8   Mapping Format Data   511

8.1	Transition from SWIFT MT to ISO 20022 MX		512
8.2	Functions and Configuration		513
	8.2.1	Map Format Data Apps Overview	514
	8.2.2	Map Format Data for Incoming Files from Banks	520
	8.2.3	Map Format Data for Advanced Payment Management	536
8.3	Summary		538

9   Bank Statements   539

9.1	Bank Statement Processing		540
	9.1.1	Overview	540
	9.1.2	Bank Statement Reconciliation	543
	9.1.3	Bank Statement Monitor	544
	9.1.4	Manage Bank Statements	548
	9.1.5	Upload Bank Statement	550
	9.1.6	Japanese Bank Statements	557
	9.1.7	Bank Statement Forwarding Using Advanced Payment Management	559
	9.1.8	Intraday Bank Statements	559
	9.1.9	Processing Rules	562
9.2	Bank Statement Configuration		565
	9.2.1	New Bank Statement Import Program: FEB_FILE_HANDLING	565
	9.2.2	Lockboxes	569
	9.2.3	Bank Statement Posting	573
	9.2.4	Search String Configuration	579
	9.2.5	BAdIs	581
9.3	Summary		582

Contents

10 The Payment Factory — 583

- 10.1 Designing a Payment Factory — 583
- 10.2 Processes and Functions — 588
    - 10.2.1 In-House Banking Processes — 590
    - 10.2.2 Payment Factory Functions — 595
    - 10.2.3 Payment Orchestration — 598
- 10.3 Quantitative and Qualitative Factors — 600
- 10.4 The Reconciliation Factory — 602
- 10.5 Summary — 605

11 Outlook on Payments and Bank Communication with SAP — 607

- 11.1 SAP Solutions — 607
    - 11.1.1 SAP Digital Payments Add-On — 608
    - 11.1.2 SAP Digital Currency Hub — 613
- 11.2 SAP Outlook by Kolja Ewering — 617
- 11.3 Summary — 620

Appendices — 621

- A ISO 20022 Transformation — 623
- B Ongoing Regulations — 639
- C The Authors — 645

Index — 647

Preface

With this book, we aim to provide an introductory guide that offers an understanding of the complexities of payment transactions. Especially in our increasingly fast-paced and complex world, we hope this work will provide guidance and ideas for future payment projects.

SAP is investing significantly in expanding its treasury and finance capabilities, with a strong focus on enhancing and optimizing payment processes. As digitalization becomes increasingly important and organizations face growing pressure to reduce manual workload, cut costs, and streamline operations, the need for modern, centralized, and automated financial solutions is more critical than ever. To meet these demands and to equip customers with advanced tools, SAP is continuously releasing new functionalities in the payments area—introducing and evolving solutions such as SAP S/4HANA Finance for advanced payment management, in-house banking for SAP S/4HANA Finance for advanced payment management, Bank Account Management (BAM) in SAP S/4HANA, SAP Multi-Bank Connectivity; updating SAP S/4HANA Finance for cash management to answer business needs; and providing many other features that support payment processes and bank statement handling. Considering these developments and the rapid pace of innovation in this space, we decided to create this book to explore and explain the latest advancements in SAP's payment landscape, providing a comprehensive guide to the tools shaping the future of financial operations and covering changes expected to occur in the future.

The main aim of this book is to provide a comprehensive, functional understanding of how to build and manage payment processes in SAP, with a strong focus on designing and implementing a payment factory. Our goal is to equip you with the knowledge of what needs to be done, why it is important, and what key factors must be considered when developing your organization's payment landscape within SAP. We begin by introducing the foundational modules used in payment processing, starting with BAM, which establishes the base for managing balances required for payments. We then move to cash management, where you gain visibility into payment impacts and overall liquidity. SAP Multi-Bank Connectivity is a central topic in this book as it's the main tool for linking your SAP system with external banks and service providers. We also delve into advanced payment management, which allows you to consolidate payments from multiple systems and sources into a unified flow, and in-house banking, which enables centralized intercompany processing. To provide a full view of the payment ecosystem, we cover related modules such as payment communication management for payment centralization and matching, as well as automation tools for format data generation and other supporting technologies.

Although we do address key aspects of configuration and underlying technology, this book primarily focuses on delivering a functional perspective, helping you understand the rationale behind each solution and how they interconnect to build a streamlined, controlled, and scalable payment infrastructure. We concentrated on the latest technologies and functionalities that are often underused or less familiar to organizations, especially those related to payments that go beyond the traditional scope of standard finance capabilities. These foundational areas have been available on the market for a long time and are already well covered in existing materials, including other SAP PRESS books.

Our main aim with this book was to give you the opportunity to fully understand how these solutions are used in practice and how end-to-end payment processes work from a functional point of view. While we are aware that certain topics could have been described in more depth, our intention was to include as much relevant information as possible in this broad and evolving area of functionality.

Target Audience

This book is intended for a wide range of readers who are involved in, or impacted by, payment processes within SAP. The target audience includes SAP consultants responsible for designing and implementing payment-related solutions; treasury department professionals who manage cash, liquidity, and bank relationships; and cash management teams working to optimize financial operations. It is equally valuable for companies exploring SAP's capabilities as for those evaluating the implementation of payments or more advanced concepts like centralized payment factories. This book is also a useful guide for organizations and individuals assessing their current bank connectivity landscape, trying to understand the requirements and possibilities offered by SAP.

In addition, it serves technology enthusiasts and SAP professionals interested in gaining insight into the latest treasury-related systems, such as BAM, advanced payment management, in-house banking, SAP Bank Communication Management, SAP Multi-Bank Connectivity, and other tools essential for modern payment processing. Finally, it is a valuable resource for anyone looking to gain a deeper understanding of the future of payments in SAP and how these evolving solutions can help streamline, automate, and centralize financial operations.

In this book, we cover a broad range of topics that intersect not only with payment processing but also with other essential areas of SAP, such as finance, associated accounting, bank account management, and cash management. While our focus remains on the functional aspects of building a payment factory, it's important to note that many of these areas are deeply integrated and foundational to running efficient and compliant financial operations. For readers seeking in-depth knowledge on specific systems—such as cash management, treasury and risk management, or liquidity planning—we recommend exploring dedicated publications on these topics that are available from Rheinwerk Publishing. These resources provide a more detailed exploration of the individual components and capabilities of SAP's financial suite and are valuable references for anyone looking to unlock the full potential of these solutions.

How To Read This Book

We recommend starting this book from the very beginning, with particular attention to Chapter 1. This chapter lays the foundations by explaining what payments are in the SAP context, which modules are involved, key process flows, types of payments, prerequisites, and the decision points related to payment execution. It provides a comprehensive functional overview to help you understand how the different parts of an SAP system support your payment strategy.

Chapter 2 focuses on connectivity options—an essential component of any payment process. It covers various ways to connect SAP with banks and service providers, such as host-to-host communication, SWIFT protocols, and SAP Multi-Bank Connectivity. This will help you evaluate which integration method best fits your organization's needs.

Chapter 3 introduces the core system: Bank Account Management. Without a proper bank account setup, executing payments and managing liquidity is not possible. This chapter explains how to establish and maintain your bank structure in SAP, which serves as the technical and functional starting point for any payment process.

After these chapters, you can read the remaining ones based on your needs. Each covers different supporting systems, which may or may not be relevant depending on your setup. However, we strongly recommend reading Chapter 11 regardless, which looks into the future of payments in SAP. Written in cooperation with SAP product owners, it highlights upcoming innovations and planned enhancements to help you stay prepared for what's next.

Throughout the book, we've also provided several elements that will help you access useful information:

> **Tips and Tricks**
> Boxes with this symbol provide you with recommendations as to how you can simplify your work.

> **Notes**
> Boxes marked with this symbol contain additional information or important contents that you should keep in mind.

Preface

> [!] **Warnings**
> Boxes with this symbol contain details worth considering. Moreover, it warns you of common errors or problems that might occur.

How This Book Is Organized

This book is structured to serve the various individuals who work on payment processes in the SAP environment. Each chapter illustrates a specific knowledge area and builds on the skills obtained in previous sections. The chapters are as follows:

- **Chapter 1**
  In this chapter, we cover the functional processes involved in executing payments. We also introduce the key SAP solutions that support payment operations, all of which will be explored in detail in subsequent chapters.

- **Chapter 2**
  This chapter outlines the various bank connectivity options available in SAP. We describe how each option functions, the business scenarios they support, and what preparations are needed to begin using them, including onboarding processes. The information provided in this chapter helps you understand both the functional and technical requirements for bank connectivity both within the context of SAP, but also for any other system or technology you may wish to connect to a bank.

- **Chapter 3**
  Here, we discuss BAM, its importance, and how it supports payment operations. We will explain how to set up bank accounts in SAP and highlight the core features and configuration options available within BAM.

- **Chapter 4**
  This chapter introduces advanced payment management, which enables centralized grouping and processing of payments. We also present the in-house banking functionality as a variant of advanced payment management, which is used to build and manage internal banking structures.

- **Chapter 5**
  We focus on SAP Multi-Bank Connectivity in this chapter, which enables direct integration between SAP and external banks. This chapter explains the technical and functional architecture of SAP Multi-Bank Connectivity, as well as its role in streamlining and securing bank communications.

- **Chapter 6**
  This chapter covers how you can effectively manage your cash position and short-term liquidity using the cash management functionality. You'll learn how to assess the real-time impact of outgoing and incoming payments on your liquidity and how to forecast cash balances to ensure financial stability and accurate planning.

- **Chapter 7**

    In this chapter, we dive into the details of SAP Bank Communication Management. We explain how payment batching works, how multilevel approval workflows can be configured, and explore additional SAP Bank Communication Management functionalities that support secure, controlled, and compliant payment execution processes.

- **Chapter 8**

    Here we introduce the Map Format Data app, which enables the transformation of outgoing and incoming payment and statement files into formats that are expected by your banks and can be successfully processed by SAP. This chapter explains how to set up mappings and transformations and thus ensure compatibility with different banking standards.

- **Chapter 9**

    This chapter focuses on processing bank statements in SAP. You will learn about the underlying mechanisms used to import, process, and post statements, as well as how these activities update your cash positions and feed into cash management for accurate and real-time reporting.

- **Chapter 10**

    In this chapter, we introduce the concept of a payment factory and explore both the qualitative and quantitative benefits of centralizing payment operations. We help you evaluate business cases for implementing more complex payment scenarios and show how consolidation into a centralized structure can streamline operations, reduce costs, and increase transparency.

- **Chapter 11**

    The final chapter provides an outlook on the future of payment processing in SAP. We review SAP's current roadmap and innovations in the payment space, helping you understand how your organization can align with upcoming technologies and stay ahead in digital treasury transformation.

Conclusion

Reading this book will equip you with comprehensive knowledge of how to manage payment processes using SAP solutions. It serves as a key reference for understanding the value and complexity of modern payment operations and provides detailed insights into how SAP modules can be leveraged to centralize, automate, and streamline payments. With a strong focus on both functional processes and system configuration, this book supports professionals in building efficient payment architectures—from basic bank integration to advanced concepts like payment factories and in-house banking. The concepts and tools covered here will empower you to make informed decisions, design effective payment processes, and fully harness SAP's capabilities in payments and treasury operations.

Acknowledgments

I would like to express my sincere gratitude to my coauthor, Jean-Michele Szczecina, for initiating the idea of writing this book—a comprehensive summary of the many projects we've delivered together over the past years. His continuous support, both professionally and personally, has been invaluable.

I am deeply thankful to the many customers and colleagues I've had the privilege of working with throughout my career. Although I cannot name everyone individually, your support, collaboration, and trust have contributed immensely to the knowledge and experience that shaped this book. Your patience and openness made it possible to grow within the SAP ecosystem and to turn practical experience into lasting expertise.

I would also like to extend my appreciation to KPMG as an organization for its commitment to innovation and for investing in cutting-edge technologies. I'm particularly grateful for access to KPMG's SAP systems and platforms, which provided me with the tools and environment to explore and test new concepts throughout the writing process.

Adrian Matys
July 2025

This book would never have come to life without the support and knowledge-sharing of the incredible people who are truly passionate about payments, treasury operations, and SAP.

First and foremost, I want to sincerely thank my coauthor, Adrian Matys. Your personal commitment—both professionally and privately—to the role of an SAP treasury consultant is not only remarkable but truly inspiring for me and for more people than you admit to yourself.

I also wish to express my gratitude to Dirk Matusall: Without your mentorship, both professionally and personally, I wouldn't be the person I am today.

Furthermore, I want to extend my thanks to Zanders as an organization. Its dedication to excellent SAP treasury consulting is embedded in its DNA. From day one, Zanders fully supported this project and provided our demo system, enabling its development. With over 25 years of partnership in SAP treasury, Zanders stands as one of the most experienced players in the field. In this regard, I would also like to acknowledge the entire Zanders SAP Center of Expertise, led by Eliane Eysackers, and thank Laura Koekkoek, Ivo Postma, Karsten Kohl, and Michal Sarnik for their valuable insights and collaboration on ongoing SAP payment projects.

A special thank you goes to Ralf Klein from SWIFT; Your sparring and discussions on payments were invaluable in shaping this book.

Finally, I want to express my deepest appreciation to my girlfriend Aleksandra Borowiecka. You kept the world running while Adrian and I were immersed in this project, often late at night and on weekends. Without your support and unwavering dedication, this book would not have been possible.

Jean-Michele Szczecina
July 2025

Last, But Not Least...

A heartfelt thank you also goes to Kolja Ewering for his guidance, for reviewing the content, and for generously sharing insights into SAP's future direction and innovations in the evolving landscape.

Special thanks to our incredibly patient editor, Meagan White, for her unwavering support and guidance throughout this journey. Emily Nicholls from Rheinwerk Publishing also played a crucial role during the early development of the book. We are grateful to both of them—as well as to the entire Rheinwerk Publishing team—for their dedication and for giving us the opportunity to share our knowledge with a wider audience.

Make the Most of This Journey

With this book, we aim to provide an introductory guide that offers readers an understanding of the complexities of payment transactions. In our increasingly fast-paced and complex world, we hope this book will provide you with guidance and ideas for your future payment projects.

Adrian Matys & **Jean-Michele Szczecina**

Chapter 1
Payments Within a Corporation

In any business, collections and sales are typically seen as the most important functions, driving revenue and growth. However, the payments side is equally, if not more, critical. Timely payments play a key role in maintaining strong relationships with vendors, ensuring the smooth receipt of goods, and managing cash flow effectively. These factors are essential for sustaining business operations and fostering trust with suppliers, which in turn supports long-term success and stability. Without a solid payments system, even the best sales and collections efforts can be undermined.

This chapter emphasizes the importance and complexity of payment processes and how corporate payments can be managed using SAP solutions. Section 1.1 describes what payments are and the imperative points and challenges that corporations might face. Section 1.2 explains how payments are executed in SAP, their evolution, and the SAP solutions used in payment processing. It also outlines the functions of the following solutions:

- Financial accounting in SAP S/4HANA Finance
- Bank relationship management using Bank Account Management (BAM) in SAP S/4HANA
- SAP Multi-Bank Connectivity
- SAP S/4HANA Finance for advanced payment management
- In-house banking for SAP S/4HANA Finance for advanced payment management (usually shortened to just *in-house banking*), the older SAP In-House Cash solution, and the differences between them
- SAP Bank Communication Management
- SAP S/4HANA Finance for cash management

Section 1.3 discusses payment processes from a functional perspective, highlighting their benefits and how they are executed in SAP. The subtopics in this section include the following:

- Payments in the name of (PINO)
- Direct debit
- Payments on behalf of (POBO)

1 Payments Within a Corporation

- Intercompany payments
- Collection on behalf of (COBO)
- Cash pooling
- Internal and external treasury payments
- Preprocesses (the master data required to run these payments) and postprocesses (other treasury activities impacted by payments)

1.1 Payments as a Business Imperative

The primary target of every company is to make a profit, which is the fundamental driver of business success. This goal is achieved by selling goods or services and generating revenue. However, the process of creating profit goes beyond simple sales. Companies typically acquire goods or services and add value to them—whether through manufacturing, improvement, or providing additional services. By enhancing the original product or service, companies can offer something more appealing to customers, thus justifying a higher price and ultimately increasing their profitability. The key to success lies in this value-added process, which differentiates a company from competitors and boosts its financial performance.

To successfully operate and achieve the previously mentioned profitability, a company must first acquire the necessary goods or services, which often involves paying suppliers or service providers. Such payments are essential for obtaining the resources needed to create or enhance the products or services offered to customers. Without timely and efficient payments, a company risks disrupting its supply chain, delaying production, or damaging relationships with vendors. Therefore, the payments process is crucial not only for maintaining operational continuity but also for the survival and growth of the business. Ensuring that payments are made promptly and accurately fosters trust and reliability with suppliers, which is foundational for long-term success.

In addition to businesses, the economy is also supported by various governmental organizations, which, unlike companies, do not aim to generate profit but instead focus on providing essential services to the public. To operate effectively and fulfill their missions, these organizations also require resources and must make payments for goods, services, and employee wages. Because they are typically funded through taxpayer contributions or government grants, it is vital for them to run efficiently, responsibly, and transparently. Proper payment management is crucial for maintaining financial integrity, ensuring that public funds are used appropriately and thus sustaining the trust of taxpayers and other stakeholders. Therefore, payments are as important for government entities as they are for businesses, ensuring the smooth delivery of services and the proper use of allocated resources.

Payment processing is a highly demanding and time-consuming task that requires significant manual effort (consuming a lot of organizational resources) to ensure accuracy and prevent costly errors. It involves verifying proper invoices, ensuring that vendors are paid on time and in the correct amounts, handling multiple currencies, and confirming that no double payments occur. This process is complex, as even small mistakes—such as incorrect amounts, missed payments, or payment duplications—can lead to financial discrepancies and significant losses, strained vendor relationships, and potential legal issues. Given the intricacies involved, the payments process can be both challenging and resource-intensive, making it crucial for organizations to allocate substantial effort to manage and monitor payments carefully to avoid expensive errors.

This underscores why significant effort and resources are devoted to developing advanced payment technologies, automation systems, and fraud-detection measures. By streamlining and automating payment processes, reducing errors, and preventing fraudulent activities, organizations can improve efficiency, maintain financial stability, safeguard public and private funds, and, most importantly, achieve economy of scale, preparing themselves for future growth and ensuring that payment processes are scalable.

As the economy becomes increasingly reliant on digital transactions, the need for robust, secure, centralized and automated payment systems is greater than ever, supporting the overall health of businesses, governments, and the economy.

As mentioned, managing payment processes is already complex because of their importance, and it becomes even more challenging when operating in the international market. This complexity increases further due to the following factors:

- **Currency management**

    When dealing with international payments, companies must manage multiple currencies, including exchange rates, foreign transaction fees, and the risk of currency fluctuations. It is important to ensure that payments are made from an account using the same currency as the payment; otherwise, the organization will incur additional costs due to foreign exchange (FX) rates.

    For certain FX payments involving exotic currencies, where transactions in that currency are infrequent, it is advisable to execute the payment from an operational account in a different currency. It is recommended to monitor such payments and associated costs to identify when there is a business that needs to open an account in the selected currency.

    In certain countries, local regulations may require that FX translations and postings be done using the exchange rates provided by the national central bank, rather than relying on rates from external sources such as Bloomberg or other market data providers. These regulations are typically put in place to ensure consistency, transparency, and compliance with local accounting standards and tax laws. By using the official exchange rates set by the national bank, organizations can align their financial practices with government guidelines, which may be particularly important for tax reporting and regulatory compliance. Failing to adhere to these requirements could result in penalties or complications during audits, making it essential for businesses operating in such regions to monitor and implement the correct rates as mandated by local authorities.

- **Multiple payment methods**
Businesses often need to manage a variety of payment methods, such as bank transfers (domestic, cross-border, urgent, normal salary, etc.), credit cards, checks, and digital wallets, each with different processing requirements, costs, and timelines.

It is recommended to monitor and use the appropriate payment methods for each transaction. For example, domestic and standard payment methods should be used for routine payments, while urgent payment methods should only be used when necessary to achieve timely payments. Otherwise, the organization may incur unnecessary costs, or the bank may even reject some payments due to the incorrect payment method being applied.

- **Timely payments**
Ensuring that vendors, employees, and contractors are paid on time is crucial for maintaining relationships and avoiding penalties or disruptions in service. Certain payments, such as high-volume payments or treasury payments, must be designated as urgent ones.

- **Invoice verification**
Ensuring that invoices are accurate, posted efficiently, match purchase orders, and legitimate requires meticulous attention to detail to prevent overpayments or errors (such as payments to a wrong vendor or an unverified bank account).

Payment disputes with vendors or customers over amounts, timing, or service delivery can delay the payment process and create friction in business relationships.

Each organization must implement strong controls to prevent duplicated payments, ideally at the invoice-posting level, to avoid unnecessary financial losses. Without these controls, there is a risk of accidentally paying the same invoice multiple times, leading to serious costs and a significant administrative burden in recovering the overpaid amounts. Duplicated payments not only strain financial resources but also consume valuable time and effort when identifying and rectifying the errors. In some cases, it may be difficult to retrieve the funds, especially if the recipient is unwilling or unable to return the excess payment. Therefore, ensuring that robust controls and reconciliation processes are in place is critical to minimizing the risk of duplicated payments and maintaining smooth financial operations.

- **Vendor management**
Managing multiple vendors with different payment terms (or the same vendor in multiple legal entities), contract stipulations, and invoicing methods increases administrative workload and the likelihood of errors.

It is recommended to implement a proper vendor verification and onboarding process from both a functional and system perspective—by following the four eyes principle when setting up vendor bank accounts, for example. In the *four eyes* approach, one person creates something, and another person must explain what the first person created. However, there can be six, eight, or even more eyes, depending on the activity. Essentially, the term represents the number of explanations or layers of understanding required to fully process or verify the original creation. This approach helps ensure that fraud is avoided and payments are not made to embargoed counterparties.

- **Fraud prevention**

  In today's digital age, fraud has become more sophisticated, frequent, and increasingly dangerous, posing a significant challenge for many organizations. Cybercriminals are constantly developing new methods to exploit vulnerabilities, making it crucial for businesses to implement stringent security measures and safeguards. To effectively mitigate these risks, organizations must establish clear and robust rules (like vendor and counterparty verifications), protocols, and monitoring systems designed to detect and prevent fraudulent activities. By taking a proactive approach, such as using advanced fraud-detection technologies, training staff, and regularly reviewing payment processes, companies can better protect themselves from the financial and reputational damage that fraud can cause.

- **Compliance and regulations**

  Payment processes must comply with a variety of legal and regulatory requirements, such as tax laws, anti-money-laundering regulations, and industry-specific standards, which can vary by country and region.

  These points require a lot of attention and checks due to various requirements. For example, certain countries may require payments to be sent with special-purpose codes that explain their nature. Failure to comply with these requirements may result in payment rejection or fines for missing other reporting regulations.

- **Payment decentralization**

  Organizations often operate through multiple legal entities, each of which may have its own payment system, potentially utilizing different online banking platforms or even separate banks. Payments may be executed by different individuals or teams within each entity, leading to varying procedures and practices. Even if multiple entities within the organization use the same bank, each one may have its own distinct method for processing payments. This decentralized approach can create complexities in managing payment processes, as it introduces a range of systems, teams, and practices that may not be standardized across the organization, making it more challenging to ensure consistency and efficiency.

  The main aim of each organization should be to streamline and automate these processes as much as possible.

- **Bank requirements**

  Each bank may use different payment formats, technologies, and connectivity requirements, making integration with payment systems more complex. Banks often have their own preferred formats for processing transactions, such as specific file types or message standards, and they may rely on various technologies for secure communication, including different protocols for data transmission. Even when the same payment format is used, there may be specific differences and variations between banks in how the format is implemented, such as unique field mappings, data requirements, or validation rules. In addition, banks may have unique connectivity options, such as direct connections, third-party intermediaries, or online banking platforms, each with distinct setup procedures and security protocols. This variability means that organizations must be adaptable and prepared to handle a range of technical and operational requirements when working with multiple banks, ensuring seamless and secure payment processing across different financial institutions.

- **Integration with accounting systems**
  Payments need to be seamlessly integrated with accounting systems for accurate financial reporting, reconciliation, and auditing purposes and for updating the general ledger (G/L) and subledgers.

- **Intermediary banks**
  Intermediary banks are typically used in international payment transactions, especially when the sending and receiving banks do not have a direct relationship or corresponding accounts with each other. They are commonly used in cross-border payments, in which the sender's bank and the recipient's bank are located in different countries and do not have direct payment channels or accounts set up for such transfers. Intermediary banks help route the transaction through a network of correspondent banks, facilitating the exchange of funds in different currencies and jurisdictions. They are also used when dealing with less common or smaller financial institutions that might not have the infrastructure or relationships to handle international payments directly. In these cases, intermediary banks ensure that payments can still be processed smoothly, albeit at a cost, and they typically charge fees for their services.

In this book, we will focus on both aspects of payments: technology, with an emphasis on how SAP solutions can assist you in managing your payment processes; and the business side, concentrating on how to run your payment processes in the most efficient way. We will elaborate further on how SAP solutions can help you address all the challenges mentioned in this chapter.

1.2 Payments with SAP

The SAP S/4HANA system provides organizations with functionalities that support optimized and efficient payment processes, helping to streamline operations and meet all payment-related requirements. It offers a range of features that cater to different aspects of payment management, from processing transactions to ensuring compliance with regulatory standards. This section will explore the various SAP solutions that assist with payments, providing an overview of their functionalities and a high-level overview of how they work. Subsequent chapters will explain more about how these solutions can be configured and how they work from a technical point of view.

Before we go into detail, it's important for you to understand that some of these solutions are specific to certain needs. Depending on your organization's requirements, only some of these solutions may be needed to best support your payment processes.

It is also important to understand that there may be technical differences in how certain solutions are implemented across different SAP S/4HANA versions, such as the public cloud, private cloud, or on-premise versions, as well as variations based on the specific SAP S/4HANA release being used. These differences can affect the functionality, configuration, and integration of BAM with other solutions or external systems. In later chapters, we will aim to capture and explain these distinctions to provide a clearer understanding of how the implementation and usage of BAM may vary depending on the deployment model and version.

1.2.1 Payments Management Within Financial Accounting

Traditionally, back in the SAP ERP days, payments were primarily managed within the financial accounting (FI) module, specifically under accounts payable (AP), where most of the activities and configuration related to payments were carried out, and they are still located there even in the latest version of SAP S/4HANA Finance. This setup was sufficient back when payment processing was simpler and less prone to fraud or technological complexity. However, as businesses grew and the importance of payments increased, so did complexity, risks, and reliance on technology. Fraud prevention, cash management, and optimizing liquidity became critical, leading to a shift in responsibilities toward treasury departments. In response to these evolving needs, SAP recognized the growing importance of these areas and developed additional solutions and functionalities, like advanced payment management, SAP Bank Communication Management, SAP Multi-Bank Connectivity, and in-house banking, which are described later in this chapter. These solutions allow for more sophisticated payment management, enhanced security, and better integration with broader treasury and risk management functions.

The most important integration point between financial accounting in SAP S/4HANA Finance and postings with payments is the *invoice*, which serves as the basis for both payment execution and the associated accounting entries. The invoice must be cleared during the payment process, ensuring that the corresponding balance sheet is updated accordingly. In addition to the invoice, the payment request acts as the second basis for payments as it provides the necessary details to initiate the transfer. It is essential to ensure that invoices and payment requests are not posted twice, as this could lead to discrepancies in financial reporting. SAP S/4HANA provides built-in controls to prevent this by checking for any outstanding or duplicate invoices or payment requests during the payment run. If an invoice or payment request has already been processed, SAP S/4HANA will ensure that no further payment is made for that same entry, effectively safeguarding against double postings and maintaining the accuracy of financial records.

Invoice

An *invoice* in SAP S/4HANA is a document that records a transaction between a business and a supplier or customer. It serves as a formal request for payment for goods or services provided. In SAP S/4HANA, invoices are captured in AP for vendor invoices and in accounts receivable (AR) for customer invoices. These invoices are posted to the vendor or customer account; the creation of master data for each vendor or customer is mandatory to ensure accurate processing. The vendor master data typically contains key information such as the vendor's name, address, bank details, payment terms, tax information, and any other relevant data for processing payments. Similarly, customer master data includes the customer name, address, contact details, payment terms, bank account information, and credit limits. This master data is crucial for ensuring that payments are routed to the correct accounts and that all transactions are accurately recorded in the system. When a vendor or customer invoice is posted, SAP S/4HANA uses the corresponding master data to identify payment terms, due dates, and payment methods, helping to automate and streamline the payment process.

When a vendor invoice is received, SAP S/4HANA creates an entry in the system, allowing the company to track liabilities and schedule payments accordingly. Similarly, when a customer invoice is issued, SAP S/4HANA records the receivable and manages the payment process. In the case of *netting*, SAP S/4HANA allows the offsetting of both vendor and customer invoices to streamline the payment process, ensuring that the outstanding amounts are reconciled efficiently. The invoice serves as the primary source for initiating payments, providing the necessary details for both accounting entries and payment requests.

Payment Request

In SAP S/4HANA, there are numerous types of payments that can be executed without the need for an invoice. These payments, which are often related to treasury activities, include bank-to-bank transfers, free-form payments, and financial instrument payments such as interest payments, loan repayments, and foreign exchange transactions. In addition, POBO (payments on behalf of) and COBO (collections on behalf of) payments are typically associated with SAP In-House Cash or in-house banking processes. In these scenarios, such as when the system generates payments within a central entity, there is no associated invoice. Instead, a payment request is generated to initiate the payment process.

The *payment request* is a key object in SAP S/4HANA that stores all the necessary information to execute the payment. This includes details about the paying entity, the paying bank account, the target bank account, the origin of the payment, and many other crucial elements. The information contained in the payment request is essential to ensure the payment is processed accurately and to generate a payment file that can be successfully accepted by the bank. The payment request serves as the foundation for initiating these transactions, enabling SAP S/4HANA to handle a wide range of non-invoice-based payments with the necessary precision and flexibility to meet banking requirements.

> **Duplicated Payments**
>
> In SAP S/4HANA, when it comes to processing payments, it is crucial to implement strict controls to avoid duplicate payments at all costs. Duplicate payments can occur due to repeated invoice postings or through parallel manual payment processes. To prevent this, organizations must plan ahead and incorporate robust change management practices. This includes implementing validation checks and system controls within SAP that ensure the same invoice cannot be posted more than once. Once an invoice is posted, approved, and validated, it will be picked up for payment automatically—making it essential that no duplicates exist at this stage. Care also must be taken to prevent manual payments from being executed for invoices already processed through the system.
>
> To prevent duplicate payments in SAP S/4HANA, it's essential to implement validations within SAP S/4HANA Finance at the invoice-processing stage. Standard checks can include invoice number and reference number validation to ensure uniqueness. Pattern checks can be introduced as well—for example, flagging payments to the same vendor in the same currency and amount that occurred within the past two weeks. These controls can help identify potentially duplicated entries before they reach the payment stage. Organizations may also benefit from leveraging third-party tools specifically designed to detect and prevent duplicate payments by running comprehensive checks across historical data.

1.2.2 Bank Account Management

The most important element for payments and overall cash operations within an organization is the bank. Every payment, whether an outgoing or incoming transaction, is executed through a bank account. The bank serves as the central point where a company manages its money, ensuring that funds are available for operational needs, investments, or settling liabilities. In other words, the bank is not just a place for storing money; it plays a crucial role in managing the flow of cash within the organization. All payments are initiated from the company's bank account and are paid to another bank account, making the relationship with the bank vital for smooth financial operations. Effective management of these banking relationships and accounts is essential for maintaining liquidity, optimizing cash flow, and ensuring the timely execution of payments across the organization.

With the current limited reliance on cash or barter systems, especially in large corporations, the role of banks has become even more crucial. As the need for digital transactions increases, banks are investing heavily in the digitalization of their services to meet the demands of modern business operations. The integration of advanced technologies, such as digital payments, real-time monitoring, and automated cash management tools, has become essential to streamline financial operations. In this digital age, technology plays a vital role in ensuring that payments are executed efficiently, securely, and in a timely manner, making technological innovation within the banking sector increasingly important for businesses.

Due to the critical role banks play in the overall payment process, SAP has also invested heavily in developing Bank Account Management in SAP S/4HANA to help companies successfully manage their bank relationships. BAM is a powerful tool that enables organizations to efficiently manage their bank accounts, monitor account balances, and ensure compliance with internal and external banking requirements. By integrating BAM, companies can streamline their interactions with financial institutions, enhance transparency, and gain real-time insights into their cash positions. This tool also facilitates the management of bank master data, the centralization of banking details, and the reconciliation of bank statements, ultimately improving cash visibility and control.

In addition to traditional banks, there are now many alternative payment solutions being developed to facilitate transactions, such as PayPal and other digital platforms. These alternatives provide companies with greater flexibility in collecting payments or making transfers, especially in global or online environments. Recognizing the growing importance of these payment methods, BAM is designed to accommodate such alternatives. With BAM, companies can manage these nonbank payment solutions by creating them as alternative accounts within the system.

In addition to traditional banks and alternative payment solutions like PayPal, BAM also allows companies to manage virtual bank accounts (VBAs) and internal accounts from SAP In-House Cash or in-house banking. These virtual or internal accounts, which are often used for centralizing cash management and optimizing liquidity across various subsidiaries or departments, can be created and maintained within BAM just like external bank accounts.

Once a bank account is created in BAM, it can be mapped to the house bank and bank account ID, along with the corresponding G/L account. This mapping is crucial for the execution of payments, as well as for posting bank statements. From a payment and bank statement execution perspective, these links are essential; without them, it would be impossible to process payments or accurately record transactions. The creation of the house bank or bank account ID is therefore a prerequisite for payment processing in SAP S/4HANA. It is important to note that while these mappings are vital for full BAM functionality, it is also possible to create house banks or bank account IDs without utilizing the full scope of BAM capabilities. This distinction will be further explored later in this book in Chapter 3.

In addition to linking payments and creating house banks and bank accounts, BAM offers many other features (depending on the SAP S/4HANA version), which are described in detail later in Chapter 3. Some of the most important features are as follows:

- **Approval process for bank account opening, adjustment, and closing**
  BAM provides the opportunity to set up various workflows for the approval of changes, as well as the opening and closing of bank accounts. You can define which fields will trigger the approval workflow, determine which account statuses will be used, and specify how approvals are given and by whom. SAP S/4HANA offers the following three types of approval processes:
  - Direct change: Changes are automatically incorporated without the need for additional approval.
  - Dual control: Another person must approve the change before it is implemented, ensuring an added layer of security.
  - Workflow: A more sophisticated approval process in which custom approval rules can be built. This allows you to define who approves specific changes and the order in which approvals are given.

  These workflows give organizations flexibility and control over their bank account management, ensuring compliance and reducing the risk of unauthorized changes.

- **Payment signatories**
  BAM also allows the maintenance of payment signatories directly under each bank account, enabling organizations to manage signatory approvals within the BAM system. This feature allows users to approve payments based on the amounts involved, providing an additional layer of control. This approach serves as an alternative to using SAP Bank Communication Management signatories, and the key differences between the two methods are described later in Chapter 7.

- **Cash position analysis on bank accounts**
  Many parameters set up in the bank account master data are crucial for later cash management processing. This data serves as the foundation for analyzing bank statements and monitoring bank balances. It provides essential details that help in reconciling bank statements, tracking cash positions, and ensuring accurate liquidity management. Properly maintaining the bank account master data ensures that cash management functions can operate smoothly and provide valuable insights into a company's financial status.

- **Bank account hierarchies**
  BAM provides functionalities that allow organizations to maintain and create bank account hierarchies tailored to their specific business needs. These hierarchies can

be organized based on various factors, such as cash pools, banks, currencies, or geographies, allowing companies to structure their bank accounts in a way that aligns with their financial operations.

- **Bank statement management**
In BAM, you can set up various parameters for processing both end-of-day and intercompany bank statements. BAM provides an overview of the bank statement data, allowing organizations to efficiently manage and reconcile these statements. By configuring the relevant parameters, businesses can ensure that the bank statements are processed accurately and in line with internal processes.

- **Bank fee analyzer**
In BAM, you can analyze billing files and the fees paid by your organization to a bank. This functionality allows you to track and review the costs associated with banking services, providing insights into the fees charged for various transactions.

- **Attachments**
In BAM, you can attach files and documents, such as bank account contracts, to each bank account. This feature allows organizations to centralize important information related to their bank accounts within the system. With the appropriate authorizations, you can easily access these attachments, ensuring that all relevant parties have the necessary documents at their fingertips for reference, review, or compliance purposes. This functionality enhances document management and ensures that important records are securely stored and accessible when needed.

- **Review process**
BAM includes a functionality to trigger a review process, which prompts you to regularly analyze your bank accounts and determine whether they are still needed or are redundant. This review process helps organizations identify and close unnecessary or unused accounts, ensuring that only essential accounts remain active. By enforcing periodic reviews, companies can streamline their bank account structures, reduce administrative overhead, and minimize potential risks associated with maintaining unnecessary accounts. This functionality ultimately supports better account management, cost savings, and enhanced operational efficiency.

- **Contact management**
In BAM, you can assign both internal and external people to each bank account, allowing you to clearly define who is responsible for the account from both the bank's perspective and within the organization. This feature helps ensure that the right individuals, whether they are bank representatives or internal staff, are easily identifiable and accountable for managing the bank account.

- **Overdraft limits**
In BAM, there is an option to maintain an overdraft limit for each bank account. This feature allows organizations to define a maximum permissible overdraft for their accounts, helping to manage cash flow and prevent unauthorized overdraft situations. By setting an overdraft limit, businesses can monitor and control their bank

balances more effectively, ensuring they stay within predefined financial boundaries and avoid potential fees or liquidity issues.

- **Cash concentration**
  In BAM, there is an option to create cash concentration structures, allowing organizations to link multiple bank accounts together and automatically trigger sweeps among these accounts. This feature eliminates the need for external cash-pooling solutions by enabling internal cash management within the system. With cash concentration, companies can efficiently consolidate funds from various accounts into a central account, optimizing liquidity and ensuring better cash flow management. This functionality is particularly useful when triggering sweeps among accounts belonging to different banks as it simplifies the process of managing cash across multiple banking relationships. It helps organizations streamline their cash management processes and reduce reliance on third-party cash pooling arrangements.

- **Bank account contracts**
  SAP has introduced a functionality called *bank account contracts* to help organizations better differentiate among various types of bank accounts, such as external bank accounts, alternative accounts, and internal, in-house banking accounts. This feature allows businesses to assign specific attributes and characteristics to each account, providing greater flexibility in managing them. With bank account contracts, organizations can decide which accounts require approval workflows, define specific activities for each account type, and set unique processing rules.

- **Power of attorney**
  The power of attorney functionality in BAM allows organizations to manage and document authorization for individuals who have the legal right to act on behalf of the company in banking matters. This feature enables companies to define and track the individuals who are authorized to make decisions or execute transactions related to bank accounts. By linking the power of attorney to specific bank accounts within BAM, organizations can ensure that only authorized individuals can perform certain actions, such as signing payments or approving changes to bank account information.

- **Bank account reporting**
  When all bank accounts are properly maintained in BAM with dedicated attributes, organizations can run various reports that provide valuable insights into their financial operations. This includes generating mandatory reports, such as the foreign bank accounting reports required in many countries. With detailed and accurate data stored in BAM, companies can quickly generate these reports, thus ensuring compliance with local regulations and improving transparency.

Based on these features, the implementation of BAM brings several key benefits to organizations. These include improved control over bank accounts, enhanced visibility into cash management, and streamlined payment processes. With BAM, you can ensure better compliance by managing and tracking bank account information, signatories, and approval workflows efficiently. It also facilitates smoother reconciliation of bank statements and enhances liquidity management through functionalities like cash concentration and overdraft limits. In addition, the ability to generate mandatory reports, such as foreign bank accounting reports, ensures that organizations meet regulatory requirements. Most importantly, BAM serves as a single point of truth, centralizing all bank account data and providing a reliable source of information for decision-making. Overall, BAM helps optimize financial operations, reduce risks, and increase efficiency, making it a vital tool for managing banking relationships and cash flow across the organization.

1.2.3 SAP Multi-Bank Connectivity and Bank Connectivity

Although all the solutions discussed here play a vital role in processing, generating, and optimizing payments, the most critical part of the payments process is connectivity with banks. Without reliable bank integration, even the most advanced payment solutions cannot function effectively. There are multiple ways to establish this connectivity, with a wide range of tools available on the market. We explore these options in detail—including their pros, cons, and implementation approaches—in Chapter 2.

SAP S/4HANA, as an enterprise resource planning (ERP) system, is responsible for recording financial transactions and running overall accounting processes. Payment processing is an integral part of SAP S/4HANA Finance, and when you create or execute payments, you essentially need to generate payment files, send them to the bank, and then receive payment statuses and bank statements in return. Performing this communication manually can be not only labor-intensive but also risky, as it introduces the potential for human error. Therefore, automating this process by connecting your system directly to the bank through various bank connectivity options is strongly recommended. Without this technology, your processes become significantly more time-consuming and expose your operations to unnecessary risks.

There are various bank connectivity options you can consider. You can opt for a direct host-to-host connection, use an EBICS connection, or choose a SWIFT bureau for SWIFT-based connections. Recently, APIs and SAP Multi-Bank Connectivity have also become popular choices. There are numerous other options and third-party vendors available as well, so selecting the best bank connectivity option is a nuanced topic that requires careful consideration. You can find detailed descriptions, guidance on what to watch out for, and everything you need to know about bank connectivity in Chapter 2.

SAP also provides its own dedicated solution for bank integration, known as SAP Multi-Bank Connectivity. This platform streamlines and secures communication between your SAP system and banking partners. SAP Multi-Bank Connectivity is a cloud-based solution that simplifies and automates communication between your SAP S/4HANA system and multiple banks worldwide. SAP Multi-Bank Connectivity enables secure, standardized connections to over 11,000 banks, providing a single point of access for exchanging payment files, retrieving bank statements in real time, and sending secure messages through SWIFT and other protocols. Its core functionalities include automated payment file processing, real-time status updates, comprehensive transaction monitoring, and robust error handling. By using SAP Multi-Bank Connectivity, you can replace multiple, fragmented bank connections with one streamlined interface, reducing manual work, enhancing security and compliance, and centralizing your bank communications—all of which contribute to greater efficiency and visibility in your financial processes. Detailed information on how SAP Multi-Bank Connectivity works, how to set it up, and its benefits can be found in Chapter 5.

1.2.4 SAP S/4HANA Finance for Advanced Payments Management

Most organizations today have a more complex landscape, with multiple ERP systems in place, not just SAP S/4HANA. Each of these systems typically has its own unique payment processes, which can create significant challenges when it comes to integrating them with banks. Connecting each ERP system individually to a bank is a time-consuming and expensive process, requiring custom configurations, ongoing maintenance, and additional resources. Moreover, managing multiple systems can make it difficult to get a clear overview of payments across the organization, as there is no single platform to consolidate data. This lack of centralization can lead to inefficiencies, duplication of effort, and suboptimal use of resources, further complicating financial operations and increasing the risk of errors or missed opportunities. As a result, many organizations are seeking more streamlined, centralized solutions to manage their bank connectivity and payments across different ERP systems.

To meet the growing complexity of managing payments across multiple ERP systems, many third-party payment tools have been developed to connect to various ERP sources and streamline payment processes. These tools are designed to integrate payments from different systems into a centralized platform, simplifying the overall payment execution and reconciliation. In addition, many corporations have developed their own custom solutions within their central SAP systems, creating payment factory models to consolidate and manage payments more efficiently. These payment factories enable organizations to process payments centrally regardless of which ERP system is used, thus improving consistency, reducing errors, and optimizing resource utilization. Although these custom solutions can be highly effective, they often require significant investments in development, maintenance, and ongoing support.

To address the growing need for streamlined payment processes across multiple ERP systems, SAP developed SAP S/4HANA Finance for advanced payment management in recent years. Advanced payment management is designed to centralize and standardize payment processing, offering a solution that integrates with various ERP systems and connects them to banks. By consolidating payment workflows into one platform, advanced payment management helps organizations reduce complexity, improve efficiency, and ensure greater control over their payments. It bridges the gap between different ERP sources, providing a unified approach to payment management that enhances visibility, minimizes errors, and allows for better resource allocation across the enterprise.

Advanced payment management was initially configured for mass payments for banks and was made available to corporations for the first time with the SAP S/4HANA 1809 release. Initially intended as a supplement to SAP In-House Cash, it now fully replaces that system with SAP S/4HANA natively developed in-house banking functions. The Map Format Data app, originating in financial accounting, has been linked with advanced payment management in SAP S/4HANA 2023, where it optimizes incoming and outgoing format exchanges via a rule-based system. Through APIs, additional SAP S/4HANA solutions and intelligent services such as watchlist screening or SAP Business Integrity Screening can be connected.

> **Advanced Payment Management License**
>
> To use advanced payment management, it is important to note that, on the top of the license for advanced payment management, you also need a valid SAP S/4HANA Finance for cash management license, as advanced payment management and SAP Bank Communication Management (for BAM accounts and SAP Bank Communication Management approval) are both part of the broader cash and liquidity management functionality within SAP S/4HANA.

Figure 1.1 demonstrates how advanced payment management helps you optimize payments.

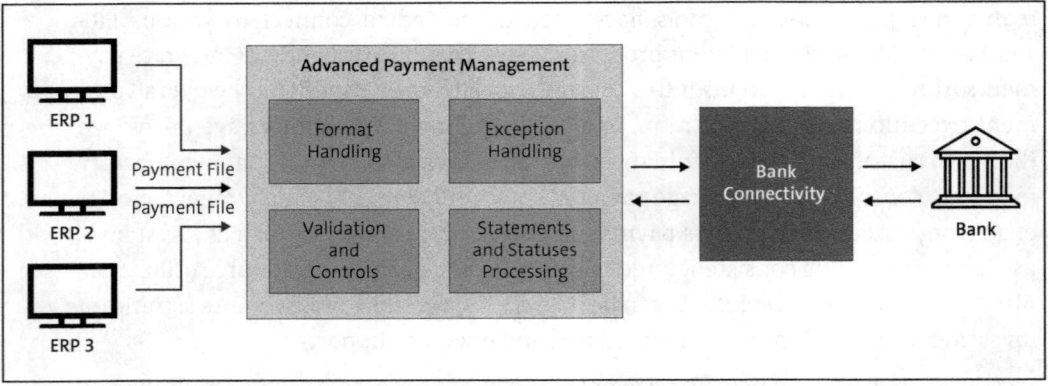

Figure 1.1 Advanced Payment Management in the Payments Landscape

In many organizations, multiple ERP systems are responsible for executing payments and generating payment files. These files are then transferred to advanced payment management, which checks, validates, and corrects them before deciding on the appropriate payment type, such as internal payments, PINO, or regular payments. Once the payment files are processed and corrected, they can be sent to the bank via a bank connectivity tool, such as SAP Multi-Bank Connectivity. The bank then processes the payments and sends back payment acknowledgments and bank statements, which are integrated into SAP S/4HANA for reconciliation and further financial processing.

Advanced payment management provides several key functionalities that streamline and optimize the payment process for organizations, including the following:

- **Centralized payment management**
  Advanced payment management consolidates payments from multiple ERP systems, enabling a centralized platform for processing and managing all payments.

- **Payment file validation and correction**
  This functionality validates and corrects payment files, ensuring that data is accurate and consistent before processing.

- **Payment categorization**
  Advanced payment management categorizes payments (e.g., internal payments, external payments, or PINO) based on predefined criteria, ensuring the correct handling of each payment type.

- **Automated payment execution**
  Advanced payment management automates the payment execution process, reducing manual intervention and minimizing errors.

- **Integration with bank connectivity**
  Once payment files are processed, advanced payment management integrates with bank connectivity tools like SAP Multi-Bank Connectivity, allowing seamless communication between SAP S/4HANA and the bank.

- **Bank acknowledgments and statements processing**
  This functionality handles the reception and processing of bank acknowledgments and statements, providing real-time updates on payment status and facilitating bank reconciliation.

- **Manual payments execution**
  There is also an option to generate payments manually directly using advanced payment management SAP Fiori apps.

- **In-house banking**
  Advanced payment management is a must-have when implementing the new in-house banking. Once advanced payment management is integrated, all payments generated within the system will be processed and routed through advanced payment management, ensuring they are handled accurately and efficiently. Advanced payment management is responsible for generating payment orders in in-house banking, as well as processing POBO and COBO payments where necessary. This centralized processing ensures that payments are correctly categorized, validated, and routed according to an organization's specific payment requirements, making it an essential tool for managing internal banking processes within SAP.

1   Payments Within a Corporation

Although advanced payment management and SAP Bank Communication Management (see Section 1.2.6) offer similar functionalities, they are typically used together within the payment process to manage different stages. Advanced payment management is primarily focused on the incoming side of the payment workflow; it handles the initial receipt, validation, and approval of incoming payment files or payment requests. On the other hand, SAP Bank Communication Management is specifically designed to manage approvals on the outgoing side once the payments have been processed and are ready to be released to the banks. This division of responsibilities ensures a clear separation of duties and enhances control and compliance in the end-to-end payment process.

Figure 1.2 illustrates the overall and basic payment process flow when using advanced payment management. This diagram provides a clear overview of the core steps involved in the payment process, highlighting key stages from invoice processing to payment execution. This figure covers the fundamental flow, but more complex scenarios and advanced use cases will be explored and described in greater detail later in this book, providing a deeper understanding of how advanced payment management can be implemented and applied in various business contexts.

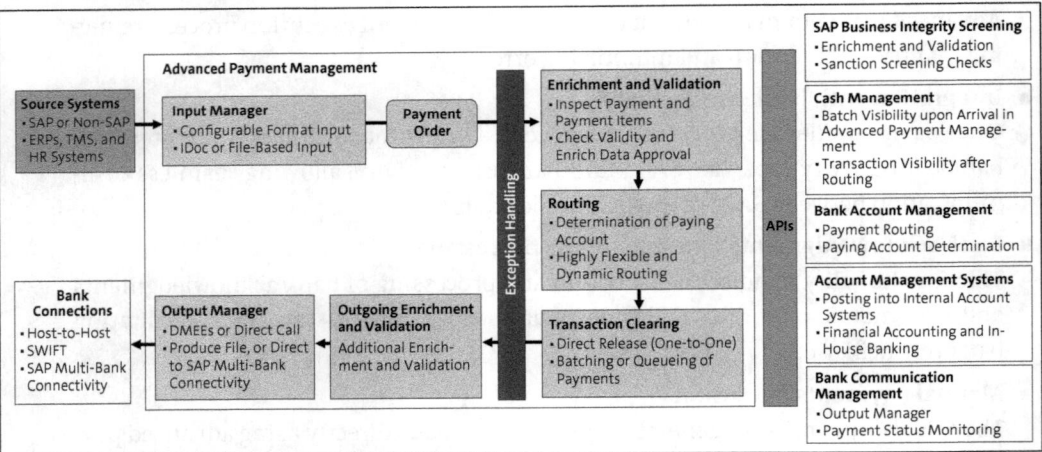

Figure 1.2 Advanced Payment Management: Solution Structure

SAP S/4HANA is built around four key processing points that ensure efficient and accurate handling of payment transactions. The first point, *input/output manager*, is responsible for handling the input and output of payment data, ensuring that payments are correctly received, processed, and sent to the appropriate parties. The second point, *enrichment and validation*, involves enhancing the payment data with additional information and validating it for accuracy and compliance before proceeding with further processing. Next, *routing* determines the optimal path for the payment based on predefined rules, ensuring that it is directed to the correct bank or payment system. Finally, *transaction clearing* handles the reconciliation of payments, ensuring that transactions are matched and cleared correctly in the system, thereby maintaining financial accuracy and integrity. Together, these four processing points enable a streamlined and error-free payment management process.

The advanced payment management processing of payments and how the features mentioned here impact flow are described in more detail ahead. It is important to understand that these flows are integrated into the broader payment processes that are explained later—specifically, internal payments, PINO payments, and POBO payments. Advanced payment management acts as the initial layer, receiving payments from various sources, then managing and structuring payments before they move into these specific scenarios. It's also crucial to note that when using in-house banking, implementing advanced payment management is mandatory. It can also help when you use SAP In-House Cash to support your payment factory. In these cases, payments are first processed through advanced payment management, where they are validated, grouped, and routed correctly, and only then are they handled by the in-house banking or SAP In-House Cash processes. Advanced payment management therefore serves as a critical starting point for efficient and compliant payment execution within the SAP landscape.

Figure 1.3 shows how a simple PINO process works in advanced payment management. You will find more information and explanations of what this looks from a technical perspective in SAP S/4HANA in Chapter 4.

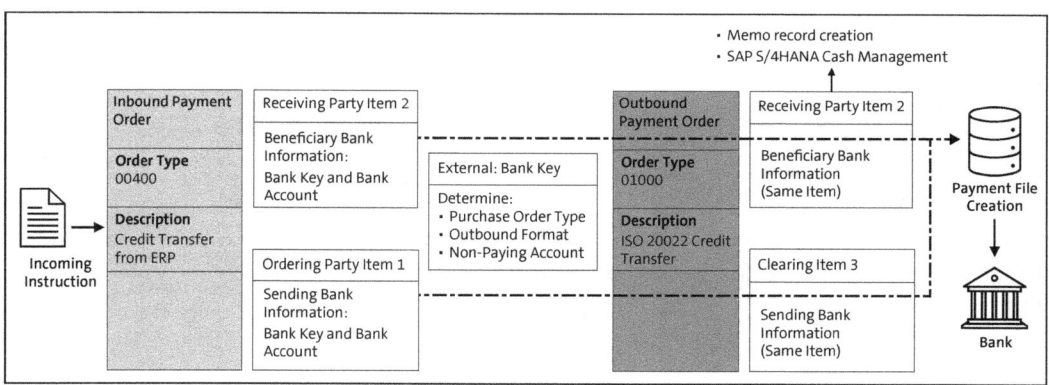

Figure 1.3 PINO Scenario in Advanced Payment Management

The process flow illustrates the basic PINO scenario. In this setup, a payment request is received from an affiliate, and advanced payment management creates a payment order containing both the recipient's bank account details and the paying company's bank account information. The payment file is then checked and, if necessary, transformed based on predefined validations or rules; however, the critical banking details—such as bank accounts—are not changed during this process. After validation, an outbound payment order is generated in advanced payment management, carrying the same banking information as provided in the original file. The final step involves creating the payment file and sending it to the bank. Throughout the process, advanced payment management does not modify the key payment details, ensuring consistency and compliance with the original payment instruction.

Figure 1.4 shows the process flow for the POBO process in advanced payment management. This is slightly different than the basic PINO flow. For POBO, the incoming payment file contains the SAP In-House Cash account of the paying entity and the beneficiary's bank account. When advanced payment management receives this file, it creates a payment order in in-house banking or SAP In-House Cash. At the same time, an outbound advanced payment management payment order is created, but this time it reflects the head bank account (the central paying account) instead of the original in-house cash account. Importantly, the beneficiary's account details are not changed during this process. In parallel, the creation of the SAP In-House Cash payment order automatically determines the payable and receivable positions between the affiliate and the in-house bank in the system's books. The detailed processing of the SAP In-House Cash payment orders and the related payments from an in-house banking perspective will be further explained in Chapter 4.

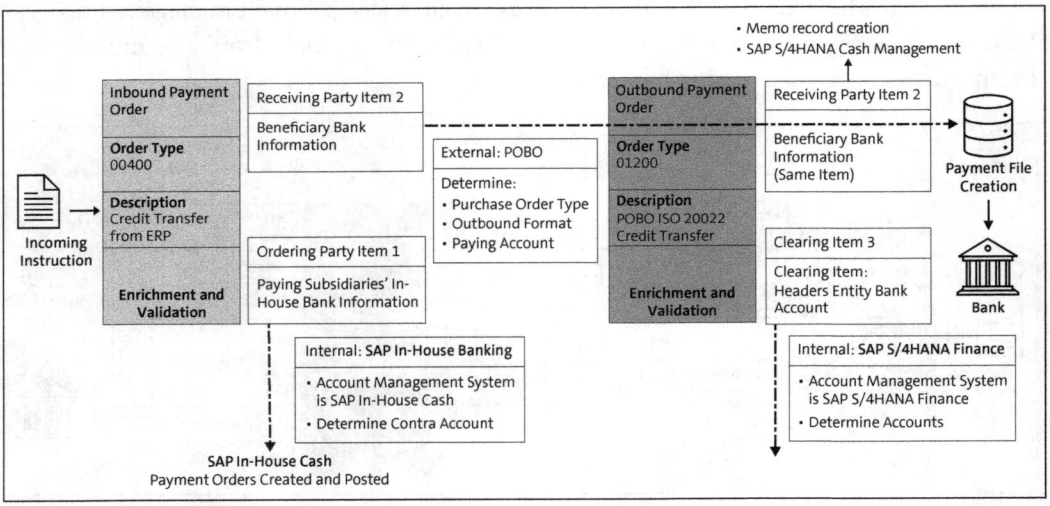

Figure 1.4 POBO Scenario in Advanced Payment Management

> **[!] Payment Order**
>
> In SAP S/4HANA, *payment order* can refer to different technical objects depending on the context. In the area of payments, a payment order can be created within SAP In-House Cash or in-house banking to represent intercompany transactions, where it determines the necessary intercompany postings in the subledger, recording debits and credits for intercompany accounts. Throughout this book, we use the terms

> payment order, SAP In-House Cash payment order, and in-house banking payment order interchangeably. Functionally, they represent the same concept—a posting to the subledger. At the same time, a payment order can also exist as an item within advanced payment management, carrying detailed payment information such as payer, beneficiary, and amounts. Although they share the same name, these payment orders are technically different objects, serving different purposes within the overall payment process.

Figure 1.5 shows the processing of internal payments in advanced payment management, which follows a specific flow designed for intercompany transactions. In this scenario, a payment request is received for an intercompany invoice, where both the payer's and beneficiary's bank accounts are internal accounts managed within SAP In-House Cash or in-house banking systems. Because both sides of the transaction are internal, this is treated as an internal payment or potentially part of a netting process. In such cases, the goal is not to send the payment out to a bank for external processing. Instead, the payment order is created and processed entirely within SAP, and the transaction is posted directly in SAP In-House Cash and in-house banking. This internal posting reflects the clearing of the payable and receivable positions between group entities.

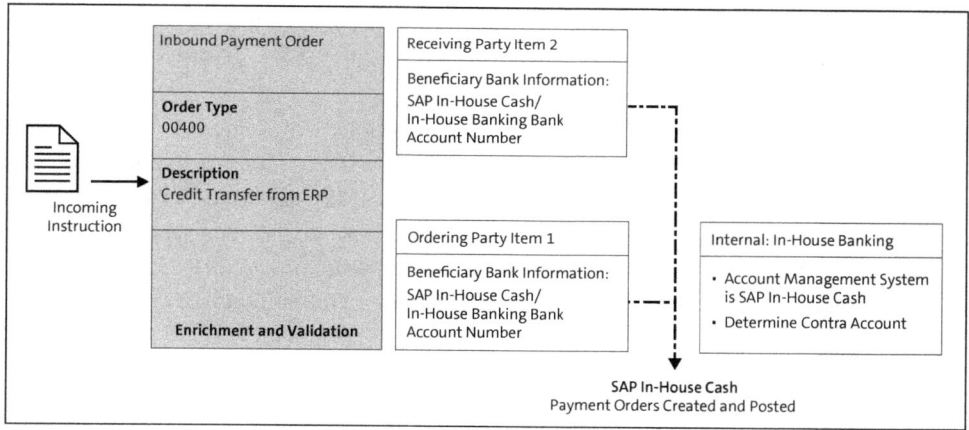

Figure 1.5 Internal Payment in Advanced Payment Management

These scenarios highlight how advanced payment management plays a crucial role in determining the type of payment scenario being processed. Whether it's facing internal, PINO, POBO, or external payments, advanced payment management is essential for identifying the correct flow and ensuring that each payment is handled properly according to its nature. This functionality is particularly important for building an efficient payment factory, in which centralized payment processing needs to accommodate different types of payments seamlessly. Advanced payment management provides the necessary structure, validation, and flexibility to manage this complexity. More information about the benefits of using advanced payment management and how it supports the design and operation of a payment factory can be found in Chapter 10.

> **Central Finance Payments Versus Advanced Payment Management**
>
> It's important to clarify that Central Finance payments are often mistaken for advanced payment management payments, but these are two entirely different concepts. Whereas advanced payment management and the other solutions described in this book are focused on managing payments within a central system after the invoice has already been posted, the payment run has been executed, and information has been passed to advanced payment management, Central Finance operates more as a central clearing factory. In a Central Finance scenario, invoices are posted in local SAP ERP or source systems and then replicated to the Central Finance system, where they are subsequently processed for payment. This means that the source systems remain the origin of the transactional data, but the actual payment execution and clearing are handled centrally in the Central Finance system.
>
> With the introduction of central payment functionality in Central Finance 1809 and 1909, businesses can now perform centralized clearing and payment execution directly in the Central Finance system. When an open item is replicated from a source system to Central Finance, it is technically cleared in the source system, and the payment is executed centrally based on defined payment terms in Central Finance. This approach solves the inconsistent issues seen in earlier releases, where open items remained open in both systems.
>
> In contrast, for advanced payment management, SAP Bank Communication Management, SAP In-House Cash, and other solutions discussed here, the process begins after the invoice has been posted into the central SAP S/4HANA system; only then can payments be processed using these tools. Therefore, Central Finance represents a separate architectural and functional strategy, where payment processing is an extension of centralized financial operations rather than a component of a traditional payment factory.

1.2.5 In-House Banking Versus SAP In-House Cash

Running payments and managing cash efficiently is a challenging and often costly task for organizations. The complexity of handling multiple payment processes, ensuring accuracy, complying with regulations, and maintaining connectivity with various banks can require significant time and resources. As a result, many organizations are actively seeking ways to reduce these costs by automating processes, centralizing payment management, and streamlining their cash management strategies. One effective approach to achieving this is by using virtual banking, which allows organizations to execute as many transactions as possible internally, thus reducing reliance on external banks. Implementing POBO and COBO payments also further minimizes the need for multiple external bank accounts, enabling organizations to optimize their payment flows while keeping costs down. In uncertain times, when organizations are under pressure to find savings and maximize efficiency, streamlining cash management through one central entity acting as an internal bank becomes even more critical. This approach allows organizations to focus on internal cash and financing before turning to external sources, reducing their dependence on external banks and lowering financing costs.

To meet the growing need for more efficient cash management within organizations, SAP introduced SAP In-House Cash many years ago. SAP In-House Cash was essentially a modified version of the much older SAP Bank Customer Account solution (which has now been retired with the transition to SAP S/4HANA), which was originally designed for banks. SAP adopted this solution to meet the unique needs of corporations, enabling them to establish and manage their own internal banking capabilities. With SAP In-House Cash, organizations could centralize their cash management processes, allowing them to handle transactions within the company, optimize liquidity, and reduce reliance on external banks. This solution quickly became quite popular, and thousands of corporations around the world are still using it today. SAP In-House Cash has provided companies with better control, transparency, and efficiency over internal cash flows, making it an essential tool for many businesses to streamline their financial operations and improve overall cash management.

However, SAP In-House Cash relies on many older technologies, such as its dependency on IDocs, and lacks integration with newer solutions like BAM. SAP also has ceased the active development of SAP In-House Cash, which makes it clear that a more modern solution is needed. To address these limitations and provide customers with more advanced technology, SAP introduced the new in-house banking in SAP S/4HANA. In-house banking leverages the latest technologies, offers seamless integration with BAM, and is designed to be more scalable and adaptable for future growth.Table 1.1 lists the differences between both solutions, as of mid-2025.

Description	SAP In-House Cash	In-House Banking
Basic prerequisite to run the solution	None	Advanced payment management
Formats supported	IDoc	XML (Pain001, CAMT, etc.)
Enabled for SAP S/4HANA Cloud Public Edition	No	Yes
Enabled for SAP S/4HANA Cloud Private Edition	Yes	Yes
Enabled for SAP S/4HANA on premise	Yes	Yes
User frontend based on SAP Fiori	No	Yes
Analytical dashboard available	No	Yes

Table 1.1 SAP In-House Cash Versus In-House Banking Functionalities

1 Payments Within a Corporation

Description	SAP In-House Cash	In-House Banking
Business user dashboards available	No	Yes
Transfer of account balance to reference account same currency	Yes	Yes
Transfer of account balance to reference account in FX currency	Yes	Yes
Transfer of account balance to an external account	Yes	Yes
Cross bank area clearing	Yes	No
Distribution and splitting of external bank statements	No	Yes
Payment scenario intercompany payments supported	Yes	Yes
Payment scenario payments on behalf supported	Yes	Yes
Payment scenario receivables on behalf/central incoming (same in-house banking)	Yes	Yes
Payment scenario receivables on behalf/central incoming (cross in-house banking)	No	Yes
Payment scenario collection on behalf supported	Yes	Yes
Payment scenario forwarding supported	No	Yes
Cross-bank-area payments (within same clearing area)	Yes	Yes
Correction of erroneous payments	No	Yes
Status notifications to subsidiaries via pain.002 (payment status sent by the bank)	No	Yes
Flexible routing configuration via master data	No	Yes
G/L transfer takes place during the end of day	Yes	Yes
G/L transfer takes place several times a day	No	Yes
G/L transfer without required shadow finance system (without full setup of SAP S/4HANA Finance in the same system; for both solutions, you need to create company codes, G/Ls used, cost centers, etc.)	No	No

Table 1.1 SAP In-House Cash Versus In-House Banking Functionalities (Cont.)

Description	SAP In-House Cash	In-House Banking
Interest calculation based on reference and fixed interest rates	Yes	Yes
Interest calculation based on new reference rates (SOFR, €STR, etc.)	Yes	Yes
Group limit besides single limits	No	Yes
Calculation of fees and bank charges	Yes	Yes
Internal cash pooling via in-house bank accounts (within same in-house banking)	Yes	Yes
Internal cash pooling via in-house bank accounts (across in-house banking)	No	No
Withholding tax calculation	Yes	Yes
Interest compensation	Yes	Yes
Support of IDoc FINSTA	Yes	No
Support of the SWIFT MT940 bank statement format	No	Yes
Support of the camt.053 bank statement format	No	Yes
Support of PDF document	Yes	Yes
Support email distribution of all format types	No	Yes
Account Statement distribution via SAP Multi-Bank Connectivity	No	Yes
Support of mass account creation process	No	Yes
Bank account management integration via semi-automatic function	No	Yes
Bank account management integration via electronic BAM (eBAM; creation of the account based on the request created in BAM)	No	Yes
Update of cash management with actual payments	Yes	Yes
Update of cash management with internal bank account balances	Yes	No
Direct integration of payments from SAP and non-SAP systems via IDoc	Yes	Yes

Table 1.1 SAP In-House Cash Versus In-House Banking Functionalities (Cont.)

Description	SAP In-House Cash	In-House Banking
Direct integration of payments from SAP and non-SAP systems via file upload	No	Yes
Direct integration of payments from SAP and non-SAP systems via SAP Multi-Bank Connectivity	No	Yes
Payment processing to external house bank without payment Transaction F111	No	Yes

Table 1.1 SAP In-House Cash Versus In-House Banking Functionalities (Cont.)

It's clear that SAP In-House Cash and in-house banking have significant differences. SAP In-House Cash, due to its long-standing presence in the market, is capable of supporting more complex scenarios, such as transactions between multiple bank areas. This has made SAP In-House Cash a choice for many over the years. In contrast, in-house banking is still in the early stages of its customer base, with its adoption numbers growing steadily—but in-house banking is under active development, with new features and functionalities being added each year, suggesting that the number of organizations adopting it will increase in the coming years. On the other hand, SAP In-House Cash is no longer actively developed, and SAP is no longer marketing or selling licenses for it, which indicates that its future in the market is limited compared to the rapidly evolving in-house banking—though in-house banking might not yet be able to support more complicated scenarios.

Please note that the comparison table and information provided reflect the current state of functionalities at the time of writing. SAP has already announced changes and enhancements planned for later in 2025, particularly for SAP S/4HANA Cloud Public Edition. We expect the number of functionalities supported by in-house banking to grow in the future, as many additional features are already on the SAP roadmap. Therefore, the comparison should be understood as a snapshot of the current capabilities, which are subject to change as the solution continues to evolve.

SAP In-House Cash/in-house banking both are comprehensive technical solutions that provide significant value by optimizing and automating key financial processes. Their main functionalities include the following:

- **Intercompany netting**
  This feature allows businesses to offset intercompany transactions, settle them cashlessly by posting results on the internal bank accounts, and hence reduce the number of payments and improve cash flow management across multiple entities within an organization.

- **Payments on behalf of**
  SAP In-House Cash/in-house banking enable one company entity to process payments on behalf of others, streamlining payment processes and allowing organizations to significantly reduce the number of bank accounts, especially those in foreign currencies, by centralizing payment processing. This reduction in the number of accounts helps lower FX costs as fewer currency conversions are required, leading to more favorable exchange rates and minimizing associated fees. POBO also enhances cash management efficiency by consolidating payments across multiple entities into a single, streamlined process.

- **Collections on behalf of and receivables on behalf of**
  This functionality allows organizations to centralize the collection of receivables for multiple entities, in the same way as POBO, improving cash flow and reducing the administrative burden of managing separate accounts receivable.

- **Cash pooling**
  This feature enables the centralization of cash from multiple accounts into a single pool, optimizing liquidity management and reducing the need for external financing by leveraging internal funds.

- **Manual payments**
  SAP In-House Cash/in-house banking allows for the processing of manual payments, giving businesses control over specific transactions that may not be automated, ensuring flexibility in payment management.

- **Cash and interest concentration (internal and external)**
  This feature consolidates cash from both internal and external accounts, improving liquidity visibility and managing interest costs more effectively.

- **Internal interest calculation**
  SAP In-House Cash/in-house banking automatically calculates internal interest on virtual bank accounts, ensuring accurate financial reporting and optimizing the management of intercompany financing.

- **Balance calculation and notification**
  This feature calculates balances across various accounts and sends notifications to all participants with bank accounts in the SAP In-House Cash/in-house banking subledger.

- **Bank charges**
  SAP In-House Cash/in-house banking also offers the functionality to calculate various charges and allocate them to the relevant affiliates. For example, it can track and charge affiliates for maintaining bank accounts, processing payments, or executing transactions. This ensures that each entity within the organization is accurately billed for the financial services they utilize, promoting transparency and cost allocation.

- **Limit management**
  The solution enables the setting and monitoring of financial limits across entities, ensuring that companies do not exceed their authorized limits and improving financial risk management.

- **Tax calculations**
  SAP In-House Cash automates tax calculation (interest withholding tax), ensuring compliance with local and international tax regulations and reducing manual intervention.

- **Internal bank account management**
  This functionality centralizes the management and creation of internal bank accounts, streamlining processes and ensuring proper reconciliation and reporting.

- **Internal bank statement generation**
  SAP In-House Cash/in-house banking generates accurate and timely bank statements, providing organizations with clear insights into their financial standing in virtual accounts and helping them track cash flows and run subledger reconciliation, thus ensuring intercompany balances are posted accurately.

- **Automatic posting**
  This feature automates accounting postings, reducing manual effort and ensuring the accuracy of financial records.

- **Integration with various areas of SAP S/4HANA, like cash management or financial accounting**
  SAP In-House Cash/in-house banking integrates seamlessly with areas of SAP S/4HANA such as cash management and financial accounting, providing a unified solution for managing financial operations across the organization.

These functionalities collectively enhance financial visibility, control, and efficiency, making SAP In-House Cash/in-house banking powerful tools for businesses looking to streamline their cash management, optimize their financial operations, and reduce their costs and accumulating gains in the central entity (bank area owner).

The SAP In-House Cash/in-house banking setup is structured as shown in Figure 1.6, where the header entity functions as the internal bank for the affiliated entities.

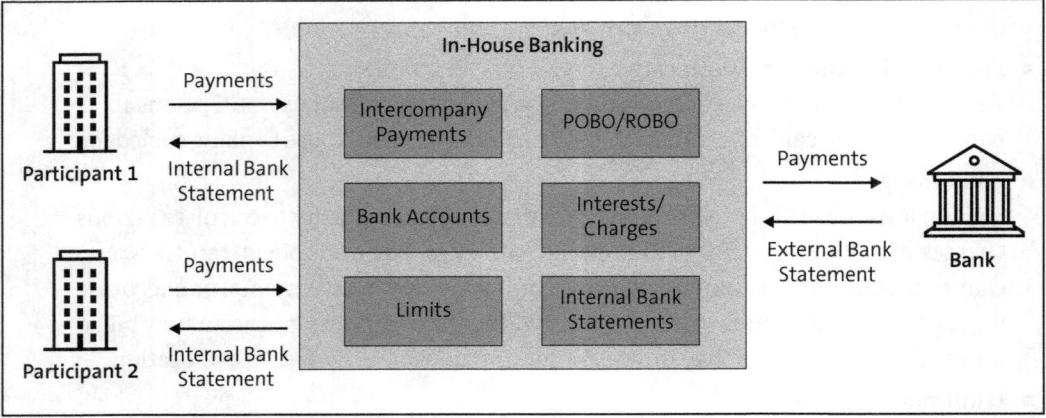

Figure 1.6 SAP In-House Cash/In-House Banking Scenario

In this setup, the header entity centrally manages cash flows and payment transactions on behalf of the subsidiaries or affiliates. This allows for efficient liquidity management, with the header entity consolidating payments and optimizing financial processes across the entire organization. By acting as an internal bank, the header entity provides services such as cash pooling, intercompany netting, and centralized payment processing, streamlining operations and reducing the need for multiple external banking relationships.

Implementing SAP In-House Cash/in-house banking is a highly complex topic due to the numerous legal constraints that must be addressed during the process. These constraints often include considerations such as transfer pricing, which involves ensuring that intercompany transactions comply with tax regulations and arm's length pricing rules. It is also essential to manage external bank accounts in such a way as to ensure that the reduction of external physical bank accounts does not lead to legal challenges, especially when multiple jurisdictions are involved. Another complexity arises from the need to handle accumulating balances on virtual bank accounts, which must be done in compliance with local regulatory requirements. Given these challenges, SAP In-House Cash/in-house banking implementation projects are usually more complex from a legal perspective than a technical one, as organizations must navigate a variety of legal frameworks and ensure compliance with financial regulations in different regions.

The selection of the header entity acting as the bank area owner is another complex aspect of SAP In-House Cash implementation. Ideally, this entity should be situated in a jurisdiction where access to cheap capital and favorable taxation policies is available, as this enables the organization to optimize its financial operations. The header entity typically accumulates gains based on the spreads and charges applied to the affiliates using the internal bank. However, there can be instances in which losses occur, such as due to FX hedging transactions. In these cases, the organization must carefully plan to offset such losses with profits from other operations. This becomes particularly challenging if the in-house bank is set up in a newly created entity, as it may lack the broader financial base or existing operations that could help mitigate these losses. Hence, careful consideration must be given to the jurisdiction and structure of the header entity to ensure that it can effectively manage both profits and potential losses.

Section 1.3 will dive deeper into specific payment scenarios that occur within the SAP In-House Cash and in-house banking setups, exploring how different transactions are handled and the processes involved. As you progress further in this book, our focus will shift to in-house banking, providing a more detailed overview of how to use and set up this newer solution. This will include guidance on its configuration and key features, ensuring that you gain a comprehensive understanding of its applications in modern financial management.

1.2.6 SAP Bank Communication Management

SAP Bank Communication Management provides a centralized platform that plays a crucial role in managing various banking relationships and transactions, ensuring that financial activities are accurately reflected across the company's accounts. Integration with SAP's ERP systems enhances this process, enabling seamless and efficient management.

It is important to mention that SAP Bank Communication Management is an older solution available in SAP ERP, and it continues to be available and recommended for monitoring payments, particularly within local systems (if you use multiple systems in your payments landscape). SAP Bank Communication Management provides organizations with the tools to manage their bank transactions, streamline bank communication, and ensure accurate reconciliation of payments. However, advanced payment management is gradually overtaking many of SAP Bank Communication Management's capabilities, offering a more centralized solution for managing payments across the organization. Advanced payment management brings the much-anticipated ability to centralize payment processes in one system, improving efficiency and providing better oversight. However, both advanced payment management and SAP Bank Communication Management are designed to work together seamlessly as part of the overall payment processing architecture in SAP.

As mentioned earlier, advanced payment management is primarily responsible for the incoming side, where it handles functional approvals, validations, and initial processing of payment files or payment instructions. Once the payments have been functionally approved and processed through advanced payment management, the responsibility is handed over to SAP Bank Communication Management, which serves as the treasury-level control point. SAP Bank Communication Management is used for final payment approvals, liquidity checks, file encryption, and the release of payment files to banks. This structured handover between advanced payment management and SAP Bank Communication Management ensures strong governance, segregation of duties, and compliance, while also supporting a smooth and secure end-to-end payment process.

At the same time, it's important to note that SAP is actively working on new payment concepts and evolving its payment architecture. In the future, the usage and integration of advanced payment management, SAP Bank Communication Management, and other solutions may change or become more unified, depending on SAP's development direction. Organizations should remain informed and adaptable as these solutions continue to evolve to meet changing business needs and industry standards. After reading these chapters, you will see how advanced payment management and SAP Bank Communication Management are integrated and complement each other within the end-to-end payment process.

The following is an overview of the concept of SAP Bank Communication Management, its main functionalities, and its role in the overall payment landscape:

- **Payment processing**
  SAP Bank Communication Management streamlines and automates the payment process by consolidating payment instructions from different SAP S/4HANA areas such as AP, AR, treasury, SAP In-House Cash/in-house banking, and payroll.

- **Payment formats**
  SAP Bank Communication Management supports a wide range of payment formats and ensures that payment batches are securely transmitted to banks through approved communication protocols, including SWIFT, host-to-host, and other secure channels (bank connectivity options are discussed in Chapter 2).

- **Payment approvals**
  The system allows for payment approval workflows, providing a robust level of control and visibility into outgoing payments.

- **Payment statuses**
  SAP Bank Communication Management offers detailed tracking and auditing features, ensuring all payment statuses—from initiation to settlement—are readily available, thus improving transparency and compliance.

- **Bank statement processing**
  SAP Bank Communication Management also facilitates the automated import and processing of bank statements. It supports multiple formats, such as MT940, BAI2, CAMT, and others, ensuring compatibility with various banking systems. Once imported, the bank statements are automatically reconciled with open items (whenever rules are configured and automatic matching could be performed) in SAP, significantly reducing manual intervention and errors. The system can identify discrepancies, allowing finance teams to resolve unmatched items promptly. This automation speeds up the month-end-closing process and provides real-time visibility into cash flow, liquidity, and overall financial health.

In today's corporate payments landscape, managing multiple banks can be highly complex, with multiple banks used (see Figure 1.7). SAP Bank Communication Management simplifies this by centralizing payment workflows, reducing errors, enhancing security, and ensuring real-time synchronization of financial data.

This decentralized approach can be unified through proper bank communication channels and SAP Bank Communication Management. One of the standout features of SAP Bank Communication Management, as with advanced payment management, is the connectivity functions, which streamline treasury operations by allowing corporations to execute payments, monitor transaction statuses, and receive bank statements through a single interface. However, connecting with multiple banks often requires various interfaces and technologies, sometimes necessitating a proprietary interface for each bank. The effort needed to maintain these interfaces can lead to high costs and manual labor, with yearly expenses for a banking interface ranging from $10,000 to

$50,000 (a rough estimate, depending on many factors). These costs encompass not only the interface itself but also ongoing maintenance, training, testing, and updates.

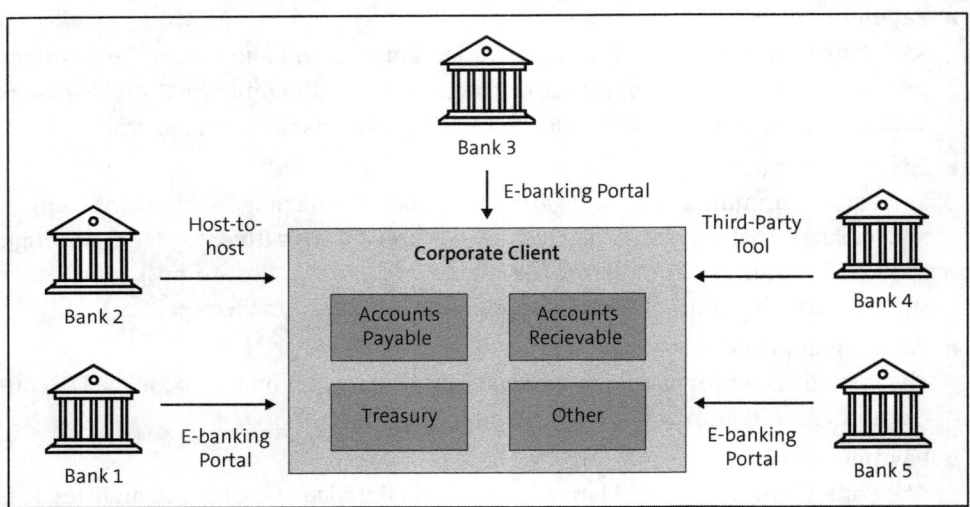

Figure 1.7 Decentralized Banking Landscape

SAP Bank Communication Management's robust payment approval process plays a critical role in maintaining control over financial operations, subjecting all outgoing payments to rigorous verification. Batch processing capabilities further enhance efficiency, allowing organizations to manage large volumes of payments, which is common for businesses with a global footprint. SAP Bank Communication Management's comprehensive reporting and audit trail features also provide the transparency and traceability needed to manage complex banking structures and ensure compliance with regulatory standards.

Within organizations, there have been instances of attempted manipulation of payment files. The process of sending these files to the bank often involves multiple people and significant manual work, with unencrypted files sometimes stored in easily accessible directories. Having multiple bank connection interfaces means that staff must either be familiar with all of them or specialize in a few, leading to inefficiencies. Furthermore, current communication methods do not allow real-time status tracking of files, causing errors to be detected too late, which can result in additional fees and post-processing costs. The signature verification process, required by banks before executing any payment order, introduces extra manual steps that delay payments. Moreover, for each bank, companies must designate a group of staff members who are authorized to send payment files, adding another layer of complexity.

Figure 1.8 shows how SAP Bank Communication Management can help you to centralize your payment landscape.

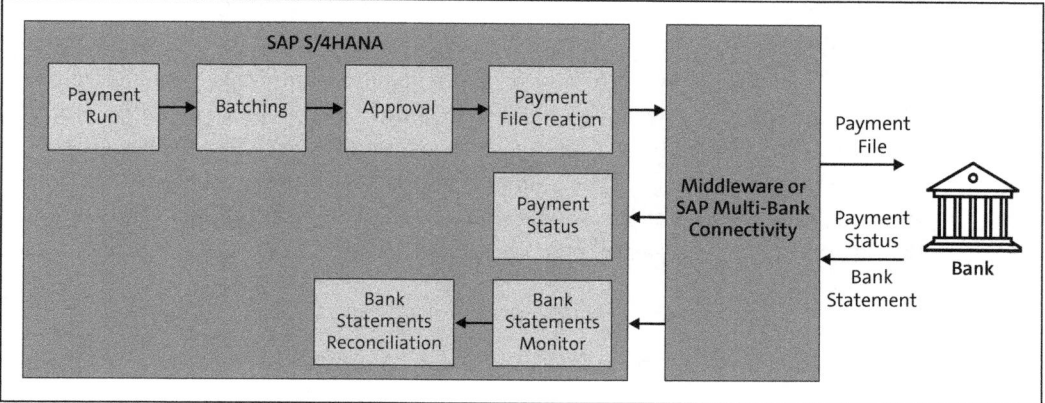

Figure 1.8 Centralized Payment Landscape Using SAP Bank Communication Management

Using both SAP ERP and SAP S/4HANA, a payment file is generated with the help of the payment program (Transaction F110 or F111). Once generated, the file is downloaded to a file server and then uploaded to a third-party system responsible for managing the transmission of files to and from the bank. Within this system, the file undergoes an authorization process, in which staff members log in to verify that an accompanying document matches the file total. If the verification is successful, they proceed to sign. Once the required number of approvers (typically two) has signed, the system sends the file to the bank. The bank then verifies that the signatures match those on file for the company's authorized approvers before executing the payment.

Banks also send files such as account statements to companies. These are received and decrypted by the third-party system, which acts as a central hub for collecting statements. In many cases, it converts the SWIFT message type 940 (SWIFT MT940) message into a format such as a statement.txt or lineitem.txt file. A staff member then retrieves the account statements and uploads them into the appropriate systems.

SAP Bank Communication Management eliminates the need for separate interfaces for each bank by providing direct access through the SWIFT network or any other bank connectivity options. SAP Process Integration or SAP Integration Suite (the newer cloud solution) play a critical role in linking the SAP S/4HANA system to the SWIFT network. SAP Bank Communication Management facilitates the transmission of a company's financial transactions with banks, including services for creating, sending, and tracking payment orders, as well as receiving bank statements. SWIFT operates a global financial messaging network that ensures secure and reliable exchanges between banks and financial institutions.

1   Payments Within a Corporation

> [!] **Bank Connectivity**
>
> Although SAP Bank Communication Management helps you to build a unified and centralized connection and to centralize the processing of all payments, it doesn't provide any bank connectivity options itself. You still need to build and choose the bank connectivity option that best fits your needs. In this context, you can refer to Chapter 2 or Chapter 5 and the detailed description of the standard SAP Multi-Bank Connectivity solution. This solution can serve as a best practice approach to establish secure and streamlined bank connections.

Figure 1.9 shows the core processes of SAP Bank Communication Management, which are as follows:

- **Payment run**
  The starting point for the SAP Bank Communication Management process flow is the execution of the payment run, which involves processing payments for open invoices using dedicated payment methods. This step ensures that all outstanding invoices are cleared and payments are made accordingly. In SAP ERP and SAP S/4HANA, payments are executed through Transaction F110 for invoices and Transaction F111 for payment requests. For cloud versions, the payment run is carried out through the Manage Automatic Payment app (F0770). This functionality is available in both the SAP GUI and SAP Fiori interfaces, depending on whether the organization is using the on-premise or cloud version of SAP S/4HANA, allowing for seamless integration and efficient processing of payments across different system environments.

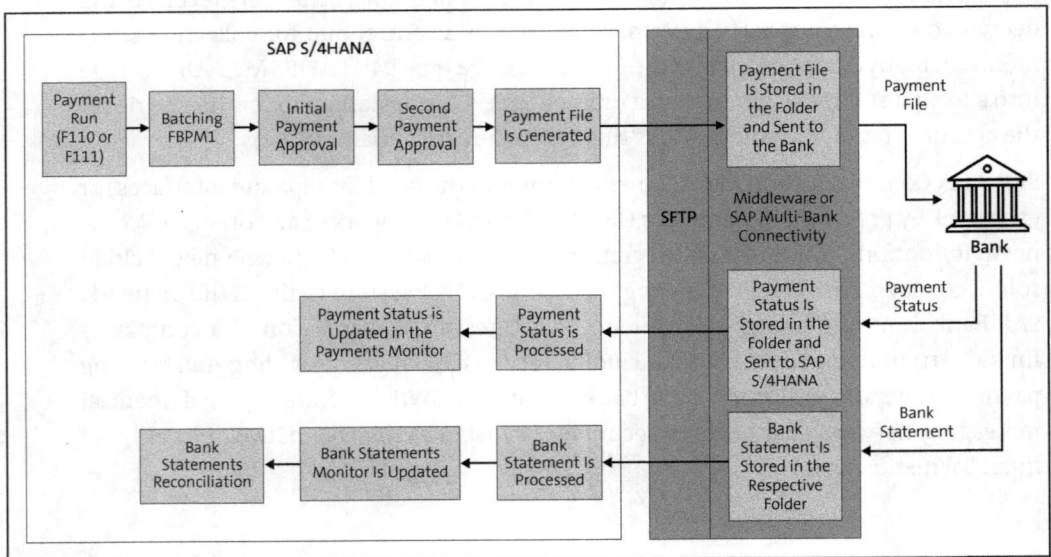

Figure 1.9 SAP Bank Communication Management Process Flow

- **Batching of payments**
  This process groups payment runs into batches, allowing for easier management. Payments can be grouped based on customizable rules, such as vendor, payment methods (to differentiate between urgent payments that need to be executed immediately and domestic payments that can be grouped into multiple runs), payment origin (such as bank-to-bank transfers, which can be autoapproved, or vendor payments requiring payment approval), or specific value ranges (e.g., to differentiate approvals based on amounts, such as autoapproval for payments up to 50,000 USD or the equivalent; other currency payments would be translated into the USD amount), with, for example, all payments above 10 million USD requiring CFO approval.

- **Approval of batches**
  Once batches are created, designated staff members approve them. Payment media are only generated after the final approval, significantly reducing the potential for unauthorized payment file manipulation. Digital signatures also support internal audit processes.

- **Payment file generation**
  Due to security reasons, the payment file (the type of payment file generated is determined by the configuration) is only created once the payment is fully approved in SAP Bank Communication Management. Once generated, it is stored in a dedicated folder and immediately sent to the bank. The main purpose of this is to ensure that no one can access or modify the file in any way.

- **Storage of files**
  All outgoing and incoming files are stored in a dedicated folder on the SFTP server, to which no one should have access. The middleware solution should pick files from this folder and send them to the bank or store files received from the bank. SAP jobs should have access to these folders to process the incoming files received from the bank.

- **Status tracking**
  SAP Bank Communication Management tracks the status of payment orders from initiation to execution by the bank, provided the bank sends status messages. This feature highlights errors and allows for the customization of automated alert workflows.

- **Bank statement monitor**
  This process gathers information from the SAP Integration Package for SWIFT and the SAP ERP or SAP S/4HANA system, allowing for real-time visibility of bank statement imports and processing.

By utilizing a single interface to connect with banks, SAP Bank Communication Management simplifies payment processes. When integrated with SWIFT, SAP Bank Communication Management also supports the Universal Financial Industry (UNIFI) standards defined by the International Organization for Standardization (ISO), with Single Euro Payments Area (SEPA) standards being a subset of UNIFI.

1.2.7 SAP S/4HANA Finance for Cash Management

Cash management is a critical foundation for executing payments effectively. One of its key roles is to ensure that there is sufficient liquidity available in the system to cover outgoing payments. This involves continuously monitoring account balances, analyzing current and expected cash inflows and outflows, and reviewing all financial messages exchanged with banks—such as bank statements and payment confirmations. Without proper liquidity planning, there is a real risk that payments may be rejected by the bank due to insufficient funds. That's why a strong cash management process isn't just about tracking money: It's about enabling smooth, reliable, and compliant payment execution. In this book, we will focus on the most important cash management functionalities to support your payment processes.

When analyzing cash management, several types of analyses are available. First, you can look at your cash position, which shows available cash on hand. Next, you can also focus on short-term liquidity forecasts, which help predict how much cash will be available in your bank accounts over the next two weeks. Long-term planning is also important for setting plans and forecasts for the coming years. Finally, from a cash management perspective, it's also valuable to analyze historical data, comparing actual and realized cash flows to previous forecasts and budgets. Figure 1.10 explains more cash management dimensions and what you can analyze.

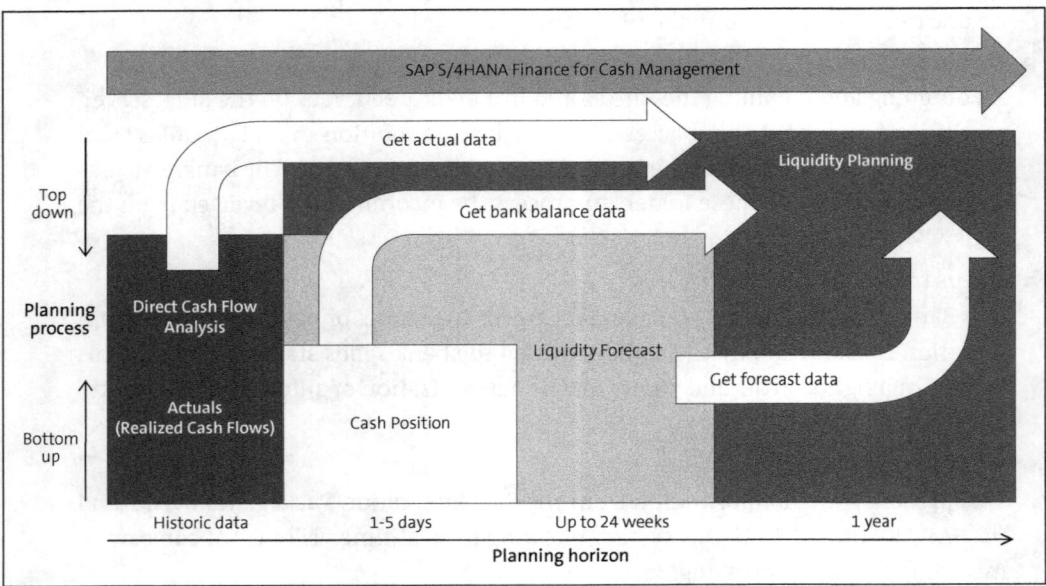

Figure 1.10 SAP S/4HANA for Cash Management

For cash management, start by analyzing historical data. This brings actual data into the cash flow analysis—data that has already occurred and been confirmed with bank statements and that is already assigned to liquidity categories. Next, focus on your cash position, which reflects your current balances and available funds on bank accounts, including a short-term horizon of up to five days (covering AP and AR). This step aims to predict what funds will be available over the next few days. After that, move on to longer-term liquidity forecasts, using information from sources like logistics, AR, AP, and other inputs. Once you have actual data, balance data, and forecast data in place, you can work on long-term liquidity planning, which covers one year or more. The further into the future you plan, the less accurate the data tends to be, which requires building certain models and assumptions to create a reliable plan.

From a cash management perspective in SAP, there are many tools, solutions, and methods available to support your planning and decision-making. However, in this book, we will focus specifically on cash position and short-term liquidity forecast. Our goal is to help your organization ensure that there is enough liquidity to meet all payment obligations and maintain financial stability in the short term. By focusing on these two key areas, we'll provide you with practical guidance and best practices to strengthen your cash management processes and keep your operations running smoothly. There are multiple books and other sources for more detailed information about all the SAP solutions that support your cash and liquidity processes.

Figure 1.11 shows what a typical and basic cash management process flow can look like in SAP S/4HANA. Chapter 6 will offer more information about how this flow can be managed using SAP solutions.

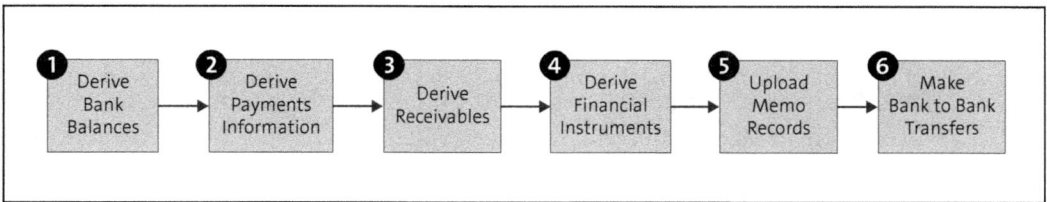

Figure 1.11 Cash Management Process

Let's walk through the process step by step:

❶ A fundamental starting point for effective cash management is knowing how much money is available across your bank accounts and portfolio. This visibility begins with the bank statement, which provides the actual account balances reported by the banks. By uploading these bank statements into the system, you can see the current cash position and available funds. This process enables accurate tracking and informed decision-making for payments and liquidity planning. Once the bank statement is in place, the system reflects the labeled balance, giving a clear picture of the organization's financial standing. For more details on how to handle and configure bank statements, see Chapter 9.

❷ The next essential step in the cash management process is to analyze payables data and identify all expected payments within a selected time frame. For example, if you're planning for the next two weeks, then you need to gather a clear view of all upcoming payments—such as vendor invoices and supplier obligations—that are scheduled for execution. This forecast allows you to anticipate the total outflows and align them with available liquidity. It's crucial to ensure that each bank account designated for these payments has sufficient funds to cover the expected transactions. If not, there's a significant risk that payments could be rejected due to insufficient funds, potentially impacting vendor relationships and operational continuity.

❸ In the next step of cash management, the focus shifts to analyzing receivables and expected inflows. It's important to review what incoming payments you anticipate from your customers over the next two weeks or another analyzed period of time. This involves gathering all available information related to open customer invoices and expected payment dates, as well as any other inflow forecasts. By identifying these expected cash receipts, you can understand what will positively impact your bank balances and contribute to maintaining sufficient liquidity. This analysis is critical for creating a complete picture of your cash position and ensuring you are prepared to meet upcoming financial obligations.

❹ Another critical step in the cash management process is incorporating data from treasury operations and financial instruments. This includes all upcoming payments related to interest on loans, debt repayments, foreign exchange settlements, and bond coupon payments—essentially, any financial obligations that must be settled within the forecast period. At the same time, you must account for incoming flows such as interest income from deposits, returns from money market instruments, or other treasury investments. These transactions often involve high-value amounts, making them highly impactful for your overall liquidity position. Therefore, integrating treasury-related flows into the cash management process is essential for building an accurate and reliable liquidity forecast that reflects both operational and financial realities.

❺ In the next step of the cash management process, it's important to manually enter any financial information that is not captured elsewhere in SAP S/4HANA or other planning and forecasting tools. This includes expected cash flows or obligations that are known to the business but not yet reflected in the system—such as upcoming one-time payments, ad hoc revenues, or other anticipated financial activities. These entries are typically recorded as memo records in cash management, serving as placeholders for expected inflows or outflows. Including this manual data ensures that the cash position and liquidity forecast are as complete and accurate as possible, thus providing full visibility into all known financial movements, even those outside of standard system processes.

❻ Once you have a clear view of your expected cash balances for the selected time period, the final step is to optimize liquidity by transferring funds between bank accounts. If certain accounts are forecasted to have a surplus while others are projected to face a shortfall, you can proactively move money from the surplus accounts to those needing additional liquidity. These bank-to-bank transfers help ensure that all accounts have sufficient funds to cover expected outflows, thus preventing payment failures due to insufficient balances. Such internal fund movements are typically managed centrally by the treasury team and represent an efficient and essential way to maintain balance across your banking landscape. This step is a key part of the overall payments process, enabling seamless execution of obligations while minimizing the need for external financing.

All of these steps form the foundation of a basic cash management process, which is essential for understanding how much money is currently available and what your organization's future liquidity and cash position will look like. This process ensures that there is enough liquidity in each bank account to support the execution of all planned payments. The exact steps may vary, depending on your organization's size, structure, and specific requirements, but the core approach remains the same. In practice, the process can become more complex, involving additional layers of forecasting, approvals, and integration with treasury tools. However, at a high level, this scenario outlines how cash management should work to support reliable and efficient payments processing. Chapter 6 provides a detailed explanation of how this entire process is managed within SAP S/4HANA, including configuration, tools, and best practices.

1.3 Payments Processing

In this section, we will focus on the various payment types and process flows related to payments, covering both the business and functional aspects. We will explore the different categories of payments, such as domestic and international transactions, as well as the steps involved in executing these payments efficiently. The discussion will address the business perspective, including how payment processes align with organizational goals, as well as the functional side, detailing the necessary procedures, controls, and systems required to manage payments effectively. By examining both sides, this section will provide a comprehensive understanding of how payment types and processes integrate within broader business operations.

1.3.1 Payments in the Name Of

Within a group or organization, there are typically multiple entities, each responsible for running their own payment processes. These entities often operate independently, using different technologies, payment systems, and processes tailored to their specific needs. Each entity may have its own team dedicated to managing payments, further contributing to the lack of standardization across the organization.

This decentralized approach can create challenges in maintaining consistency, ensuring compliance, and managing operational efficiency, adding to the lack of global visibility and inefficient use of resources as each entity follows its own procedures and leverages its own technological solutions to process payments.

Centralizing all payments within a globally dedicated AP team or treasury team (depending on the size of the organization) brings numerous benefits to an organization. By consolidating payment processes, companies can streamline operations, reduce redundancies, and ensure consistent payment practices across all entities. A centralized AP team can standardize payment procedures, improving efficiency, accuracy, and compliance with company policies and regulations. This centralization also allows for better control over cash flow as the team can more effectively manage payment schedules, identify opportunities for early payment discounts, handle potential errors faster, address fraud management, enact better controls, improve visibility, and optimize working capital.

We refer to the process in which a central team executes payments using a centralized platform as *payments in the name of*. In this model, the central team processes payments on behalf of each entity, while still utilizing the bank accounts that belong to each individual entity. The team is responsible for paying invoices and ensuring that the relevant general ledger and subledger accounts are updated accordingly. This approach allows for streamlined payment execution, greater efficiency, and cost savings, while maintaining the integrity and independence of each entity's financial records.

SAP S/4HANA is an excellent tool for implementing the PINO process. It offers flexibility in managing payments across multiple entities by consolidating all these entities within a single SAP system, thus allowing the AP team to execute payment runs in the name of all entities.

When using multiple ERP systems, you can connect them to a central SAP S/4HANA system, where advanced payment management will assist with all necessary processing and transformations.

The process flow shown in Figure 1.12 demonstrates how the PINO process works in a single SAP S/4HANA environment. The process flow changes slightly when different solutions are used.

Ultimately, all payments are routed through SAP Bank Communication Management and/or SAP Multi-Bank Connectivity, ensuring secure, streamlined payment execution and transmission to the bank. This centralized approach provides enhanced control, transparency, and efficiency in managing global payments.

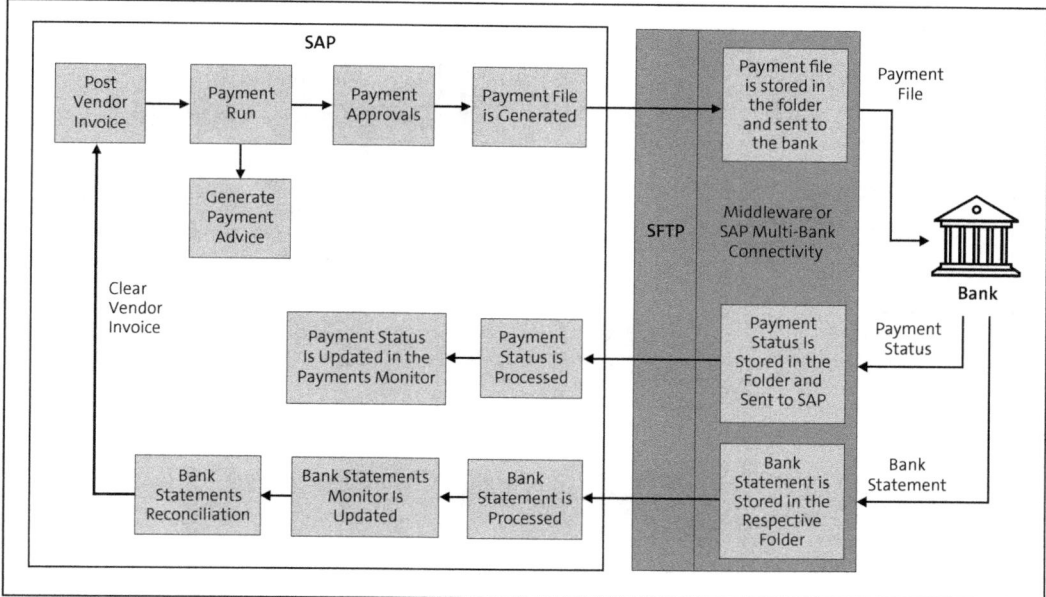

Figure 1.12 PINO Process Flow

1.3.2 Direct Debit

Direct debit is a widely used payment method that allows businesses to collect payments directly from their customers' bank accounts. However, it is important to note that direct debit is not technically a payment in the traditional sense: Unlike typical payments, in which customers voluntarily transfer money to businesses, in direct debit, businesses initiate the transaction by demanding money from their customers.

To put it another way, unlike other vendor payment methods, the customer does not actively initiate the transfer of funds. Instead, after obtaining proper authorization, the business demands payment. This makes direct debit a more controlled form of transaction for the organization, but it also places responsibility on businesses to manage the process accurately to avoid disputes.

An organization can trigger direct debits for customers by generating payment runs for their open invoices. In this process, you create a specific payment file, typically in the pain.008 format, which contains the necessary details to initiate the direct debit transactions. This file is then sent to the bank in the same way as it is for normal payments. The difference lies in the fact that with direct debits, you are requesting payment from the customer's account rather than receiving payment; however, the process of generating the payment file and transmitting it to the bank is the same as in standard payment runs. This ensures consistency and automation in handling both payments and direct debits (collections).

Figure 1.13 illustrates what the direct debit process flow can look like.

1   Payments Within a Corporation

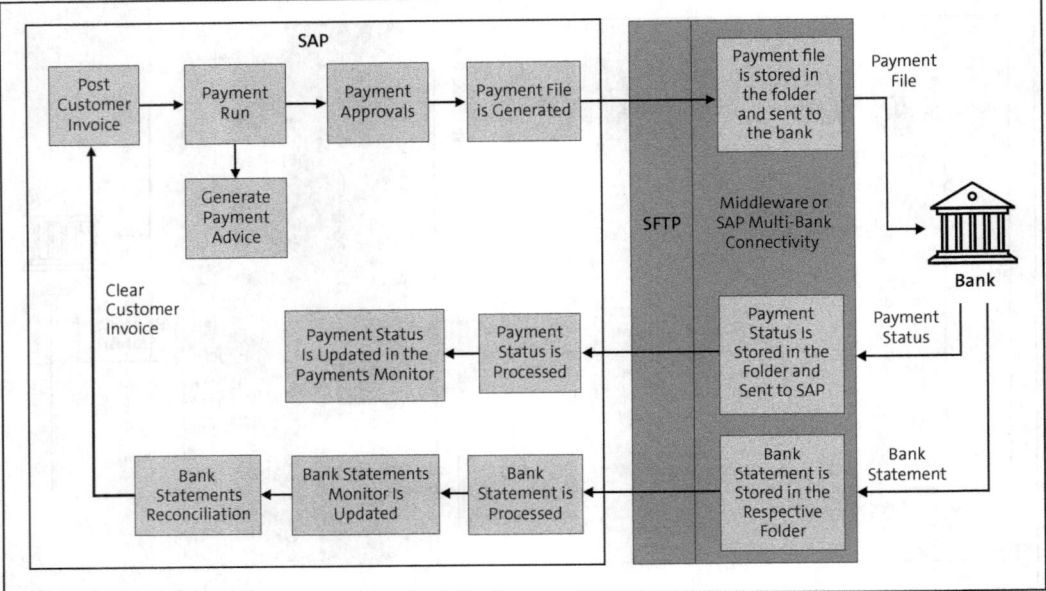

Figure 1.13 Direct Debit Process Flow

From the perspective of the overall process flow, direct debit is not significantly different from a standard PINO payment. Both processes involve similar steps and follow a comparable payment procedure. The primary difference lies in the fact that direct debit is executed based on the customer invoice, which links the payment directly to the outstanding amount. In addition, a different file is sent to the bank for processing, and this file contains the necessary information for the bank to complete the payment. This distinction ensures that the payment is accurately matched to the customer's invoice, making direct debit a more specific form of payment processing within the broader PINO system.

From a technical perspective, however, there are more notable differences. For direct debit, you need to create a specific incoming payment method and configure the entire incoming payments setup. You must also enable direct debit functionality for your customers and ensure that you store the relevant direct debit signatories. This is crucial so that the bank can make the payment correctly. These additional configuration steps are necessary to facilitate the direct debit process within the broader payment system.

Finally, monitoring incoming payments is particularly important in the direct debit process. It is essential to track the payments closely as there are instances in which a customer might not have sufficient funds to cover the payment. In such cases, the payment may not be made, or it could be delayed until the funds are available. It is quite common for a direct debit to fail due to a lack of funds in the customer's account. In such cases, it is crucial to have a proper process in place to monitor and address the issue promptly. This is especially important if the invoice is cleared during a payment run, as it might take some time to notice that the payment was not actually received. Banks typically send information such as the camt.054 statement, which provides details about direct debits that were not executed. Once a payment has failed, it is important to initiate a retry process, but having established procedures to collect the outstanding money is equally important. These procedures might include reinitiating the direct debit, contacting the customer for payment, or pursuing other collection methods to ensure the funds are recovered in a timely manner. Proper monitoring and well-defined processes help minimize the risk of missed payments and ensure financial accuracy.

1.3.3 Intercompany Payments

Paying external counterparties is undeniably an important aspect of any organization's financial operations, but intercompany payments often represent a significant portion of the total payment flow. Depending on the size and structure of the organization, as well as the number of legal entities within the corporate group, intercompany transactions can make up a substantial part of all payments. These payments are typically made between different branches, subsidiaries, or affiliates within the same organization, and they play a crucial role in maintaining liquidity and ensuring proper financial management across the entire group. In large organizations with multiple legal entities operating in different regions or business sectors, intercompany payments can be quite frequent, and their effective management is essential to ensure smooth operations and compliance with financial regulations.

In the case of intercompany payments, organizations typically have greater control over how and when these transactions are executed compared to external payments. Unlike external payments, which are governed by strict contractual agreements and third-party obligations, intercompany payments do not require adherence to the same level of external regulations or terms. This allows for a more streamlined and efficient process, as internal policies and agreements can be tailored to the needs of the organization, making it easier to manage the financial relationships between different entities within the group and gain significant efficiencies and cost savings.

Physical Bank Transfers

Each intercompany payment can be executed directly as a normal physical payment, similar to how external payments are processed. However, the key difference lies in the level of control an organization has over these payments. With intercompany payments, the timing and execution can be more flexible. For instance, the organization can delay certain payments, treat them with lower priority, or wait until there is enough liquidity before processing the payment. This added control allows for better cash flow management and enables the organization to optimize its financial operations by making strategic decisions about when and how to execute these payments based on internal needs, rather than being constrained by external contractual obligations or deadlines.

However, it is still recommended to settle intercompany payments within a reasonable timeframe to maintain good financial practices and avoid potential disruptions within the organization. Having more control over the timing provides flexibility, but relying on delayed or prioritized payments too frequently is not advisable. This approach can become inefficient and lead to higher operational efforts and costs (processing bank charges for executed payments), as each payment requires processing, tracking, and administration. Over time, the accumulated costs associated with managing these payments can outweigh the benefits of delay or prioritization. Therefore, it is generally better to aim for timely execution while leveraging flexibility in cases of urgent need of liquidity.

It is recommended to use intercompany payments as physical transfers only when legally required and no other options are available. From a processing perspective, netting and cashless settlements will deliver more value, better controls, and greater cost savings compared to physical transfers.

Netting

Intercompany netting is a financial process that involves gathering invoices for all affiliates within a corporate group and calculating the final amount that needs to be paid or received for a selected period. Instead of making individual payments for each transaction between subsidiaries, intercompany netting consolidates all the outstanding amounts, allowing the organization to offset receivables and payables across the various entities. This process simplifies the payment flow by reducing the number of transactions and ensures that only the net balance—either a payment or receipt—needs to be settled. By centralizing the payments in this way, organizations can improve cash flow management, reduce administrative costs, and minimize the risk of errors associated with processing multiple intercompany transactions.

Traditionally, intercompany netting can be either payable-driven or receivable-driven. In a *payable-driven approach*, the paying entity selects which invoices to pay, while in a *receivable-driven approach*, the collecting entity starts the process by selecting which invoices to collect. Both methods offer different perspectives on how the netting process is managed. However, it is crucial to set up this process to be either fully payable-driven or fully receivable-driven. Mixing the two approaches can lead to confusion and complications, particularly in reconciliations, as it may result in discrepancies or mismatched transactions between entities. Consistent setup ensures smooth reconciliation, accurate financial reporting, and the proper alignment of intercompany balances.

Netting can be run in the following two ways:

- **Decentralized**
  In decentralized netting, the process is run independently among all participating affiliates. Each entity calculates and determines its own net position—whether an amount to be paid or received—from the other entities in the group. Each affiliate is

responsible for gathering and offsetting invoices with all other participants, ensuring that only the net balance, rather than the total sum of individual transactions, is paid or received (see Figure 1.14).

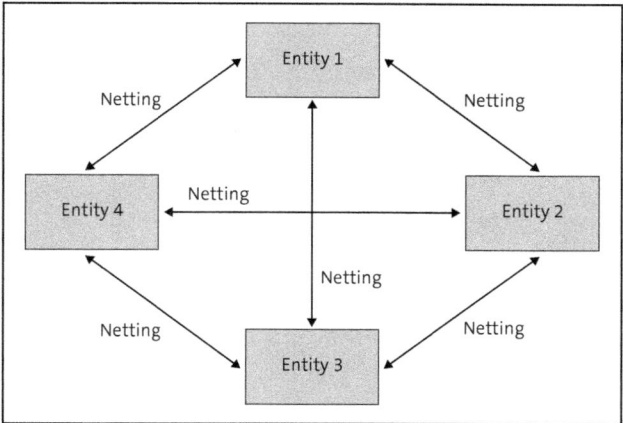

Figure 1.14 Decentralized Netting

Standard SAP supports decentralized netting through the Transaction F110 payment program, which allows for the execution of payments between selected invoices between affiliates. In this process, each affiliate independently processes its own payments and invoices, selecting the appropriate transactions to be included in the netting run. Using Transaction F110, the system can automatically identify and offset the receivables and payables between the participating entities, calculating the net amount to be paid or received.

This decentralized approach allows for greater flexibility and autonomy for each entity in managing its own intercompany transactions, but it requires careful coordination to ensure that the netting process is accurately executed and reconciled across all entities involved. It provides more distributed control over payments, with each participating entity handling its own settlements.

- **Centralized**
  In centralized netting (see Figure 1.15), a designated header entity, often referred to as the *netting center*, takes responsibility for calculating the netting on behalf of all participating affiliates. The netting center consolidates all invoices from the various entities, offsets the receivables and payables between them, and calculates a single net amount for each participant. This final amount represents what each entity needs to pay or receive, reducing the number of individual transactions that need to be settled. The netting center acts as the central point of control, ensuring that all intercompany balances are accurately netted and that only one payment or receipt is made between the netting center and each participant. This centralized approach simplifies the reconciliation process, improves efficiency, and reduces the administrative burden for individual entities.

1 Payments Within a Corporation

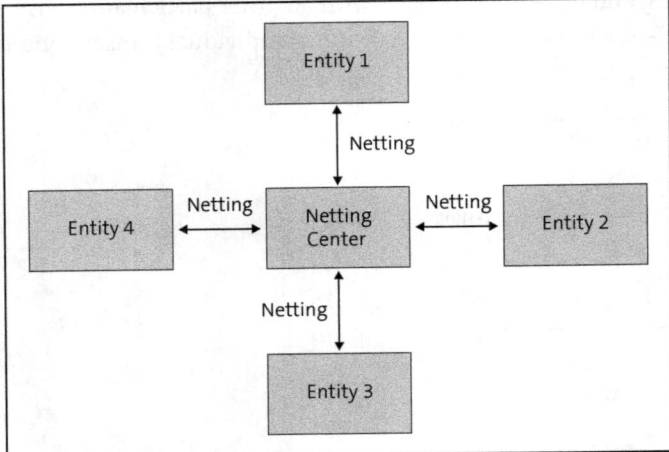

Figure 1.15 Centralized Netting

Standard SAP S/4HANA provides limited functionality for centralized netting as it primarily focuses on cashless settlements through SAP In-House Cash or in-house banking. These tools are designed to facilitate the process of managing intercompany transactions but do not fully support the complex netting of receivables and payables between multiple affiliates in a centralized manner. To execute a more robust centralized netting process, organizations typically require enhancements to the standard SAP system, or the use of a dedicated tool specifically designed for centralized netting.

When you design or plan the netting process, you should consider the following points, in addition to whether netting should be payable- or receivable-driven:

- **Legal constraints**
  Before implementing intercompany netting, it is important to verify whether netting is legally allowed in the specific country or countries where your affiliates operate. Different jurisdictions have varying regulations regarding the offsetting of receivables and payables between entities, and in some regions, netting may be restricted or subject to specific legal requirements. For instance, certain countries may require formal agreements or impose limitations on the ability to net transactions across legal entities, especially if it involves cross-border payments or different currencies. Ensuring compliance with local regulations is crucial to avoid legal and tax-related issues. Therefore, before setting up a netting process, it is recommended to consult with legal and tax advisors to confirm that netting is permissible and to understand any country-specific restrictions or obligations.

- **Netting currencies**
  When deciding on the netting approach, it is crucial to determine the appropriate currency strategy. Netting can be conducted using the transactional currency of each

affiliate, the functional currency of each entity, or the group currency. Using the transactional currency may help preserve consistency with the invoicing currencies, but it can expose the organization to foreign exchange risks. Opting for each entity's functional currency can reduce FX exposure, as it aligns with the local currency in which the entity operates, mitigating the impact of currency fluctuations. Alternatively, using the group currency for netting provides a consolidated view and simplifies accounting, but it may result in more complex FX conversions. In addition, the approach for handling FX payments must be addressed, especially in cases in which an entity does not have a bank account in the required currency. This may involve converting funds to the necessary currency or using a centralized treasury or netting center to manage the currency exchange. Determining the right currency approach is vital for reducing FX risk and ensuring efficient and cost-effective netting operations across the organization.

- **Timing**
  The choice of timing for intercompany netting is a crucial factor that significantly impacts the organization's operations and payment processing. Netting can be run at different intervals, such as quarterly, monthly, or even more frequently, depending on the needs of the organization. Running netting less frequently, like on a quarterly basis, may reduce the administrative burden but can lead to larger, less manageable transactions and potentially more significant cash flow fluctuations. On the other hand, conducting netting on a monthly or more frequent basis provides more regular cash flow management and quicker reconciliation of intercompany balances, but it also requires more operational effort, coordination, and system resources to process transactions efficiently. The timing decision should align with the organization's financial strategy, liquidity management needs, and the complexity of its intercompany relationships. Frequent netting offers tighter control over cash flow, while less frequent netting may reduce operational overhead but could increase exposure to currency and liquidity risks.

Figure 1.16 demonstrates the netting process, which runs as follows:

❶ **Select invoices to pay**
The first step in the netting process is to identify the invoices that need to be paid. This involves reviewing all outstanding invoices to determine which ones are due for payment and should be included in the current netting cycle.

❷ **Netting cutoff**
For the netting process to be effective, all included invoices must be submitted before the established netting cut-off time. Once the cut-off is reached, any newly submitted invoices will be excluded from the current round and will instead be carried over to the next netting cycle.

❸ **Determine exchange rates**
The exchange rate used for netting and cross-currency calculations must be carefully

determined to ensure accurate financial settlements. Typically, the exchange rate applied in the netting process may include a spread compared to the standard bank rate. This spread represents the netting center's gain and acts as a markup for the services they provide, including facilitating the currency conversion and managing the associated risks.

❹ **Execute netting**
Once all invoices are submitted, the necessary rules are set up, and the exchange rate is determined, the netting process can be executed. During this stage, the netting center consolidates the financial data, considering all payables, receivables, and any applicable exchange rates. As a result, each participant will receive the final amount they are either required to pay or entitled to receive, either to or from the netting center. This streamlined settlement ensures that all participants fulfill their obligations while optimizing cash flow and reducing the need for multiple individual transactions, thus simplifying the overall financial settlement process.

❺ **Execute payments**
Once the netting results are received, the next step is to execute the payments through physical transfers. This involves transferring the netted amounts between the participants and the netting center based on the calculated obligations. The payments are typically made via bank transfers, ensuring that the correct amounts are settled in the appropriate currencies.

❻ **Run reconciliation**
The final step in the netting process involves comprehensive reconciliations to ensure that all intercompany invoices are correctly cleared. This step is critical not only for the invoices that participated in the netting cycle but also for those from other affiliates, whether they are customer or vendor invoices, depending on the triggering side. In addition to verifying the settlement of amounts, proper cross-currency clearings must be performed using the agreed-upon exchange rates to ensure accuracy in currency conversions. These reconciliations verify that all outstanding amounts, both within the netting system and outside of it, are properly accounted for and cleared. By performing these checks, businesses can confirm that no discrepancies remain, and all financial transactions are accurately settled across all entities involved, ensuring the integrity of the entire intercompany reconciliation process.

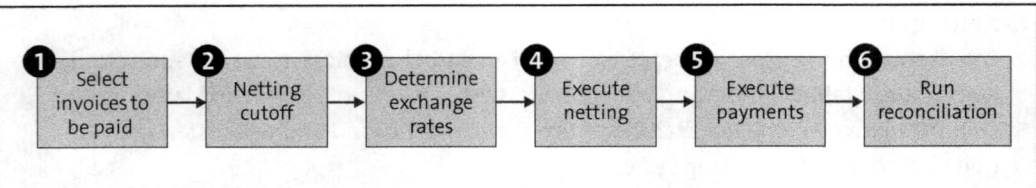

Figure 1.16 Netting Process

Cashless Settlements

Intercompany payments are typically supported by SAP In-House Cash or in-house banking and are processed as cashless settlements. This means that each intercompany payment is executed within the system through a standard payment run. However, instead of generating physical payments, the amounts are posted to the intercompany virtual accounts managed in SAP In-House Cash or in-house banking. This approach allows for the seamless transfer of funds between entities without the need for actual money transfers, streamlining the reconciliation process and improving the efficiency of intercompany settlements. By using virtual accounts, businesses can simplify financial flows while maintaining accurate records within the system.

As a result, all transactions processed as intercompany payments are posted to the SAP In-House Cash or in-house banking accounts, where they are stored as balances. These balances represent the outstanding amounts among the entities involved in the intercompany settlement. Over time, interest is calculated on these balances, which reflects the financial implications of holding or owing funds within the system. The interest is typically determined based on pre-agreed-upon rates and serves to compensate for the time value of money.

Thanks to this process, no physical transfers are needed, allowing for a more streamlined and efficient way of managing intercompany transactions. This provides better control over intercompany finances by centralizing and automating the settlement process, ensuring that all intercompany payments and balances are accurately recorded in the system. This approach also results in significant time savings, as the entire process is automated, reducing manual intervention. You only need to review the results in the reports, which simplifies your role and minimizes the risk of errors.

Technically, the intercompany payments process differs slightly between SAP In-House Cash and in-house banking. The technical differences between the two were described earlier in Section 1.2.5.

Figure 1.17 shows how intercompany payment is executed in SAP In-House Cash. It shows a payable-driven process flow, which generates a PAYEXT IDoc based on the vendor invoice payment. For a receivable-driven process, there is a DIRDEB IDoc generated.

Figure 1.18 shows what the intercompany payments process flows look like for in-house banking.

1 Payments Within a Corporation

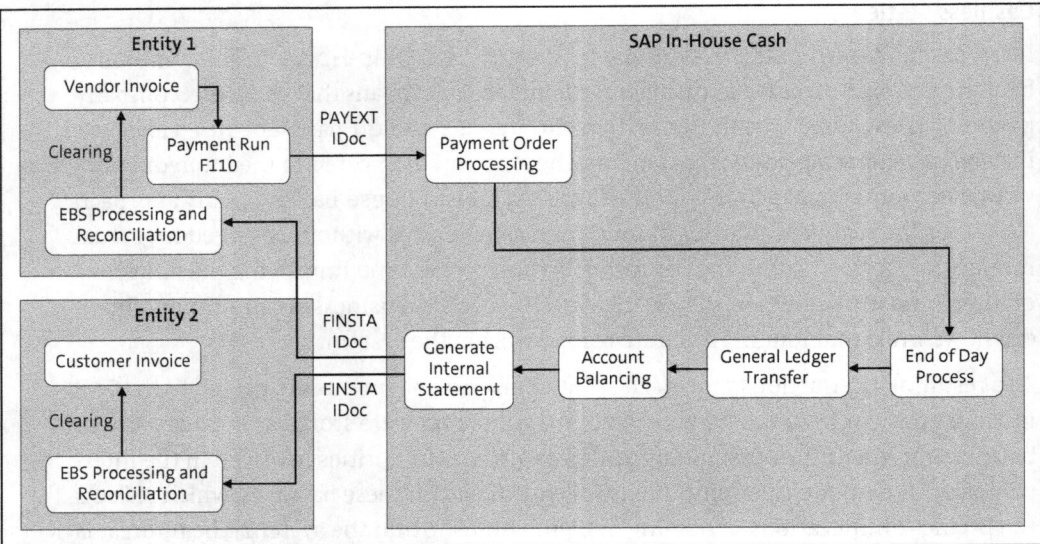

Figure 1.17 Intercompany Payment Using SAP In-House Cash

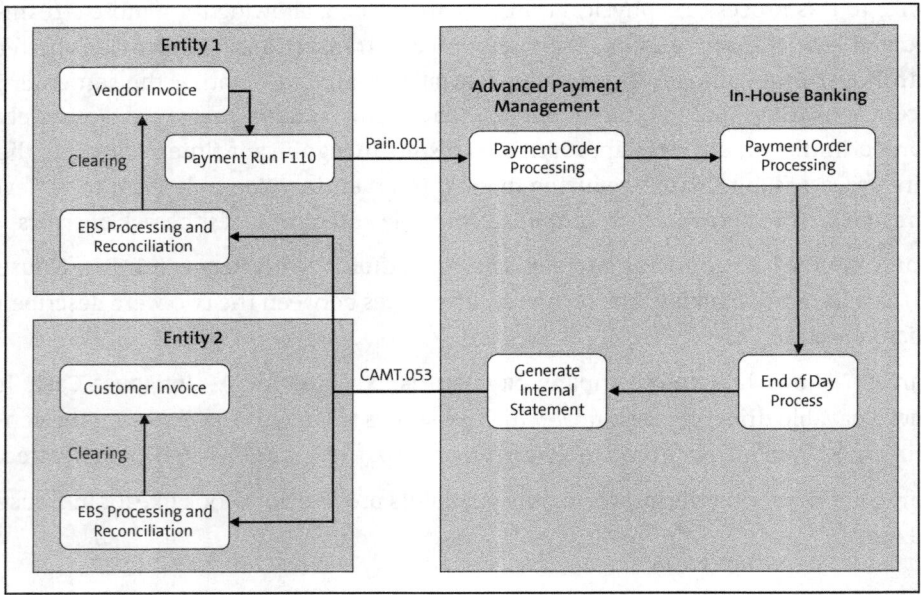

Figure 1.18 Intercompany Payment Using In-House Banking

From a business perspective, the processes in both SAP In-House Cash and in-house banking are essentially the same, with identical key inputs and outputs involved in managing payments and cash flows. However, the main differences arise from a technical standpoint, as highlighted earlier in Table 1.1. SAP In-House Cash utilizes the IDoc format for communication and data exchange, while in-house banking adopts newer

XML formats for these processes. In addition, in-house banking requires the integration of advanced payment management to process payments, introducing an extra layer of functionality compared to SAP In-House Cash. These technical distinctions allow in-house banking to support more advanced payment features, but the core business processes remain similar across both solutions.

1.3.4 Payments on Behalf Of

Payments on behalf of is an advanced and highly efficient payment process in which a central entity executes payments on behalf of its affiliates. Instead of payments being made by the individual affiliate's accounts, all payments are processed through the bank accounts of the central entity. This POBO approach centralizes the payment system, streamlining cash management and simplifying the banking structure.

Once the central entity executes a payment on behalf of an affiliate, a posting is made in the SAP In-House Cash or in-house banking subledger to reflect the intercompany payable between the affiliate and the central (header) entity. This posting ensures that the transaction is accurately recorded in the financial systems, establishing a liability from the affiliate to the central entity. The subledger entry helps track the intercompany balance, ensuring that the payment is properly allocated and that both entities' financial records remain aligned.

Before implementing the POBO process, it is crucial to conduct a thorough legal analysis to determine if the procedure is legally allowed in the countries or currencies involved. Different jurisdictions may have specific regulations governing intercompany transactions, payment processing, or the management of funds on behalf of others. In addition, certain currencies may have restrictions or regulatory requirements that must be considered before centralizing payments. Performing a legal analysis helps identify any potential limitations, such as local banking laws, tax implications, or compliance obligations, that could impact the implementation of POBO.

The main benefits of using POBO are as follows:

- **Reduced number of bank accounts**
  When POBO is implemented, the central entity opens bank accounts in each necessary currency, and payments are executed directly from these accounts. This eliminates the need for affiliates to maintain individual accounts in those currencies. Instead, the central entity opens one bank account for each currency they operate in.

  Typically, affiliates are required to maintain at least one bank account in their functional currency for legal reasons, often related to solvency requirements. However, with the implementation of POBO and in-house banking, it is sometimes possible for entities to close all their other bank accounts if the central entity handles all payments and collections for the affiliate. This arrangement can lead to significant cost savings, as the central entity manages the payables and collections transactions and eliminates the need for each affiliate to maintain multiple bank accounts. By consolidating banking activities into fewer accounts, affiliates can reduce bank fees, minimize administrative overhead, and streamline their financial operations.

- **Reduced FX costs**
Significant cost savings are achieved, particularly for FX payments. In a traditional setup, affiliates would incur costs related to currency conversions, including FX translation fees. However, with POBO, as payments are made from the central entity's multicurrency accounts, there is no need to pay for FX translations, streamlining the process and reducing the associated costs.

- **Cash management**
Once the POBO process is in place and as many external payments as possible are routed through a single central paying account, managing cash becomes significantly easier. With only one central account to monitor, liquidity management becomes more streamlined as there is no need to track multiple bank accounts across various affiliates and currencies. This centralized approach allows for a clearer overview of the organization's cash position, making it simpler to assess available liquidity and plan for future cash needs. In addition, with all payments being processed through the same account, forecasting cash flow, optimizing working capital, and ensuring sufficient funds are readily available for transactions become much more efficient tasks.

- **Counterparty risk management**
When implementing POBO, the number of bank accounts is significantly reduced, as all payments are funneled through a central entity's accounts. This reduction in accounts leads to a more focused banking setup, requiring fewer banking partners to manage. As a result, counterparty management becomes easier and more streamlined. With fewer banks involved, it is simpler to maintain relationships, negotiate better terms, and manage the overall banking network.

However, there is a trade-off: Relying on just one banking partner increases the risk of dependency. Should any issues arise with the primary banking partner, such as operational disruptions or unfavorable terms, the organization could face challenges. On the other hand, once the risk of dependency becomes too significant, it is easier to migrate to another partner due to the centralized nature of the POBO setup, thus allowing for the flexibility to adapt and manage risks more effectively.

- **Cross-border payments**
When planning the implementation of POBO, one key consideration to optimize the process further is the potential opening of local bank accounts in the countries where your vendors are or where your currency is used locally. This strategy allows payments to be made domestically rather than relying on cross-border transactions, offering several advantages in terms of cost, efficiency, and speed. When payments are made domestically, an organization can benefit from lower transaction costs as most domestic transfers are less expensive than international ones.

Figure 1.19 shows how the POBO process flow is executed in SAP S/4HANA Finance using SAP In-House Cash, which proceeds as follows:

❶ **Post vendor invoice**
The prerequisite for running the payment run is posting the vendor invoice.

❷ **Execute payment run**
This is done in Transaction F110. During the payment run, the vendor invoice is paid, and a special IDoc in PAYEXT format is generated as the payment medium. The affiliate's SAP In-House Cash account is used as the paying account. Depending on the configuration, payment advice can also be generated to inform the vendor about the executed payment.

❸ **IDoc generates posting in the SAP In-House Cash subledger**
The IDoc generates a posting in the SAP In-House Cash subledger and posts the payment order to both the affiliate and header entity accounts.

❹ **Payment request generation**
During posting, a payment request is generated, containing the central entity's bank account as the paying account, compared to the SAP In-House Cash account number belonging to the affiliate.

❺ **Payment request paid via Transaction F111**
The payment request is processed using Transaction F111, thus generating the payment file. It is highly recommended to use SAP Bank Communication Management to batch the payment first and go through the payment approval process. The system also can process acknowledgment files and update the payments monitor.

❻ **Payment file generation**
The payment file is generated based on the executed payment run and sent to the bank for payment execution.

❼ **End-of-day process in subledger**
Every day, an end-of-day process is executed in the subledger. During this process, account balancing is performed, G/L transfers (intercompany payables and receivables) are posted, and an internal bank statement in FINSTA format is generated.

❽ **Account balancing**
Interest, closing balance, and bank charges are calculated.

❾ **General ledger transfer**
In this step, intercompany payables and receivables are posted in the balance sheet of the header entity.

❿ **Generate internal bank statement**
An internal bank statement is generated for the selected SAP In-House Cash bank accounts.

1 Payments Within a Corporation

⓫ **FINSTA posting**
The FINSTA statement is sent and received in SAP S/4HANA Finance and is used for clearing the vendor payment run.

⓬ **External bank statement**
For external payments, an external bank statement is received and should be cleared in the Transaction F111 payment run.

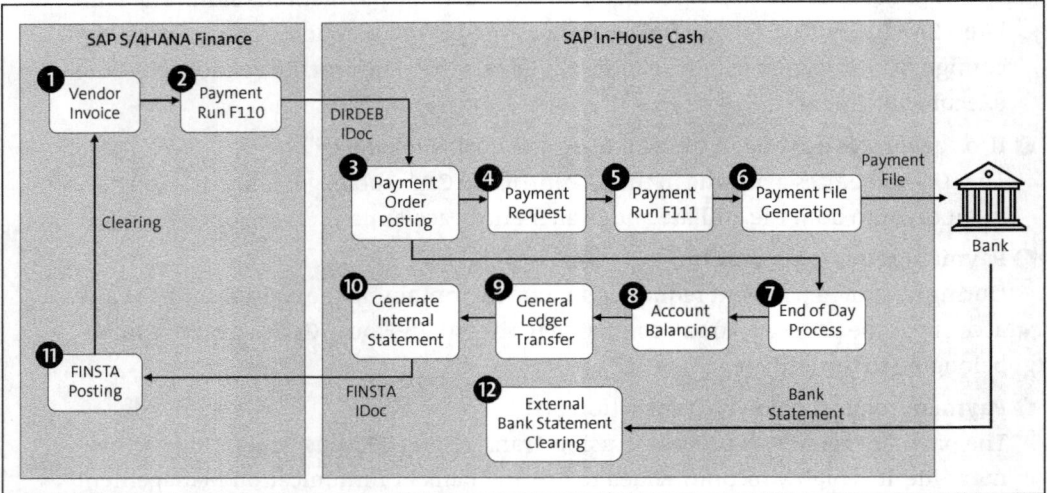

Figure 1.19 POBO Process Flow Using SAP In-House Cash

Functionally, the POBO process flow for in-house banking (see Figure 1.20) achieves the same results, but there are some technical differences in how it is executed:

❶ **Post vendor invoice**
The prerequisite for running the payment run is posting the vendor invoice.

❷ **Execute payment run**
This is done in Transaction F110. During the payment run, the vendor invoice is paid, and a pain.001 format file is generated as the payment medium (exactly like for all external bank payments). The affiliate's in-house banking account is used as the paying account. Depending on the configuration, payment advice can also be generated to inform the vendor about the executed payment.

❸ **Advanced payment management generates incoming payment order**
An incoming payment order is created in advanced payment management. The posting in the in-house banking subledger is completed.

❹ **Outgoing payment order is generated**
Advanced payment management creates an outgoing payment order containing the relevant payment information, including the header bank account details for the paying bank account.

❺ Payment file generation
The payment file is generated based on the executed payment run and sent to the bank for payment execution. Advanced payment management also generates the pain.002 file to inform local subsidiaries that the payment was executed or rejected.

❻ End-of-day process in subledger
Every day, an end-of-day process is executed in the subledger. During this process, account balancing is performed, G/L transfers (intercompany payables and receivables) are posted, and an internal bank statement in camt.053 format is generated.

❼ General ledger transfer
In this step, intercompany payables and receivables are posted in the balance sheet of the header entity.

❽ Generate internal bank statement
An internal bank statement is generated for the selected in-house banking bank accounts.

❾ Internal bank statement posting
The camt.053 statement is sent and received in SAP S/4HANA Finance and is used for clearing the vendor payment run.

❿ External bank statement
For external payments, an external bank statement is received and creates postings in the header's entity balance sheet.

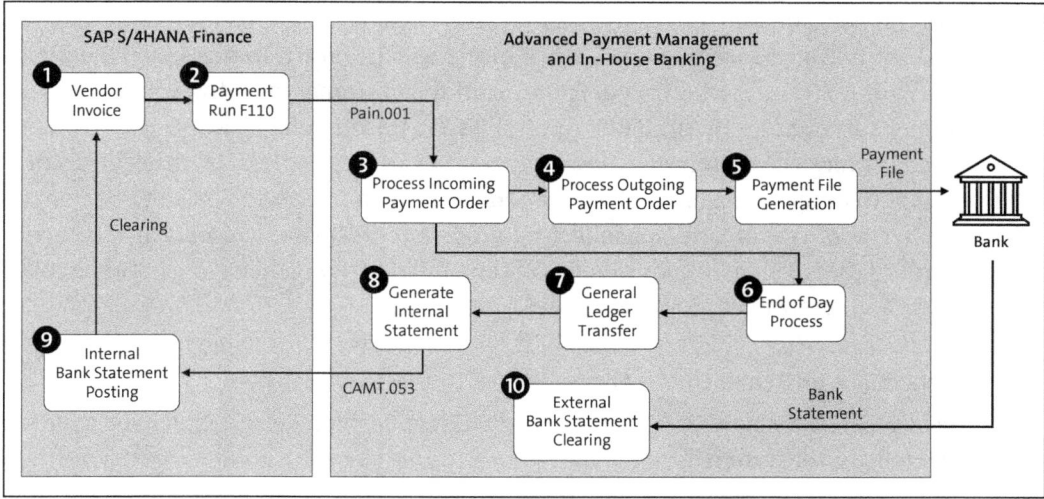

Figure 1.20 POBO Process Flow Using In-House Banking

1.3.5 Collections/Receivables on Behalf Of

Collections on behalf of (COBO), sometimes referred to as *receivables on behalf of* (ROBO), is a payment process in which the central entity collects customer payments on behalf of its affiliates. Instead of each affiliate handling its own customer payments, the central entity manages the entire collections process, consolidating payments into its own bank accounts. This approach simplifies cash management, reduces administrative overhead, and ensures that all payments are centralized, making it easier to track and reconcile accounts. By handling collections centrally, the organization can also optimize liquidity management as all customer payments are pooled into one account. This pooled cash allows for more efficient management of funds, making it easier to allocate resources, plan investments, and optimize returns (interests). COBO/ROBO also enables the central entity to potentially negotiate better banking terms and improve the efficiency of cash flow across the business.

It is only when both COBO (for receivables) and POBO (for payables) are fully implemented—covering both the payables and receivables sides—that an organization can truly conclude that affiliates no longer need their own separate bank accounts. With the central entity managing both payments and collections, affiliates can rely on the central system to handle all financial transactions, thus eliminating the need for individual accounts in each affiliate. This centralization further enhances liquidity management, as the funds are already pooled into one account, making it easier to optimize investments and cash flow on a global scale.

For this book, we will examine COBO as a direct debit payment process in which the central entity triggers payments on behalf of the affiliate entity. In this setup, the central entity initiates the collection of funds from the customer, directly debiting the customer's account for the affiliate's receivables. On the other hand, ROBO refers to a process in which the customer makes a payment to an account held by the central entity rather than the affiliate. In ROBO, the central entity manages the receipt of payments on behalf of the affiliate, consolidating funds into its own accounts. This distinction clarifies the roles and responsibilities of the central entity in managing both the collections and receivables processes.

Collections on Behalf Of

Figure 1.21 shows how COBO works (central direct debit on behalf of) in the classic SAP In-House Cash system.

The COBO process flow is very similar to the one used in POBO, and it has the following steps:

❶ **Post customer invoice**
   The prerequisite for running the payment run is posting the customer invoice.

❷ **Execute payment run**
   This is done in Transaction F110. During the payment run, the vendor invoice is paid,

and a special IDoc in DIRDEB format is generated as the payment medium. The affiliate's SAP In-House Cash account is used as the paying account. Depending on the configuration, payment advice can also be generated to inform the vendor about the executed payment.

❸ **IDoc generates posting in the SAP In-House Cash subledger**
The IDoc generates a posting in the SAP In-House Cash subledger and posts the payment order to both the affiliate and header entity accounts.

❹ **Payment request generation**
During posting, a payment request is generated, containing the central entity's bank account as the paying account, compared to the SAP In-House Cash account number belonging to the affiliate.

❺ **Payment request paid via Transaction F111**
The payment request is processed using Transaction F111, generating the payment file. It is highly recommended to use SAP Bank Communication Management to batch the payment first and go through the payment approval process. The system can also process acknowledgment files and update the payments monitor.

❻ **Payment file generation**
The payment file is generated based on the executed payment run and sent to the bank for payments execution.

❼ **End-of-day process in subledger**
Every day, an end-of-day process is executed in the subledger. During this process, account balancing is performed, G/L transfers (intercompany payables and receivables) are posted, and an internal bank statement in FINSTA format is generated.

❽ **Account balancing**
Interest, closing balance, and bank charges are calculated.

❾ **General ledger transfer**
In this step, intercompany payables and receivables are posted in the balance sheet of the header entity.

❿ **Generate internal bank statement**
The internal bank statement is generated for the selected SAP In-House Cash bank accounts.

⓫ **FINSTA posting**
The FINSTA statement is sent and received in SAP S/4HANA Finance and is used for clearing the customer payment run.

⓬ **External bank statement**
For external payments, an external bank statement is received and should be cleared in the Transaction F111 payment run.

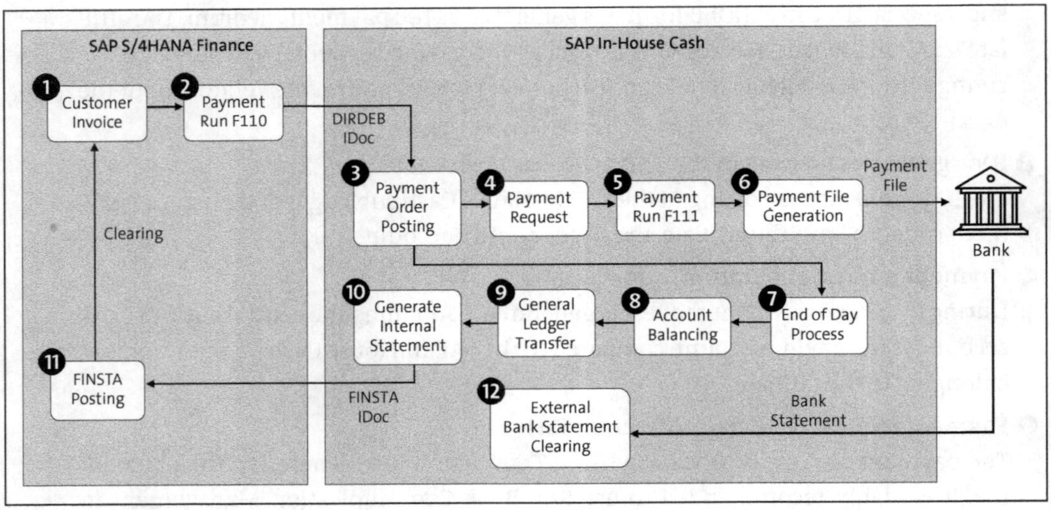

1 Payments Within a Corporation

Figure 1.21 COBO Process Flow Using SAP In-House Cash

Figure 1.22 shows how the COBO process flow works using the newer in-house banking.

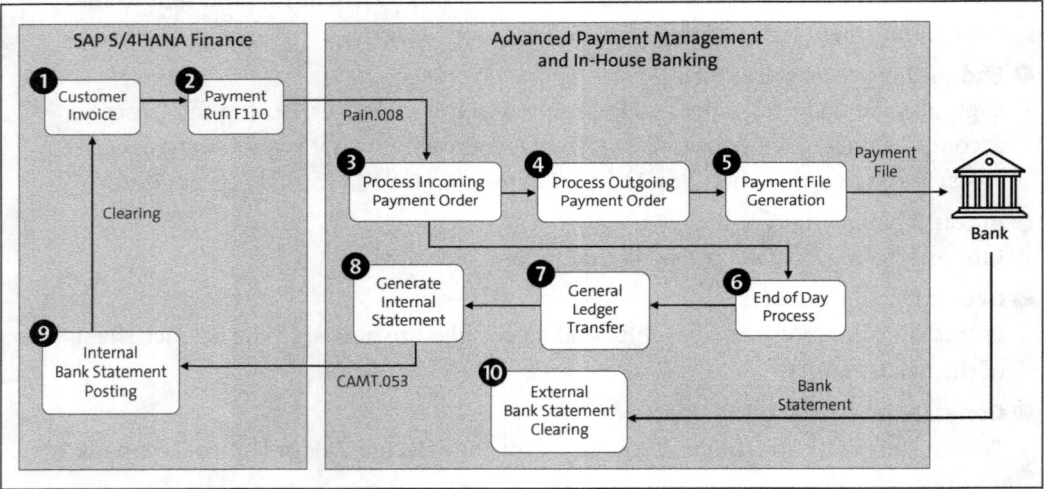

Figure 1.22 COBO Process Flow Using In-House Banking

The process flow is very similar to the one used in POBO, with the differences that an XML file is generated in pain.008 format, and you collect money from your customers.

Receivables on Behalf Of

ROBO is a specific process flow in which customers make payments directly to a central bank account held by the central entity. The prerequisite for this process is a special agreement with the bank to generate a virtual International Bank Account Number

(IBAN) for each affiliate. These virtual IBANs are unique identifiers assigned to each affiliate but linked to the central entity's main bank account. The virtual IBANs are then provided to the affiliate's customers, who use them to make payments. When a company makes a payment, the bank recognizes which virtual IBAN was used and automatically transfers the funds to the central entity's physical bank account.

Once the payment is made, the bank generates a statement and sends it to the central entity. Each receivable on the statement will contain the corresponding virtual IBAN used by the customer, which is crucial for reconciliation. The virtual IBAN is referenced to ensure that the correct affiliate's receivables are matched to the funds received. This information is then used to properly post the transaction in the SAP In-House Cash or in-house banking subledger, where it is recorded as an intercompany payable or receivable. This process enables efficient tracking, reconciliation, and posting of intercompany transactions, ensuring that all payments are correctly allocated and thus improving overall cash flow management.

Figure 1.23 shows how ROBO is executed in the classic SAP In-House Cash setup.

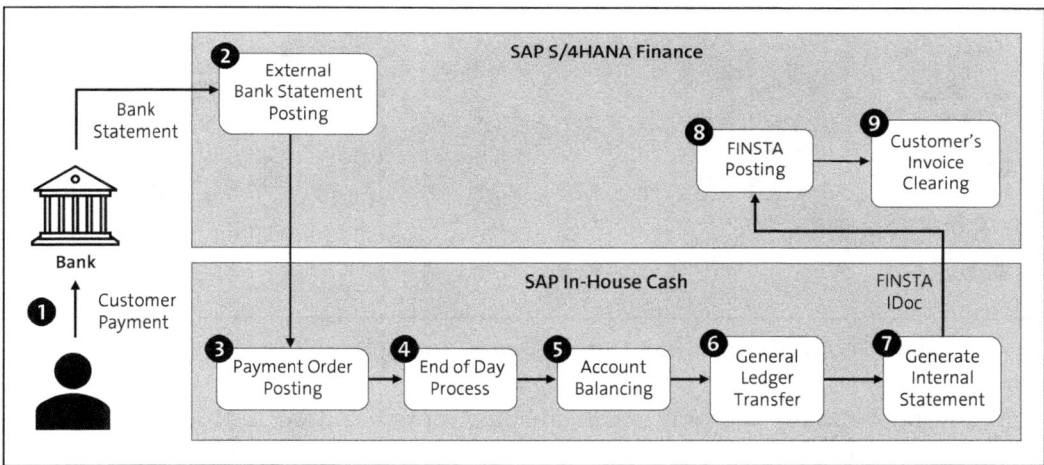

Figure 1.23 ROBO Process Flow Using SAP In-House Cash

The ROBO process flow differs from COBO and POBO as it is triggered by the customer's payment. It proceeds as follows:

❶ The customer first executes the payment for their invoice, paying via the designated virtual IBAN. The bank processes the payment and deposits the funds into the central bank account belonging to the header entity.

❷ The bank generates the bank statement and sends it to the central entity.

❸ The bank statement is posted in the financial accounting system, updating the central entity's general ledger. Based on the provided virtual IBAN, SAP In-House Cash creates a posting in the SAP In-House Cash subledger, recording the payment in the accounts of both the central entity and the affiliate.

1 Payments Within a Corporation

❹ Every day, an end-of-day process is executed in the subledger. During this process, account balancing is performed, G/L transfers (intercompany payables and receivables) are posted, and an internal bank statement in FINSTA format is generated.

❺ Account balancing is performed.

❻ The G/L transfer is executed.

❼ The internal bank statement is generated.

❽ The FINSTA statement is sent and received in SAP S/4HANA Finance and is used to clear the payment in the central entity's subledger.

❾ The customer's invoice is cleared in the affiliate's subledger.

Figure 1.24 shows how ROBO is executed in the newer in-house banking solution.

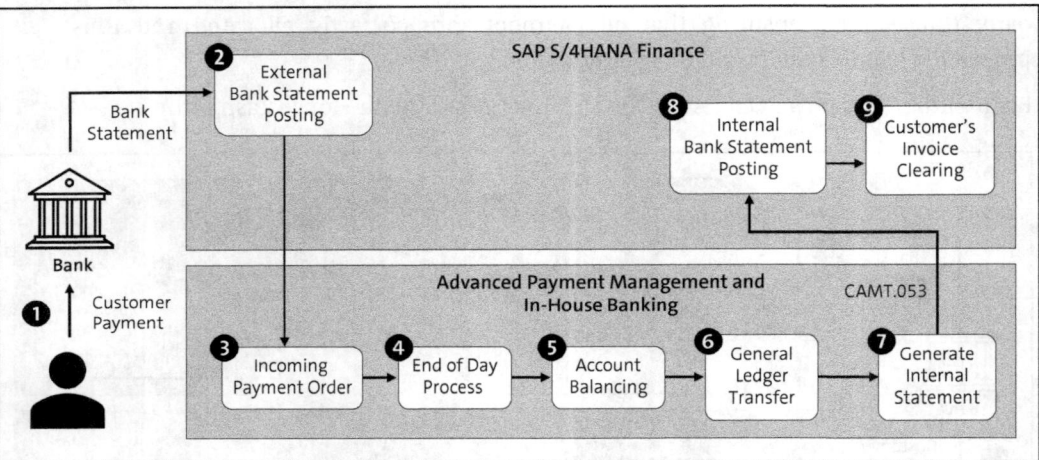

Figure 1.24 ROBO Process Flow Using In-House Banking

ROBO works the same way in both in-house banking and SAP In-House Cash; the only differences are in the statement format and the choice of solution.

1.3.6 Cash Pooling

Cash pooling is a process used to optimize liquidity management within a group of companies, where affiliates' accounts are typically zero balance accounts (ZBAs). In this setup, the affiliates' accounts are automatically funded from a central bank account, allowing for the centralized management of cash. The central entity transfers funds into the affiliate accounts as needed, ensuring that each affiliate has the liquidity required to operate, while excess cash from affiliates is pooled back into the central account. This approach eliminates the need to manage cash management and liquidity forecasting for all the individual affiliate accounts as these tasks are centralized and can be handled solely for the central bank account.

Figure 1.25 shows how cash pooling works in an organization. Some specific details you should be aware of are as follows:

- USD and EUR cash pools are in place, both of which are set up as zero balance cash pools. This means that excess cash or negative balance is transferred by the bank, ensuring you have a balance of zero.
- Both cash pools fall under one central legal entity.

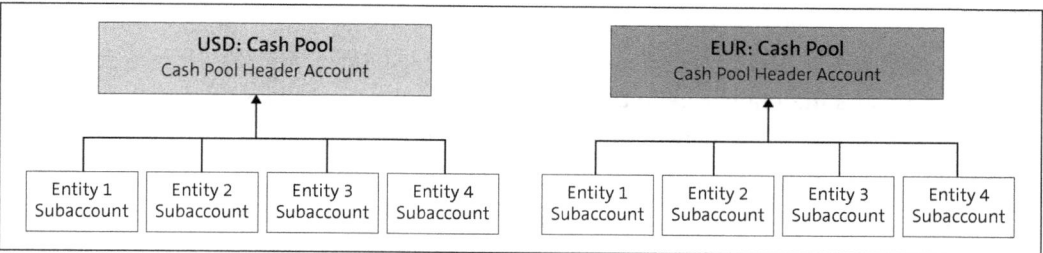

Figure 1.25 Cash Pooling Hierarchy Example

This process acts as an intermediary solution for managing cash flow before more advanced processes like POBO and COBO are fully implemented. Once both POBO and COBO are in place, cash pooling becomes redundant, as these processes centralize both payables and receivables, eliminating the need for separate cash management at the affiliate level.

This process refers to *physical cash pooling*, where the bank physically sweeps funds between the central account and the affiliate accounts daily. This means that any excess cash in the affiliate accounts is transferred to the central entity's account, while funds are transferred from the central account to the affiliates' accounts when liquidity is needed. Performing this daily sweep helps ensure that liquidity is effectively managed across the group, and all cash is centralized.

An alternative to physical cash pooling is *notional cash pooling*, which does not involve actual movement of funds between accounts. Instead, notional cash pooling aggregates the balances of multiple accounts at the end of each day for the purpose of calculating interest or managing liquidity, without physically transferring the funds. In this model, the affiliates retain control over their individual accounts, and no money is physically swept into the central account. Instead, a notional balance is used for internal purposes, allowing the group to benefit from consolidated liquidity management without the need for cash transfers. Some banks even offer notional cash pooling at the cross-currency level, enabling organizations to consolidate liquidity and manage interest across accounts in different currencies. This feature allows companies to optimize liquidity management on a global scale, improving financial flexibility while avoiding the need for FX transactions.

However, it's important to note that notional cash pooling is quite limited in many jurisdictions due to regulatory restrictions. Some countries or regions may have rules that restrict or prohibit notional pooling arrangements, particularly when dealing with cross-border or cross-currency transactions. As a result, companies must carefully evaluate local regulations before implementing this method of liquidity management.

Furthermore, standard SAP S/4HANA does not have a clear solution for notional cash pooling, and organizations often need to implement workarounds using cash management, SAP In-House Cash, or in-house banking. These workarounds are required to simulate the notional pooling functionality in SAP S/4HANA, making it more complex to set up and maintain. Due to these complexities, this book does not focus on notional cash pooling; it is outside the scope of standard SAP S/4HANA functionality and requires significant customization.

Figure 1.26 shows how physical cash pooling is executed in the standard SAP In-House Cash setup. The cash pooling process resembles the ROBO payment process from a technical standpoint. Functionally, it is completely different as payment is initiated by the bank to move funds between cash pool accounts.

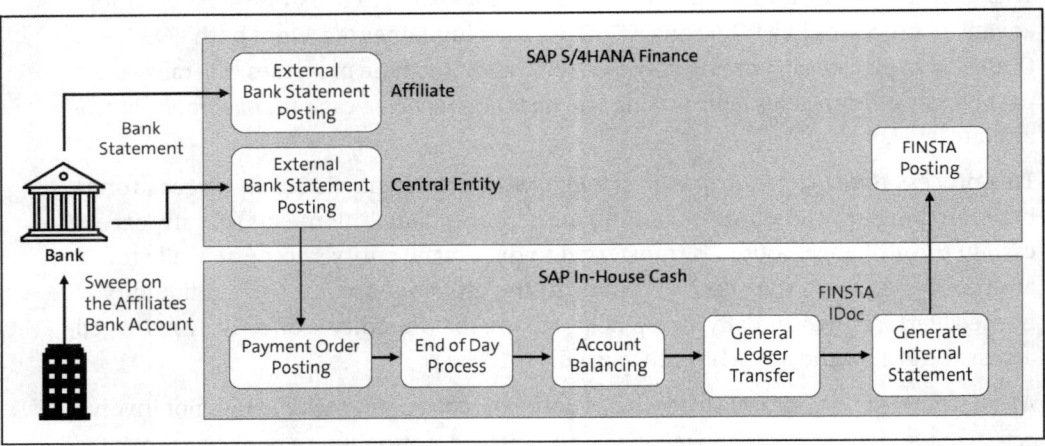

Figure 1.26 Cash Pooling Using SAP In-House Cash

Figure 1.27 shows how cash pooling is executed in the newer in-house banking setup:

1. The bank executes a sweep on your behalf, either funding the negative balance on the account or sweeping excess cash. Then it generates a bank statement and sends it to both the affiliate and the header entity.

2. The bank generates the bank statement and sends it to the central entity.

3. The header entity receives the statement and, based on the sweep (with the system looking for an identifier specified in the configuration—usually a virtual bank account number), it creates an SAP In-House Cash payment order that represents the update in the virtual bank account along with the corresponding receivables and payables.

1.3 Payments Processing

4. Every day, an end-of-day process is executed in the subledger. During this process, account balancing is performed, G/L transfers (intercompany payables and receivables) are posted, and an internal bank statement in FINSTA format is generated.
5. Account balancing is performed.
6. The G/L transfer is executed.
7. The internal bank statement is generated.
8. The FINSTA statement is sent and received in SAP S/4HANA Finance in both the affiliate and central entity and is used to update financial accounting.

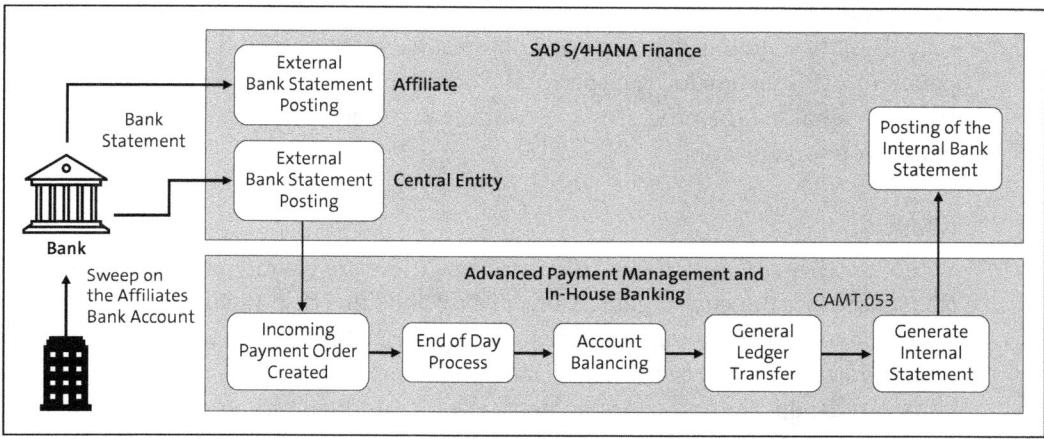

Figure 1.27 Cash Pooling Using In-House Banking

Processing cash pooling via in-house banking is very similar to how it's done in SAP In-House Cash. The key difference is that the incoming payment order is first created in advanced payment management, and the corresponding postings are executed in the in-house bank accounts. These transactions are then processed through end-of-day processing. In addition, instead of using a FINSTA IDoc, there is either a camt.053 or MT940 statement generated, depending on your system setup.

Central Incoming Payments
In in-house banking, the solution for handling the cash pooling and ROBO processes is called *central incoming payments*.

1.3.7 Treasury Payments

In addition to the various vendor and customer payments mentioned earlier, there are multiple other payments that can be made by the treasury department. These payments may include internal transfers between company accounts, debt servicing, interest payments, tax-related payments, or payments to financial institutions for services rendered. The treasury department is responsible for managing these payments, ensuring that the company's liquidity needs are met while maintaining control over cash flow. The treasury also monitors and manages the overall cash position, ensuring that payments are made efficiently and in line with the company's financial strategy. This centralized approach helps to streamline financial operations and improve the company's overall financial management.

These payments can generally be divided into internal and external payments. *Internal payments* involve transactions within the organization, such as transfers between company accounts, funding affiliates, or settling intercompany balances. *External payments*, on the other hand, refer to payments made to third parties outside of an organization, including debt servicing, interest payments, tax obligations, and vendor payments. This section focuses specifically on these types of payments executed by the treasury, detailing the processes and controls involved in managing them effectively within your organization.

Internal

When it comes to internal treasury payments, these are payments executed between affiliates within the same organization or into a bond that belongs to your organization. These transactions typically represent internal transfers of funds that help manage liquidity and ensure that the organization's financial obligations are met within the corporate group. The following payment types are internal treasury payments:

- **Bank-to-bank transfers**
  The treasury team is responsible for managing liquidity across an organization's bank accounts. Their primary task is to ensure that sufficient funds are available in each account to meet the company's daily financial needs. To achieve this, they execute bank transfers to move funds between accounts, optimizing liquidity and ensuring that there is enough cash available in the right accounts at the right time. This includes transferring excess funds from operational accounts to centralized accounts, funding affiliates, or ensuring that accounts with specific obligations (such as debt servicing or tax payments) are adequately funded.

  These bank transfers are executed directly in SAP S/4HANA, where payment files are generated and sent to the bank for processing. The SAP S/4HANA system ensures that the transfers are recorded in real time and updates the relevant financial data across the company's accounts.

- **Cash concentration**
  This is a functionality that allows organizations to set up and manage cash pools, enabling the centralization of liquidity across various accounts. With this feature, you can create cash pools and trigger them manually, ensuring efficient management of funds across different bank accounts. Cash concentration in SAP S/4HANA facilitates cross-bank transfers, enabling the movement of funds between accounts held at different banks. In addition, you can define target balances for each pool,

which helps to optimize liquidity across the organization. The system can propose and execute sweeps based on these target balances, automatically transferring funds between accounts to maintain the desired liquidity levels. This functionality enhances cash management by automating cash pooling processes and improving cash visibility.

- **Intercompany financial transactions**
  Intercompany financial instruments, such as interest payments, FX transactions, and loan repayments, can also trigger treasury payments. These transactions often arise from intercompany agreements and require careful management to ensure accurate cash flow and risk management data. Ideally, these financial instruments should be captured and managed within SAP Treasury and Risk Management, which allows for effective tracking and management of liquidity and associated risks. When a payment is due for these intercompany financial instruments, the system generates a payment request, which is then processed through the payment run using Transaction F111. This process ensures that the necessary payments are executed efficiently, and the corresponding financial transactions are reflected in the relevant accounts.

  There are two ways to settle intercompany financial instruments: cashless settlement or external settlement. In *cashless settlement*, payments are handled within the system by posting them to the SAP In-House Cash subledger, which creates corresponding intercompany payables or receivables between the entities. This process does not involve physical cash movement and is ideal for managing internal cash flow and balances. Alternatively, the intercompany financial instruments can be settled *externally* (exactly as it is done with external counterparties), meaning that a payment file is generated and the payment is routed to the bank for a physical transfer. This approach involves actual movement of funds and is used when cash settlement between entities is required. By providing both settlement options, SAP allows organizations to manage intercompany payments in the most efficient and flexible way according to their financial needs.

Figure 1.28 shows how intercompany FX transactions can be settled via SAP In-House Cash or the in-house banking subledger via the following steps:

❶ First, you capture the financial instrument in SAP Treasury and Risk Management. This example is showing an intercompany FX deal.

❷ You post the deal using Transaction TBB1 at maturity, when both legs of the deal are exchanged and posted in the G/L. A payment request is generated.

❸ You need to execute the payment run using Transaction TBB1 and process the payment request. Based on the master data setup for the counterparty, the system posts the payment order in the SAP In-House Cash subledger and posts both legs to the affiliate accounts, as in the intercompany payments process flow.

1 Payments Within a Corporation

❹ A payment order is created in either SAP In-House Cash or advanced payment management, and corresponding postings are handled in the subledger.

❺ End-of-day processing is executed as in all other cases.

❻ Account balancing is executed.

❼ The G/L transfer is executed.

❽ The internal bank statement is generated (FINSTA for SAP In-House Cash; camt.053 for in-house banking).

❾ The internal bank statement clears the postings generated by Transaction TBB1.

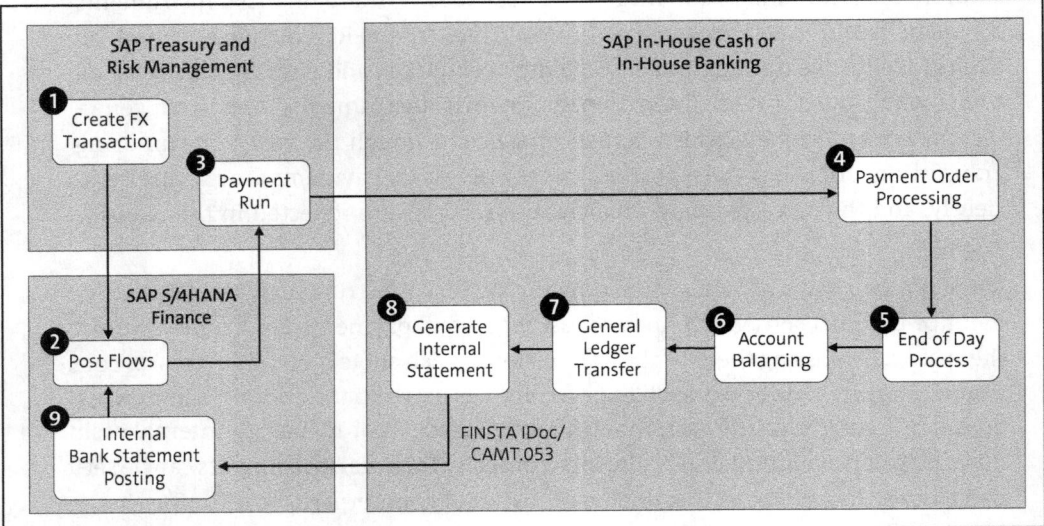

Figure 1.28 Settle Intercompany FX Transaction Using SAP In-House Cash or In-House Banking

External

Similarly, the treasury can also execute certain external payments, which are payments to bank accounts belonging to different counterparties outside of the organization. These include payments to suppliers, counterparties, banks, and others. You can classify these external treasury payments as follows:

- **Free-form payment**

  SAP S/4HANA offers functionality for executing free-form payments, which allows for manual payments when there is no invoice involved, but funds need to be transferred. These payments are typically used for nonstandard transactions, such as paying for the acquisition of a company or another special payment. To initiate a free-form payment, manually select the paying account and enter the partner's bank details, specifying the recipient and payment amount. Once these details are entered, the system generates a payment request, which then needs to be processed through the payment run using Transaction F111.

If the payment is expected to recur, it can be saved in the system as a template. This allows for quicker processing of similar payments in the future as the details (such as the paying account and partner bank information) can be reused without needing to be manually entered each time.

However, free-form payments should be executed only as exceptions. This is because they provide less security compared to regular automated payments. SAP S/4HANA is designed to run automatic payments with a high level of control and validation, which ensures accuracy and reduces the risk of fraud or errors. For free-form payments, you manually enter payment details, bypassing many of the built-in controls of the automated processes. Therefore, it is essential to implement additional controls and approval workflows for such payments to minimize risk and ensure that they are processed securely and accurately.

- **External financial transactions**
  Like intercompany financial transactions, external financial instruments such as FX transactions, money market instruments, derivatives, and other similar instruments can also be managed within SAP Treasury and Risk Management. These instruments, which are traded with external counterparties, often have associated cash flows that require payment on specific due dates. SAP Treasury and Risk Management allows organizations to track and manage these external financial transactions efficiently. When payment is due for any of these instruments, the system can generate a payment request and initiate the payment process. This payment request is then routed through the SAP S/4HANA system for approval and execution, after which it is sent to the bank for physical processing.

Figure 1.29 shows the process is of settling external financial instruments for external loan interest, which has the following steps:

1. First you conclude a contract with an external bank—in this case, an external bank loan.
2. You settle the interest payments based on the agreed-upon timeline.
3. When the due date is reached, the system posts the interest using Transaction TBB1 and generates a payment request.
4. The payment request needs to be paid using Transaction F111. If using SAP Bank Communication Management, the payment should be batched and sent through the payment approval process.
5. A payment file is generated and sent to the bank for execution.
6. Once the payment is made, the bank should include this payment in the next day's bank statement.
7. The statement is received and posted, and the clearing of the posting is handled via Transaction TBB1.

1 Payments Within a Corporation

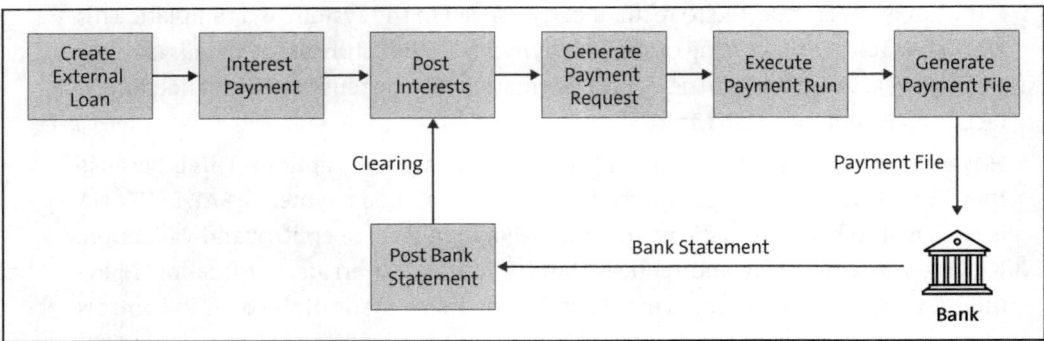

Figure 1.29 Settle External Transaction Using SAP TRM

1.3.8 Preprocessing and Postprocessing

In this section, we will explain the requirements for the master data necessary to execute payments as preprocesses. Master data plays a critical role in ensuring that payments are processed accurately and efficiently within SAP S/4HANA. This includes setting up key entities, such as vendor and customer accounts, bank details, payment methods, and counterparty information. Properly configuring master data ensures that the payment process runs smoothly, with the correct payment instructions being generated and routed to the appropriate bank accounts. We also will list some of the steps required as postprocesses for payments, such as clearing the payment postings, reconciling bank statements, and ensuring proper accounting in the G/L. By addressing both pre- and postprocesses, you can streamline the payment workflow, minimize errors, and ensure compliance with organizational and regulatory requirements.

Preprocesses

In SAP S/4HANA, several master data objects are crucial for the efficient processing of payment scenarios. These objects need to be created and maintained to ensure smooth transaction processing and compliance with financial regulations. The key master data objects include the following:

- **Bank accounts**
  Bank account data must be set up for both internal and external accounts. This includes details such as the bank's name, account number, IBAN, and country. For intercompany payments, it is essential to maintain accurate bank account information for affiliates to facilitate smooth transactions.

- **Business partners**
  Business partner master data refers to the entities involved in payment transactions, such as suppliers, customers, external and internal treasury counterparties, or internal entities. These records need to include key details like the business partner's

name, address, and contact information, as well as payment terms, banking details, and tax information. Proper maintenance of business partner data ensures accurate and efficient processing of payments.

- **Payment methods**
  Payment methods define how payments are made, whether via bank transfer, check, or other methods. This data must be properly configured to align with the company's payment practices and ensure accurate processing in line with external regulations and business needs.

- **Payment terms**
  Payment terms specify the conditions under which payments are due, including discount periods, due dates, and payment frequency. Proper maintenance of payment terms ensures that the organization can track when payments are due and manage cash flows effectively.

- **Payment program**
  The payment program is used to manage the execution of payments, selecting the invoices due and determining the payment method. This configuration ensures that the right invoices are paid according to the defined criteria.

- **Bank determination**
  Bank determination involves specifying which bank accounts should be used for various payment types and which banks are eligible for outgoing and incoming payments. It also includes prioritizing which bank account will be used in case of multiple options, and this data is necessary for processing payments effectively.

- **General ledger accounts**
  G/L accounts need to be set up to manage financial transactions related to payments, such as cash accounts and clearing accounts. Accurate G/L account configuration is crucial for financial reconciliation and reporting.

- **Bank chains**
  In SAP S/4HANA, *bank chains* is SAP's term for intermediary banks. Intermediary banks are used when direct payment connections between the sender and recipient's banks are not available, ensuring that payments can be routed through a network of correspondent banks for cross-border transactions. Proper configuration of bank chains and intermediary banks is essential for organizations with complex payment requirements or multiple bank accounts.

Each of these master data objects plays a vital role in ensuring accurate, compliant, and efficient processing of payments. Maintaining up-to-date and accurate data for all these objects is key to minimizing errors, optimizing cash flow, and streamlining payment processing within SAP S/4HANA. Maintenance and usage of these master data objects will be described in the following chapters.

Postprocesses

After the payment processes are completed in SAP S/4HANA, several important postprocesses need to be executed to ensure accurate financial tracking and management. These include the following:

- **Processing of bank statements**
  Bank statements need to be processed to update the system with the latest transactions, ensuring that all outgoing and incoming payments are reconciled with the actual bank data. This process helps maintain up-to-date cash balances and supports the reconciliation of financial records.

- **Intraday bank statements**
  Intraday bank statements are processed to capture real-time or daily updates on bank transactions, allowing for more timely monitoring of cash positions and quick identification of discrepancies. This is especially useful in businesses with high transaction volumes that need frequent updates.

- **Payment monitoring**
  This process involves monitoring the status of all payments to ensure that they have been executed successfully and within the expected timeframes. Payment monitoring helps detect any delays, failures, or discrepancies in payments, enabling quick corrective actions to be taken when necessary.

- **Collections of failed direct debits**
  When direct debit payments fail due to insufficient funds, incorrect account details, or technical issues, the collections process is initiated to either retry the payment or take other actions to recover the outstanding amount.

- **Payment rejections**
  Payment rejections occur when payments cannot be processed due to issues such as incorrect payment details or blocked accounts. This process ensures that rejected payments are properly identified and corrective actions are taken to resolve the issue, preventing further payment delays.

  Rejection of payments can occur at two stages in the payment process: before the payment file is sent to the bank during the payment approval process; or after the file has been submitted and the payment is rejected by the bank. Rejections during the approval process may happen due to incorrect payment details, such as missing or incorrect bank account numbers, invalid payment terms, or insufficient funds in the payer's account. Once a payment reaches the bank, it can be rejected due to reasons like issues with the recipient's account, mismatched payment instructions, or failure to meet regulatory requirements. When a payment is rejected, it is essential to quickly identify the cause of the rejection by reviewing the error messages or reports provided by the bank, such as the pain.002 message or payment rejection notices. After identifying the reason for the rejection, corrective actions need to be taken, such as correcting the payment details, updating the recipient's bank account information, or ensuring there are sufficient funds in the payer's account. Once the issue is resolved, the payment should be resubmitted, and proper procedures should be followed to ensure that all parties involved are notified and the payment is processed successfully.

- **Refunds**
  It is also possible that the paying bank might accept a payment and send a notification about it, but a rejection may then occur on the recipient's bank side. This can happen for various reasons, such as missing information on the intermediary bank side or the target bank account being closed. In such cases, the money is sent back and refunded, a process known as a *rebounded payment*. Typically, organizations learn about this situation through their bank statements, where the amount received does not match the original payment amount, as the recipient's bank may have deducted transfer fees. Once a payment is rebounded, it is essential to identify and properly allocate the refunded amount to the correct invoice, which can be challenging. It may require extra effort to correct these errors, such as obtaining the correct vendor bank account information, especially if the issue was due to incorrect or outdated details. Once the correct information is secured, the payment must be reprocessed, which can be time-consuming and may cause delays in completing the payment cycle. Proper reconciliation and prompt follow-up are critical to ensure the payment is accurately reflected in the financial records and that any discrepancies due to bank charges are addressed appropriately.

- **Reconciliation of payment transactions**
  After payments are processed, the reconciliation of payment transactions is essential to ensure that all payments have been correctly recorded and matched against bank statements and accounting records. This step helps identify any discrepancies between the payment and the G/L.

- **Clearing open items**
  Open items in accounts need to be cleared once payments are successfully processed and matched to invoices. This ensures that AP and AR are up to date and that outstanding liabilities are properly accounted for.

These postprocesses ensure that payments are effectively managed, discrepancies are promptly addressed, and financial records are kept accurate, contributing to smooth financial operations and compliance with internal controls and external regulations.

1.4 Summary

You can see that payments are an extremely complex topic from a business perspective, with many factors to consider, such as compliance with tax regulations, handling foreign currency transactions, managing liquidity across different entities, and ensuring accurate reconciliation of bank statements. If not managed efficiently, these complexities can lead to serious problems, such as delayed payments, errors in financial reporting, increased operational costs, or even regulatory penalties. Inefficient payment processes can also cause disruptions in cash flow, affect relationships with suppliers and customers, and create a lack of transparency in financial operations. This emphasizes why technology is essential for automating and streamlining payment management. We introduced you to the SAP solutions that play a key role in managing these payments in this chapter. The following chapters will delve into the technical aspects of setting up these solutions and demonstrate how they can optimize payment management and support smooth financial operations.

Chapter 2
Bank Connectivity

This chapter provides an overview of common connectivity types and message formats for transmitting payments, account statements, and bank protocols. We also highlight quick wins and offer ideas for a cost-effective and secure transformation in the payments sector.

In this chapter, we offer a comprehensive overview of the technical requirements, functionalities, and standards in payment transactions. We delve into the essential global and local standards, offering ideas and insights. Furthermore, you will gain insights into interbank payment transactions. In short, this chapter introduces the topic of payments for corporations and equips you for your next payments project.

We begin by describing largely vendor-agnostic bank connectivity topics: First, we discuss the connectivity standards (EBICS, SWIFT, etc.), then we cover how to secure and harmonize payment processing, and then we walk through the process of transmitting and receiving electronic bank details. In addition to bank connectivity, this chapter also describes different formats, payment systems, and payment methods. You will gain insight into payment instruments, as well as a behind-the-scenes look at how interbank payments work. This chapter serves as an introduction to the payment system as a whole.

2.1 Connectivity Standards and Security in Payments

Not only banks but also corporations are facing massive transformations in the payments sector. Although banks must address strategic questions such as how to handle the growing influence of payment service providers, central banks are discussing how to regulate payments through distributed ledger technologies and developing their own central bank digital currencies.

The evolution of bank connectivity has been particularly intriguing over the past decade. Established providers like SWIFT have responded to current challenges and emerging competition, such as Ripple, or local connectivity forms like EBICS in the Eurozone, by developing their own API solutions and advanced services. These include format validation services, and products like SWIFT GPI. The key challenges of the banking world and the rapid development of alternative payment methods are directly reflected in the treasury departments of multinational corporations. There are growing demands from the e-commerce sector to offer current payment methods and to increase technological and regulatory requirements.

Overall, there are increasing demands in the area of payment security, an ever-growing heterogeneity of technologies and payment methods, and a concurrent transformation of payment standards and their impact on ERP and treasury management systems.

The current requirements for payment processing can be categorized into the following overarching demands, which complement and depend on each other:

- **Technology requirements**
  There are key technological requirements for modern payment management. In particular, the demands for security in payment processes, such as the avoidance of media breaks (when information has to be transferred from one media to another) through straight-through processing (STP), increasingly stringent encryption requirements, and fraud prevention, are essential in the age of artificial intelligence. In this chapter, you will find ideas and suggestions for how to enhance payment security within your corporation, and we will specifically address the issue of media breaks across different systems.

- **Trends and formats**
  For automating processing and sending payment requests, protocols, and bank statements, standards are indispensable. This applies to the integration of banks as well as payment service providers. This topic is constantly evolving. This is partly due to local legal requirements; technological demands, such as protecting against fraud; and technological innovations and trends. As authors, we are particularly eager to see how central bank digital currencies will develop, to name just one of many examples. In this chapter, we introduce the common standards and processes in payment transactions. For dealing with payment service providers, you will find suggestions for your system architecture in Chapter 11.

- **Legal requirements**
  Whether you're working with domestic payment transactions or cross-border-managed cash pools, legal requirements and restrictions, particularly regarding global capital management, are constantly changing. This includes legal nuances and developments that lead to ongoing adjustments in payment processes, including new purpose codes in payment instruments. Legal requirements, such as verification of payee, also directly influence the functionalities and architecture of payment systems. We'll discuss these requirements.

The following sections delve into specific connection methods. We'll introduce key standards and protocols such as SWIFT and EBICS, alongside modern approaches for direct bank integration via APIs and host-to-host connections. These solutions enable efficient and secure communication between businesses and financial institutions.

2.1.1 SWIFT

In the intricate world of global finance, where trillions of dollars are transacted daily, seamless communication between financial institutions is paramount. At the heart of this complex network is the Society for Worldwide Interbank Financial Telecommunication (SWIFT). Established in the early 1970s, SWIFT revolutionized the way financial information was exchanged across borders, setting a unified standard that enabled robust, secure, and rapid transactions.

SWIFT is not a direct money transfer service; rather, it is an international messaging network used by banks and financial institutions to securely transmit information and instructions through a standardized system of codes. With over 11,000 members in more than 200 countries, the network facilitates an average of over 42 million messages per day, making it an indispensable component of the modern banking system.

In short, SWIFT serves as the backbone of the financial industry, providing the ideal platform for exchanging payment instructions, account statements, and protocols with your bank. In the following sections, we will not only provide an overview of SWIFT but also offer valuable insights and food for thought for designing your system infrastructure in the area of payment transactions. By understanding these elements, you can effectively design your payments infrastructure, taking all circumstances into account to create a sustainable, cost-effective, and secure system.

In the following sections, we outline the key points that will help you successfully get started with SWIFT, not only from a technical perspective but also from a business standpoint. We also explore the various possibilities and options available to you. We'll cover the following topics:

- **Connectivity options**

  We'll present technical connection options for the SWIFT network and delve into their characteristics and functionalities.

- **Contracting and membership**

  We'll explore contractual and business aspects of SWIFT and outline the available options for corporations to join the SWIFT network and exchange messages with banks.

- **Key transmission services**

  We'll discuss the different ways messages can be exchanged within the SWIFT network, explaining their characteristics and specific features. We'll also define and discuss the terms *FIN*, *FileAct*, and *InterAct*.

- **Bank identifier code**

  The bank identifier code (BIC) has become an essential part of everyday financial transactions. Nearly every financial institution can be uniquely identified through its BIC. If a corporate entity wishes to join the SWIFT network, it will receive its own BIC. We'll delve into the details and examine the structure of the BIC.

2   Bank Connectivity

- **Onboarding**
  After outlining the technical and business fundamentals of SWIFT, we'll provide a step-by-step guide to the SWIFT onboarding process.

- **Further SWIFT services**
  Beyond its core messaging service, SWIFT has significantly expanded its product and service portfolio—not only for banks but also for businesses. We explore the key SWIFT products and services, such as the SWIFT Tracker from SWIFT GPI and the SWIFTRef Bank Directory, along with the advantages they offer within the SAP system.

Connectivity Options

Due to the increasing requirements for payment transactions and the exponentially growing security requirements for payment management, SWIFT is offering additional programs, such as the SWIFT Customer Security Program and an expanded cloud service offering. In addition to new connectivity options, additional services such as sanction and embargo screening and reporting and payments visibility services such as SWIFT GPI are being launched and enhanced. Figure 2.1 shows the connectivity options for SWIFT, which we'll cover in the following sections.

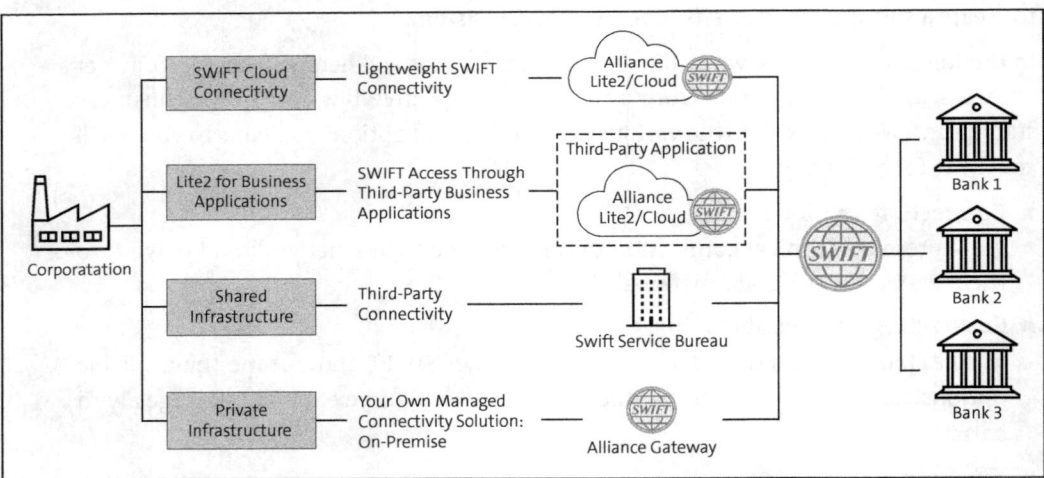

Figure 2.1 SWIFT Connectivity Options

Private Infrastructure

SWIFT's private infrastructure option (known as the SWIFT Alliance Gateway option) is outdated and arguably the most expensive way to connect to the SWIFT network. It involves a physical server that hosts all necessary components to connect to the SWIFT network and exchange data over it. We are currently aware of only a few companies that continue to operate their own SWIFT infrastructure, and they are either in the process of transitioning or have already switched to an alternative connectivity method.

2.1 Connectivity Standards and Security in Payments

Shared Infrastructure

Because managing your own SWIFT infrastructure is not only IT cost-intensive but also potentially personnel-intensive, requiring at least two security officers, a shared infrastructure option is available. This is typically offered by SWIFT Service Bureaus for banks and for corporations. A direct connection, usually via host-to-host communication, is established between the respective treasury management system or ERP system and the SWIFT Service Bureau. The data carriers are transmitted to the SWIFT bureau over a secure path, and the bureau manages the bank connection via SWIFT as a service. Well-established bureaus offer their own software for monitoring payment files, as well as an overview of transmitted payment files.

Figure 2.2 illustrates a system with shared SWIFT infrastructure. You can see that nearly all components required for connecting to the SWIFT network are hosted by the Shared Service Bureau (SSB). The SSB has the flexibility to choose which infrastructure it uses to connect to the SWIFT network.

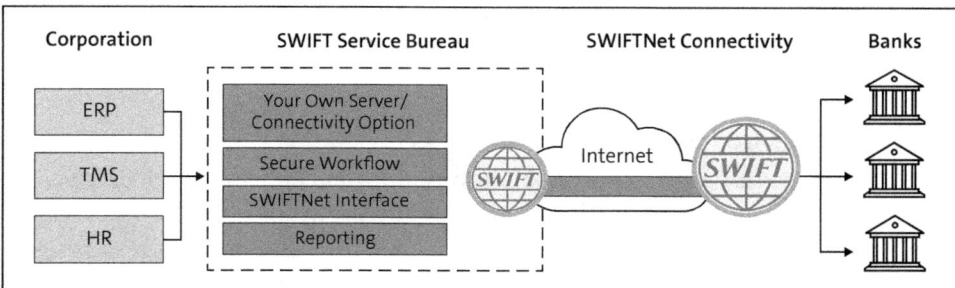

Figure 2.2 SWIFT Service Bureau

Alliance Lite2

Alliance Lite2 offers a simplified, file-based method for transferring payments to banks. It is crucial that access to the transfer software, AutoClient, is secured appropriately. Through AutoClient, folder structures can be created for incoming and outgoing messages. Payment files can be placed directly in these structures using Alliance Lite2–compatible software. Companion files, which contain essential information about the sender and recipient, accompany these payments. Payments are job-scheduled and transmitted through AutoClient into the SWIFT network and then forwarded to the correct bank.

The SWIFT network generates acknowledgement (ACK) and no acknowledgement (NACK) protocols to confirm or reject a successful transfer, although these are only for the transfer to the SWIFT network itself. The receiving bank provides additional confirmation in the form of a pain.002 bank confirmation message, verifying receipt and initiation of the payment process.

Note that the final confirmation of payment execution is reflected on the account statement, which is made available on the next banking business day.

Figure 2.3 illustrates the SWIFT Alliance Lite2 system in a schematic IT infrastructure. Presystems such as an ERP system or treasury management system (TMS) send file-based messages to the server, where AutoClient operates, or it retrieves the messages via a file picker. In practice, a middleware solution is often implemented to facilitate the transport of messages between the pr-systems and AutoClient.

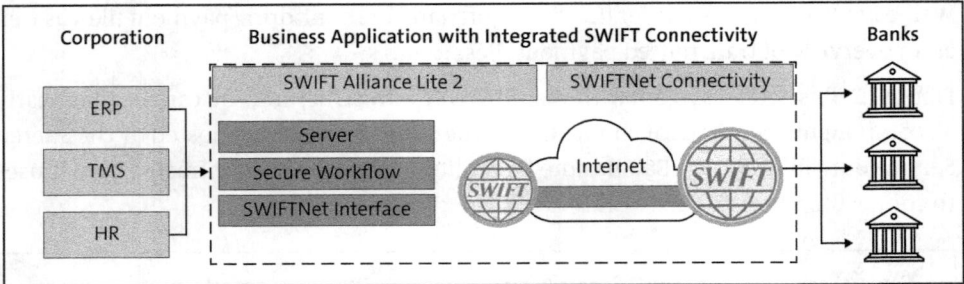

Figure 2.3 Alliance Lite2

Lite2 for Business Applications

SWIFT's Alliance Lite2 for Business Applications provides companies with direct access to the SWIFT network without the need to build their own complex infrastructure. It combines the benefits of the standard version of Alliance Lite2 with features designed specifically for business applications. All necessary components for SWIFT connectivity are integrated directly with the TMS or online banking solution. Essentially, SWIFT's AutoClient, which handles file management, is embedded within the software. Potential providers for this solution include SAP, with its cloud-based multi-bank connector, as well as standalone providers like TIS or Kyriba, which have integrated Alliance Lite2 for Business Applications into their solutions.

Figure 2.4 illustrates how the components of Alliance Lite2 can be integrated into a third-party application, such as a cash management system. In this way, payment messages can be transmitted directly from an ERP or TMS system to the third-party application and then forwarded via the SWIFT network to the bank.

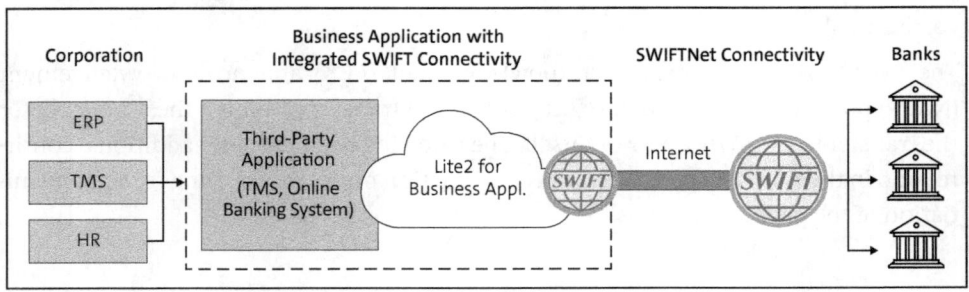

Figure 2.4 Alliance Lite2 for Business Applications

2.1 Connectivity Standards and Security in Payments

SWIFT Cloud Connectivity

Since 2018, SWIFT has offered a fully cloud-based connectivity option called *Alliance Cloud*. In this solution, all necessary components are hosted by SWIFT, relieving businesses of the need to manage their own infrastructure. Companies only need to establish an API connection with Alliance Cloud, allowing them to control all required components through the SWIFT-hosted platform.

Alliance Cloud is considered a successor to the Alliance Lite2 solution, providing a more streamlined experience as it eliminates the need for organizations to rely on their own SWIFT infrastructure. This solution is particularly appealing to businesses looking to benefit from a fully integrated and low-maintenance SWIFT connection.

Figure 2.5 depicts the Alliance Cloud workflow. Unlike Alliance Lite2, all components, including servers, are hosted on SWIFT's own cloud infrastructure. In this case, the customer does not need to maintain a dedicated server for applications such as AutoClient.

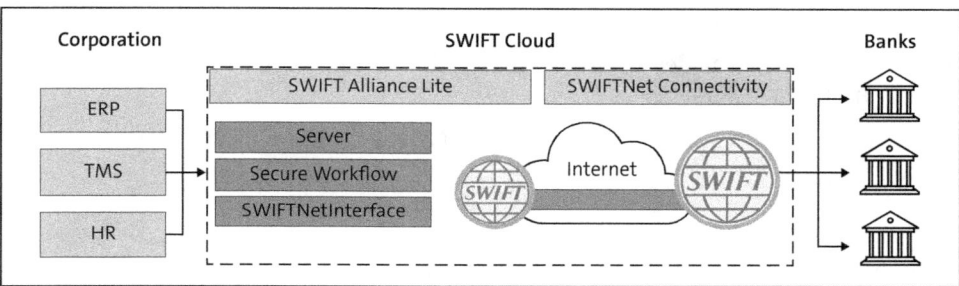

Figure 2.5 Alliance Cloud

Contracting and Membership

SWIFT is considered one of the world's most secure networks, with top-tier stability. To protect this network, SWIFT, as an institution, implements strict criteria for user admission; it does not allow just anyone to join. In the following sections, we present the contract and membership options available for joining the SWIFT network.

Standardized Corporate Environment

In the Standardized Corporate Environment (SCORE) framework, a contractual relationship is established between SWIFT and a corporate entity. A comprehensive *know your customer* process is essential and can be demanding. In general, a nonlisted corporation willing to use SCORE has to be recommended by a financial institution that participates in SCORE and that is located in a Financial Action Task Force (FATF) member country. To initiate this process, a recommendation letter from a member bank is required. Member banks are administratively registered banks that are authorized to facilitate the know your customer process. These banks are designated by SWIFT as Supervised Financial Institutions. You can usually find out if your bank supports SCORE by checking its website or consulting your financial advisor.

Once you have successfully completed a SCORE contract, you can exchange messages with any bank that is SWIFT-enabled and supports the SCORE framework, allowing seamless integration into your system environment, as illustrated in Figure 2.6.

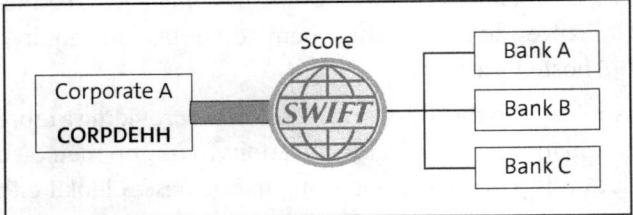

Figure 2.6 SWIFT SCORE Contract

There is an alternative path for SCORE, approached through a recommendation letter from your bank. SWIFT requires the following criteria to be fulfilled:

- The corporation must be located in an FATF member country.
- The bank that provides the recommendation letter must be a participant in the SCORE closed user group.
- The bank must confirm its willingness to exchange traffic with the corporation over SWIFTNet.

Member-Administered Closed User Group

A Member-Administered Closed User Group (MA-CUG) is essentially a contractual relationship between the administrating bank and your company, as illustrated in Figure 2.7. You can have multiple MA-CUG agreements in place simultaneously to exchange messages with your bank via the SWIFT network. The key difference here is that the admission rules and know your customer processes are conducted by the administrating bank. In short, for companies for which SCORE is not feasible, MA-CUG offers a suitable alternative. The contract is made directly with the bank in this scenario, and the terms are also tailored individually. Typically, your bank will determine which message types can be exchanged over the network and which cannot. Therefore, it is important to review the formatting strategy in advance and align it with the services offered by the bank to establish a sustainable banking connection from the outset.

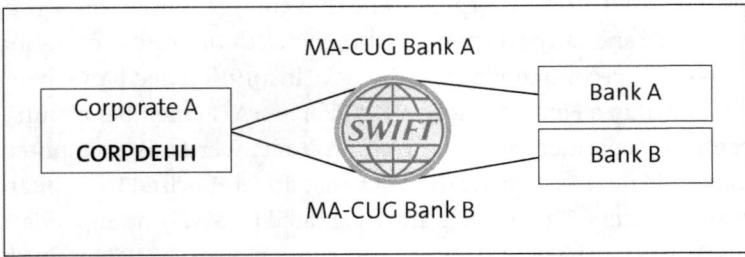

Figure 2.7 SWIFT MA-CUG

To participate in a closed user group as a corporate entity under the SWIFT qualification criteria, the following prerequisites are essential:

- **Stock exchange listing**
  The corporate must be listed on a regulated stock exchange in a country that is a member of the FATF.

- **Majority ownership**
  If not directly listed, the corporate can be majority-owned by an entity that meets the criteria for being listed within a stock exchange, provided a declaration of ownership is submitted to SWIFT from the eligible parent entity.

The following requirements also apply:

- **Legal status**
  The entity must be duly incorporated, validly existing, and duly organized as a legal entity.

- **Financial and compliance standing**
  The entity needs to be in good financial standing and compliant with applicable laws and regulations.

- **Audit standards**
  The entity should be subject to regular audits following internationally recognized accounting standards by an independent audit firm.

Treasury Counterparty

The Treasury Counterparty (TRCO) was introduced in 1998 when SWIFT opened its network to corporations. Through the member group TRCO, companies can exchange financial messages in the third SWIFT message category (Treasury Markets), MT3*, including treasury deal confirmations, spots, forwards, money market transactions, and FX deals. The requirements for participation are as follows:

- The company must engage in a large volume of treasury activities incidental to its core business with multiple bank counterparties.
- It must not qualify as a SWIFT member, submember, or any other category of participant.
- The company must be in good standing financially and comply with applicable laws and regulations.
- It must be regularly audited according to international accounting standards by a recognized audit firm.

Key Transmission Services

The FIN, FileAct, and InterAct SWIFT services each serve distinct functions in banking communications in order to transmit messages. The individual bank, similar to the type of contract, determines which of these services are offered to corporations. In this section, we highlight their differences and categorize these services.

FIN

Financial Network (FIN) is a messaging service provided by SWIFT for the exchange of structured financial messages in the MT and ISO 15022 formats. It is used worldwide by financial institutions and supports secure, standardized communication. The main advantage of FIN is the high quality of data transmission, as the defined message types are validated formats. A service called *FinCopy* (also known as *Store and Forward*) also can be activated; it automatically creates a copy of each message for fallback scenarios during every transfer, providing additional stability. Following is an overview of messages that are transferred via SWIFT for corporations through the FIN network:

- Request for Transfer: MT101
- Account Statement: MT940
- Intraday Statement: MT942
- Notice to Receive: MT210
- Return Messages: MT195
- Free Form Messages: MT199
- Deal Confirmations: MT3xx

> **Format Transformation: MT to ISO20022 (CBPR+ and HVPS+)**
>
> An important aspect of the MT to XML-based ISO 20022 (MX) migration is the development of the Cross Border Payments (CBPR+) and Domestic High Value Payments (HVPS+) specifications within the ISO 20022 XML standard. These specifications dictate how an XML message should be populated in terms of data and field requirements for CBPR+ and HVPS+. Simplified, both programs deal with the format transformation from MT messages to XML-based ISO 20022 messages. This primarily affects settlement systems and financial institutions. However, it also has a direct impact on your ERP or treasury system. We will discuss the specific implications of ISO 20022 in Appendix A. At this point, it's simply important that you have encountered these terms and can roughly categorize them.
>
> Note that HVPS+ refers to domestic RTGS clearing systems, and several countries are in the process of making their domestic clearing systems natively ISO 20022 XML compliant.

InterAct

SWIFT InterAct is a service designed for the transfer of structured ISO 20022 messages. This service validates the messages before forwarding them to the respective receiving financial institution.

> **Store and Forward with InterAct**
> Similar to FIN, the Store and Forward service can create a copy of the message, enhancing security in case a fallback is needed. Note that SWIFT is one of the most stable systems globally and has been categorized as having *five nines availability*, which indicates an availability rate of 99.999%.

You can transmit the following messages via the InterAct service:

- Payment Initiation: pain.001
- Direct Debit: pain.008
- Bank Protocol: pain.002
- Account Statement: camt.53
- Intraday Account Statement: camt.52
- Credit/Debit Notification: camt.54

FileAct

The SWIFT FileAct service is arguably the most flexible and the most cost-effective service, making it ideal for high-volume payment processing. FileAct, as a service from SWIFT, transmits messages and files without format specifications. The format specifications are provided by the financial institution itself and are not validated by SWIFT. This means that files, for example, can be transmitted zipped through FileAct. In addition to common financial messages like payment initiation, account statements, and bank protocols, documents in PDF form, such as PDF account statements or IDocs, can also be received through the service.

> **Adapting to Current Challenges**
> The SWIFT network is one of the most secure closed networks in the world, with extremely high availability, serving as the backbone of the financial industry. It is responding to current market developments by investing in advanced technologies such as distributed ledger technology (DLT) for authorization and payment transactions. Although SWIFT solely transports messages and does not actively conduct payments, we are witnessing a steadily growing range of services that support credit institutions and banks as they navigate an increasingly regulated market.
>
> Furthermore, due to the ISO 20022 migration, SWIFT has planned a transformation of the FIN channel. Since March 2023, SWIFT has taken a major step in the MT/MX migration; it now supports the exchange of ISO 20022 XML messages over the FINplus network. Previously, this was only possible via the FileAct or Interact channel. In parallel, legacy MT format messages continue to be exchanged over the FIN network; the MT flow for message categories 1 (customer payments), 2 (FI transfer), and 9 (statements) through the FIN network will be decommissioned in November 2025.

> As such, between March 2023 and November 2025, financial institutions need to be able to receive and process MX messages through FINplus on the inbound side and optionally send MX messages or MT messages for outbound messaging. After that period, only MX will be allowed.

Bank Identifier Code

The SWIFT bank identifier code (BIC) is used for the unique identification of SWIFT participants. In general, a participant can apply for multiple SWIFT codes. In practice, however, it should be at least two BICs: one for the production environment and one for the test environment. In the banking world, we see both 8-character and 11-character BIC codes. The 8-character BIC codes are typically used by the headquarters, while the 11-character BICs are used by subsidiaries. The BIC is structured as shown in Table 2.1.

CORP	US	01	XXXX
A four-letter code identifies the SWIFT participant	Country code (two letters)	Location code (number or letter)	Optional brand code (4 letters)

Table 2.1 Bank Identifier Code

Onboarding

Each bank connection is unique and should be meticulously planned to save costs and ensure a sustainable and secure connection. The underlying reasons vary, including increasing regulatory requirements, bank-specific services, and constant updates to formats, all contributing to a heterogeneity in banking connections. In this section, we present a sample project approach and provide some food for thought on what you should consider when structuring your next bank connection via SWIFT.

We have divided the sample project approach into five phases, as shown in Figure 2.8, which we will introduce step by step. For the initial pilot connection, we recommend allocating four to six months, taking into account both internal and external availabilities.

Let's look at the five phases in more detail:

- **Prepare**
  The better prepared you are for the project, the lower the chance of failure. During the prepare phase, you will examine the overall architecture and identify relevant stakeholders for your project. In addition to the treasury team, your in-house IT should be involved, as well as all relevant contacts from the AP and AR areas for each bank connection. A new bank connection may also be associated with new formats and processing steps. Create a realistic project plan that takes availability into consideration.

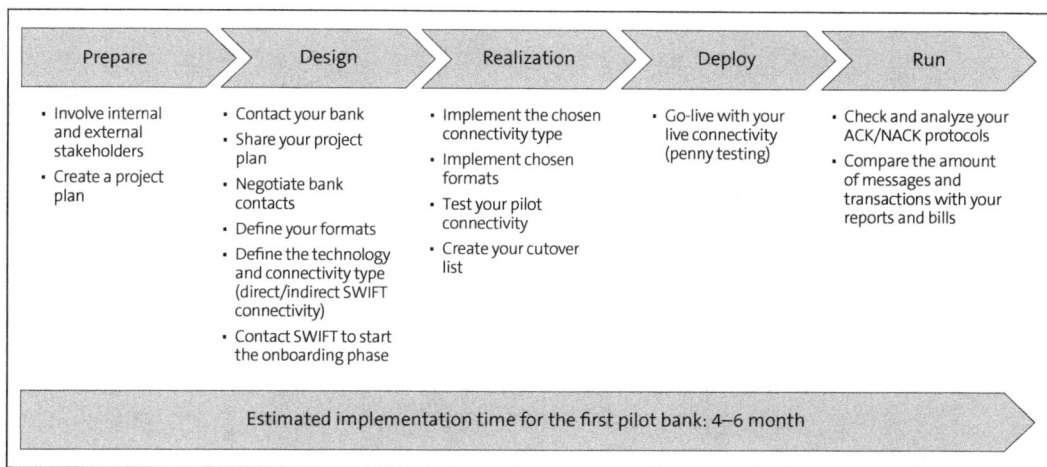

Figure 2.8 SWIFT Onboarding

- **Design**

  In the design phase, you focus on the details. During this phase, it's crucial to understand all available options and decide on the type of connection to establish (direct or indirect SWIFT connection), as well as negotiate the contractual specifics with the bank. What options does the bank offer to join the SWIFT network? Is SCORE or MA-CUG the appropriate connection type? Which formats fall within the scope? You also should contact SWIFT through its website. Both your bank and SWIFT will guide you through the process and assist you with any documents needed for the know your customer process. We recommend conducting a preliminary study before establishing the bank connection to thoroughly understand and consider all requirements related to accounts payable, accounts receivable, human resources, IT, and treasury. Once all contractual matters are resolved and the connection design is finalized, you can proceed with the implementation.

- **Realization**

  In the realization phase, all requirements from the involved departments should be defined to connect your bank via SWIFT using the chosen connector. For a FIN connection using classic MT files, header information is required, which reflects the specific MT format. For a FileAct connection, you must be able to create what are known as *companion files*. These companion files include the sender and receiver of the SWIFT file and act as guides for the SWIFT network. Aside from the BIC of the sender and receiver, the file also specifies whether the transmission is inbound or outbound.

  Typically, the bank provides a test or pilot connection along with the specific details for the technical setup. We recommend actively using the test connection to thoroughly evaluate your setup in the system from front to end.

- **Deploy**
  During the deployment phase, you execute the go live. We highly recommend meticulously planning this process using a cutover plan and blocking out the necessary time. Ideally, start with small transactions, known as *penny tests*, such as low-value invoices or account transfers. Schedule several penny tests and conduct as many low-value scenarios as possible to prevent any surprises during large-scale payment runs.

- **Run**
  The run phase is designed to stabilize payment processes. Be prepared for processing errors, rejected payments, or errors in bank statement imports. Although a properly executed testing phase and penny tests minimize risks, issues in master data can still lead to errors. To be prepared for such scenarios, it is beneficial to allocate sufficient time for the run phase to secure necessary support from the bank, your IT department, or, potentially, external consultants.

  Use the run phase to intensively verify reports. Well-structured SWIFT service bureaus regularly provide reports on sent and received messages, which you can reconcile with your SWIFT billing.

Further SWIFT Services

Over the years, SWIFT has evolved from an infrastructure provider to a service provider in the areas of payments and securities settlement. In addition to basic connectivity, SWIFT maintains a standardized banking directory with global master data and services in the know your customer area. In the following sections, we present services relevant for corporations that go beyond basic bank connectivity and can support and enhance processes within your ERP system and treasury department.

SWIFT GPI

The world of payments is undergoing an unprecedented transformation. SWIFT is responding to competitors like Ripple with new innovations that ensure security while also enhancing speed and transparency in transactions. SWIFT GPI is an initiative aimed at improving cross-border payments by making them faster, more transparent, and traceable. This effort introduced a new field: the unique end-to-end transaction reference (UETR), which allows payments to be tracked in real time, offering greater transparency for both senders and recipients—similar to package tracking. Banks participating in the SWIFT GPI program send a pain.002 message to SWIFT, providing information on the current status of the payment. This is particularly a game-changer for foreign payments. For treasurers, SWIFT GPI provides transparency into which correspondent bank the payment is with and when. Banks also can opt to include payment information that further enhances transparency.

Moreover, SWIFT GPI enables immediate confirmation of payment receipt by the beneficiary, reducing uncertainties. Figure 2.9 provides a schematic representation of the SWIFT GPI Tracker.

2.1 Connectivity Standards and Security in Payments

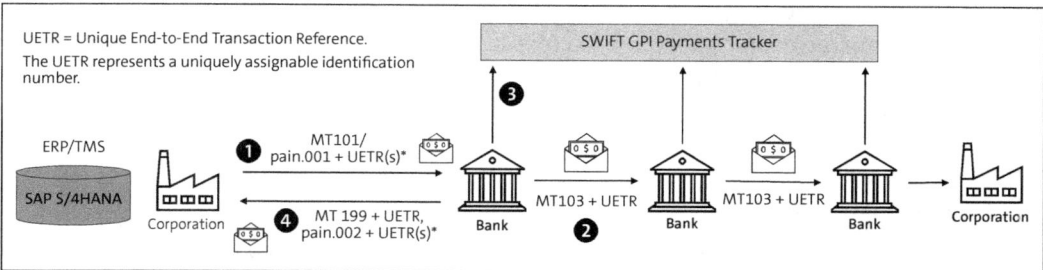

Figure 2.9 SWIFT GPI Tracker

In Figure 2.9, we illustrate the SWIFT GPI Tracker process using the transmission of a payment request via MT101 or pain.001:

❶ **Corporate client creates a payment order with a UETR number**
This is done via MT101 or pain.001 format for single or batch payment orders.

❷ **Bank creates an MT103 interbank payment order**
The bank adopts the UETR number from the corporate payment order.

❸ **Information sent to the tracker**
Details such as costs, cost allocation (SHARE, OUR, etc.), the amount, and so on are transmitted to the tracker.

❹ **Initiating bank sends updates to the corporate client**
Updates are transmitted via an MT199 or pain.002 message, providing the corporate client with the current status of the payment order.

SWIFTRef

Master data is the foundation for the successful execution of payments as well as for the processing of account statements. Especially with the latest format updates from 2019 and new developments in the PSD3 environment, such as the verification of a payee, an organized and fully reliable master database is key.

SWIFT and providers like Accurity provide regularly maintained databases that can be imported into the SAP S/4HANA system via standard transactions like BIC2 or BIC2N or SAP Fiori apps like Transfer BIC Data (BIC2) or Import Bank Directories (BIC2S), thereby populating table BNKA in the SAP S/4HANA system.

SWIFTRef is a comprehensive database provided by SWIFT. It serves as a pivotal resource for accessing reference data about financial institutions worldwide. This includes essential information such as bank identifier codes (BICs), International Bank Account Number (IBAN) structures, national bank codes, and other critical data needed for efficient international payment processing.

The primary purpose of SWIFTRef is to enhance the accuracy and efficiency of global financial transactions. By offering access to standardized and reliable data, the database helps financial institutions, corporations, and various organizations minimize errors in payment processing. This not only reduces the risk of transactional failures but also speeds up the overall execution of cross-border payments. One alternative provider for bank master data is Accurity, which offers a processable file that can be imported into the system via Transaction BIC2.

2.1.2 Host to Host

Via host-to-host connection, backend systems can be connected with each other, establishing a bilateral connection to the bank server, following the rules set by each bank. This means the bank determines the technical requirements for the connection, such as which protocol (e.g., HTTPS, SFTP) should be used and which formats should be transmitted via the protocol. Consequently, a host-to-host connection is tailored for each bank.

In practice, after successfully completing the online banking agreement and defining the framework conditions, you receive host-to-host parameters from the bank, such as server information, credentials, and links. Depending on the type of connection, a certificate exchange might occur. The certificates are used to encrypt the transmission path and the message itself. Note that these certificates usually have an expiration date—and they should. You should review and update your host-to-host connections approximately every two years to comply with current security standards, in coordination with your bank.

Be aware that the security requirements for a host-to-host connection may vary from bank to bank. After successfully setting up a host-to-host connection, messages such as payment instructions, account statements, and bank reports can be exchanged over the connection. All data sent via the host-to-host connection is considered signed and approved and is executed by the bank accordingly.

2.1.3 API Usage

Application programming interfaces (APIs) are templates of protocols and tools designed to enable communication between different software applications. Essentially, you can think of an API as a blueprint that sets out the rules for connecting to a bank's systems. When these rules are followed, systems can be directly integrated and communicate seamlessly. In the banking sector, APIs are developed by banks to provide programmatic access to their services, allowing for smooth integration and interaction.

One of the biggest challenges with APIs is their heterogeneity. In simple terms, each bank can create its APIs according to its own rules. Standards like ISO 20022 are inherently based on XML, which standardizes certain functionalities like the verification of a payee. However, this does not yet apply to the exchange of cash management information, such as intraday items and transactions, or for triggering banking services like over-the-counter transactions. Moreover, cross-bank authentication is not yet fully resolved.

Ideally, a bank should offer an API feature that interacts directly with a treasury management system (TMS) or an ERP system, allowing payments—especially instant payments—to be triggered directly via the API and bank account updates to be conducted via the API. A successful pilot project has been run by J. P. Morgan in collaboration with SAP Multi-Bank Connectivity.

2.1.4 EBICS

Electronic Banking Internet Communication Standard (EBICS), as a communication standard, comes with three layers of encryption based on Hypertext Transfer Protocol Secure (HTTPS). In addition to having a public and private key, EBICS users are initialized, which can present a significant advantage over alternative connection forms.

> **EBICS User Initialization**
>
> An EBICS user is an authorized participant in the EBICS system who securely exchanges transactions with banks. Each user is initialized with unique keys and certificates to ensure authentication and encryption. We describe EBICS users and the initialization process in detail in later in this section.

Unlike host-to-host connection and SWIFT, which are pure communication forms, EBICS has an intelligent signature process integrated into its logic, following the signing process logic in the Germany, Switzerland, Austria (GSA) region. EBICS, developed by the German banking industry in 2006, is gaining increasing popularity as a standardized communication protocol between banks and corporations. The reason for this is simply the unbeatable price-performance ratio achieved through high standardization.

Furthermore, EBICS offers user-specific signature logic. Primary and secondary user signatures can be designated and stored in an EBICS contract. Following are some key aspects for choosing an EBICS connection:

- **Open standard**
  The EBICS standard is open and can therefore be used flexibly. Its functionality and current schemes are continuously published on the EBICS website, which facilitates its adoption. As a result, leading treasury management providers support the EBICS standard and have established it as an integral component for bank integration.
- **State-of-the-art technology**
  The EBICS standard is continuously evolving to ensure its encryption meets current requirements. It supports various data formats, including XML and ZIP, making it a robust and adaptable protocol for secure file transfers in banking.
- **Security**
  Continuous optimization ensures that the EBICS protocol remains a secure protocol. With individual EBICS users, permission concepts can be implemented flexibly.

- **Flexibility**
  The variety of configuration options with EBICS makes it a highly flexible protocol for bank integration. Depending on the capabilities of the bank, there are virtually no limits to data exchange. Even files in ZIP format can be sent via EBICS.

In the next sections, we describe the core components of the EBICS system in detail. First, we'll explain the technical specifications and structure of the EBICS environment. This is followed by a look at the practical application of EBICS, focusing on its effective use and the initialization process. Finally, we'll introduce special features such as the distributed digital signature and corporate seals, providing additional security and efficiency.

Entities

In the realm of EBICS, there are key entities that structure the system's functionality. Understanding these entities is essential for implementing and managing EBICS effectively:

- **Host (EBICS bank server)**
  The host is the EBICS bank computer system, uniquely identified by the host ID. Some banks may have multiple host IDs to account for their various EBICS servers. The bank communicates the EBICS host ID along with the URL necessary for customer access.

- **Partner (or customer)**
  This organizational unit, whether a company or an individual, enters into a contractual agreement with the bank. The contract outlines which administrative order types and business transaction format (BTF) combinations will be used, specifies the relevant accounts, and identifies which users (subscribers) at the company will communicate with the EBICS bank server, along with their authorized permissions. The partner ID uniquely identifies the partner.

- **User (or subscriber)**
  Users, whether human or technical systems, are linked to a customer and identified by a combination of user ID and partner ID on the EBICS bank server. A technical subscriber facilitates data exchange between the customer and the financial institution but does not authorize orders. Human users have the ability to authorize orders.

- **Signature class**
  Each EBICS user (subscriber) receives at least one signature class related to the electronic signature for authorization. The signature class defines the quality of the subscriber's electronic signature. Banks can design detailed authorization models that assign different signature classes based on the BTF, amount limit, or accounts used.

  Table 2.2 provides an overview of the different signature types.

User Type	Description	Usage
A-signature	First signature	First signature; valid only in conjunction with an additional A-signature, a B-signature, or an E-signature.
B-signature	Second signature	Secondary signature; only valid in conjunction with a primary signature (A-signature) or a single signature (E-signature). Two B-signatures, or a combination of a B-signature and a T-signature, are insufficient to authorize a payment document. In such cases, the bank server will reject the transaction.
E-signature	Single signature	The E-signature is solely authorized to sign and can thus authorize payments without additional signatures.
T-signature	Transport signature	The signature serves for transmission of files and is used for errands such as collecting account statements or bank protocols. Payments can also be transmitted to the bank with a T-signature for a distributed signature process.

Table 2.2 EBICS User Types

- **BTF identifier**
  Within EBICS, BTFs structure and categorize the different types of messages and transactions that can be processed through EBICS. Each BTF is designed for specific types of payments or transactions, such as the following:
  - SEPA credit transfers
  - Direct debits
  - Account statement retrieval
  - Payment transaction files in various formats

  By using BTFs, companies can ensure that their transactions are processed correctly, making them a central component in the implementation of EBICS for electronic payments.

- **Order types**
  Order types are used to categorize transactions and transmitted payment instruments. They are divided into general order types, which are oriented based on incoming and outgoing file formats and administrative order types. Administrative order types are specific technical orders that facilitate various functions, including downloading technical information, initialization, key management, order cancellation, and using electronic distribution standards.

  Table 2.3 and Table 2.4 provide an overview of the common general and administrative order types, respectively.

2 Bank Connectivity

Identifier	Transmission Direction	Description	Format
AEU	U	Send Free Text Message Export Letters of Credit	DTAEA
AIA	U	Send Import Letters of Credit	DTALC
AID	U	Send Import Letters of Credit Document Filing	DTALCA
AKA	D	Retrieve Import Letters of Credit	DTALCR
AKD	D	Retrieve Import Letters of Credit Settlement	DTALCD
AZV	U	Send AZV in Diskette Format	DTAZV
CBC	D	Retrieve Payment Status Report for Direct Debit via XML-Container	XML container with pain.002.002.03
CCT	U	Send Credit Transfer Initiation (ZKA/EPC Specification of SEPA Transfer)	pain.001.002.03
CCU	U	Send Urgent Euro Transfer	pain.001.001.03
CIP	U	Send a Collector with (Terminated) SEPA Real-Time Transfers: Instant Payment	pain.001.001.03
CD1	U	Send Direct Debit Initiation (SEPA Core Direct Debit CORE 1)	pain.008.001.02
CDB	U	Send Direct Debit Initiation (SEPA Business Direct Debit)	pain.008.001.02
CDD	U	Send Direct Debit Initiation (SEPA Core Direct Debit)	pain.008.002.02 message (Chap. 2 of Annex 3 of the DFÜ Agreement)
CDZ	D	Retrieve Payment Status Report for Direct Debit	ZIP file with 1-n messages pain.002.002.03 (Chap. 2 of Annex 3 of the DFÜ Agreement)

Table 2.3 General Order Types

2.1 Connectivity Standards and Security in Payments

Identifier	Transmission Direction	Description	Format
C52	D	Retrieve Bank to Customer Account Report	camt.052.001.02
C53	D	Retrieve Bank to Customer Statement Report	camt.053.001.02
C54	D	Retrieve Bank to Customer Debit Credit Notification	camt.054.001.02
XBS	D	Retrieve Electronic Bank Fee Statement in camt.086 Format	camt.086
DDG	D	Retrieve Foreign Exchange Trade Confirmation	MT300
DHB	U	Send Foreign Exchange Trade Confirmation	MT300
FTD	U / D	Send or Retrieve Free Text File	ASCII
GAB	D	Retrieve Guarantees	MT760
GAK	U	Send Guarantees	MT760
GFB	D	Retrieve Aval Follow-up Messages (Inquiry on Extend or Pay)	Notice of Use
GUB	D	Retrieve Aval Messages (Creation)	Amendment
GUK	U	Send Aval Messages (Creation)	Amendment
INT		Send International Payment Transactions	MT101
PTK	D	Retrieve Protocol File	Text-based EBICS protocol
RFT	U	Send Request for Transfer	MT101
STA	D	Retrieve Swift Daily Statements	MT940
VMK	D	Retrieve Short-term Reservations	MT942
WPA	D	Retrieve Securities Settlement	MT510
WPC	D	Retrieve Depository Statement	MT571

Table 2.3 General Order Types (Cont.)

Identifier	Transmission Direction	Description	Format
DKI	D	Retrieve Foreign Exchange Rate Information (Euro)	CSV
DMI	D	Retrieve Foreign Exchange Market Information	CSV
DSW	D	Retrieve Foreign Exchange Swap Information	CSV
XAZ	U	AZV Submission to Third Party Bank	Any agreed-upon format
XCT	U	SEPA Credit Transfer ISO20022	pain.001.001.03

Table 2.3 General Order Types (Cont.)

Identifier	Transmission Direction	Description
HAA	D	Retrieve Retrievable Order Data
HAC	D	Retrieve XML Protocol File
HCS	U	Send Public Keys
HIA	U	Send Initial Public Key
HKB	D	Retrieve Bank Keys
HKD	D	Retrieve Customer and Participant Information of the Customer
HPD	D	Retrieve Bank Parameter File
HTD	D	Retrieve Customer and Participant Information of the EBICS Participant
HVD	D	Retrieve VEU Status
HVE	U	Submit VEU Signature
HVS	U	Cancel VEU Signature
HVT	U	Retrieve VEU Order Details
PUB	U	Send Public Key
SPR	U	Block Access Authorization

Table 2.4 Administrative Order Types

2.1 Connectivity Standards and Security in Payments

> **Order Types Are Limited Under EBICS 3.0**
>
> Starting with EBICS version 3.0, the protocol has been streamlined to use two order types primarily: file upload (FUL) and file download (FDL). All other order types are included within the payment data as business file formats (BFFs). To facilitate this, mapping tables and specifications are available for download directly from the website of the German banking industry. These resources provide the necessary guidelines for integrating and handling various transaction types effectively. For more information, visit *https://www.ebics.org/en/technical-information/btf-mapping*.

This foundational understanding of EBICS entities is crucial for navigating and utilizing the EBICS system effectively in professional settings. EBICS is a broadly adopted standard with an open design that has gradually established itself across Europe and continues to gain momentum. It is highly recommended to take a closer look at this standard and evaluate whether it might be preferable to alternatives like SWIFT or host-to-host connections, especially considering its lower costs.

Onboarding Process

Despite its many technical facets and details, EBICS stands out as a straightforward yet highly secure connectivity method, especially in implementation. Figure 2.10 illustrates the following onboarding steps, which are required for onboarding and establishing a connection between your bank and EBICS:

❶ **Preparation**
Before implementing EBICS, check how EBICS connectivity will be established. To send or receive messages via EBICS directly from the SAP system, an EBICS connector is required. This can be developed in house or purchased externally, or you can use SAP Multi-Bank Connectivity. You also should establish an online banking agreement with your bank beforehand, specifying the EBICS version, the transaction types used, and the assigned users. Before setting up an EBICS connection, all relevant information must be gathered to ensure a smooth implementation. This includes defining order types, assigning users, and collecting essential server details. Banks typically provide these details, which can also be found in the EBICS contract. Proper preparation of these elements lays the groundwork for a secure and efficient connection.

❷ **User creation**
As already described in the previous section, the authorization of payment carriers within EBICS works with EBICS users. These must be stored in the e-banking system or EBICS connector as a prerequisite for step 4.

❸ **Key generation and exchange**
EBICS users must be initialized once at the bank. This process essentially registers the EBICS user with the bank. After successful initialization, payments can be signed and

thus authorized by the EBICS user. Technically, this requires a certificate exchange. For each user, the system generates a cryptographic key (private and public key). Using technical order type INI, keys are exchanged with the bank.

❹ **Receipt of bank parameters**
- There is an exchange of data with the bank:
- Bank acknowledgment: The bank processes the received keys and user information. It authenticates the details and prepares for secure communication.
- Bank keys: The bank sends its own public keys back to the company. This exchange completes the encryption setup, enabling two-way secure communication.

❺ **Signed initialization letter**
To ensure that the initialization is carried out by authorized individuals, the EBICS connection and EBICS users must be activated by the bank. This requires a physical initialization letter to be printed, signed, and sent to the bank. The initialization letter displays the received key as a hash value, which must be signed by the EBICS user holder.

❻ **Activation and initialization of EBICS user**
After both parties—the company and the bank—have exchanged and verified keys, the bank activates the EBICS user. Often, this activation requires additional verification, which could include confirmation emails, phone calls, or physical document verification, to ensure all parties are who they claim to be.

❼ **Testing and go live**
Conduct thorough testing to confirm the setup is correct. This involves sending low-value transactions, via penny tests, through the system to ensure that all parts of the communication channel are operational and secure.

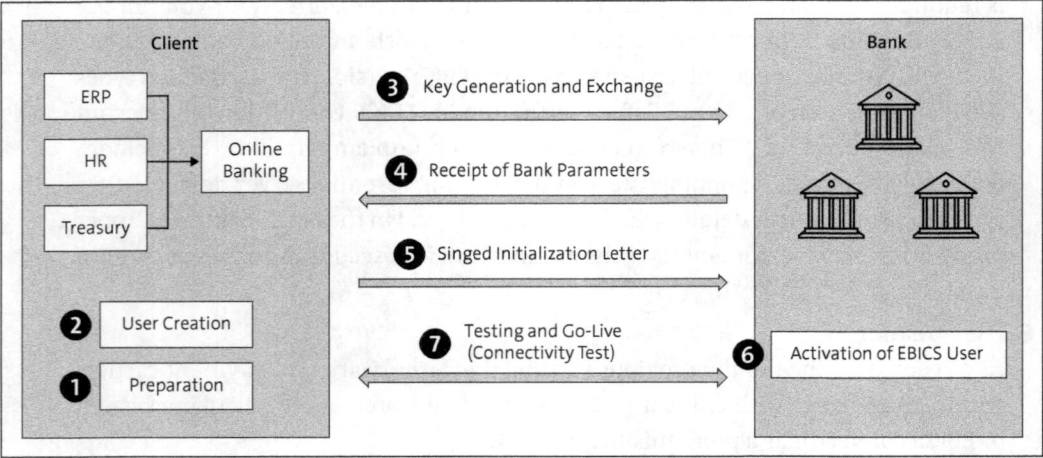

Figure 2.10 EBICS Version 2 Initialization

Data Transfer

Data transfer via EBICS is conducted through a multiencrypted HTTPS connection, as described earlier. The prerequisites for this process are that the recipient bank supports EBICS and that you use an EBICS-enabled software or connector, such as SAP Multi-Bank Connectivity. Once EBICS users are successfully initialized, data carriers like payment files can be transmitted to the bank. Protocols and account statements also can be retrieved from the bank server via pull requests with the appropriate order type. Figure 2.11 illustrates a typical Single Euro Payments Area (SEPA) credit transfer, as well as the ISO 20022 protocol and account statement retrieval using EBICS 2.5:

❶ **Credit transfer**
The payment file is transmitted to the bank server using order type CCT.

❷ **Protocol retrieval**
The pain.002 bank protocol is fetched from the bank server using order type CRZ. This protocol contains a four-character code and the corresponding message ID of the previously sent payment file for identification purposes. It indicates whether the payment file was accepted for processing or rejected.

❸ **Account statement retrieval**
End-of-day account statements are retrieved from the bank server using order type C53, while intraday account statements are fetched using order type C52.

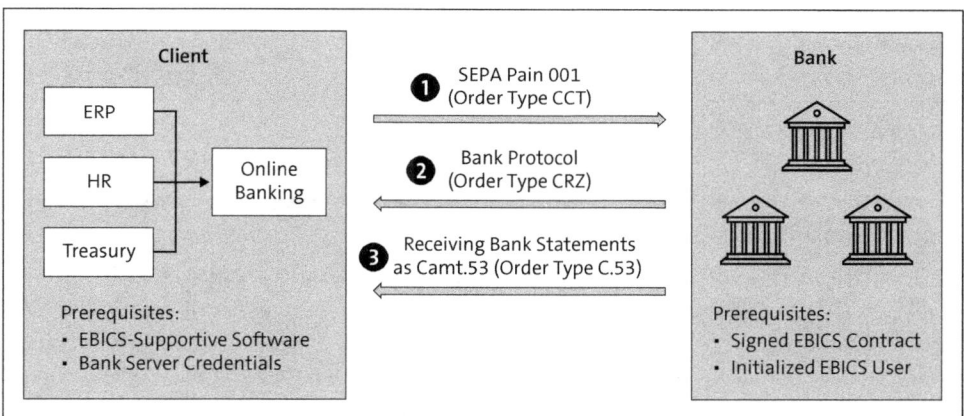

Figure 2.11 EBICS Prerequisites

Distributed Electronic Signature

The principle of a distributed electronic signature (DES) allows authorized signatories from different locations to place electronic signatures on a payment instrument submitted to the bank's computer. The connection between both files is established in the system via an order number or order ID. Authorized employees of a company can access payment instruments available on the bank's server, inquire about the pending signature of payment orders, and sign and thus approve the payment via electronic signature. DES is used exclusively with EBICS. Payment orders, for example, are initially transmitted from the finance or treasury department to the bank's computer but are not booked there immediately. Authorized signatories can retrieve DES status data and verify which payment orders are pending approval and which signatures have already been provided or are still missing. The orders to be signed are fetched from the bank's computer, electronically signed, and sent back. The bank only executes the orders once all required signatures are in place.

This principle is also often used in collaboration with HR service providers. For instance, they deposit payment instruments for salary payments on a bank server, which can then be signed by an authorized signatory. Figure 2.12 illustrates EBICS DES process sending a pain.001 credit transfer payment medium to the bank as an example.

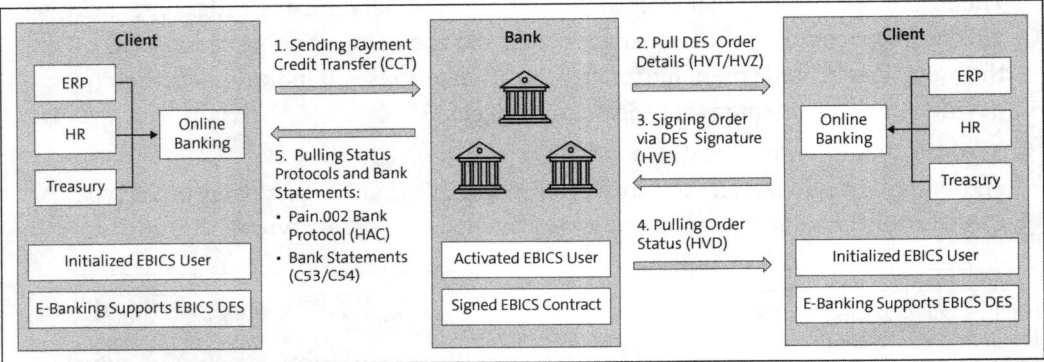

Figure 2.12 EBICS Distributed Electronic Signature

Bank-Specific Details
Order types may vary depending on the bank. As an example, consider a specific feature of German bank Commerzbank, which uses individual order types for distributed signature. At Commerzbank, there is a difference in order types under EBICS 2.5 when fetching and sending the second signature. Commerzbank uses order type ESG to fetch electronic second signatures and order type ESP to send electronic second signatures. Neither is supported by payments, but order types HVZ (Retrieving Additional Information), HVT (Retrieving DES Order Details), and HVE (Sending Signature) are supported in parallel during individual retrieval.

Corporate Seal with EBICS
The use of a corporate seal or chief financial officer (CFO) user in conjunction with the EBICS protocol offers an intriguing solution for companies looking to reduce administrative tasks while increasing flexibility in internal approval processes. Here are several key points to consider:

- **Reduced administrative burden**
  By employing a static e-signature user, the need for continuous management and registration of individual EBICS users with the bank is eliminated. This can significantly lessen administrative load, particularly for large organizations with frequent changes in authorized signatories.

- **Flexibility in internal approval processes**
  Companies can manage payment approval rights internally by configuring them within the online banking system. This eliminates the need for bank intervention during personnel changes, allowing for a more rapid adaptation to organizational shifts.

- **Security considerations**
  It's crucial to ensure that the online banking system supporting this solution has robust security measures in place. Access to the e-signature user should be strictly controlled to prevent misuse.

- **Supported systems**
  Not all online banking systems support this kind of solution. During implementation, it's important to verify that the existing infrastructure can accommodate this type of user management.

- **Legal and compliance requirements**
  Companies should check whether the use of a corporate seal or CFO user complies with legal and regulatory requirements in their jurisdiction. This includes adherence to electronic signature regulations and data protection laws.

This solution offers substantial benefits, particularly for larger organizations with complex approval structures and frequent personnel changes. However, implementation should be carefully planned to ensure that all aspects, such as security, compliance, and technical prerequisites, are adequately addressed.

EBICS 3.0

Since November 2023, banks have been offering EBICS 3.0 as the most recent and up-to-date version. This version is binding in the GSA region until approximately November 2025. Here is a summary of the most important changes:

- Increased standardization: Local EBICS "flavors" are unified to simplify implementation.
- Enhanced encryption: Since version 2.5, the minimum encryption level has been 2048 bits. This is continuously increased with the EBICS 3.X version.
- XML for EBICS: Text-based EBICS protocols like PTK are migrated to the XML-based pain.002 HAC.

The new version of EBICS increases the security of the communication standard and makes it more attractive in the EU. In addition to Germany, Austria, Switzerland, and France, the communication standard is increasingly offered by banks in Spain, Portugal, and the Netherlands, as well as in the Nordic countries. Recently, the first banks in Poland have started to offer this communication standard, marking a rising trend.

For companies that mainly use host-to-host connections or SWIFT for bank connectivity within the Eurozone, it may be worthwhile to look at EBICS and consider switching their connectivity method, provided their banking partner offers EBICS.

2.1.5 Bank Collection and Forwarding Service

Direct bank connections or regional standards like EBICS can be used to connect with local banks and transmit or collect messages over the bank's internal SWIFT network via *request for forwarding*, also referred to as *European Gateway* or *SWIFT forwarding*. This service allows for cost-efficient corporate banking without the need to connect each bank directly via host-to-host communication or SWIFT.

A SWIFT forwarding agreement is established and signed with each specific bank, and payment files are sent to the bank through a designated order type solely meant for forwarding. In this arrangement, the bank serves merely as a transmitter of the message. This approach also can be applied to account statements. Numerous banks in the GSA region and France actively promote this service as an additional cash management option for their corporate clients. Account statements are centrally gathered through the bank's SWIFT network and delivered to corporations through the existing EBICS channel. This method reduces implementation tasks and simplifies maintenance. Figure 2.13 illustrates SWIFT forwarding via an EBICS channel.

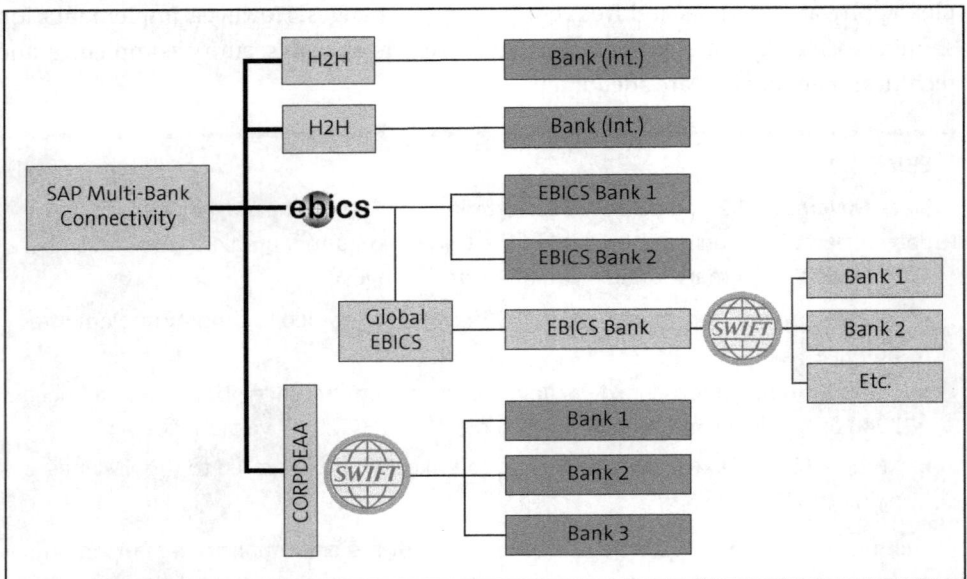

Figure 2.13 Forwarding via EBICS

2.2 Financial Standards

In the previous section, we discussed the topic of bank connectivity in detail. Now you might ask, What data can be exchanged via these interfaces? How are standards in the financial sector defined? You will find the answers to these questions in this chapter. We will examine global message types (MTs) and classify ISO 20022 messages. In Appendix A, you can find additional details on ISO 20022 and its impact on your SAP S/4HANA system. In this section, we'll focus on the basics of incoming and outgoing payments, providing an overview of which message types should be within the scope of your project. Specifically, we've structured the section to discuss different payment instruments, payment schemas, payment systems, and payment formats for incoming and outgoing payments and bank fee statements. We'll also offer some inspiration for process enhancements and potential automation opportunities within your company.

2.2.1 Payment Instruments

The European Central Bank describes an *electronic payment instrument* as a personalized device (or a set of devices), piece of software, and/or set of procedures agreed upon between the end user and the payment service provider to request the execution of an electronic transfer of value. Typical examples of electronic payment instruments include payment cards, credit transfers, direct debits, e-money transfers, and digital payment tokens. In this section, we'll focus on the payment instruments most commonly used in the B2B sector: credit transfers, instant payments, direct debits, and checks.

Credit Transfer

According to the Bank for International Settlements, a *credit transfer* is an essential payment instrument characterized by the movement of funds initiated by the payer, or originator, with the ultimate aim of making those funds available to the beneficiary. This process involves a payment order in which the instructions move from the payer's bank to the beneficiary's bank. In many cases, this transfer may involve multiple intermediary banks or systems, illustrating the complex network that supports global financial transactions. In the following sections, we'll discuss both domestic transfers and cross-border transfers.

Domestic Credit Transfer

Domestic payments within a country have the essential advantage of often being settled directly between the sender's and recipient's bank. This usually takes place through local clearing systems, such as ACH in the United States or Faster Payments in the UK. These systems facilitate fast, efficient processing, allowing transactions to be completed within the same day or even instantly. As a result, this method of payment is particularly cost-effective as it minimizes intermediaries and reduces fees associated with cross-border transactions.

Cross-Border Credit Transfer

Cross-border payments are financial transactions in which the payer and the beneficiary are located in different national jurisdictions. Figure 2.14 shows the settlement of cross-border payments. Banks operate within local clearing systems; they rely on correspondent banks for cross-border payments if they do not have a direct connection to the respective foreign bank. This structure allows financial institutions to facilitate international transactions even without direct relationships, though it can introduce additional complexity, costs, and settlement time.

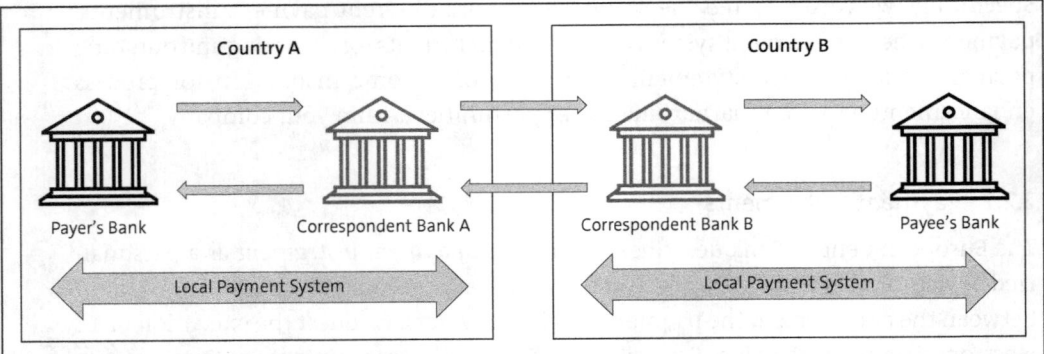

Figure 2.14 Cross-Border Payments

From a technical perspective, cross-border payments involve several key components and challenges that distinguish them from domestic payments.

Currency conversion is a fundamental aspect, requiring the exchange of one currency for another. This process is subject to exchange rate risk, which can impact the value received by the beneficiary due to fluctuations in currency values. As such, financial institutions must employ strategies to manage this risk effectively.

A network of intermediaries, including correspondent banks, must facilitate the transfer of funds across different countries. This multilayered process can introduce delays and increase transaction costs. Each intermediary may levy processing fees, contributing to the overall expense of the transaction.

Regulatory compliance is another critical factor in cross-border payments. Transactions must comply with international regulations, as well as the specific legal frameworks of the countries involved. This includes adhering to anti-money-laundering directives, counterterrorist financing measures, and know your customer requirements. These regulations are designed to prevent illicit activities but can also create additional documentation and processing requirements, thus impacting transaction speed.

Recent advancements in financial technologies are beginning to mitigate some inefficiencies in cross-border payments. Innovations such as blockchain technology and the implementation of standardized protocols like SWIFT GPI are enhancing transaction transparency, reducing processing times, and lowering costs by minimizing the need for traditional intermediaries.

One of the key aspects to consider with cross-border payments is the cost of transactions and conversions incurred during local clearing with the sending bank, the correspondent bank, and the recipient bank via local clearing networks.

It is crucial to determine who bears the costs—whether the sender, the recipient, or both—before executing the transaction. Banks also may charge fees as a percentage of the payment amount, which can significantly impact margins and must be considered in advance. The method of fee allocation is governed by payment instruction keys, which we introduce in the following box.

> **Payment Instruction Keys**
>
> Payment instruction keys are integral in defining how costs associated with a payment transaction are allocated among the parties involved. They play a significant role in the cost-sharing aspect of payment processing. Here are the essential payment instruction keys used for cost allocation in payment transactions:
>
> - OUR: The sender bears all the transfer costs. This means that the sender's bank fees and any intermediary or recipient bank fees are covered by the sender, ensuring that the full amount is received by the beneficiary.
> - SHA: The costs are shared between the sender and the beneficiary. The sender pays their bank fees, and the beneficiary covers the fees charged by their bank. Any intermediary bank fees are typically deducted from the transfer amount.
> - BEN: The beneficiary bears all the transfer costs. The sender does not cover any fees, so all costs, including those from intermediary and recipient banks, are deducted from the transfer amount, meaning the beneficiary receives a reduced amount.

Instant Payments

The speed of payment transactions has significantly increased due to pressure from alternative payment service providers. This has raised concerns about regulatory compliance, such as money laundering controls, and whether banking systems can handle mass payments instantly. In response, banks have adapted their settlement systems.

Instant Payments Within the US

The US Faster Payments Council has projected significant growth in instant payments by 2028. The US has two main systems for instant payments: FedNow and Real-Time Payments (RTP). Both enable 24/7 real-time transaction processing, but they differ in governance, access, clearing mechanisms, and transaction limits.

FedNow is a real-time payment system introduced in July 2023 by the US Federal Reserve. It operates round the clock, every day of the year, and is available to all banks and credit unions in the country. Participation is voluntary, and the system is directly embedded within the Federal Reserve's infrastructure, enabling instant processing without intermediary clearinghouses. The default transaction limit for FedNow is $100,000 per payment, but banks can increase this up to $500,000.

RTP, launched in 2017, is operated by the Clearing House, a consortium of major US banks. Like FedNow, RTP is available 24/7 and enables instant settlement. Participation is also voluntary, with around 353 banks currently connected. In February 2025, RTP raised its transaction limit to $10 million per payment, making it particularly attractive for businesses that handle large real-time transactions. Unlike FedNow, RTP relies on the Clearing House's clearing network for settlement.

Neither FedNow nor RTP is mandatory for US banks; that is, there is no legal requirement forcing banks to adopt either system. The decision to participate depends on a bank's strategy, infrastructure, and target customers. Although the US lacks specific laws governing instant payments, financial institutions must comply with general banking regulations.

One key challenge is the lack of interoperability between FedNow and RTP. Payments initiated via FedNow cannot be automatically received by banks using only RTP and vice versa. This fragmentation creates difficulties for businesses that work with multiple banks and payment partners. However, there have been discussions about potential interoperability. The American Bankers Association (ABA) has urged the Federal Reserve to work toward technical interoperability between FedNow and RTP. However, although both systems use similar messaging standards, their payment processing setups differ, making direct interoperability complex.

Some financial institutions are adopting both systems to maximize reach and flexibility. Over time, industry collaboration and regulatory efforts may lead to greater compatibility, but as of now, they remain distinct networks. The fact that both systems use the ISO 20022 standard is already a significant step toward the harmonization of the clearing and payment system landscape.

Payments Within the European Union

Instant payments in the SEPA area are electronic transfers that make funds available in the recipient's account within ten seconds, operating continuously. They are facilitated by the SEPA Instant Credit Transfer (SCT Inst) scheme, introduced in 2017, ensuring immediate availability and real-time confirmation. The adoption of instant payments varies across SEPA regions.

To standardize and encourage the use of instant payments, the Instant Payments Regulation (IPR) was adopted in March 2024. It requires payment service providers to offer instant transfers, mandates that costs for instant transfers will not be higher than standard ones, and provides payers with a free service to verify payee identities.

The European Central Bank supports these initiatives as part of its strategy to promote instant payments and monitors their adoption across the euro area.

Here are the key facts about instant payments, based on EU regulations:

- **Mandatory implementation**
  Payment service providers are required to offer real-time transfers 24/7 to their customers.
- **Cost parity**
  Instant payments must not be more expensive than traditional SEPA or SWIFT transactions.
- **Security requirements**
  Customers are screened daily against sanction lists. Their assets must be frozen immediately in the case of matches.
- **Transaction limits**
  Currently, real-time transfers are capped at €100,000, with the possibility of lifting this limit in the future.

Some additional rules are as follows:

- **Data validation**
  Payment service providers are required to perform thorough checks to ensure the accuracy of transaction details. For example, the International Bank Account Number (IBAN) must be matched with the recipient's name. If inconsistencies are detected, the payer must be alerted, ensuring a high level of fraud prevention and reducing errors.
- **Support for bulk file settlements**
  Businesses can process multiple payments simultaneously through bulk file submissions. This is particularly beneficial for payrolls, supplier payments, and other high-volume transaction needs as it allows efficient management of large-scale payments within the instant payment framework.
- **Cross-border payment enablement**
  Instant payments are designed to work seamlessly across EU member states, facilitating real-time cross-border transactions. This ensures businesses and individuals can transfer funds across borders as quickly and conveniently as domestic transfers.
- **Settlement timeframe**
  Every instant payment must be processed and completed within 10 seconds, including fund transfer, validation, and confirmation. This strict time frame ensures rapid fund availability for recipients, improving financial efficiency for users.

Direct Debit

The SEPA direct debit system standardizes direct debit transactions across participating European countries, facilitating both business-to-consumer (B2C) and business-to-business (B2B) payments, as you'll see in the following sections.

SEPA Core Direct Debit

SEPA Core Direct Debit, a B2C setup, is designed for transactions between businesses and consumers. It allows creditors, such as companies or service providers, to collect payments from consumers' bank accounts, provided a valid mandate has been authorized by the debtor. The key features are as follows:

- **Mandate authorization**
  The debtor (consumer) provides a signed mandate to the creditor, authorizing the collection of payments from their account.

- **Refund rights**
  Consumers have the right to request a refund for any authorized transaction within eight weeks without providing a reason. For unauthorized transactions, the refund period extends up to 13 months.

- **Processing timeline**
  The creditor must submit the direct debit request to their payment service provider at least one business day before the due date.

As shown in Figure 2.15, direct debits have the following process steps:

❶ **Mandate issuance**
The payer (debtor) grants the payee (creditor) a written mandate. This mandate authorizes the creditor to collect payments from the debtor's account and instructs the debtor's bank to execute these payments. The mandate must include specific details:

- The debtor's name and address
- The creditor's identification number (creditor identifier [CI])
- A unique mandate reference number to identify each direct debit

❷ **Advance notification (prenotification)**
The creditor informs the debtor in advance about the upcoming debit. The timelines are as follows:

- Typically, notification should occur at least 14 calendar days before the debit date.
- Shorter notice periods can be agreed upon between the parties.
- The notification specifies the amount, the debit date, and other relevant details.

❸ **Submission of the direct debit**
The creditor submits the direct debit request to their bank (the first collecting bank). The submission is made in XML format in line with SEPA specifications. The deadlines for direct debits are as follows:

- First-time debits: At least five business days before the due date
- Recurring debits: At least two business days before the due date

2.2 Financial Standards

❹ Processing/clearing by banks

The following tasks take place in this step:

- The creditor's bank forwards the direct debit to the debtor's bank via the SEPA clearing system.
- Banks verify the validity of the details, such as the IBAN and CI.

❺ Debit of the debtor's account

On the due date, the debtor's bank debits their account and transfers the amount to the creditor's bank. There is a condition: The debtor's account must have sufficient funds. Otherwise, the direct debit may be rejected.

❻ Crediting the creditor's account

The collected amount is credited to the creditor's account, typically on the same day or the day after the debtor's account is debited.

❼ Reversal (optional)

If issues arise, different reversal mechanisms apply. The debtor's right to refund can be as follows:

- B2C: Refunds can be requested within eight weeks without needing to provide a reason.
- B2B: No refund is allowed for authorized transactions.
- Unauthorized Transactions: Refunds can be requested within 13 months.

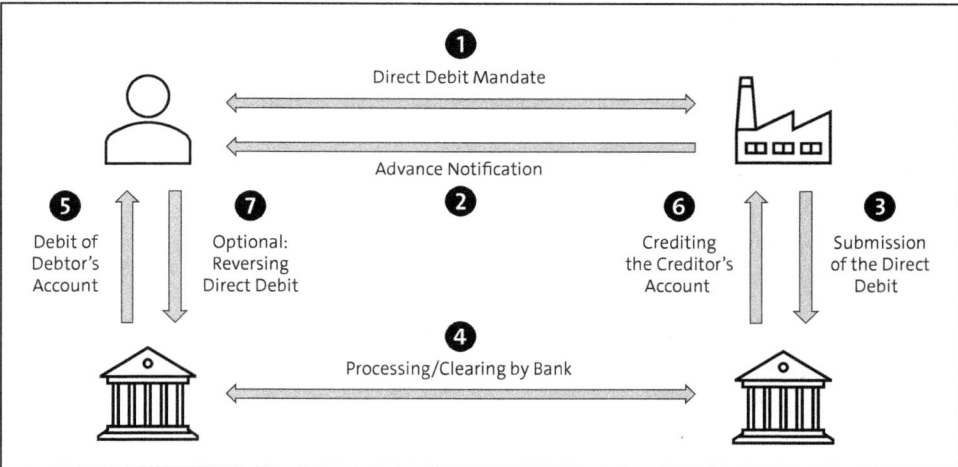

Figure 2.15 Direct Debits

> **Archiving Requirements**
>
> The mandate must be archived by the creditor for at least 14 months after the last direct debit as it may be needed for future verification.

SEPA B2B Direct Debit

The SEPA B2B Direct Debit scheme is tailored exclusively for transactions between businesses. It enables creditors to collect payments from debtors' business accounts, given that a valid mandate is in place. The key features are as follows:

- **Mandate authorization**
  The debtor (business) provides a signed mandate to the creditor. The debtor's bank also must verify the existence of this mandate before processing any direct debit requests.

- **No refund rights**
  Unlike the B2C scheme, B2B Direct Debit does not offer refund rights for authorized transactions. Once the payment is processed, it is considered final.

- **Processing timeline**
  The creditor is required to submit the direct debit request to their PSP at least one business day before the due date, like the B2C scheme.

The key differences between the B2C and B2B direct debit options are as follows:

- **Target audience**
  B2C is intended for consumer transactions, whereas B2B is exclusively for business-to-business transactions.

- **Refund policy**
  B2C allows consumers to request refunds within specified timeframes, whereas B2B does not permit refunds for authorized transactions.

- **Mandate verification**
  In the B2B scheme, the debtor's bank is obligated to verify the mandate before processing payments, adding an extra layer of security. This verification step is not required in the B2C scheme.

- **Participation requirement**
  All banks within the SEPA region are mandated to support the B2C scheme. In contrast, participation in the B2B scheme is optional for banks.

Understanding these distinctions is crucial for businesses to select the appropriate direct debit scheme that aligns with their transaction needs and to ensure compliance with SEPA regulations.

Check Payments

Check payment is a traditional method in which a written, signed document authorizes a bank to transfer money from the payer's account to the recipient. Common in personal and business transactions, checks are useful when electronic payments are unavailable or unsuitable. Businesses often use checks for large payments, for vendor settlements, or to maintain a paper trail. Although check payments are less common in today's digital world, they remain quite popular in certain countries—particularly in the US and UK, where checks continue to be a common method of payment for businesses and individuals alike. Checks are also used in some other countries, although their popularity is generally declining.

As a result, whenever you implement payment systems or solutions, it's important to ensure that checks are included in the scope of your payment processing capabilities, especially if you operate in regions where checks are still widely used. This ensures that businesses can efficiently handle check payments alongside more modern electronic methods. However, it is essential to note that check payments generally take longer to process compared to electronic methods and can involve additional administrative efforts for tracking and clearing.

To execute a check payment in SAP S/4HANA, follow these steps:

1. Maintain bank master data in the system, including the details of the bank account from which the check will be issued.
2. Configure the payment method as *check* within the payment program (table FPAYH). Specify the check format, whether it will be printed manually or automatically, and the relevant printing parameters.
3. Create the check lot, which is a unique identifier for a group of checks issued in a specific payment run.
4. Configure the payment run to include checks as part of the payment method, along with parameters such as check number assignment and vendor data.
5. Use SAP S/4HANA's functionality to print checks through the check print program, formatting the check with necessary information such as the payee, amount, and bank details.
6. Properly reconcile issued checks against bank statements, ensuring that cleared checks are accurately reflected in SAP S/4HANA.

In the US, the escheatment process complicates check payments, requiring organizations to manage unclaimed funds per state laws. If a payee does not cash a check within three to five years, depending on state regulations, the payment is deemed *abandoned*. The issuer must then report and remit these funds to the state's unclaimed property office. This adds an administrative burden as organizations need to track uncashed checks, identify which ones require escheatment, and comply with various state requirements. Handling unclaimed funds and managing claims from rightful owners is time-consuming and complex, necessitating meticulous record-keeping and adherence to legal standards.

Lockboxes are specialized services provided by banks to help businesses efficiently manage incoming payments, including checks and other forms of remittance. They allow companies to have their customers send payments directly to a designated post office box, where the bank retrieves the payments, processes them, and deposits the funds into the company's account. This service is particularly common in the United States.

The Bank Administration Institute version 2 (BAI2) standardized electronic file format is used by banks to transmit detailed payment and remittance information from lockbox processing. The BAI2 file includes essential data such as payment amounts, customer details, and any associated invoices or payment references, allowing businesses to automate the reconciliation of payments with their accounts receivable records. Lockboxes and BAI2 files are commonly used by businesses with high volumes of payments, such as utilities, insurance companies, and large retailers, to streamline payment collection and reduce manual processing time.

By using lockboxes and BAI2 files, organizations can enhance cash flow management, improve operational efficiency, and ensure more accurate and timely reconciliation of payments.

In SAP S/4HANA, the lockbox functionality in accounts receivable (AR) simplifies processing and reconciling incoming payments. When a company uses a lockbox service, the bank processes payments and sends remittance data in a format like BAI2. This data is imported into SAP through lockbox processing to automatically post customer payments. The system matches payments with outstanding invoices, reducing manual work and improving cash flow management. SAP S/4HANA also allows for automatic reconciliation and generates detailed reports for tracking payments. Using lockbox processing reduces administrative efforts and enhances the speed and accuracy of cash applications.

2.2.2 Payment Schemes

In simplified terms, a *payment scheme* is a set of rules for processing payment transactions using tools like payment cards, mobile apps, and e-wallets. These rules cover issuing payment tools, offering services to accept payments, and processing transactions. The organization running the payment scheme creates these rules, may carry out these activities itself in some models, and oversees the scheme's operation and decision-making.

Payment schemes are available across countries, such as SEPA, but there are also local "flavors" like the SEPA scheme from SIX in Switzerland or STUZZA from Austria. In a payment traffic implementation, software providers often offer templates that focus on country-specific requirements and align with organizations like the German Banking Industry Committee (DK), SIX, or STUZZA. Nevertheless, experience shows that a country-specific customization of formats is not sufficient. Despite the unification of formats to a small extent, the heterogeneous system landscape of the banking industry requires not only country-specific but also bank-specific adjustments.

Single European Payment Area

The SEPA system has its roots with the introduction of the euro. On January 1, 1999, 11 EU member states decided to adopt the euro: Belgium, Germany, Ireland, Spain, France,

Italy, Luxembourg, the Netherlands, Austria, Portugal, and Finland. The unified currency faced the challenge of a heterogeneous settlement, format, and system landscape. This issue laid the foundation for SEPA.

SEPA was introduced 2008 for credit transfers and 2009 for direct debits. It establishes a harmonized environment in which individuals, businesses, and organizations can make electronic payments under the same conditions, rights, and obligations, regardless of national borders. This initiative supports the EU's broader goal of creating a unified financial market by eliminating barriers to cross-border transactions. SEPA is applicable to euro payments and covers credit transfers, direct debits, and card payments.

Today, SEPA is not limited to the EU member states. In addition to the 27 EU countries, it also includes several non-EU participants, such as Norway, Iceland, Liechtenstein, Switzerland, and Monaco. This broad participation ensures that SEPA's benefits extend to a wide range of economic activities across Europe.

The key payment schemes under SEPA include the following:

- SEPA Credit Transfer: A standardized framework for transferring funds between bank accounts across SEPA countries
- SEPA Direct Debit: A mechanism allowing businesses to collect recurring payments directly from customers' accounts (discussed in the previous section)
- SEPA Instant Credit Transfer: A real-time payment system enabling transactions to be completed within seconds, available 24/7

Although SEPA has significantly streamlined payment processes, challenges remain. Adoption of SEPA standards can vary across countries and sectors, and ongoing efforts are required to ensure full compliance and utilization. As financial technology evolves, SEPA also must continue to adapt to incorporate innovations such as blockchain and digital currencies. Overall, SEPA has significantly contributed to the transparency, efficiency, and cost optimization of European payments. The processing speed for mass payments has improved from the previous D+2 to D+1. On the business day following a bank's acceptance of a payment order (subject to sufficient funds and cutoff times), the recipient must have received the liquidity.

In the following sections, we explore the developments and innovations within SEPA, such as IBAN only, SEPA Request-to-Pay (SRTP), and verification of a payee via SEPA.

IBAN Only

With the introduction of SEPA, local account numbers were replaced by IBANs and BICs. Since February 1, 2016, *IBAN only* applies throughout the SEPA area. This means that customers can initiate payments by providing only the beneficiary's IBAN, without needing to include the associated BIC. The originating bank must add the BIC, however, as it's essential for interbank clearing.

For the originating bank, this means that they must allow manual entry of SEPA payments in online banking with only the mandatory IBAN and optionally without a BIC. The IBAN only functionality does not check whether the payer is a consumer, meaning it can be used by both consumers and other types of payers.

> **Note**
> Even after the introduction of IBAN only, a BIC is still required, for example, for payment transactions outside the SEPA area.

SEPA Request to Pay

SRTP, officially launched in June 2021, is gaining popularity, particularly as the transaction limits for instant payments decrease. Instead of the traditional approach of sending an invoice and awaiting payment, billers can now use SRTP to request payment directly from the recipient. The recipient receives a prefilled digital payment request within their banking environment, with the invoice document attached. This process requires only authorization of the payment, eliminating the need to manually enter details like the payment purpose and the recipient's IBAN.

The process is particularly appealing for companies with the following attributes:

- A high volume of invoices
- Significant manual allocation effort
- Customers who are self-payers

This process also aids in optimizing working capital, reduces the need for paper invoices, and mitigates the risk of direct debit chargebacks (in case of rejection). Therefore, it is especially beneficial when dealing with customers who have a negative credit rating.

Here are the key characteristics, functions, and processes of SRTP:

❶ **Payer identification**
After the payer has opted for SRTP, they typically need to identify themselves to the payee via a website or app. The mandatory data required is agreed upon between the SRTP provider and the parties involved, the payer and payee, and they generate a token. This token is stored in an SRTP identification register accessible by all banks, which contains a mapping to the payer's bank. Only the payer's own bank has access to the underlying identity and account information. This setup allows the payee's bank to identify the payer's bank using the token, eliminating the need for the payer to share their account details with merchants or other banks.

❷ **Request to pay initiation by payee**
If no additional payment service provider is placed between the payee and the executing bank, a pain.013 request to pay initiation file can be created from the available

data. This file is then transmitted through the communication channel agreed upon with the bank (such as EBICS, SWIFT, host to host, or API), thereby initiating the process at the payee's bank. The following information has to be provided to the request to pay provider as per the SRTP scheme rulebook (EOC014-20/2023 Version 3.2):

- E001: Name of the payee (mandatory)
- E004: Address of the payee (mandatory)
- E002: Trade name of the payee (optional)
- E006: Merchant category code of the payee (mandatory)
- B001: IBAN(s) used by the payee in the SRTP messages (mandatory)
- B002: Enrollment start date (mandatory)
- B003: Enrollment end date (optional)
- B004: Enrollment contract reference type (optional)
- E012: Ultimate payee (optional)
- B006: Payee logo (optional)
- B008: Visibility (optional)
- B009: Service activation allowed (mandatory)
- B010: Service description link (optional)
- B011: Payee service activation link (optional)
- B013: Limited presentment indicator (optional)
- B014: Counterpart identification type (optional)
- B015: Contract format type (optional)
- B016: Contract reference type (optional)
- B017: Payee instruction (optional)
- B019: Activation request delivery party (mandatory)

❸ **Request to pay settlement**
The payee's bank receives the pain.013 RTP message and identifies the payer's bank using the RTP identification register. The payee's bank then forwards the pain.013 RTP message to the payer's bank. Upon receiving the pain.013 RTP message, the payer's bank uses the RTP identification token to identify the payer and retrieve their account details.

❹ **Request for confirmation**
The payer's bank sends the request to pay to the payer's banking app. The payer has several options to choose from:
- Accept the payment and execute it immediately.
- Defer the payment to pay it later.
- Reject the payment.

❺ **Payment confirmation**
After the payer responds to the message, their feedback is forwarded to the payee's bank via pain.014: a payment activation request status report. The payee's bank then informs the payee of the payer's response.

❻ **Settlement of the request to pay over interbank direct debit**
If the payer accepts the payment settlement, the amount is collected via interbank direct debit. For this, the payee's bank sends a pain.008 direct debit message to the payer's bank to withdraw the amount from the payer's account. Subsequently, the payee receives an intraday account statement (camt.052) or an end-of-day account statement (camt.053) the following day with the credited amount.

You can see an overview of this process in Figure 2.16.

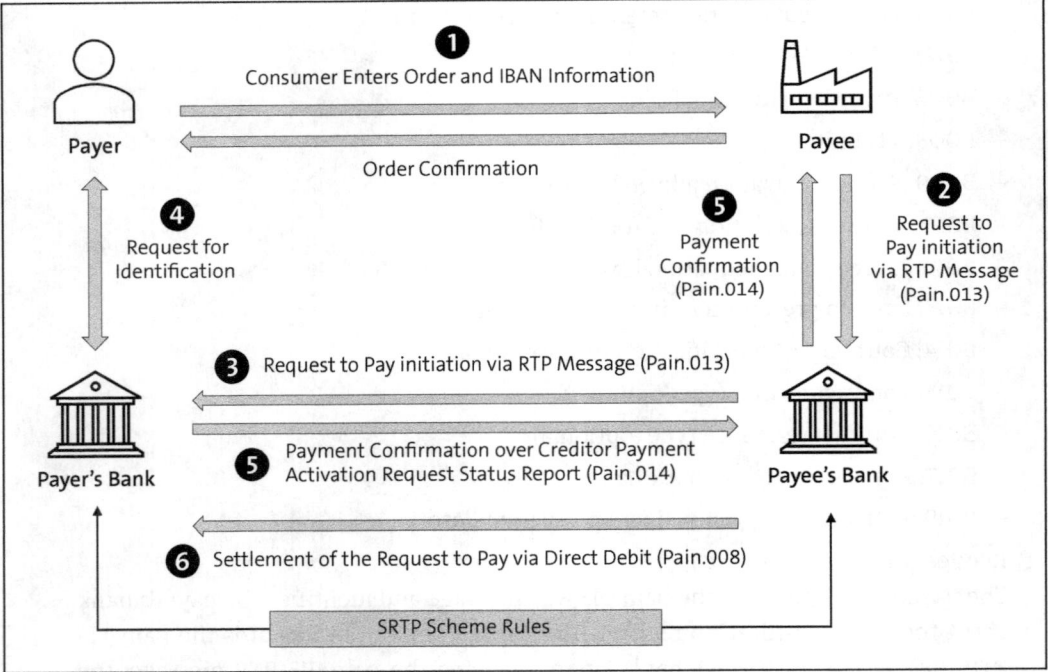

Figure 2.16 Request to Pay

Overall, the SRTP process, especially when combined with instant payments and electronic invoices, is a valuable tool for optimizing the working capital and cash management processes. The flexible structuring and customization of payment terms make cash inflows more predictable. It remains to be seen whether the process will establish itself in the B2B sector. However, companies heavily involved in e-commerce or those with a strong B2C focus looking to digitize their operations are likely to benefit significantly.

2.2 Financial Standards

Verification of Payee

Originally defined as an additional security mechanism for real-time transfers (or *instant payments*), the European Commission is expanding the SEPA scheme for verification of a payee to include regular SEPA transfers in its current draft of the Payment Services Regulation (PSR) in the SEPA area. The idea behind the verification of a payee is quite simple and definitely sensible: It aims to verify whether the specified account actually belongs to the named recipient, thereby enhancing security in payment transactions and making life more difficult for fraudsters.

The scheme rulebook was first published on October 10, 2024, and the implementation of and support for verification of payee will be mandatory for banks starting from October 5, 2025 based in the current publications of the European Payments Council. The rulebook is based on the API and XML-based ISO 20022 standard.

To achieve this, the European Payments Council (EPC), in collaboration with SWIFT, has launched a project that uses an anonymized generated identifier to uniquely identify recipients. The verification of the payee will be carried out in conjunction with the bank's local master data.

Figure 2.17 shows a simplified overview of the process, which has the following steps:

❶ **Payer sends a payment file to the bank**
The payer sends a payment order to the executing bank. In our case, for example, this could be a SEPA credit transfer via pain.001.

❷ **Instant verification of payee request**
Upon receiving the payment order, the bank initiates a verification process to confirm that the provided IBAN matches the designated payee. This involves utilizing a routing and verification mechanism, which acts as a gateway to facilitate coordination between the bank's master database and the Electronic Directory Service of the EPC. The routing and verification mechanism handles the communication of verification of payee requests and responses. Once the master data is successfully matched, a response is generated.

❸ **Instant verification payee response**
The payer receives a pain.002 report from their bank, which communicates the result of the verification. The possible outcomes are as follows:

– Match: The requested IBAN matches the provided recipient's name.
– Close match: The sent name partially matches the name held by the recipient's bank.
– No match: The sent name does not match the name held by the recipient's bank.
– No check: This may happen, for example, if a timeout occurred during the communication process.

Upon receiving the response, the payer can decide to authorize or cancel the payment. It is possible to proceed even with a "no match" result, but it would be at the payer's own risk.

2 Bank Connectivity

④ Electronic signature
After reviewing the result, the payer can electronically sign the payment document and authorize the payment. With EBICS, this process works through distributed signatures.

⑤ Credit notification
Once the payment has been executed, the recipient receives a credit, which can be viewed through the intraday account statement (camt.52) or on the next banking business day via the end-of-day account statement (camt.53).

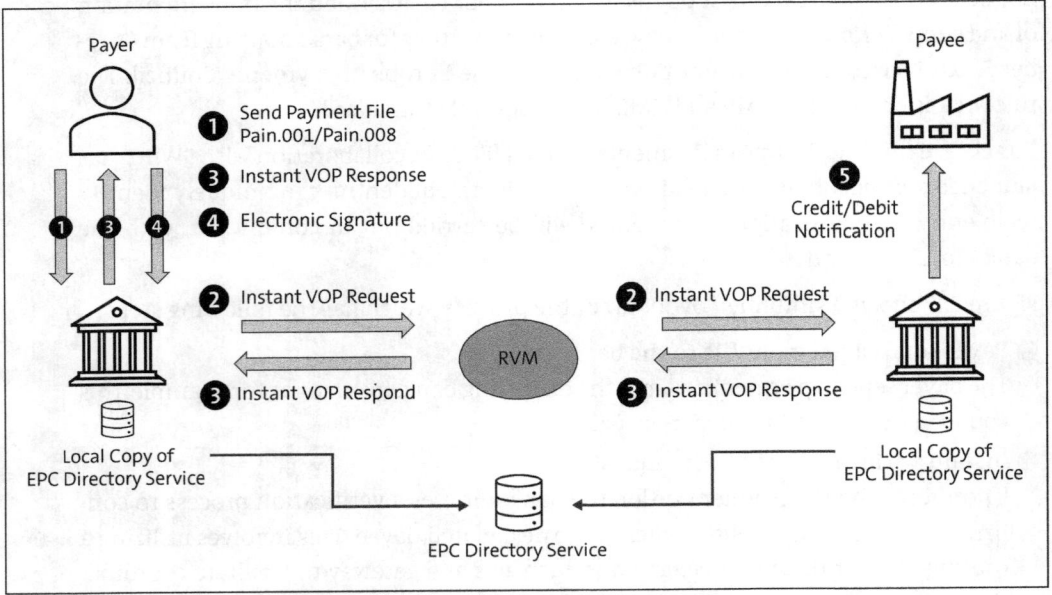

Figure 2.17 Verification of Payee

The Instant Payment Regulation stipulates that corporate customers (specifically non-consumers) can voluntarily participate in the verification of payee process. If they opt in, the customer can have their data verified; opting out skips the verification of payee process and sends the payments without verification. This allows for process control. The EU requires banks to offer the verification of payee process free of charge when it is related to a payment. Although the verification of payee process is technically separate from the actual payment process, it should not be used without an associated payment. Overall, it is currently expected that there will be complications in payment transactions. In particular, a heterogeneous system landscape and different databases can lead to complications.

For more information about this process, visit *https://www.europeanpaymentscouncil.eu/what-we-do/other-schemes/verification-payee*.

> **Check Your Master Data**
>
> It is recommended to reconcile the master data with your bank to ensure how it is stored and whether all the master data is indeed correct. Furthermore, your system should support the processing bank protocols of pain.002 messages in version pain.002.001.10.

ISO 20022 CGI

At the beginning of the new millennium, efforts were made to harmonize local and diverse formats under a unified technological framework. Extended Markup Language (XML) was chosen as the basis for this unified format. XML offers the advantage of being platform-independent, allowing information to be transported across systems through defined rules and thus providing the necessary flexibility. These rules were encapsulated in the ISO 20022 standard. ISO stands for the International Organization for Standardization, an independent, nongovernmental, international organization that defines global standards. The management of the ISO 20022 standard has now been transferred to SWIFT, which supports the global implementation of the standard.

The ISO 20022 standard facilitates interbank communication as well as corporate-to-bank and bank-to-corporate message exchange. It is continuously developed by the Common Global Implementation (CGI) group, which consists of leading financial institutions, consulting firms such as Zanders, and corporations.

The ISO 20022 standard or Universal Financial Industry Message Scheme (UNIFI) for XML is the basis for today's payment transactions and for all SEPA data formats. In future, these formats will also serve as the basis for SWIFT formats. All SEPA message types used for customers and banks are based on ISO 20022.

2.2.3 Payment Systems

Payment systems are crucial frameworks that enable the transfer of funds between participants via established instruments, procedures, and rules. These systems, managed by set agreements, include participants and an operational entity. Historically, payment systems date back to early central banks such as the Bank of Amsterdam, which established structured mechanisms for merchants and governments. Beyond traditional systems, payments can also be conducted using methods like banknotes or digital tokens. Modern innovations, including platforms like PayPal and Satispay, provide alternative payment methods that require more sophisticated systems compared to traditional approaches.

Figure 2.18 illustrates the key steps of a transaction settlement in a payment system.

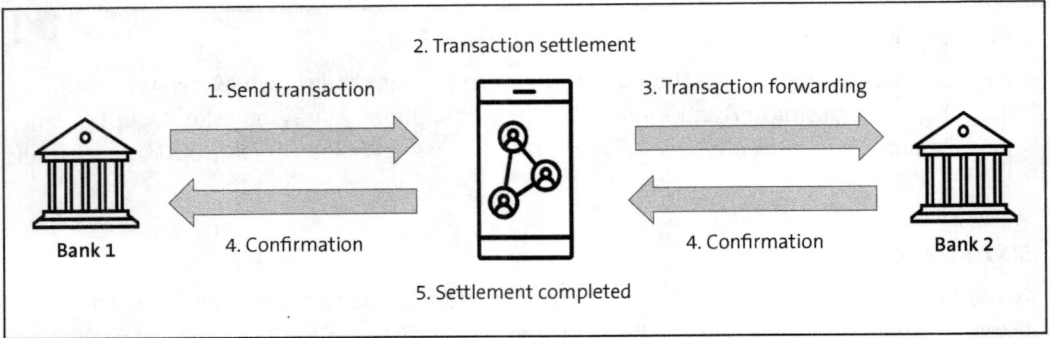

Figure 2.18 Payment System: Transaction Settlement

Let's consider some common types of payment systems:

- **Real-time gross settlement system**

  In a real-time gross settlement (RTGS) payment system, transactions between financial institutions are settled individually, in real time, without netting or delay. This means that each transaction is processed and settled immediately, and funds are transferred on a gross basis, rather than being aggregated with other transactions.

  In such a system, payments are final and irrevocable once processed, eliminating the settlement risk that could arise if transactions were delayed. The system operates continuously throughout the day, allowing for the rapid and secure transfer of large-value payments.

  For example, TARGET2, operated by the European Central Bank, is an RTGS system that processes euro-denominated transactions across the Eurozone, ensuring that funds are transferred without delay and with immediate finality.

  Unlike systems that use netting (where transactions are grouped and settled at the end of the day), RTGS systems settle payments individually, which increases the transparency of payments and reduces the risk of defaults or settlement failures. However, because payments are settled in real time, participants typically need to have sufficient liquidity to cover their payments as they occur. This requires effective management of available funds and liquidity to avoid delays or failure to settle payments.

- **Net settlement system**

  A net settlement is an interbank payment settlement system in which banks collect transaction data throughout the day and exchange this information with a clearinghouse and the central bank to settle outstanding amounts. Banks track their electronic and physical credit and debit transactions, and at the end of the day, the net difference is settled between the participating banks. There are different types of net settlement systems, as follows:

- Bilateral net settlement system: Payments are settled for each bilateral combination of banks. Banks with a positive net balance are credited, and those with a negative balance pay the difference.
- Multilateral net settlement system: Received transfers are offset against sent transfers. A positive balance indicates a net credit position, while a negative balance indicates a net debit position.
- Deferred net settlement system: Payment obligations can be deferred to a later time based on agreements between the involved parties.

In contrast, RTGS settles transactions immediately. In net settlements, the risk of default is higher as payments are not made instantly.

Now that we've looked at the types of settlement systems, let's examine some common clearing and settlement systems:

- **CHIPS**

   CHIPS is a privately operated large-value payment system in the US, run by the Clearing House, a company owned by banks. It processes over 400,000 payment orders per day, with a total value of approximately $1.8 trillion.

   The system employs a patented liquidity-saving algorithm that matches and nets payments either on a bilateral or multilateral basis, thus enhancing liquidity efficiency.

   Although not strictly an RTGS system, CHIPS processes some payments individually. With more than 40 direct participants, CHIPS provides a private alternative to the Federal Reserve's Fedwire system.

- **CHAPS**

   The Clearing House Automated Payment System (CHAPS) is the UK's RTGS system that has been operated by the Bank of England since 2017. It facilitates high-value and time-critical payments, offering efficiency and reliability.

   The system has over 30 direct participants and more than 5,000 financial institutions that access CHAPS through these direct participants. CHAPS employs a liquidity-saving netting algorithm to optimize payment processing and reduce liquidity demands. The Bank of England also provides intraday liquidity on a collateralized basis to ensure smooth transaction flows. This combination of advanced algorithms and liquidity support makes CHAPS a robust and secure platform for critical payment operations. For more information, visit *https://www.bankofengland.co.uk/payment-and-settlement/chaps*.

- **TARGET2**

   TARGET2 is the RTGS system of the EU, facilitating the rapid and secure transfer of funds between banks across the European Union. It enables banks to process payments in real time, ensuring immediate settlement of transactions.

As of March 2023, TARGET2 has been replaced by T2, which continues to provide these essential payment services within the EU. The system is integral to the smooth functioning of the EU's financial markets, supporting monetary policy implementation and the stability of the financial system. By providing a reliable platform for high-value and time-critical payments, TARGET2 enhances the efficiency and security of financial transactions across Europe. For more information, visit *https://www.ecb.europa.eu/ecb-and-you/explainers/tell-me/html/target2.en.html*.

- **TIPS**

In November 2018, the European Central Bank (ECB) launched a new instant payment service: TARGET Instant Payment Settlement (TIPS). This service enables both private and commercial bank customers to process payments across Europe within seconds, 24/7, throughout the entire year. With TIPS, the ECB facilitates cross-border transfers in less than ten seconds. TIPS is considered part of TARGET2, discussed previously. Its key unique feature is the ability to transfer funds instantly in central bank money, 365 days a year. To encourage bank participation, the ECB has adopted an attractive pricing strategy. The price per instant payment transaction is fixed at 0.2 cents (€0.002), split equally between the sending and receiving participants in TIPS (i.e., €0.001 for the originator and €0.001 for the beneficiary). Monthly fees also apply for PSPs and automated clearing houses (ACHs) connected to TIPS for holding accounts on the platform. TIPS can be used by both individuals and businesses. For more information, visit *https://www.ecb.europa.eu/paym/target/tips/html/index.en.html*.

- **EURO1**

EURO1 is a pan-European large-value payment system (LVPS) operated by EBA CLEARING and launched in 1999 with the euro. It serves as a private alternative to the ECB's TARGET2, processing high-priority and large-value euro transactions.

EURO1 operates on a multilateral net settlement basis for liquidity efficiency while providing immediate payment finality. To manage credit risks, it employs bilateral and multilateral limits, with finality ensured only within these constraints. Discretionary limits also determine loss-sharing in case of participant default.

With 51 direct participants and over 5,000 indirect users, EURO1 is a key infrastructure for secure and efficient euro transactions across Europe. For more information, visit *https://www.ebaclearing.eu/services-single-payments/euro1/*.

Zengin

The Zengin payment system is a unique and specialized payment network in Japan, designed to handle domestic interbank transfers securely and efficiently. What makes Zengin special compared to other payment methods is its focus on real-time payment processing for a wide range of transaction types, including credit transfers, direct debits, and batch processing of payments. Zengin enables the seamless and rapid exchange of payment data between participating financial institutions within Japan, ensuring that funds can be transferred quickly between banks, whether for individual or corporate payments.

Zengin uses a specific text-based file format, commonly known as the Zengin format, for payment processing. These Zengin text files contain detailed information about a transaction, including payer and payee bank account numbers, payment amounts, and any associated remittance information. The format adheres to strict standards to ensure smooth communication between banks and minimize errors during processing. However, Japan is currently in the process of transitioning to XML-based messages as part of the global shift to the ISO 20022 standard. This change is expected to bring more structured, flexible, and interoperable messaging for payment transactions. As a result, we can anticipate growing adoption of XML-based Zengin formats in the future, allowing for easier integration with international systems and enhanced support for more complex financial transactions.

Payments processed via Zengin can be made in real-time or in batch mode, depending on the nature of the transaction. For real-time payments, funds are typically transferred and available in the recipient's account almost instantly, making it ideal for time-sensitive transactions. Batch processing allows multiple payments to be bundled together and processed at once, which is useful for bulk payments, such as payroll or vendor settlements.

One of the key features of Zengin is its high level of security, which ensures that payment data is encrypted and transmitted securely between banks. In addition, the system is tightly regulated by the Japanese Bankers Association (JBA), ensuring compliance with domestic financial regulations and safeguarding against fraudulent activities.

Another distinctive feature of the Zengin system is its bank statement generation process. Zengin generates multiple bank statements per day for each participating bank, ensuring that transactions are promptly recorded and reconciled. These statements are often available in real time or at regular intervals throughout the day, providing businesses and individuals with up-to-date information on their account activity. These frequent updates help ensure transparency and allow users to quickly detect and resolve any discrepancies in their payments. Compared to international systems, Zengin's ability to issue frequent and timely statements is a valuable feature, supporting efficient cash flow management and payment tracking within Japan's financial ecosystem.

Compared to international payment systems, Zengin is tailored specifically for the Japanese market, with banks in Japan using it for routine, low-cost, and high-efficiency transactions. Its ability to process both real-time and batch payments, along with its robust file formats, security measures, and daily bank statement generation, makes it a crucial component of Japan's financial infrastructure. As the system transitions to XML formats aligned with ISO 20022, Zengin will become even more adaptable and integrated into global payment networks.

In SAP S/4HANA, the processing of Zengin formats is integrated into financial accounting (FI) and AR, allowing businesses to efficiently manage domestic payment transactions in Japan. Outgoing payments can be generated by program RFFOJP_T.

Zengin files, typically in text format, are imported into SAP S/4HANA using the electronic bank statement functionality by using the IBS_JP app and the IBSX_JP app (for XML files). SAP S/4HANA's electronic bank statement processing allows the system to read and interpret the payment data contained within the Zengin text files, such as payment amounts, payer and payee account details, and remittance information. The system then automatically processes this data, matching payments to open invoices and updating the relevant customer accounts.

India UPI

The Unified Payments Interface (UPI), introduced in 2016 by the National Payment Corporation of India (NPCI), is a cutting-edge, real-time payment system designed for seamless and instant credit transfers. Sponsored by NPCI—a not-for-profit organization owned by Indian banks and Fintechs—UPI provides a simple way to link your bank accounts to a mobile application. This integration allows you to conduct fund transfers and merchant payments efficiently from a single platform.

UPI's open protocol fosters innovation and scalability, supporting over 350 bank accounts with a rapidly growing user base exceeding 100 million monthly active users and handling more than 7 billion transactions per month. The ecosystem thrives due to its open architecture, enabling complementary solutions like Bharat QR for QR-based payments, Bharat Billpay for secure and efficient bill settlements, and BHIM Aadhaar Pay, which facilitates biometric-based payments for merchants.

In addition to UPI, NPCI manages RuPay, a domestic card scheme, and extends UPI's capabilities globally by collaborating with international operators. These partnerships enable merchants abroad to accept payments via UPI and RuPay, broadening its utility for Indian tourists and driving global financial integration.

2.2.4 Payment Formats

Formats in bank communication are designed to convey information for executing and confirming bank transactions consistently across the payment schemes. To facilitate automation, standardized formats have been defined by institutions such as SWIFT, as well as by banking associations like the German Banking Industry Committee and SIX in Switzerland, in addition to central bank systems like Zengin in Japan.

Over the past decades, a wide range of standards have emerged in the market. These are gradually being replaced by the global XML-based format standard ISO 20022.

The following sections provide an overview of the currently prevalent formats, their structure, and their advantages and disadvantages.

CSV Formats

Comma-separated value (CSV) files are text files used to store and exchange structured data easily. As the name suggests, these files use commas to separate data fields or columns within a dataset. Depending on the software and settings, other delimiters such as colons, semicolons, spaces, or tabs may also be used. Overall, the format is used for payment instructions as well as account statements and transaction information. Older formats like those used by Multicash (e.g., Auszug/Umsatz.txt) are often in this style.

Interestingly, in the payment sector, this type of file is experiencing a resurgence. Payment service providers are increasingly using the CSV format to transmit transaction data. Due to their straightforward structure and wide interoperability, CSV files offer a quick way to establish interfaces, even if they are not the most cutting-edge solution. However, there is a broader trend toward XML for data exchange.

Fixed-Length Formats

Fixed-length formats are a method of storing or transmitting data where each field has a fixed, predetermined length. A well-known example of this is the Data Carrier Exchange with Foreign Countries in Payment Transactions (DTAZV) format. In this specific format, each piece of information in a record is stored in a precisely defined number of characters.

For example, a DTAZV record might include fields such as account number, amount, and payment description, with each field having a fixed length: the account number is 10 characters, the amount is 12 characters, and the payment description is 30 characters. If the account number is "123456789," it is stored in the fixed 10-character field as "123456789 ", with a space appended to the number to fill the length. The amount "1500.50" might be stored in the 12-character field as "000000150050" to reach the required length. A description like "Rent" would be stored in the 30-character field as "Rent ", with additional spaces to ensure the fixed length.

One significant advantage of this format is the ease of reading its data: Because the position and length of each field are predefined, data can be processed automatically and efficiently. Additionally, the overall length of a record remains constant, simplifying the management of large datasets.

However, there are also disadvantages, such as space wastage: When the actual information is shorter than the reserved length, the unused space is still occupied. Moreover, the format offers a rigid structure; changes to the data structure require extensive adjustments to the systems processing the data.

Despite these limitations, formats like DTAZV are useful in certain areas, especially where quick and consistent data processing is needed, although today more flexible formats are often preferred.

In addition to DTAZV, other payment transaction formats created in fixed-length formats include DTACH (used in Switzerland), MT940/MT942 (SWIFT formats for electronic

2 Bank Connectivity

banking statements), and EFT files in some regions for electronic funds transfer. These formats similarly employ fixed-length records to ensure data integrity and processing efficiency.

SWIFT Message Types

The SWIFT message type format, abbreviated as SWIFT MT, historically replaced the transmission of data via fax and teletype in international correspondent banking. With SWIFT's FIN, the MT format was defined.

An MT101 message consists of multiple structured blocks, with the basic header, application header, and text block containing the key information:

- The basic header block (block 1) carries essential routing details, such as the bank's sender address and an indication that it is a financial message. This block defines the message format and direction.
- The application header block (block 2) specifies the type of message being sent—for MT101, this is a payment instruction—and identifies the intended recipient. It also includes information about the message priority and the role of the recipient (e.g., the FIN receiver in SWIFT).
- The text block (block 4) is the most critical section, as it contains the actual payment details. This block includes all relevant fields, such as the payment reference number (:20:), ordering customer information (:50K:), beneficiary details (:59:), amount and currency (:32B:), value date (:30:), and expense allocation details (:71A:). Optional fields may be used as well for the payment purpose or intermediary bank details. The text block can contain multiple individual payments within a single message, making it the core of the MT101 format.

Listing 2.1 shows an example of an MT101 payment.

```
MT101 - Request for Forwarding
{1:F01BANKBEBBAXXX0000000000}{2:01011205030605BANKDEFFFXXX-
30074748530306051205N}{3:{113:ROMF}}{4:
:20:REFERENCE12345
:28D:1/1
:50A:/12345678901234567890
BANKDEFFXXX
:30:231022
:21:RELATEDREF1
:32B:EUR100000,
:59:/DE09876543210987654321
BENEFICIARY NAME
:70:Invoice 1234 Payment
:71A:SHA
-}{5:{CHK:123456789ABC}}
```

2.2 Financial Standards

Header (Blocks 1 & 2):
{1:F01BANKBEBBAXXX0000000000}: Sender BIC and other identifiers.
{2:O101...}: Message type (MT101) and processing information.

Application Header (Block 3):
{3:{113:ROMF}}: Optional application header with additional information.

Text Block (Block 4):
:20:REFERENCE12345: Transaction reference number.
:28D:1/1: Page number and continuation indicator.
:50A:/12345678901234567890\nBANKDEFFXXX: Ordering customer's account and BIC.
:30:231022: Requested execution date in YYMMDD format.
:21:RELATEDREF1: Related reference, optional for reconciliation purposes.
:32B:EUR100000,: Currency and amount of the transaction.
:59:/DE09876543210987654321\nBENEFICIARY NAME: Beneficiary's account number and name.
:70:Invoice 1234 Payment: Remittance information for the beneficiary.
:71A:SHA: Details of charges (SHA means shared).

Trailer (Block 5):
}{5:{CHK:123456789ABC}}: Checksum for message validation and integrity.

Listing 2.1 MT101 Message: Customer Payment Example

SWIFT MT formats are generally intended for individual transactions and are categorized as follows (note that the transactions use British English spelling):

- MT1xx: Customer Payments and Cheques
- MT2xx: Financial Institution Transfers
- MT3xx: Treasury Markets—Foreign Exchange, Money Markets, and Derivatives
- MT4xx: Collection and Cash Letters
- MT5xx: Securities Markets
- MT6xx: Reference Data
- MT7xx: Documentary Credits and Guarantees/Standby Letters of Credit
- MT8xx: Travellers Cheques
- MT9xx: Cash Management and Customer Status

The following are the important SWIFT MT formats for corporate payments and account statement processing in categories 1, 2, and 9:

- **Category 1: Customer Payments and Cheques**
  - MT101: Request for Transfer
  - MT103: Single Customer Credit Transfer

- MT104: Direct Debit and Request for Debit Transfer Message
- MT110: Advice of Cheque(s)

- **Category 2: Financial Institution Transfers**
  - MT200: Financial Institution Transfer for its Own Account
  - MT202: General Financial Institution Transfer
  - MT202 COV: Cover Payment Message (used alongside MT103)
  - MT203: Multiple General Financial Institution Transfer
- **Category 9: Cash Management and Customer Status**
  - MT940: Customer Statement Message
  - MT941: Balance Report
  - MT942: Interim Transaction Report

These formats are particularly relevant for corporations as they facilitate transactions and streamline financial communication between banks.

Migration of Major MT Formats to the ISO 20022 Standard

SWIFT initiated a project in 2018 to gradually transition from the traditional MT formats to the ISO 20022 XML standard following a consultation with its members. Currently, there is a coexistence phase between the conventional message types and the new ISO 20022 XML formats. This transition primarily affects financial institutions. However, in the market, banks are passing these format changes on to end customers and, in some cases, are already phasing out the classic MT format. The deadline for the transition is currently November 2025. However, since the end of 2024, it has become evident that many financial institutions will miss this deadline as they have not yet fully updated their systems. We recommend proactively contacting your bank to plan your own format transition if you have not already done so.

ISO 20022 CGI Formats

ISO 20022 is an international standard for the electronic exchange of financial messages, providing a uniform, structured language for various financial transactions. It is based on XML and enables standardized processing and automation of payment clearing, securities transactions, and other financial services. There are three types of XML-based messages in the ISO 20022 standard that are relevant for payment transactions, as follows:

- The pain format (payment initiation): Used for initiating credit transfers or direct debits.
- The pacs format (payments clearing and settlement): Used for the exchange of payments between banks.

- The camt format (cash management): Used for reporting account information between participants in payment transactions, such as account statements and intraday account activity.

> **Note**
> The pain and camt types of messages are particularly relevant for corporations.

Table 2.5 shows how the versioning of a typical ISO 20022 file is structured.

Business Area Payment Initiation	Message (001 = Credit Transfer)	Variant/Subformat	Version Number
pain	001	001	09

Table 2.5 Structure of pain.001 Message

Figure 2.19 illustrates an end-to-end payment order process, starting from the transmission of a pain message, followed by the steps leading to the receipt of a payment status update, and concluding with an end-of-day bank account statement.

The banking industry is undergoing a major transformation with ISO 20022, prompting a fundamental shift across financial institutions. We have dedicated an entire section of Appendix A to this topic, providing detailed insights into ISO 20022, the global format transition from MT to MX, and its impact on ERP systems and TMS.

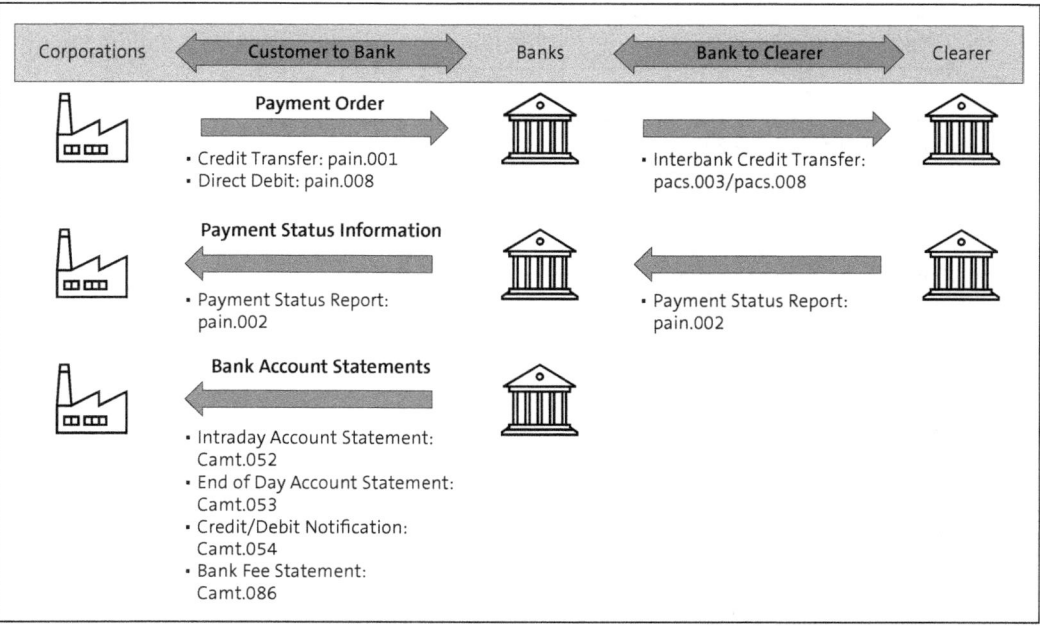

Figure 2.19 Payment Processes

2.2.5 Bank Fee Analysis

An analysis of bank transaction fees is essential for companies with high transaction volumes to actively manage banking policies. For the first time since 2006, there is a unified and standardized electronic format for bank fee information known as the TWIST Bank Service Billing (BSB) Standard. Since 2008, this standard has been further developed under ISO 20022 in the format of camt.086, is a message type under ISO 20022 that falls within the category of statements and notifications (advisories and reporting).

Using the XML-based message and notification format camt.086, banks provide their customers with a detailed breakdown of fees and costs for banking services such as account management, transaction fees, cash management, and more.

If a bank offers this service and message type, the reporting can generally be requested and obtained on a monthly or quarterly basis, either separately for each account or consolidated. The elements listed in Table 2.6 are included in a typical camt.086 message.

Element	XML Tag
Host currency	\<HstCcyCd\>
Currency Exchange	\<CcyXchng\>
Currency Exchange	\<CcyXchng\>
Ledger Balance—Avg Net	"LBAN"
Collected Balance—Avg Net	"CBAN"
Taxable Service Charges	"TXSC"
Settle Charge Ttl—Current Period	"SCCP"
Service Charges Due Before Tax	"SCBT"
Tax Total Sum	"TXTS"
Tax Total Sum	"TXTS"
Charges & Taxes Due This Stmnt	"CTND"
Non Balance Compensable SVCs	"NBCS"
Total Services	\<Svc\>
Exempt Services	"XMPT"
Taxable Services	"TAXE"
Tax Region Number	\<TaxRgn\>\<Rgnb\>

Table 2.6 Camt.086 Elements

Element	XML Tag
Tax Region Name	<TaxRgn><Nm>
Taxes Per Service	<TaxId>
Tax Due to Region	<TaxDueToRgn>
Tax Region Number	<TaxRgn><Rgnb>
Tax Region Name	<TaxRgn><Nm>
Taxes per service	<TaxId>
Tax Due to Region	<TaxDueToRgn>

Table 2.6 Camt.086 Elements (Cont.)

2.3 Summary

The payments landscape is becoming faster with the advent of instant payments. The shift to new formats and XML brings increased opportunities to enhance matching and automation rates, although it also adds complexity to payment processes. Increasing regulatory requirements, such as verification of payee, enhance security but also increase complexity and demand significantly more resources than before. New technologies, including digital central bank currencies and alternative payment methods, along with payment service providers, are leading to rapid and exciting developments in payments that companies should consider in their architecture. The flexibility to incorporate new payment methods and respond to current market demands should be a key part of the initial system architecture planning. AI in payments also supports the accelerated speed and increasing complexity of system structures, aiding in error resolution, customization, and tracking where a payment is stalled if urgent action is needed.

Chapter 3
Bank Account Management

Bank accounts are the cornerstone of payment processing, and managing them effectively is crucial for smooth financial operations. A variety of activities, including managing bank relationships, revolve around maintaining and utilizing these accounts.

As described in Chapter 1, Bank Account Management (BAM) in SAP S/4HANA serves as the foundation for many other treasury functions. The bank account is the central master data element and the starting point for all treasury-related activities. Without a properly maintained and approved bank account, it is not possible to execute payments, upload or process electronic bank statements, or perform any meaningful cash or liquidity checks. In essence, the entire treasury process landscape depends on the availability and accuracy of bank account data managed within BAM.

In this chapter, we will explore the usage of BAM and its integration within the SAP S/4HANA system, focusing particularly on how SAP Fiori apps are utilized and their impact from a functional perspective. This chapter will highlight the key configuration steps necessary to enable core functionalities, offering insight into the most critical elements that support payment execution and help manage your bank account data.

3.1 Functions and Processes

In this section, we will explain how bank accounts are created and updated in the system, covering the main functionalities and how to configure subaccounts. Although SAP S/4HANA offers many additional features related to BAM, our focus will be on the core functionalities essential for setup and maintenance. This will provide a clear understanding of how to manage bank accounts effectively within the SAP S/4HANA environment.

In this section, we will explore how BAM serves as a central repository for managing all bank accounts in your SAP system and how you can use it as a single point of truth for all your banking relationships. You will gain a clear understanding of the concept of a bank key, its significance within SAP, and how it is essential for uniquely identifying each bank. We will walk you through the process of creating banks for your suppliers and customers, covering the key data and requirements you need to consider. Finally, we'll explain how to manage and establish all your bank relationships, including both external and internal banks, to ensure streamlined and accurate handling of financial transactions within your SAP S/4HANA environment.

> **[!] Bank Account Management Versus Electronic Bank Account Management**
>
> Electronic BAM (eBAM) represents the automation of key activities between banks and their corporate customers using software. These activities include opening bank accounts, maintaining them—such as changing account signatories or spending limits—closing them, and generating reports required by law or regulation. This automation typically involves generating and sending files to the bank to execute these activities. BAM is a core solution in SAP for managing bank accounts, but it can also be used as an eBAM solution when connected to a bank, providing an integrated platform for automating and managing these processes seamlessly.
>
> In this chapter, we will briefly touch on the eBAM functionalities available within BAM. However, it is not a very popular feature to use. With the centralization of BAM and the implementation of the latest SAP modules, the goal is to reduce the number of bank accounts being used. When there are fewer accounts to manage, implementing an automatic exchange of messages for creating and closing bank accounts becomes less practical and cost-effective.

3.1.1 Bank Keys

In SAP S/4HANA, the bank key is a unique identifier used to represent a bank within the system. It stores essential details about a bank, such as its name, address, SWIFT/BIC code, and clearing information. The bank key is critical for payment processing as it ensures transactions are routed to the correct financial institution. In many countries, it corresponds to a standardized bank code—for example, the sort code in the UK or routing number in the US, or the Zengin code in Japan. The bank key is a critical identifier that represents a specific bank branch, not just the bank itself. This means that the same bank operating in different cities—or even different branches within the same city—can have separate bank keys.

The bank key is a required element when entering any bank account data in SAP S/4HANA. It must be provided not only for bank accounts created in BAM, but also for bank accounts assigned to business partners such as vendors and customers. Without a valid bank key, it is impossible to create or maintain any bank account in the system as it forms the foundation for identifying the banking institution associated with the account.

> **[!] Creation of Master Data**
>
> Preparing master data in SAP S/4HANA, especially data related to bank accounts, can be a challenging and time-consuming task. One of the main difficulties is identifying and assigning the correct bank key for each bank account number. Although the bank key is

often embedded within the bank account number—particularly in countries where an IBAN is used—SAP still requires the bank key to be entered separately in the master data for vendors, customers, and internal accounts.

This requirement remains even though, from a payment execution perspective, the IBAN alone is sufficient in many countries. In SAP S/4HANA, however, the bank key must be explicitly maintained and linked to detailed bank master data, including the address. This means that before you can even load vendor or customer bank accounts, you must first prepare and upload all necessary bank keys, defining them in the system as banks with their corresponding information.

To facilitate this, companies often choose to import SWIFT files or purchase access to external bank master databases to prepopulate their systems. Alternatively, when the actual bank key is unknown, using the SWIFT code as the bank key is a common and accepted workaround, especially in global implementations. Nevertheless, the preparation of accurate and complete bank master data remains a key dependency and potential bottleneck in system setup.

3.1.2 Creating Banks

Banks can be created in SAP S/4HANA using several different methods, depending on the system version and preferred approach. Here are the main options:

- **The Manage Banks for Master Data app (F6437)**
  Available in SAP S/4HANA 2023 and SAP S/4HANA Cloud Public Edition. This app is used to create simple banks used for master data.

- **The Manage Banks app (F1574)**
  Available for lower versions of SAP S/4HANA (prior to SAP S/4HANA 2023). Offers similar functionality to the Manage Banks for Master Data app but provides more information.

- **Transaction FI01**
  This classic SAP GUI transaction is used primarily in SAP ERP systems or where SAP Fiori apps are not available. It is not recommended to use this transaction for creating bank keys in SAP S/4HANA.

To open a new bank account, you can choose between two primary approaches in SAP: creating a new bank account directly in the Manage Bank Accounts app or requesting a new bank account using the Submit Bank Account Applications feature. With the first option, you can implement an approval process to manage the creation of new bank accounts effectively, providing better control and governance. Alternatively, the second approach allows you to submit a bank account application for opening a new account using the Submit Bank Account Applications app. After the application is approved through the Approve Bank Account Applications app, an inactive bank account is created and can then be fully processed in the Manage Bank Accounts app.

3 Bank Account Management

Which method you choose depends on your organizational requirements, but using approval workflows is recommended to enhance visibility and oversight when opening new bank accounts. In this section, we will describe both approaches and explain how to manage your bank keys, as well as how to prepare for a bank migration.

Submit Bank Accounts

Starting with SAP S/4HANA 2023, there is an option to use the Submit Bank Account Applications app (F5861), which allows users to initiate a structured request for creating a new bank account before the account is created in the system. Historically, you would create a bank account in SAP only after you had contracted with the bank, as the system did not have built-in program processing for account creation. However, SAP has evolved, and it now offers options to automate and streamline this process through BAM. This is especially useful in organizations with multiple teams involved in the bank account lifecycle, such as treasury, accounting, compliance, or local finance departments. Using this app, users can fill in preliminary data and submit the application for review and approval. Once approved, the system can automatically generate an inactive bank account based on the submitted information. When opening a bank account and creating it in the system, there are several prerequisites and activities that need to be completed, including securing internal approvals and finalizing arrangements with the bank. The Submit Bank Account Applications app helps manage this process by guiding you through the necessary steps and facilitating the submission of bank approvals. Only after these approvals are in place can a bank account be officially created and configured in the system, ensuring compliance and proper setup before any financial transactions can take place.

To submit a new bank account request, you can use the Submit Bank Account Applications app. When you open the app, you'll see a list of all existing bank account requests. To create a new request, simply click **Create**. Figure 3.1 shows the main screen where you can submit bank applications. In the new application form, you need to provide an **Application Title** for the request and an **Application Reason** for opening the account. You will also need to specify key details such as the **Company Code**, the **Currency** of the new account, the **Bank Country/Region** where the account will be opened, and the **Contract Type**. There's also an option to upload attachments, which can be used to support the request; for example, you can upload a document explaining the business need for the new account. Once all the required data is entered, click **Create** to submit the request. In this example, the request is for opening a new INR (Indian Rupee) bank account in India.

Once you create the bank account application, it is initially saved as a draft. To initiate the approval process, you also need to submit the bank application for approval.

3.1 Functions and Processes

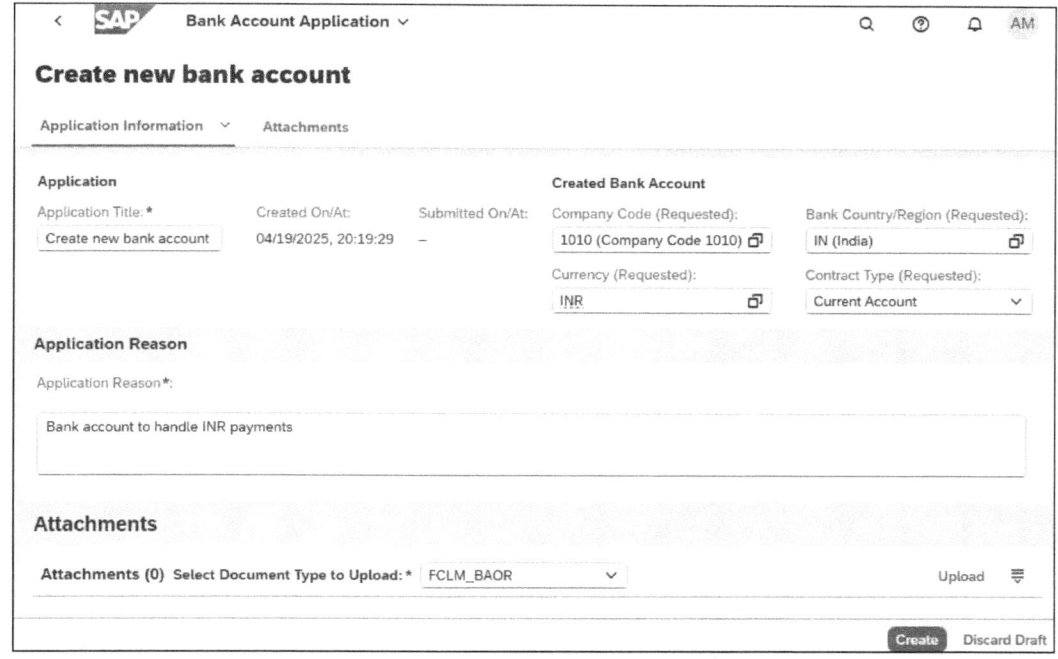

Figure 3.1 Submit Bank Account Application

Approve Bank Account Applications

To approve bank accounts via the approval workflow, use the Approve Bank Account Applications app (F5859). When you open this app, you will see only those requests that are assigned to you for approval, meaning that the system filters the requests based on the workflow configuration and routing logic. You won't see all bank account applications, but only the ones you're authorized to approve in your current workflow step. The system is designed so that depending on the workload setup and user roles, you should not be able to approve the bank account application that you previously submitted. This ensures proper segregation of duties and prevents conflicts of interest. Instead, you should only be able to view bank account applications created by other users, ensuring that approval processes are handled by designated personnel to maintain security and compliance.

Once inside the app, click any visible request to open its details. As shown in Figure 3.2, you will have the following options:

- **Approve** the request to confirm the creation or update of the bank account.
- **Reject** the request to stop the process and mark it as rejected.
- **Edit** the request. This is often used when additional input is required before the account can be finalized. This is particularly useful if approvers need to add data such as the general ledger (G/L) account, validity dates, or other key attributes discussed earlier.

155

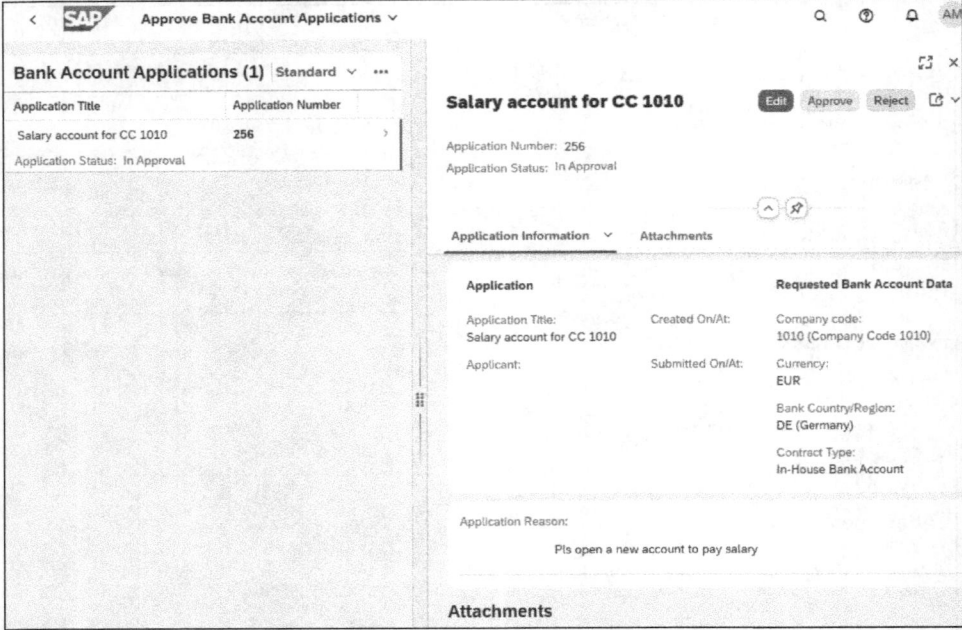

Figure 3.2 Approve Bank Account Application

This **Edit** option is quite common in more complex workflows, where the final approver is also responsible for enriching the bank account with technical or financial details before it's officially created in the system. Once the bank account application is approved, you can proceed with the creation of the new bank and bank account in the system. It's crucial to select the correct **Contract Type** during this process as it determines many further functionalities. As you'll see in Chapter 4, this step is particularly important because it helps differentiate between in-house bank accounts and external bank accounts, ensuring that the appropriate configurations and functionalities are applied to each type.

Once the application is approved, you can proceed with the creation of the bank accounts in the system. This step allows you to set up the necessary bank details and configure the account according to the specified contract type and other relevant parameters.

Create Bank

Let's now walkthrough how to actually create a bank. We'll examine three options for doing so:

- Manage Banks—Master Data app
- Manage Banks app
- Transaction FI01

Manage Banks—Master Data App

Starting with SAP S/4HANA 2023, you should use the Manage Banks—Master Data app (F6437) to create master data for the banks your company, your business partners, your customers, and your suppliers use to conduct business. When you open this app (see Figure 3.3), you can either check the banks created for a certain country or click **Create** and create a new bank by entering the **Bank Country/Region** and the **Bank Key**.

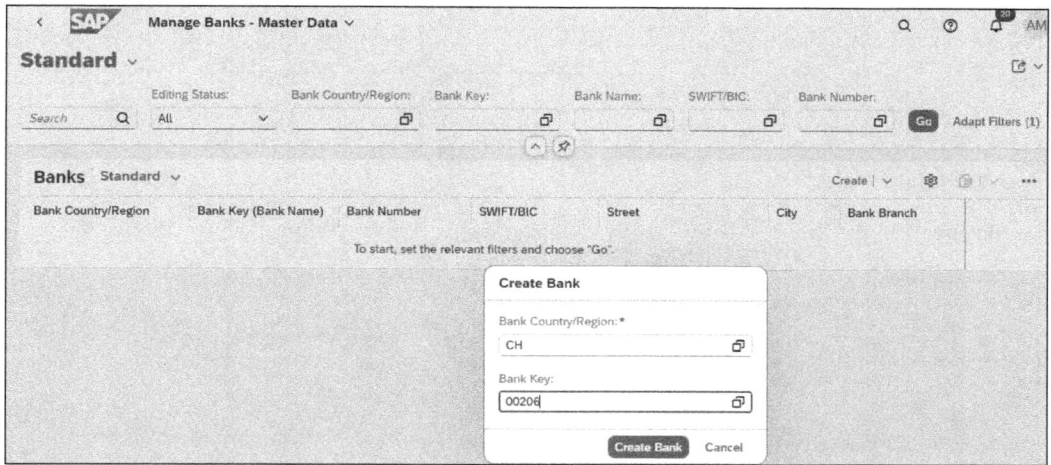

Figure 3.3 Create Bank

In this example (see Figure 3.4), we will create the UBS bank located on the Paradeplatz in the center of Zurich. After you click **Create Bank** in Figure 3.3, you will need to fill in other information. When creating a bank in SAP S/4HANA, it is important to enter the **SWIFT/BIC** code and bank **Address**, both of which may be required for generating payment files and ensuring accurate payment processing. The SWIFT code ensures international compatibility, while the address (including international versions, e.g., to create addresses in different languages when different alphabets are used) can be used to fulfill country-specific formatting or compliance requirements. Once all the necessary details are entered, you can create the bank.

Upon creation, the system will generate the bank record, and you will be able to view its full details and history (see Figure 3.5), including any changes or updates made over time. This ensures transparency and traceability for audit or reconciliation purposes.

Once the bank is successfully created and saved, it becomes available for use in your counterparty master data, such as vendors, customers, or business partners. This allows you to link bank details to transactions like payments or collections. If you intend to create a house bank (a bank account owned by your company), you can do so using the Manage Banks—Cash Management app (F1574A). This app enables you to define house banks and assign the bank accounts necessary for treasury operations and payment processing; the functionality is similar to that of the classic Manage Banks app.

3   Bank Account Management

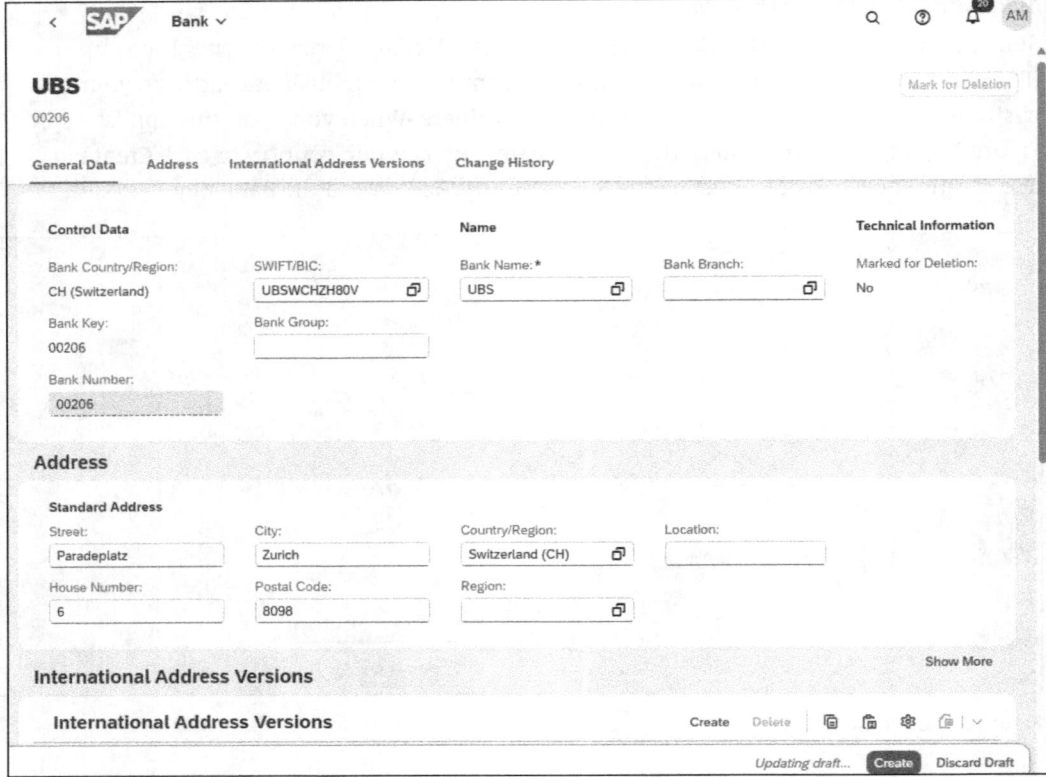

Figure 3.4 Create Bank Address

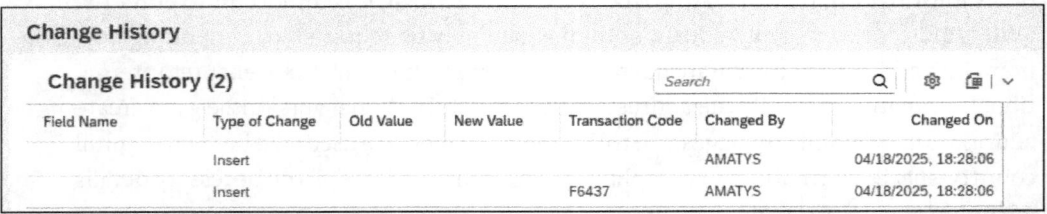

Figure 3.5 Change History Create Bank

Manage Banks App

An alternative option for creating banks in SAP S/4HANA is the Manage Banks app (F1574) (see Figure 3.6), which, although deprecated, is still available in versions after SAP S/4HANA 2023 (and you need to use it if you have a lower SAP S/4HANA version). This app not only allows you to create and maintain banks but also provides extended capabilities for managing house banks. It also enables you to enter more detailed information, such as contract types. Despite being phased out in favor of newer apps, it remains available and a useful tool.

3.1 Functions and Processes

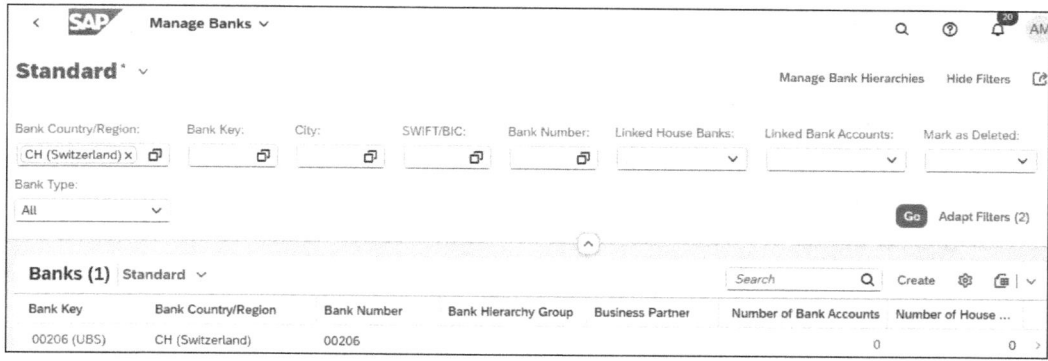

Figure 3.6 Manage Banks

This app functions similarly to the one discussed previously for bank creation; however, as Figure 3.7 shows, when you click **Create**, you'll see more detailed fields to be filled in.

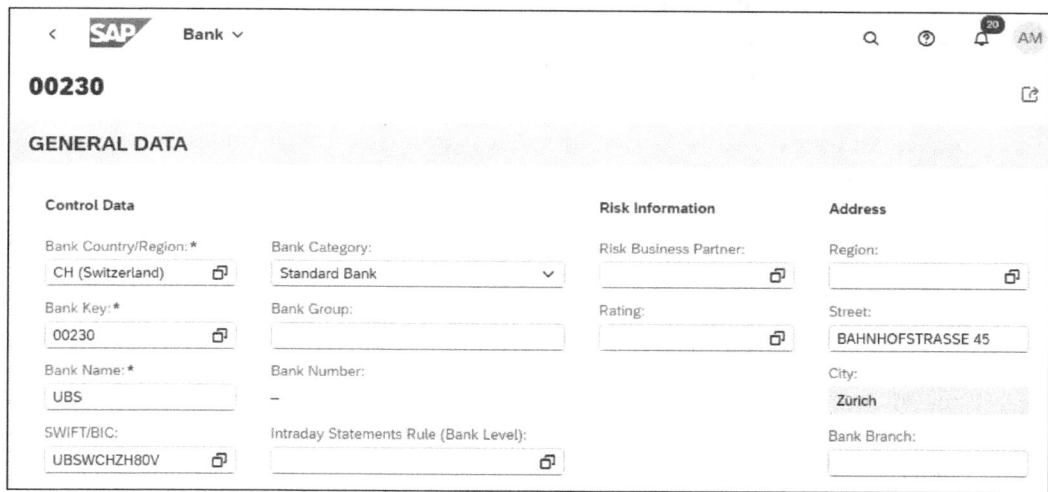

Figure 3.7 Manage Banks: General Data

In this app, you enter the same core information as in other bank creation tools—such as the **Address** and **SWIFT/BIC** code—but you also gain access to several additional, important fields, like the following:

- **Bank Category**
  This defines the type of bank being created. The default option is **Standard Bank**, but you can also select categories like **In-House Bank Account** for internal banking operations or **Technical Bank Account** for system-internal postings and testing. The selection here impacts how the bank accounts behave in later processes, which will be explained further in Section 3.2.

- **Intraday Bank Statement Rule**
  This setting determines whether intraday bank statements are expected from this bank, and how frequently. Its function and configuration will be detailed later in Chapter 9.
- **Risk Business Partner**
  This is used for integration with the Bank Risk app, allowing risk exposure to be tracked against specific banks.
- **Bank Rating**
  This field captures the rating of the bank, which is useful for treasury and risk management evaluations.

These additional options allow for more precise and business-aligned bank master data management.

Once you save the bank record, you will be able to provide more information, like **House Banks**, **Contact Info**, **Related Branches**, **Bank Service Mapping**, and **Netting Business Partner**, and you will be able to see the **Change History** information.

One of the most important options available in this app is the ability to create a house bank (see Figure 3.8). When creating a house bank, you must first assign it to a **Company Code** and define a **House Bank ID**/code. The house bank ID will be used later when you create bank accounts. After this, you can provide additional key information that enhances how the bank is used in various processes. This includes **Communication Data**, which specifies how the company will communicate with the bank and who the relevant contact person is. You can also maintain general data, as well as configuration for data medium exchange to enable electronic payment file processing.

Furthermore, you can define the account used for bank charges (**Charges Account** section) and execution (**Execution** section), and assign an **Instruction Key** that indicates, for example, whether the payer or payee will cover the transaction charges. You can also maintain the **Reporting Data** needed for central bank reports, ensuring compliance with regulatory requirements. There is also an option to make the house bank relevant for **EDI Partner Profiles** (relevant for SAP In-House Cash house banks; without this setting enabled, you won't be able to run SAP In-House Cash payments or post SAP In-House Cash statements).

In most cases, only the company code and house bank ID are commonly used and required when creating a house bank. These are the essential details needed to define and use a house bank in standard payment- and bank-related processes. Additional functionalities—such as communication data, instruction keys, reporting details, or EDI partner profiles—are optional and being slowly deprecated by SAP; they're typically used only for very specific business or regulatory requirements.

In the classic SAP In-House Cash solution, the bank accounts created in BAM and used for payments (such as those generating PAYEXT or DIRDEB IDocs) must be marked as EDI-relevant, and the associated payment methods must also be flagged as EDI-relevant. If these settings are not maintained, IDoc generation in SAP In-House Cash is not possible. In contrast, with in-house banking, these EDI-relevant flags are no longer required, simplifying the configuration and execution of payments.

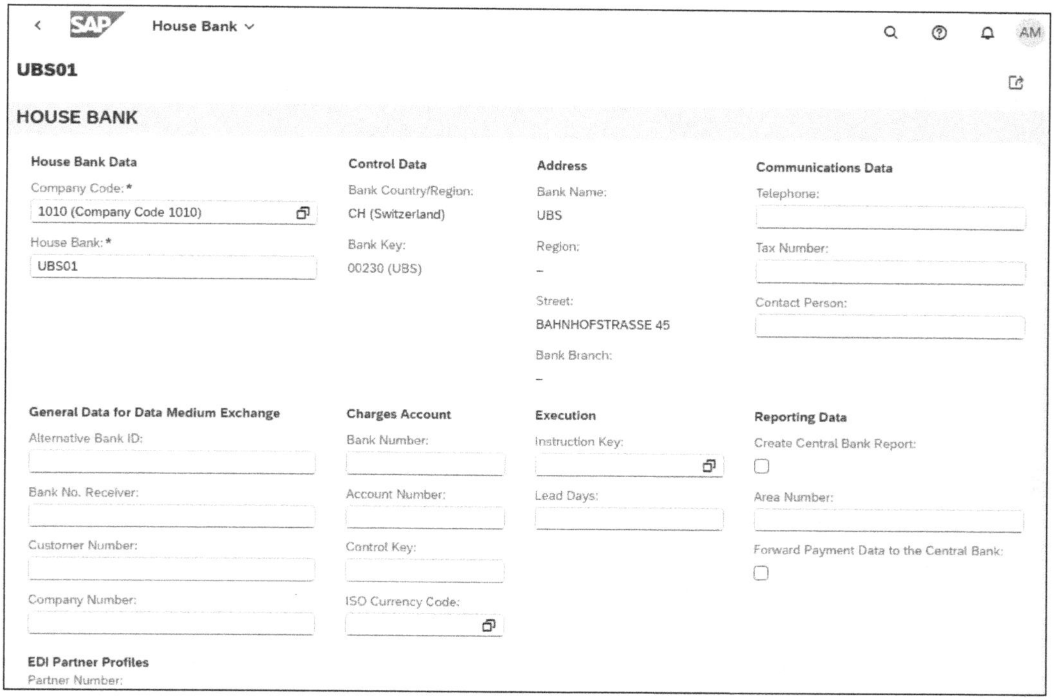

Figure 3.8 Create House Bank

In addition, you can create contact information within the house bank setup to indicate who is responsible for managing the bank account on the bank's side, as well as who serves as the internal contact person from your organization. However, it's important to note that the contact person must first be created as a business partner in the system before they can be assigned to the house bank.

In many cases, bank address data is not straightforward; banks may have multiple addresses—such as branch-specific addresses, headquarters addresses, or addresses in different languages using special characters. SAP S/4HANA provides advanced functionalities to handle these scenarios, allowing users to define detailed and flexible address data. This is especially important because accurate and complete address information is often required to ensure smooth payment processing on the bank's side.

Such functionalities are available when creating a bank in SAP S/4HANA via the advanced address feature (as shown in Figure 3.9). To access this option, you first need to create a bank, then open it for editing and click the **Advanced Address** button. This option allows you to enter more detailed and structured address information beyond the standard fields. With advanced address, you can capture additional data such as the building name, floor number, district, PO box details, and more granular location specifics.

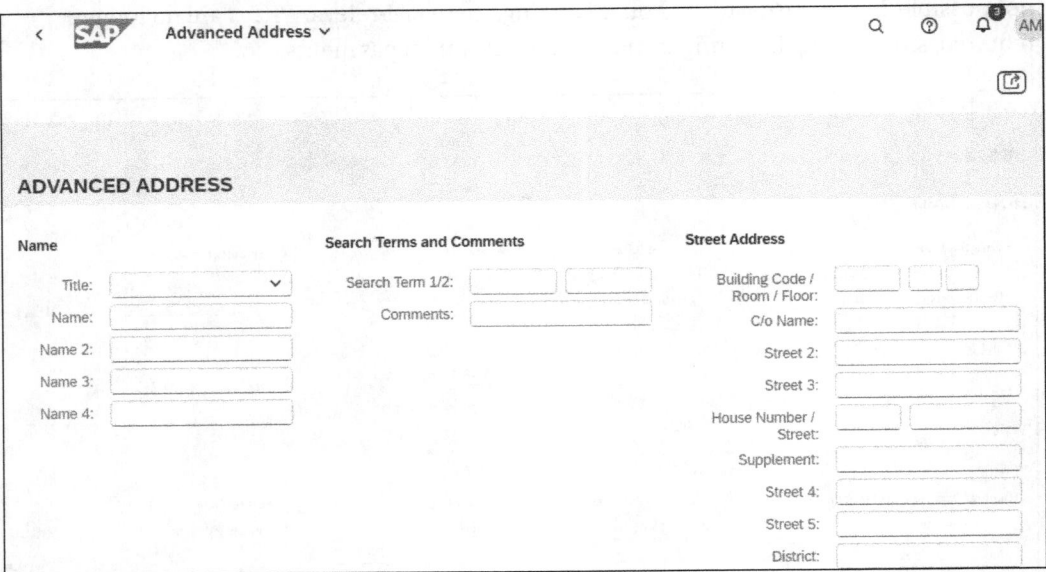

Figure 3.9 Advanced Address

This can be especially helpful in scenarios such as the following:

- **Payment file generation**
  Some formats or local regulations may require extended address data to ensure the payment is accepted by the banking system.

- **Avoiding delays**
  Providing a full and accurate address helps avoid rejections or delays due to incomplete information.

- **Compliance and reporting**
  For certain jurisdictions, especially in regulated industries, full address details might be required for audit, compliance, or central bank reporting purposes.

- **Intercompany or in-house bank structures**
  When managing a large number of banks and bank accounts, the advanced address structure helps to clearly differentiate and organize data.

Using advanced address information enhances data quality and reduces the likelihood of errors or missing fields during payment processing and reporting.

In the Manage Banks app (F1574), you also can view all the bank keys and banks that have been created in the system, along with their corresponding bank category. This information is particularly important when working with SAP In-House Cash, in-house banking for SAP S/4HANA Finance for advanced payment management, or any other technical accounts that will be used for specific purposes within your organization. The bank category helps determine how each bank account behaves in the system and ensures that the correct category is assigned, which is essential for proper configuration and usage.

Figure 3.10 shows how the Manage Banks app can be used for reporting functionalities. In this view, you also can see how many house banks and bank accounts have been created for the selected bank.

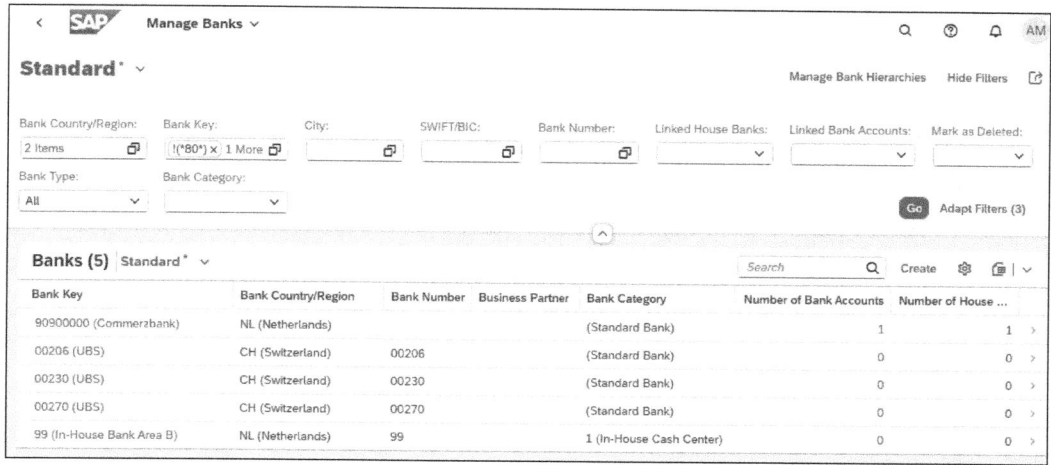

Figure 3.10 Manage Banks

Transaction FI01

Finally, as an alternative to using the SAP Fiori apps for bank creation, you can use the classic SAP GUI Transaction FI01 (see Figure 3.11). This transaction allows you to manually create banks by entering the required information such as bank key, name, address, and SWIFT code. If you are still working in an SAP ERP system, Transaction FI01 is the only option to create a bank, but the transaction is also still available in SAP S/4HANA.

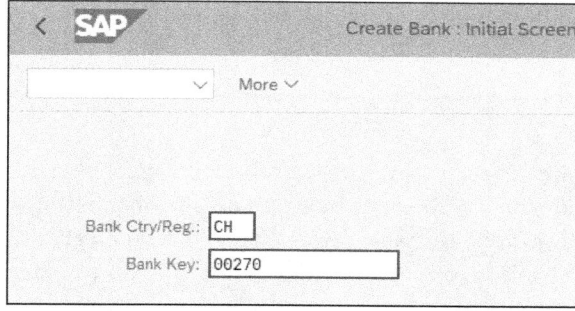

Figure 3.11 Transaction FI01: Create Bank

Transaction FI01 is very straightforward to use. You simply need to enter the **Bank Ctry/Reg.** and the **Bank Key** to begin the bank creation process. Once these two fields are filled in, the system will allow you to proceed with entering the remaining bank master data, such as the **Bank Name**, **Address**, and **SWIFT/BIC** code, as shown in Figure 3.12.

3   Bank Account Management

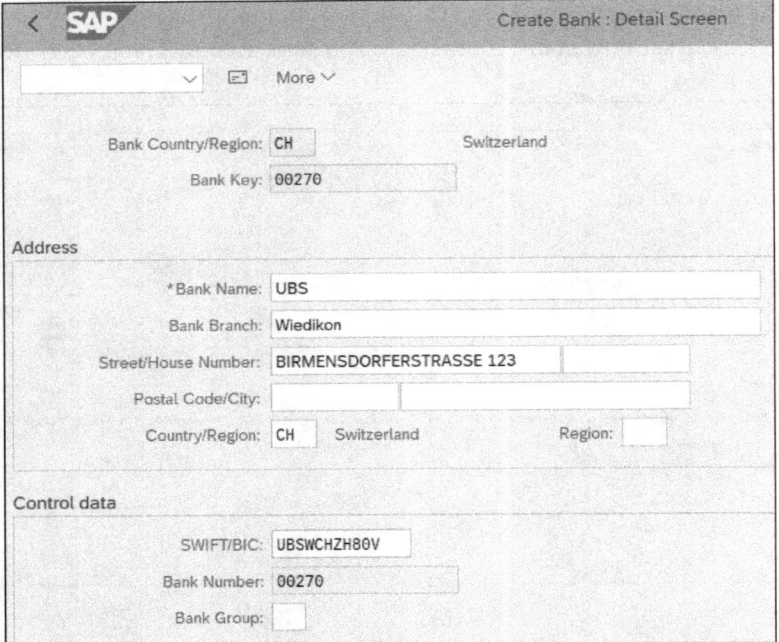

Figure 3.12 Transaction FI01: Additional Bank Data

When using SAP S/4HANA version 2022 or higher, the system automatically creates all banks in the background as standard banks, meaning that the bank category (called the *contract type* in previous versions) is set to **Standard Bank**. This default behavior has important implications for future configuration activities. The bank category determines what kind of functionality and configuration options are available for that bank (such as the ability to switch off workflow approvals for technical or SAP In-House Cash accounts).

Change of Contract Type

It is not possible to change the bank category after the bank key has been created, so it is essential to choose the correct category from the beginning, especially when setting up house banks for SAP In-House Cash/in-house banking (see Chapter 4 to learn about the link between BAM and in-house banking accounts) or other technical purposes. This means that when you use SAP S/4HANA 2023 and up, you cannot use these functionalities and will need to use the newest apps. Otherwise, it won't be possible to determine the correct contract type.

Bank Migration

As you can see, when implementing a new SAP S/4HANA system, you may have to migrate hundreds of thousands or even millions of bank records, which is virtually impossible to manage if you need to enter them one by one in the SAP Fiori app. However, there are several options available to streamline the migration process:

- **Using Transaction FI01**
  This transaction can be used for uploading bank master data in older systems or even in SAP S/4HANA. With SAP Legacy System Migration Workbench, you can map the fields from your source system to SAP S/4HANA and upload large sets of bank master data. However, it is recommended to use SAP Legacy System Migration Workbench only in SAP ERP systems as it may limit the ability to enter a lot of information, such as certain bank configurations or additional fields required in SAP S/4HANA. Using SAP Legacy System Migration Workbench in newer systems can restrict the full use of features and fields that are available in the SAP S/4HANA environment.

- **Using the SAP S/4HANA migration cockpit for bank master data**
  In SAP S/4HANA systems, you can use the migration cockpit to load bank data into your new system in bulk, without having to manually create each bank key or bank record. You can prepare a data file and then use the standard migration objects provided by SAP.

- **Using a BAPI**
  Alternatively, you can write a custom program using the BAPI_BANK_CREATE BAPI to automate the creation of banks and bank keys in the new system. This BAPI provides a standardized way to create banks and their details programmatically. This approach is highly beneficial when dealing with large volumes of data and complex migrations.

These options provide flexibility depending on your specific needs and the tools available in your SAP environment, ensuring that large-scale migrations can be performed efficiently.

Manage Bank Accounts

As mentioned previously, banks (and their corresponding bank keys) must be created in the SAP S/4HANA system for all relevant financial institutions, including those related to vendors, customers, and any other counterparties your organization interacts with.

The next step is to create house banks and corresponding bank account IDs for the banks your company uses for its own financial operations. These house banks and account IDs are critical because they will be used for executing payments, receiving funds, performing bank reconciliations, and supporting overall cash and liquidity management within the SAP S/4HANA system. Creation of, changes to, and closure of bank accounts is handled in the Manage Bank Accounts app (F1366A).

Figure 3.13 shows how you can display and manage your organization's own bank accounts, maintain bank relationships, and adjust their associated attributes and settings. This app serves as a central point for viewing and updating details such as account numbers, currency, bank assignment, and usage information. To work efficiently within the app, you need to define selection criteria—such as **Company Code**, **House Bank**, or bank **Account Number** or, in the example shown, the **Bank Country/Region** information for where the bank accounts are located. This lets you filter the results and display the relevant bank accounts already created in the system.

Figure 3.13 Manage Bank Accounts View

From the Manage Bank Accounts app, you have access to a wide range of functionalities beyond just displaying account details. You can import and export bank accounts, manage bank account hierarchies, create and attach documents, and initiate review and approval workflows. In addition, you can directly navigate to related SAP Fiori apps such as Manage Bank Statements, Cash Flow Analyzer, Manage Bank Account Balances, and Monitor Bank Account Balances. You also can create new bank accounts or delete existing ones directly from this interface. We will describe all these SAP Fiori apps and features in more detail in Chapter 6; for now, we will focus specifically on the creation of new bank accounts within this app.

Figure 3.14 shows what happens when you click the **Create** button. The system will start the creation of the new bank account and will ask you first to select the **Company Code** for which you will create the bank account and the **Account Type**. In this example, the account type is **01**, for a current bank account (representing an external bank account; the values in this dropdown list are from the configuration described in Section 3.2.4). The account type is created in the configuration and determines the fields available and the behavior of the account.

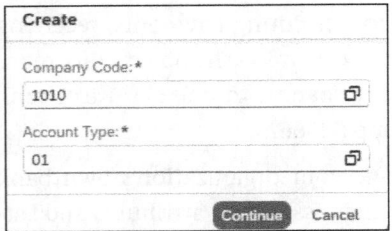

Figure 3.14 Create Bank Account

When creating a new bank account (see Figure 3.15), the first step is to fill in the **Header** data. This includes entering the **Account Description** and bank **Account Number**, and linking the account to the **Bank Country/Region** and **Bank Key** that you previously created. You'll also need to enter the **Bank Control Key** and account **Currency**. If the bank account is for a country that uses IBAN (i.e., an "IBANized" country), you can simply click **Generate IBAN**, and the system will automatically propose an IBAN based on the entered data. If the system is unable to generate the IBAN—for example, due to missing formatting rules—then you can always enter the IBAN manually.

Figure 3.15 Create Bank Account Header Data

> **Creation of the Bank Account**
>
> As previously mentioned, the contract type (bank category) cannot be selected or changed at this stage. It is automatically derived from the bank creation step completed earlier. Once a bank is created with a specific contract type, it is fixed and cannot be modified later, so it's important to ensure that the correct category is chosen during the initial bank setup.
>
> Also, note that in SAP S/4HANA you can assign only one currency to each bank account. In practice, many banks offer multicurrency accounts, allowing multiple currencies to be handled under a single account number—but SAP S/4HANA does not support this functionality directly.
>
> To manage multicurrency accounts in SAP S/44HANA, the common approach is to create one house bank and then multiple bank accounts under that house bank, each with the same account number but a different currency assigned. This allows for accurate processing and reconciliation (however, you will need to set up payments and configuration for each bank account separately) across various currencies while maintaining alignment with how SAP S/4HANA structures bank account data.

The next pieces of information to be checked or entered is under **General Data** (as shown in Figure 3.16). Here, the system automatically derives certain information from the company code selected earlier. You can also specify the **Account Holder**, which is a free-text field that allows you to enter the name of the entity owning the account. This can be either an SAP-registered entity or any external party. You also can define the **Profit Center** and **Segment**, which are particularly useful for reporting and can also be automatically derived during bank statement reconciliation processes. You will also select the **Account Type**, which determines how the bank account behaves within the system (whether it's a current account, loan account, etc.). Another key field is **Opening Date**. This date is crucial because no transactions can be carried out before it (there would be a lot of errors in various places in SAP S/4HANA: payments, statements, treasury and risk management, and cash management). Once the account is created and approved, this opening date, along with other key data, usually cannot be changed, thus ensuring the consistency and auditability of bank master data.

Figure 3.16 Create Bank Account General Data

The **Technical ID** is automatically generated by the system once the bank account is created and serves as a unique identifier for internal system processes. The **Financial Object Number** is particularly relevant for SAP Treasury and Risk Management; it's used to link the bank account to financial objects for risk analysis and reporting. If this functionality is configured in your system, the number will be derived automatically. All of the information under **Bank Details**—such as the bank name, address, SWIFT code, and bank key—are pulled directly from the bank record that was created in the previous step, ensuring consistency and avoiding redundant data entry.

Figure 3.17 shows the next tab in the bank account, **Bank Relationship**, which is very important and affects many functionalities in SAP S/4HANA.

What you select at this stage significantly impacts various functionalities later in the SAP S/4HANA system. Each field and value chosen here—**Account Type**, **Intraday Bank Statement**, **Balance Update Method**, and so on—can influence how the bank account behaves across different modules, from cash management and payment processing to reconciliation and reporting. In Chapter 6 and Chapter 9, we'll take a closer look at each of these settings and how they are derived, and we'll explain how they affect system behavior.

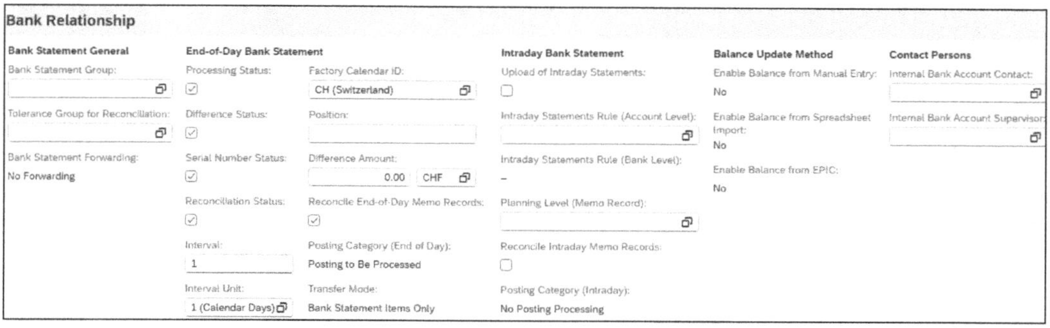

3.1 Functions and Processes

Figure 3.17 Bank Account: Bank Relationship

In the next step, shown in Figure 3.18, you need to select the **Bank Statement Group**, which is defined in the system configuration. This grouping is used to organize bank statements, allowing them to be displayed and managed at an aggregated level. You must also specify the **Tolerance Group for Reconciliation**, which determines the acceptable tolerance levels when reconciling line items during the bank statement reconciliation process. This helps streamline reconciliation and ensures consistent handling of small discrepancies.

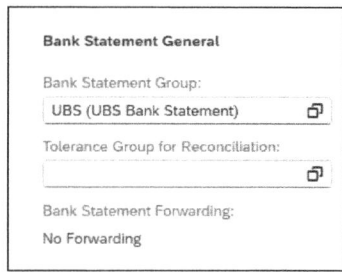

Figure 3.18 Bank Account: Bank Statement General

In addition, there is an option for **Bank Statement Forwarding**, which defines whether the imported bank statement files—both end-of-day and intraday statements—should be automatically forwarded to subsequent systems (e.g., other SAP systems, legacy environments, or third-party applications). This functionality is particularly relevant in complex system landscapes and is governed by configuration settings.

The **End-of-Day Bank Statement** section (see Figure 3.19) is a critical component for leveraging the Bank Statement Monitor app, which is explained in detail in Chapter 9. In this section, you define how frequently you expect to receive the end-of-day bank statements by specifying the **Interval Unit** (e.g., daily, weekly), and this frequency is monitored against the selected **Factory Calendar**. This ensures that expected receipt dates align with working days and holidays based on the relevant calendar. This expectation forms the basis for monitoring whether statements are received on time and in accordance with your reconciliation processes.

169

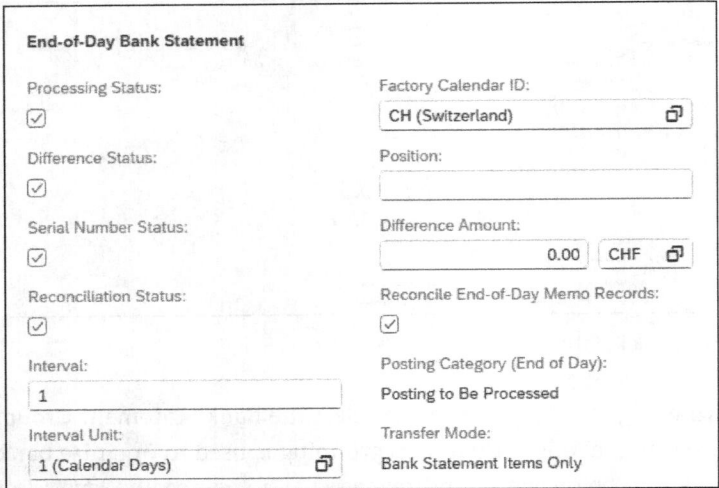

Figure 3.19 Bank Account: End-of-Day Bank Statement

This setup enables the system to track several key statuses within the Bank Statement Monitor app, including the following:

- **Processing Status**
  Whether the bank statement has been imported and processed successfully.
- **Difference Status**
  Reflects any discrepancies between the G/L account balance and the balance reported in the bank statement.
- **Serial Number Status**
  Checks for gaps or duplication in the sequence of incoming statements.
- **Reconciliation Status**
  Indicates whether the items in the statement have been matched and cleared properly against the corresponding G/L entries.
- **Reconciliation Differences**
  These define the acceptable thresholds for automatic reconciliation of items.
- **Reconciliation of End-of-Day Memo Records**
  This option, used by certain apps, enables you to also reconcile memo records that appear at the end of the day with the forecasted flows. This is purely a treasury forecasting activity and is independent of the accounting reconciliation.

Moreover, you define the **Posting Category**, which indicates whether the statement should be posted automatically or kept for review. This influences the overall automation level of the bank statement processing.

Finally, the **Transfer Mode** setting determines how cash flows are updated in cash management. Both the posting category and the transfer mode come from the cash management setup and are not editable directly at the BAM level. These settings play a crucial role in determining how data flows through SAP's treasury and cash management functions and directly impact downstream processes and visibility into cash positions.

Under the **Intraday Bank Statement** section, shown in Figure 3.20, you first indicate that intraday bank statements are expected and will be uploaded for the specific bank account by selecting the **Upload of Intraday Statements** checkbox.

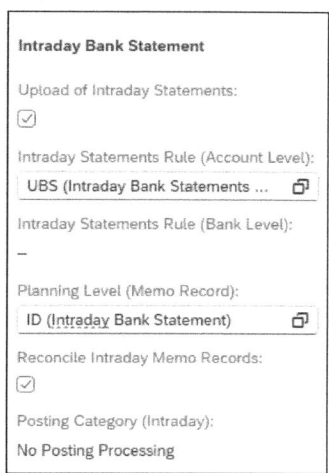

Figure 3.20 Bank Account Intraday Bank Statement

If enabled, several critical configurations follow:

- **Intraday Statement Rule**
  This defines how frequently you expect to receive intraday statements throughout the day (e.g., every hour, multiple times per day). It helps structure your expectations and set up monitoring properly in Bank Statement Monitor and related apps. This rule is created in Transaction FCLM_BRM_RULE or the Define Monitoring Rules – Intraday Statements app.

- **Planning Level**
  This determines the level at which line items from the intraday statements are posted into cash management. It affects how cash positions are visualized and analyzed across planning hierarchies.

- **Intraday Memo Records Reconciliation**
  Here you specify whether you will reconcile memo records from intraday statements using the dedicated Reconciliation of Memo Records app. This improves cash visibility and helps identify temporary entries or holds.

- **Posting Indicator**
  This setting specifies whether actual postings should be made in the system based on the content of intraday statements. This is especially relevant when real-time updates to cash positions or ledger postings are needed.

If you select and configure these entries correctly and proceed with creating them, you will enable the full functionality related to intraday bank statements. This includes the ability to receive and process intraday updates from your banks, as well as to use the Bank Statement Monitor—Intraday app (F3671) and to perform memo record reconciliation in dedicated SAP Fiori apps.

The balance update methods for bank accounts (see Figure 3.21) are predefined in the Define Bank Account Settings—Bank Statements app (F5488) and cannot be changed or selected directly within the Manage Bank Accounts app. These methods determine how bank account balances can be updated in the system and are typically assigned based on company code requirements.

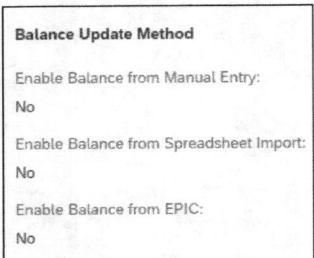

Figure 3.21 Bank Account: Intraday Balance Update Methods

The available balance update methods include the following options:

- **Enable Balance from Manual Entry**
  Allows users to manually input balances for bank accounts using the Manage Bank Account Balances app. This is useful for quick updates or ad hoc adjustments.
- **Enable Balance from Spreadsheet Import**
  Enables users to upload a spreadsheet to update bank account balances in bulk via the same app. This method is efficient for handling multiple accounts at once.
- **Enable Balance from EPIC**
  Allows automatic updating of bank account balances using an integration specific to the Chinese market—the Electronic Payment Integration for China (EPIC) functionality.

In the **Contact Person** section of the bank account (see Figure 3.22), you can define both internal and external contacts related to the selected account:

- For external contacts, you should assign the person responsible for the bank account on the bank's side. This person must be created in the system as a business partner.
- For internal bank account supervisors, you need to define relevant system users who will be responsible for managing the bank account internally. To do this, create business partners with the role category BUP003 (Employee), and then assign the system user IDs to these business partners. Some older SAP S/4HANA versions might require employees to be created with HR roles.

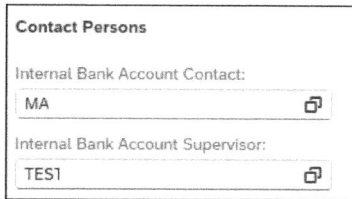

Figure 3.22 Bank Account: Contact Person

Now, scroll down to the **House Bank Account Connectivity** section, shown in Figure 3.23. Here you create a **House Bank** and **House Bank Account** ID and assign these to the selected bank account (by entering a value here in **House Bank Account**; or can you add this later by clicking the line for the correct account). A house bank in SAP represents your company's own bank within the system. It acts as an internal structure that organizes and manages your external bank accounts. The house bank is created at the bank level and is always linked to a specific bank key, which uniquely identifies a bank branch (via a routing number, sort code, Zengin code, etc.).

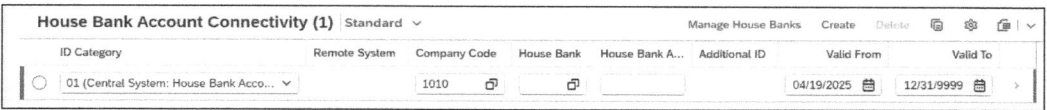

Figure 3.23 Bank Account House Bank Account Connectivity

The house bank is not a bank account itself, but a logical container that groups one or more bank accounts (bank account IDs) for a particular company code. It helps with account management, payment processing, and reconciliation. You create the house bank in the Manage Banks—Cash Management app or the Manage Banks app during the earlier setup phase, when you also provide the company code and define the house bank ID. Once the house bank is created, you can proceed with assigning it to the relevant bank account ID, finalizing the structure that links your SAP S/4HANA bank account to an actual bank institution for operational use.

Figure 3.24 shows how to create a new bank account after clicking the **Create** button.

Once you click **Create**, you'll be prompted to select an option for **ID Category**, which defines how the bank account is managed:

- **00 Central System: G/L Account**
  Used when you have only a G/L account centrally and the account is managed in the local system; this is not a popular option.

- **01 Central System: House Bank Account**
  The bank account is fully managed in the current system. This is the most common setup when all postings and bank-related processes happen in the same SAP S/4HANA system and should be used for most scenarios.

- **02 Remote System: House Bank Account**
  The bank account is managed in a different system; you're only bringing the account details into this system for central visibility or consolidation purposes (e.g., for centralized cash management or reporting).

- **03 Remote System: G/L Account**
  You store the bank account centrally for reporting purposes, but the G/L account is created only in the local system, together with corresponding postings.

- **04 Others**
  Any other specific scenario not covered by the other options.

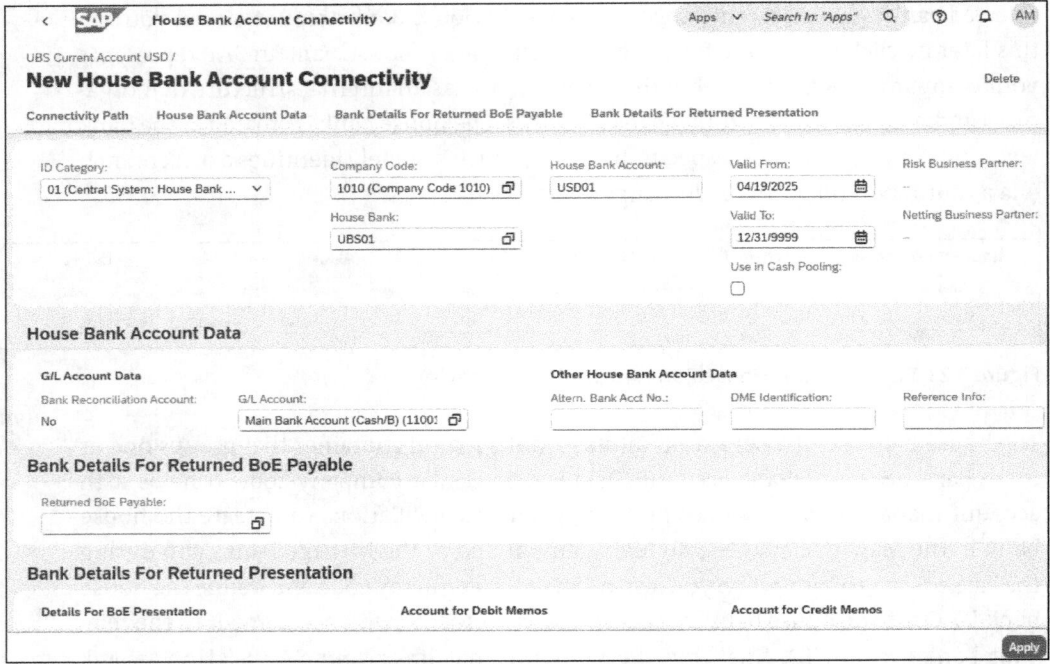

Figure 3.24 New House Bank Account Connectivity

House Bank is automatically derived based on the bank and bank key you selected earlier. You'll then need to do the following:

- Assign a **House Bank Account** ID, which uniquely identifies the specific bank account within the house bank.
- Define the validity period by entering the **Valid From** and **Valid To** dates.
- Select **Use in Cash Pooling** if you want this account used in cash pooling.
- Select the **G/L Account** that the bank account will post to. This G/L account will be used for all financial postings related to this specific bank account.

Other house bank account fields such as **Alternative Bank Account Number**, **DME Identification**, or **Reference Information** can be used to support specific bank statement processing scenarios. These fields become especially useful when you have one of the following situations:

- The bank sends additional identifiers in the bank statement (e.g., a different account number than the one maintained as the main account number).
- You need to use data medium exchange (DME) identification to differentiate between various payment files or bank accounts.
- The **Reference Info** can help with matching incoming bank statement lines to the correct internal bank account, particularly when banks include reference data that doesn't directly match your primary account settings.

> **House Bank and Bank Account ID**
>
> It's important to note that both the **House Bank Account** ID and the **Bank Account** ID can contain a maximum of five characters each. Because of this limitation, it's highly recommended to establish a consistent naming convention across the entire system to maintain clarity and avoid confusion.
>
> This logic should be simple yet informative—ideally allowing users to identify key details about the account at a glance. A common and effective practice is to include abbreviations for the bank name and the currency. For example, an account at HSBC in USD could be named HSUSD, while one at Citi in EUR might be CTEUR.
>
> Defining and adhering to a system-wide standard ensures better governance and simplifies ongoing bank account maintenance, reporting, and integration activities across different areas of SAP S/4HANA and company codes.

> **G/L Accounts Used for Bank Accounts**
>
> In the classic approach, typically, each bank account is associated with its own G/L account, and at least one bank clearing account is created for each bank. However, with the new approach introduced by SAP S/4HANA, there is now the concept of a single G/L account for all bank accounts within a particular company code. This allows for greater simplicity in the chart of accounts by reducing the number of G/L accounts needed.
>
> Before moving forward, it is important to verify if your system supports this functionality and whether it will enable you to perform bank reconciliations with a single G/L account. If this feature is available and you can reconcile your bank accounts using just one G/L, it is highly recommended as it simplifies the overall financial structure and reduces the complexity of managing multiple G/L accounts for different bank accounts.

When you further scroll down your screen in the Manage Bank Accounts app, you will see the **Payment Approvers** tab (see Figure 3.25), where you can assign payment approvers and define which groups or individuals are authorized to approve payments based on specific amounts. If the approval process is not dependent on the amount, you can select the **Unlimited Amount** option. Here, you will also specify the validity period for the payment approver and define which person or group is responsible for each step in the approval process.

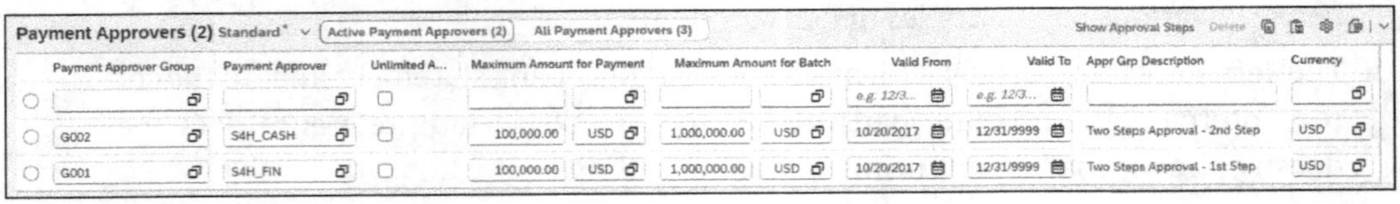

Figure 3.25 Bank Account Payment Approver

By clicking **Show Approval Steps**, you will be able to view the configured approval flow, including who is responsible for which step of the payment approval process and the corresponding approval patterns. This allows for clear visibility into and control over the payment approval workflow, ensuring that the right people are involved at the right stages of the process.

Note that from the BAM perspective, the payment approval rules can only be created based on amounts. In addition, you need to maintain the individuals and groups responsible for approvals for each bank account. If your approval process requires more complex rules or logic beyond simple amount-based conditions, you would either need to implement custom development or utilize SAP Bank Communication Management or advanced payment management for more advanced payment approval workflows and processing. These solutions provide greater flexibility and customization for handling payment approvals and related processes.

> **Maintain Payment Approvals for Multiple Bank Accounts App**
>
> There is also an alternative SAP Fiori app called Maintain Payment Approvals for Multiple Bank Accounts (F1372), shown in Figure 3.26, which allows you to create and maintain payment approval rules for more than one bank account at the same time.
>
> Once you open the app, you can select the bank accounts you want to modify. Then, click **Add Payment Approver** to assign a new approver. You will be able to select the **Approver** and **Approval Group**, define the **Amounts** and **Validity**, and assign the **Payment Approval Group**. Alternatively, you can select multiple bank accounts and use the **Change Payment Approver** option to update the approvers for the selected accounts. You also can update the overdraft amount limits, or remove approvers entirely by clicking **Revoke Authorization**.

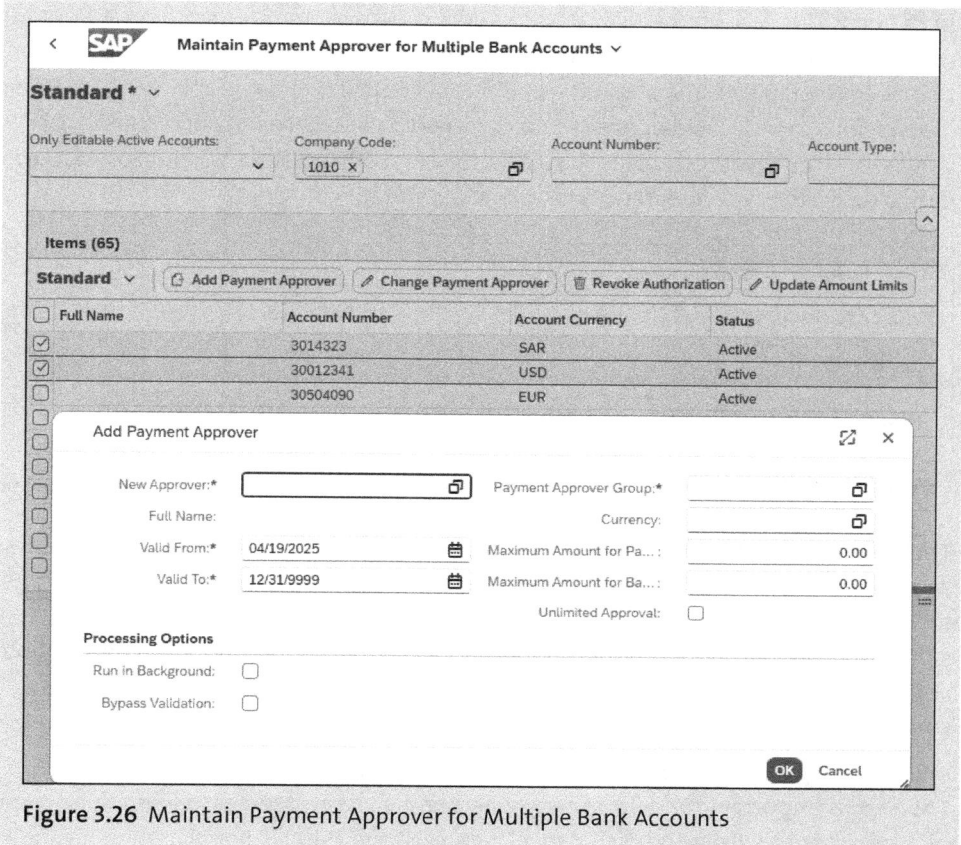

Figure 3.26 Maintain Payment Approver for Multiple Bank Accounts

The next step in maintaining the bank account, shown in Figure 3.27, is defining the overdraft limits. Here, you can specify the overdraft limit for a particular bank account, including the validity period, the overdraft amount, and the currency. These overdraft limits are checked and utilized in various applications within cash management to ensure proper monitoring and management of cash flow.

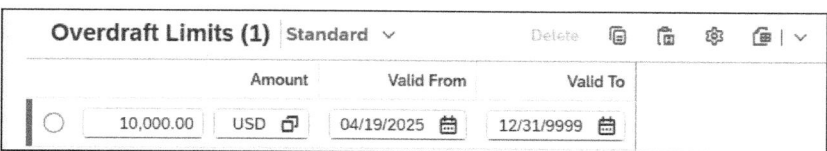

Figure 3.27 Bank Account Overdraft Limits

Overdraft limits are available and used in a few SAP S/4HANA solutions, primarily to monitor and control cash availability. However, the overall functionality around overdraft limits is still limited, and they are not yet widely integrated into broader payment or liquidity management processes. Their use remains basic, mainly serving as a reference point rather than a dynamic control mechanism.

In the next step (see Figure 3.28), you can create a description for the bank account in multiple languages. This feature is particularly useful for multinational organizations that need to manage bank accounts in different countries and regions. By maintaining descriptions in local languages as well as centrally in English or any other standard language, you can ensure consistency and clarity across all locations while also accommodating local language preferences for users in different countries.

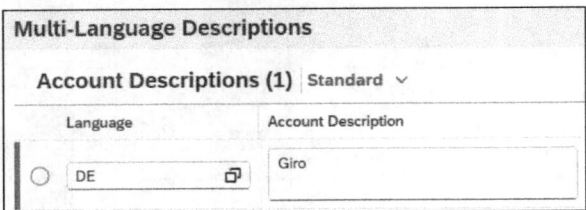

Figure 3.28 Bank Account: Multilanguage Descriptions

This helps with efficient communication and streamlines the management of bank accounts globally. This functionality allows you to conduct thorough research and meet international reporting requirements across various countries, especially when it comes to payment processing and variant generation. It also ensures that descriptions are clear and understandable for all stakeholders involved, supporting consistent and compliant communication across different regions and systems.

The next step involves entering **Cash Pool** data for the bank account (see Figure 3.29). This step is important for determining whether the account serves as a header account or a subaccount within the cash pool structure. If the value is populated under the **As Header Account** area, it means that the account is a header account for the selected cash pool. In this case, you also need to specify a payment method that will be used for sweeps in case of any transfers made from this header account. If the value is populated under the **As Subaccount** area, it means that the account is designated as a subaccount for the selected cash pool. For subaccounts, you also need to specify the payment method used for the sweeps from this account, the target balance—which indicates the expected balance on this account after the sweep is completed—and the maximum and minimum transfer amounts to be executed.

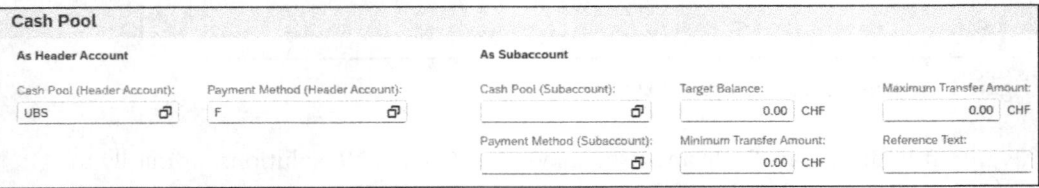

Figure 3.29 Cash Pool Data

Cash pooling, also known as *cash concentration* in previous versions of SAP S/4HANA, allows companies to manage liquidity more efficiently by consolidating balances from multiple accounts into one central account. This functionality is highly relevant for payment processing and cash management. However, note that in SAP S/4HANA 2023, new dedicated functionalities have been introduced for this specific solution. Cash pools are no longer managed within a single master data object but are now handled centrally through the Manage Cash Pools (Version 2) app (F3266A), provided as part of SAP S/4HANA Finance for cash management. This means that if you are using at least SAP S/4HANA 2023, you maintain participating accounts in the Manage Cash Pools apps. If you have an older version, you need to specify and assign them individually in each bank account.

Also note that SAP plans to remove this functionality from the Manage Bank Accounts app in the future and move it fully to dedicated apps. It is already deprecated as of SAP S/4HANA Cloud Public Edition 2025, but it can still be used for now.

The bank correspondence functionality (see Figure 3.30), available just below the **Cash Pool** tab in Manage Bank Accounts app, adds features that enable automatic correspondence with the bank, making it easier to manage communications electronically. One key feature is the ability to create e-banking functionality, allowing your company to automate the exchange of electronic messages and documents with the bank. We will cover the e-banking functionality in Chapter 4—specifically, during the process of creating an in-house bank account from a BAM account.

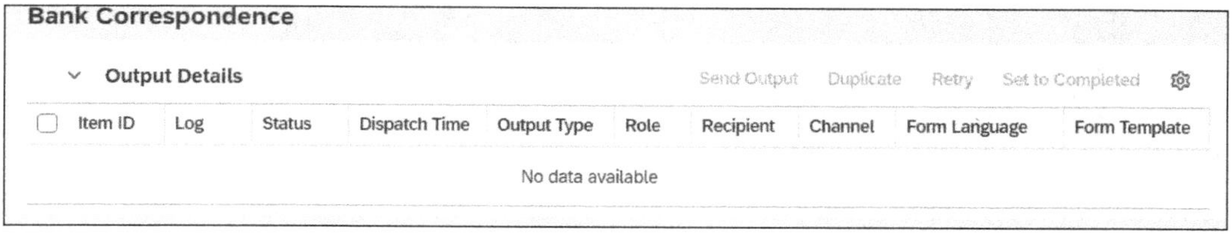

Figure 3.30 Bank Correspondence

This capability enhances your bank communication by enabling you to automatically generate and send correspondence, reducing the manual effort involved in creating and managing letters. For instance, automatic notifications can be sent to the bank when key account changes occur, such as account openings or updates to payment approvers. E-banking functionality also can integrate with digital banking services to facilitate smoother, faster transactions and communication between your company and the bank. However, the main idea behind using centralized payment systems is to reduce the number of banks. Using and paying for e-banking services, which require technical support, can be too costly to justify reducing the number of banks. Therefore, it is generally recommended to manage this manually.

Once you have created a bank account, and if you have configured the SAP Document Management service in your SAP S/4HANA system, you can easily attach relevant documents to a bank account. To do this, go to edit mode, click **Manage Attachments** (see

Figure 3.31), and the system will open the Manage Attachments app (F4812; see Figure 3.32), in which you can create and attach documents related to the bank account. This app is used for all attachments in the system, so you need to select **Bank Attachments** from the list.

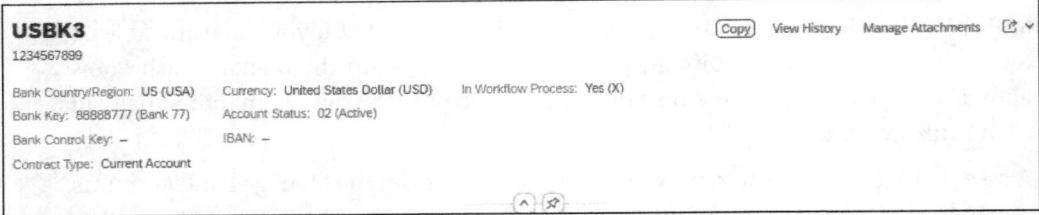

Figure 3.31 Manage Attachments

This functionality allows you to store and organize all related documents, such as contracts, agreements, or any other files, in a centralized location, making it easier to access and manage them when needed.

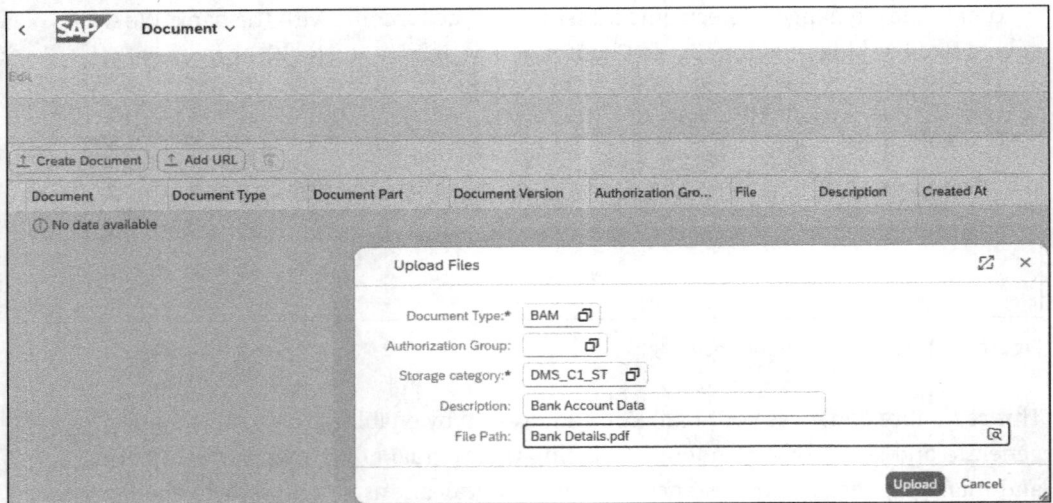

Figure 3.32 Add Attachment to Bank Account

Once the documents are attached, you will be able to see all the relevant files linked to a particular bank account (see Figure 3.33). The Manage Attachments app (F4812) allows you to store and organize all information coming from the banks, such as agreements, bank statements, contracts, and correspondence. By centralizing these documents, you ensure easy access and efficient management, making it simpler to track and reference important information associated with each bank account.

Once you have populated all the required information for the bank account, you can save it. At this stage, the account will be saved as an inactive account. The next steps in the process will depend on the configuration settings, but typically, bank accounts should go through an approval process. The approval process, including who approves the account and the approval steps, will be determined by the configuration setup. This ensures that all necessary checks and validations are performed before the bank account is fully activated and ready for use. We will discuss approvals in Section 3.1.4, but first we'll talk about options for mass import/migration of bank accounts.

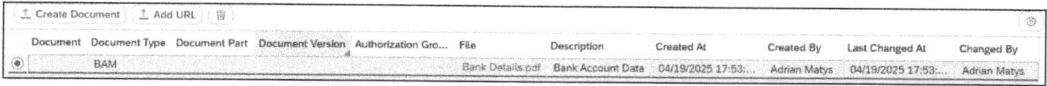

Figure 3.33 Manage Bank Account Attachments

3.1.3 Uploading, Migrating, and Replicating Bank Accounts

When you have multiple ERP systems across your organization, it's highly beneficial to establish a central bank account repository to manage all bank accounts in one place and replicate them into connected systems. This approach minimizes manual work, reduces inconsistencies, and eliminates redundant updates across environments.

BAM is an excellent tool to serve as this central repository. It allows for centralized creation, maintenance, and governance of bank account data, while also supporting automatic replication to other ERP systems via integration. This ensures that all systems are always aligned with up-to-date and accurate bank account information, leading to operational efficiency, data integrity, and compliance benefits.

Replication options for bank accounts are as follows:

- XML-based replication (import/export via SAP Fiori app)
- IDoc replication
- Web services/APIs

Currently, SAP is also developing some additional features for bank account replication in SAP S/4HANA Cloud Public Edition.

Import and Export Bank Accounts App

Creating all bank accounts one by one in the Manage Bank Accounts app can be extremely time-consuming, especially during system go lives, large-scale migrations, or major system updates. To streamline this process, SAP provides the Import and Export Bank Accounts app, which is a much more efficient alternative. You can also open this app from the Manage Bank Accounts app by clicking **Import and Export Bank Accounts**.

In this application (see Figure 3.34), you can first download the predefined Excel template that includes all the necessary fields expected by the system by clicking **Download XML Spreadsheet Template**. Once you've populated the template with the required bank account data, you can upload it directly into the system, and SAP S/4HANA will automatically create the bank accounts based on the information provided in the file.

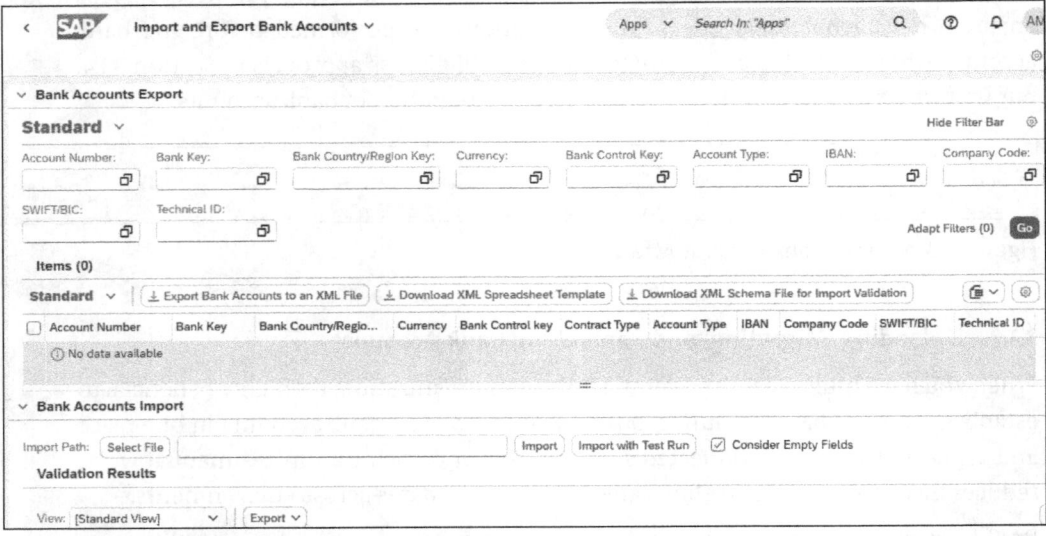

Figure 3.34 Import and Export Bank Accounts

The Import and Export Bank Accounts app processes files in XML format. This standardized format ensures compatibility and allows for structured data exchange between systems.

You can download the current bank account data from the system in XML format, make the necessary changes or additions, and then upload it back—either to the same system or to another system, such as quality or development. This makes it particularly useful for keeping environments synchronized or regularly updated.

In addition, during project testing phases or cutover activities, you can prepare an XML file with all the finalized bank account information and upload it into the production system as part of the data migration process. This helps avoid manual data entry, ensures consistency, and speeds up the transition to the new environment.

IDoc Bank Accounts Replication

SAP has provided standard message types to facilitate the replication of house banks, house bank accounts, and bank accounts using the data replication framework (DRF). This framework is especially useful when setting up a centralized BAM approach or during system landscape transitions, such as implementing a new SAP S/4HANA instance. We only want to briefly mention this functionality here; there are many other sources for detailed information on how it works technically and how it integrates into various SAP processes. For example, you can check out SAP Note 3524248.

Transaction DRFOUT triggers the replication; here you can select the appropriate replication model and filter criteria to define which bank data should be sent to which system. Transaction DRFOUT is a standard SAP transaction used for triggering the replication of master data, including bank accounts, business partners, materials, and more. Its functionality is highly dependent on the DRF configuration. Within Transaction DRFOUT, you define replication models, filter criteria, and target systems, which control how and where the data is sent.

When it comes to bank account replication, Transaction DRFOUT supports sending data via IDoc message types like BAMMAST (for bank accounts) and HBHBMAST (for house banks and house bank accounts). The flexibility of this transaction makes it a powerful tool not only for bank-related data but also for a wide range of master data objects, if the appropriate replication models and message types are configured. Proper setup in Transaction DRFIMG, along with maintaining partner profiles in Transaction WE20, is essential for enabling and controlling these replications

Web Services/APIs

Replication of bank accounts using web services or APIs with custom code is the most flexible and robust option for integrating systems, but it does not use a standard tool. It allows you to write custom logic using OData services or SOAP/REST APIs, enabling real-time replication across multiple systems. This approach is particularly well-suited for organizations adopting microservices architecture or requiring on-demand updates as it offers scalability and adaptability for complex enterprise environments.

3.1.4 Bank Account Approvals

Once a bank account is created in SAP, there are three main options for how it can be approved, depending on the configuration settings:

- The first option is *no approval*, in which the account becomes active immediately after creation without requiring any further validation.

- The second option is *dual control*, in which one user creates the bank account and another user must approve it using the Approve Bank Account Changes—Two-Person Verification app (F6264). This provides a simple layer of control and can be performed by any authorized user.

- The third and most flexible option is the use of a *workflow*, which allows for building more complex approval processes. In a workflow setup, predefined users must approve the account based on specific rules, and in some steps, they may even be required to input additional information. For example, the first person may create the bank account, and another designated user can later assign the house bank account ID or other attributes. This level of flexibility in workflows ensures proper governance and accountability, tailored to the organization's internal control requirements.

3 Bank Account Management

Two-person verification is shown in Figure 3.35. When you open the Approve Bank Account Changes—Two-Person Verification app, you will see a list of bank accounts requiring two-step verification. The process is straightforward: Simply click **Request** to view the details of the account, including any attached documents. From there, all you need to do is click **Approve** or **Reject**.

This app does not support workflows; instead, it follows a simple dual control mechanism. Any user with the appropriate role can approve any bank account, regardless of the specific content. However, the only restriction is that you cannot approve changes, openings, or closures for accounts that you yourself have requested. This ensures a basic level of control and separation of duties in the approval process.

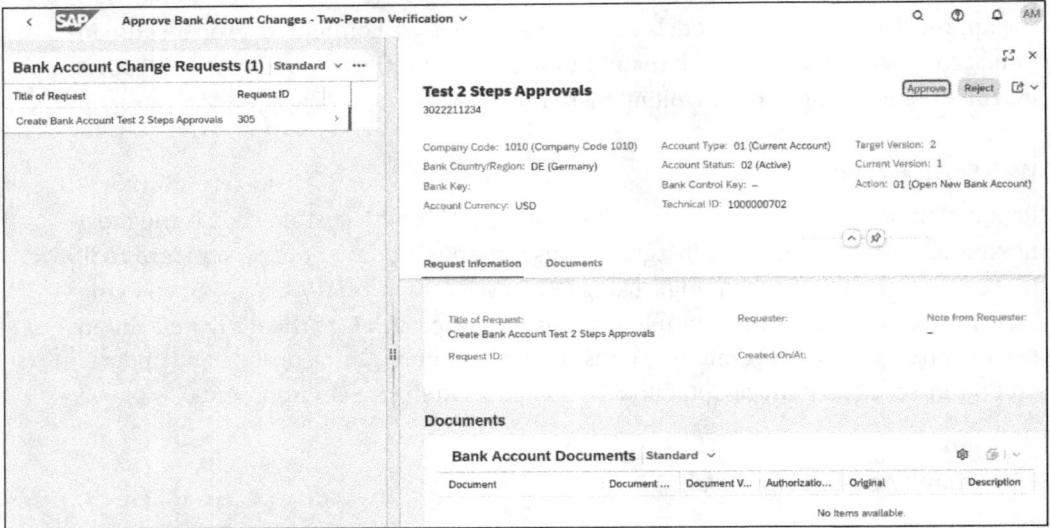

Figure 3.35 Approve Bank Account Changes—Two-Person Verification

The Bank Account Change Requests app (F6265) is designed to facilitate the management and monitoring of bank account change requests within the two-person verification process. This app (see Figure 3.36) allows users to view all bank account change requests that require approval, providing a centralized overview of their statuses. Users can filter and check the details of each request, including whether they have been approved, are pending approval, have been rejected, or have been withdrawn. This functionality ensures transparency and control over bank account modifications, aligning with organizational compliance requirements.

The status of all submitted bank account applications (no matter the type of approval selected) can be tracked in this app. It provides a centralized overview of all requests created for opening, changing, or closing bank accounts. You can view detailed information such as the status (e.g., **Draft**, **Submitted**, **In Approval**, **Approved**, **Rejected**, or **Withdrawn**), the applicant, the submission date, and other key details. It's especially useful for monitoring the progress of applications in organizations with multiple teams involved in BAM, ensuring transparency and timely follow-up.

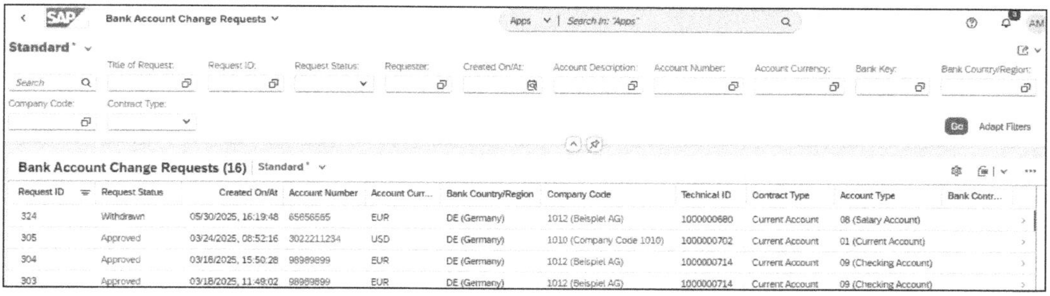

Figure 3.36 Bank Account Change Request

If you are approving bank accounts using a workflow (instead of the dual control approach), you will use the Approve Bank Account Applications app (F5859).

This app allows you to manage the approval process for bank account creation or changes in a more controlled and flexible way. It is especially useful in organizations that require more structured approval flows, possibly involving multiple steps or designated approvers. Key features of this app include the following:

- **Approve or reject bank account applications**
  You can review the application details and either approve or reject the request directly within the app.

- **Automatic creation of inactive bank account**
  Once approved, the system automatically creates an inactive bank account based on the details provided in the application.

- **Email notifications**
  The system automatically sends an email to the applicant to inform them of the approval or rejection status.

- **Withdraw application**
  You can confirm the withdrawal of a bank account application. If the application is withdrawn, the process of creating the bank account is stopped. If an inactive bank account was already created before the withdrawal, it will be automatically marked as **Withdrawn** in the system.

This app supports structured and auditable workflows, making it a strong alternative to the simpler two-step, dual control method.

Withdrawal of a request can be done directly from the Manage Bank Accounts app, regardless of whether you're using dual control or a workflow-based approval. To withdraw a request, follow these steps:

1. Open the Manage Bank Accounts app.
2. Find and select the bank account for which you've submitted a request (this could be a creation, change, or closure).
3. If the request is still pending approval, you'll see an option to withdraw the request. Simply click **Withdraw**, and the system will cancel the approval process.

Once a bank account is approved, its status changes to **Active** and it becomes available for use. At this point, the house bank and bank account ID are also activated. It's important to note that once a bank account reaches the **Active** status, it can no longer be deleted. Only **Inactive** accounts can be deleted. If an account has already been activated, the only option is to close it. This ensures that all activated accounts remain in the BAM system for audit, monitoring, and reporting purposes.

BAM workflows can be maintained using the Manage Workflows—For Bank Accounts app (F2796). This app (see Figure 3.37) enables business process specialists to define and manage approval workflows tailored to bank account processes within a line of business or across the entire organization. With this app, you can review predefined workflows or create new ones by copying existing templates. You can customize the workflow logic based on your company's requirements by specifying preconditions, step sequences, and recipients. It's also possible to configure automatic approvals for certain steps by setting specific criteria such as the change request creator, account type, or company code. The app also allows you to activate, deactivate, and delete workflows and to define the order in which workflows are evaluated, ensuring that the correct workflow is triggered for various bank account scenarios.

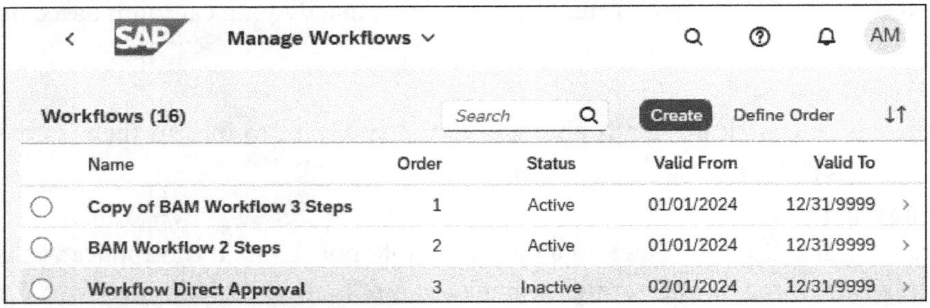

Name	Order	Status	Valid From	Valid To
Copy of BAM Workflow 3 Steps	1	Active	01/01/2024	12/31/9999
BAM Workflow 2 Steps	2	Active	01/01/2024	12/31/9999
Workflow Direct Approval	3	Inactive	02/01/2024	12/31/9999

Figure 3.37 Manage Workflow for BAM Applications

When creating a workflow in the Manage Workflows—For Bank Accounts app, you have a wide range of possibilities to tailor the process to your organization's needs. You can define the triggering events that initiate the workflow, such as creation, modification, or closure of a bank account. You can also specify rules and conditions, such as which company codes, account types, or change initiators the workflow should apply to. Within the workflow, you can build multistep approval processes, assigning different recipients for each step and even enabling or skipping steps based on certain criteria. In addition, you can decide which fields should trigger the workflow when changed and

3.1 Functions and Processes

whether a step requires user input or just approval. This flexibility allows you to create complex and dynamic workflows that reflect real organizational policies and approval hierarchies.

3.1.5 Manage Bank Account Hierarchies App

Once your bank accounts are created and approved, it's a common practice to group them into bank account hierarchies for better organization, reporting, and operational management. These hierarchies help structure accounts based on various criteria, such as cash pooling structures, geographical locations, currencies, or other custom-defined reporting needs. This is managed through the Manage Bank Account Hierarchies app (F4973). Within this app (see Figure 3.38), you can define and maintain flexible hierarchical structures by assigning bank accounts to different nodes, allowing you to visualize and analyze your account structure more effectively. These hierarchies can also be leveraged in reporting tools, cash concentration processes, and cash management analysis across your organization. When you open the Manage Bank Account Hierarchies app, the system will display the existing bank account hierarchies. To create a new hierarchy, simply click the **Create** button.

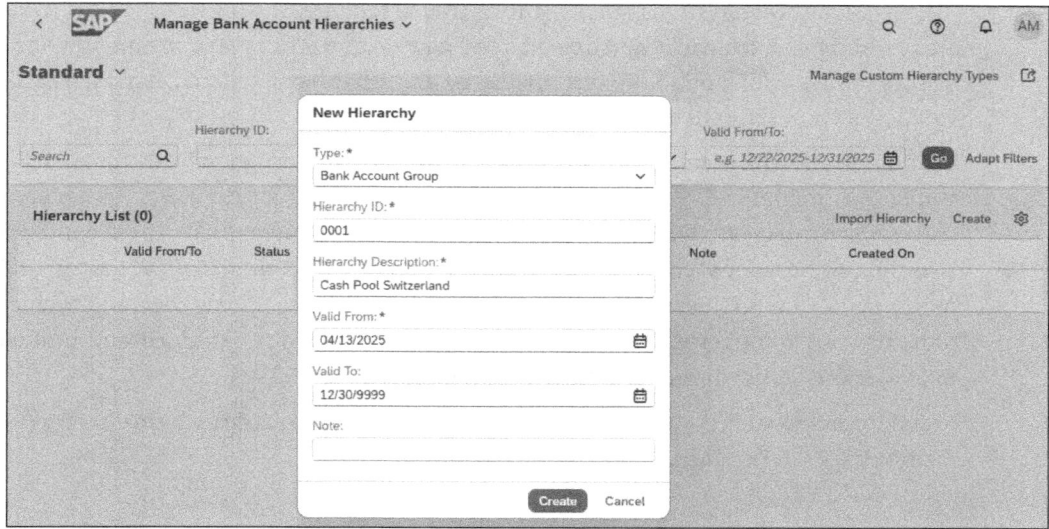

Figure 3.38 Manage Bank Account Hierarchies

You will then need to select the **Type** for the hierarchy, enter the **Hierarchy ID**, provide a **Hierarchy Description**, specify the validity period (**Valid From** and **Valid To**), and click **Create**. The app also allows you to import hierarchies from an Excel file, making it easier to set up multiple hierarchies at once, especially for large-scale environments or migrations. Then you can simply start populating your hierarchies by creating headers and assigning bank accounts, as shown in Figure 3.39.

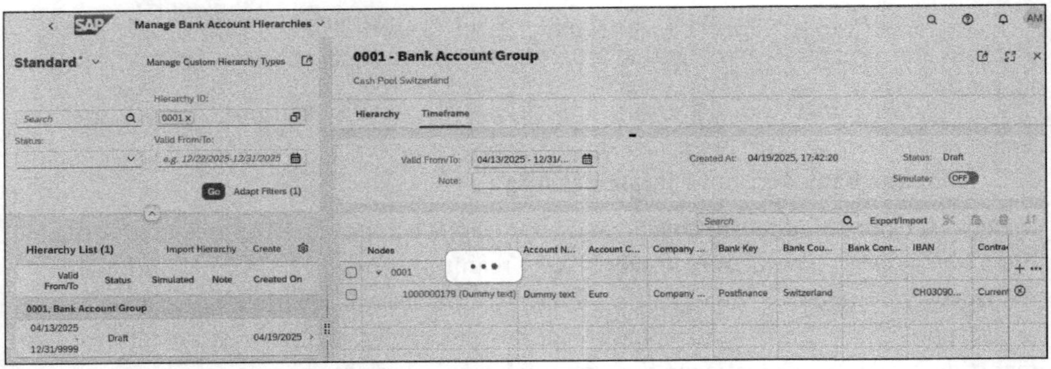

3 Bank Account Management

Figure 3.39 Create Bank Account Hierarchy

You can update existing bank account hierarchies by selecting the desired hierarchy, opening a new view, and clicking **Edit**. Once you're in edit mode, you can click the **+** icon to add new nodes or bank accounts to the hierarchy. A node represents a grouping level; for example, you can have a **Europe** hierarchy containing nodes that represent countries in Europe. The app also provides an export/import feature, which allows you to download and upload hierarchy files. This is especially useful when you need to transfer hierarchies between systems, such as during user acceptance testing (UAT): You can create a hierarchy, download it, and upload it to a new system. After you make any changes, be sure to save your work. Note that once a hierarchy has been updated, it must be activated by another user to ensure proper approval.

In addition to bank account hierarchies, there are multiple apps available for creating similar hierarchies in SAP S/4HANA. In Chapter 6, we will explore liquidity item hierarchies, which are created in the same way as bank account hierarchies. These concepts also tie into broader accounting structures such as profit centers, G/L accounts, and other types of hierarchies. The approach to managing and organizing these hierarchies remains consistent across various functionalities, ensuring streamlined and unified management across different accounting and financial processes.

Hierarchies created in this app are used in a couple of reports we'll describe in Chapter 6—mainly, the Cash Flow Analyzer app.

3.1.6 Manage Bank Account Reviews App

The Manage Bank Account Reviews app (F7510) enables you to initiate, manage, and monitor periodic reviews of bank account master data to ensure its accuracy and completeness. This app allows you to organize reviews by reference date and divide the process into specific review areas, each tailored to different responsible reviewers. For each area, the system takes a snapshot of the bank account master data at the time of review initiation. This snapshot serves as the basis for the review and is preserved throughout the process. Once a review is started, a workflow is triggered to guide the assigned reviewers, who complete their tasks via the My Inbox—All Items app. You can continuously track the progress of each review area in the Manage Bank Account Reviews app, ensuring the process stays on schedule and facilitating issue resolution if needed. Once all checks are finalized, you can mark the review as **Completed**. This app is a key component for enforcing governance and control over bank account data in a structured and auditable manner.

Prior to SAP S/4HANA 2024, initiating the bank account revision process required starting directly from the Manage Bank Accounts app or from the individual bank account itself. Users would then switch to a dedicated bank account revision app to validate and manage the revisions they initiated. However, SAP S/4HANA is now shifting toward a more centralized and structured approach with the Manage Bank Account Reviews app, which streamlines the process and offers better oversight and control. It is considered best practice to perform regular reviews of bank accounts to ensure that all accounts are still required and properly maintained. This is particularly important for large organizations, where it's common to find unused or outdated bank accounts that have not been formally closed. A structured revision process helps identify such cases, improves centralization of BAM, and supports clean and efficient financial structures across the organization.

3.1.7 Closing Bank Accounts

The closure of a bank account in SAP S/4HANA is performed through the Manage Bank Accounts app. To initiate the process, open the relevant bank account and click **Mark for Closing**. The system will open the popup shown in Figure 3.40, in which you need to select a date for the **Planned Closing Date** and add a **Note**.

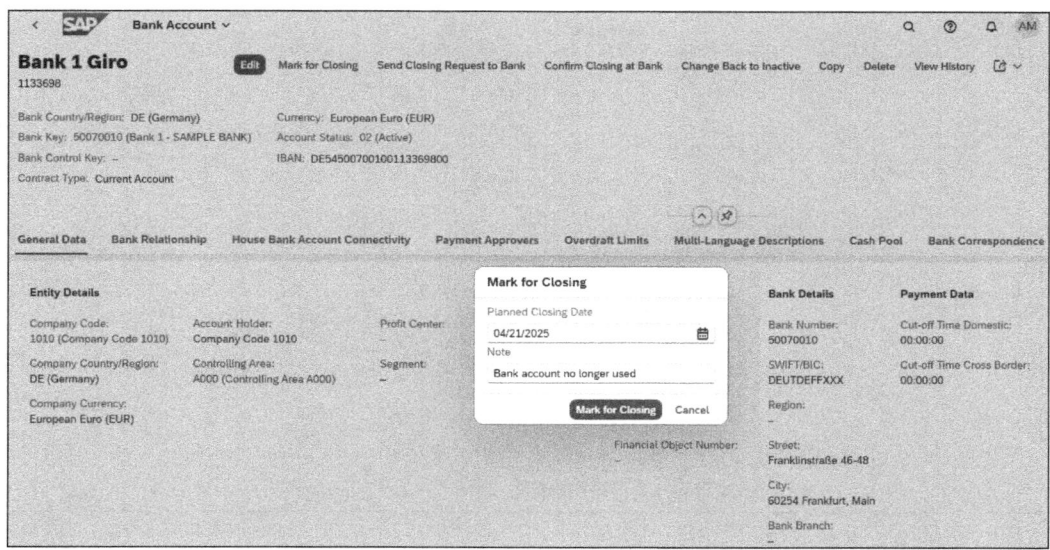

Figure 3.40 Close Bank Account

Once this action is taken, the system—depending on your configuration—will trigger a workflow requiring approval of the closure request. After approval, there may be additional steps in the process, such as sending a closure request to the bank, confirming the closure at the bank, and finalizing the account closure in the system. These steps vary based on how the workflow is set up, which will be covered in more detail later in Section 3.2. It's crucial to note that once a bank account has been activated, it cannot be deleted from the system. It can only be closed, and this closure will always follow a structured approval and workflow process to ensure compliance, traceability, and proper recordkeeping for audit purposes.

3.1.8 Manage Bank Chains App

Bank chains in SAP S/4HANA are used for managing and processing intermediary bank accounts, which are often required in international or complex payment scenarios. When a direct connection to the beneficiary bank is not possible or practical, intermediary banks are involved in routing the payment through one or more additional financial institutions. These intermediary relationships are maintained as bank chains in SAP S/4HANA using the Manage Bank Chains app (F4004). In this app, you can define the sequence of intermediary banks that should be used for specific transactions, ensuring accurate and efficient payment processing. This setup is particularly important in global organizations in which multiple banking relationships exist and payments need to be routed through a network of banks.

Once you open the Manage Bank Chains app and click **Create**, you'll arrive at the screen shown in Figure 3.41, where you can create a new bank chain and intermediary bank.

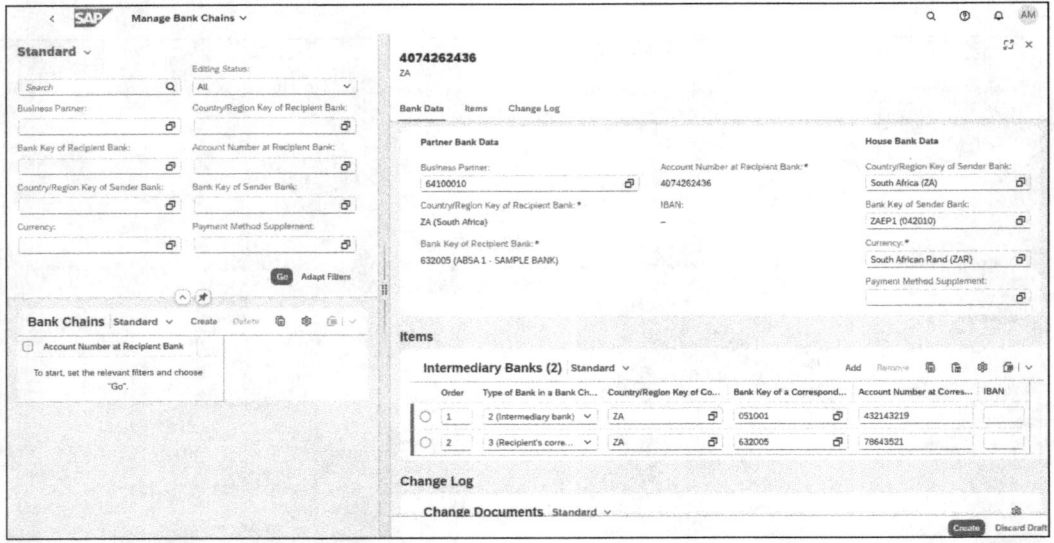

Figure 3.41 Create New Bank Chain

3.1 Functions and Processes

After you click **Create**, select the business partner for whom you want to execute the payment. Then choose the account from which the payment will be made, the currency in which you would like to execute the payment, and the payment method to supplement it. Then click **Add**, select the intermediary bank, and specify its country, bank key, and bank account. Next, select the recipient's bank account (the ultimate beneficiary account), including its country, bank key, and bank account.

In the example shown in Figure 3.42, we processed payments through a bank chain by utilizing an intermediary bank. Specifically, we initiated payments from a bank account located in South Africa, denominated in South African Rand, and routed the funds through an intermediary bank account—also based in South Africa—before they reached the final beneficiary's local bank account.

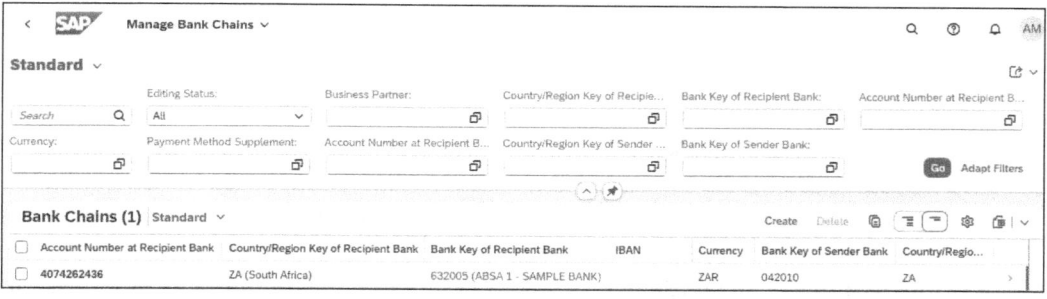

Figure 3.42 Manage Bank Chain

3.1.9 Foreign Bank Accounts App

In many countries, there is a regulatory requirement to register foreign bank accounts held by entities located within that country, along with their respective balances. To support compliance with these regulations, SAP S/4HANA offers the Foreign Bank Accounts app (F1575). This app enables you to identify and manage foreign bank accounts owned by your company and to track employees who hold the power of attorney over these accounts. The collected information can be used to file mandatory reports—such as the Report of Foreign Bank and Financial Accounts (FBAR) required for US persons—or for internal analytical purposes, depending on your country or region's legal requirements.

When you open the app (see Figure 3.43), you'll notice that the system has already predefined certain countries, but you can easily add more as needed; these are the countries where the entity is located. Simply select the timeline for which you want to review the accounts (in the **Time Range** field), choose the foreign accounts created for the relevant entities by selecting values for **Home Country/Region**, and click **Go** to view the results.

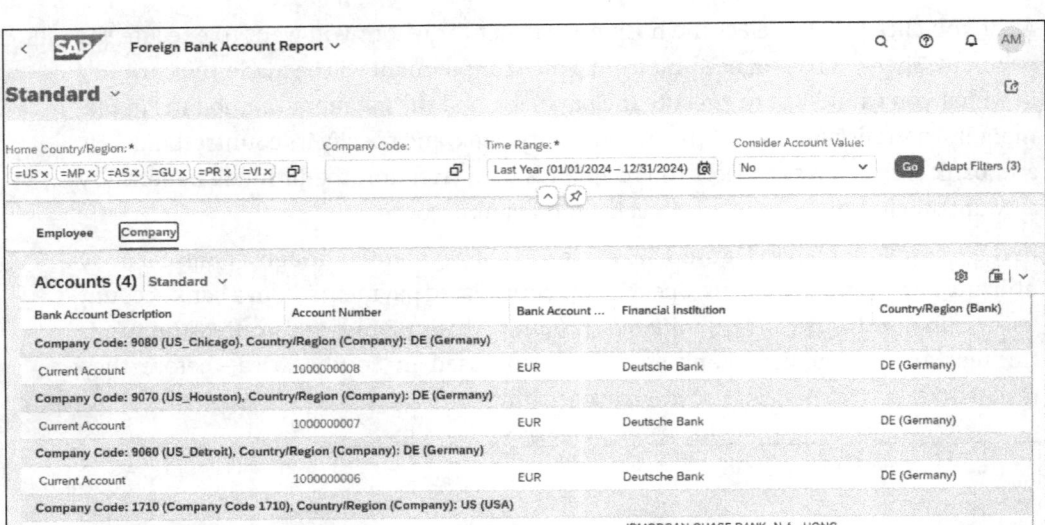

Figure 3.43 Foreign Bank Account Report

As a result, the system will display all foreign bank accounts associated with entities in the selected countries that are valid within the chosen time period. By switching to the **Employee** tab, you can also view all employees who hold power of attorney and are authorized to receive payments in the selected bank accounts.

This report is designed to provide much of the information needed for various regulatory reporting requirements. Some countries, such as France, may also require that you report the balances of these foreign accounts. In such cases, you'll need to refer to other reports as well, which we'll explore later in this section. You can then combine the relevant information and manually prepare the necessary regulatory submissions. Although the app supports multiple use cases, its primary focus is to fulfill the requirements of US authorities.

3.1.10 Bank Relationship Overview App

The Bank Relationship Overview app (F3775) provides a wide range of valuable information about your organization's banking landscape. It offers insights into both incoming and outgoing payments—showing their volumes and amounts—based on selected bank accounts. You can also access detailed bank profiles, including how many banks are associated with specific bank keys, the number of bank accounts, and how many company codes are using them. In addition, the app displays a breakdown of bank fees for selected accounts, account statuses (**Active**, **Closed**), workflow statuses (such as open requests), and revision history. In essence, this app consolidates all key data previously available across various apps and bank statements, offering a comprehensive view of your entire bank relationship setup within the system.

When you open this app (see Figure 3.44), you can select the banks you would like to analyze via the following selection parameters: **Display Currency**, **Exchange Rate Type** and **Date**, **Bank Key**, and **Bank Country/Region**.

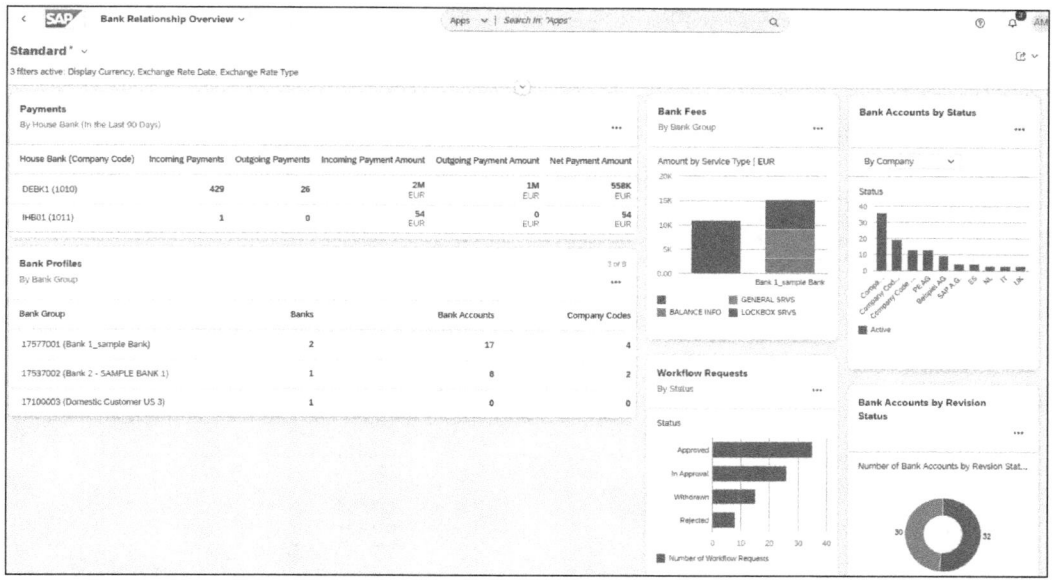

Figure 3.44 Bank Relationship Overview

Once you click **Go**, the system will generate all the necessary information and display a comprehensive overview for the selected bank accounts, giving you immediate access to key data and insights.

3.1.11 Power of Attorney for Banking Transactions

In every corporate organization, the process of granting authorizations and determining who is authorized to represent the company in its relationships and transactions with banks is critically important. In many countries, these authorizations must also be officially registered in commercial registers and approved by the CFO. Furthermore, they need to be regularly reviewed and audited to ensure compliance. Traditionally, managing this information across multiple locations and systems has been cumbersome and complex, making regulatory reporting a challenge. To support and simplify this process, SAP introduced the power of attorney functionality (as of SAP S/4HANA 2022), allowing companies to centrally maintain authorizations and responsible persons directly within SAP S/4HANA, ensuring better governance, easier access, and streamlined reporting.

The *power of attorney* defines the legal authorization granted by a company to one or more individuals to perform banking transactions on their behalf. It outlines the principals, authorized representatives, associated bank accounts, payment approval rules, and permitted activities assigned to each representative.

You can create a power of attorney record using the Manage Powers of Attorney for Banking Transactions app (F5742). Once you open this app, you will see the screen shown in Figure 3.45.

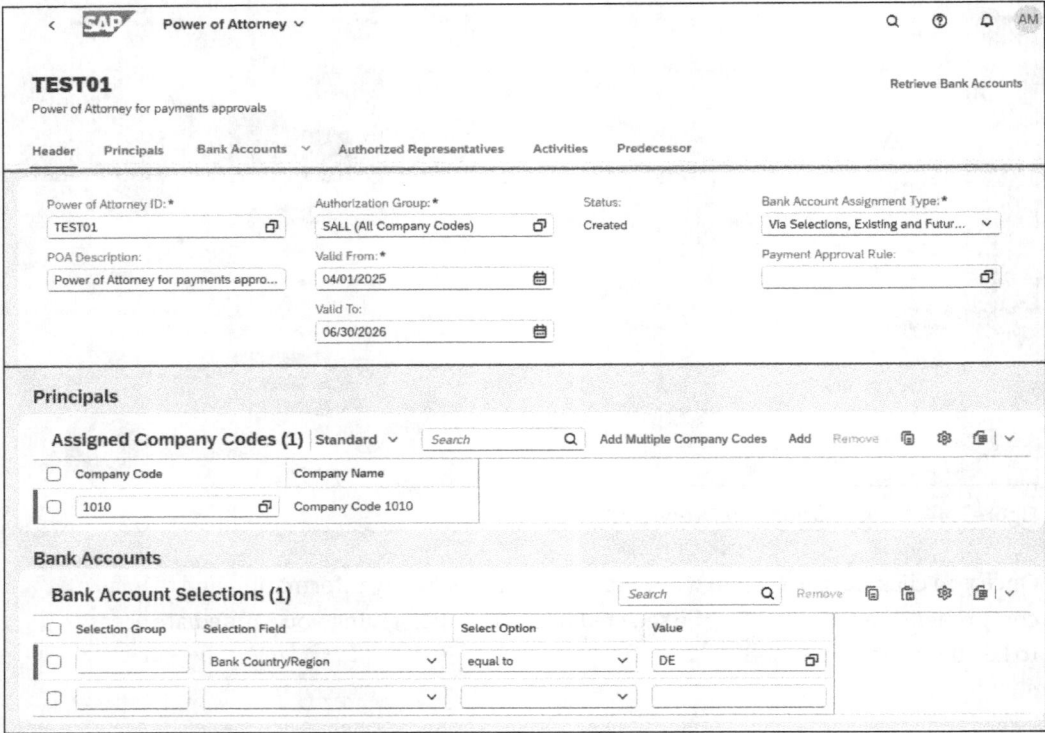

Figure 3.45 Create Power of Attorney

When creating a power of attorney record, you first need to assign a **Power of Attorney ID** and provide a **POA Description**. In this example, we are creating a power of attorney specifically for the approval of payments. Here, select the relevant **Authorization Group** and define the validity period by specifying the start and end dates. Next, you must indicate whether the power of attorney applies to a specific **Payment Approval Rule** or if it should cover all existing bank accounts in the system, as well as any bank accounts that will be created in the future.

To select the *principals*—the companies that grant the power of attorney to one or more individuals (authorized representatives)—you need to choose the appropriate **Company Code**s in the system; these represent the principals of the power of attorney holders. After selecting the relevant company code, you then define the bank accounts to which the power of attorney will apply. In this example, we are selecting bank accounts created within a specific country, but you can also search by bank keys, specific bank accounts, currencies, technical IDs, or any other relevant criteria. Once you have made your selections under **Bank Account Selections**, click the **Retrieve Bank Accounts** button. The system will then automatically retrieve all bank accounts that meet the specified criteria, as shown in Figure 3.46.

Technical ID	Bank Key	Account Number	Bank Country/Region	Currency	Bank Control Key	Contract Type	Account Type	Bank Account Assigned At
1000000000	50070010	1133698	DE (Germany)	EUR			01 (Current Account)	
1000000004	99999911	99999911	DE (Germany)	EUR			02 (Current Account (Internal))	
1000000003	88888886	1133888	DE (Germany)	EUR			01 (Current Account)	
1000000002	50070010	40414243	DE (Germany)	USD			01 (Current Account)	
1000000001	82080000	2580061	DE (Germany)	EUR			01 (Current Account)	

Figure 3.46 Power of Attorney: Assigned Bank Accounts

Next, you need to select *authorized representatives*; these are individuals (typically employees) to whom your company has granted the power of attorney.

Authorized Representative

For your authorized representatives, you must create a business partner and assign the business partner role **TR1000 Authorized Representative** (see Figure 3.47). Only business partners with this role can be assigned as authorized representatives in a power of attorney record.

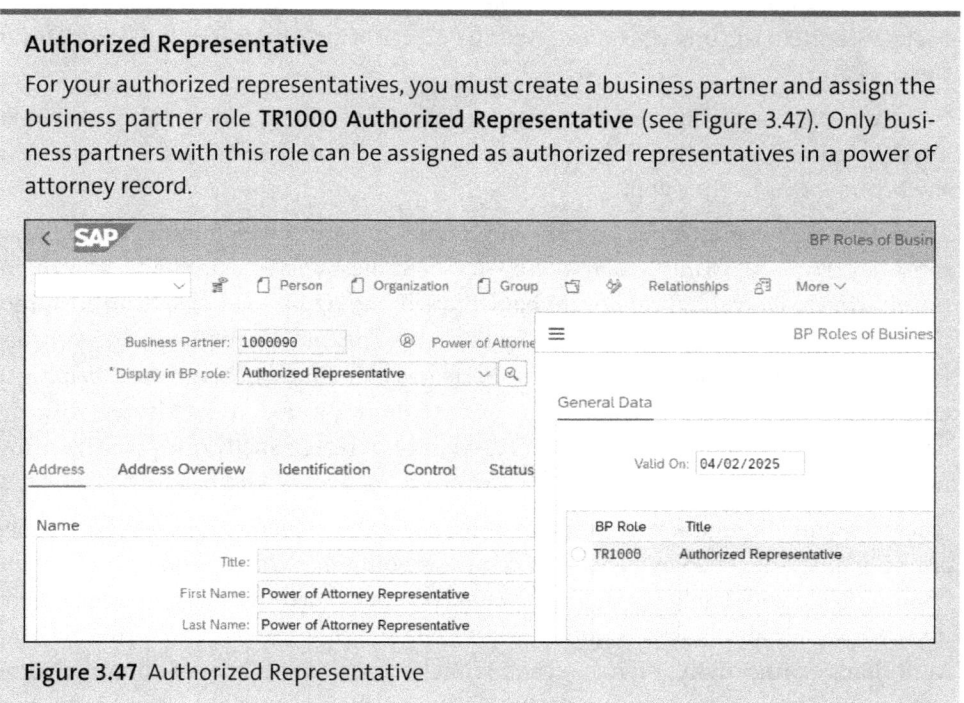

Figure 3.47 Authorized Representative

After selecting the representatives, you must define *activities* (see Figure 3.48), which are the banking transactions for which authorization has been granted. The system offers a set of predefined activities, but you can also define custom activities as needed.

Once you have selected all the required information, click **Create**. After the power of attorney record has been created, you must submit it for approval.

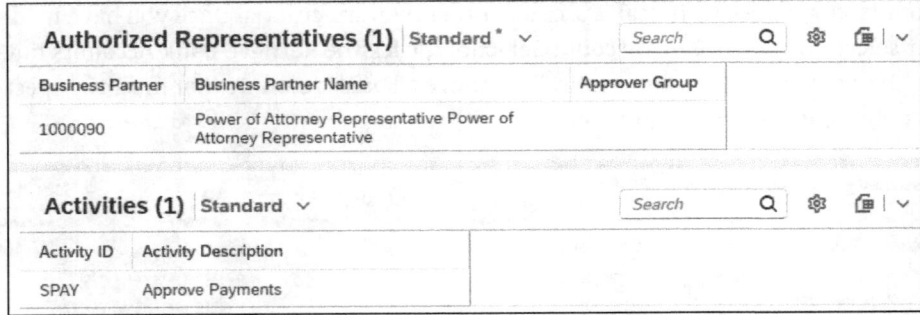

Figure 3.48 Create Power of Attorney Representatives and Activities

When a power of attorney record is submitted for approval using the Manage Powers of Attorney for Banking Transactions app, and an approval workflow is enabled, a workflow task is automatically created in the My Inbox—All Items app for all users in the responsible team who have the authorization to approve powers of attorney. To ensure compliance with the dual control principle, the workflow only assigns approval tasks to users other than the one who submitted the power of attorney record (the initiator of the workflow).

If the workflow is configured with email notifications, the approvers also receive an email informing them that the approval process has started, including a direct link to the My Inbox—All Items app.

After the powers of attorney have been granted and are active, administrative tasks must be carried out to implement them—for example, by setting up changes and communication to go through a relevant bank branch (see Figure 3.49) or setting up authorizations for online banking managed by the bank. The Implement Powers of Attorney for Banking Transactions app (F6374) supports you in managing this process, helping to ensure that the granted powers are properly communicated and activated with the banks. For each power of attorney record, you must create exactly one power of attorney implementation.

Once your implementation is complete, you can use the Implement Powers of Attorney app to manage several key tasks. You can create outgoing correspondence objects to communicate with banks, as well as register incoming correspondence objects. If you do not have an automated connection or if eBAM is not implemented through SAP Multi-Bank Connectivity, then you can manually update the status of outgoing correspondence objects. The app also allows you to replace an existing power of attorney record or revoke one when necessary, helping you maintain accurate and up-to-date records of your banking authorizations.

The power of attorney and automatic exchange of information features are not very widely used. In most cases, only power of attorney documents are maintained to clearly identify who is authorized to use the accounts and who has access to them in the online banking platform. This is primarily done for audit and reporting purposes, ensuring that there is a clear record of account access and usage rights.

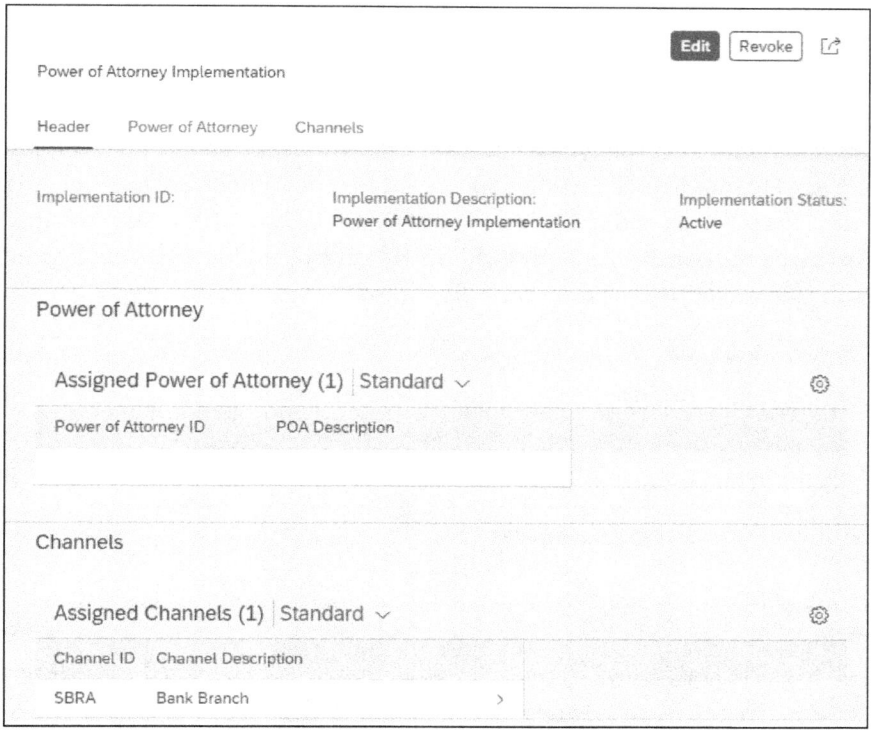

Figure 3.49 Implement Power of Attorney

3.1.12 Bank Fees

Bank fees are an essential aspect of managing bank accounts, and they are typically incurred for each transaction. Maintaining bank accounts is associated with various costs, including account maintenance fees, transaction fees, and other charges depending on the services provided by the bank. These fees are closely tied to the number and type of transactions processed, as well as the specific bank account terms. Identifying and tracking the exact cost of these bank relationships can be quite complex, as fees can vary based on account activity, transaction volume, and other factors. Often, the detailed breakdown of these costs is not immediately clear, making it challenging to have full visibility into how much is being paid for the services provided by the bank.

Banks usually send this information about bank fees as part of the camt.86 file, which provides detailed fee breakdowns. SAP S/4HANA offers tools designed to help with the reconciliation of these bank fees, making it easier to match the bank's statements with your internal records and ensure accurate accounting. However, the process still requires careful review and reconciliation to ensure all costs are accounted for correctly.

3   Bank Account Management

The camt files you receive from your bank can be uploaded into SAP S/4HANA for further analysis using the Import Bank Services Billing Files app (F3002). This app allows you to import and process the bank fee data from the camt files, enabling you to perform detailed analysis, reconcile fees, and ensure that the bank charges align with your internal records.

When you open the Import Bank Services Billing Files app (F3002), you will see a single option to import your file. Simply select the file you wish to upload and import it into the system (see Figure 3.50) via the **Select Source File** field. It is important to ensure that the bank account reflected in the file already exists in the system. If the bank account is not created in the system, the file import will fail. However, if the bank account is set up correctly, the file will be uploaded successfully without any issues.

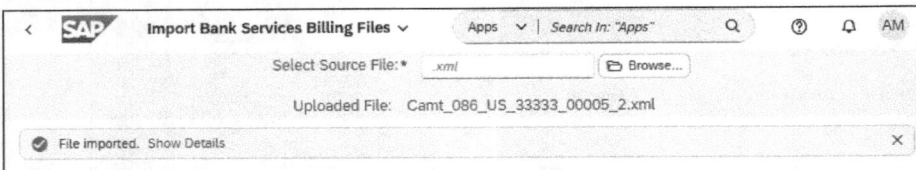

Figure 3.50 Import Bank Services Billing Files

> [!]
> **File Version**
> Note that banks may send various versions of the camt86.xml files. Older versions of SAP S/4HANA prior to version 2023 might not be able to process newer versions of these XML files. In such cases, you may need to build a transformation to convert the file into a compatible format, or you can check SAP Notes (SAP's online service system) for any necessary updates. SAP typically releases updates through SAP Notes that can be uploaded to your system, allowing newer versions of the files to be processed correctly.

The Manage Bank Fee Conditions app (F3185) allows you to create, edit, and delete bank fee conditions. The condition defines how bank fees should be charged for specific bank services, as shown in Figure 3.51. The system uses these conditions to validate the imported bank fee data, checking for any errors or improper charges. The main features of this app are as follows:

- **Create, edit, and delete bank fee conditions**
  You can define conditions for how fees should be applied to various banking services.
- **Assign conditions to bank services**
  Once conditions are defined, you can assign them to relevant bank services and use them to validate imported bank fee data in the Monitor Bank Fees app.

To create a new bank fee condition, you first need to enter the **Condition ID** and specify the **Valid From** date. Then, provide a **Condition Description**, select the **Pricing Type**, and choose the **Condition Currency**. Once these fields are completed, click **Create**. Now enter the conditions that are applied by the bank, which should come from your contract. These include thresholds, unit price, annual rates (as a percentage), base charge, minimum charge, and maximum charge. After filling in all the required information, click **Create** again to finalize the setup.

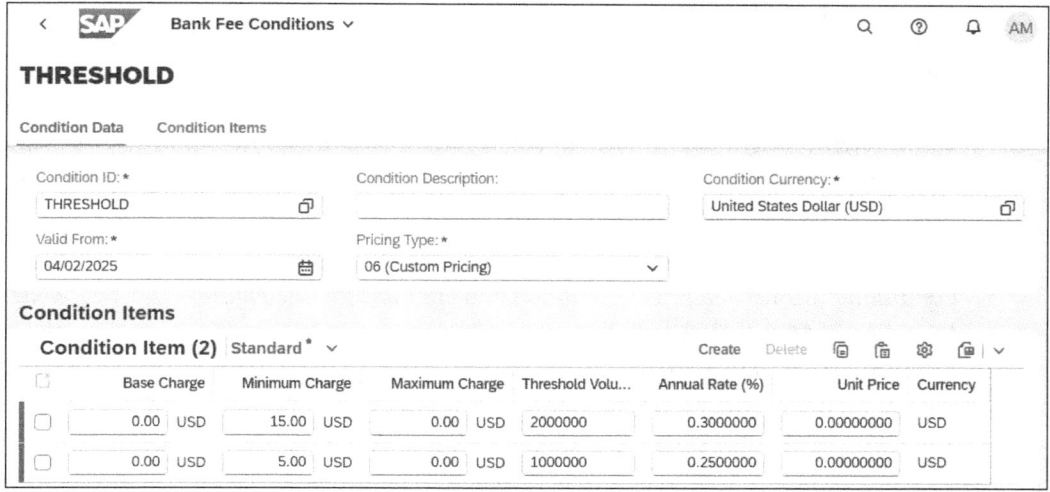

Figure 3.51 Manage Bank Fee Conditions: Create

In this example, we are creating conditions and charges that the bank can apply based on certain thresholds. For instance, if there is an excessive balance in your bank account, then the bank may charge fees. These conditions allow you to define how such charges are applied depending on the account balance or other predefined criteria.

Once you've created the bank fee conditions, you can analyze the fees uploaded from the statements using the Monitor Bank Fees app (see Figure 3.52).

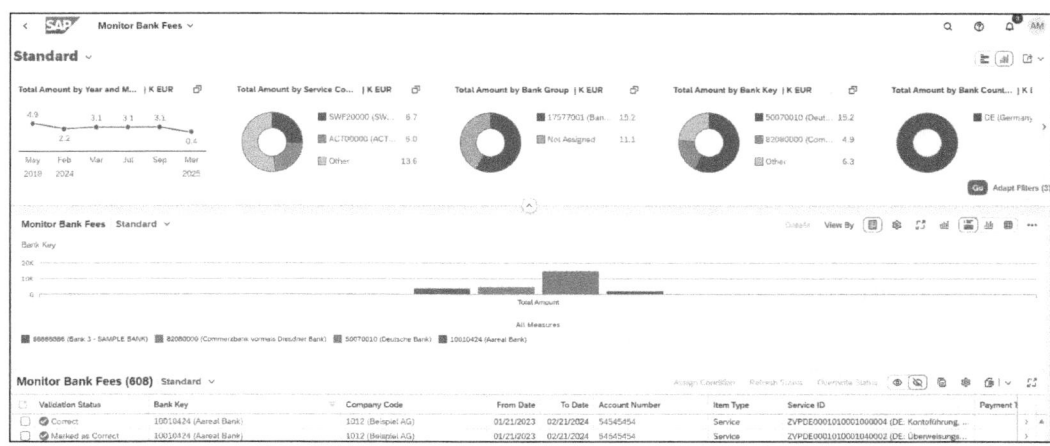

Figure 3.52 Monitor Bank Fees

3 Bank Account Management

This app allows you to view a detailed breakdown of your fees per bank account, including the online items that come from the billing services file. It provides a comprehensive overview, enabling you to ensure that the fees charged align with the conditions you've set and helping you to identify any discrepancies or unexpected charges.

The app helps you analyze how much you've paid per selected bank account. You can adjust the selection criteria to focus on specific bank accounts; under **Monitor Bank Fees**, you'll see all the line items from the bank services file. The app shows whether these line items have been assigned to any condition type. If a line item hasn't been automatically assigned a condition, you can manually assign it. Simply click the line item, and the system will open a new view in which you can assign the appropriate condition type to that line item.

From the line item in the **Bank Billing Services** file, select a line item and click the **Assign Condition** button, then select **Condition** and click **Assign** (see Figure 3.53) or click **New Condition** (system will move you to the Manage Bank Fee Conditions app). Once the condition is assigned correctly, it will be considered in the analytics and statistics for your bank fees, allowing for more accurate tracking and reporting. This ensures that your bank fees are properly categorized and integrated into future analyses.

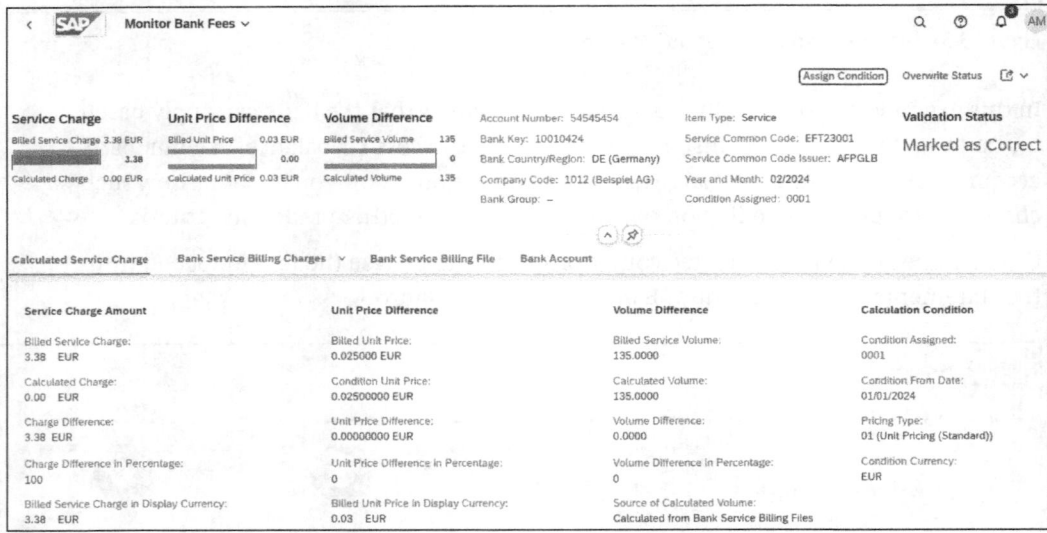

Figure 3.53 Assign Conditions to Bank Fee

Once all the relevant conditions have been assigned, you can begin analyzing and monitoring bank fees. The system will automatically track these fees and update them across various reporting sections, ensuring consistent and accurate visibility into your financial data.

3.2 Configuration

As you've learned throughout this book, in cash and bank activities, SAP S/4HANA is increasingly shifting functionalities to SAP Fiori apps, allowing business users to manage many tasks directly through the applications without needing technical support. This move supports a more user-friendly and accessible approach to daily operations. However, there are still certain activities that require backend configuration, particularly those related to defining the behavior of accounts or objects used in cash management. These configurations ensure that the system processes and interprets financial data correctly in line with organizational requirements.

In this section, we'll first examine two configuration activities that are also necessary for SAP S/4HANA Finance for cash management (Chapter 6) and SAP Bank Communication Management (Chapter 7): defining basic settings and enabling payment approvals. From there, we'll get into BAM-specific configurations for number ranges, bank account master data, bank account contract types, and field status groups.

3.2.1 Define Basic Settings

Defining basic settings is one of the most critical configuration activities as it determines how cash and bank relationship management and SAP Bank Communication Management will function within the SAP S/4HANA system. Follow menu path **SAP Financial Supply Chain Management • Cash and Liquidity Management • General Settings • Define Basic Settings** to reach the screen shown in Figure 3.54. This activity lays the foundation for the behavior of cash and liquidity management by establishing essential parameters such as the **Cash Scope** of functionality (**Basic Scope** or **Full Scope**, depending on your license and the associated cost).

You will also find change request grouping, bank account control modes, snapshot settings, and the availability of cash request features. These settings shape how the system processes cash flows, manages liquidity data, and interacts with other areas of SAP S/4HANA, like accounting and BAM. This activity also includes key technical setup areas such as defining the source application for accounting data, runtime parameters for the flow builder, and default liquidity items for liquidity analysis. Choosing the right configuration here is vital as it directly influences how cash operations are structured and monitored across the organization. Let's now describe each function a bit more:

- **Enable snapshot**
  The enable snapshot feature in SAP S/4HANA Finance for cash management allows you to activate the snapshot function within the full scope of cash management. When the **Enable Snapshot** field is enabled, cash flows are segmented by the time they are generated in One Exposure from Operations. This allows you to take snapshots of cash flows at different times and compare various versions of cash management data, such as cash positions or forecasts for a particular day. For example, you

can use snapshots to trace cash position reports for the previous month and assess how they have changed over time. It's important to note that the snapshot functionality is not supported in the basic scope. Even in the full scope, this feature is not enabled by default; it must be activated manually. Note that in this book, we are not covering snapshots in cash management in detail.

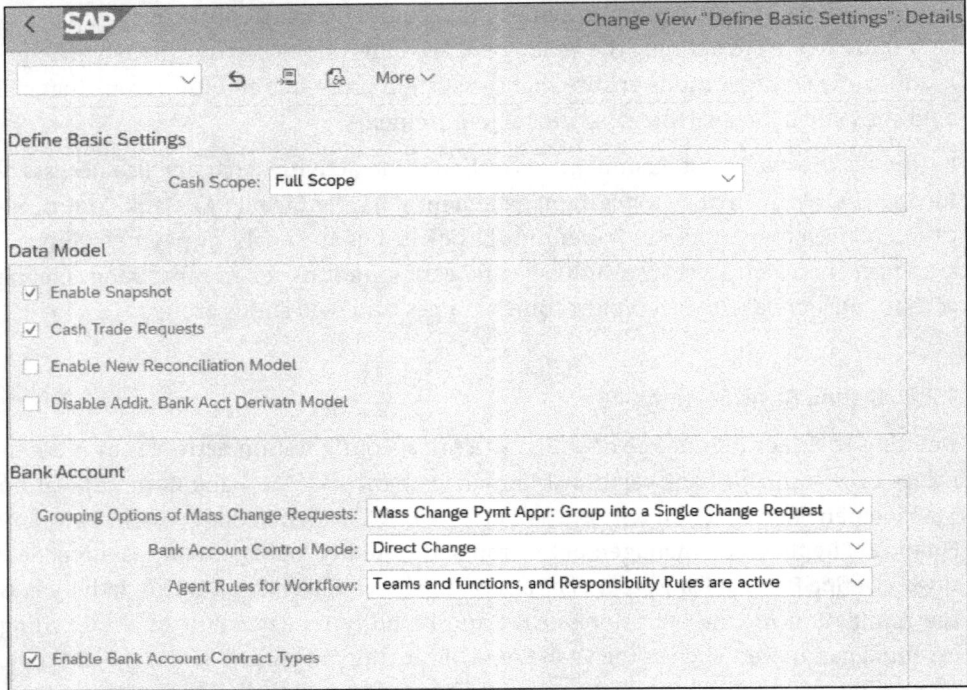

Figure 3.54 Define Basic Settings

- **Cash trade request**
  A cash trade request in cash management is a feature that allows you to create, track, and manage short-term investments and borrowing activities within your organization's cash position management processes. It can be triggered directly from the cash management module and can also be used to send trade requests to external trading platforms to create deals. This capability helps treasurers to manage liquidity more effectively, ensuring that surplus funds are invested efficiently or short-term borrowing needs are met in a timely manner. It's worth noting that cash trade requests fall within the broader topic of treasury and risk management in SAP, which we do not cover in detail in this book.

- **New reconciliation model**
  The **Enable New Reconciliation Model** option activates the new cash flow reconciliation model within cash management, providing enhanced capabilities for reconciling cash flows. This new model offers more functionalities and can be used to

reconcile cash flows against a broader set of data compared to the old model. If you have this functionality available in your system, you should use the new reconciliation model to take advantage of its advanced features. When the new reconciliation model is enabled, you should use the Reconcile Cash Flows—Intraday app for your daily cash flow reconciliation tasks. If the new reconciliation model is not enabled, you will need to use the Reconcile Cash Flows—Intraday Memo Records app instead. This flexibility allows you to tailor your reconciliation processes based on your specific requirements and system configuration.

- **Additional bank account derivation model**
The **Disable Additional Bank Account Derivation Model** option in cash management allows you to control how the system fills in missing bank account information during accounting flows. By default, the system applies an additional bank account derivation logic when key details such as the bank account, house bank, or house bank account information is absent. This additional logic is executed before saving the data. However, if you activate the **Disable Additional Bank Account Derivation Model** option, the system suspends this additional derivation process, allowing you to manage and control bank account assignments manually if needed. It's important to note that in cash management, having all flows assigned to the correct bank accounts is critical for maintaining an accurate cash flow forecast. Therefore, if you choose to disable this option, you must implement an alternative method to assign flows to bank accounts to ensure accuracy. For more information on these alternatives, see Chapter 6.

- **Grouping options for mass change requests**
The grouping options for mass change requests in BAM allows you to control how change requests for updating payment approvers in multiple bank accounts are grouped together for workflow processing. By default, when you initiate a mass change to update payment approvers across several accounts, a change request is generated for each bank account, and these requests are grouped into a single mass change request. However, you can customize how these requests are grouped: into a single mass change request, or into multiple mass change requests by company code, account type, or a combination of both. This flexibility ensures that workflows align with your organization's structure and responsibilities. It's important to note that this setting is only relevant when workflows are enabled for BAM and is not applicable to basic cash management. If you choose **Group into a Single Change Request**, all change requests are combined into one, requiring approvers to be authorized for all company codes and account types. **Group by Company Code** creates separate mass change requests for each company code, so approvers need responsibility for specific company codes. **Group by Account Type** creates separate requests by account type, so approvers must be responsible for specific account types. Lastly, **Group by Company Code and Account Type** generates highly specific mass change requests for each company code and account type pair, giving you the most granular control over approval workflows.

- **Bank account control mode**
  Via the bank account control mode, you can select how you would like to approve bank changes within BAM. By choosing the appropriate grouping option, you determine how change requests are organized and routed to approvers based on your company's structure and workflow needs, as described in Section 3.1.4.

- **Agent rules for workflow**
  An agent in BAM is a user who can receive workflow tasks (work items) in their inbox and perform necessary actions on them. In the **Agent Rules for Workflow** section, you can define whether the workflow should consider teams and functions only, or also include responsibility rules. Teams and functions group users by role or organizational structure, whereas responsibility rules allow for more granular assignments based on criteria such as company code, account type, or other attributes. Typically, you would select both teams and functions and activate responsibility rules to ensure that the right users—whether organized by teams or by functions or assigned via responsibility rules—receive and process workflow tasks efficiently.

- **Bank account contract types**
  The **Enable Bank Account Contract Types** setting in BAM allows you to activate the use of contract types for bank accounts within your system. By enabling this option, you can classify and categorize bank accounts based on different contract types, such as checking, savings, or term deposit accounts, depending on how they are managed within your organization. This feature helps ensure that bank accounts are appropriately categorized and managed according to your internal standards and contract-specific requirements. The use of bank account contract types and when you select them is described in Section 3.1.2.

3.2.2 Enable Payment Approval

In the configuration activity for enabling payment approvals, you will define which payment approval mechanism will be used: payment approvers maintained in the BAM master data, or via the SAP Bank Communication Management configuration, which is described in detail in Chapter 7. This decision is crucial, as it determines where and how approval rules are managed and executed. As described earlier, using BAM allows for a simpler, payment-based approval setup, while SAP Bank Communication Management provides more flexibility and control, enabling complex approval workflows based on various parameters such as amount, payment method, or company code.

You can find this configuration under **SAP Financial Supply Chain Management • Cash and Liquidity Management • Bank Account Management • Enable Payment Approval**.

To enable the payment approval process in BAM, add the following two entries to table TPS34 (however, we recommended you maintain them via the menu path) as shown in Figure 3.55:

- **Process** "OBANK002": **Function Module** "FCLM_BAM_BCM_AGT_PRESEL": **Product** "BAM"

- **Process** "OBANK004": **Function Module** "FCLM_BAM_BCM_REL_PROC_CTRL": **Product** "BAM"

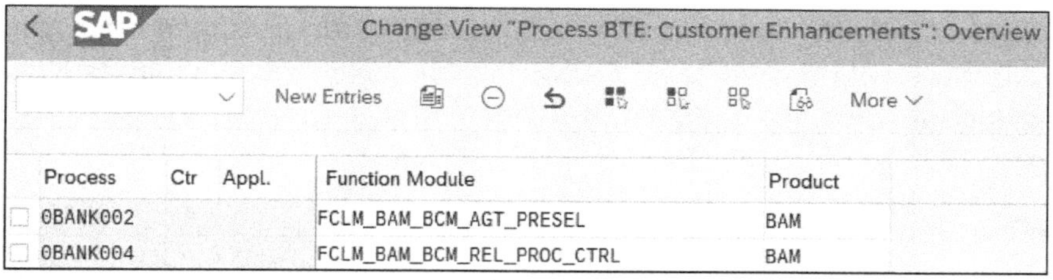

Process	Ctr	Appl.	Function Module	Product
OBANK002			FCLM_BAM_BCM_AGT_PRESEL	BAM
OBANK004			FCLM_BAM_BCM_REL_PROC_CTRL	BAM

Figure 3.55 Enable Payment Approver

If you want to continue working with the SAP Bank Communication Management setup, you do not need to do anything here; leave the blank values, and the system will pick up rules and workflow assignments from the SAP Bank Communication Management customizing, as described in Chapter 7.

3.2.3 Number Ranges for Technical Accounts ID

The first thing you always need to set up when working with BAM in SAP S/4HANA is the number range. SAP S/4HANA uses various number ranges for different purposes, and several of them are specifically related to BAM. Among these, one of the most important is the number range for technical IDs (see Figure 3.56), which represents the internal numbering of bank accounts in the system. As shown in the earlier screenshots, these technical IDs play a crucial role in identifying and managing bank accounts. Before you can use BAM effectively, it's essential to first create and assign the appropriate number range for these technical IDs.

You can find the setup for the number ranges under **SAP Financial Supply Chain Management** · **Bank Account Management** · **Basic Settings** · **Define Number Ranges for Bank Account Technical IDs**.

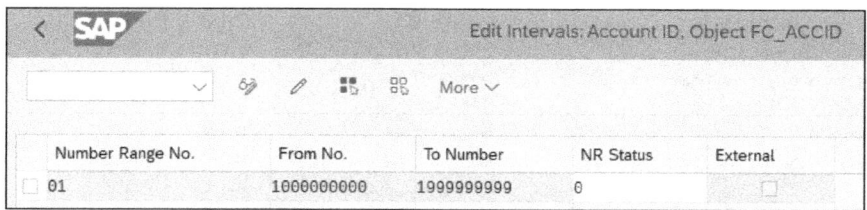

Number Range No.	From No.	To Number	NR Status	External
01	1000000000	1999999999	0	

Figure 3.56 Number Range for Bank Account Technical ID

Once you open this customizing activity, you can click **Edit** and provide the desired number range.

Here, you need to specify the range used for BAM accounts. By default, this range can typically be set from 1 to 9999999999, providing a wide scope for assigning internal technical IDs to bank accounts. However, for specific needs, you can use any other number range; these numbers are used only for internal purposes. The example shown here also presents a different number range. This number range ensures that each bank account receives a unique internal identifier, which is crucial for tracking and managing accounts effectively within SAP S/4HANA. Make sure this range is defined and assigned before creating any bank accounts in BAM.

3.2.4 Define Settings for Bank Account Master Data

In this customizing activity, you are specifying the most important settings for BAM, including the account types and how they are going to behave within the system. This activity can be found under **SAP Financial Supply Chain Management • Bank Account Management • Basic Settings • Define Settings for Bank Account Master Data**.

First you need to specify a bank account **Type** (see Figure 3.57) used for BAM and an **Account Type Description**; an underlying **Contract Type—Current Account** represents external bank accounts, whereas a **Payment Service Provider Account** can be used for technical accounts provided by third-party payment providers such as PayPal. **Technical Account** is for any other technical or interim accounts you may want to use. **In-House Bank Account** is used for accounts required in in-house bank processing, and **Other** can cover any remaining accounts. It's good practice to use different contract types for various accounts to differentiate them clearly and give yourself the flexibility to use different functionalities.

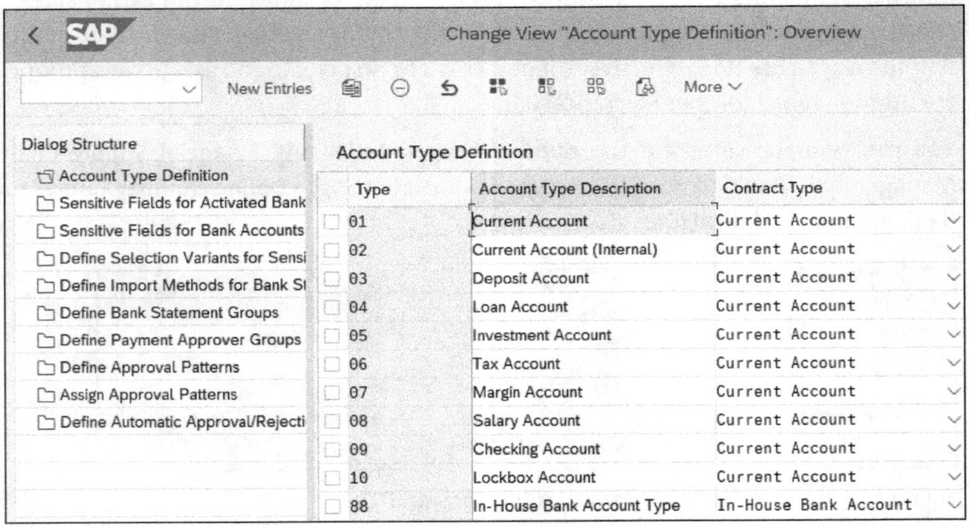

Type	Account Type Description	Contract Type
01	Current Account	Current Account
02	Current Account (Internal)	Current Account
03	Deposit Account	Current Account
04	Loan Account	Current Account
05	Investment Account	Current Account
06	Tax Account	Current Account
07	Margin Account	Current Account
08	Salary Account	Current Account
09	Checking Account	Current Account
10	Lockbox Account	Current Account
88	In-House Bank Account Type	In-House Bank Account

Figure 3.57 Bank Account Types

3.2 Configuration

Next, click **Sensitive Fields for Activated Banks**. Here you need to specify which fields are going to trigger approval workflows and which can be changed without triggering a workflow (see Figure 3.58). To specify which fields are sensitive fields and workflow fields in BAM, simply click **New Entries**, then select the object that represents the tab where the field is located in the bank account (for example, **General Data** or **Payment Approver**), and choose the field name you want to include. Once you've made your selections, click **Save** to confirm your entries. This setup ensures that only the designated fields are monitored for sensitive changes or included in workflow triggers. Fields specified here will trigger the approval workflow.

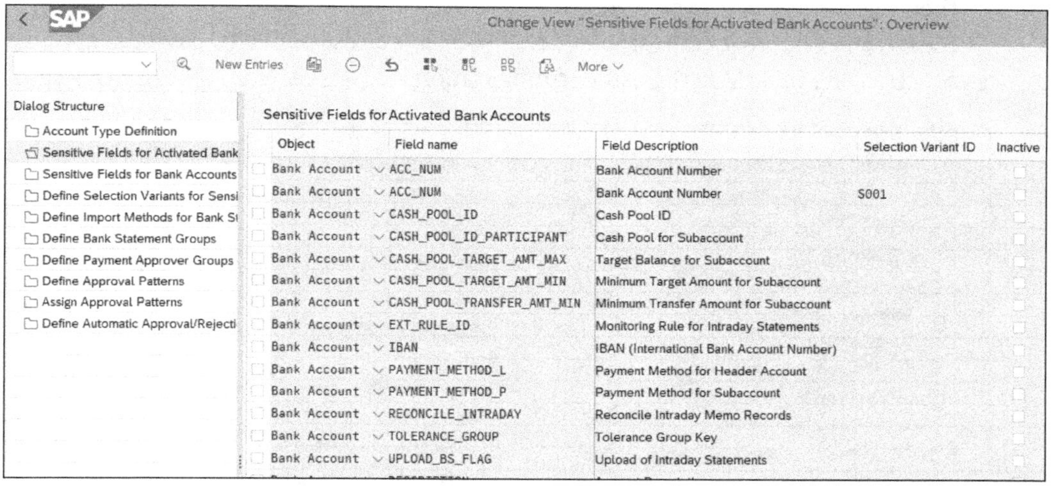

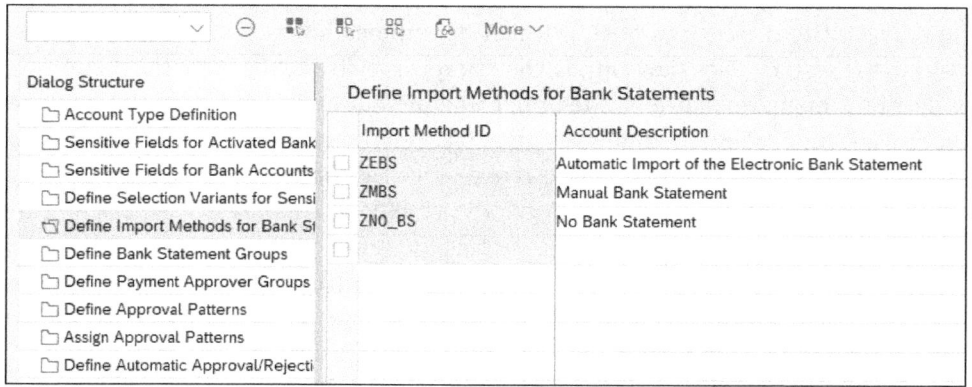

Figure 3.58 Sensitive Fields for Approval Workflow

In the next step, click **Define Import Methods for Bank Statements**, where you need to specify the import methods for the selected bank statements (see Figure 3.59).

Figure 3.59 Import Methods for Bank Statements

This is a setting you define within the BAM account, and it serves to inform you and others about how a particular bank statement is being uploaded into the system—for example, via manual upload, file interface, or an automated connection with the bank. (There are no predefined options in the system; these are text fields, so you can enter any values here.) Although this setting does not have a significant impact on subsequent processing or system behavior, it plays an informative role, helping to describe the source and method of statement delivery.

In the next step, click **Define Payment Approver Groups** (see Figure 3.60). Here you can specify any values; enter the approver group (**Appr. Grp.**) and its **Description**. You will then select these approver groups in the Manage Bank Accounts app—specifically, on the **Payment Approver** tab (provided that you have already enabled payment approver functionality in BAM in the earlier configuration steps).

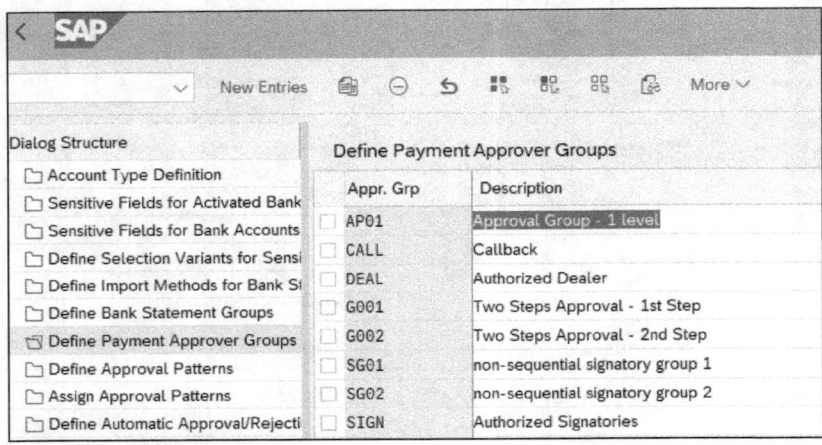

Figure 3.60 Define Payment Approvers

In the next configuration activity, **Define Approval Patterns**, you need to define the approval pattern (see Figure 3.61), which determines how payments are approved within the system. Click **New Entries**, then simply specify the name for the pattern and select how many sequences are required. SAP provides up to four levels of payment approvals, supporting the eight-eyes principle. You then assign each approval level to an approver group and specify the currency, as well as the minimum amount for both payments and batches that will trigger the workflow. This involves specifying the sequence and structure of approvals, such as the number of approval levels and the roles responsible for each step. Once the approval pattern is defined, it must be assigned to the relevant company codes and bank codes to ensure that the correct approval process is applied during payment processing for those entities.

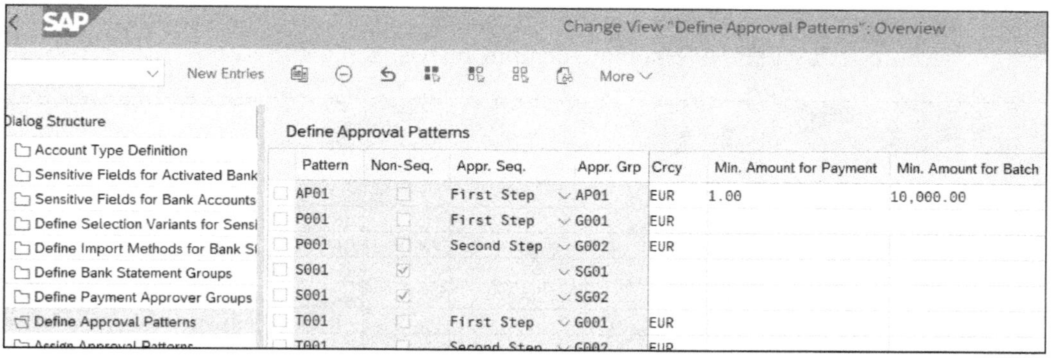

Pattern	Non-Seq.	Appr. Seq.	Appr. Grp	Crcy	Min. Amount for Payment	Min. Amount for Batch
AP01	☐	First Step	AP01	EUR	1.00	10,000.00
P001	☐	First Step	G001	EUR		
P001	☐	Second Step	G002	EUR		
S001	☑		SG01			
S001	☑		SG02			
T001	☐	First Step	G001	EUR		
T001	☐	Second Step	G002	EUR		

Figure 3.61 Define Approval Patterns

3.2.5 Define Settings for Bank Account Contract Types

As described earlier, this configuration activity is used to set up additional settings related to the behavior and capabilities of bank accounts—specifically, when bank account contract types are enabled. Here you define how bank accounts behave throughout their lifecycle—for example, which statuses are relevant, how accounts are activated in each status, and what repair actions can be taken. You can also control the visibility of specific tabs in the bank account master data, depending on the type of bank account. These settings can be tailored based on contract type, account type, and company code, allowing for a more flexible and structured approach to managing bank account data.

Note that this activity is only applicable if you have already enabled bank account contract types in your configuration. The prerequisites include activating the contract type feature in the **Define Basic Settings** activity (see Section 3.2.1) and defining the contract types themselves in the **Define Settings for Bank Account Master Data** activity (see Section 3.2.4).

You can find this configuration activity under **SAP Financial Supply Chain Management • Bank Account Management • Basic Settings • Define Settings for Bank Account Contract Types**.

First, click **Define Repair Function for Bank Accounts**. In this activity (see Figure 3.62), you define repair actions for bank accounts to manage corrections during their lifecycle. Two options are available:

- **Return to Preceding Status**
  Allows for reverting a bank account to its previous status; enabled by default.
- **Restore Predecessor Version**
  Restores the account to an earlier version; useful for both status and data changes. A version is created when the account is created, its status changes, or sensitive data is updated.

Both options have an impact on the next configuration steps, in which you define how account statuses should change depending on user actions.

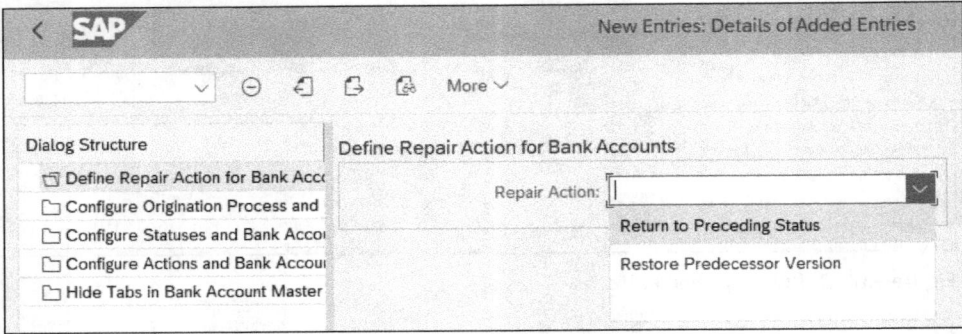

Figure 3.62 Repair Action for Bank Accounts

Next, click **Configure Origination Process and Manual Creation** (see Figure 3.63). When you click **New Entries** in this configuration step, you can select a contract type and decide whether to enable the originating status, which requires you to submit a bank account for creation. You can also choose to disable manual creation for accounts that are generated automatically—such as those created through the eBAM interface or the default BAM—thus restricting manual entries and ensuring consistency.

When the bank account contract type is enabled but the bank account origination process is not activated in the **Define Settings for Bank Account Contract Types** configuration, the system uses the default opening process, which includes only two statuses: **Inactive** and **Active**. If you enable the bank account origination process, additional statuses become available—**In Opening Process**, **Opened at Bank**, and **Active**. This allows for the use of eBAM correspondence and the possibility to design more advanced workflow scenarios. When manual creation is disabled, new bank accounts can only be created through the bank account application process. This means that a formal application must be submitted using the Submit Bank Account Applications app to request the creation of a new account.

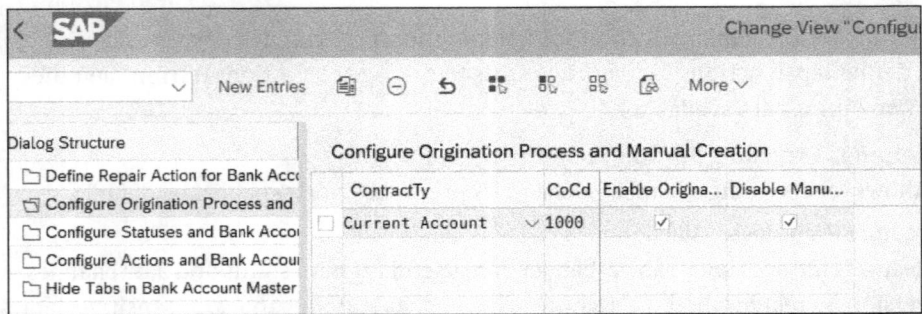

Figure 3.63 Configure Origination Process and Manual Creation

In the next step, click **Configure Statuses and Bank Account Control Modes**. In this activity, you configure the statuses and control modes for the bank account lifecycle. SAP S/4HANA provides predefined mandatory and optional statuses. Mandatory statuses are always enabled, whereas optional ones are enabled by default, unless you create an entry and deselect the **Enable Status** option.

The statuses include **Inactive**, **Active**, **Closed**, **Application Approved**, and others depending on whether the bank account origination process is enabled. For each status, you define the relevant bank account contract type, account type, and company code. Two control modes are defined for each status:

- **Control Mode to Next Status**
  Controls progression to the next status.
- **Control Mode to Preceding Status**
  Controls regression to the previous status.

Once a bank account reaches the **Pre-Opening Preparation** status, it cannot be reverted to **Application Approved**.

In our example (see Figure 3.64), you can, for example, set up **Direct Change** for **In-House Bank Accounts**, meaning that no workflow will be triggered for changing statuses in in-house bank accounts. In this way, you can differentiate the setup for the external and internal bank accounts.

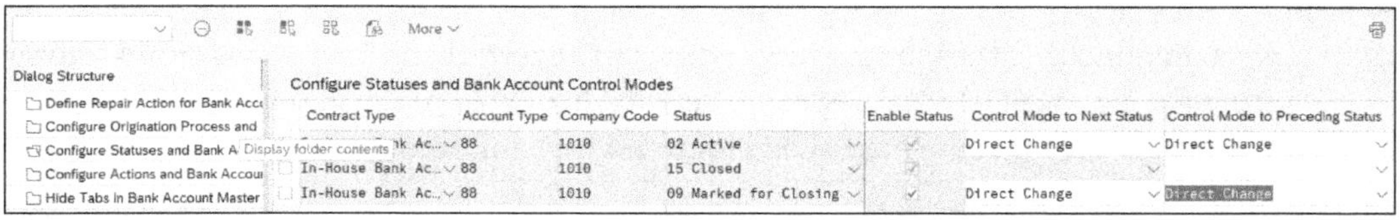

Figure 3.64 Configure Statuses and Bank Account Control Modes

In the next activity (see Figure 3.65), click **Configure Action and Bank Account Control Modes**. Here you can configure the bank account control modes for actions performed on bank account master data, such as modifying or deleting a bank account. In this configuration step, select the contract type and the company code, then choose the action—whether it applies to changing the status or deleting the account. You also specify if it is a direct change, requires two-person verification, or should go through a workflow. Finally, you set the control mode for restoring the account, which applies only to tracing and managing attending bank accounts—for example, transitioning an account from inactive to active or closing a bank and its associated accounts completely. This setup is mainly required when you're using eBAM functionality as normally you only need the **Active**, **Inactive**, and **Closed** bank account statuses. SAP provides these additional statuses for more complex scenarios if needed.

3   Bank Account Management

If the **Restore Predecessor Version** repair action is enabled, you will also need to specify the **Control Mode for Restore**. This setting determines how the bank account is restored to its previous version. For our example, we want to have autoupdating turned on for in-house bank accounts.

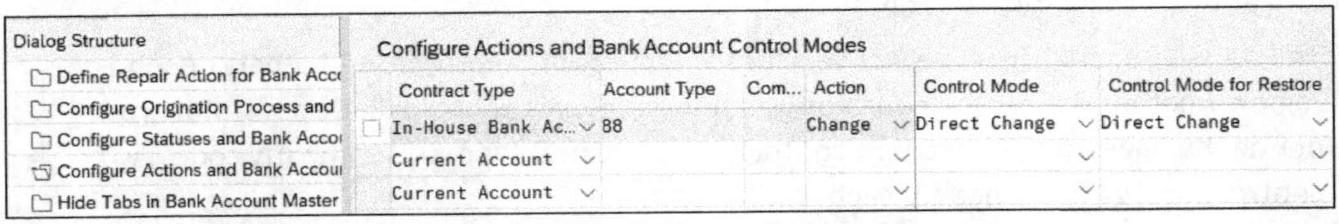

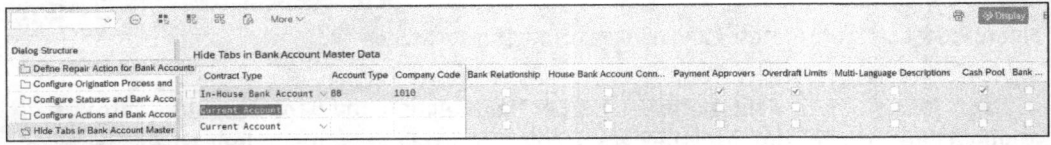

Figure 3.65 Configure Actions and Bank Account Control Modes

In the **Hide Tabs in Bank Account Master Data** activity, you define which tabs are visible in the bank account master data. For example, the **Cash Pools** tab may not be needed for technical bank accounts, so you can deselect this option for bank accounts with the technical bank contract type.

You can specify the bank account **Contract Type**, **Account Type**, and **Company Code** for each entry. The combinations you can use, in order of priority, are as follows:

1. Contract type + account type + company code
2. Contract type + account type
3. Contract type

Keep in mind that the tab visibility settings will override the field status group settings from the **Manage Field Status Groups** activity. In our example (see Figure 3.66), we have hidden the tabs that are not needed for in-house bank accounts.

Figure 3.66 Hide Tabs in Bank Account Master Data

3.2.6   Define Field Status Groups

This customizing activity (see Figure 3.67) allows you to define field status groups and specific rules that control how fields appear and behave in the Manage Bank Accounts app. You can set fields to be mandatory, hidden, editable, or read-only on the user interface. The process involves creating field status groups, selecting variants, and assigning them to specific activities and account types. In this activity, you first need to specify the **Define Field Status Group** and then select **Define UI Field Status**. The system will then display a long list of all the fields available in BAM; currently, there are 235 fields. If you have any custom fields, you also need to assign them here. For each field that is relevant to you, you can choose whether the field status should be read-only, hidden, editable, or mandatory. Note that in most cases, you don't need to change anything, because SAP already provides a wide range of field status groups that you can use. You should only create or modify entries here for specific cases that truly require it.

In most cases, the default settings are sufficient, but you may adjust them based on your needs—such as by making the **IBAN** field mandatory for certain account types or hiding fields that are not relevant (e.g., **Intraday**).

You can find this configuration activity under **SAP Financial Supply Chain Management • Bank Account Management • Manage Field Status Groups**.

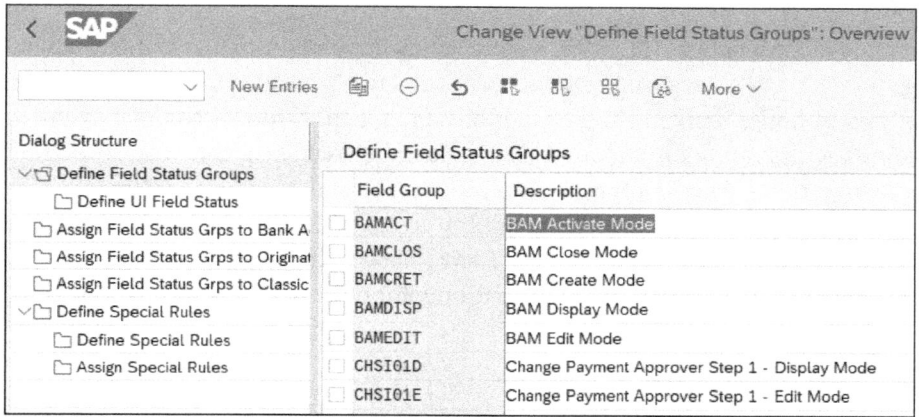

Figure 3.67 Define Field Status Groups

If you create new field status groups, you need to assign them to bank account master data (see Figure 3.68). However, SAP has predefined these values, so changes are only required if you have specific needs. If you need to change anything, click **New Entries**, select the **Contract Type** and **Scenario**, and then assign your newly created field status group.

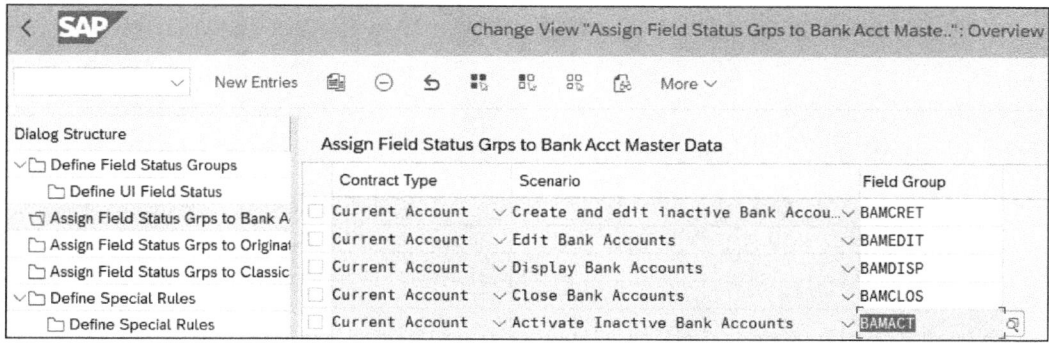

Figure 3.68 Assign Field Status to Bank Account

3.3 Summary

In this chapter, we explored the core functionalities of BAM within SAP S/4HANA, focusing on its essential role in maintaining a centralized and controlled structure for managing bank accounts. We walked through the prerequisites and configuration steps needed to set up BAM and highlighted the key apps and tools provided by SAP to simplify and standardize your banking processes. From creating and managing bank accounts to understanding their integration into payment workflows, we laid the foundation for more advanced treasury operations. We also clarified that in-house bank accounts are created and managed within BAM, further reinforcing its central role in your treasury setup.

With BAM accounts now in place, you're ready to move on to the next stage—implementing more advanced payment scenarios using advanced payment management and in-house banking. These solutions allow you to model complex payment workflows across your internal and external landscape, enabling greater control and efficiency. From there, you can progress into cash management, where you'll manage balances, assign cash flows, and forecast liquidity across your organization. At this point, you should have a solid understanding of BAM's place within the broader SAP S/4HANA ecosystem and be well prepared to continue building a comprehensive, integrated treasury framework.

Chapter 4
Advanced Payment Management and In-House Banking

In an increasingly complex world, the payment industry is evolving at an unprecedented speed. Therefore, it is all the more important to design IT infrastructure in a sustainable and modular way to be able to respond to future developments. In this chapter, we build upon the foundational technical sections and provide insight into both the functionality and technical implementation of advanced payment management and in-house banking.

This chapter outlines the processes of both in-house banking for SAP S/4HANA Finance for advanced payment management (the successor to SAP In-House Cash) and SAP S/4HANA Finance for advanced payment management. It gives click-by-click instructions for performing the corresponding configuration activities, for which there is significant overlap in the SAP S/4HANA system. It also explains how to integrate with two governance and risk solutions—SAP Business Integrity Screening and SAP Watch List Screening—to conduct fraud detection and sanction and embargo screening.

4.1 Advanced Payment Management Functions and Processes

Although advanced payment management only entered the market with SAP S/4HANA 1809 FPS 2, it has been recognized as a proven solution for processing high-volume payments from the beginning. The foundation for its development was the Payment Engine (FS-PE), which was already available in SAP ERP for the banking industry. Advanced payment management continues to benefit from the functionalities and expertise of the banking industry, particularly when it comes to performance in processing large volumes of payment transactions.

However, there are not only advantages; some banking industry requirements differ significantly in certain aspects, and the flexibility and design freedom that advanced payment management offers should be approached with caution. A well-thought-out and sustainable concept is key if advanced payment management is to be used as a platform for payment processing.

Since the transformation of the Payment Engine into advanced payment management, SAP has continuously worked on simplifying and expanding its payment-processing capabilities. More and more functionalities that previously required transportable customizations can now be configured via SAP Fiori apps by key users or even end users.

In Chapter 1, we explained business functionalities and processes such as payments in the name of (PINO), payments on behalf of (POBO), and collection on behalf of (COBO). In this section, we will guide you step by step through how these functionalities and processes are implemented in advanced payment management and provide an in-depth exploration of its capabilities.

We'll cover the following specific topics:

- **Key components**
  Section 4.1.1 covers the components and functions of advanced payment management, providing insight into the integration of third-party systems via SAP Multi-Bank Connectivity in conjunction with the input manager. It explores payment routing and the conversion of target formats.

- **Functions and apps**
  How is advanced payment management managed from both key user and end user perspectives? Section 4.1.2 offers insights into the SAP Fiori apps and transactions used to control advanced payment management.

- **Master data**
  Master data plays a critical role in advanced payment management as well. In Section 4.1.3, we examine key master data elements such as service-level agreements with payment factory participants, routing and clearing agreements, and their functions.

Prerequisites

Advanced payment management requires SAP S/4HANA 1809 FPS 02 or higher. It is subject to licensing, along with in-house banking. However, for a complete representation of processes with integrated payment approval, SAP Bank Communication Management is also required. To use SAP Bank Communication Management, SAP S/4HANA Finance for cash management must also be licensed.

The connector part of SAP Multi-Bank Connectivity is included with the license. However, bank connectivity via SAP Multi-Bank Connectivity requires an additional license. In general, we recommend contacting your SAP key account manager for any licensing-related inquiries.

4.1.1 Key Components

Advanced payment management consists of individual components that each contribute to the automated processing of payment carriers and bank messages. Figure 4.1 provides an overview of the components within advanced payment management, as well as those that can be directly linked to advanced payment management. We will examine the key components of advanced payment management—from the integration of third-party SAP and non-SAP systems via SAP Multi-Bank Connectivity, through the input manager, to the output manager. In addition, we will explore the functionalities and characteristics of each component.

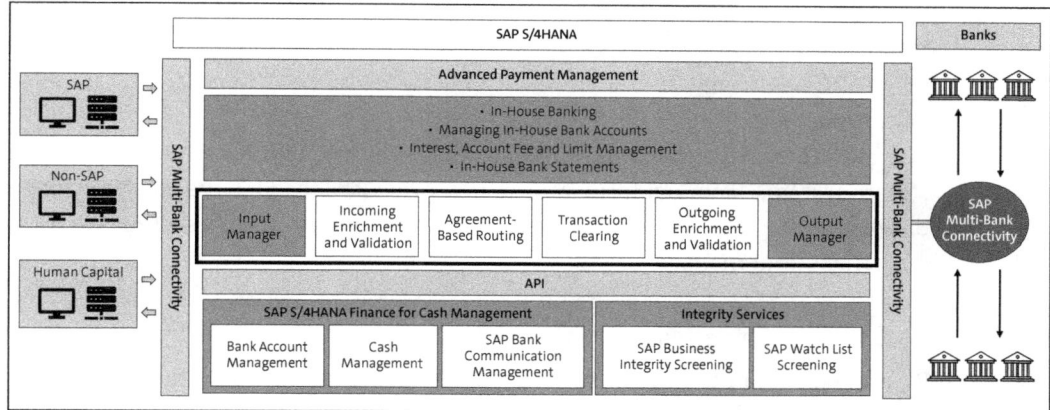

Figure 4.1 Advanced Payment Management

We'll cover the following topics in this section:

- **SAP Multi-Bank Connectivity**
  We have dedicated an entire chapter (Chapter 5) in this book to SAP Multi-Bank Connectivity and its connector. In this section, we will simply put SAP Multi-Bank Connectivity into context for this chapter.

- **Input manager**
  All incoming messages are transferred to the advanced payment management via the input manager, which is specifically designed for managing and orchestrating messages related to mass payments. We introduce this component here and detail its tasks.

- **Incoming enrichment and validation**
  From the enrichment of payment messages to screening against predefined rules, as well as the management and integration of third-party systems, enrichment and validation handle a wide range of tasks. In this section, we introduce these key components.

- **Exception handling and control**
  Along with security, stability plays a crucial role in payment transactions. But what happens when there are exceptions or errors? For this, advanced payment management has a centrally positioned component that handles transactions that require special attention or correction.

- **Routing and transaction clearing**
  Advanced payment management serves as a central hub in payment transactions, particularly due to the routing and clearing agreements that govern payment processing. In this section, we highlight these key functionalities.

- **Outgoing enrichment and validation**
  During processing, payments can be batched, corrected, or reentered. In this section, we introduce not only the incoming tasks but also the responsibilities of outgoing enrichment and validation.

- **Output manager**
  The output manager handles outgoing communication with banks as well as upstream systems. Payment formats are either processed through the output manager or entirely recreated based on predefined target formats. In this section, we introduce the output manager.

SAP Multi-Bank Connectivity

SAP Multi-Bank Connectivity is capable of connecting not only banks but also external third-party SAP and non-SAP systems. It consists of two components: the integrated connector within the SAP system and SAP Multi-Bank Connectivity on Cloud Foundry. Detailed information about SAP Multi-Bank Connectivity functionalities can be found in Chapter 5.

However, SAP Multi-Bank Connectivity has two key functions. The first is connecting third-party SAP systems and routing incoming and outgoing messages. The second is routing messages within the SAP system itself. Through local routing, payment messages and account statements are internally transferred from one module to another.

Thus, SAP Multi-Bank Connectivity is not only responsible for connecting external systems and banks but also plays a crucial role in handling and routing messages within the internal in-house bank.

Input Manager

The input manager is responsible for converting externally or internally created payment media into the internal advanced payment management meta format, stored in the file handler database. To achieve this, the input manager is equipped with a variety of input converters that allow you to process payment media in XML, SWIFT Message Type (MT) format, or flat-file formats such as DTAZV by importing them into advanced payment management's structured format.

Converters are essential for mapping and storing payment data using an internal metadata format. They are defined by the unique combination of format, medium, and channel, which can be configured in customizing. How the input converters are configured is explained in more detail in Section 4.3.1. If needed, converters can be expanded or customized to meet specific requirements.

During this process, the incoming file is also saved in its original format within the file handler database.

> **Control of the Input Manager**
>
> In general, all payment media imported via the import manager are displayed directly in the Manage Payment Items app. However, the backend still provides transactions, primarily originating from the Payment Engine, that allow for the selection and retrieval of payment media.
>
> The file handler database can be accessed via Transaction /PF1/FH_SHOW_DB.

There are several ways to manage the file handler. Import converters from the file handler can be linked to process customizing in SAP Multi-Bank Connectivity for automated imports or used via the Manage Incoming Payment Files app (F1680). However, there is also the option to use the file handler manually—for example, to import data carriers via a specified file path. In the following sections, we present both options.

Import File (Expert)

For manual import in the backend via the expert mode, use Transaction /PF1/FH_IPM_EXPERT. This can be accessed through SAP GUI or found in the end user SAP menu under **Accounting** • **Financial Supply Chain Management** • **Advanced Payment Management** • **Processing** • **Import** • **/PF1/FH_IPM_EXPERT**. After calling the transaction, the control panel for the input manager appears as shown in Figure 4.2.

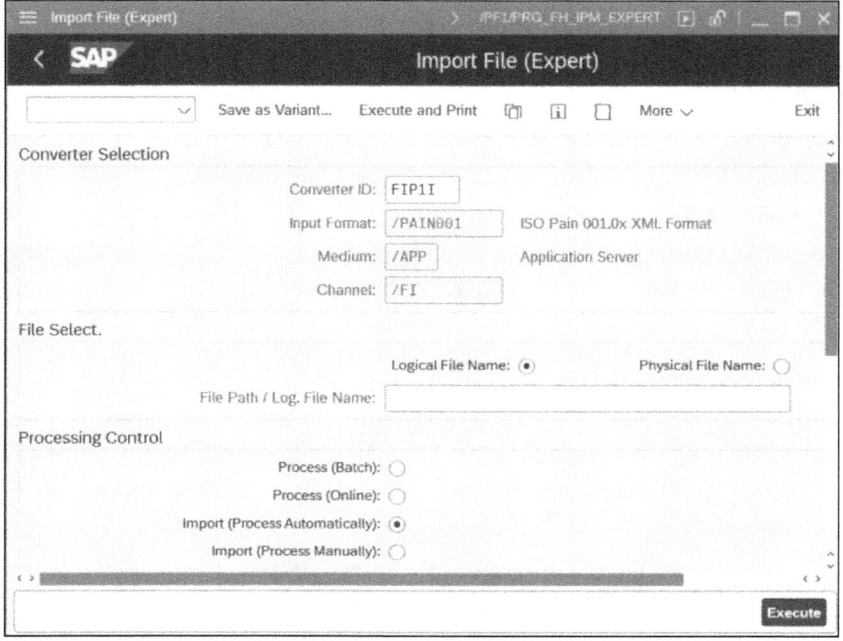

Figure 4.2 Import File (Expert)

The important areas on this screen are as follows:

- **Converter Selection**
  After you enter a value for the **Converter ID**, you can select the desired import converter by either multiple selection or direct entry. Note that the converter must be configured for the file type being imported.

> **Import Path Set in Advance**
> In customizing, the converter for manual import can be configured in advance, meaning that the import path will be automatically filled in upon selection. This is mandatory if job-controlled files are to be imported using the import converter.

- **File Select**
  Through file selection, the physical path can be specified using a selection window or by entering it manually. Alternatively, an internal SAP path can be selected via the **Logical File Name** field using Transaction FILE.

- **Processing Control**
  The processing of the imported file is controlled via processing control, with the following options available for selection:
  - **Process (Batch)**
    Batch mode is used for processing large amounts of data and executing tasks in the background, rather than in real time.
  - **Process (Online)**
    Data will be imported directly within a dialog mode and processed immediately.
  - **Import (Process Automatically)**
    The input manager reads the file directly and converts it into the internal database, the file handler database. With this option, the file can be processed further automatically. For this purpose, use Transaction /PF1/POLLER.
  - **Import (Process Manually)**
    Through this option, the file is read into and converted in the file handler database, but it must be manually initiated for further processing. This can be done in the backend via Transaction /PF1/PO_EXPERT.

Manage Incoming Payment Files

The Manage Incoming Payment Files app (F1680) integrates with advanced payment management, facilitating the manual import of payment files into the SAP S/4HANA system. Detailed explanations of its functionalities and customization options can be found in Chapter 9.

To start the import process, click the **Import** button. This action will open a popup window in which you can trigger the import process via the input manager by clicking the **Payment File for Advanced Payment Management** option, as shown in Figure 4.3.

4.1 Advanced Payment Management Functions and Processes

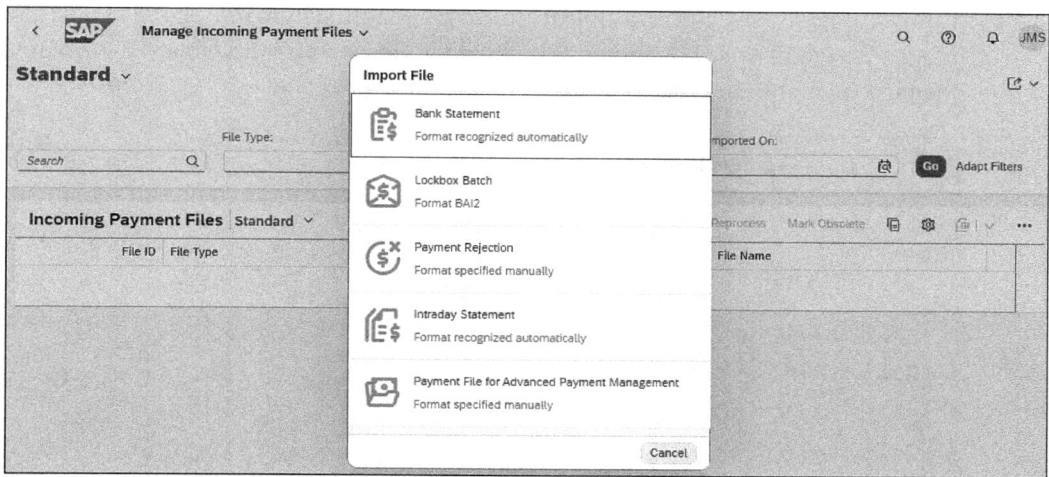

Figure 4.3 Import File

A popup window will then open in which import payment files using drag-and-drop functionality. Within this window, you can choose the suitable input converter for processing the payment files using the dropdown menu provided under **Converter ID** (see Figure 4.4).

Clicking the **Process** button initiates the import, and the payment file is forwarded to advanced payment management for further processing.

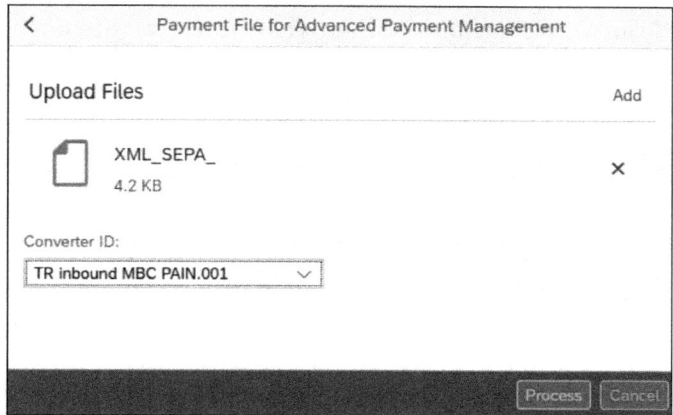

Figure 4.4 Payment File for Advanced Payment Management

After a successful import, the user receives an overview of the imported item. Details such as attachments and log information are only visible within the app to the user who performed the import. From this overview, a direct jump into advanced payment management is possible. By clicking the number shown for **Imported Records**, a popup window opens that provides links to advanced payment management applications.

Figure 4.5 shows how you can navigate from the imported item into the advanced payment management SAP Fiori apps via links for **Repair Payments**, **Analyze Payments**, **Manage Payments**, or **Create Payments**.

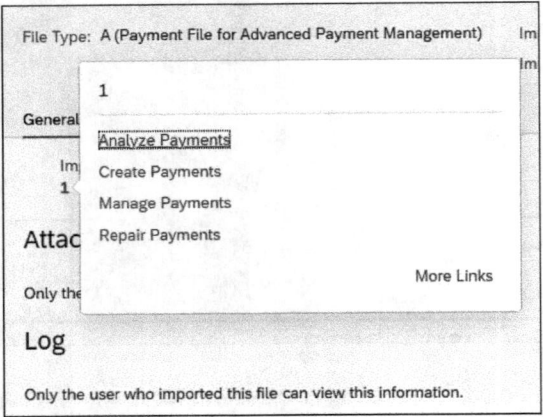

Figure 4.5 Links for Advanced Payment Management Apps

Enrichment and Validation

The enrichment and validation functions in advanced payment management provide comprehensive verification mechanisms to ensure payment data quality. They support individual validation rules depending on the transaction or payment order type and can be seamlessly integrated with sanction screenings, such as via the SAP Business Integrity Screening or SAP Watch List Screening solution. These functions also offer a development framework for creating custom validations and for enriching or adjusting payment data. Enrichment and validation is the first processing step after the payment files have been successfully imported, converted, and mapped to their corresponding payment order and payment item types, as follows:

- **Payment order**

    The payment order represents an internal, external, or manual payment request. This payment request can be either an individual or a batch payment request. Payment orders are created automatically during inbound or outbound processing within the system. Every time a payment instrument is imported into advanced payment management for manual processing that was created using the Create Payments app (F3648) for manual payment requests, a payment order will be created. Each payment order is uniquely defined by the following business key fields:

    – **Clearing Area**
      Specifies the legally independent organizational unit where a payment order was created or imported from.

    – **Payment Order Number**
      Number range.

- **Payment Order Date**

  The system retrieves the payment order date based on the reconciliation date defined in advanced payment management.

Payment orders essentially represent the header information of each payment. All payment orders created in the system are stored and listed in a dedicated database table (table /PF1/DB_ORDER).

- **Payment items**

  Within advanced payment management, the payment item serves as a central business object, representing a distinct stage of a transaction. It consolidates structured data essential for execution, tracking, compliance, and reporting. To streamline complex payment processes, payment items are categorized into different types, ensuring accurate processing, posting, and monitoring within the system. The following are some of the payment types available in the system:

  - **Recipient Party Item (RCP)**

    This item represents the payment to the recipient, such as a supplier or customer. It contains specific information like invoice numbers or payment purposes. These items originate from incoming payment orders and are crucial for the correct allocation and posting of payments.

  - **Originator Item (ORP)**

    This item stands for the paying party, typically the company initiating the payment. It serves as the counterentry to one or more recipient party items and ensures that the payment is correctly recorded in the system.

  - **Clearing Item (CLR)**

    Clearing items are generated during the clearing process, either at the completion of a payment batch or when forwarding an individual payment order. They are posted to the bank clearing account (Nostro account) or customer account and are used for the internal reconciliation of payments.

  - **Turnover Item (TOV)**

    Turnover items are special payment items that are not assigned to either an originator or a recipient. They typically arise from internal accounting systems, such as handling securities transactions or calculating interest.

  - **Statement Entry**

    A statement entry represents a single transaction from an electronic bank statement, such as a credit, a debit, fees, or interest. These items are imported and processed within the system as part of the electronic bank statement (MT940, camt.053).

Payment items are stored in a separate database (table /PF1/DB_ITEM). Key components stored within a payment item include the following:

- Business partner information (e.g., vendor, customer)
- Invoice or remittance data

- Address details
- Banking and account details
- Regulatory reporting attributes (e.g., purpose codes, tax references)
- Status and control information related to lifecycle processing (e.g., validation, enrichment, posting)

The interaction between a payment order and a payment item gives advanced payment management the necessary flexibility to function as a central hub for payment transactions, even in companies with complex payment processes.

Figure 4.6 presents a schematic representation of a payment order. For example, this could be a credit transfer consisting of an originator item (ORP, the payer) and multiple payment recipients (RCP, payment receivers). It is important to note that a payment order amount is always balanced between ORP and RCP.

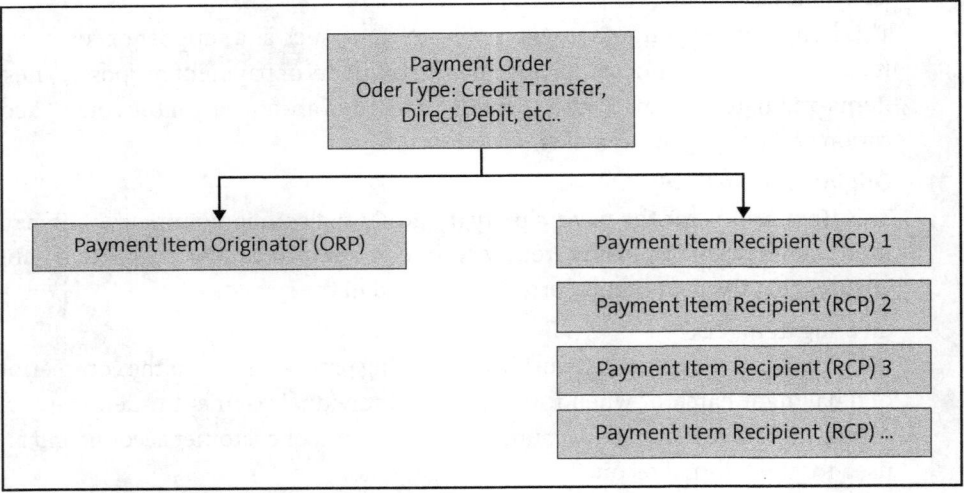

Figure 4.6 Payment Orders and Payment Items

The enrichment and validation process validates the attributes of payment orders and payment items and enriches them where necessary. Checks can also reach out to external applications like SAP Business Integrity Screening or additional non-SAP systems. Each payment order and payment item has its own dedicated enrichment and validation checks; a cross-check on the whole payment also takes place (e.g., checksums throughout the payment order).

Checks are clustered in *check sets* for payment orders and payment items (ORP and RCP sets). These check sets can be modified in terms of adding or deleting certain checks. Table 4.1 provides an excerpt of the existing checks available in the SAP S/4HANA system.

Check ID	Description	Check ID Name
23	Mandatory fields check—based on configuration, multiple elementary field checks are be executed.	Mandatory Fields Check
148	Check cut-off time of the processing channel against SLA (inbound) and clearing agreement (outbound).	Due Date—Cut-Off Time Check
214	Status notification registration for template ID with SLA.	Status Notification (SLA required)
24	Mandatory fields check—based on configuration, multiple elementary field checks are be executed.	Mandatory Fields Check
145	Validates account information. Checks that at least one of the following fields or combinations of fields is filled in: ■ Bank country, bank key, bank account number ■ IBAN ■ BIC, account number	Account Details
430	Checks whether ordering account exists in BAM.	House Bank Account Check
275	Checks whether BIC and IBAN are consistent.	BIC-IBAN Consistency
187	Checks if the debit item total equals the credit item total.	Debit Credit Consistency Check
90	Duplicate check.	Duplicate Check
261	Status notification registration for template ID with SLA.	Status Notification (SLA Required)
26	Mandatory fields check—based on configuration, multiple elementary field checks are executed.	Mandatory Fields Check

Table 4.1 Existing Checks in Advanced Payment Management

Due to the fact that many checks within enrichment and validation are strongly connected to the service-level agreements (SLAs) defined in the master data (refer to Section 4.1.3), these two topics must therefore be considered together.

In addition, advanced payment management offers an outgoing enrichment and validation process that works in exactly the same way as enrichment and validation during payment processing. Outgoing enrichment and validation allows you to carry out last-minute checks on the payments—for example, last-minute sanction and embargo list screening in case a payment that was parked within advanced payment management ended up in an exception-handling process.

> **Enrichment and Validation Integration Usage Is Not Mandatory**
>
> An approval via SAP Bank Communication Management can be triggered during outgoing enrichment and validation. Only after final approval can the payment be forwarded to the output manager. In case of rejection of a single or batch transaction, the entire transaction is passed to exception handling, where a reaction to the rejection is determined. Usage of outgoing enrichment and validation is not mandatory.

After enrichment and validation, processing continues through routing control, which basically routes payments toward banks and bank accounts based on defined business rules and clearing scenarios. Those business rules can use any attributes of a payment: payment type (HR, treasury, travel etc.), amount, priority, currency, and so on.

Routing is followed by clearing processing, where the rules determined in the routing are executed. These two processing steps are configured through master data and therefore described in Section 4.1.3.

Exception Handling and Control

Whenever an error occurs in processing or a hit is detected in relation to screening (blacklist, whitelist, or sanction and embargo), exception handling is involved. The end-to-end payment processing is divided into a sequence of logical processing steps. If the system detects errors in any of these steps that prevent further processing of a payment, then an appropriate reaction is required. Exception handling addresses all situations in payment order and payment item processing that deviate from straight-through-processing. It is embedded in the payment order processing and can be directly invoked by straight-through-processing components such as the file handler, payment processing (especially enrichment and validation), routing control, and clearing.

Exception handling filters out irregular payment transactions from the process and directs them to postprocessing. For all payment transactions that cannot be corrected based on predefined rules, they can be manually reviewed and, if necessary, corrected in the Repair Payments app (F3651).

Figure 4.7 illustrates the exception handling process, including the correction process.

If an error occurs in one of the components connected to exception control, the system immediately stops the ongoing process and transfers the error to exception control. Exception handling is used to set up rules that define response types for error situations that arise during processing. For each error, exception handling can determine and execute a configurable reaction. The preferred responses in a given error situation are controlled by response types, which can be precisely defined within the error handling framework.

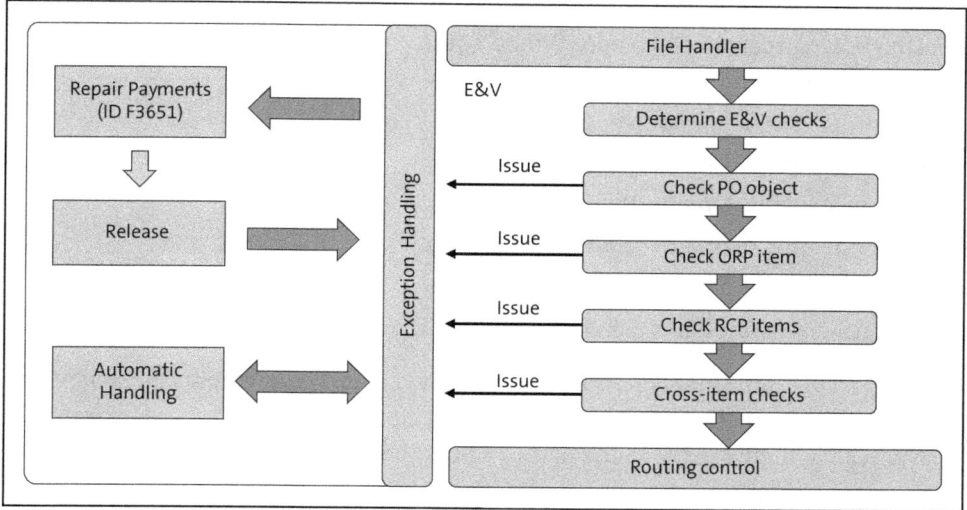

Figure 4.7 Exception Handling

Exception handling is designed to establish rules that define response types for errors that occur during processing. It is based on freely definable rules that determine which reaction is favored in a specific error situation. The reaction is configurable and depends on various parameters related to processing steps in advanced payment management, such as the process, the object category, the check phase, the process step that caused the error, and specific payment characteristics. Exception handling rules are maintained within master data, while exception handling response types, response type priority, and so on must be configured in customizing. Exception handling rules must be created for a processing step that is associated with an object category and an error phase.

Exception handling is based on two key aspects: customizing and master data. In Section 4.1.3 on master data, as well as in Section 4.3.1 in the discussion of enrichment and validation, we will take a closer look at the integration of exception handling and exception control.

Routing and Transaction Clearing

Routing control receives payment items that have been enriched and validated and delivers them to clearing processing, referenced to a specific route and clearing agreement.

Routing controls the further processing of a payment and is directly connected to exception handling. Each route results in a *clearing agreement*, which is part of the clearing process and states how advanced payment management operates for internal and external payments—that is, which outgoing channel, format, and medium should be used. The clearing agreement, found during the routing control process, defines which clearing scenario is preferred for a specific payment and thereby offers many possibilities, such as the following:

- Batching of, for example, HR payments
- Grouping of payments based on bank, account, currency, or the like, until a certain time or specified amount is reached
- Direct clearing of urgent or real-time payments
- Definition of cutoff times
- Adjustment of due dates using, for example, markups

Figure 4.8 illustrates the structure of routing and clearing agreements for the internal processing of payments, such as intercompany transactions, as well as the external processing of payments made in the name of or on behalf of an payment factory participant.

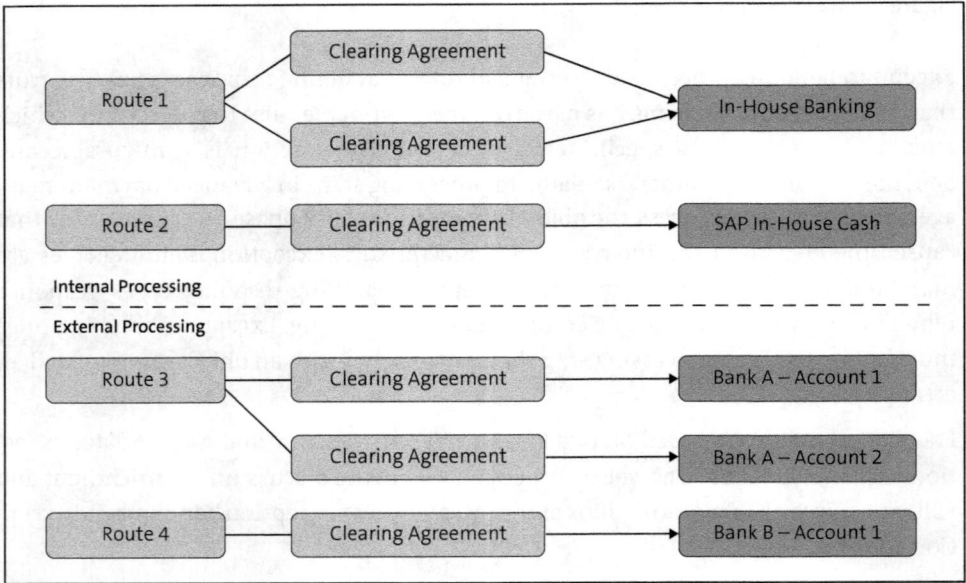

Figure 4.8 Routing and Transaction Clearing

Output Enrichment and Validation

Within payment processing through advanced payment management, corrections and enrichments are applied to individual payments. Therefore, advanced payment management provides a final outgoing output enrichment and validation. Payment transactions leaving the system can undergo a format and content check shortly before transmission, ensuring they are screened against blacklists, whitelists, and sanctions and embargo lists.

Output Manager

The output manager consolidates all processed payment information and generates an external target format using output converters. Therefore, having an overview of all required output formats is essential for designing the output manager effectively.

Output converters transform the internal format into the desired external target format, which does not necessarily have to be limited to payment mediums. In addition to traditional payment files such as pain.001 and pain.008, advanced payment management can generate account statements in formats like camt.53 and camt.54, as well as pain.002 status messages.

The output converter can be connected directly to SAP Multi-Bank Connectivity. Alternatively, external files can be written either to the internal temporary storage or to a path specified in the customizing settings.

Figure 4.9 illustrates the function of the output converter. You can see how the external format is transformed into the internal file handler database based on ISO 20022 standards. If a converter-specific mapping has been customized and is required, then the file is enriched or converted into the desired output format. The external format can then be forwarded either directly to the SAP Multi-Bank Connectivity connector, stored in the internal temporary storage (TemSe), or written to an external path.

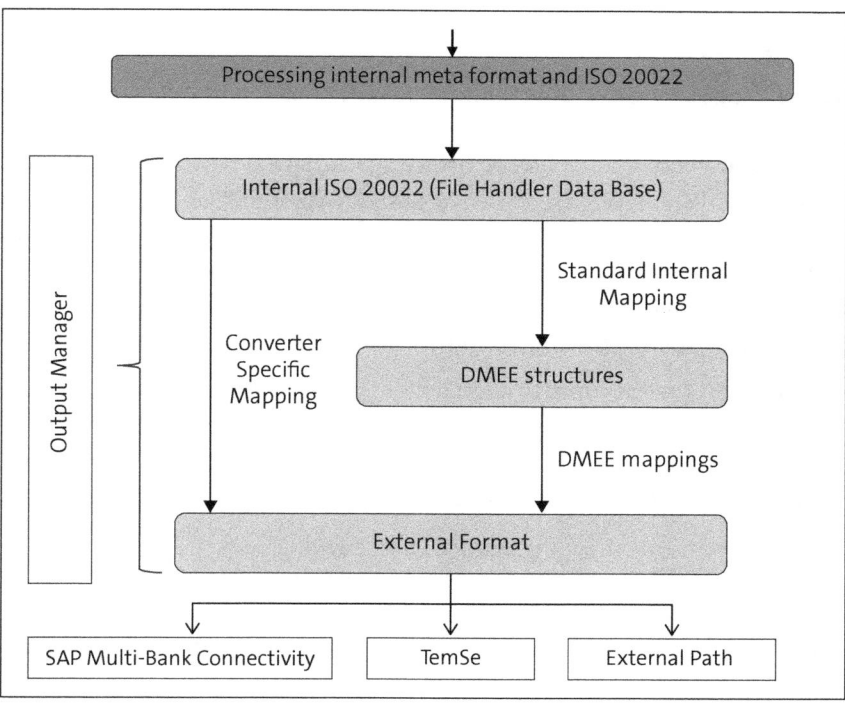

Figure 4.9 Output Manager

4.1.2 Functions and Apps

In the following sections, we will discuss the advanced payment management SAP Fiori apps and their functions. These apps serve to both monitor and actively manage and correct payments in case of errors. The following apps are presented in this section:

- **Create Payments (F3648)**
  This app is used for the manual entry of payments.

- **Repair Payments (F3651)**
  This app is part of exception handling and is used for reviewing and correcting payments.

- **Manage Payments (F3647) and Manage Payment Batches (F4039)**
  These apps are used for monitoring and managing payment batches. Individual payments can be analyzed, and payment items can be added to batches if needed.

- **Manage Payment Items (F7051)**
  This app provides users with a detailed view of individual payment items, including both payment originator and receiver items.

- **Payments Analyzer (F4040)**
  Here we delve into the reporting functions of advanced payment management, showcasing the structure and capabilities of the app.

Create Payments

Manual payments can be created directly from advanced payment management. The unique aspect of this functionality is that no direct booking is required. This enables the centralized management of manual payments within the SAP system, even for companies that are not directly integrated into the SAP S/4HANA system. For this purpose, SAP provides the Create Payments app (F3648). The Create Payments app provides SAP with a significant expansion beyond the traditional manual payment request entry in Transaction FIBLFFP (Free Form Payment) under SAP ERP. The fields are directly aligned with ISO 20022, offering significantly more options.

When opening the app, the appropriate payment order must first be selected, for which a manual payment needs to be created. Before entering the SAP Fiori app, a filter selection screen appears (see Figure 4.10). You can choose whether to create a manual payment from scratch (**New Payment Order**) or use a predefined template (**Payment Order from Template**).

The selected **Clearing Area** must be specified as a mandatory field. In addition to the clearing area, you also have the option to select the **Payment Order Type** directly. The selection result is displayed in the lower section, showing all available payment order types that can be used for manual payment processing.

4.1 Advanced Payment Management Functions and Processes

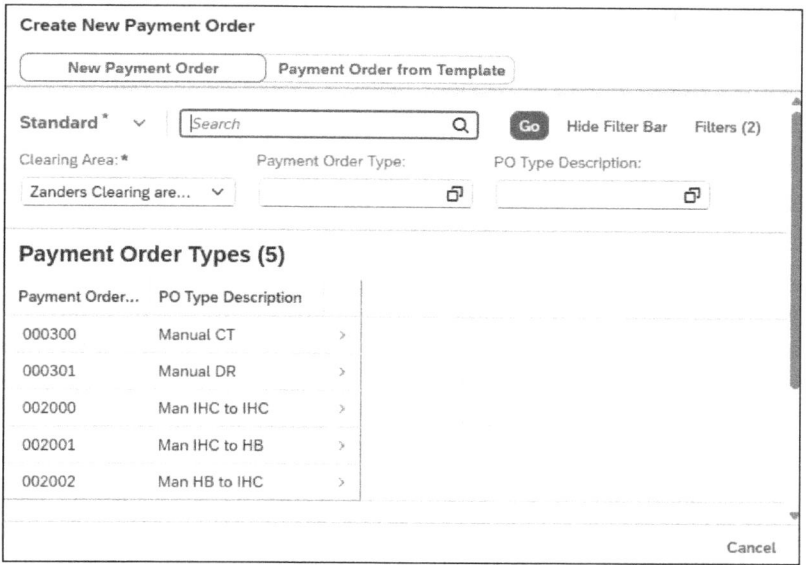

Figure 4.10 Create Payments: Selection

> **Payment Orders and Manual Payments**
> Before a payment order can be used for manual payments, it must be enabled in customizing. This customizing step is covered in detail in Section 4.3.1.

After selecting the desired order type, an entry screen opens, as shown in Figure 4.11. In the **Order & Originator** section, you must enter the payment details of the originator from which the payment was initiated.

The **Order & Originator** tab is divided into the following field groups:

- **General**
  In this area, you select the transaction type corresponding to the order type. In this case, we have chosen an order type for a credit transfer, which is linked to order type: manual entry credit transfer in the customizing settings for processing. These fields are noneditable as they are automatically populated and derived from the customizing configuration. The **General** area also holds the following fields:
  - **Payment Priority**
    Under this field, you can select the processing priority, allowing you to prioritize a manual payment for processing. Depending on the master data settings (clearing and routing) and provided that advanced payment management is not experiencing an unusually high data load, the selected priority does not significantly impact processing. In general, manual payments are processed immediately by advanced payment management after pressing the **Process** button.

- **Payment Information ID**
  This field can optionally be assigned manually and serves as a unique identifier for a group of payment items processed as part of a common payment order.

- **Category Purpose Code**
  In this field, you can specify the type of the payment. This is particularly important for reporting and processing purposes at the receiving bank. HR payments with the purpose code SALA are handled and processed differently from, for example, treasury payments (TREA).

- **Service Level Code**
  This field's value can also be assigned at the payment level and determines how the payment should be processed by the recipient bank. For example, **URGP (Urgent Payment)** indicates a certain prioritization in processing.

- **Account**
  Under the **Account** tab, you select the executing bank connection. Via the **Account Number** field, you can jump directly to BAM, where you can select the desired bank connection and the account to be debited. All other fields, such as **Bank Number**, **BIC**, **IBAN**, and so on, are automatically populated after selecting the bank account.

- **Dates & Process Control**
  Through this field, you can control the execution of a payment by setting the execution date.

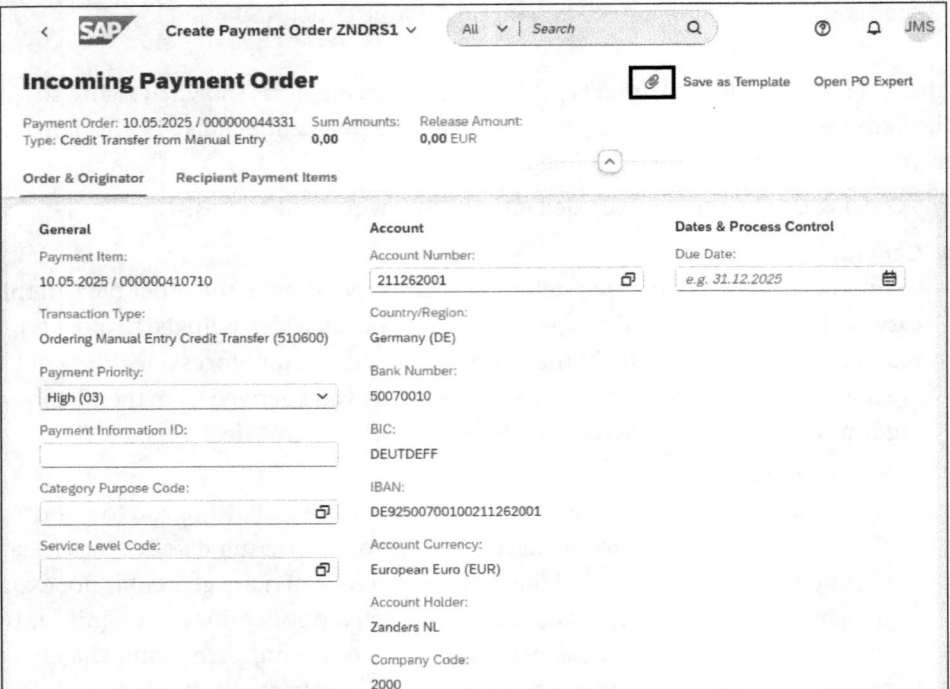

Figure 4.11 Create Payments Originator

Moving to the top right of the screen, if you click the **Add Attachment** button (highlighted in Figure 4.11), then you can attach a file—such as an invoice in PDF format—to the payment order as an attachment.

With the **Save as Template** button, the completed manual payment can be saved as a template directly, allowing for future reuse. A popup window, as shown in Figure 4.12, will appear. The template requires you to provide both a short and a long description. By selecting the **This Template Is Private** flag, you can define that a template should be available only to your private user account; otherwise, it will be available to all users. If you click the **OK** button, the template will be saved and made available for the next payment entry.

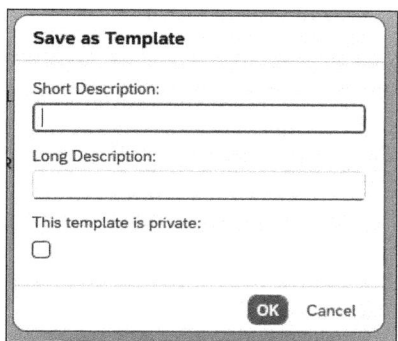

Figure 4.12 Save as Template

Finally, the **Open PO Expert** button moves you to Transaction /PF1/PO_EXPERT in advanced payment management, an expert tool for the detailed display and analysis of payment orders. It allows you to examine individual orders along with their associated payment items, status information, and potential errors in depth. In addition, manual actions such as rechecking or approval for further processing can be initiated here.

This transaction is designed for experienced users in the treasury or in IT. It primarily serves as a tool for error analysis and process monitoring. Transaction /PF1/PO_EXPERT in advanced payment management provides the same detailed functionality, enabling you to analyze payment orders, payment items, and system errors while allowing manual interventions when necessary.

Moving to the **Recipient Payment Items** tab, you can see that the payment item is an individual transaction sent to the recipient. Using the **Add** button, as shown in Figure 4.13, one or more payments can be linked to the payment order.

After clicking the **Add** button, the **Recipient Item** window opens (see Figure 4.14). The input mask is based on ISO 20022 and allows you to capture mandatory fields such as the recipient's account number and IBAN/BIC. Structured and unstructured address details for the recipient can be entered as well. Extra identifiers, such as the UETR for the SWIFT Tracker, are automatically generated.

4   Advanced Payment Management and In-House Banking

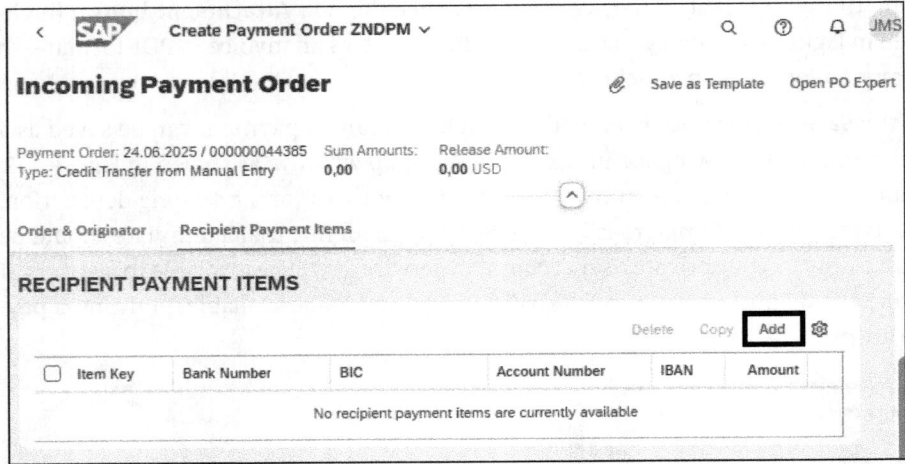

Figure 4.13 Recipient Payment Items

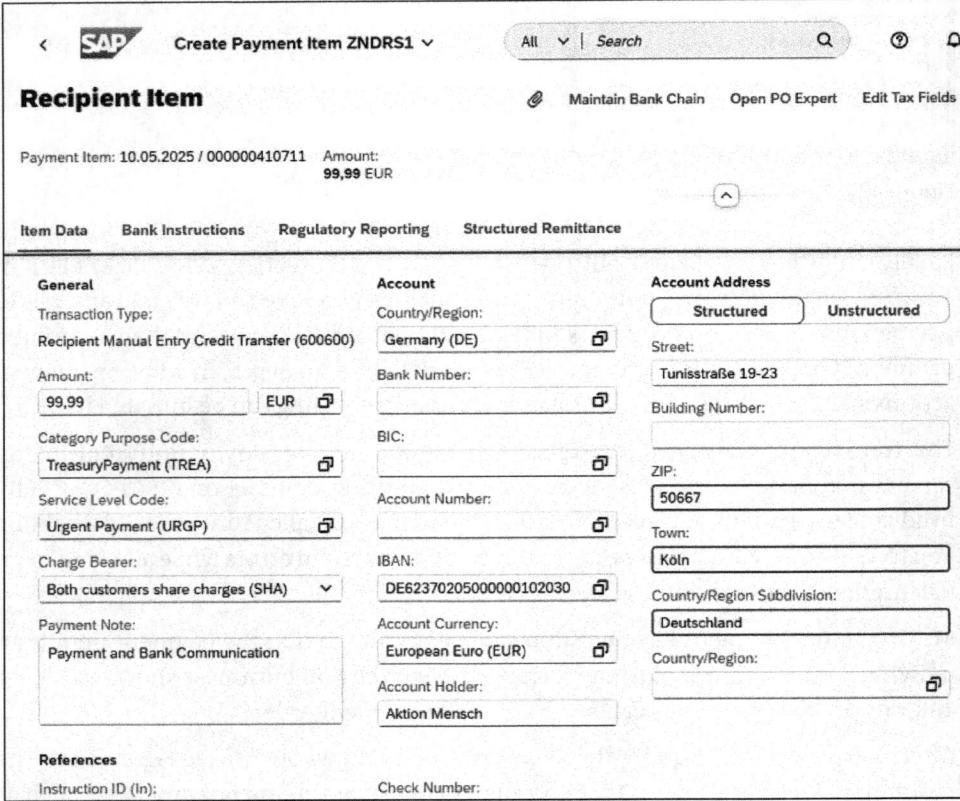

Figure 4.14 Create Payment Item

4.1 Advanced Payment Management Functions and Processes

Once all relevant data for the payment instruction has been entered, the item can be added to the payment order using the **Apply** button, located in the bottom-right corner of the Create Payments app (see Figure 4.15).

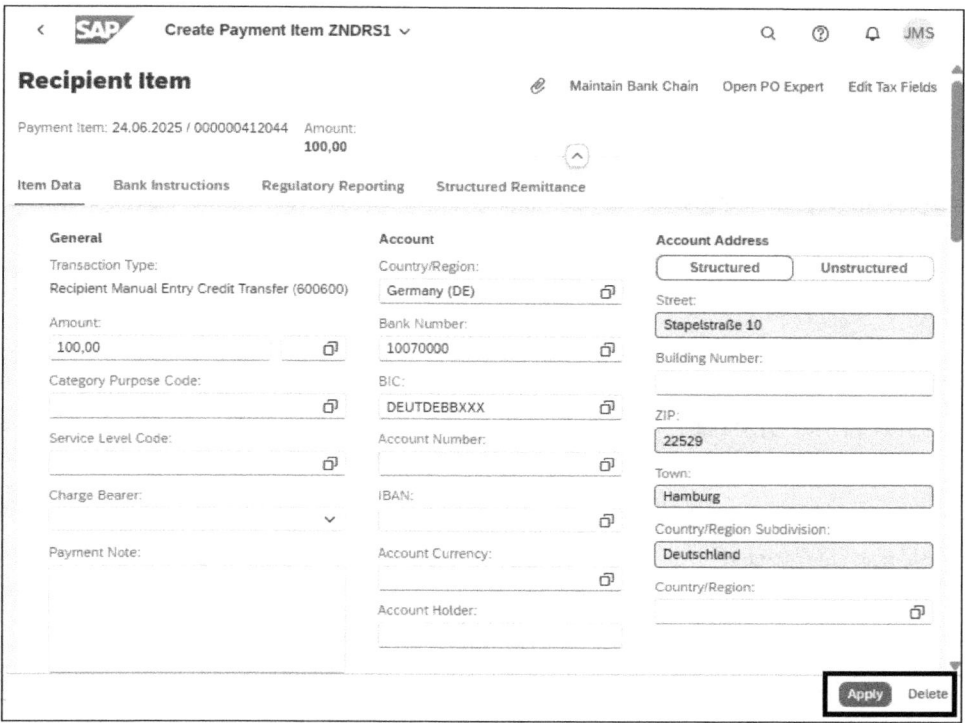

Figure 4.15 Payment Order

After the payment item has been added to the payment order, several options become available in the bottom-right corner of the app (see Figure 4.16), as follows:

❶ **Simulate Processing**

Clicking this button simulates the processing of the payment order. This function helps verify whether all relevant fields have been sufficiently filled to ensure successful processing through advanced payment management.

❷ **Save**

Clicking this button generates a payment order number and saves the payment in **Status 103: Draft**. In the Manage Payments app (ID F3647), the payment order can be accessed, processed, or further edited if necessary.

❸ **Process and New**

By clicking this button, the payment order is saved and processed through advanced payment management. At the same time, a new window opens for entering a new manual payment.

235

4   Advanced Payment Management and In-House Banking

❹ **Process**

By clicking this button, the payment order is processed through advanced payment management and the SAP Fiori app is closed. The user is then redirected back to the SAP Fiori launchpad.

❺ **Delete**

By clicking this button, all previous entries are deleted.

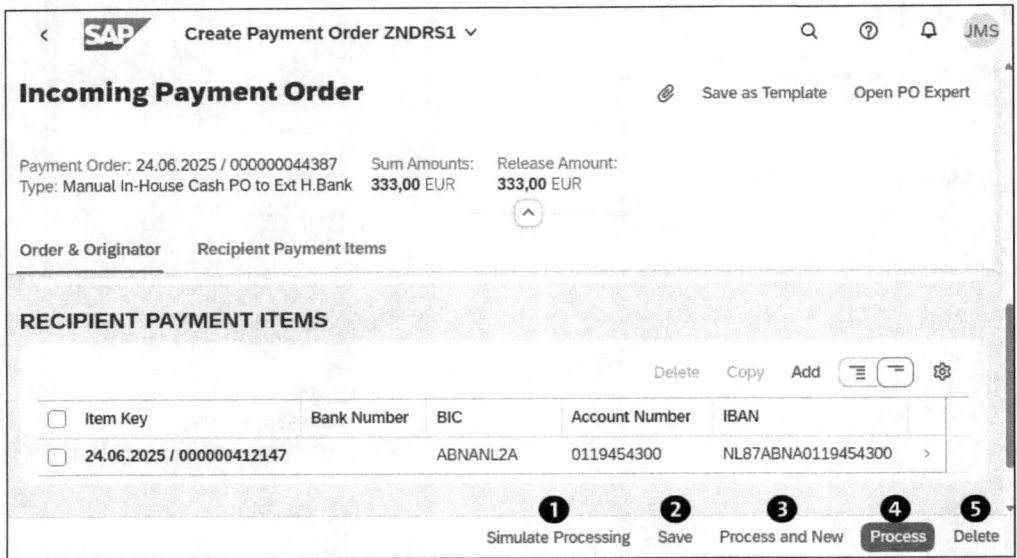

Figure 4.16 Recipient Payment Items App

> **Approvals**
>
> The approval of manual payments can be managed through a dedicated advanced payment management workflow. SAP provides a standard workflow template, which can be configured for payment orders (release object /PF1/ORDER) as well as payment items (release object /PF1/ITEM).

Repair Payments

The Repair Payments app (F3651; see Figure 4.17) allows you to repair payments that cannot be processed due to an error and that have therefore been transferred to exception control. You can correct the erroneous data yourself or trigger automatic reactions from the app, such as returning the payment to the sender.

When you enter the Repair Payments app, you will see an overview of faulty payments that have been parked for review. The reasons for these errors can vary—from missed cutover times to missing information to even lacking customization required for further processing. The errors may have either technical or business-related origins.

4.1 Advanced Payment Management Functions and Processes

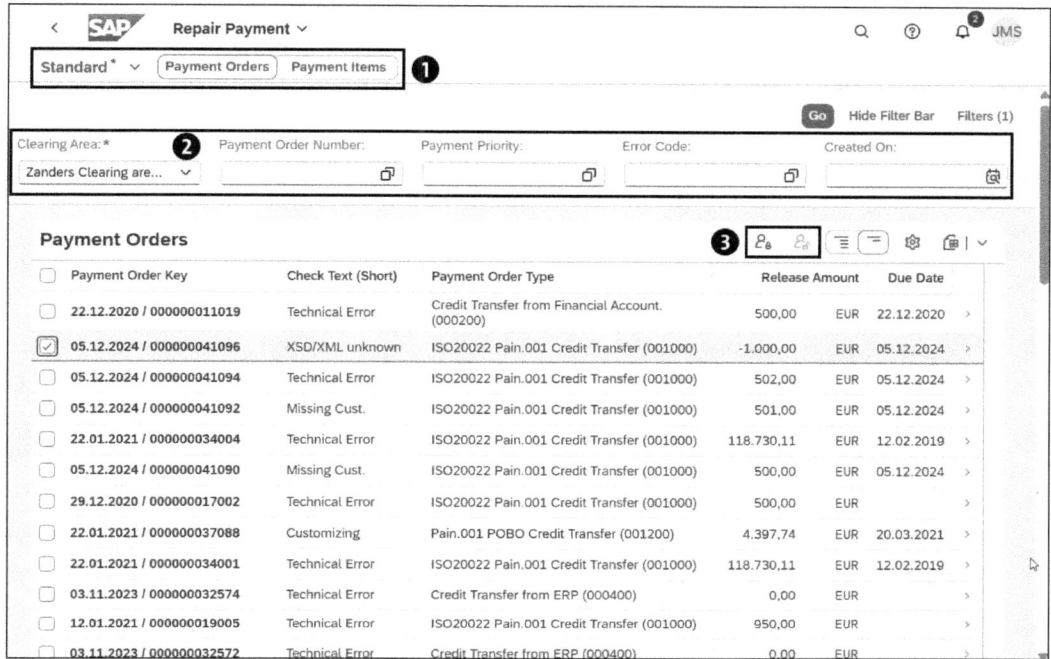

Figure 4.17 Repair Payments App

Let's look at the key functions of this app:

❶ **Manage variant and selection**
Not only can you manage the variants within the app, but you can also switch between selection at the payment order and payment item levels. By clicking the drilldown button, you can access the variant management options in the app. Selections and filter settings can be saved through variant management.

❷ **Filter selection**
The app offers a filter selection within the SAP Fiori standard design, which can be individually expanded using the **Filter** button located in the top-right corner of the app. An important note: The **Clearing Area** filter is mandatory and must be selected in advance.

❸ **Manage work basket**
The erroneous transactions displayed depend on the individual work basket you are assigned. Erroneous objects can also be reassigned to another work basket. If you want to reassign a work basket to your user, you can do so using the **Set** and **Remove Reservation** buttons.

Figure 4.18 shows the selected work basket assigned to the logged-in user. Within the app, there are several functions, as follows:

- **Exception handling**
  Through exception handling, the repair process can be initiated directly if possible, with the system providing support in finding solutions. If the error message originates from customizing, then the user must decide whether the issue can be ignored and processed further (e.g., in the case of a warning) or if deeper support, such as assistance from IT, is required.

- **Payment processing**
  The payment can be processed again using the **Process** button. Alternatively, the item can also be rejected.

- **Simulate processing**
  This functionality is present throughout nearly all SAP Fiori features in advanced payment management. The **Simulate Processing** button allows you to perform a preliminary simulation of the processing to verify whether it proceeds as expected. However, in the Repair Payments app, this option is currently available only for incoming orders.

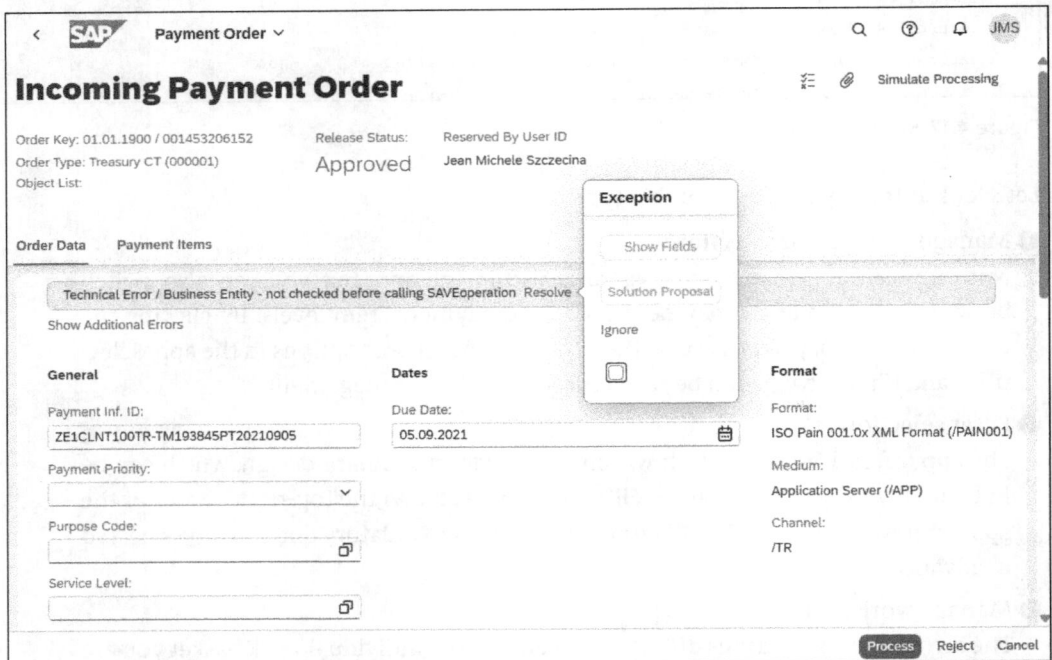

Figure 4.18 Repair Payments: Work Basket

Overall, the Repair Payments app represents an essential component in the exception handling process, providing the ability to continue the payment process without having to retroactively cancel the entire payment run. Nevertheless, downstream processes should also be established within the company to prevent manual repair operations on payments from recurring.

4.1 Advanced Payment Management Functions and Processes

Manage Payments

The Manage Payments app (F3647) serves as the central hub for managing incoming and outgoing liquidity flows within advanced payment management. In this app, the status of each incoming and outgoing payment order and payment item can be monitored.

After opening the app, you will see the overview (see Figure 4.19), which is divided into the filter area ❶ and the filter result area ❷.

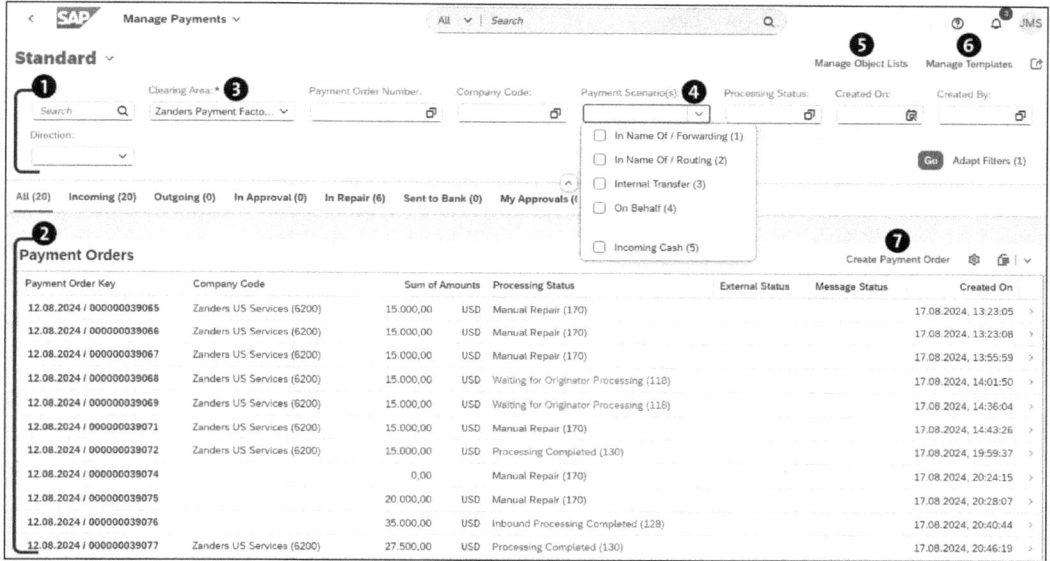

Figure 4.19 Key Filters and Navigation Points

Let's look at the key filters and navigation points within the app:

❸ **Clearing Area**
The **Clearing Area** field is marked as mandatory in almost all advanced payment management apps and must therefore be selected.

❹ **Payment Scenario**
In addition to filtering by processing status and by incoming or outgoing payments, advanced payment management provides the option to select payment orders and items based on the payment process.

❺ **Manage Object Lists**
An object list represents all payment media that are either loaded into or created by the advanced payment management system. It provides an overview and contains information related to payment data, such as format details, file name, references (payment run ID, SAP Multi-Bank Connectivity message ID, data media reference), sender and receiver details, and file handler data, including the original payment media.

239

❻ Manage Templates
This button navigates you to the **Template Management** section of the Create Payments app. There you will see an overview per clearing area and have the option to select and delete individual templates.

❼ Create Payment Order
This button navigates you to the Create Payments app (F3648), which was described in detail in the earlier section on that app.

Manage Payment Batches

With the Manage Payment Batches app (F4039), payment batches can be reprocessed and reviewed. The app allows you to monitor the status of batch-processed payments and assign unallocated payment items to payment batches.

The app also provides a direct link to the Manage Payment Items app, where payment items can be reallocated as needed. This reallocation enables individual payments to be redirected when necessary, such as when changing the sender bank. This may be required, for example, to process foreign payments under the best conditions or to modify the sending bank due to liquidity considerations. Figure 4.20 shows the Manage Payment Batches app.

The Manage Payment Batches app overview opens with a predefined filter set to the statuses **Open (02)** and **Final Processing Started (05)**. You will see a cross-clearing area view of the payment batches and have the option to manually process a selected batch by choosing the respective payment batch and clicking the **Process** button.

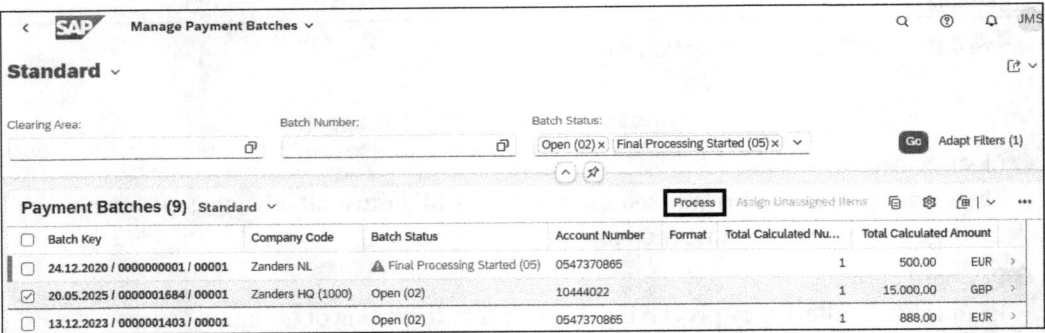

Figure 4.20 Manage Payment Batches

Manage Payment Items

The Manage Payment Items app is directly linked from many applications and provides a detailed overview of all payments processed within advanced payment management. It not only offers insight into processed payment items but also enables drilldown functionality and links to all relevant processing reports. Unlike most other advanced payment management apps, there are no mandatory filters in the Manage Payment Items app. The app provides a cross-company area overview of all processed payment items.

4.1 Advanced Payment Management Functions and Processes

Figure 4.21 shows an overview of the processed payment items. You can filter individual payment items using the filter bar with criteria such as **Clearing Area**, **Company Code**, **Bank Country**, and the account information. Alternatively, the standard search functionality within the filter is also available. After you click **Go**, the filtered results appear in the lower half of the app.

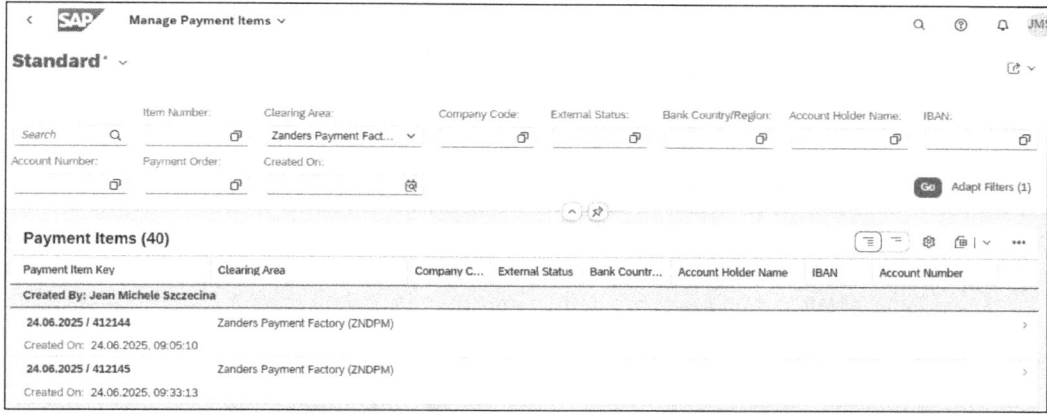

Figure 4.21 Manage Payment Items App

By clicking the **Detail** > button, you are directed to the detailed view of the respective payment item (see Figure 4.22).

There are several key functionalities and links within the Manage Payment Items app:

❶ **Display Log**

The **Display Log** button allows you to access the processing protocol for the payment item. When opened, the log file is usually empty. Log entries are generated when the payment item undergoes manual processing via the **Process** button or is subjected to a recall.

❷ **Add Note Attachment**

You can attach a file—such as a PDF document—to the payment item via drag and drop using this button. It also allows you to add a note to the attachment.

❸ **Flow**

The **Flow** button offers a visual representation of the processing status. As shown in Figure 4.23, the process chain is displayed, allowing you to see which steps have already been completed and which are still pending. Alternatively, you can access this overview by clicking the **Flow** ❾ button.

❹ **Recall**

The **Recall** button moves the payment item back to the **In Repair** or **Draft** status, making it editable again. The item can then be modified as needed.

4   Advanced Payment Management and In-House Banking

❺ Process

The **Process** button allows for manual processing of the payment item. This button triggers the next system-defined action, transferring the payment item—depending on its processing status—either into the approval workflow or directly into outgoing validation.

❻ Reassign

This button enables reassignment of the payment item to a new route and clearing agreement. For example, an individual payment can be redirected if it needs to be processed through a different bank account than was originally intended.

❼ Display Internal XML

This button displays the internal XML structure of the payment item.

❽ Display Tax Fields

You can view additional tax-related fields such as **Tax ID Number**, **Registration ID**, **Tax Type/Method**, and **Amount** via this option.

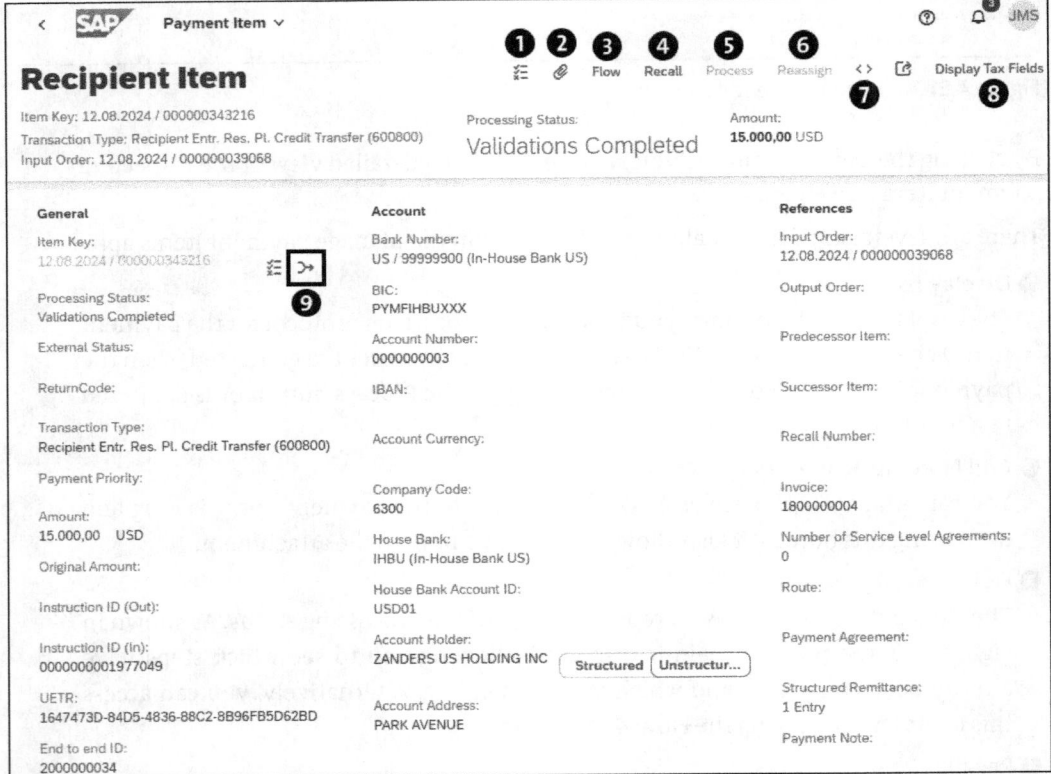

Figure 4.22 Payment Items Details

Figure 4.23 shows the process flow for a single payment item. The entire process chain is displayed, along with an indication of the current processing stage of the item.

4.1 Advanced Payment Management Functions and Processes

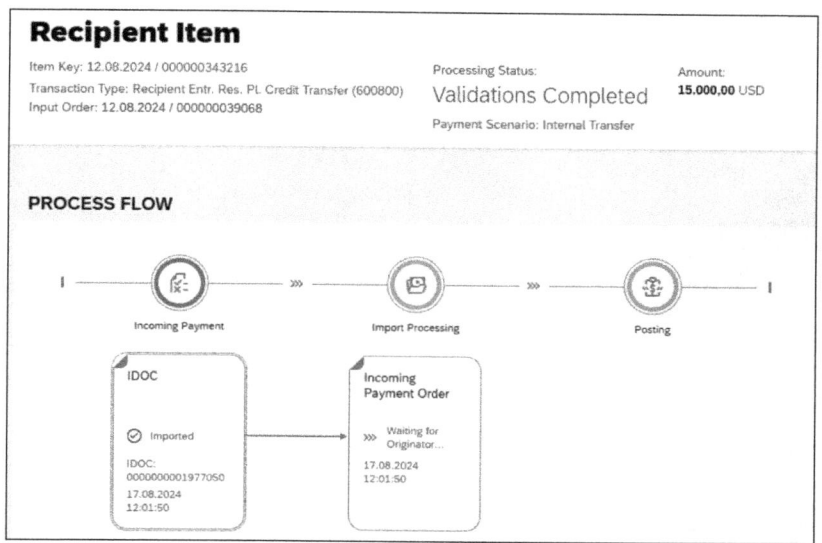

Figure 4.23 Process Flow for Single Payment Item

Payments Analyzer

The SAP Fiori apps in the Payments Analyzer series within advanced payment management are designed for monitoring and analyzing payment processes. The Payments Analyzer consists of analytical apps that are currently available only for on-premise and SAP S/4HANA Cloud Private Edition users. In total, SAP S/4HANA provides six analytical SAP Fiori app variants for advanced payment management, together offering an overview of processed and unprocessed payments. Let's look at the structure of the Payments Analyzer app:

- **Filters**

  In the upper section (see Figure 4.24), filters can be set for display currency, creation date, and payment status. Saved statuses can also be stored as variants within the standard SAP Fiori functionality.

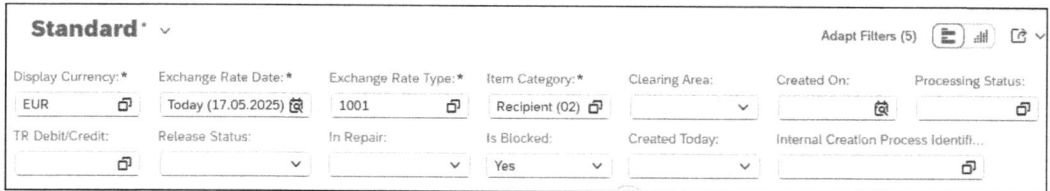

Figure 4.24 Analyze Payments: Filters

- **Processing status**

  In the middle section, the selection results are visualized (see Figure 4.25). As part of the standard SAP analytical apps, various display options are available, ranging from a tabular view to a pie chart representation.

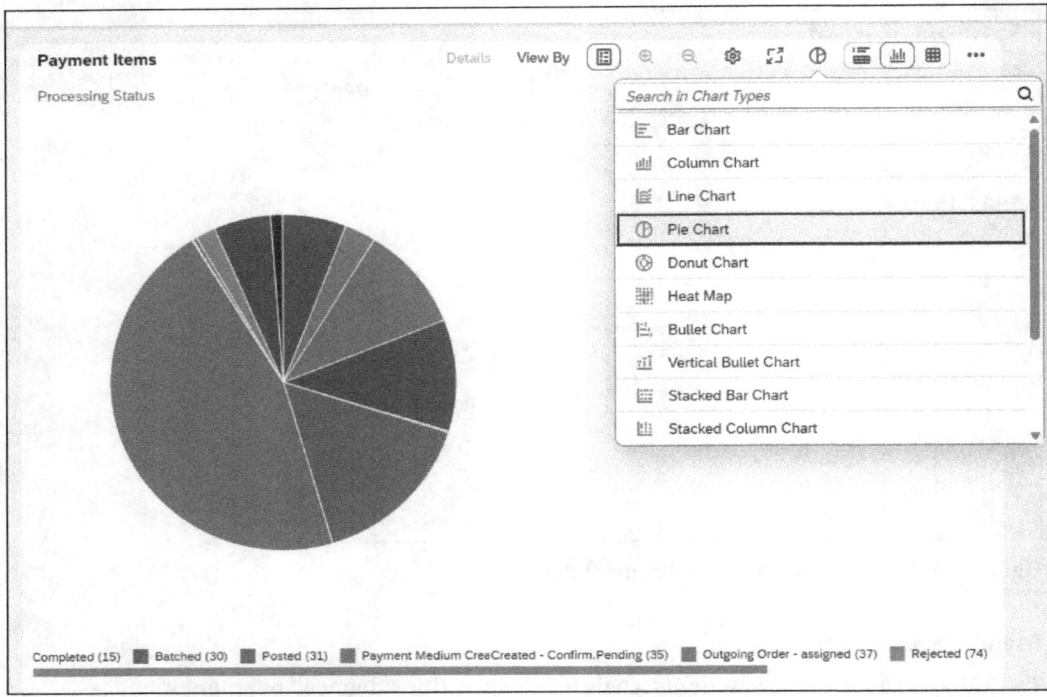

Figure 4.25 Payment Analyzer: Analytical Report

- **Payment items**
  In the lower section of the app, the selected payment items are displayed (see Figure 4.26). The app allows you to navigate to the Manage Payment Items app by clicking the detail icon › for a payment item of your choice.

Payment Items (1.107)

Item Key	Created On	Processing Status	Transaction Amount in Displa...		Release Status	Internal Creatio...	TR Debit/Credit
19000101 /000604291457	07.09.2021	Validations Completed (15)	14.877,49	EUR	Automatic Processing		Credit (C)
19000101 /000820791540	07.09.2021	Outgoing Order - conversion successfull ...	14.880,46	EUR	Automatic Processing		Credit (C)
19000101 /001194470272	05.09.2021	Validations Completed (15)	21.276,60	EUR	Automatic Processing		Credit (C)
19000101 /001287766233	06.09.2021	Validations Completed (15)	2.119,15	EUR	Automatic Processing		Credit (C)
19000101 /001617068934	07.09.2021	Validations Completed (15)	14.876,00	EUR	Automatic Processing		Credit (C)
19000101 /001762853295	05.09.2021	Rejected (74)	21.276,60	EUR	Automatic Processing		Credit (C)
19000101 /001840079407	05.09.2021	Validations Completed (15)	21.276,60	EUR	Automatic Processing		Credit (C)
			201.481.665,36	EUR			

Figure 4.26 Selected Payment Items

Advanced payment management offers a range of Payments Analyzer apps with various preselected filters, as follows:

- **Payment Analyzer—Items in Process**
  Displays all payments currently being processed by the system. These are transactions that are underway but not yet completed.

- **Payments Analyzer—Items for Release**
  Shows payment transactions that are awaiting user authorization or release. This app supports compliance processes like dual control or approval workflows.
- **Payments Analyzer—Items Received**
  Lists payments that have been received into the system but have not yet entered further processing. Useful for tracking incoming payment transactions in real time.
- **Payments Analyzer—Items in Repair**
  Displays payments that contain errors or require correction before they can be processed further. This might include missing master data or failed validations.
- **Payments Analyzer—Blocked Items**
  Shows payments that have been automatically or manually blocked due to risk checks, policy rules, compliance violations, or missing approvals.
- **Payments Analyzer—Items Completed**
  Lists all successfully processed and completed payment transactions. Useful for auditing, reporting, or reconciliation.

4.1.3 Master Data

Master data in advanced payment management serves as a fundamental cornerstone that significantly influences the efficient control and processing of payment transactions. These essential data points determine whether payments should be processed in batches, undergo detailed validation, and be handled according to internal cutoff times. Their precise configuration is therefore crucial for optimizing payment flows within the system.

However, there are three types of master data elements that are key for payment processing within advanced payment management and in-house banking:

- Service-level agreement
- Clearing agreement
- Route

All three master data types will be explored in detail in the following sections. In advanced payment management, master data can be configured via both SAP GUI and the SAP Fiori launchpad, ensuring flexibility and user-friendliness in customization and maintenance. In addition, modification processes can be controlled through an integrated approval workflow, which supports compliance with internal regulatory policies and ensures transparency.

The following sections will explore the specific applications and functionalities within advanced payments management that are critical for managing and optimizing payment processes.

In addition, we provide insight into the Manage Payment Agreements app, which serves as the successor to the Maintain Route and Clearing Agreement app. We also offer an overview of the master data apps in the enrichment and validations area, such as Exception Control. Through the Exception Control app, you can see how status management in advanced payment management can be controlled.

Last, but not least, the Maintain Payment Blocks app from the enrichment and validations area enables the integration of payment blocks in advanced payment management—such as on a currency or country level.

Service-Level Agreement

The SLA originates from the banking sector as an agreement between a bank and its customer. Within the framework of advanced payment management, the SLA defines how transactions from connected systems and subsidiaries should be processed. It establishes the procedures for handling incoming and outgoing payments for each participating entity.

The SLA is usually maintained by key users in the treasury department or payment factory and is available as master data in SAP GUI under Transaction /PF1/SLA or in the Service-Level Agreement app (/PF1/SLA). Each SLA is established at the level of the clearing area.

> **Clearing Area: Definition**
>
> A *clearing area* is an organizational unit within advanced payment management and, in a broader sense, serves as the hub for transaction processing. The clearing area is assigned its own business master data and business objects and features an independent end-of-day processing system.
>
> The clearing area links not only master data but also transactional data such as payment orders, payment items, and transactions. It also functions as a cornerstone for authorization management within advanced payment management.

As previously described, SLAs regulate the agreed-upon processing of transactions. The following functionalities are among those supported by an SLA:

- **Managing payment processing and clearing**
  The SLA ensures the orchestration of processing—for example, by determining which participants can use which channels and which products or payment types. Furthermore, the processing is managed to decide whether payments should be batched based on predefined criteria, grouped via a queue and subsequently batched, or processed directly. Internal cutoff times for processing can be defined as well.

- **Payment validations**
  Participating entities can utilize central services, such as the verification of payments for both technical and content-related accuracy, within advanced payment management. Checks—such as duplication detection—can be performed as well to ensure reliable transaction processing.

- **Correspondence with participating entities**
  Correspondence with participating entities can be structured through SLAs to ensure smooth communication and transparency in payment processing. For instance, status reports such as pain.002 messages can be sent to participants, providing updates on transaction processing stages

Multiple SLAs can be configured per clearing area, each at different levels. Let's examine the various service-level types:

- **Clearing area SLA**
  The clearing area SLA is the general SLA for an entire clearing area. This SLA serves as the standard regulation when no more specific SLAs—such as segment-, group-, or customer-specific SLAs—are defined for a payment.

- **Segment SLA**
  A segment SLA refers to an organizational subunit within the clearing area, such as one based on geographic regions or business units. This type of SLA allows for differentiated management of business requirements for individual market segments.

- **Group SLA**
  A group SLA applies to a group of business partners. Such groups can be defined based on industry, revenue volume, risk class, or contract status. The purpose is to centrally manage SLAs for specific internal customer groups without needing to configure them individually for each customer.

- **Customer SLA**
  The customer SLA is the most individualized type of SLA and applies to a single business partner. Once a customer SLA is defined, it overrides all other SLA types. This type of SLA allows for the precise representation of contractually agreed-upon or operationally relevant specifics in payment transactions.

After opening the Service-Level Agreement app, a web GUI interface opens (see Figure 4.27) and offers a dynamic selection. Before you can create or select an SLA, you must first choose a clearing area. On the right side, you can double-click selection fields to select them and add them to the selection mask. Clicking the **Execute** button brings you to the overview of the selected clearing area.

Click the **Create SLA** button shown in Figure 4.27 to create a new SLA. Each SLA must be assigned a unique ID. After the creation of the clearing area, it will appear in the tree on the left side under the respective node (see Figure 4.28). Clearing areas with the **Clearing Area** SLA type will be displayed under the **Clearing SLA** node.

4 Advanced Payment Management and In-House Banking

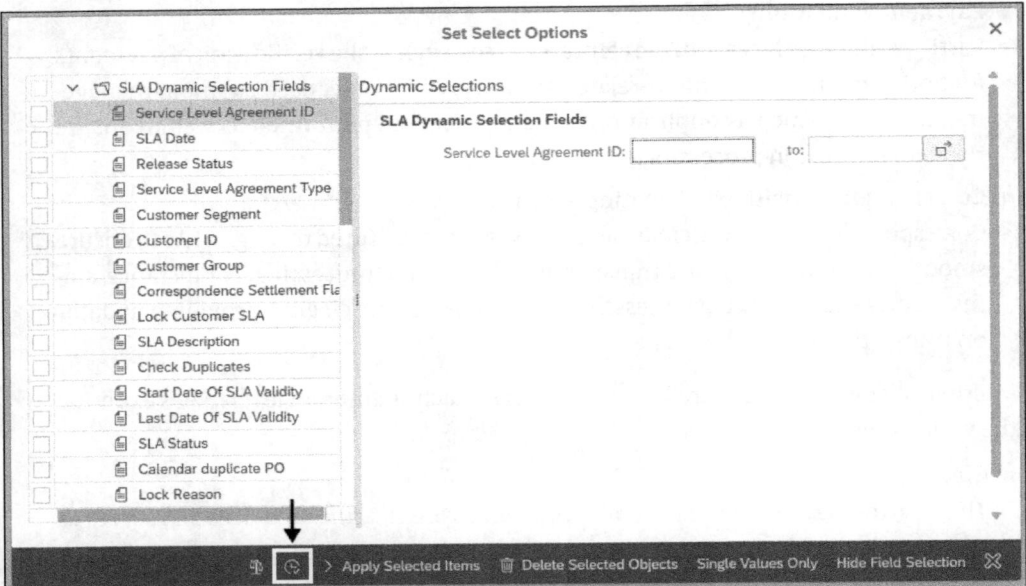

Figure 4.27 Service-Level Agreement: Selection

Let's examine the areas of the **Basic Data** tab shown in Figure 4.28 ❶:

❷ **SLA Information**

SLA status management can be controlled via the **SLA Information** area. Before an SLA can go into live operation, it must be in status **Active**. This must be activated manually via the **Activate** button. If the SLA should be locked for live operation, this can also be done through SLA status management using the **Lock** button. The SLA type allows the selection of the appropriate type. As previously described, the orchestration level can be chosen here.

❸ **Options**

Through the options, individual services can be activated as needed. You can enable services such as enrichment, validations, or notifications. These checks are performed based on the corresponding set of rules (see Figure 4.29).

❹ **Assigned Customer**

Here, the SLA can be restricted to selected entities. If no selection is made, then the SLA applies to all incoming transactions.

If the **Enrichment and Validation** box has been checked under the **Basic Data** tab, then the conditions for the examinations can be set on the **Enrichment and Validation** tab using the **Set of Rules** button. The selection conditions can include, for example, the payment type, country, and bank account, as well as the respective channel, as shown in Figure 4.29. Enrichment and validation do not refer only to purely content-related examinations but also include the validation of compliance with cutoff times under this point. For this purpose, custom selections or rulesets can be configured.

4.1 Advanced Payment Management Functions and Processes

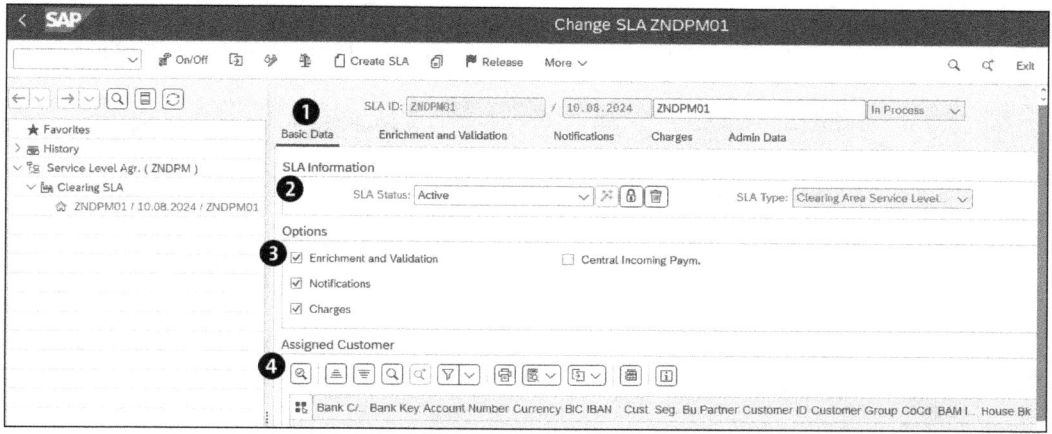

Figure 4.28 Manage Service-Level Agreements: Create Agreement

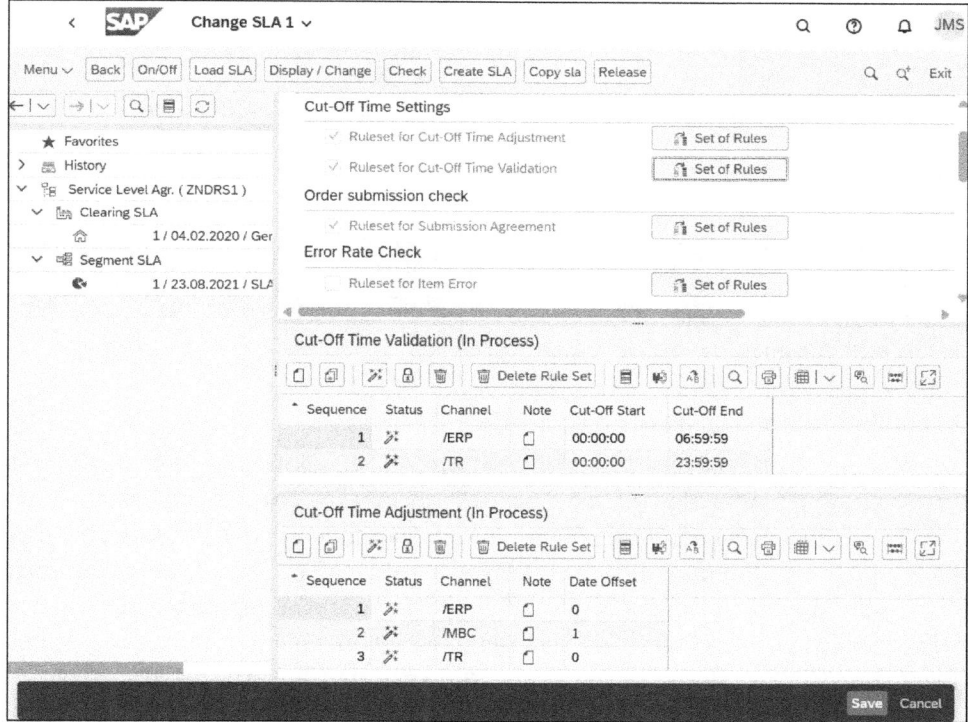

Figure 4.29 Display Service-Level Agreement

The **Notification** tab is used for correspondence with participant entities. The setting displays the correspondence options and rules, such as correspondence type, role, and recipient. Correspondence recipients can be specified or locked. The different types of correspondence are triggered by the specific processes in advanced payment management. Figure 4.30 shows the settings for payment feedback through the ERP channel.

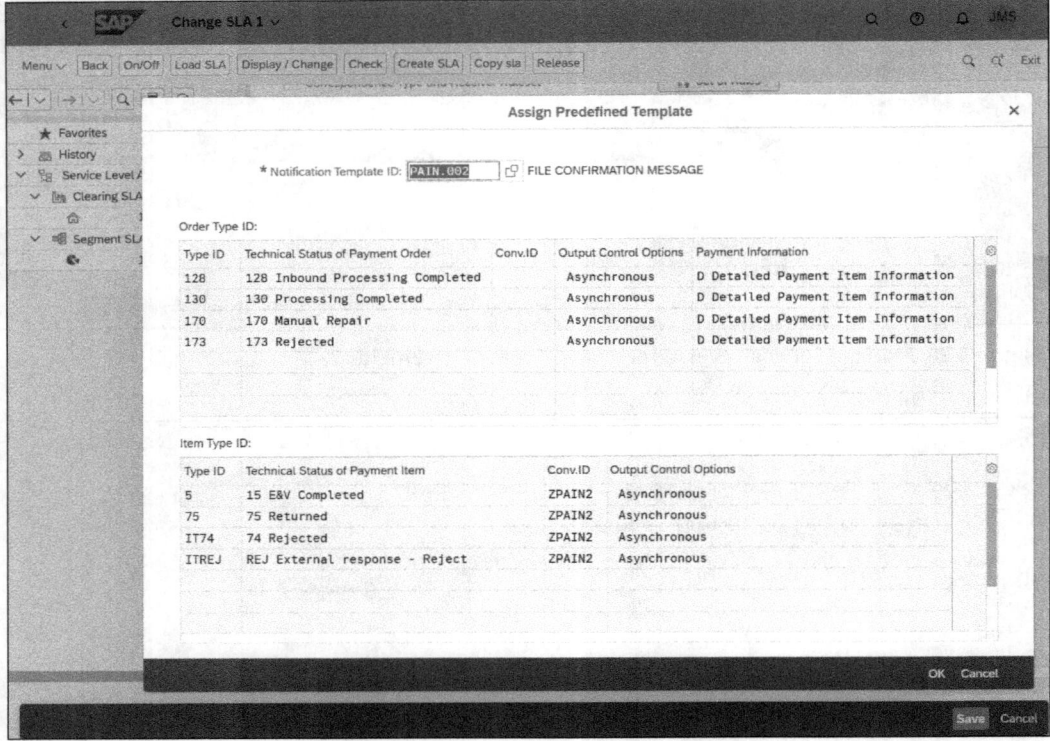

Figure 4.30 Assign Predefined Template

Maintain Route and Clearing Agreement

The Maintain Route and Clearing Agreement app (/PF1/RN) is used for routing and transaction clearing (see Section 4.1.1). It essentially focuses on two components: the route and the clearing agreement. These components determine whether an item is an internal payment (e.g., internal payment, payment on behalf of [POBO], or external payment for forwarding as in a payment in the name of [PINO]) transaction. The functionality can be accessed via Transaction /PF1/RN, as well as the Maintain Route and Clearing Agreement app (/PF1/RN).

> **Maintain Route and Clearing Agreement App**
>
> The Maintain Route and Clearing Agreement app is only available for on-premise and private cloud systems. For public cloud customers, this app has been replaced by the Manage Payment Agreements app (F4629). Since the 2020 FPS 0 release, the Manage Payment Agreements app has been available for on-premise systems as well as private cloud systems.

4.1 Advanced Payment Management Functions and Processes

Route and clearing agreement master data is mandatory for this functionality; at least one agreement must be present in the system to process payments. You can create a new route or clearing agreement by right-clicking **Routes** and selecting **Create Route** or **Create Clearing Agreement**, as shown in Figure 4.31.

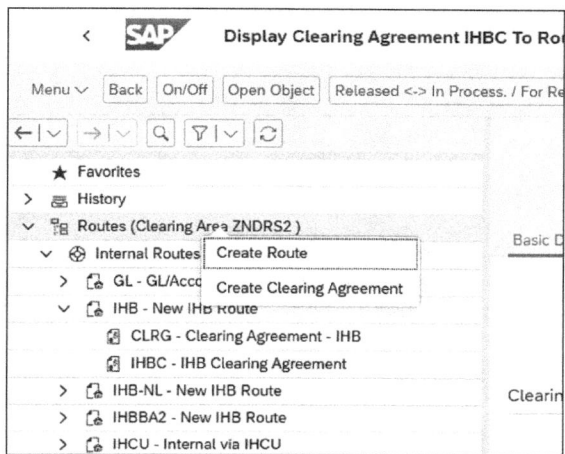

Figure 4.31 Create New Route or Clearing Agreement

Figure 4.32 provides examples of internal and external route IDs and clearing agreements. Ahead, we present an example of the setup for an internal route. Within a route, multiple clearing agreements can be organized and structured hierarchically. A route can be assigned to a company code, a bank account, or both simultaneously.

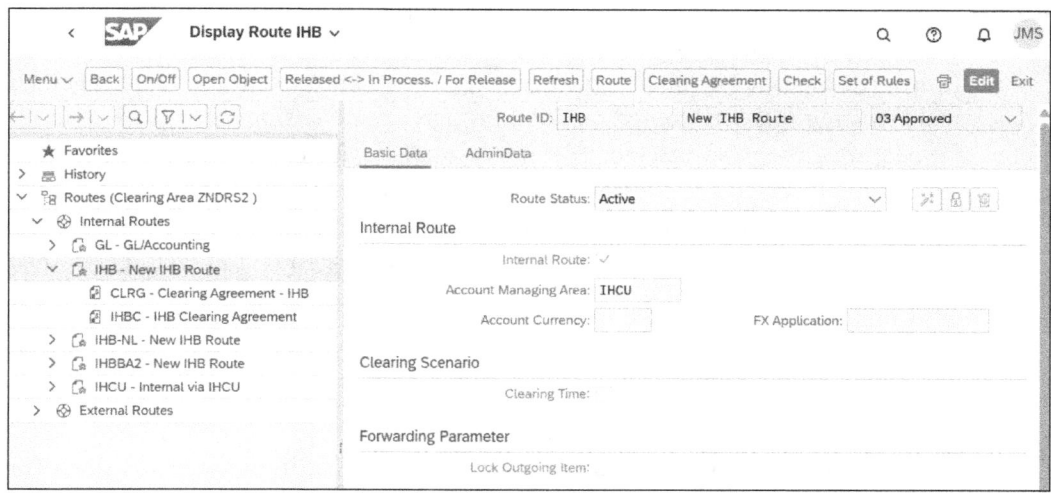

Figure 4.32 Internal Route

4   Advanced Payment Management and In-House Banking

Figure 4.33 provides an example of a clearing agreement. Within the clearing agreement, you can define how the payment should be processed in detail. For instance, you can specify whether the payment should be handled within a batch or processed directly online, along with the respective conditions.

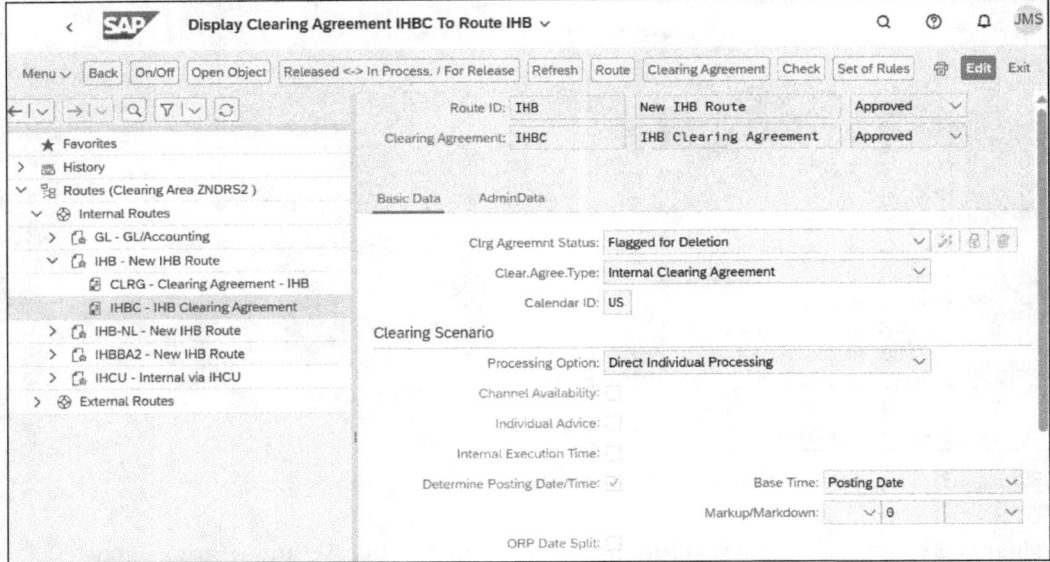

Figure 4.33 Display Clearing Agreement

Manage Payment Agreements

The Manage Payment Agreements app (F4629) is used for clearing payments and essentially fulfils the same tasks that the Maintain Route and Clearing Agreement app performs. The Manage Payment Agreements app allows for the centralized management of all existing rule sets, routes, and clearing agreements. Using the app, these can be viewed, modified, and deleted. The app also allows for controlling how payments should be processed and is essential for routing. The payment agreements primarily define processes such as internal payments, POBO, and pure forwarding—that is, PINO.

The Manage Payment Agreements app (see Figure 4.34) is divided into three tabs:

- **Payment Rule Sets**
  A payment rule set is created for each company code, and all associated payment rules are attached under the payment rule set.

- **Payment Agreements**
  Under the payment agreement, the conditions for clearing are established. This determines whether the payment should be processed individually or as a batch payment. Cutoff times for processing can be set as well.

4.1 Advanced Payment Management Functions and Processes

- **Payment Rules**

  Under the payment rules, the detailed selection criteria are defined—for example, specifying which bank and bank account should be used to process a POBO transaction.

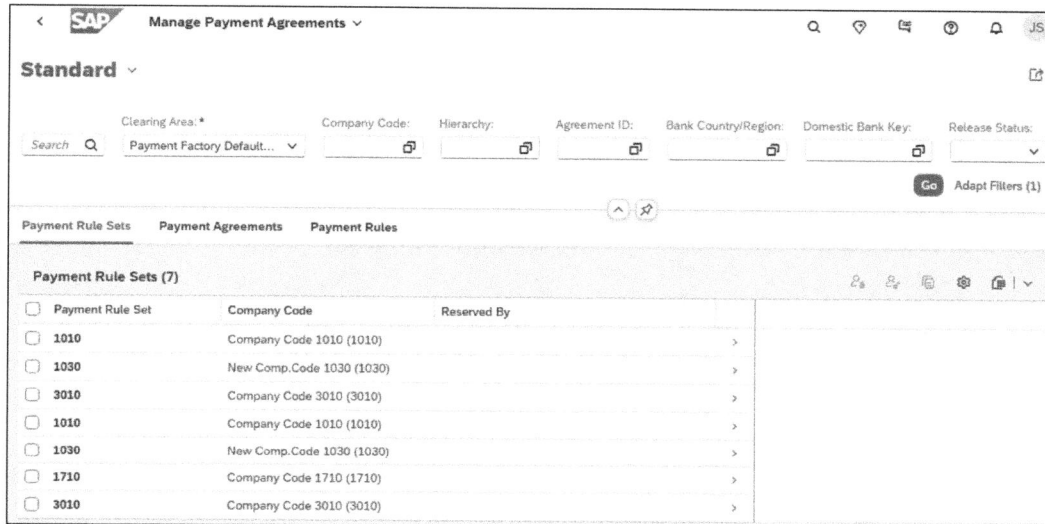

Figure 4.34 Payment Rule Set

> **[!] Information Loss Within the Manage Payment Agreements App**
>
> Information loss may occur between the Maintain Route and Clearing Agreement and Manage Payment Agreements apps. It may occur if existing data—for example, data created in the backend—is modified here. For instance, payment rule sets created from the transaction or the Manage Route and Clearing Agreement app do not have an entry under **General Company Code**.

Figure 4.35 shows a POBO payment agreement. Payment orders that meet the selection criteria of the associated payment rules are selected and cleared under the conditions defined here.

Figure 4.36 shows a payment rule with the respective conditions for clearing. A rule is set for the selection of POBO payments based on the stored IBAN. **PO_IBAN** (payment order IBAN) is selected under **Characteristic**, with **Inclusive** as the selection type and equal to (**EQ**) as the option. Payments for which the selected IBAN is the beneficiary will be processed as POBO payments.

253

4   Advanced Payment Management and In-House Banking

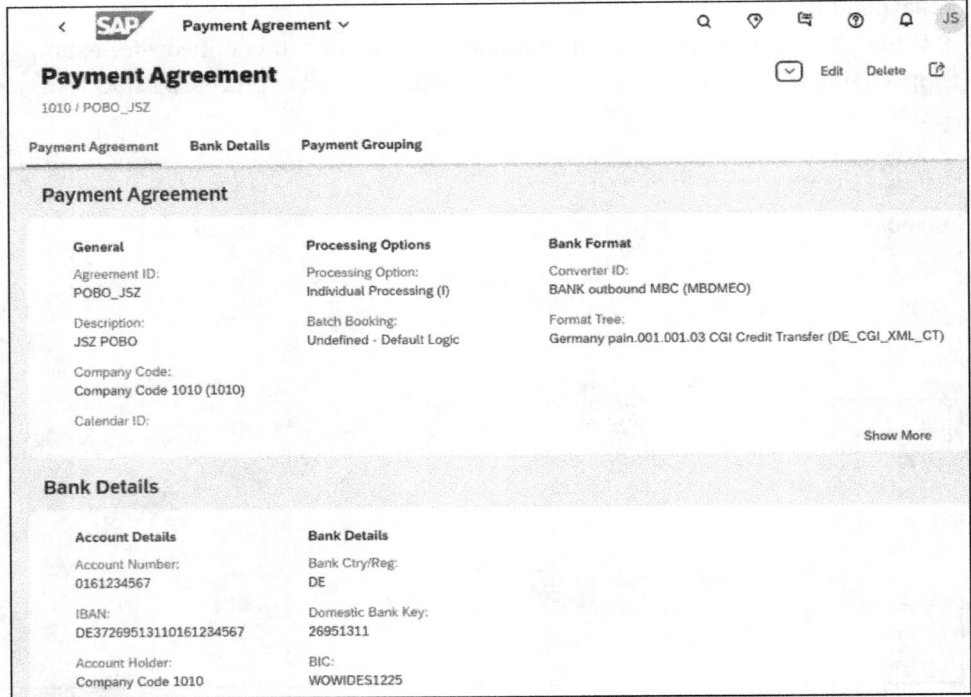

Figure 4.35 Payment Agreement

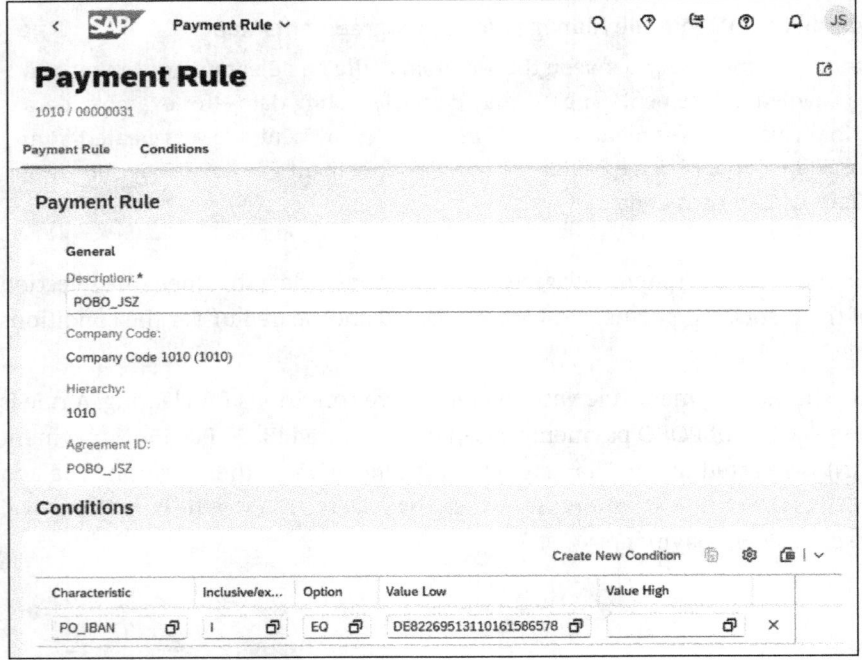

Figure 4.36 Payment Rule

Exception Control

Ideally, advanced payment management is configured such that the rules are finely tuned and defined to allow payments to be fully automated and processed via straight-through processing. But what happens when deviations occur? When information is missing in the payment medium, or a payment fails the validation checks? In such cases, exception handling takes effect. Exception handling is closely linked to every process step and is directly invoked whenever discrepancies arise during format processing, validation, or routing and clearing. The reasons for this can be varied, ranging from new, unknown bank formats to erroneous information or hits on sanction and embargo lists. These errors will result in error codes. Through exception control, the status can be managed actively to determine how the data should be processed.

Table 4.2 and Table 4.3 show some standard SAP error codes and possible processing options that can be configured via the Exception Control app (/PF1/EH).

Error Code	Error Description	Processing
290000	An unknown error during the XSD validation.	Reject
290001	The XSD validation cannot process this invalid XML document.	Reject
290002	The XSD definition for this XML document is not available. Check Transaction /PE1/XSD for XSD definitions.	Reject
290003	For each XML tag, the child nodes must respect the sequence definition given by the XSD definition.	Reject

Table 4.2 Syntax Errors

Error Code	Error Description	Processing
13	Check cut-off time of the processing channel against the SLA (inbound) and clearing agreement (outbound).	Ignore
155	Duplicate due to internal checksum logic. Checksum logic is dependent on configuration.	Reject
156	Duplicate due to external checksum.	Reject
157	Duplicate check—concurrent lock exists.	Reject

Table 4.3 Inbound Payment Order Checks

Exception control can be accessed through the Exception Control app (/PF1/EH) or via Transaction /PF1/EH. When you click on the SAP Fiori tile for the Exception Control app, a Web Dynpro application will launch, opening Transaction /PF1/EH in the background. Prior to configuring the exception control, it's essential to select the relevant **Clearing**

Area for the app, as shown in Figure 4.37. The popup window shown appears when the Exception Control app is opened for the first time.

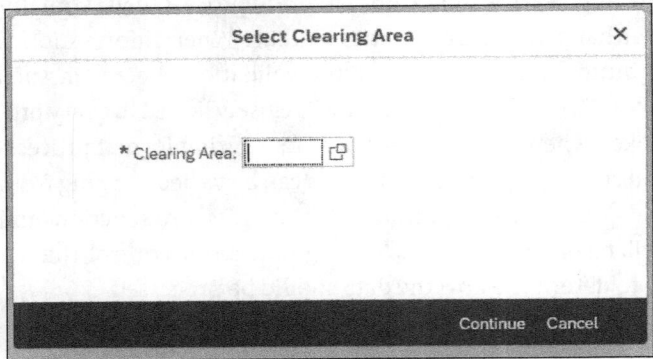

Figure 4.37 Exception Control: Select Clearing Area

After selecting the **Clearing Area**, you can choose a specific process step in the system, as depicted in Figure 4.38 ❶, where you wish to establish active control. This can be configured for both incoming and outgoing processing at the payment level or the payment item level. At the payment item level, you can decide whether to apply this to the originator, recipient, turnover, or clearing item level. By double-clicking the object, you can access the detailed view ❷ and then customize the standard response type (**Stand. Response Type**). The first step is to select the type via the multiple selection option ❸.

Figure 4.39 shows the **Executed Checks** tab. These error codes describe the checks performed during this process step and the possible error codes for the business object. They can also be used to establish specific response types.

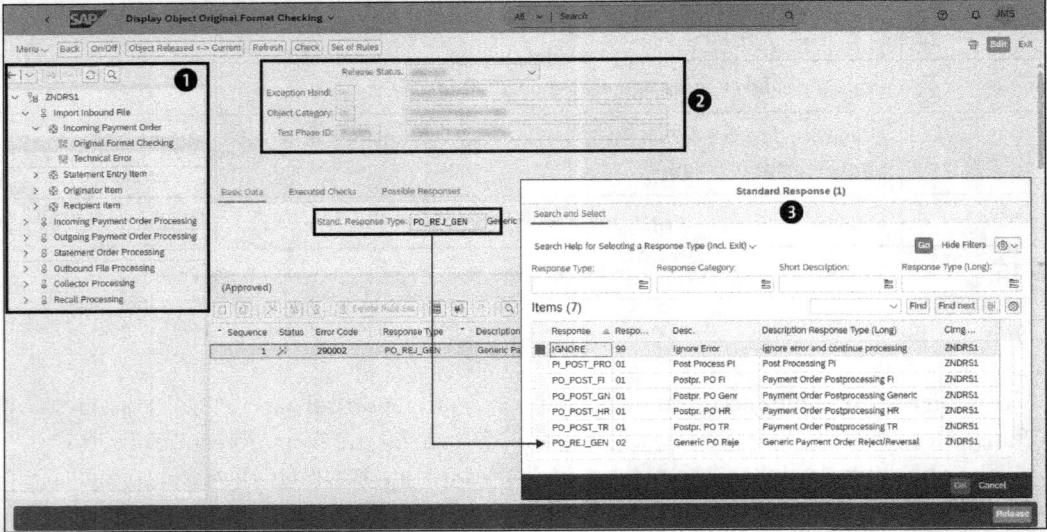

Figure 4.38 Exception Control: Basic Data

4.1 Advanced Payment Management Functions and Processes

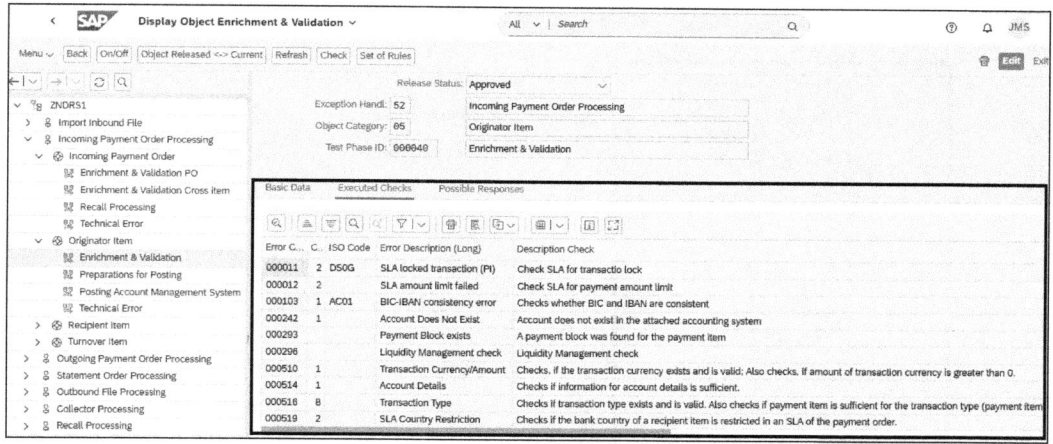

Figure 4.39 Exception Control: Executed Checks

In the **Possible Responses** tab (see Figure 4.40), you will see the types of responses available for this process step and the business object. SAP makes a logical preselection, providing different types of responses depending on the process step, business object, and other factors.

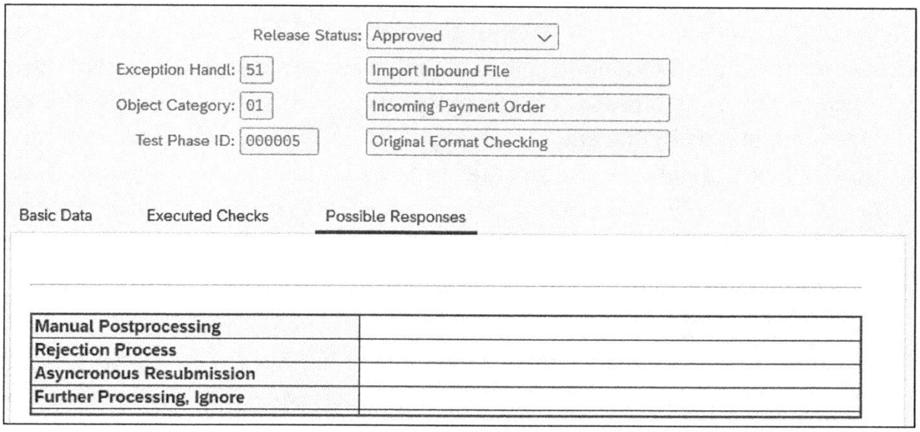

Figure 4.40 Exception Control: Possible Responses

If the default reaction type should not always be executed, then specific scenarios can be defined using rule creation, allowing deviations from the standard. For example, there may be a particular focus on treasury or HR payments. Figure 4.41 shows the selection process used if the executed checks need to be restricted based on specific criteria such as format, medium, and so on. If a criterion needs to be added, simply double-click a characteristic and enter the appropriate value.

257

4   Advanced Payment Management and In-House Banking

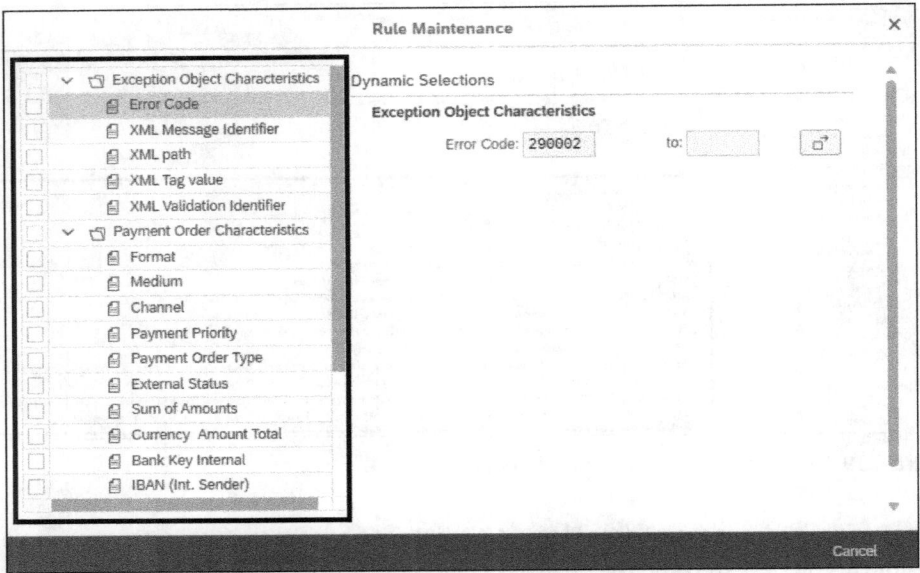

Figure 4.41 Exception Control: Rule Maintenance

Maintain Payment Blocks

The Maintain Payment Blocks app (F3650) is part of the enrichment and validation process, designed to set payment blocks. With this app, payment blocks can be configured based on countries or regions, bank correspondence partners, or currencies. The app can be operated directly by the end user or key user. Once the validations are set and saved, they receive the status **Active** and are immediately ready for use. Payments that meet the set criteria, such as a blocked payment currency, are automatically blocked and forwarded to the Repair Payments app (F3651). In this section, we'll introduce this app and its configuration options.

Upon launching the app, you will arrive at the overview screen. Before configuring the payment blocks, ensure that the correct clearing area has been selected. This can be done using the **Select Clearing Area** button located in the lower-right corner of the app, as shown in Figure 4.42.

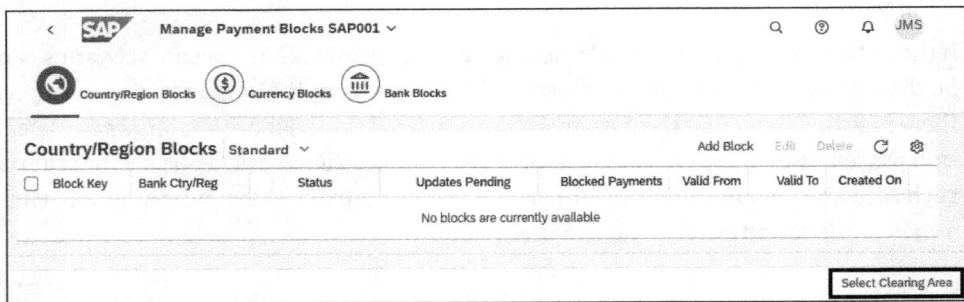

Figure 4.42 Manage Payment Blocks

4.1 Advanced Payment Management Functions and Processes

After selecting the target clearing area, a new block can be added by using the **Add Block** button, as shown in Figure 4.43.

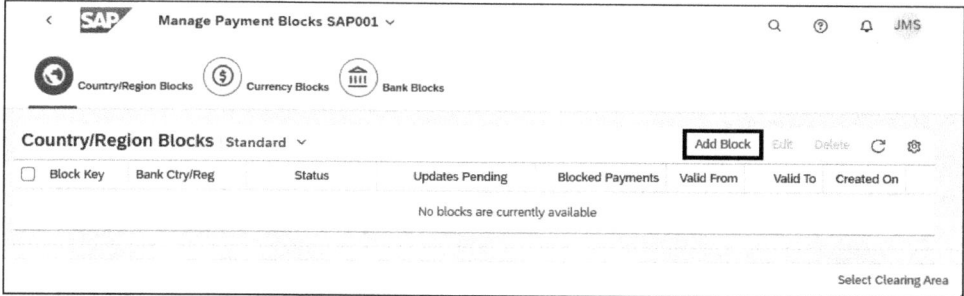

Figure 4.43 Manage Payment Blocks: Add Block Button

After clicking the **Add Block** button, a popup window opens, in which a new block can be added. It is important to set the value date correctly to ensure the block is activated within the desired timeframe. Figure 4.44 shows an example. The block can be set based on **Country/Region**, **Currency**, or **Bank**, as maintained in the system.

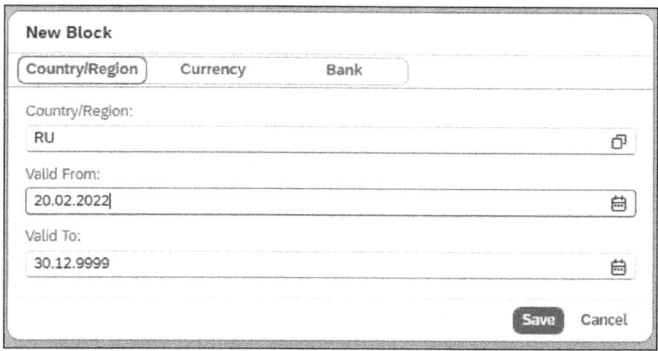

Figure 4.44 Add New Payment Block

> **Selected Blocks Need to Be Configured**
> If a block is to be set on a bank, for example, the bank must also be maintained in the system with a bank key in table BNKA to be selectable.

Once you click **Save**, the payment block appears in the app overview and is immediately available in the system with status **Active**. Subsequent payments that are selected for this specific block during the enrichment and validation process are directly forwarded to the Repair Payments app. In the Repair Payments app, decisions can be made about further processing of the payment.

259

4.2 In-House Banking Functions and Processes

In Chapter 1, we introduced the concept of in-house banking, providing an overview of its purpose, its key features, and how it supports centralized cash management and external and internal payment processing. In this section, we will take a deeper dive into its practical aspects, focusing on the core apps and functionalities used within SAP S/4HANA to execute the processes and payment flows described earlier. We will walk through how these processes are managed from a functional perspective using in-house banking and explain the essential configuration settings and master data required to successfully operate an in-house banking environment. This section aims to bridge the theoretical foundations with hands-on execution within SAP S/4HANA.

Traditionally, all operations related to virtual banking in SAP were handled through the classic SAP In-House Cash solution, which was available in SAP ERP. SAP In-House Cash is still available in SAP S/4HANA; however, SAP is gradually limiting its availability to new customers. Organizations implementing higher versions of SAP S/4HANA may be required to transition to advanced payment management and in-house banking, which have been introduced as the new options and are being actively developed by SAP. Because the classic SAP In-House Cash solution is no longer being enhanced, SAP encourages organizations to adopt advanced payment management and in-house banking for future-proof solutions. The key differences between SAP In-House Cash and the new advanced payment management/in-house banking functionalities were explained earlier in Chapter 1.

Because SAP In-House Cash is no longer actively supported by SAP and does not make use of SAP Fiori apps—SAP hasn't even developed any SAP Fiori interfaces for SAP In-House Cash—we will only occasionally reference it in this section. Instead, our primary focus will be on the capabilities and processes within the new in-house banking, which has been designed to fully integrate with the modern SAP Fiori experience and is being continuously developed by SAP as the strategic direction for in-house banking operations. In-house banking is a new innovation: an internal banking solution that allows subsidiaries to maintain in-house bank accounts with their headquarters. It became available with the SAP S/4HANA 2022 release.

To begin this section, we will focus on the functional usage of in-house banking. We'll start with a brief overview of the most important topics, highlighting only the key prerequisites required to use in-house banking effectively. After that, we'll cover the creation of in-house banking accounts and their lifecycle management. Next, we'll move on to discuss system limits, the setting up of bank charges, and how conditions and interest conditions are configured. Finally, we'll quickly walk through the main payment processes that take place within in-house banking, providing a high-level understanding of how these operations are handled.

4.2.1 Prerequisites

Before you can start using in-house banking, there are several important prerequisites that must be met. Some of these prerequisites involve arrangements outside of SAP, such as agreements with external banks and third parties to set up virtual accounts, cash pooling structures, or automated transaction handling. On the internal side, you must ensure that the necessary configurations are completed within other areas of SAP S/4HANA—primarily, in SAP S/4HANA Finance:

- **Virtual banks and agreements with the banks**

  If you want to run internal payments and manage your internal banking processes in SAP S/4HANA, the first step is to establish contracts with your external banks for the use of virtual banking services. These agreements allow you to process receivables on behalf of (ROBO) transactions, cash pooling, and other automated transactions through the banks. In practice, this means that special virtual accounts must be created to handle these types of operations. *Virtual accounts* are unique identifiers provided by banks that help to centrally manage and track incoming payments. Instead of setting up separate real bank accounts for every internal entity or customer, virtual accounts allow you to easily distinguish and allocate incoming funds, simplifying the cash management, reconciliation, and centralization of your payment processes.

- **Intercompany agreements**

  Before executing internal payments such as netting or intercompany payments between affiliates, it is essential to have formal agreements established between the entities involved. These agreements define the conditions under which payments will be processed, including who will initiate the transactions, whether the process will be driven by payables or receivables, and which party is responsible for starting the payment. They must also outline settlement rules, such as the currencies to be used for balancing the accounts, payment terms, cutoff dates, and any additional conditions specific to the relationship between the entities. Once the agreements are finalized, you can simply assign them to the relevant in-house bank accounts within SAP S/4HANA.

- **Setup in SAP S/4HANA Finance**

  Another important prerequisite for using SAP In-House Bank is setting up the necessary general ledger (G/L) accounts. You need to have specific G/L accounts created and ready, especially intercompany payable and receivable accounts, which will be used for posting internal bank statements and managing internal transactions. In addition, cash-in-transit accounts are required to clear payments during the movement of funds, along with G/L accounts for posting bank charges and interest related to in-house banking activities.

It's also crucial to ensure that the required cost centers are created and assigned to profit and loss accounts so that charges and income are properly tracked. Furthermore, you must maintain trading partners in the system to correctly represent intercompany relationships and balances in receivables and payables.

In addition, configuration for payment methods is required, with dedicated payment methods created specifically for intercompany payments. You will also need to maintain G/L accounts to act as clearing accounts for the execution of payments, ensuring that the payment flows are accurately recorded. Further information regarding the processing of statements and how to configure them will be covered in Chapter 9.

- **Advanced payment management and SAP Multi-Bank Connectivity setup**
  Proper setup of advanced payment management and SAP Multi-Bank Connectivity, as described in Chapter 5, is essential for the successful functioning of in-house banking. These configurations provide the necessary processing rules and ensure the correct route of internal payments. For in-house banking to operate effectively, both advanced payment management and SAP Multi-Bank Connectivity must be set up properly. Without these configurations, the processing and routing of internal payments would not function correctly, impacting the smooth execution of in-house banking transactions. Therefore, ensuring that advanced payment management and SAP Multi-Bank Connectivity are properly configured is a critical prerequisite for the seamless operation of in-house banking.

4.2.2 Account Management Lifecycle

The most important component of an in-house bank, like any external bank, is the bank account. Each affiliate participating in the internal bank must have designated bank accounts created within the system. These accounts serve as the central location where all transactions related to that affiliate will be posted and stored. The balances for each affiliate within the internal bank also are accumulated and tracked in these accounts. Without properly established bank accounts, it would be impossible to accurately manage, monitor, or reconcile the financial flows between the different entities within the internal bank. Therefore, creating and maintaining these bank accounts is essential to the smooth operation and accurate financial reporting of the in-house banking system.

In-house banking is typically located within the payment factory, which acts as the central unit for managing internal payments across affiliates. When setting up accounts in in-house banking, it's essential to ensure that each internal bank account created in in-house banking is also reflected in BAM. This step is necessary for operational purposes such as payment execution and bank statement processing. If the affiliate's operational system is the same as the one hosting in-house banking, then the account is created in BAM within that same system. However, if the affiliate is using a different SAP system, then the in-house banking account must still be created in the central in-house banking system, and a corresponding internal bank account must be created in BAM within the affiliate's system (to create the house bank and bank account ID). This ensures that local payment processes and reconciliations can function properly while staying connected to the central in-house banking structure.

Before setting up an in-house bank in SAP S/4HANA, it's essential to carefully consider its geographical location. The choice of location can significantly impact on your company from a tax, legal, and operational standpoint. Common countries selected for in-house banks include the United Kingdom, Ireland, the Netherlands, Switzerland, and Luxembourg, or tax friendly districts in the US, or Hong Kong or Singapore in Asia, often due to favorable tax frameworks or financial infrastructure.

To streamline operations and effectively manage global financial workflows, organizations often establish multiple in-house banking centers strategically located across different regions, such as the US, Europe, and Asia. This approach addresses the challenges posed by time zone differences, ensuring round-the-clock operational efficiency and timely execution of transactions. By having regional in-house bank centers, companies can simplify financial processes by creating single, consolidated transactions within each region and maintaining a clear balance between these centers. This not only enhances transaction accuracy but also facilitates compliance with local regulations and reduces the complexity of cross-regional financial management. Quite a common scenario for US-based companies is to build one center in the US and one in Europe or/and Asia.

However, the decision should not be based solely on tax considerations. It is equally important to locate the in-house bank in a country where your organization has substantial business activity, operational presence, and access to capital. A thorough analysis of your company's global structure, regulatory environment, and liquidity needs should guide the final decision.

Once you have decided where to locate your in-house bank, the next step is to create it in the system, following the same process as for setting up any other bank account. You will need to define a bank key that represents your in-house bank. Because this is not an external financial institution, you have the flexibility to define the bank key as needed. It's common practice to use a distinctive and easy-to-identify key, such as 99999 for an in-house bank located in Switzerland. This helps clearly differentiate internal banks from external ones in your configuration and reporting.

When setting up the bank key for your in-house bank, follow the procedure outlined in Chapter 3. If you are using SAP S/4HANA 2022 or higher, you need to use the Submit Bank Account app to create and submit the bank account. If you are using a lower version, then you can create it directly in the Manage Banks app. During the submission process, make sure to set the **Contract Type** as **In-House Bank Account** to correctly classify it in the system as an internal bank entity (if you are using a lower version than 2108, then the contract type is not available).

Figure 4.45 shows an example of the bank key for one of the in-house bank's bank area. In SAP S/4HANA, the bank area is the internal bank within your organization, and it serves as a centralized entity for managing internal transactions and financial processes. Essentially, setting up a bank area means creating your own internal bank to efficiently handle financial operations across affiliates. As you learned in Chapter 3, a crucial step in establishing this internal bank is the configuration of a bank key for each bank in the system. The bank key acts as a unique identifier, enabling the system to distinguish between different banking entities and ensuring streamlined operations. Properly defining and configuring these bank keys is essential for maintaining order and accuracy in the in-house banking structure, aligning it with both organizational needs and country-specific requirements.

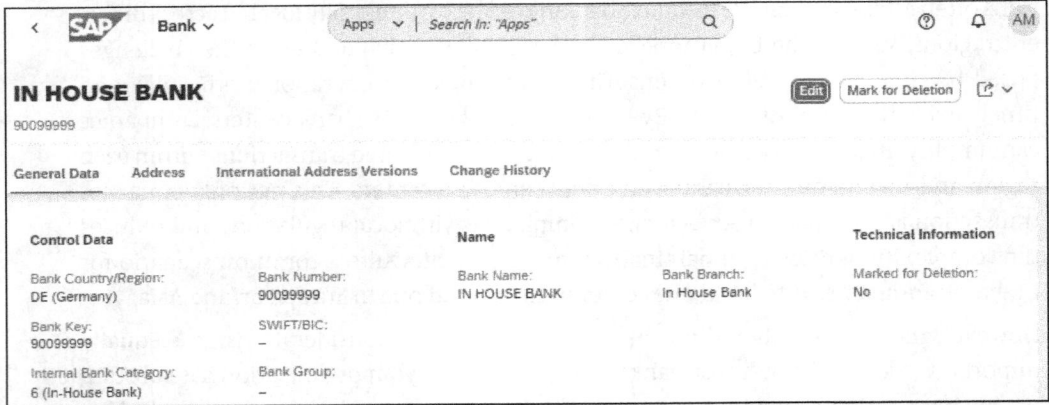

Figure 4.45 In-House Bank Key

> **Bank Key for In-House Bank**
>
> It's important to note that when creating a bank key for your in-house bank, the format of the bank key must follow the country-specific rules defined by SAP. For example, if you are setting up a bank key for Germany, it must consist of exactly eight characters; for Switzerland, the bank key must have five characters; and for the US, it must be nine characters. Each country has its own requirements, and you must adhere to these guidelines as outlined in the system's country-specific configuration. If you attempt to enter a bank key that does not meet the required format for the selected country, then the system will return an error and prevent you from proceeding. Always consult the local configuration guide to ensure the correct setup.

4.2.3 General Functions

As we've discussed, the most important part of any bank, whether external or internal, is the bank account. In Chapter 3, you discovered how external bank accounts are created and managed using BAM. For internal bank accounts, SAP also provides dedicated solutions that allow companies to manage their internal banking activities efficiently. These solutions are specifically designed to handle the creation, maintenance, and operation of internal bank accounts, separate from the management of external bank accounts. They support the definition of account conditions, transaction rules, and intercompany settlement procedures, ensuring that internal payments and cash management processes are properly structured and compliant. Having dedicated tools for internal bank accounts is essential in order to mirror real banking operations within the organization and maintain full control over internal liquidity movements.

In the following sections, we'll delve into the intricacies of master data, emphasizing the creation and management of in-house bank accounts. These discussions will also encompass key elements such as defining bank conditions and interest, calculating bank fees, and setting bank limits. By exploring these foundational aspects, we aim to outline a comprehensive framework that supports the efficient operation and governance of internal banking processes within SAP S/4HANA systems.

Create In-House Banking Accounts

Traditionally, in SAP In-House Cash, internal bank accounts (SAP In-House Cash accounts) were created and maintained using Transaction F9K1 in SAP GUI. This method was widely used for many years; however, over time, it became increasingly problematic. One of the key challenges was the lack of integration between SAP In-House Cash accounts and BAM, meaning that internal accounts had to be created separately and maintained independently from the central bank account repository in SAP S/4HANA. To address these limitations, SAP introduced a new in-house banking solution with dedicated SAP Fiori apps that allow users to create, manage, and link internal bank accounts in a streamlined and integrated way, fully aligned with BAM.

There are three main ways to create in-house bank accounts in SAP S/4HANA:

- First, you can request their creation through the standard electronic Bank Account Management (eBAM) process, in which accounts are initiated in BAM and automatically created as in-house bank accounts using predefined configurations. This method is efficient and aligns with automated workflows.
- Second, you can upload in-house bank accounts from an XML file, which is particularly useful for mass creation or during data migration activities, such as moving from test to production environments.
- Third, you can manually create in-house bank accounts using the Manage In-House Bank Accounts app, which allows for full control over the account setup and customization. Each method serves different purposes depending on the business context and system landscape.

The recommended approach (which works only if you have SAP S/4HANA 2408 or higher; otherwise, you need to create accounts manually) is to use the internal eBAM connection to link your bank accounts created in BAM directly with your in-house bank accounts. This integration ensures consistency and eliminates manual effort by automatically creating and synchronizing the bank accounts across both applications. Without this setup, you would need to manually create and maintain the same bank account data separately in both BAM and in-house banking / SAP In-House Cash, increasing the risk of errors and misalignment.

The creation of an in-house bank account can be initiated from the Manage Bank Accounts app, which is also used for maintaining external bank accounts and is described in detail in Chapter 3. To begin, open the app and click **Create**. Then, select the appropriate **Company Code** where you want the in-house account to be created. In the **Account Type** field, type "88", which specifically designates the account as an in-house bank account. This setup ensures that the account is correctly recognized and managed within your in-house banking environment. Figure 4.46 shows how you can create an in-house banking account within the Manage Bank Accounts app.

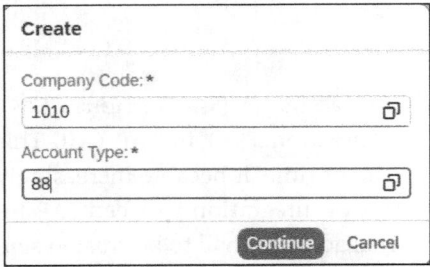

Figure 4.46 Create In-House Banking Account Using Manage Bank Accounts

After setting account type 88 for the in-house bank account, proceed by entering the account number, selecting the appropriate bank key that represents your in-house bank, and providing the country, currency, and the opening date. Also select the relevant in-house bank service to link the account to the correct internal banking process. Once all mandatory fields are filled in, save the account, and then click **Start Opening Account**, which is available as a status option next to the **Edit** button. This action transitions the account into the opening process. After that, navigate to the **Bank Correspondence** section to continue with the configuration and communication settings related to the account. Figure 4.47 shows the required details for the in-house banking account.

In the **Bank Correspondence** section, you need to select the correspondence type that has been preconfigured for use with eBAM; this will be the specific type used to trigger the request for creating an in-house bank. Once selected, enter the required details, such as recipient information or internal routing data, and add a meaningful description to clearly identify the purpose of the request. After filling in all necessary fields (see Chapter 3 and the external bank account creation; we follow this procedure during the internal bank account creation), click **Send Account Opening Request** to initiate the formal creation process through eBAM (see Figure 4.48). This step ensures that the in-house banking account is properly logged and processed within the system's internal banking structure.

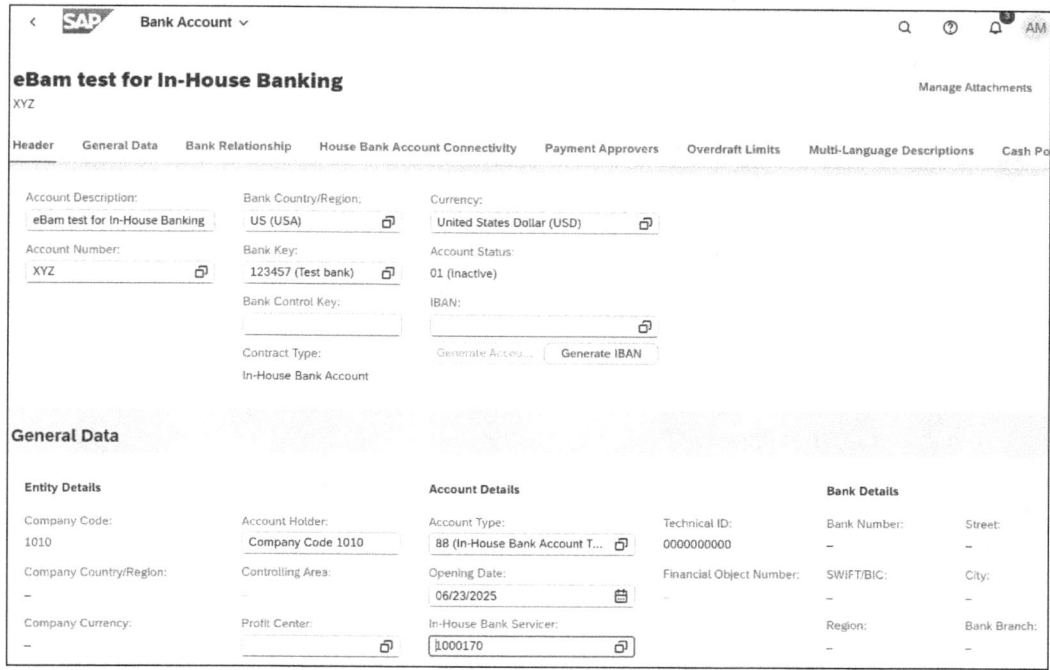

Figure 4.47 Account Details for In-House Banking Account

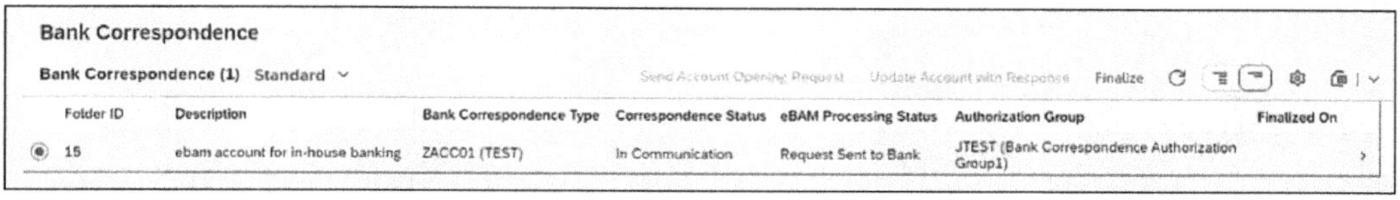

Figure 4.48 eBAM Bank Messages

Once the account opening message is sent, navigate to the Manage Bank Messages app to monitor the status of both outgoing and incoming messages. This app allows you to track the communication flow used within the internal eBAM process. Here, you can verify whether the outgoing message for creating the in-house bank account was successfully processed and whether a corresponding incoming confirmation message has been received. Monitoring these messages ensures that the internal eBAM process is functioning correctly and that the new in-house banking account has been created and acknowledged within the system.

Once the message has been successfully processed, the final step is to enter the **House Bank Account ID** and assign the appropriate G/L account for the BAM account; see

Chapter 3 for more information. After entering this data, simply activate the account to complete the setup. You can then verify that the account has been created correctly in the in-house bank by using the Manage In-House Bank Accounts app. This confirms that the internal eBAM process has worked as expected and that the in-house banking account is now fully operational and ready for use within your SAP S/4HANA environment.

> **Bank Account Number**
> Note that the in-house banking account does not currently use the bank account number created in BAM. Instead, it uses a number from the number range specified in the in-house banking configuration.

Alternatively, you can use Manage In-House Bank Accounts app (F5942) to create in-house banking accounts manually directly in the application. Figure 4.49 shows how an account can be created.

Figure 4.49 Create In-House Bank Account

When you open the Manage In-House Bank Accounts app and click **Go**, the system displays a list of existing in-house bank accounts based on predefined selection criteria and filters. To create a new bank account, you can click the **Create** button, which opens a popup window prompting you to enter the key details required for setup. These include the **Bank Area**, **Account Currency**, the **Business Partner** for whom the account is being created, and the **Opening Date**. You also have the option to select an **Account Template**, which automatically fills predefined fields based on a previously created account setup, streamlining the process and ensuring consistency across similar accounts.

> **Note**
>
> When creating a new in-house bank account, there are a couple of important checks that need to be considered. First, the opening date of the bank account cannot be earlier than the current posting date maintained for the selected bank area. Second, the business partners used during the creation process should represent affiliates or entities holding the bank account. In-house banking, just like SAP In-House Cash, is designed in a way that allows the inclusion of participants and account holders that may not be part of the same SAP system as the one hosting in-house banking, providing flexibility for cross-system or external entity integration (including entities in different ERP systems than SAP S/4HANA).

To use business partners for the creation of in-house bank accounts in SAP, they must first be properly created and maintained as business partners within the system. Specifically, they need to have the appropriate business partner roles assigned—BKK010 (Account Holder) and BKK030 (Correspondence Recipient). Figure 4.50 shows these roles created for the business partner in Transaction BP or using the Manage Business Partner Master Data app (F3163), which are essential for identifying which entities are authorized to hold accounts and to receive relevant bank correspondence. Once these roles are correctly maintained and active, the business partners can be selected when creating internal bank accounts. It's also crucial to ensure that the business partner master data is complete, particularly address data and the **Trading Partner** field, which is required for generating correct accounting postings during G/L transfers and reconciliation. The same requirements applied to the old SAP In-House Cash solution.

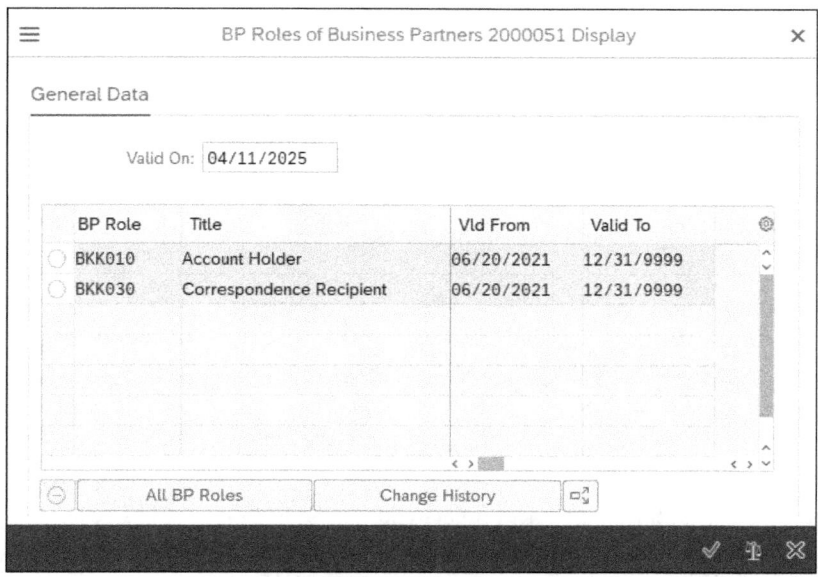

Figure 4.50 Business Partner Roles

While creating an in-house bank account, you need to provide several key pieces of information that define how the account will operate. First, you must specify the **Account Number**. Whether you can enter it manually depends on the configuration of number ranges: You can set up either internal numbering, where the system automatically assigns a number from a predefined range, or external numbering, where the user enters the number manually. You will also need to enter an **IBAN**, which could be a virtual account number provided by your bank for specific purposes such as ROBO transactions or other virtual banking setups. You also must define whether this account should be marked as the preferred account for payments; this is important when multiple accounts exist for the same currency. You should also provide a meaningful **Account Description**, indicate whether payment transactions are allowed on the account, and select the account status. Note that the statuses available in the dropdown list are aligned with those used for external bank accounts managed in BAM. Figure 4.51 shows this information created in the BAM account.

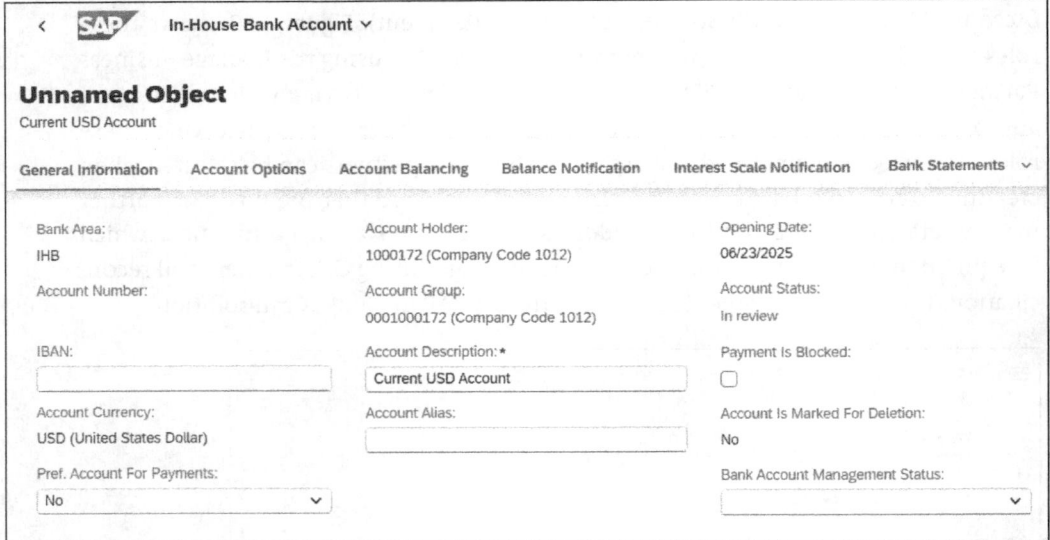

Figure 4.51 In-House Banking Account General Information

Numbering of Accounts

When it comes to defining number ranges for in-house bank accounts, it's important to consider what will make account identification and reconciliation easiest for your organization. If you're working with several affiliates, it is often beneficial to use external numbering, where you manually assign account numbers that follow a logical, recognizable pattern. For example, account numbers could include elements like currency code (e.g., USD), account type, and company code. This approach simplifies reconciliation processes and makes it easier to identify transactions linked to specific accounts.

> However, currently in in-house banking, you can change the bank account number to your selected number only by using a BAdI. Otherwise, the system will just use incremental numbers.

When creating an internal bank account in in-house banking, one of the key steps is defining the **Account Options**. These options determine how the account behaves in terms of reporting, communication with affiliates, and financial processing. You need to specify the following:

- **Account Balance Needed**
  Account balances should be created and tracked.
- **Balance Notification Needed**
  Balance notifications should be sent to affiliates to inform them about their current balances.
- **Interest Scale Notification Needed**
  Interest scale notifications reports are needed. These reports contain detailed information about interest calculations.
- **Bank Statement Needed**
  Bank statements should be generated, which consolidate all relevant information (balances, interests, fees), and serve as the basis for financial postings.

In addition, you must define how frequently **Account Balancing** should occur (e.g., monthly, quarterly), including the calculation of interest and bank fees. In most cases, such as in monthly cycles, this is set to the last working day of the month. Once account balancing begins, further information becomes available, such as withholding tax details and the next expected account balancing date. This setup ensures that the internal bank operates with the same discipline and transparency as an external financial institution.

If you select **Account Balance Needed** and **Interest Scale Notification Needed**, you will need to specify who will receive the notifications, what format will be used (based on the configuration), and to which address they should be sent. Figure 4.52 shows how it can be done in the system by specifying **Business Partner**, **Output Format ID**, and **Email Address**. Usually, the notifications for bank balances and interest scales are generated in PDF format and contain detailed information about the current balances, interest amounts, and the logic or rules applied to calculate them. These documents are typically sent to the participants—meaning the affiliates who hold internal bank accounts—who can then review them, use them for internal validation, and keep them as part of their documentation for audit and compliance purposes.

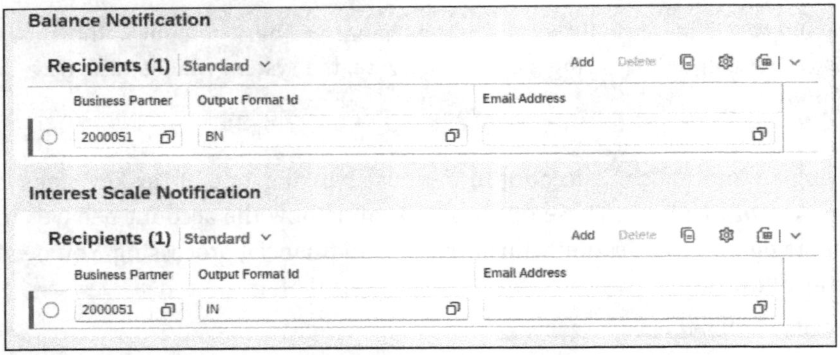

4   Advanced Payment Management and In-House Banking

Figure 4.52 Balance and Interests Notification

Figure 4.53 shows that in the next step of setting up an internal bank account, you need to define the **Bank Statement Frequency** for generating bank statements. You can specify whether statements should be created multiple times per day or just once, depending on the transaction volume and reporting needs. You also have the option to decide if statements should still be generated even when there are no postings; this is controlled by selecting or not selecting the **Suppress Without Postings** flag. After that, you need to provide the business partner holding the account and assign a correspondence recipient, in the **Business Partner** field, who will receive the statements. You must also select the output format (PDF, IDoc, etc.) and determine how it will be delivered—for instance, via email or sent directly to an integrated system. Furthermore, you should indicate whether the bank account number or its alias should be shown in the statement.

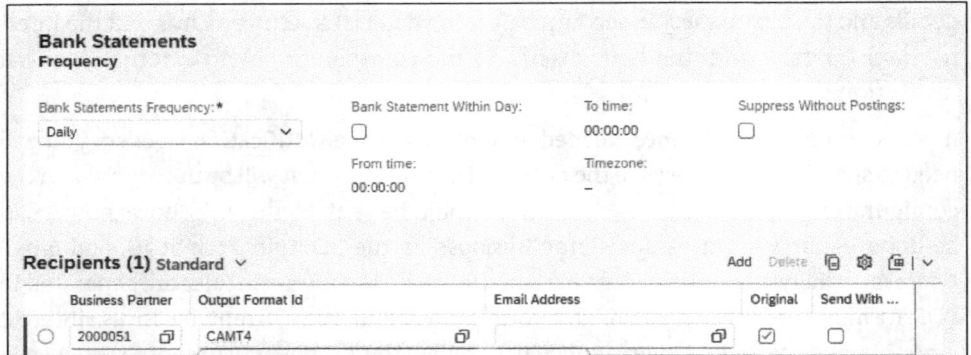

Figure 4.53 Create In-House Bank Statements: Details

Once all the necessary details are filled out, you can proceed to create the bank account, which will then trigger the approval workflow, if such a process is configured. Otherwise, the bank account can be activated immediately.

To streamline the creation of internal bank accounts, you can define a template for the **Account Options** using the Manage In-House Bank Account Templates app (F6039). Figure 4.54 shows how it can be done.

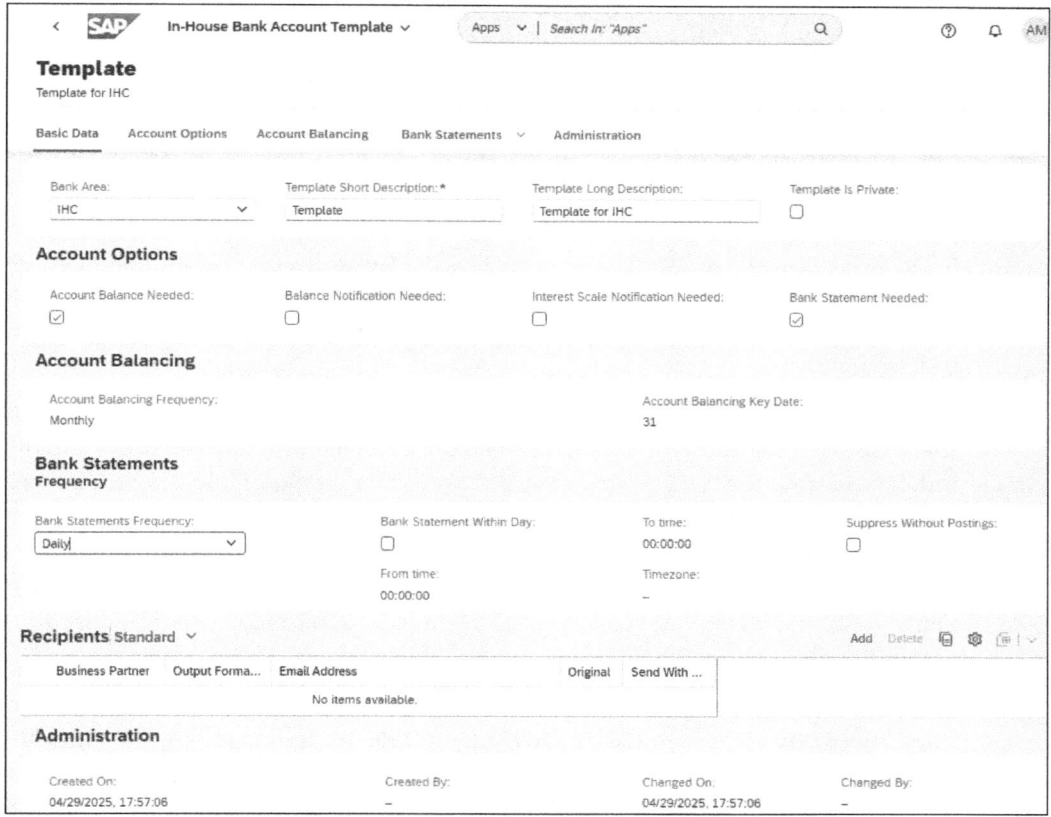

Figure 4.54 Create In-House Bank Template

This template allows you to preselect which checkboxes should be enabled—such as account balance creation, balance notifications, interest scale reports, or bank statement generation—and specify the notification formats to be used. Once this template is saved, it can be easily reused during the creation of new in-house bank accounts, ensuring consistency and saving time by eliminating the need to manually input the same information for each new account.

An important aspect to consider is the recipient for the bank statements. It's often recommended to create a dedicated recipient for each business partner who is also an account holder. To streamline the process, you can create a business partner with the recipient role defined as the target system and include this setup in a template. Then, when creating new in-house bank accounts, you simply select the template linked to the relevant system, significantly simplifying and standardizing the account creation process across your internal bank as you do not need to select a corresponding business partner for each bank account.

Create Bank Accounts Using Templates

In addition to manually creating in-house bank accounts or using eBAM via internal SAP Multi-Bank Connectivity processes, SAP also provides a convenient functionality to create bank accounts using templates. This includes the ability to download existing in-house bank accounts into an XML file and subsequently upload them from an XML file into another system. This feature is particularly useful for data migration scenarios. For example, during user acceptance testing (UAT), once all in-house bank accounts have been successfully created and validated, you can export them to an XML file and then upload this file into the production environment. This significantly reduces manual effort, ensures consistency across environments, and speeds up the transition from testing to go live.

If you haven't used the eBAM process and instead created an in-house bank account manually, it's important to remember that you also need to create the corresponding in-house bank in the system where the affiliate (account holder) is located. This is necessary to define the house bank and bank account ID locally. These elements are essential for setting up the payment configuration and for enabling the processing of bank statements in the affiliate system. Without this setup, it would not be possible to correctly route payments or reconcile bank transactions related to the in-house bank account.

When you are in the Manage In-House Bank Accounts app, you have the option to click **Download Template**, which allows you to export selected in-house bank accounts into an XML file. This functionality is particularly useful when you want to back up account data, prepare for data migration between environments (such as from testing to production), or simply reuse a predefined setup. The system will generate an XML file containing all the details of the selected accounts. Once downloaded, the XML file will be automatically saved to your browser's default download folder.

In-house bank accounts stored in an XML file can be created in the system by using the **Upload Accounts** option available within the Manage In-House Bank Accounts app, or created by directly opening the dedicated Upload In-House Bank Accounts app (F6038), as shown in Figure 4.55. Both options allow you to import multiple bank accounts efficiently, which is especially useful during data migration or mass creation scenarios. To upload the accounts, simply select your XML file from your local system and click the **Upload** button. The system will then process the file and create the corresponding in-house bank accounts based on the data contained in XML.

Once you upload the XML file, the system will process the data and allow you to create in-house bank accounts based on the uploaded information. The uploaded details—such as account numbers, currencies, business partners, and configuration settings—

will be displayed for review. You can then proceed to confirm and finalize the creation of each account directly within the app. This functionality simplifies the mass creation process and ensures consistency with previously defined account structures.

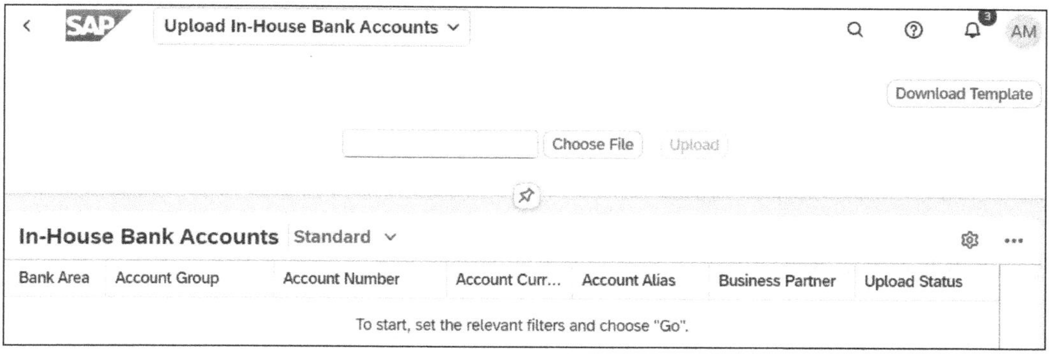

Figure 4.55 Upload In-House Bank Accounts

Manage In-House Bank Conditions

Once you have created an in-house bank account, the next critical step is defining the conditions, which refer to the interest rates that will be applied to credit balances, debit balances, and potential overdrafts, the latter being tied to the limits that will be defined in the following step. Interest calculation is a fundamental part of the in-house banking process as one of the core goals of in-house banking is to reduce reliance on external financing and instead enable internal funding. In this setup, the header entity—typically the treasury or finance institution acting as the bank area owner—functions as an internal bank. It applies interest charges to affiliates based on their balances, with a spread between credit and debit rates. The difference between these rates represents a gain for the internal treasury.

Normally, the bank area owner, as the treasury center, aims to source cheap financing within the group—by using the excess liquidity of cash-rich affiliates—and redirect it to entities needing funding. This internal funding model is not only more efficient but also cheaper than seeking financing from external counterparties, making in-house banking a strategic tool for optimizing group-wide liquidity. You can use the Manage In-House Bank Conditions app (F5941), as shown in Figure 4.56, to maintain your internal interest conditions.

Once you open the Manage In-House Bank Conditions app, you can simply click **Go** to view all existing financial conditions defined for in-house bank accounts. These conditions determine how interest is calculated for debit and credit balances. If needed, you can create a new condition or define a condition group, which allows you to group multiple bank accounts under the same set of financial rules. To create a new condition, click **Create**, and a popup window will appear prompting you to enter the key details required to define the condition, such as the group (**Group**), the type of balance it applies to (credit or debit) (**Interest Type**), and the interest terms (**Interest Calculation Type** and **Interest Calculation Method**).

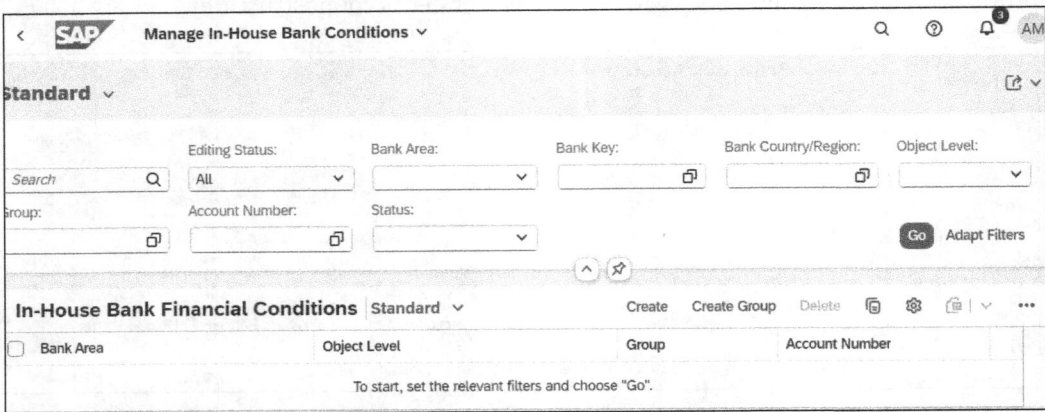

Figure 4.56 Manage In-House Bank Conditions

Once you click **Create** in the Manage In-House Bank Conditions app, the next step (see Figure 4.57) is to select the **Bank Area** for which the condition will apply.

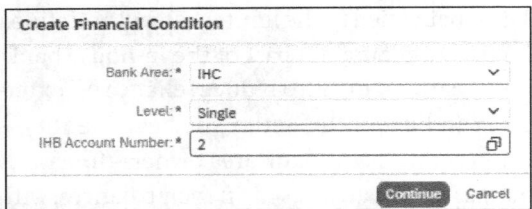

Figure 4.57 Create Financial Condition

Then, you need to define the **Level** at which the condition is being created. You have the flexibility to set it at a global level for the entire bank area, for a specific bank group, or at the individual level for a selected in-house bank account (in-house banking account). In our example, we're creating a condition for a specific **IHB Account Number**. However, it's generally recommended to define conditions at the bank area level to maintain consistency across multiple accounts. Alternatively, you can organize accounts into bank groups—for instance, based on country-specific risks, currency zones, or other strategic criteria—and apply interest conditions at that level.

Once you confirm the initial setup, the system will take you to a new view where you need to define the interest conditions. Start by clicking **Create**, and specify the **Interest Type**; typically, you will set up interest for **Credit**, **Debit**, and potentially for **Overdraft**, which will apply when an account exceeds its assigned limit (limits will be defined in the next step). Next, define the **Interest Calculation Type** and the **Interest Calculation Method**, along with the validity period for the condition. It's important to note that you can define multiple validity periods, allowing you to adjust interest terms over time as needed. Then, determine whether the interest will be calculated using a fixed rate or based on a **Reference Interest Rate**. If you're using a reference rate, you can also apply a spread to adjust the final rate under **Reference Rate Details**. Figure 4.58 shows a filled-in interest screen.

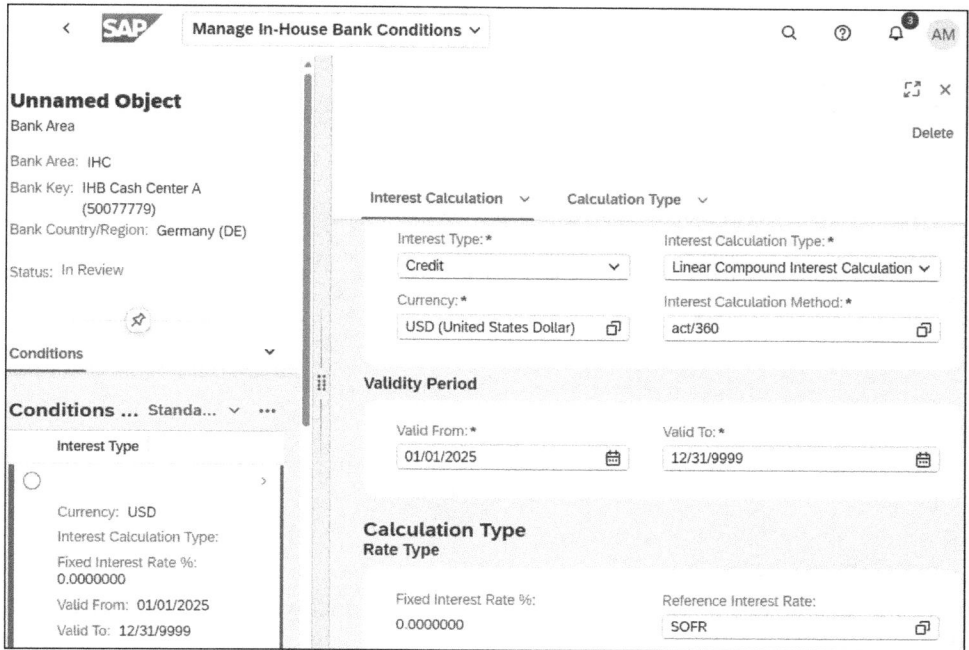

Figure 4.58 Create Interest Conditions

You'll then continue creating conditions for all required interest types. Once you are ready, save your conditions. Figure 4.59 shows an overview of the created interest conditions.

Once you create a condition, it will initially be in **In Review** status. To proceed, click the **Submit** button. If an approval workflow is defined in the system, the condition will go through the necessary approval steps. If no workflow is configured, the system will automatically activate the interest condition upon submission. Once activated, the defined interest terms will start applying to your in-house bank account as specified.

You can always review your interest conditions by navigating to the Manage In-House Bank Accounts app. Within this app, you also have the option to simulate interest calculations for a selected bank account over a specific period, which helps validate if the conditions are working as expected. Note that balances and interest calculations will only apply if the corresponding in-house bank accounts are created in BAM. This is essential because only then can you create payment items and postings—which require

4   Advanced Payment Management and In-House Banking

a valid house bank and bank account ID—to reflect transactions accurately within the system; without postings, it will not be possible to add them.

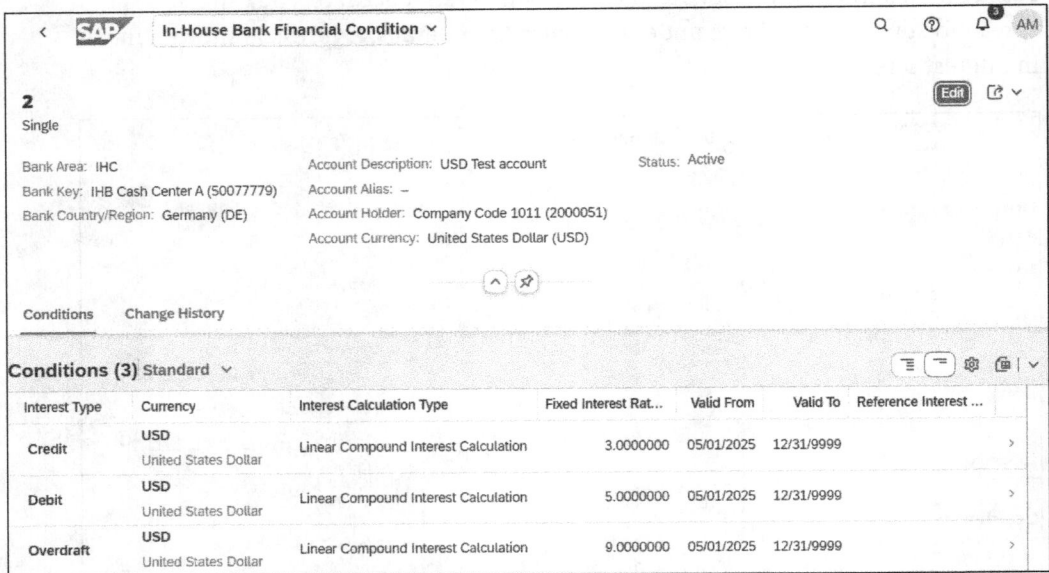

Figure 4.59 In-House Bank Financial Condition Overview

Manage In-House Bank Limits

The next important step after creating interest conditions is to define limits for your in-house bank accounts. These limits allow you to control how much selected affiliates can transact, helping to manage risk and ensure financial discipline. For example, you may want to restrict certain entities from initiating too many transactions or from exceeding specific transaction volumes or amounts. This setup is done using the Manage In-House Bank Limits app (F5940), where you can assign and monitor limits. Once you open the app, you can display and check all the limits created or you can create a new group or a new limit.

Once you click **Create** (Figure 4.60 shows the popup you will see), you will need to select the bank area for which you want to set the limit. Then, you can choose a group that was previously created, such as a risk group or country group, or you can opt to select a single account for the limit. If you select a group, on the next screen, while creating the limit, you will also need to specify the accounts that the selected limit will apply to. This flexibility allows you to tailor the limits either to specific accounts or broader groups based on the business requirements or risks associated with the transactions.

Once you click **Continue**, on the next screen (see Figure 4.61), you will need to specify the **Account Limit**, the currency of the limit, and the validity period for the particular limit. If you previously selected a group, you will also be required to provide the accounts to which this limit will apply. This ensures that the limit is accurately allocated to the correct accounts or group of accounts, helping manage the risk and control the number of transactions affiliates can create within the defined parameters.

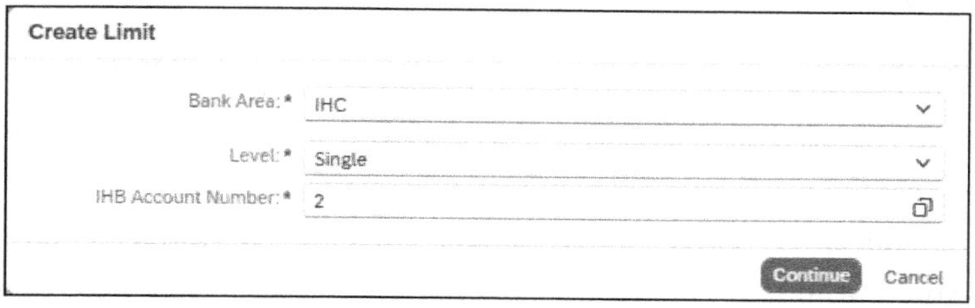

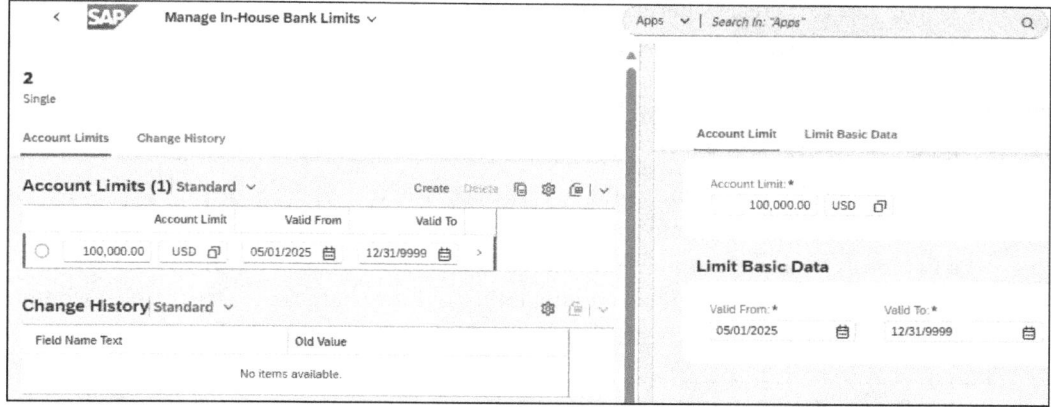

Figure 4.60 Create Limit

Figure 4.61 Create Limits Data

To create a limit, click **Save**; the system will then register the new limit in **Review** status. Next, click **Submit**. If an approval workflow is configured, the limit will go through the defined approval steps; otherwise, it will be automatically activated. Once the limit is active, it will immediately begin to apply to the specified in-house bank account(s), enforcing the transaction thresholds you've set.

Manage In-House Bank Fees

Another way to generate results for both your affiliate entity and the in-house bank owner is by creating and applying bank fees to the transactions processed on the account. You can achieve this by defining fee rules in the Manage Bank Fee Conditions app (F6335).

When defining how to recharge costs to your affiliates, whether through interest spreads or bank fees, it's essential to coordinate closely with your tax and legal teams and plan your ssstrategy in advance. You'll need to evaluate which approach works best given potential regulatory or audit constraints; applying large spreads to balance-based interest can result in cash-rich entities paying disproportionately more than high-transaction, low-balance affiliates, whereas transaction-based bank charges might offer a fairer allocation. Conversely, combining both methods can drive costs too high and may conflict with legal limits or transfer pricing rules. By engaging your tax and legal advisors up front, you can determine the optimal mix of interest and fee structures that aligns with your treasury objectives while remaining compliant and equitable across all entities.

Creating bank charges follows a similar process as creating limits and interest conditions. To begin, open the app and click **Create**. You can set up charges for the entire bank area, a specific bank, or a selected group. If you choose a group, the next screen will prompt you to select the individual accounts to which the charge will apply. Alternatively, you can create charges for a single bank account. Once you've selected your entries, the system will move to the next screen, where you can click **Create** and choose the specific charges that you want to apply to the selected bank account. This process allows you to customize charges based on your chosen criteria and apply them accordingly.

Once you click **Create**, you will need to select the fee type (see Figure 4.62). You can choose to charge an affiliate based on transaction volumes or, as in the example, apply a flat fee. For example, you might charge a flat fee of $10 per monthly subscription for each bank account. This gives you flexibility in how charges are structured—whether they are based on activity within the accounts or a fixed periodic fee.

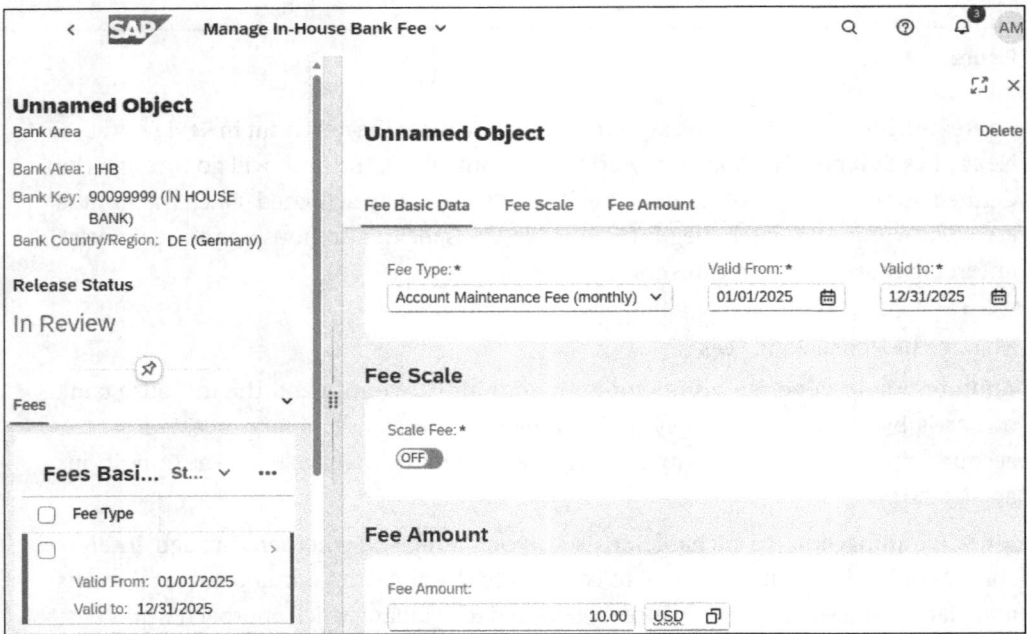

Figure 4.62 Create Bank Fees

You can also define additional fee structures—for example, per-transaction charges for domestic, international, or cross-border payments. These fees can be set as a flat rate per transaction or configured as a scale-based fee. For instance, as shown in Figure 4.63, you might specify that for a given period, affiliates pay 0.50 USD per transaction for the first 1–99 transactions, and 1.00 USD per transaction thereafter. You can apply multiple fee types to each bank account, allowing the header entity to recover costs incurred—such as those for external POBO transactions—by recharging affiliates accordingly. This flexibility ensures that transaction fees reflect actual processing costs and can be tailored to your organization's treasury strategy.

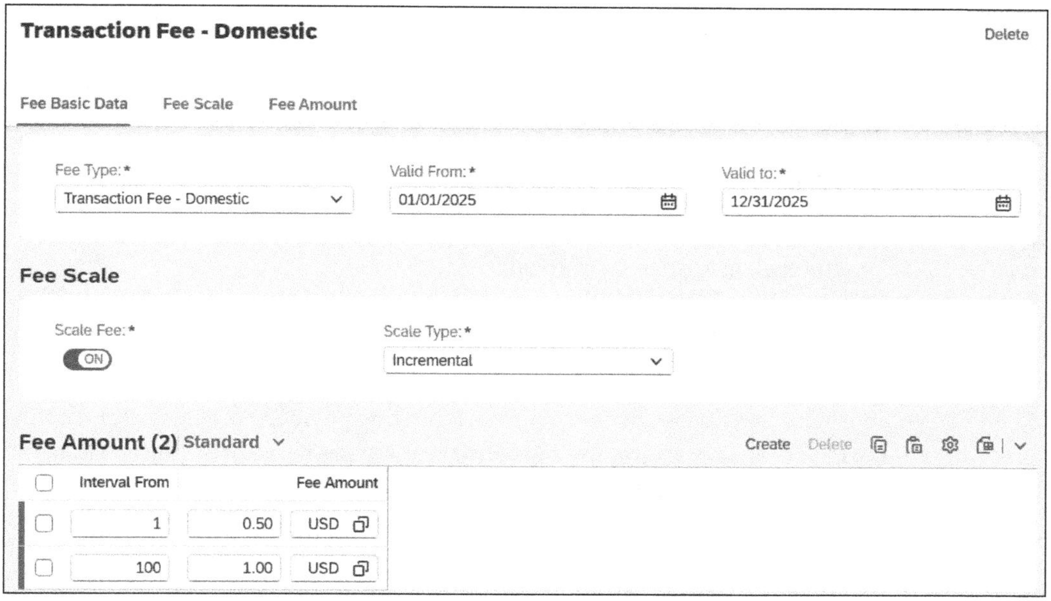

Figure 4.63 Create Incremental Fee

Once you have defined all your fee conditions, click **Create**. The new conditions and bank fees will appear in **Review** status. Next, click **Submit**. If an approval workflow is configured, the system will initiate the workflow; otherwise, the fees will be automatically activated. Once activated (see Figure 4.64), the fees will be applied to your in-house bank accounts during the end-of-day processing, ensuring they're included in that day's financial calculations.

Once you have completed your master data setup—meaning your in-house bank accounts are created, the corresponding BAM accounts with house bank and bank account IDs are in place, and all your conditions (limits, interest conditions, and bank fees) are defined—you're ready to start using those in-house banking accounts to generate and process payment transactions.

4 Advanced Payment Management and In-House Banking

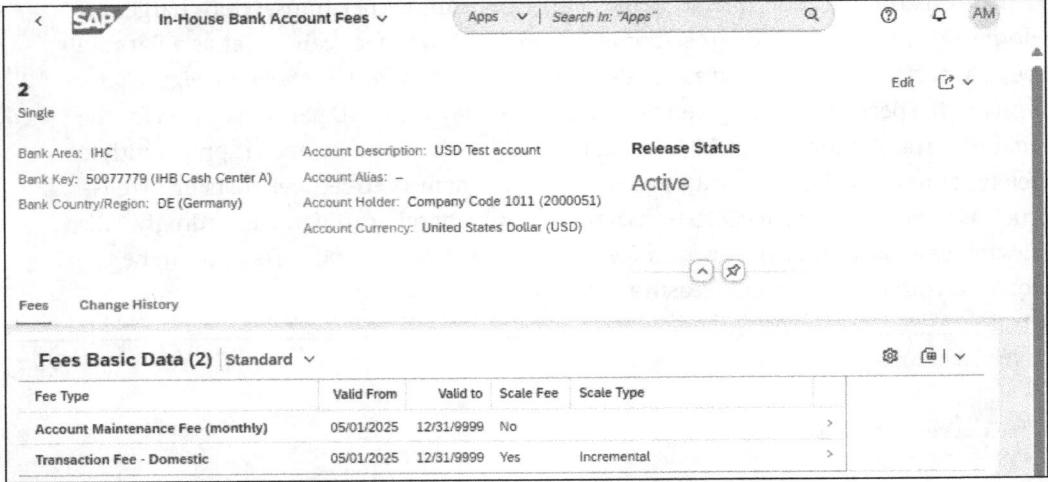

Figure 4.64 Bank Account Fees Activated

4.2.4 Payments: In-House Operations

In this section, we will delve into some of the most widely used payment scenarios within in-house banking. Our focus will be on internal payments, which streamline intercompany transactions without relying on external payment channels, and POBO payments, which enhance efficiency by consolidating payment processes for multiple entities. These scenarios exemplify the versatility and precision of in-house banking in managing complex cash operations across a global organization.

Payments on Behalf of

In the case of POBO, it is essential to have the external vendor properly maintained in the system. This includes creating the vendor master data and ensuring that their bank account information is correctly set up in the vendor master record. However, the payment itself must be executed from the in-house banking account that belongs to the affiliated company, while the target bank account remains the vendor's account.

After the POBO payment is processed within advanced payment management, the system automatically replaces the paying bank account with the one belonging to the treasury entity that is executing the payment on behalf of the affiliate. This allows for centralized payment processing, while maintaining full transparency of the original business transaction and preserving the correct accounting between entities.

We recommend running payments daily in the background or running them on scheduled timelines (e.g., once per week for the whole of the next week) to optimize and streamline payments processing. However, to execute a payment manually, open the Manage Automatic Payments app (F0770) and create parameters (see Figure 4.65). Enter the **Posting Date** (date on which payment document will be posted), **Docs Entered Up**

4.2 In-House Banking Functions and Processes

To (documents posted after this date will not be considered), **Company Code** (paying entity), **Next Payment Date** (date on which you expect the next payment to be executed), **Payment Method**, and other criteria, like suppliers and document numbers (not recommended for the automatic run). Everything should be paid for.

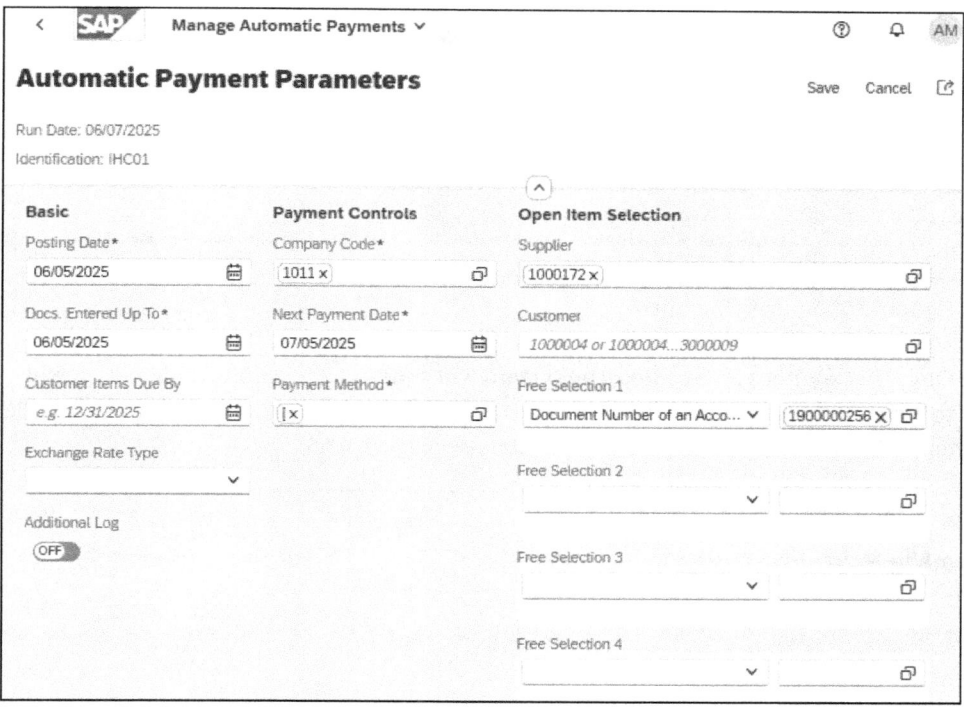

Figure 4.65 Create Parameters for POBO Payment Run

Once you click **Save**, the system will create parameters, and you will be able to create a proposal (for more info, see Chapter 7 on the payment run). Once you have parameters or proposals created, highlight a line and click **Schedule Payment** (see Figure 4.66).

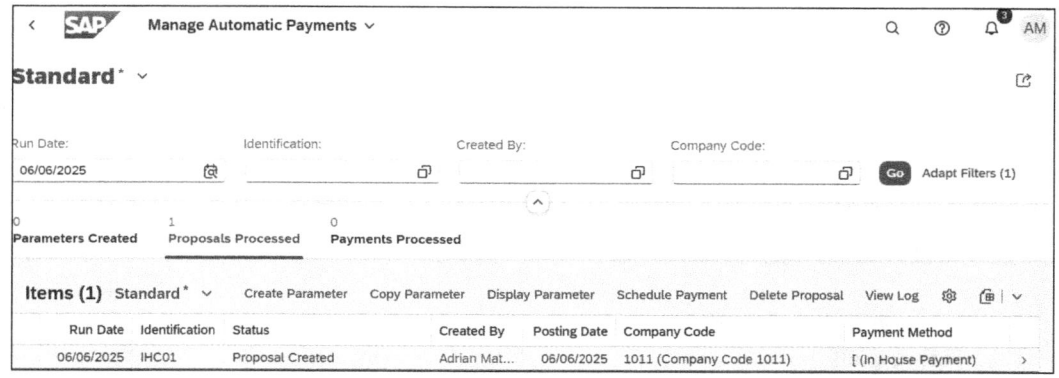

Figure 4.66 Schedule POBO Payment

283

Once you schedule a payment, the system will execute it. You will see an update in the payment status and will be able to see details of your payment (see Figure 4.67) by clicking the payment line.

Figure 4.67 Payment Executed

Once the payment is executed, the system will generate the pain.001 file, which will go to advanced payment management. You can open the Manage Incoming Payment Files app (F1680) to see your payment executed (see Figure 4.68).

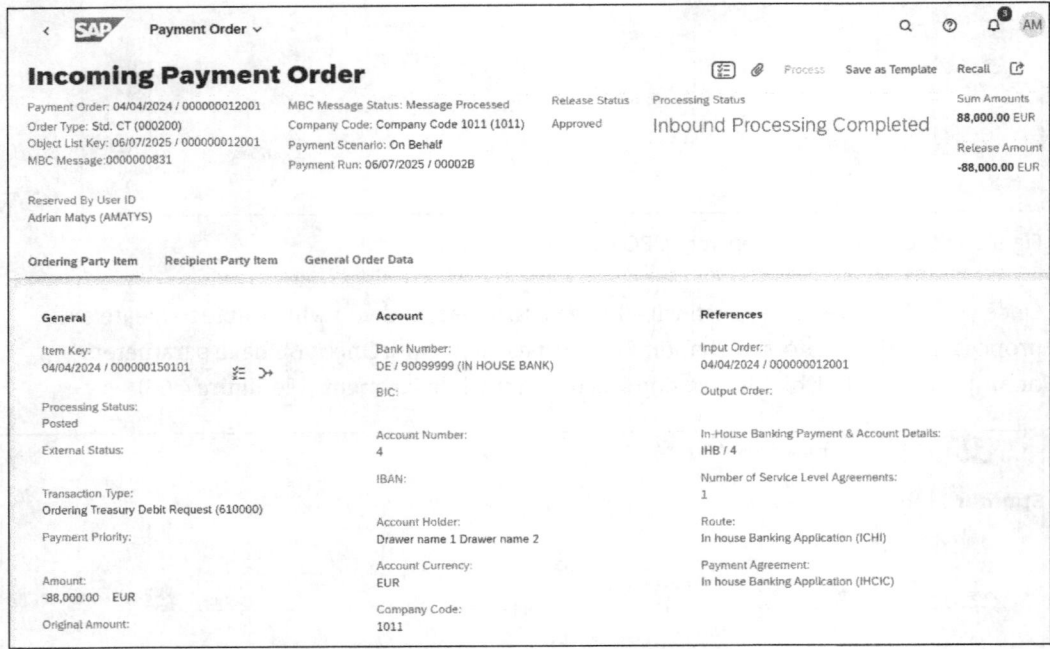

Figure 4.68 Payment File Received in Advanced Payment Management

When you click the payment line, you can display your incoming payment order (see Figure 4.69).

4.2 In-House Banking Functions and Processes

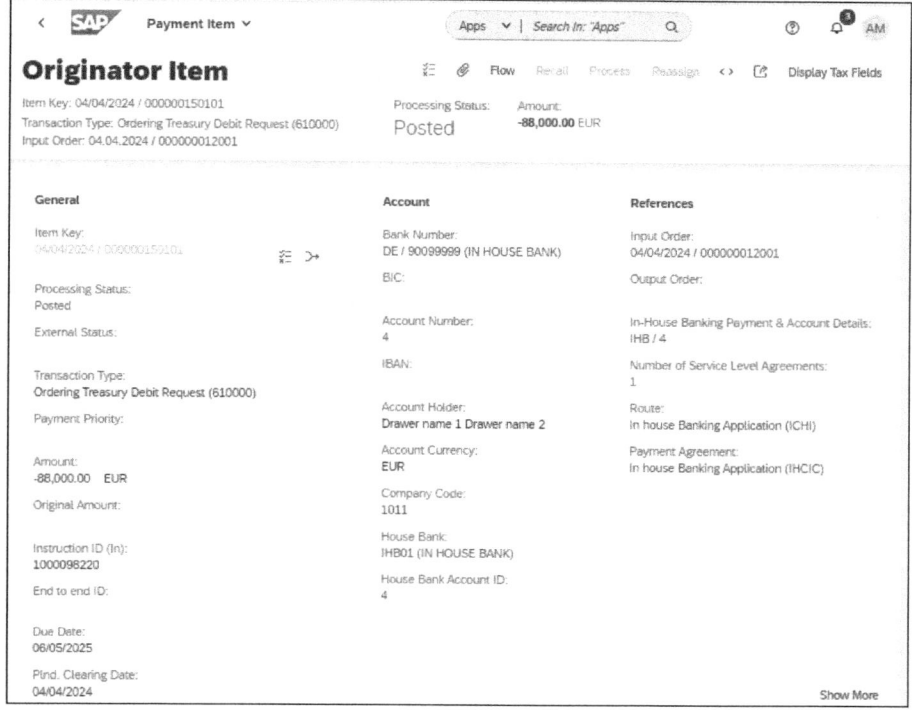

Figure 4.69 Incoming Payment Order

Note the bank number below the account, showing that payment was processed using an in-house bank account and thus that the posting will be done in the in-house bank subledger.

If you click the blue number below the item key, the system will show you details about the item key and the information passed to in-house banking (see Figure 4.70).

Figure 4.70 Advanced Payment Management Payment Item

4   Advanced Payment Management and In-House Banking

You can go back and check the SAP Multi-Banking Connectivity message generated for processing this payment (see Figure 4.71). Note that, as will be discussed in Chapter 5, all payments, even internal ones relevant to in-house banking, must be processed through SAP Multi-Bank Connectivity. Even when you do not use it as a connection option to your bank, it is still used in the background.

Figure 4.71 Manage Bank Messages: Processed POBO Payment

Based on the configuration, the system knew that this is a payment and already prepared an outgoing payment order and associated payment batch. In the Manage Payment Batches app (F4039), you can select the batch created for your POBO transaction (see Figure 4.72).

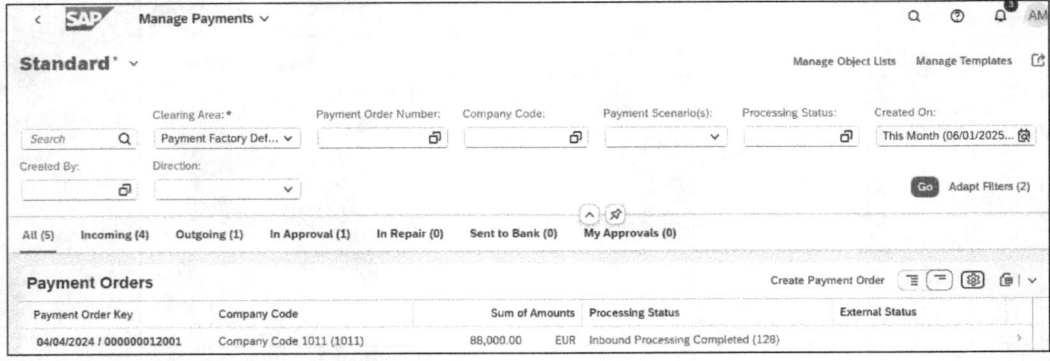

Figure 4.72 POBO Payment Batch Created

When you click the **Process** button, a payment batch will be created (see Figure 4.73) and will go through the payment approval process (see Chapter 7 for more information about payment approvals). Note that in the example, more than one payment was batched together. This is the idea of the batching process, to reduce the number of payments sent to the bank.

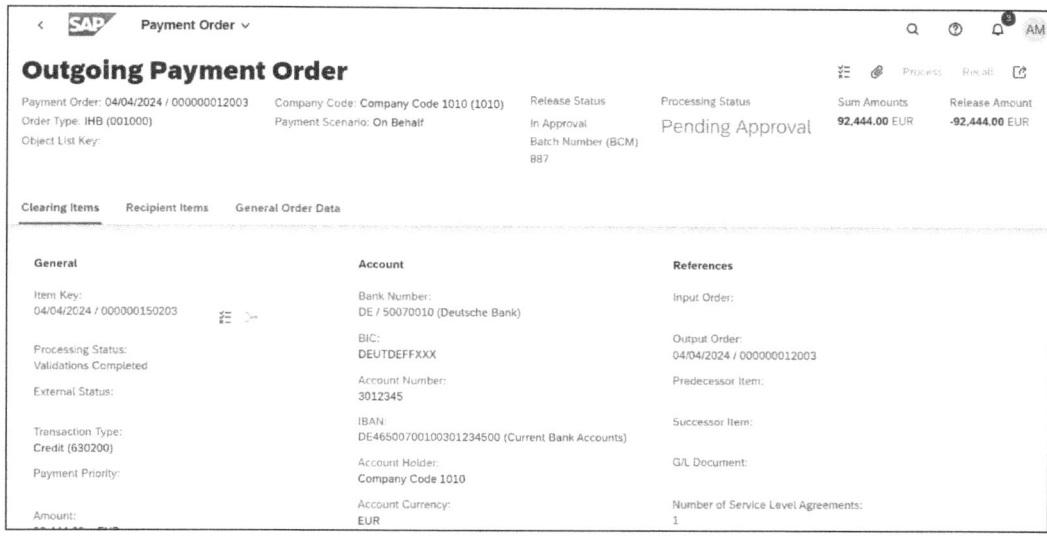

Figure 4.73 SAP Bank Communication Management: Outgoing Payment Approval

This time, the payment file includes the paying account of the header entity—the bank area owner—who is executing the payment on behalf of the other entities (see **Bank Number**). Once payment is approved and a payment file is generated, the payment file is sent to the bank to initiate and execute the actual payment. As part of this process, a POBO posting is created to reflect the resulting intercompany payables and receivables, ensuring that the financial obligations between the involved entities are properly accounted for.

In the COBO scenario, when you initiate a direct debit on behalf of another entity, the process functions similarly to the POBO model. However, instead of using a payment file like pain.001, you must generate a pain.008 file, which is specifically designed for direct debit transactions. This file format ensures that the direct debit is correctly processed and aligned with SEPA or other applicable standards.

Figure 4.74 shows in-house banking postings in the in-house banking account. You can check it in the Manage In-House Bank Payment Items and Manage In-House Bank Balances apps.

Overall, POBO processing can be fully automated, but you can check postings made in the in-house banking subledger and granted payment approvals.

4   Advanced Payment Management and In-House Banking

Figure 4.74 POBO Payments in In-House Banking Subledger

Internal Payments

Internal payment processing is very similar to POBO processing. For internal payments, it is necessary first to create the intercompany entities, customers, and vendors as business partners, then create their payment details in an internal in-house banking account. This applies to both internal counterparties and external counterparties for POBO or COBO payments; advanced payment management, once it receives the payment file (like POBO, you need to post the invoice first and execute the payment run), will know how to interpret the payment and which transaction should be used. Figure 4.75 shows which payment details you need to set up for the business partner; you can set up the payment details using the Manage Business Partner app. Like other processes, under **Payment Transactions**, you need to specify a unique ID (in the **ID** field), the country where bank account is created (**C/R** field), a **Bank Key**, a bank account (**Bank acct** field) that represents the in-house bank account number, and potentially an **IBAN**; control keys are not used for internal bank accounts.

Once this setup is complete, you can post an invoice to the intercompany entity and proceed with the payment run, exactly as you would with any external invoice. The key difference lies in the payment process: Instead of sending the payment externally to a bank, it is processed internally using the configured in-house banking accounts, ensuring that the internal transfer is handled entirely within the system without involving any external payment channels.

4.2 In-House Banking Functions and Processes

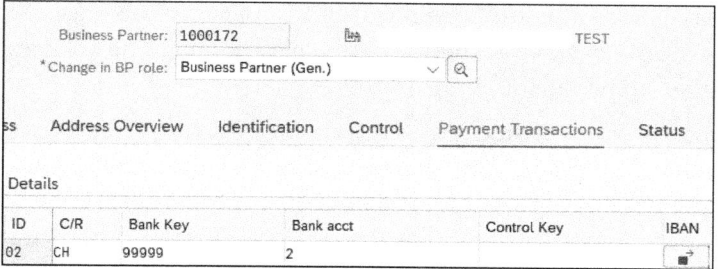

Figure 4.75 Business Partner Payment Details

There's a key difference to note here between the legacy SAP In-House Cash system and the current in-house banking setup. In the past, with SAP In-House Cash, it was necessary to configure special payment methods involving EDI to generate IDocs, set up partner profiles, and process the IDocs through the system. In contrast, with advanced payment management, you generate a standard XML file (pain.001), just as you would for external payments. The only difference is that for internal payments, the paying account must be an SAP In-House Cash account, and the receiving account must also be an SAP In-House Cash account. Based on this configuration, advanced payment management will automatically identify, upon receiving the payment file in the Manage Incoming Payments app, that this is an internal payment and will process it accordingly.

Once generated, the pain.001 file will be received in the Manage Payments app within advanced payment management, including the internal payment scenario (see Figure 4.76).

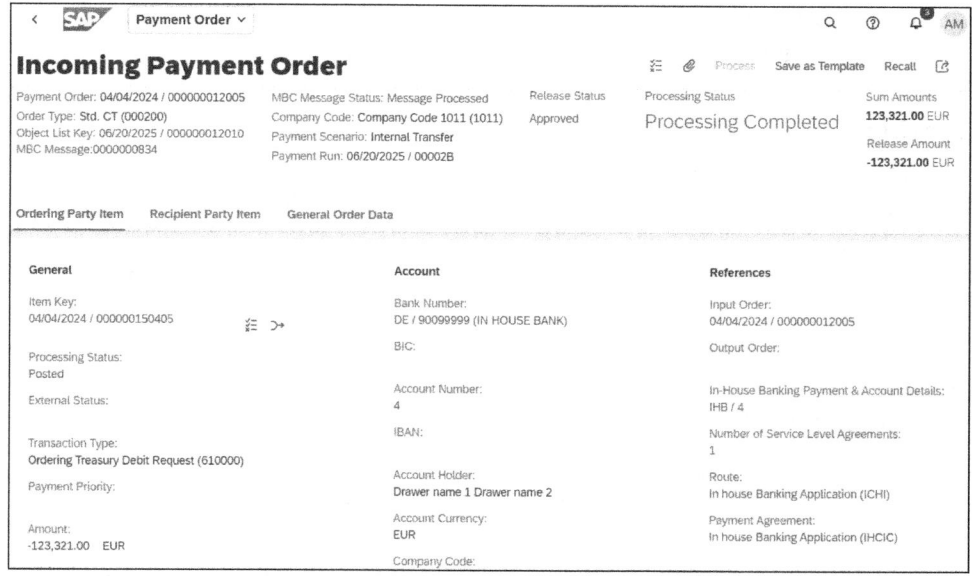

Figure 4.76 Internal Transfer Payment Order in Advanced Payment Management

289

This will trigger the creation of an in-house banking payment order and the corresponding postings on the in-house banking accounts—both the incoming and outgoing sides—exactly as specified in the pain.001 transfer details.

The defined instructions and payment order will also result in postings in the respective in-house banking account. These postings will later be used to update account balances. They will also be reflected in the in-house banking statements and will be used for interest calculation and the generation of derivative transactions. Figure 4.77 shows an example of the internal payment in the in-house banking subledger.

Figure 4.77 Internal Payment in In-House Banking

Internal payments, as previously mentioned, are used only for in-house banking accounts. The main idea is not to reduce the number of internal payments, but rather to better manage internal liquidity and optimize internal cash operations.

Check In-House Banking Balances

Once all in-house banking payments have been generated, the Manage In-House Bank Account Balances app (F6341) can be used to view and monitor the balances of the in-house banking accounts. This app provides a clear overview of all internal bank account activity, allowing users to verify that the in-house banking balances (see Figure 4.78) are accurate and up to date. These balances should align with the intercompany payables and receivables recorded in the balance sheet, which are reflected through G/L transfers and in-house banking internal bank statements. Ensuring this consistency is essential for maintaining accurate intercompany reconciliation and financial reporting.

The report in the Manage In-House Bank Account Balances app offers drilldown functionality, allowing users to navigate through different levels of account hierarchy. Initially, the report displays balances by **Bank Area**, followed by the **Account Groups** that were defined earlier, and then the individual bank accounts. Like the Cash Flow Analyzer app, you can customize the view by adjusting settings and changing the grouping of

4.2 In-House Banking Functions and Processes

accounts—for example, to create a drilldown based on currency. These custom views can be saved as variants for quick access in the future.

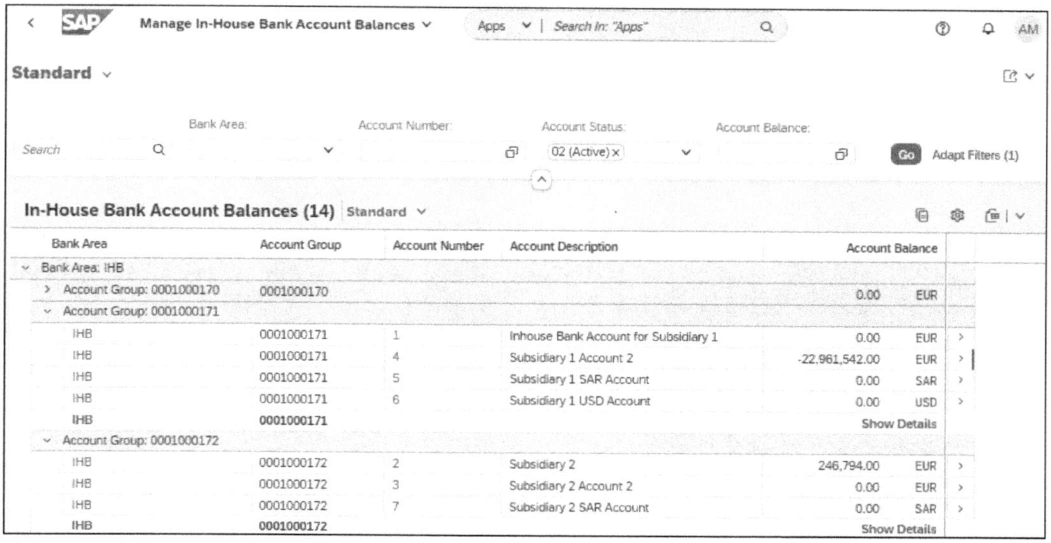

Bank Area	Account Group	Account Number	Account Description	Account Balance	
Bank Area: IHB					
> Account Group: 0001000170	0001000170			0.00	EUR
∨ Account Group: 0001000171					
IHB	0001000171	1	Inhouse Bank Account for Subsidiary 1	0.00	EUR >
IHB	0001000171	4	Subsidiary 1 Account 2	-22,961,542.00	EUR >
IHB	0001000171	5	Subsidiary 1 SAR Account	0.00	SAR >
IHB	0001000171	6	Subsidiary 1 USD Account	0.00	USD >
IHB	0001000171			Show Details	
∨ Account Group: 0001000172					
IHB	0001000172	2	Subsidiary 2	246,794.00	EUR >
IHB	0001000172	3	Subsidiary 2 Account 2	0.00	EUR >
IHB	0001000172	7	Subsidiary 2 SAR Account	0.00	SAR >
IHB	0001000172			Show Details	

Figure 4.78 Manage In-House Bank Account Balances

In addition, when a specific account is selected (see Figure 4.79), the report provides detailed information about the balance, including underlying line items and data on the internal bank statements that have been generated. This enables you to analyze in-house banking activity comprehensively and efficiently.

In the detailed view of the Manage In-House Bank Account Balances report, you can interact with the data by clicking the blue-highlighted amounts to explore transaction-level details. By clicking the **Item Number** within a transaction, you can access the specific payment items that were created in advanced payment management. From this view, it is also possible to display the corresponding journal entries and review the internal bank statements that were generated. This level of transparency and traceability supports effective reconciliation and provides a complete audit trail for each in-house bank transaction.

The corresponding in-house banking payment items can be accessed directly from the transactions in the detailed view or, alternatively, by using the Manage In-House Bank Payment Items app (F6339; see Figure 4.80). This app allows you to display all in-house banking payment items that have been posted. After you enter the relevant selection parameters—such as **Bank Area, Account**, or **Date Range**—and click **GO**, the system retrieves and displays all matching payment items. You can then select any individual item to view its full details, providing a clear and structured way to analyze and trace internal payment activity.

4 Advanced Payment Management and In-House Banking

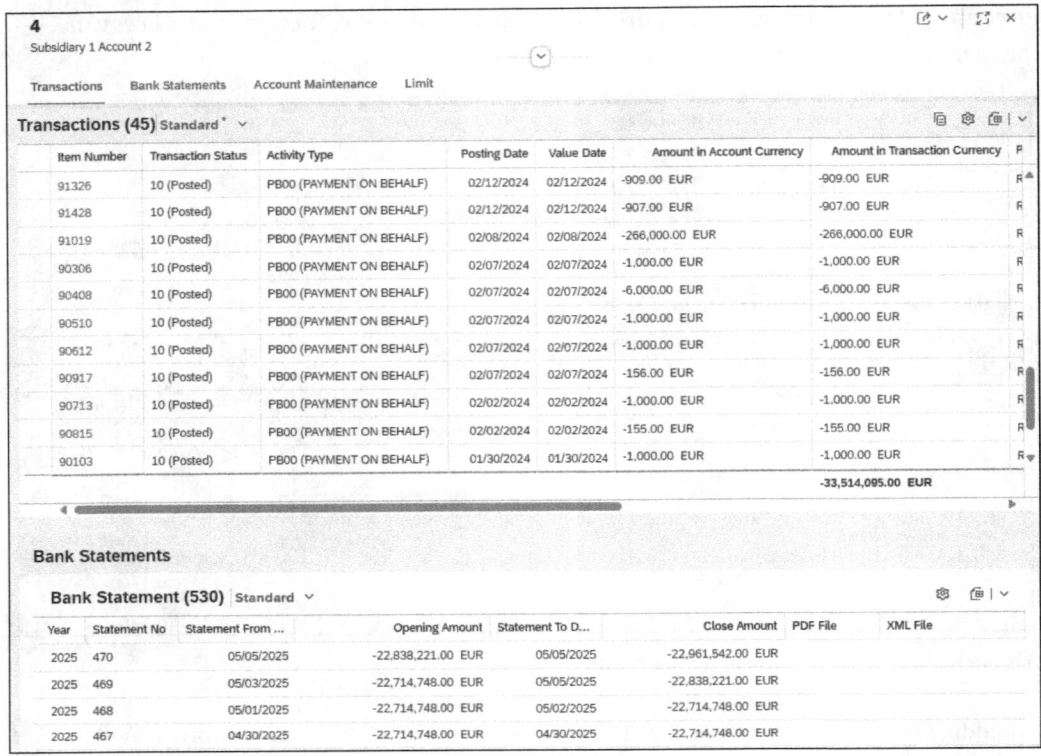

Figure 4.79 In-House Banking Balance Detailed View

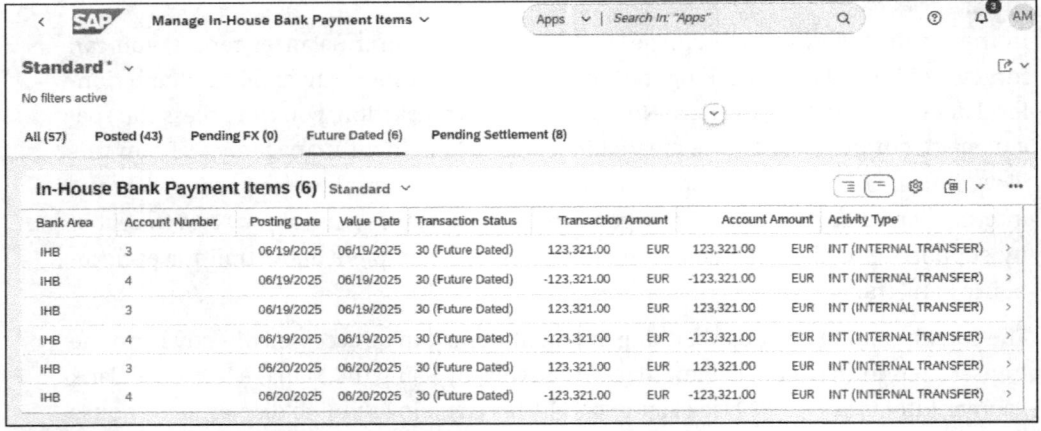

Figure 4.80 Manage In-House Banking Payment Items

End-of-Day Process

The in-house banking end-of-day process in SAP S/4HANA is a critical step to ensure accurate and complete financial processing within the in-house bank framework. It consolidates all internal payment activities, updates account balances, and ensures that intercompany payables and receivables are properly recorded through internal bank statements and corresponding journal entries. This process is essential for maintaining the integrity of internal financial data, supporting timely reconciliation, and ensuring that in-house banking accounts reflect the true financial position at the close of each day. Without the end-of-day process, key transactions may remain unposted, leading to inconsistencies between actual cash positions and reported balances in the G/L.

Historically, in SAP In-House Cash, it was only possible to run one end-of-day process per day. This limitation posed challenges for global organizations operating across different time zones as the timing of internal bank statements often didn't align with regional processing and cutoffs. The new in-house banking functionality in SAP S/4HANA addresses this issue by allowing multiple end-of-day processes and the generation of multiple internal bank statements throughout the day. This enhancement significantly improves the flexibility and timeliness of internal reconciliations and accounting, enabling more accurate and efficient financial processing across global operations.

You can use the Manage In-House Bank Application Jobs app (F6653) to schedule and monitor various in-house banking-related jobs, including the end-of-day process. This application provides a centralized and user-friendly interface to define job parameters, set execution times, and automate the regular processing of in-house banking functions.

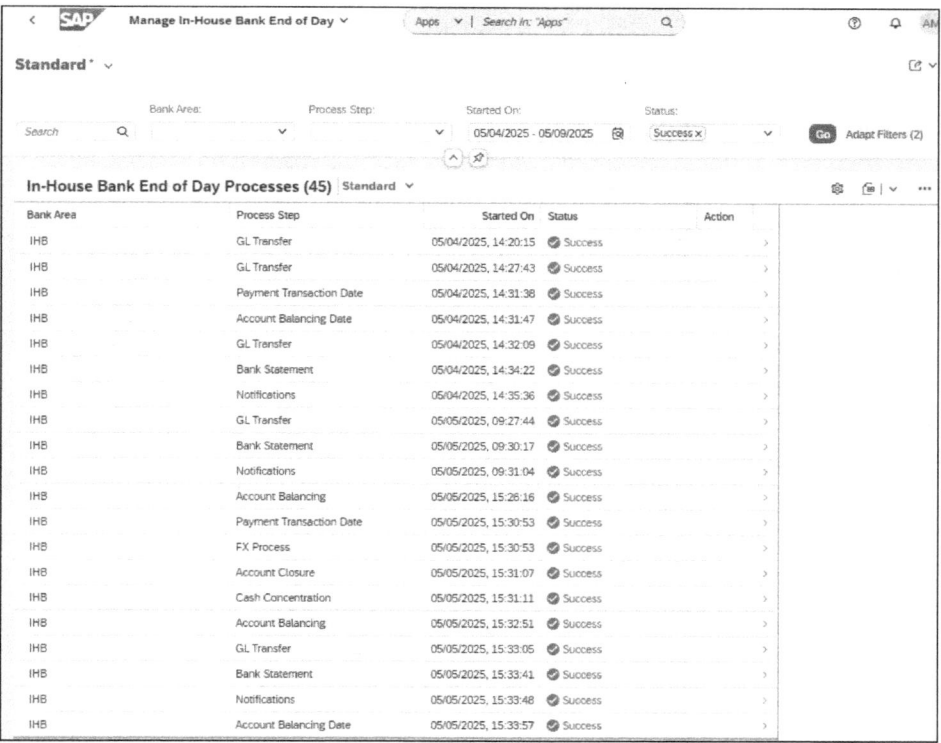

Figure 4.81 Manage In-House Banking: End of Day

The Manage In-House Bank End of Day app (F6338) allows you to monitor the status and processing steps of in-house banking end-of-day jobs (see Figure 4.81). Within this app, you can select various parameters—most commonly, the date on which the end-of-day was run or the status of the jobs (e.g., successful, in progress, or failed). After entering your selection criteria and executing the search, the system will display a list of all executed end-of-day jobs that match your filters. This provides a clear overview of job execution, helping you track progress and troubleshoot any issues that may arise during the end-of-day processing.

When you click any job in the Manage In-House Bank End of Day app, you will see detailed information about the status of the job and the associated program that was executed. This view helps you understand which specific step was performed, whether it was completed successfully, and if any errors occurred during processing. It provides transparency into the execution flow of the in-house banking end-of-day process and supports efficient monitoring and troubleshooting.

Programs associated with and required for in-house banking that are triggered during the end-of-day run are listed in Table 4.4.

Transactions	Description	Usage
PF1/IHB_ACC_CLOSE	Close Account	Used for account closure
PF1/IHB_ACC_FX	Process FX	Process foreign exchange (FX) differences coming from the in-house banking balances
PF1/IHB_ACCBAL	Account Balancing	Calculate limits, interests, bank charges and withholding tax
PF1/IHB_BK_STMNT	Create Bank Statement	Creates bank statements
PF1/IHB_FI	Transfer Payment Items to FI-GL	Post intercompany balances in the G/L
PF1/IHB_INIT_DATE	Initialize IHB EOD Dates	Initialize date for the in-house banking end-of-day
PF1/IHB_NOTIFI	Create Notification	Create and send balance notifications
PF1/IHB_BK_STMNT	Create Bank Statement	Create and send bank statement
PF1/IHB_SET_CLOSE_D	Set Account Balancing Date	Set up date for the end-of-day balance
PF1/IHB_SET_TRANS_D	Set Payment Transaction Date	Set up payment transaction date

Table 4.4 Programs Triggered by In-House Banking End-of-Day Processing

4.3 Configuration

In this section, you will gain a comprehensive understanding of the necessary configuration steps relevant for advanced payment management and in-house banking. As highlighted in previous sections, the design principle underlying this functionality is that many activities are maintained directly in SAP Fiori apps as master data. However, in addition to these SAP Fiori–based master data maintenance activities, there are also multiple configuration steps that must be carefully carried out to tailor the system to your organization's needs. This section aims to provide clear guidance on these configuration steps, ensuring a thorough understanding of how to effectively enable and integrate advanced payment processing and in-house banking structures within your SAP S/4HANA system.

4.3.1 Advanced Payment Management

Overall, SAP aims to significantly reduce customizing effort and make the configuration of its solutions as user-friendly as possible for both key and end users. This is particularly true for advanced payment management. With each new release, SAP offers new functionalities, especially in the public cloud, including SAP Fiori apps like Manage Payment Blocks and Exception Control, as well as the workflow control, which has gradually transitioned from traditional customizing to an SAP Fiori–based solution that can be easily operated by key users.

This section introduces customizing for advanced payment management. We will focus on the essential elements for deploying the functionality and offer insights into the underlying mechanisms.

> **Preconfigured Settings**
>
> Unlike most areas of SAP S/4HANA, most settings for advanced payment management are already preconfigured by SAP. This provides the advantage of being able to refer directly to SAP's initial settings and adapt or simply copy them for your own purposes. As a result, the implementation time for advanced payment management is significantly reduced.

In the following sections, we will delve into the most relevant basic settings for advanced payment management, found under menu path **Financial Supply Chain Management • Advanced Payment Management**:

- Basic Configuration
- Payment Processing
  - Enrichment & Validations

- **External Interfaces**
  - Bank Communication Management (BCM)
  - File Handler
    - Inbound Converter
    - Outbound Converter

Basic Configuration

The basic settings come with a preconfigured setup that can almost be used directly as is. However, within the basic configuration, there are additional options, such as setting up load balancing. This is a significant advantage over advanced payment management's predecessor, SAP In-House Cash; with advanced payment management, you have far more options to control performance technically. We'll cover the following topics in this section:

- **Clearing area**
  Here, we introduce one of the key customization points in advanced payment management. The clearing area serves as the central hub for payment transactions in the system and is linked to a variety of settings. We present the essential configuration aspects here.

- **Assign internal bank keys to clearing area**
  This transaction links the virtual bank, which must also be maintained within BAM, with the clearing area.

- **Integration of cash management reporting**
  External systems can be connected to advanced payment management; this also applies to internal areas, such as cash management.

- **Maintain generic number ranges**
  In advanced payment management, various processes are executed, and each is assigned its own number range. In this section, we introduce how to maintain these ranges and link them to your clearing circle.

Clearing Area

Setting up the clearing area is the first and most fundamental step in the configuration of advanced payment management. It can be defined in the IMG under **Financial Supply Chain Management • Advanced Payment Management • Basic Configuration • Organization • Define Clearing Areas**.

All transactions and structured data processed within advanced payment management are inherently linked to a clearing area; this applies not only to processing but also to authorizations and reporting. Much like the company code in SAP S/4HANA Finance or the bank area in in-house banking, the clearing area serves as the core framework for transaction processing, ensuring data separation and organization.

Each clearing area is governed by specific settings to define the transaction processing. You must define the six-character abbreviation for the clearing area, which can consist of both letters and numbers. This abbreviation serves as a key field and is required in most subsequent customizing steps. Figure 4.82 shows the configuration of a clearing area for the US market.

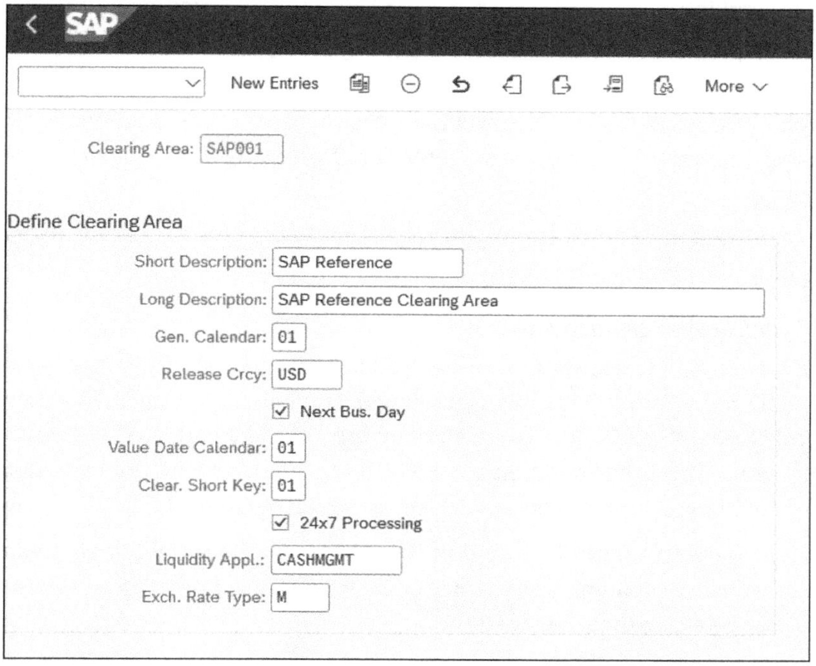

Figure 4.82 Define Clearing Area

Here you can see the following essential fields:

- **Short Description**
  The short description can be up to 15 characters long and can be freely defined. This field is primarily used for display purposes and for reporting.
- **Long Description**
  The long description can be up to 40 characters long and can be freely defined. This field is primarily used for display purposes and for reporting.
- **Gen. Calendar**
  This calendar defines the processing days, as well as the booking and value dates of the clearing area.
- **Release Crcy**
  Approvals in advanced payment management, particularly approval limits, are converted into this currency. This setting determines which currency takes precedence in the approval processes.

- **Next Bus. Day**
  Checkbox to determine whether the previous or the next business day is used to update the processing date when the processing date is not a valid working day.

- **24x7 Processing**
  This flag is used to determine whether the clearing area should generally process payments directly in online mode or operate in batch mode. If you need direct or online processing, check this box.

- **Liquidity Appl**
  This setting is optional and can be used when the clearing area needs to be connected to another application, such as cash management reporting.

- **Exch. Rate Type**
  This function is used to define a standard exchange rate type for foreign exchange translation into the release currency.

Assign Internal Bank Keys to Clearing Areas

For each bank area in SAP In-House Cash as well as in-house banking, just like a real bank, a national ID and a unique bank key are assigned. This bank key, listed in the **Bank Key** column in Figure 4.83, is linked to the clearing area (**Clrng Area**) and an account proxy category within this transaction. To connect the clearing area with the bank area, the **AMCategory** field must be set to **Inhouse Cash**, as shown in Figure 4.83.

The configuration transaction can be found in the IMG under **Financial Supply Chain Management · Advanced Payment Management · Basic Configuration · Organization · Assign Internal Bank Keys to Clearing Areas**.

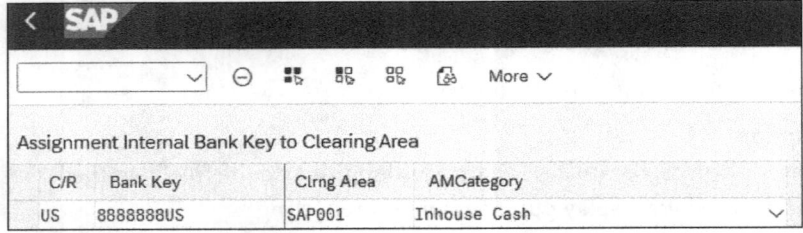

Figure 4.83 Assign Bank Key to Clearing Area

Integration of Cash Management Reporting System

Because advanced payment management can function as a central hub for group-wide payment transactions, it also has access to relevant datasets, which are particularly useful for cash management. Cash management–relevant movements, accurate to the value date, can be transferred to SAP S/4HANA Finance for cash management. Alternatively, external cash management systems also can be integrated.

To integrate SAP S/4HANA Finance for cash management with advanced payment management, the application must be defined in the customizing linked with the clearing area. This can be configured via the IMG at **Financial Supply Chain Management** • **Advanced Payment Management** • **Basic Configuration** • **Application Configuration** • **Define Local Application Management Areas**.

> **Other Systems**
>
> Through the **Define Local Application Management Areas** activity, settings are configured not only for cash management but also for SAP Business Integrity Screening or other systems you wish to connect to advanced payment management. At this point, these external applications are linked with the clearing area.

Figure 4.84 illustrates a sample configuration of the cash management application, showing, in this example, US clearing area **SAP001**.

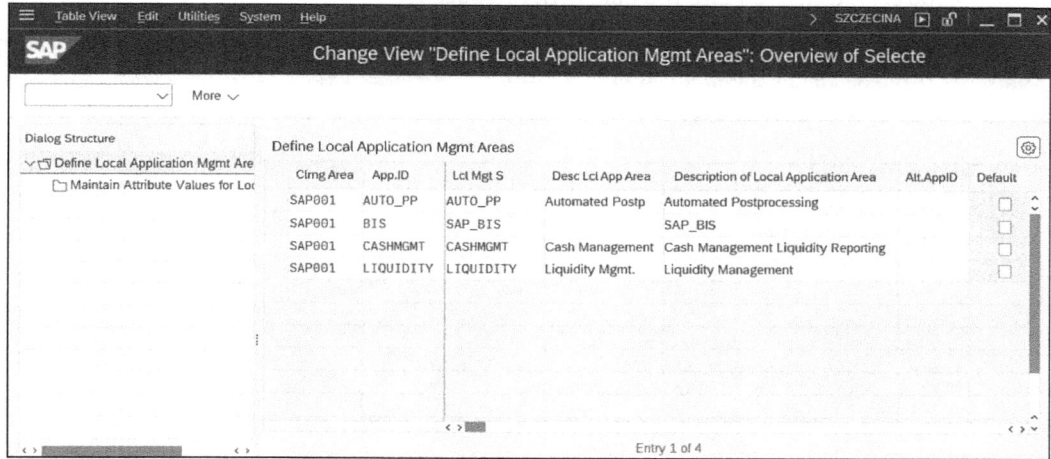

Figure 4.84 Define Local Application Management Areas

You can see the following relevant fields:

- **Clrng Area**
  In this field, the clearing area is stored, linking it to the application.
- **Application ID (App.ID)**
  This field is freely configurable; here, an abbreviation for the local application can be defined.
- **Local management system (Lcl Mgt S)**
  In this field, SAP Cash Management (**CASHMGMT**) and Liquidity Management (**LIQUIDITY**) can be selected for integration.
- **Short description of local application area (Desc. Lcl App Area)**
  This field is used for a short description (15-character max) of the local application area.

- **Description of Local Application Area**
  This field provides up to 40 characters for a description.
- **Alternative local application area ID (Alt.AppID)**
  This field is optional and only needs to be filled in if you need an alternative ID for this application.

Through the **Maintain Attribute Values for Local Application Areas** dialog, you can define detailed values for cash management (see Figure 4.85). This involves the planning type for creating memo records in cash management for the designated **Clearing Area**. Double-clicking on **Maintain Attribute Values for Local Application Areas** will take you to the parameter settings. Here you have the option to add cash management–relevant attributes, such as planning type for cash movements that are generated from advanced payment management. For this purpose, the two-character planning type code can be entered into the field **Attr.Val. Post. Item**.

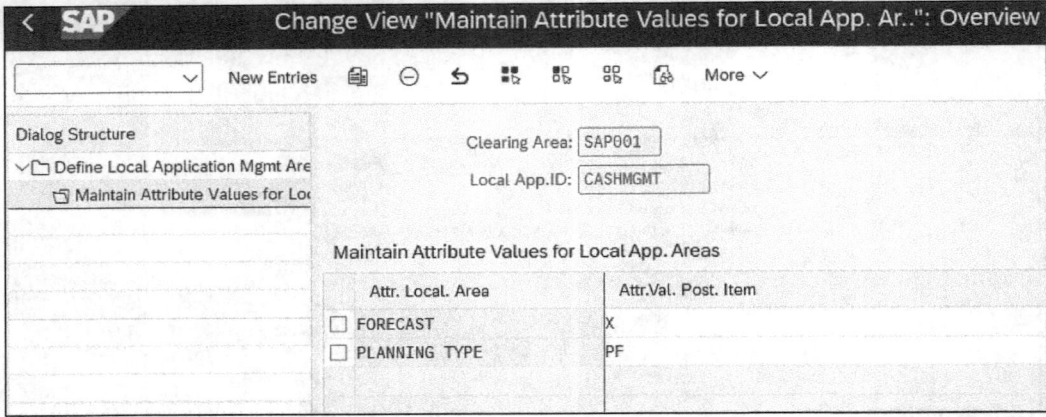

Figure 4.85 Maintain Attribute Values for Cash Management

After the correct configuration, memo records are generated in the local cash management system—for example, when payment instruments are imported and processed from external systems. Information such as the amount, value date, payer, and payee is transmitted to cash management.

Maintain Generic Number Ranges

The number range numbers are assigned to number range objects in advanced payment management, which must be maintained in Transaction SNRO (Maintaining Number Range Objects). A number range object is uniquely identified by its name and a two-digit number. You can associate number ranges with a clearing area via IMG path **Financial Supply Chain Management** • **Advanced Payment Management** • **Basic Configuration** • **Maintain Generic Number Ranges**.

The following objects can be linked with the clearing area:

- /PF1/COLL: Number range object for collectors
- /PF1/CSTID: Customer IDs for service level agreements
- /PF1/FH_OL: Number range object list number (secondary key)
- /PF1/PO_PI: Number range for payment transactions (secondary key)
- /PF1/PO_PO: Number range for payment order (secondary key)
- /PF1/PO_RE: Number range object for recalls

Figure 4.86 shows the configuration of the generic number ranges for the sample clearing area. The **No** field contains the linked number range, which you maintain within Transaction SNRO (Maintain Number Ranges).

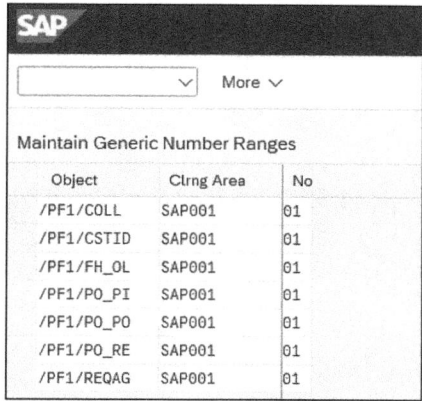

Figure 4.86 Maintain Generic Number Ranges

Payment Processing

In the following sections, we will delve into processing and examine the configuration settings for payment orders, payment items, relevant number ranges in this area, and the available options for manual payments in greater detail.

Payment Order

As initially described in Section 4.1.1, the payment order represents an internal, external, or manual payment request. For processing the payment, a number range is drawn from the input manager for each payment order and is populated in nearly all SAP Fiori apps, transactions, and reports. When it comes to payment orders, there are two configuration activities that we will discuss: maintaining number ranges and defining payment order types.

The number range for payment orders can be maintained directly via Transaction SNRO or through IMG path **Financial Supply Chain Management · Advanced Payment Management · Payment Processing · Business Objects · Payment Order · Maintain Number Ranges for Payment Orders**.

To maintain a number range, you need to enter the transaction and click the **Edit Intervals** button, as shown in Figure 4.87.

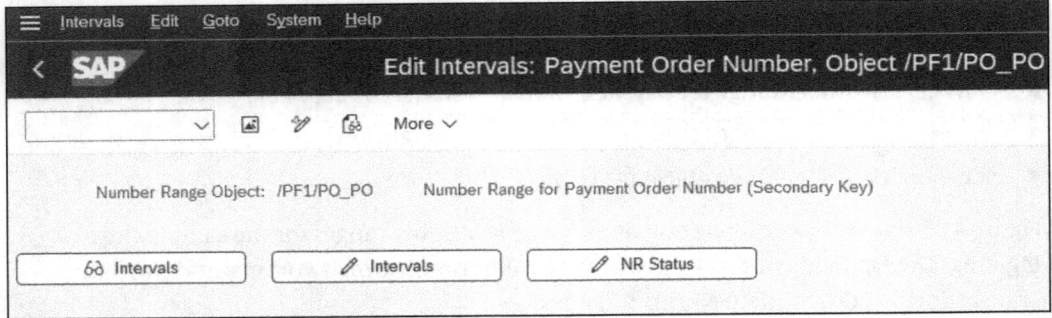

Figure 4.87 Maintain Number Range Intervals

After you click the **Edit Intervals** button, you will be taken to the number range interval maintenance area. If you click the **Insert Inset Line** button ⊕, as shown in Figure 4.88, you can add a new number range.

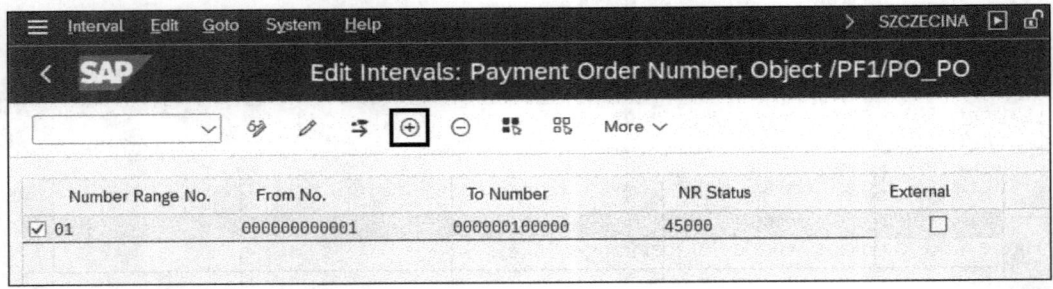

Figure 4.88 Maintain Number Range: Insert Line

Figure 4.89 shows the number range maintenance area for object /PF1/PO_PO.

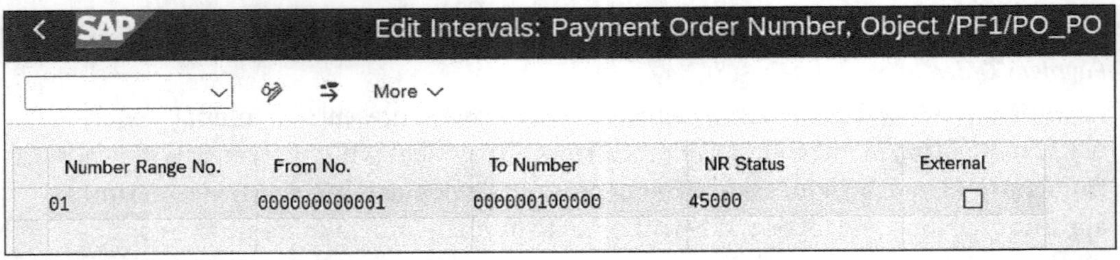

Figure 4.89 Maintain Payment Order Number Range

Payment order types are linked to the clearing area and serve as a key control element in the system by systematically classifying and managing payment orders. They enable clear differentiation and individual processing of various payment methods, such as domestic transfers, SEPA payments, international transfers, or check payments. Each payment order type defines specific processing rules, including formats (e.g., ISO20022, MT101), approval processes, payment channels, and associated banks. In addition, these types help determine which payments should be grouped, prioritized, or undergo special review.

Figure 4.90 shows an example of a payment order type configuration tailored for manual credit transfer processing. The payment order types can be maintained directly via Transaction SNRO or through IMG path **Financial Supply Chain Management** • **Advanced Payment Management** • **Payment Processing** • **Business Objects** • **Payment Order** • **Define Payment Order Types**.

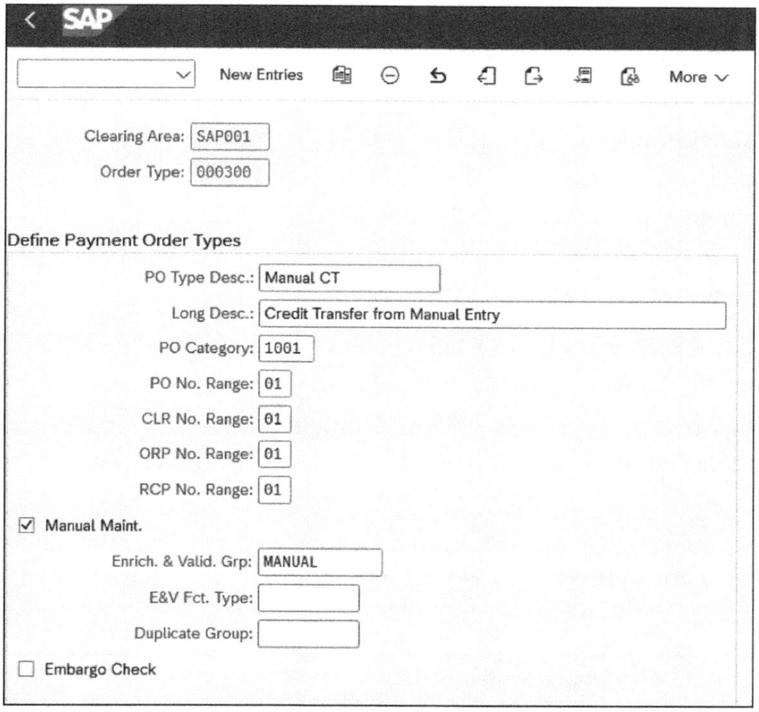

Figure 4.90 Define Payment Order Types

The following fields are relevant for payment order type configuration:

- **Clearing Area**
  The order type is always linked to a clearing circle, which can be defined in the **Clearing Area** field.
- **Order Type**
  In this field, you can define the order type number. It is generally advisable to consider a conceptual approach to the design. Here, you can refer to the standard order types provided in the sample clearing area, **SAP001**.

- **PO Type Desc**
  This field is used for a short description of the payment order type.
- **Long Desc**
  This field provides more characters and can be used for the long description.
- **PO Category**
  The payment order categories are used to classify payment orders based on their purpose, such as incoming or outgoing payments.

For the order type, you need to assign a number range to each respective item. In the next four fields, you can define which number range should be used for each type of payment order or item linked to this payment order type:

- **PO No. Range**
  Link number range with general payment order
- **CLR No. Range**
  Define number range number for clearing items
- **ORP No. Range**
  Define number range number for originating party items
- **RCP No. Range**
  Define number range number for recipient items

The **Manual Maint.** checkbox specifies which payment order types are allowed in the Create Payments app for manual payment entry.

In the **Define Payment Order Types** area, payment orders can be linked directly to enrichment and validation processes. You can use the **Enrich. & Valid. Grp.** field to achieve this.

Table 4.5 provides an overview of the predelivered payment order types from SAP for sample clearing area **SAP001**. These can be copied within the system and used for your own clearing area.

Clearing Area	Payment Order Type	Short Description	Long Description
SAP001	000001	Treasury CT	Credit Transfer from Treasury
SAP001	000002	Treasury DR	Debit Request from Treasury
SAP001	000100	Human Res. CT	Credit Transfer from Human Resources
SAP001	000101	Human Res. DR	Debit Request from Human Resources
SAP001	000200	Finan. Acct. CT	Credit Transfer from Financial Account.

Table 4.5 Standard Content: Payment Order Types

Clearing Area	Payment Order Type	Short Description	Long Description
SAP001	000201	Finan. Acct. DR	Debit Request from Financial Account.
SAP001	000300	Manual CT	Credit Transfer from Manual Entry
SAP001	000301	Manual DR	Debit Request from Manual Entry
SAP001	000400	ERP CT	Credit Transfer from ERP
SAP001	000401	ERP DR	Debit Request from ERP
SAP001	000500	Accrual EoD	Accrual Payment Order (EoD)
SAP001	001000	ISO PAIN CT	ISO20022 pain.001 Credit Transfer
SAP001	001001	ISO PAIN DR	ISO20022 pain.008 Debit Request

Table 4.5 Standard Content: Payment Order Types (Cont.)

Payment Item

The payment item represents individual entries at both the ORP (originator parts = the payer) and RCP (recipient party) levels. A fundamental setting for the payment item is the number range configuration. Figure 4.91 illustrates the maintenance of the number range (/PF1/PO_PI) for the payment item.

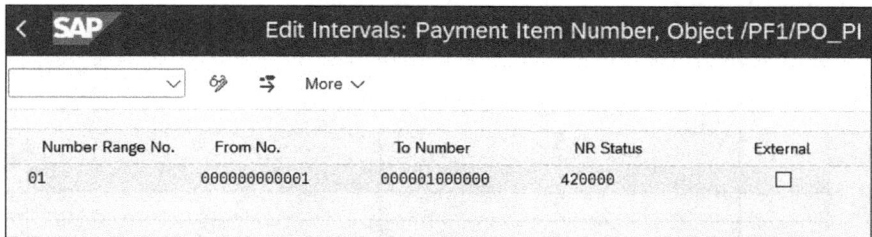

Figure 4.91 Define Number Range for Payment Item

There you can maintain the number range object (/PF1/PO_PI) directly via Transaction SNRO. Alternatively, you can follow IMG path **Financial Supply Chain Management • Advanced Payment Management • Payment Processing • Business Objects • Payment Item • Maintain Number Ranges for Payment Items**. The process is the same as that for the number range you maintained for the payment order.

Transaction types classify and process payment items, determining the handling in the system, posting direction, validation rules, and processing steps. Each defines attributes for debit or credit posting and validations during posting, with options for additional logic like simulation, prenotes, or mandate checks, ensuring accurate processing per company policies. Commonly used transaction types include the following:

- **Originator**
  Refers to payment items associated with the paying entity (e.g., company code).

- **Recipient**
  Refers to payment items associated with the payee (e.g., vendor or customer).

- **Clearing**
  Used for internal postings, especially during the clearing process when a payment run is finalized or an individual payment order is forwarded.

- **Turnover**
  Used for special payment items not directly linked to either payer or recipient, such as internal settlements or interest calculations.

- **Statement Entry**
  This is a single transaction from electronic bank statements, processed in the system.

For more information on the use of transaction types, refer to Section 4.1.1, where we discussed transaction types in detail. To configure transaction types, navigate to IMG path **Financial Supply Chain Management • Advanced Payment Management • Payment Processing • Business Objects • Payment Item • Define Transaction Types**.

Figure 4.92 shows the transactions and settings for the SAP sample clearing area.

> **Checkbox Functionality**
>
> On the screen shown in Figure 4.92, the following checkboxes and fields are available:
>
> - **Credit transaction type (Credit Trans. Type)**
>   This checkbox determines whether the transaction type is a debit or a credit.
>
> - **Transaction type group for enrichment and validation (TTGrp E&V)**
>   In this field, you can add enrichment and validation checks for a specific payment item based on the determined transaction type.
>
> - **Acct. Holder Required**
>   This checkbox indicates whether an account holder is required for a particular transaction type.
>
> - **Invalid Trans.**
>   This transaction type validation check is used to determine if the transaction type is valid. Set the checkbox to mark a transaction type as invalid. This means it can no longer be used in payment transaction processing.
>
> - **Manual Maint.**
>   Once this checkbox is selected, this transaction type is authorized for manual payments and can be used in the Create Payments app.

4.3 Configuration

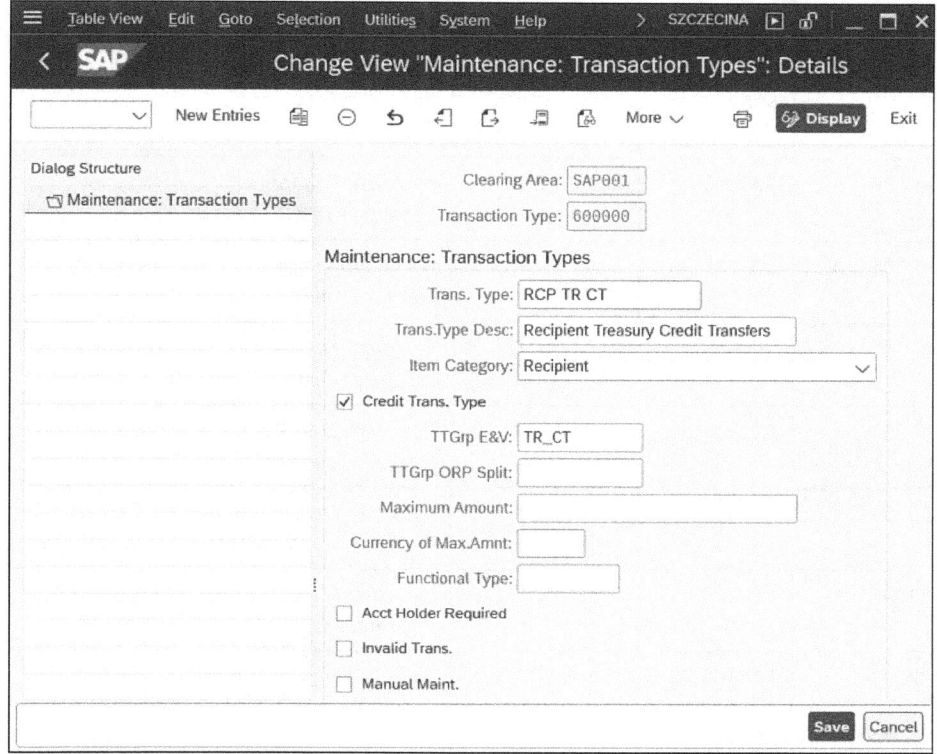

Figure 4.92 Maintenance of Transaction Types

There are several other fields available here as well:

- **TTGrp ORP Split**
  Payment items can be grouped via this field; for this, groups must be defined in table /PF1/T_TTGR_SPLT.
- **Maximum Amount**
  Specifies the maximum amount allowed for a transaction for the specific transaction type.
- **Currency of Max.Amnt**
  Specifies the currency for the maximum allowed currency for the specific transaction type.
- **Functional Type**
  The functional type maintained in enrichment and validation can be stored here. This defines specific processing rules and validations.

SAP delivers a set of predefined transaction types in the standard system, which can be used directly or adapted to meet specific business needs, as shown in Table 4.6.

Clearing Area	Transaction Type	Short Description	Long Description	Item Category
SAP001	500000	RCP TR DR	Recipient Treasury Debit Request	Recipient
SAP001	500200	RCP HR DR	Recipient Human Resources Debit Request	Recipient
SAP001	500400	RCP FI DR	Recipient Financial Acct. Debit Request	Recipient
SAP001	500600	RCP MAN DR	Recipient Manual Entry Debit Request	Recipient
SAP001	500800	RCP ERP DR	Recipient Entr. Res. Pl. Debit Request	Recipient
SAP001	510000	ORP TR CT	Ordering Treasury Credit Transfers	Originator
SAP001	510200	ORP HR CT	Ordering Human Resources Credit Transf.	Originator
SAP001	510400	ORP FI CT	Ordering Financial Acct. Credit Transf.	Originator
SAP001	510600	ORP MAN CT	Ordering Manual Entry Credit Transfer	Originator
SAP001	510800	ORP ERP CT	Ordering Entr. Res. Pl. Credit Transfer	Originator
SAP001	530000	CLR IHC DR	Clearing IHC Debit Request	Clearing
SAP001	530200	CLR FI DR	Clearing FI Debit Request	Clearing
SAP001	540000	RCP Accr. Dbt.	Recipient Accrual Debit	Recipient
SAP001	550000	ORP Acc. Dbt.	Ordering Accrual Debit	Originator
SAP001	600000	RCP TR CT	Recipient Treasury Credit Transfers	Recipient
SAP001	600200	RCP HR CT	Recipient Human Resources Credit Transf.	Recipient
SAP001	600400	RCP FI CT	Recipient Financial Acct. Credit Transf.	Recipient

Table 4.6 Standard Content: Payment Transaction Types

Clearing Area	Transaction Type	Short Description	Long Description	Item Category
SAP001	600600	RCP MAN CT	Recipient Manual Entry Credit Transfer	Recipient
SAP001	600800	RCP ERP CT	Recipient Entr. Res. Pl. Credit Transfer	Recipient
SAP001	610000	ORP TR DR	Ordering Treasury Debit Request	Originator
SAP001	610200	ORP HR DR	Ordering Human Resources Debit Request	Originator
SAP001	610400	ORP FI DR	Ordering Financial Acct. Debit Request	Originator
SAP001	610600	ORP MAN DR	Ordering Manual Entry Debit Request	Originator
SAP001	610800	ORP ERP DR	Ordering Entr. Res. Pl. Debit Request	Originator
SAP001	620400	RCP POBO FI CT	RCP POBO FI Credit Transfer	Recipient

Table 4.6 Standard Content: Payment Transaction Types (Cont.)

Once the transaction types have been defined, they must be assigned to the corresponding order types. This is done in IMG under the following path: **Financial Supply Chain Management • Advanced Payment Management • Payment Processing • Business Objects • Payment Item • Assign Transaction Types to Order Types**.

Figure 4.93 shows the configuration for the sample clearing area delivered as part of the standard SAP setup.

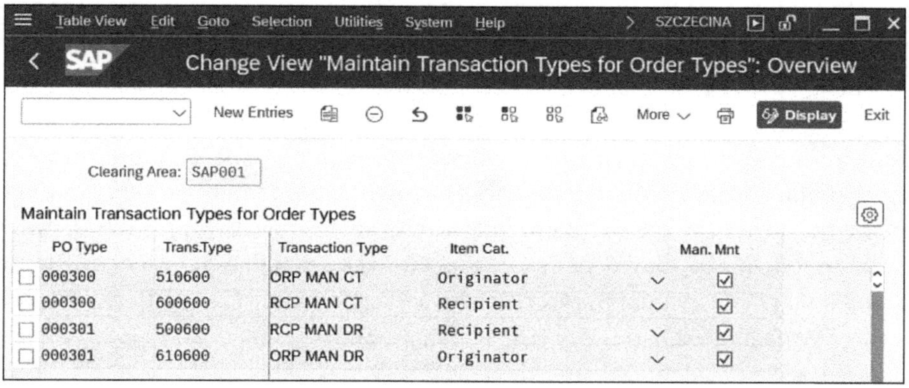

Figure 4.93 Maintain Transaction Types for Order Types

In the **PO Type** field, the payment order type is selected and linked to the transaction type (**Trans.Type**). Using the drilldown field for the item category (**Item Cat.**), you can set the categorization. You must specify whether this is an **Originator**, **Recipient**, **Clearing**, **Turnover**, or **Statement Entry**.

External Interfaces: SAP Bank Communication Management

During the payment processing in advanced payment management, a payment must be transferred to SAP Bank Communication Management for payment approval. This process step is part of the transaction clearing that takes place in advanced payment management before a payment is sent through the output manager to SAP Multi-Bank Connectivity. For this reason, an integration between advanced payment management and SAP Bank Communication Management needs to be configured via advanced payment management. For the integration, two configuration settings need to be implemented:

- **Standard BAdI for integration**
  To integrate advanced payment management with SAP Bank Communication Management, advanced payment management provides enhancement spot /PF1/BADIIMPL_OPO_PROCESS_BCM, which can be utilized directly by the responsible user or serve as a foundation for developing a custom implementation to ensure seamless integration with SAP Bank Communication Management.

- **External interface configuration**
  In addition to the BAdI implementation, standard customizing is required for the integration of advanced payment management and SAP Bank Communication Management. Within the IMG customizing, the rules for SAP Bank Communication Management must be maintained to ensure the integration functions according to business requirements. In the customizing settings, rule IDs must be assigned to a clearing area and a payment order type. This assignment is essential for SAP Bank Communication Management to technically retrieve the necessary data for payment approval.

  To do this, follow the IMG path for configuration: **Financial Supply Chain Management • Advanced Payment Management • External Interfaces • Bank Communication Management (BCM) • Define Rules for Bank Communication Management**.

  Figure 4.94 shows the **Define Rules for Bank Communication Management** customizing transaction, which must be maintained to integrate SAP Bank Communication Management approvals into the advanced payment management process. Here you can see that you have the option to link criteria such as clearing area (**Clrng Area**), company code (**CoCd**), payment order type (**PO Type**), as well as house bank (**House Bk**) and house bank account ID (**Acct ID**) with a rule maintained in SAP Bank Communication Management.

4.3 Configuration

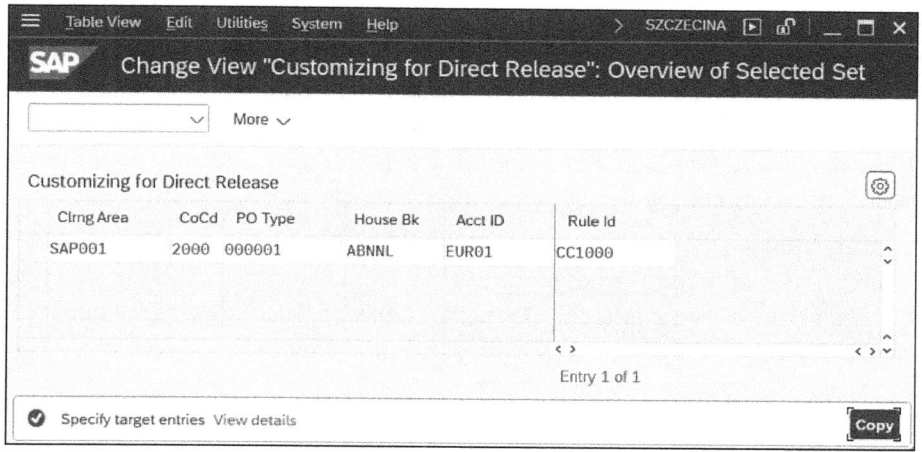

Figure 4.94 Assign Approval Rules for SAP Bank Communication Management

External Interfaces: File Handler

The file handler is an internal advanced payment management–specific database built on the ISO 20022 XML structure. It is utilized by both the input and output manager. To write messages into the file handler or read messages from it, the inbound and outbound converters are used.

The system differentiates transactions and messages based on their direction. *Inbound* messages originate from an external application or system and are sent to advanced payment management, while *outbound* messages are, in general, generated or forwarded by advanced payment management and delivered to an external application or system:

- Inbound messages come from an external system, module, or bank and can be directly forwarded to the respective target system or transformed into the advanced payment management data model, known as the file handler.

- Outbound messages are messages that are either forwarded by external systems or directly issued by advanced payment management and sent to an external application or system. They can carry payment transactions—typically sent to a service bureau or house bank—or status messages and advisories that inform connected systems or subsidiaries about the processing status.

Within advanced payment management, each external application or system is uniquely defined by a combination of format, media, channel, and direction. This set of attributes is identified by a unique converter ID.

In this section, we'll walk through configuration that is necessary for both the input and output managers; in the subsequent sections, we'll walk through settings that are unique to each of them.

Define Formats for Handler

The format is generally a freely definable field, but it serves to identify message and payment formats, such as pain.001 XML payment instructions or camt.53 end-of-day account statement formats. The first field, **Format**, serves as a key field and cannot be modified after saving. The **Desc. Format Short** field is used for a brief description, while the **Description Format Long** field provides more characters for a detailed description.

SAP initially provides a table with common formats, which can be expanded as needed. Figure 4.95 shows an example configuration of various formats, which can be individually defined in the following IMG area: **Financial Supply Chain Management · Advanced Payment Management · Payment Processing · External Interfaces · File Handler · Basic Configuration · Define Formats for Handler**.

Define Formats		
Format	Desc. Format Short	Description Format Long
/CAMT029	ISO CAMT 029	ISO camt.029.001.08 - Cancellation Respo
/CAMT053	ISO PAIN 053	ISO Pain 053 XML Format
/CAMT0547	ISO PAIN 054.07	ISO Pain 054.07 XML Format
/CAMT0557	ISO CAMT 055.07	ISO camt.055.001.07 - Cancellation
/DMEE	DMEE Format	DMEE Format
/GTSC	GTSCPayment	GTSCPaymentOrder
/GTSCCONCR	GTSC Concur	GTSC Concur
/IDOC	IDOC Format	IDOC XML Format
/IDOC2	IDOC Format	IDOC XML Format
/IDOCBLK	IDOC BLK Format	IDOC XML Bulk Format (PEXR2003)
/IDOCNEW	IDOC Format New	IDOC XML Format
/IDOCSGL	IDOC Sgl Format	IDOC XML Single Format (PEXR2003)
/INT	Internal Format	Internal Format
/INTFORM	Interest Scale	Interest Scale
/MT940	SWIFT MT940	SWIFT MT940
/PAIN001	ISO PAIN 001.0x	ISO Pain 001.0x XML Format
/PAIN0013	ISO PAIN 001.03	ISO Pain 001.03 XML Format
/PAIN0015	ISO PAIN 001.05	ISO Pain 001.05 XML Format
/PAIN0018	ISO PAIN 001.08	ISO Pain 001.08 XML Format

Figure 4.95 Define Format

Define Media

The medium typically represents the transport channel, such as /MBC for SAP Multi-Bank Connectivity, and can be customized as needed. SAP provides an initial dataset that can be used and expanded. Figure 4.96 shows an example media configuration, which can be defined in the IMG under **Financial Supply Chain Management · Advanced Payment Management · Payment Processing · External Interfaces · File Handler · Basic Configuration · Define Media**. **Medium** is a key field here. The **Descr. Medium Short** and **Desc. Medium (Long)** fields are used to describe the medium.

4.3 Configuration

> **Application Flag**
>
> The application flag (**AP flg.**) checkbox indicates if a media source of the defined medium must be picked up from a presentation server or application server.

Define Media			
Medium	Descr. Medium Short	Descr. Medium (Long)	AP flg
☐ /APP	Appl. Server	Application Server	☑
☐ /MBC	MBC	MBC	☑
☐ /PRS	Present. Server	Presentation Server (local)	☐
☐ MAIL	Mail	Mail	☐

Figure 4.96 Define Media

Define Channels for File Handler

The channel typically refers to the communication channel, such as SWIFT or EBICS. This field is also freely definable and is used to organize inbound and outbound file transactions.

Figure 4.97 illustrates the transaction for defining the channel. Before configuring the channel, the corresponding clearing area must be selected. **Clrng Area** and **Channel** are key fields, while the **Channel Descr. Short** and **Channel Description (Long)** fields are used for the description of the channel. As with the definition of formats and media for the file handler, channels should also be conceptually adapted to the implementation requirements. In contrast to channels, formats and media can be used across clearing areas. You can define channels via the following IMG path: **Financial Supply Chain Management • Advanced Payment Management • External Interfaces • File Handler • Basic Configuration • Define Channels for File Handler**.

Define Channels and Assign them to Channel Groups for FM			
Clrng Area	Channel	Channel Descr. Short	Channel Description (Long)
☐ SAP001	/BCM	Bank Com. Mgmt.	Channel for Bank Communication Mgmt.
☐ SAP001	/ERP	Entr. Res. Pla.	Channel for Enterprise Resource Planning
☐ SAP001	/FI	Financial Acct.	Channel for Financial Accounting Paymnt.
☐ SAP001	/HR	Human Resources	Channel for Human Resources Paymnt.
☐ SAP001	/INT	Internal	Channel for Internal Paymnt.
☐ SAP001	/MBC	Multi Bank Com.	Channel for Multi Bank Communication
☐ SAP001	/TR	Treasury	Channel for Treasury Paymnt.

Figure 4.97 Define Channels

External Interfaces: Define Inbound Converter

In the **Define Converter (New)** transaction, all previous settings—format, medium, and channel—are consolidated and supplemented with the direction, the necessary conversion program, and other relevant settings. These configurations depend on the specific process. In the example in Figure 4.98, we show the settings for a standard XML import with forwarding functionality.

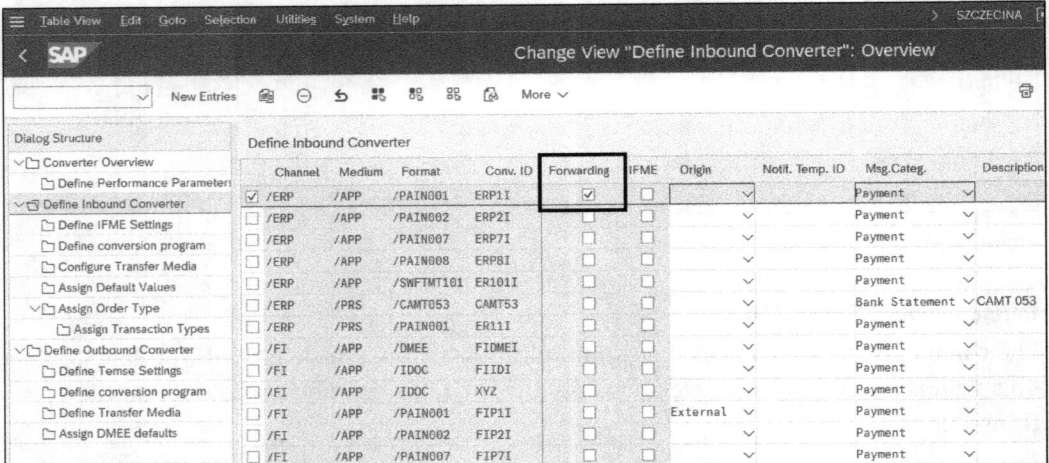

Figure 4.98 Define Inbound Converter

Figure 4.98 shows the **Define Inbound Converter** page. The configuration can be found in IMG under the following path: **Financial Supply Chain Management • Advanced Payment Management • External Interfaces • File Handler • Basic Configuration • Define Converter (New)**.

When you open the **Define Converter (New)** transaction, you will first see an overview of all incoming and outgoing converters, along with their settings.

In the following sections, we will discuss the customization of incoming and outgoing converters, but first, let's examine the functions and settings from the configuration transaction field and function overview:

- **Direction (Directn)**
  This field determines whether it is an incoming or outgoing converter.

- **Converter ID (Conv. ID)**
  A converter is defined by the combination of format, medium, and channel. This field allows a unique ID to be assigned to the converter, with up to six characters available. The **Conv. ID** field, along with the **Channel**, **Medium**, and **Format** fields, serve as key fields.

- **Class Name**

  The class serves as an interface to the file handler and manages both incoming and outgoing conversions to or from the file handler database. Incoming formats are transformed into the file handler via the selected class, while outgoing formats use converters to transform data from the file handler to DMEEX or the Map Format Data app.

- **Default converter checkbox (df.Conv.Id)**

  Using this checkbox, the respective converter can be selected as the default converter. The file handler determines the input converter based on channel, media, and format. If no converter can be determined, the default converter is selected.

- **Forwarding checkbox**

  If no validation of the format is required during forwarding, this flag can be set.

- **Origin: Local/Extern**

  This field is relevant for the processing of IDocs. The available options for the origin are **Local** and **External**. If the IDoc (e.g., PAYEXT) was created within the same system, then you should select **Local**. In this case, the reference IDoc EUPEXR, which is provided, will be checked.

- **Directory File Name**

  This field stores the logical directory, which is configured for advanced payment management via Transaction FILE.

- **Logical File Name**

  This field contains the designation of the logical path.

- **Bank Confirmation Indicator**

  This field defines whether a bank status message is expected for the sent file.

- **Code Page**

  Within this field, the code page for the file is defined.

- **Encryption checkbox**

  This field is flagged if certificates configured via Transaction STRUST are to be used for data encryption. When this flag is set, incoming messages are decrypted and outgoing messages are encrypted.

- **DMEE checkbox**

  This checkbox defines whether the DMEE is used for conversion by the converter.

- **Output Stream Class**

  This field is maintained in the outbound converter within the **Define Transfer Media** dialog step and determines how the file is output. The field can define the transfer to SAP Multi-Bank Connectivity via class /PF1/CL_OPM_OUTPUT_STREAM_MBC. In addition to this class, class /PF1/CL_OPM_OUTPUT_STREAM_FILE is available for file output via a path, and class /PF1/CL_OPM_OUTPUT_STREAM_MAIL for sending files via email.

- **Format Mapping Name**
  This field is maintained in the inbound converter within the **Define IFME Settings** dialog step. Here, the Map Format Data converter app can be linked to the inbound converter.

- **Mapping Type ID**
  This field is maintained in the inbound converter within the **Define IFME Settings** dialog step. Here, the mapping type for the Map Format Data app is determined via selection.

- **TemSe checkbox**
  This checkbox determines whether the file should be stored in the temporary storage TemSe via the outbound converter.

- **Type of Payment Medium: File or XML**
  This field determines whether the data is in XML format or another format, such as a file format like MT101.

- **Payment medium format (Pmnt Medium Format)**
  This field holds the format designation. This is particularly relevant for the outbound converter and is maintained in the **Define Temse Settings** dialog field.

- **Message Category**
  This field defines the categorization of the converter's message. The following categories are available for selection:

  – Payment
    This is a payment medium in file or XML format.

  – Bank Statement
    This is a bank statement, which can be an end-of-day bank statement.

  – Balance Notification
    This option is used if pure balance notifications from an in-house bank account are to be transmitted via advanced payment management.

  – Interest Notification
    This option is used when sending interest information from in-house banking.

  – Payment Status Notification
    If a status report on the processing of the payment is to be sent from advanced payment management, this message category is used. This is typically applied when sending a pain.002 message to an external system.

  – Converter description (Description)
    This field is used to label the converter and can be freely defined.

In the following sections, we'll discuss the setup of the input manager, from connecting it with the Map Format Data app to defining the conversion program, configuring transfer media, assigning default values, and assigning order types.

4.3 Configuration

Define Map Format Data Settings

If you want to connect the Map Format Data app (F3906) with the input manager, you can do so using the **Define IFME Settings** configuration. As shown in Figure 4.99, the Map Format Data app can be selected and integrated via the multiple selection option.

Here is an overview of the most important fields in the configuration:

- **Format Mapping Name**
  Here, the map format data, which was previously created in the Map Format Data app, can be linked to the inbound converter.

- **Mapping Type ID**
  The mapping type of the map format data is determined via selection.

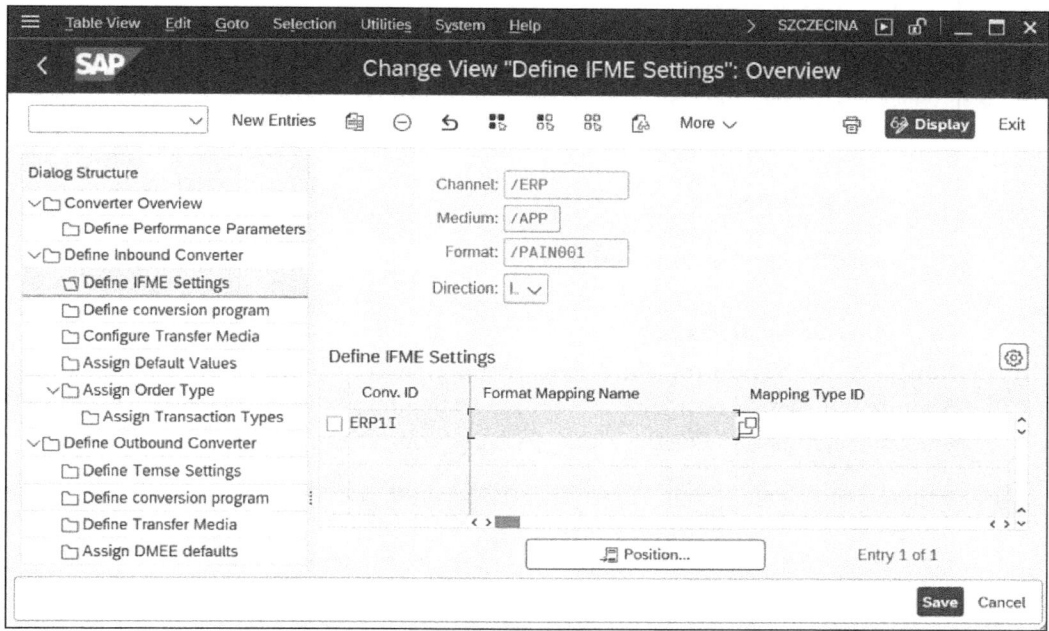

Figure 4.99 Define Map Format Data Settings

Define Input Conversion Program

Next, click **Define conversion program**. This step of the conversion program exists for both the input manager and the output manager. Files imported via the input manager are written into internal advanced paymesnt management tables using this class. For pure forwarding, if the format is already correct, fully maintained, and properly configured, and no adjustments need to be made by advanced payment management, it is sufficient to set the **Forwarding** flag in the converter. In this case (see Figure 4.100), there is no need to maintain class /PF1/CL_IPM_P_XML_FORMAT_CONV, which is responsible for converting incoming XML messages for the file handler.

317

4   Advanced Payment Management and In-House Banking

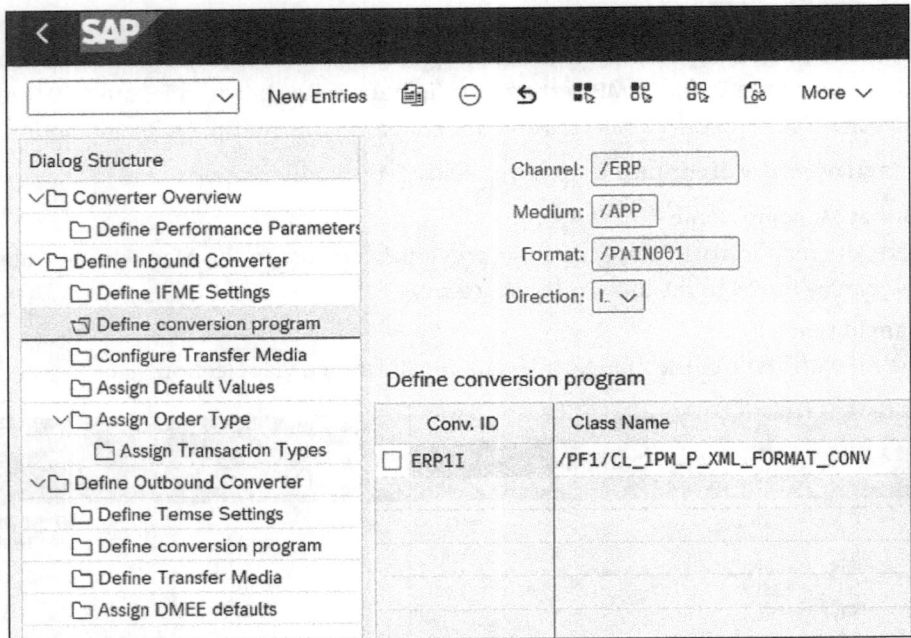

Figure 4.100 Define Conversion Program

Configure Transfer Media

Now, select **Configure Transfer Media**. Through this customizing, the import path can be stored. If files or payment carriers are to be imported into the system via a path upload, you must provide the logical path and code page.

> **Logical Path Must Be Configured**
>
> The logical path must be linked to a physical path. It must be configured for import via Transaction FILE (Cross Client File Names/Paths). For advanced payment management, the following initial logical file paths are available for configuration:
>
> - /PF1/BCM_FILE_OUT
> - /PF1/MBC_FILE_OUT
> - /PF1/XML_DISPLAY_INPUT_PATH
> - /PF1/XML_DISPLAY_INPUT_PATH

After the logical path is maintained in the converter, it is automatically populated in the import transactions of the input manager—for example, in Transaction /PF1/FH_IMPORT_DIR (File Directory Import), where the import path is automatically filled in. Figure 4.101 shows how you can link the converter ID with the logical path, which you can maintain via Transaction FILE as follows:

4.3 Configuration

- **Directory Logical File Name**
  This field stores the logical directory, which is configured for advanced payment management via Transaction FILE.
- **Logical file**
  This field contains the designation of the logical path.

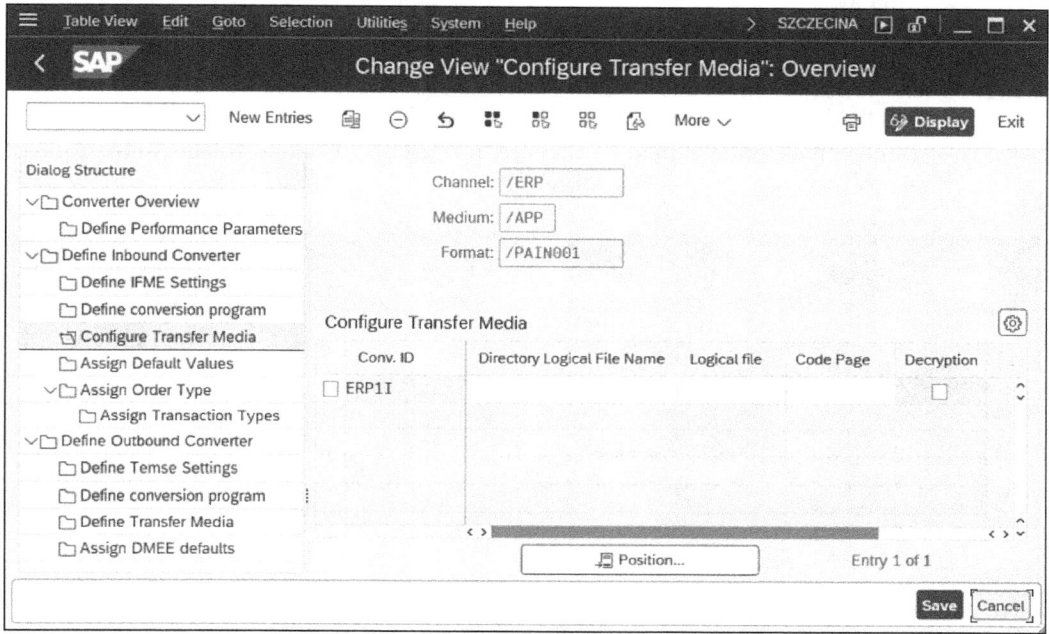

Figure 4.101 Configure Transfer Media

Assign Default Values

Now proceed to **Assign Default Values**. In this customization step, you can define default values for the converter, such as the default payment order type or the processing priority. Setting default values is optional as they are derived from both the customization settings and the master data.

>
> **Transaction Type Symbol**
>
> The transaction type symbol field (**TranTypSym**; see Figure 4.102) is a remnant from the original Payment Engine and represents the payment method used for the transaction. However, this field is generally not used in payment processing and can be considered redundant.

Figure 4.102 shows how default values for the selected converter ID can be defined. The following fields are available for definition within the default values:

- **PO Type**
  Within this field, select the payment or type that you want to set as the default.
- **Paym.Prio.**
  This field is used to define the default priority for processing within advanced payment management.
- **Clrng Area**
  This field is used to define the default clearing area.
- **TranTypSym**
  This field is used to define the default transaction type symbol that you want to link to the converter. This field is currently not used and can be considered redundant (see Transaction Type Symbol box).

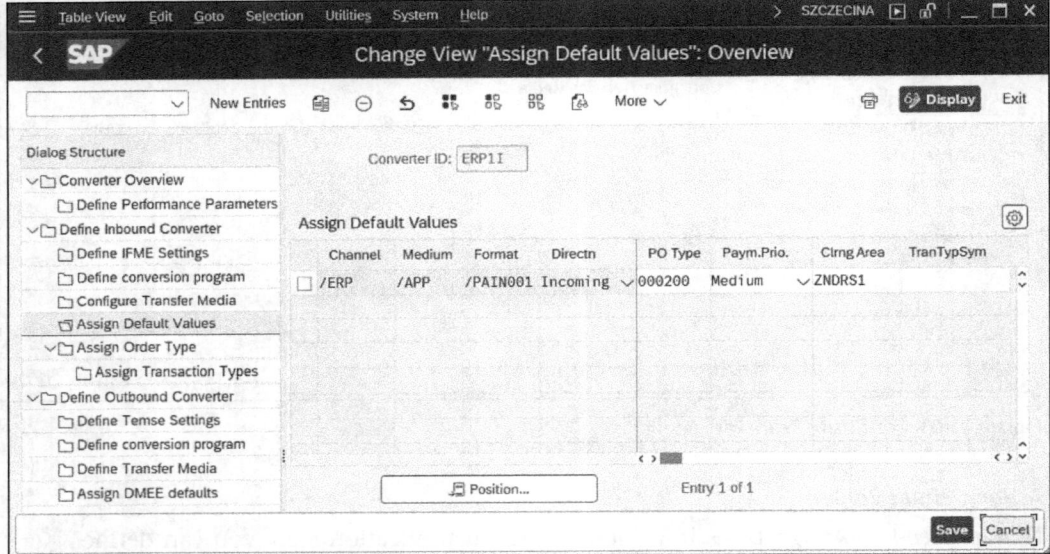

Figure 4.102 Assign Default Value

Assign Order Type

Through the **Assign Order Type** activity, the inbound converter is linked to the payment order type at the clearing area level. This is necessary for the categorization and process classification of the transaction. Figure 4.103 shows how to link the inbound converter at the clearing area level with the respective payment order types. Here, the respective payment order type is assigned to a combination of clearing area, XML message identifier, local instrument, and payment method.

In the next step, this payment order type is linked to the corresponding transaction types. These transaction types are crucial for the categorization of originator (ORP) and recipient party (RCP) payment items. Figure 4.104 shows that in the next dialog step, you can assign the respective item categories (**Item Cat.**) and the associated transaction types (**Trans.Type**) to the payment order type (**PO Type**).

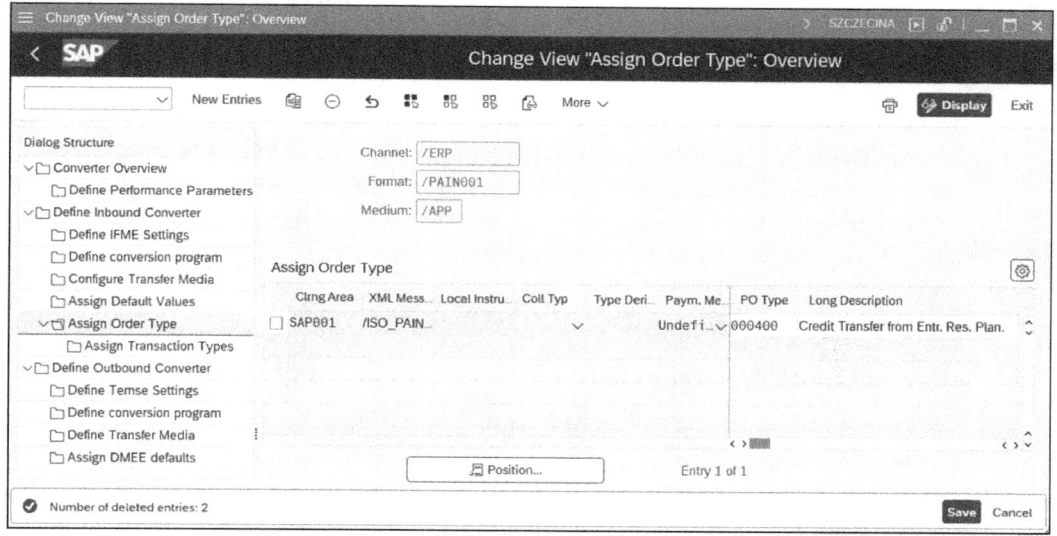

Figure 4.103 Assign Order Type

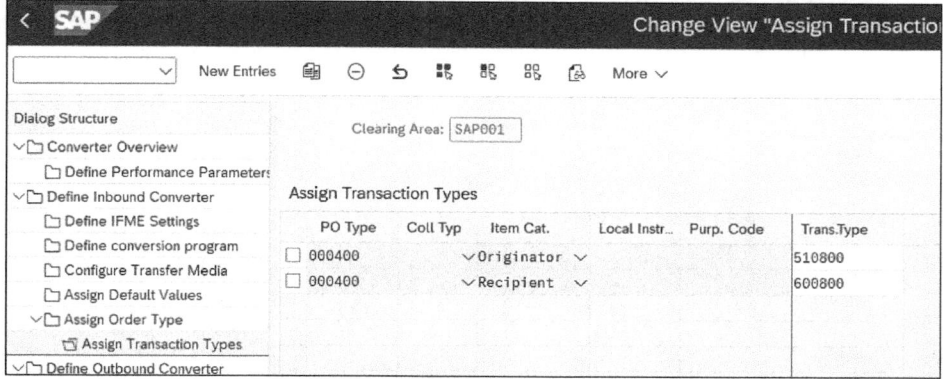

Figure 4.104 Assign Transaction Types

External Interfaces: Define Outbound Converter

The advanced payment management output manager is a central tool for controlling and managing the export of data within the advanced payment management infrastructure. Its primary function is to format data, reports, and notifications processed by advanced payment management and efficiently transfer them to defined destinations, such as external target systems, treasury management systems, online banking systems, or banks connected via SAP Multi-Bank Connectivity. The routing and transaction clearing component forwards the data to the output manager, which can then transmit it to external systems.

Following the same logic as the structure of the input converter, the output converter is responsible for exporting transactions and formats. It consists of a combination of channel, medium, format, and direction. Data and formats enriched through processing in advanced payment management within the file handler database meta format are converted into the desired target format via the output converter.

The configuration of the output converter, like the input converter, is carried out in IMG via customization path **Financial Supply Chain Management** • **Advanced Payment Management** • **External Interfaces** • **File Handler** • **Basic Configuration** • **Define Converter (New)**. In the following sections, we will explore the available configuration options.

Figure 4.105 shows an overview of output converters. Using the toolbar, you can create new converters or copy existing ones and adapt them to specific requirements. Each converter is assigned a unique ID.

Conv. ID is a six-character field that is available for free definition. To maintain clarity and organization, it is advisable to establish a sustainable naming convention early in the project. Thoughtful planning can help ensure consistency, facilitate identification, and improve overall project manageability.

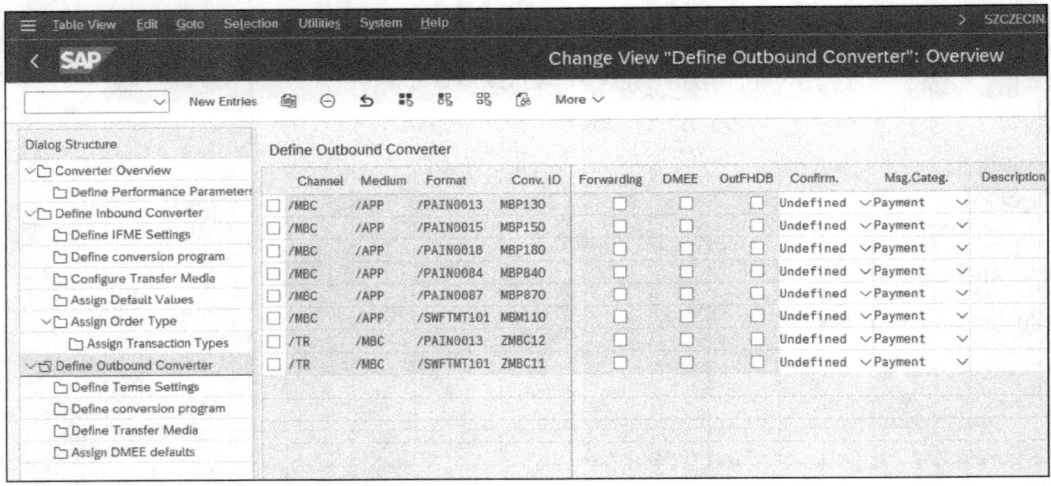

Figure 4.105 Output Converter

The following are the configuration options for outbound converters in the first configuration step:

- **Forwarding**
  This checkbox is used for the PINO procedure when a pure forwarding process is desired for the payment file.

- **DMEE**
  In the output manager, there are fundamentally two ways to generate the target format:

either through the XML conversion framework or alternatively via DMEE. If DMEE is to be used, this checkbox must be selected.

- **Use outgoing file handler database (OutFHDB)**
  Data is read from the file handler database.
- **Bank confirmation indicator (Confirm.)**
  Advanced payment management is capable of processing status messages from the bank. Here, we define whether a bank protocol is expected from the bank.
- **Message category (Msg.Categ.)**
  Here, the type of message is defined. Available message categories include payment, account statement, protocol, balance, or interest notification.

Define TemSe Settings

Now select **Define Temse Settings**. This setting allows you to control the writing and storage of data in temporary storage (TemSe), SAP's internal storage system.

Figure 4.106 illustrates the configuration screen for the output converter, which is used for storing data in TemSe. Once this dialog is maintained, the file is temporarily stored in TemSe during the creation of the target file and can be further processed from there.

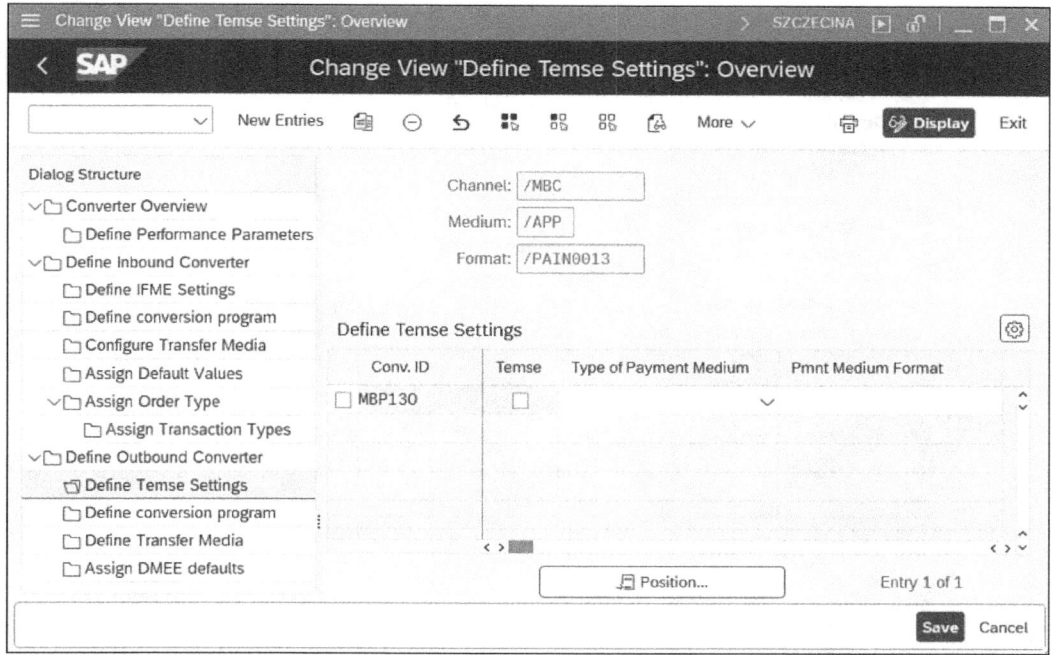

Figure 4.106 Define TemSe Settings

Figure 4.106 shows the maintenance dialog. The converter ID can be directly maintained along with the type of medium—XML (for pain.001/008 messages) or files (e.g., MT101 messages)—along with the respective format used.

> **Storage Capacity Is Limited**
>
> When storing or temporarily caching data in TemSe, it is important to note that its storage space is significantly restricted—to two gigabytes. It is designed solely for temporary data storage and should not be used for long-term data retention.

Define Output Conversion Program

If not DMEE but instead the XML conversion framework of the output manager is to be used for data carrier creation, then the designated class can be linked to the output converter via **Define conversion program**. For standard ISO2022 XML payment output messages, class /PF1/CL_OPM_P_XML_FORMAT_CONV is sufficient.

Figure 4.107 illustrates the customization step for linking the XML conversion framework with the converter. The converter class (**Class Name**) is directly assigned to the converter ID via the **Conv. ID** field.

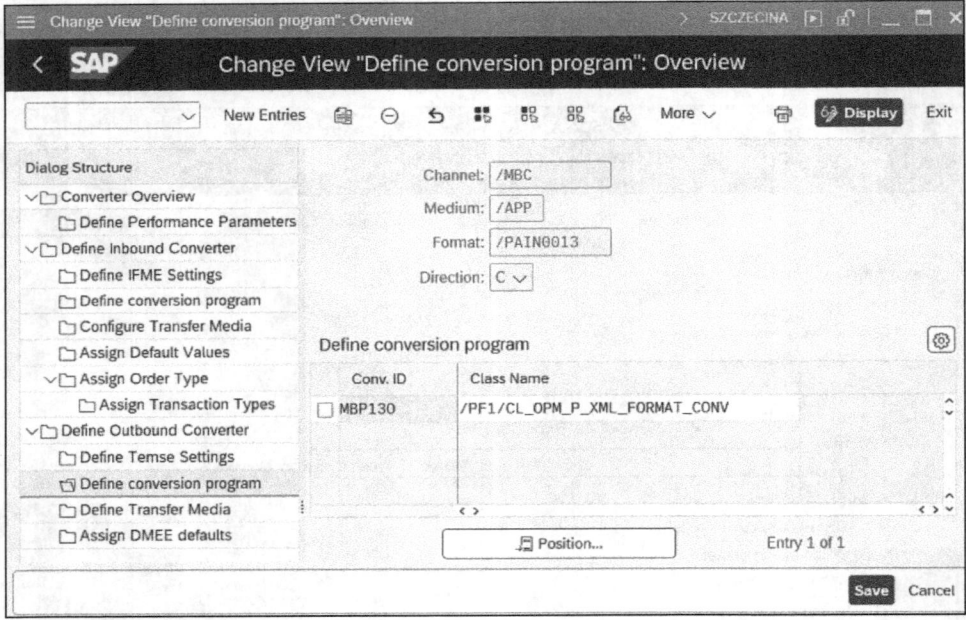

Figure 4.107 Define Conversion Program

Define Transfer Media

Next, select **Define Transfer Media**. This transaction is key when data needs to be forwarded to SAP Multi-Bank Connectivity for transmission to the bank. Class /PF1/CL_OPM_OUTPUT_STREAM_MBC is specifically used for this purpose. Figure 4.108 illustrates the customization for linking with SAP Multi-Bank Connectivity, including the following fields:

4.3 Configuration

- **Directory Logical File Name**
  This field stores the logical directory, which is configured for advanced payment management via Transaction FILE.
- **Logical file**
  This field contains the designation of the logical path.

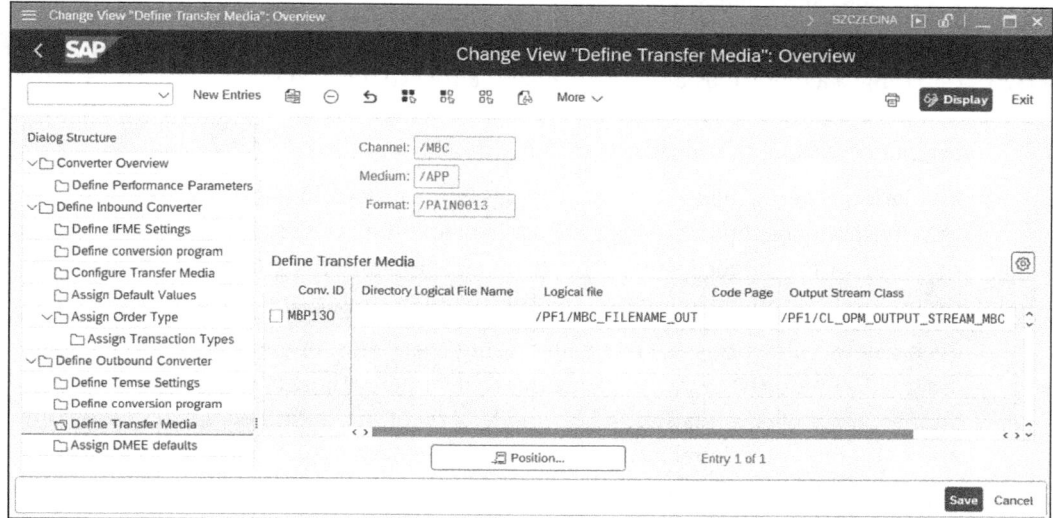

Figure 4.108 Define Transfer Media

Define DMEE Defaults

Finally, select **Assign DMEE defaults**. Figure 4.109 provides an example of the configuration settings for linking the output converter to a DMEE(X) format tree, when the XML framework is not being used.

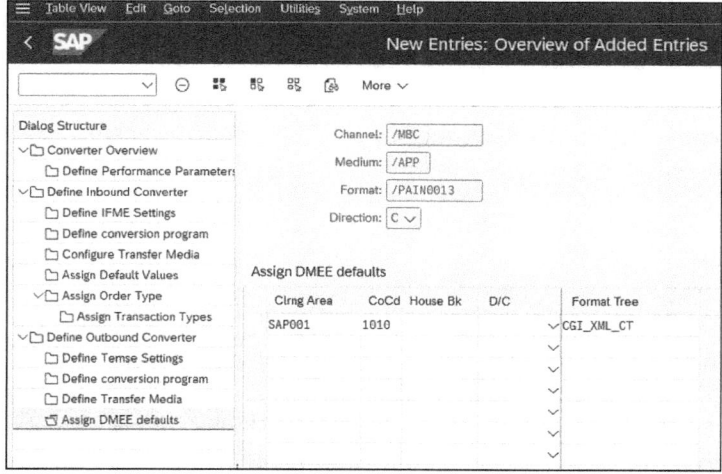

Figure 4.109 Assign DMEE Defaults

The figure illustrates how the clearing area (**Clrng Area**) and company code (**CoCd**) have been set as criteria for determining DMEE(X) format tree **CGI_XML_CT**.

The **House Bk** column will specify the house bank, while the **D/C** field allows for the inclusion of a credit or debit indicator.

Enrichment and Validations

As described in Section 4.1.1, enrichment and validation are used to check incoming and outgoing payment files and, if necessary, enrich their formats. These checks can be applied at both the payment and payment item levels, covering elements such as the originator, recipient, and cross-item consistency.

Validation is performed at two main stages in the processing flow: once after the file is imported into the input manager and again before the payment order leaves the output manager in the target format. This logic also applies to manually created payment orders, such as those generated using the Create Payments app (F3648).

Technically, validations are controlled using *check sets*, predefined groups of checks assigned a unique ID in the customizing environment. These check sets can range from simple to complex validations. SAP provides a set of standard checks that can be copied and tailored to individual business needs. If any check requires an SLA, it must also be explicitly defined. Validation logic can be grouped into four core categories:

- **Formal accuracy**
  Ensures data is structured correctly, such as by verifying that a bank account number is valid for the specified country format.
- **Referential accuracy**
  Validates the presence and correctness of linked data, such as by confirming that an IBAN exists and is properly formatted.
- **Material errors**
  Validates value thresholds, such as checking whether a payment amount exceeds defined limits for a given payment order type or group.
- **Consistency**
  Ensures logical integrity within the data, such as a debit/credit consistency check that verifies that the sender and recipient items balance properly.

The following sections will walk you through the essential customizing steps for implementing these functions.

Maintain Enrichment and Validation Check Sets

The system offers a multitude of standard checks that can be bundled into check sets and processed in a predetermined sequence. Figure 4.110 shows the customizing transaction **Maintain Enrichment and Validation Check Sets**, which can be accessed in Transaction SPRO under **Financial Supply Chain Management** • **Advanced Payment**

Management • **Payment Processing** • **Enrichment & Validations** • **Maintain Enrichment and Validation Check Sets**.

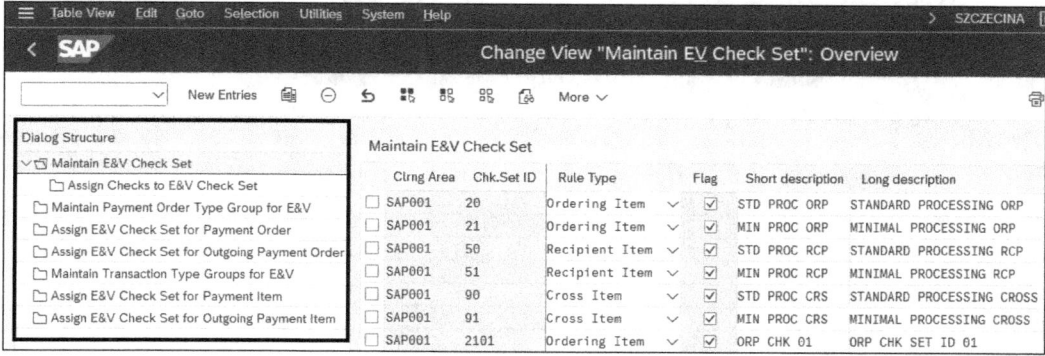

Figure 4.110 Enrichment and Validation Check Set Dialog

Check sets can be configured at both the payment order level and the payment item level. A new check set can be created using the **New Entries** button, and individual checks can be assigned via the dialog structure on the right side, under the **Assign Checks to E&V Check Set** subfolder. For example, you could have a cross-item check set, which examines the entire payment file. It would check the payment for consistency and check individual items against sanctions and embargo lists using an external application, SAP Business Integrity Screening.

> **[«]** **Sanctions and Embargo Checks with Advanced Payment Management**
>
> For a sanctions and embargo check, SAP Business Integrity Screening must be connected with advanced payment management. Alternatively, SAP Watch List Screening can also be used for this purpose. An overview of the integration is provided in Section 4.4.

> **[«]** **Maintain Individual Check ID Functions**
>
> SAP provides, by default, a rather large selection of check IDs that can be directly adopted. If individual checks beyond the standard SAP selection are required, they can be added through the cross-client customizing transaction **Define Enrichment and Validation Check Functions**. This transaction can be found in Transaction SPRO under **Financial Supply Chain Management** • **Advanced Payment Management** • **Payment Processing** • **Enrichment & Validations** • **Define Enrichment and Validation Check Functions**.

Maintain Payment Order Type Group for Enrichment and Validation
Through the next dialog step, you can define payment order type groups at the payment order level and link them with the clearing area. The defined payment order type

groups can also be directly associated with the payment order type (see our earlier discussion of payment processing configuration). Figure 4.111 shows the payment order type groups delivered by SAP, which can also be copied and assigned to your own clearing area if needed.

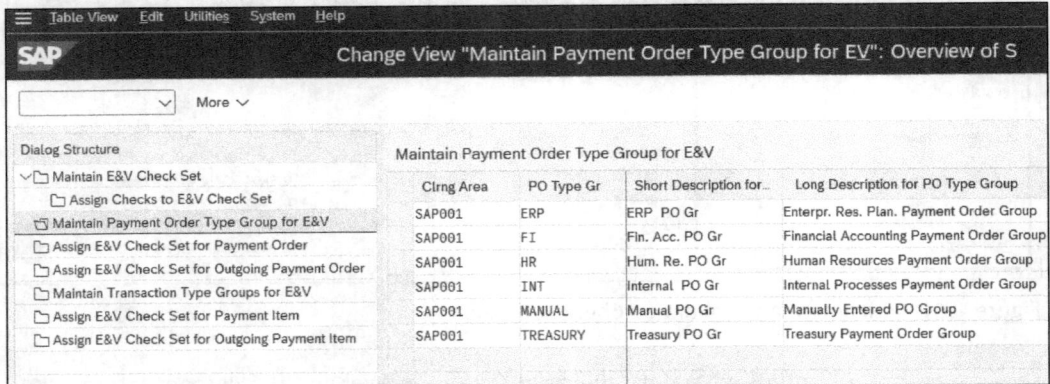

Figure 4.111 Maintain Payment Order Type Group for Enrichment and Validation

Assign Enrichment and Validation Check Set for Payment Order

The previously defined payment order type groups can be linked to the clearing area with the enrichment and validation check set in the next step. The **EV Set Type** field allows you to specify whether the check should be carried out at the payment order level or at the cross-item check level.

Execution Time

In the **Exec Time** field, you have several options to choose from, which determine when the validation should occur:

- **E: Execution Date**
  The check is not performed immediately. The system waits until the execution date before carrying out the validation.
- **S: Submission Date**
  The validation is performed on the day the order is submitted to the system.
- **X: Subm./Exec.**
  The values for execution date and submission date were taken into account.

In Figure 4.112, check sets are assigned at the payment order level. Here, the enrichment and validation group is linked to the clearing area (**Clrng. Area**) and the enrichment and validation check set type is assigned an execution time. For further identification, the channel and format can also be used as determination criteria.

Through the **Assign E&V Check Set for Outgoing Payment Order** dialog step, you can specifically define enrichment and validations for outgoing payment orders.

4.3 Configuration

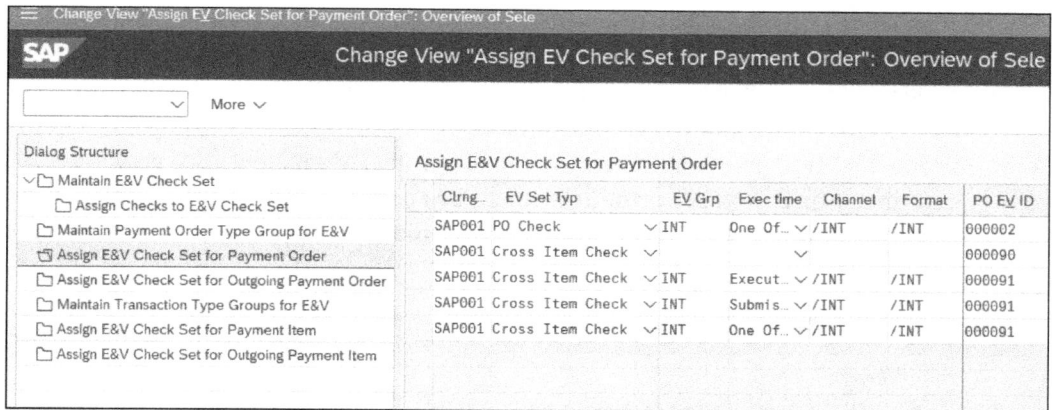

Figure 4.112 Assign Enrichment and Validation Check Set for Payment Order

Maintain Enrichment and Validation for Payment Items

In the next three dialog steps, enrichment and validation checks can be defined at the payment item level for ORP or RCP levels.

The first step is the definition of a transaction type, which is then linked with the clearing area. In Figure 4.113, the transaction types initially delivered by SAP are depicted, which can be linked with check sets at the payment item level in the subsequent dialog steps.

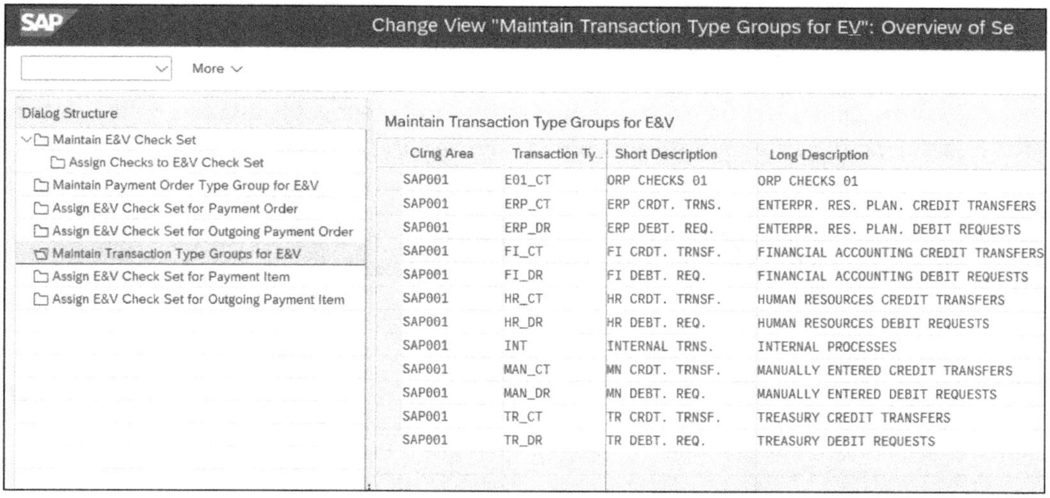

Clrng Area	Transaction Ty...	Short Description	Long Description
SAP001	E01_CT	ORP CHECKS 01	ORP CHECKS 01
SAP001	ERP_CT	ERP CRDT. TRNS.	ENTERPR. RES. PLAN. CREDIT TRANSFERS
SAP001	ERP_DR	ERP DEBT. REQ.	ENTERPR. RES. PLAN. DEBIT REQUESTS
SAP001	FI_CT	FI CRDT. TRNSF.	FINANCIAL ACCOUNTING CREDIT TRANSFERS
SAP001	FI_DR	FI DEBT. REQ.	FINANCIAL ACCOUNTING DEBIT REQUESTS
SAP001	HR_CT	HR CRDT. TRNSF.	HUMAN RESOURCES CREDIT TRANSFERS
SAP001	HR_DR	HR DEBT. REQ.	HUMAN RESOURCES DEBIT REQUESTS
SAP001	INT	INTERNAL TRNS.	INTERNAL PROCESSES
SAP001	MAN_CT	MN CRDT. TRNSF.	MANUALLY ENTERED CREDIT TRANSFERS
SAP001	MAN_DR	MN DEBT. REQ.	MANUALLY ENTERED DEBIT REQUESTS
SAP001	TR_CT	TR CRDT. TRNSF.	TREASURY CREDIT TRANSFERS
SAP001	TR_DR	TR DEBT. REQ.	TREASURY DEBIT REQUESTS

Figure 4.113 Maintain Transaction Type Groups for Enrichment and Validation

Figure 4.114 shows the **Assign E&V Check Set for Payment Item** dialog step. In this dialog step, the check set is linked with the transaction type and associated with the clearing

4   Advanced Payment Management and In-House Banking

area. Through the **EV S Type** field, you can specify whether this is an RCP, ORP, or cross-item type check. Similar to the payment order checks, you can also determine the execution time here. You can also refine the granularity at the channel and format level. Figure 4.114 shows how the clearing area and check set type criteria are assigned to a transaction type group. The execution time, format, and channel also can be added as determination criteria. Once the determination and processing criteria are selected, the corresponding check set can be assigned, which is determined based on the established criteria.

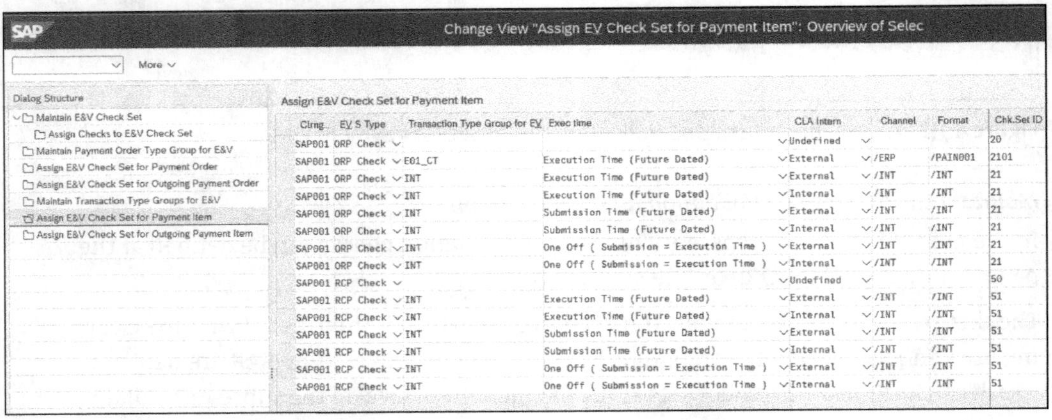

Figure 4.114 Assign Enrichment and Validation Check Set for Payment Item

In parallel with the outgoing payment order check, you can also implement check sets at the payment item level. Figure 4.115 shows an example that is not provided by the standard SAP content. Here you can see how the check set is assigned to the clearing area, transaction type, and enrichment and verification check set type determination criteria. The channel and format can be included as criteria as well.

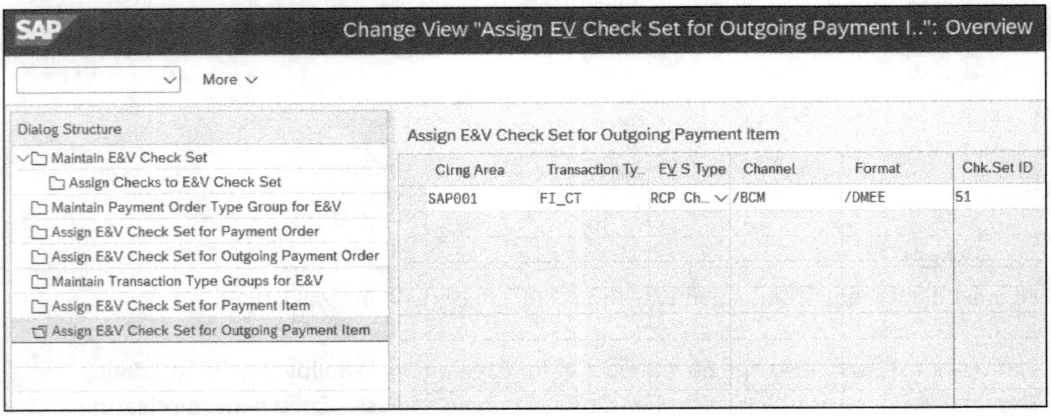

Figure 4.115 Assign Enrichment and Verification Check Set for Outgoing Payment Item

330

4.3 Configuration

4.3.2 In-House Banking

Compared to SAP In-House Cash, the in-house banking configuration is notably more limited. This limitation primarily stems from the requirement to use advanced payment management as a prerequisite for implementing in-house banking. As a result, most of the available scenarios in in-house banking are directly derived from or tied to advanced payment management, restricting flexibility. In addition, in-house banking relies more heavily on item setup for its master data, requiring a more detailed and structured approach to data maintenance. In contrast, SAP In-House Cash offered a broader range of configuration options, allowing greater customization and adaptability to various business needs without such strict dependencies.

In this section, we will delve into the comprehensive configuration required to update in-house banking. Our primary focus will be on defining the bank area, which serves as the central entity for managing key financial operations. In addition, we will explore the activity types and criteria that are essential for effective account management, alongside the creation of specific number ranges for in-house bank accounts and payment items. Finally, this section will address the integration of accounting processes, ensuring a seamless connection between financial transactions and the organization's broader financial ecosystem.

Define Bank Areas for the In-House Bank

The first essential step in setting up in-house banking is the configuration of a bank area. The bank area serves as the header entity and acts as the central structure for managing centralized netting, as well as for executing POBO and ROBO transactions. When defining bank areas, it's important to carefully consider whether you need a single bank area or multiple ones. This decision depends on several factors, such as legal and regulatory requirements, as well as your preferred approach to handling intercompany payables and receivables. By evaluating these aspects up front, you can determine the appropriate number of bank areas needed to support your organization's financial operations effectively.

You can find create in-house banking bank areas under **Financial Supply Chain Management** • **Advanced Payment Management** • **In-House Banking** • **Define Bank Areas for Account Management** (see Figure 4.116).

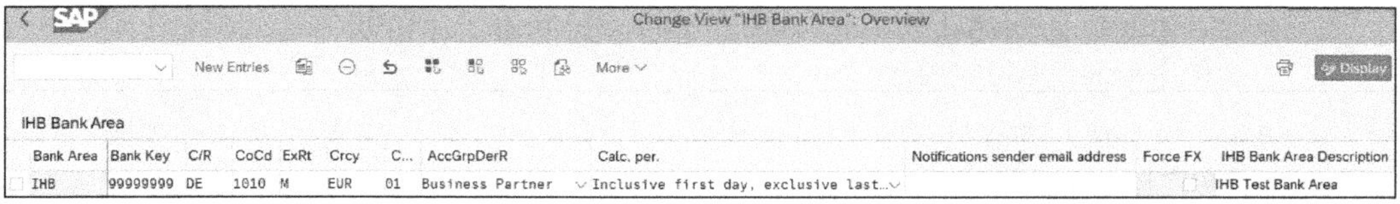

Figure 4.116 Create In-House Banking Bank Area

When defining a bank area, the following key fields must be maintained in the configuration table:

- **Bank Key** and country/region (**C/R**) are used as reference points in routing setup.
- Company code (**CoCd**) identifies the legal entity that owns the in-house bank.
- The exchange rate type (**ExRt**) determines how foreign currency amounts are converted into account currency.
- Currency (**Crcy**) defines the default currency of the bank area.
- **Calendar** specifies bank working days and holidays.
- The account group derivation rule (**AccGrpDerR**) determines the level at which account grouping is defined—either by a business partner or a trading partner.
- The account interest calculation period (**Calc. per.**) defines treatment of the first and last days in balance period interest calculation. The options are as follows:
  - Inclusive first day, exclusive last day
  - Exclusive first day, inclusive last day
- **Notification sender email address** is the email address from which system notifications will be sent.
- Direct FX (**Force FX** checkbox) controls the timing of FX conversion:
  - If selected, the FX rate at posting time is used; payment item status is set to **POSTED**.
  - If not selected, the FX rate is deferred to end-of-day processing; item is **PARKED** until finalized with the latest FX rates (useful for transactions in foreign currency).
- **IHB Bank Area Description** provides a descriptive label for identification.

Some recommendations for filling out these fields are as follows:

- It is generally recommended to create one bank area per organization that has its own bank key.
- When determining the number of bank areas, consider legal structures (e.g., independent accounting units) and geographical or political factors (e.g., regional separation).
- Each in-house bank account must be linked to a bank area at the time of creation; this assignment is permanent and cannot be changed later.
- If required, external number assignment (e.g., based on SWIFT codes) can be enabled for alternative identification.
- Ensure that the bank data related to the bank area has already been created and a bank key is available.

Define Activity Types for Account Management

In this activity type, you define all potential posting scenarios that may occur within the in-house bank. It serves as the foundation for specifying which types of transactions can be generated as payment orders and subsequently posted to in-house banking accounts. By configuring this activity type, you are effectively outlining the full range of possible financial movements within the in-house bank, ensuring that all relevant posting types are accounted for and can be processed accurately within the in-house banking framework.

You can find this activity under **Financial Supply Chain Management** • **Advanced Payment Management** • **In-House Banking** • **Define Activity Types for Account Management**.

Once you open this activity, you can specify all scenarios and transactions (see Figure 4.117).

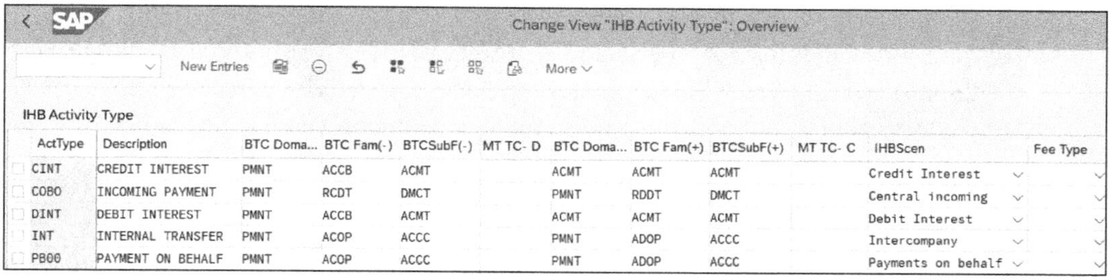

ActType	Description	BTC Doma...	BTC Fam(-)	BTCSubF(-)	MT TC- D	BTC Doma...	BTC Fam(+)	BTCSubF(+)	MT TC- C	IHBScen	Fee Type
CINT	CREDIT INTEREST	PMNT	ACCB	ACMT		ACMT	ACMT	ACMT		Credit Interest	
COBO	INCOMING PAYMENT	PMNT	RCDT	DMCT		PMNT	RDDT	DMCT		Central incoming	
DINT	DEBIT INTEREST	PMNT	ACCB	ACMT		ACMT	ACMT	ACMT		Debit Interest	
INT	INTERNAL TRANSFER	PMNT	ACOP	ACCC		PMNT	ADOP	ACCC		Intercompany	
PBOB	PAYMENT ON BEHALF	PMNT	ACOP	ACCC		PMNT	ADOP	ACCC		Payments on behalf	

Figure 4.117 Define Activity Types for Account Management

In this activity, you define activity types that categorize postings within in-house banking. These activity types play a central role in how transactions are classified, processed, and reported.

Each activity type (**ActType**) is characterized by a combination of attributes such as **Description**, credit/debit indicator (**IHBScen**), domain (**BTC Domain**), family (**BTC Fam**), and subfamily (**BTCSubF+**). Based on these attributes, activity types can be mapped to different business transaction codes. In some countries or regions, specific activity types are also tied to reporting requirements in accordance with applicable foreign trade regulations. By defining activity types, the system can accurately perform account determination to identify the appropriate G/L accounts for postings.

In addition, each activity type contains critical information needed for further processing. For example, the domain, family, and subfamily values are essential for generating bank statements. For G/L transfers, the corresponding in-house banking scenario is derived from this configuration.

Maintain Activity Type Criteria for Account Management

This configuration step allows you to define the criteria used to link specific activity types to relevant payment items within in-house banking. These criteria serve as conditions that must be met for a payment item to be correctly associated with an activity type during processing. You can find this configuration activity under **Financial Supply Chain Management • Advanced Payment Management • In-House Banking • Maintain Activity Type Criteria for Account Management** (see Figure 4.118).

The delivered configuration includes a predefined set of activity type criteria, offering a structured breakdown of how activities can be differentiated. However, you can enhance this setup by adding new activity types with more specific filter conditions as needed. This allows for greater granularity and flexibility in transaction processing.

The linkage between activity types and payment items is based on values from different parts of the payment structure:

- Header-level fields (e.g., payment method, scenario), entered under **Field Name**.
- Item-level fields (e.g., payment kind, debit/credit indicator), entered as a range (**Filter Condition**) from **Value From** to **Value To**. So, in this activity you can remove certain payment methods, countries, and so on from being processed.

The system checks all configured conditions. Only when all conditions are satisfied does it assign the specified activity type.

As already mentioned, values here are predefined; you only need to change them when necessary.

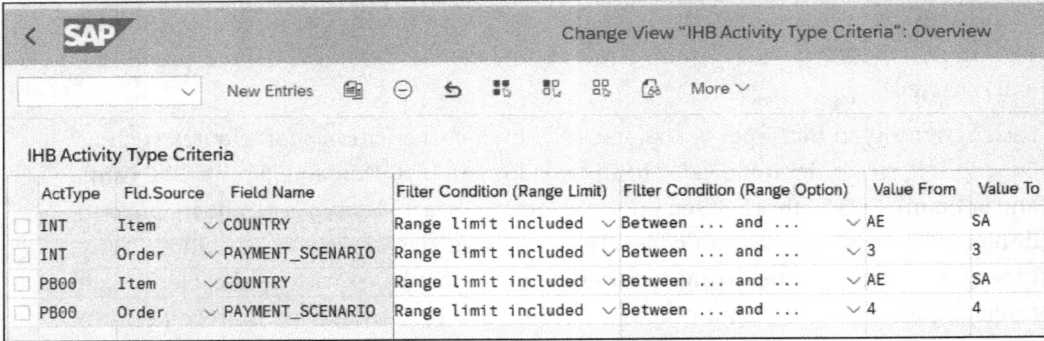

Figure 4.118 Maintain Activity Type Criteria for Account Management

Maintain Number Ranges for Payment Item Bundle

In this activity, you define number ranges for payment item bundles. A payment item bundle serves as a reference for a group of payment items that are transferred together to SAP S/4HANA Finance. This reference is stored in the financial accounting journal entry—specifically, in the AWREF (Reference Document Number) or AWKEY (Reference

Key) fields—ensuring traceability between in-house banking transactions and corresponding accounting entries.

You can create the number range under **Financial Supply Chain Management** • **Advanced Payment Management** • **In-House Banking** • **Maintain Number Ranges for Payment Item Bundle** (see Figure 4.119).

To create a new number range, click the **Edit NR** button (pencil icon), then provide the **Number Range No.** and its value in **From No.** and **To Number**. Click **Save** to close the form.

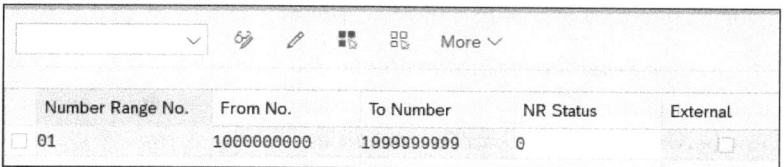

Figure 4.119 Maintain Number Ranges for Payment Item Bundle

Maintain Number Ranges for In-House Bank Account

In this activity, you can define the number ranges for in-house bank account numbers. However, note that the in-house bank account number is currently not editable. Even if you select an external number assignment in this configuration, the system will still automatically generate the next incremental number during account creation. This ensures consistency and control over account numbering, and manual entry or modification of the account number is not supported.

You can create the number range under **Financial Supply Chain Management** • **Advanced Payment Management** • **In-House Banking** • **Maintain Number Ranges for In-house Bank Account** (see Figure 4.120).

You need to create a separate number range for each in-house bank area—meaning that even if you have multiple bank areas, you might have an account with the same number in each of them.

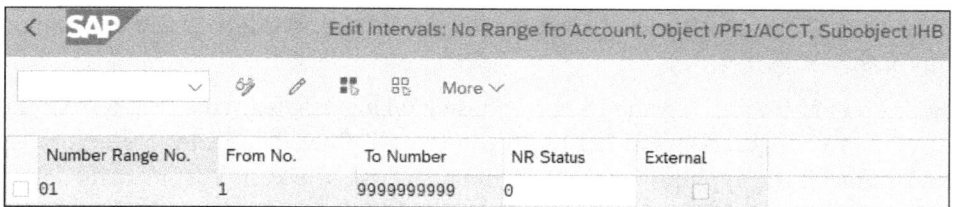

Figure 4.120 Maintain Number Ranges for In-House Bank Accounts

Accounting Integration

One of the most critical roles of in-house banking is to ensure that all financial activities and postings within the in-house bank subledger are accurately reflected in the company's financial statements. Every transaction processed through the in-house bank must ultimately be posted correctly to the G/L, impacting either the balance sheet, the profit and loss statement, or both. This alignment is essential for accurate financial reporting and compliance. The primary mechanism for achieving this is through automated bank statement postings, which transfer the summarized subledger activity into the G/L. However, in certain scenarios, direct postings from the in-house bank area to accounting are also possible, allowing for more immediate or specific financial entries. Together, these processes ensure that the in-house banking operations are fully integrated into the organization's overall financial landscape. Setting up the accounting configuration is a prerequisite for running in-house banking.

Maintain Settings for Posting Expenses and Incomes for In-House Banking

The accounting integration described here involves several critical steps to ensure a seamless alignment between in-house banking activities and the company's overall financial framework. One vital aspect of this integration is determining how controlling postings are handled, including the proper recognition of costs and revenues within the financial system. This process requires thorough configuration to categorize and allocate financial movements accurately, reflecting them in the G/L and supporting detailed internal reporting. Proper assignment of cost centers and profitability metrics plays a significant role in achieving transparency and control in financial postings, serving as the foundation for effective management and compliance.

You can use Transaction FJEPVC_IHB_DERIV_CO to maintain settings for posting expenses and incomes. In most scenarios, it is advisable to use the **Post as Costs** option (see Figure 4.121) when defining how expenses and incomes from in-house banking should be posted. This approach ensures that all relevant financial movements are recorded against a cost center, providing better visibility and control over internal cost allocations. When using this option, expenses are posted as costs, while interest revenues are treated as negative costs. This method aligns well with internal controlling practices and supports detailed reporting. Although the system also offers options to post as revenue—directly to profitability accounting—or without any cost object, these alternatives are typically used in more specific cases. For standard internal transactions and interest flows, posting to a cost center remains the most effective and transparent approach.

To post expenses and income from in-house banking as costs, you need to follow a structured configuration process to ensure that all financial entries are accurately recorded with the appropriate cost assignments. First, select the **Post as costs** option and then navigate to **Expense—Assignment Steps** within the dialog structure. Define a step description, and, if necessary, set a validity period using the **Valid From** and **Valid To** fields. You can also choose whether the in-house banking scenario should influence account determination; if activated, the scenario will become a filterable field in the next step.

4.3 Configuration

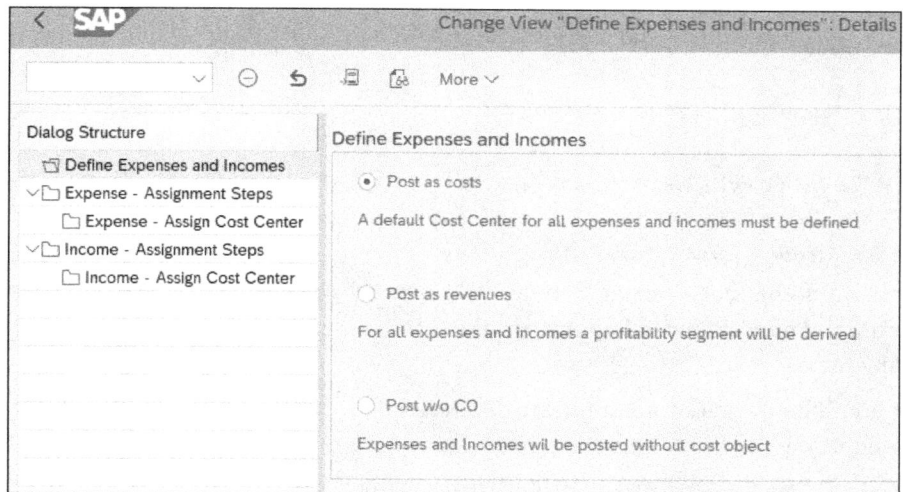

Figure 4.121 Maintain Settings for Posting Expenses and Incomes for In-House Banking

Next, go to **Expense—Assign Cost Center** and input key data such as the in-house banking scenario, company code, profit center, business area, and cost center. These values work together to determine the correct posting destination in your accounting system.

Repeat a similar procedure for income postings by selecting **Income—Assignment Steps**, again entering a description, optionally entering validity dates, and deciding on the use of the in-house banking scenario. Then proceed to **Income—Assign Cost Center** and complete the same set of organizational data fields. This ensures that both expenses and income related to in-house banking are properly allocated and reflected in your controlling and financial accounting systems.

If you select **Post w/o CO**, no further setup is required.

Assign G/L Objects to Bank Area

You can find this customizing activity under **Financial Accounting • General Ledger Accounting • Periodic Processing • Integration • In-House Banking • Assign GL Objects to Bank Area**. Alternatively, you can use Transaction FJEPV_IHBASGLOBJ and maintain the default **Profit Center** for each **Bank Area** used in in-house banking (see Figure 4.122). This assignment is essential for ensuring that financial postings are correctly integrated into your internal reporting and controlling structures. If your system is not configured to use profit centers or business areas, then attempting to access this configuration will trigger an error message as these elements are required for the setup. When profit center accounting is enabled, the system automatically derives the profit center from the bank area during journal entry postings for payment items. This helps ensure consistent and accurate financial reporting across all in-house bank transactions.

337

4   Advanced Payment Management and In-House Banking

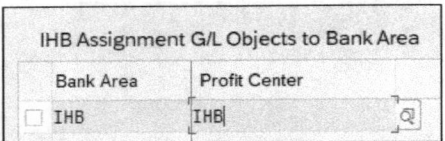

Figure 4.122 Assign G/L Objects to Bank Area

Define G/L Accounts for In-House Banking

You can find this customizing activity at **Financial Accounting • General Ledger Accounting • Periodic Processing • Integration • In-House Banking • Define GL Objects to Bank Area**.

For in-house banking, G/L accounts are determined through a set of predefined rules that guide how payment items are posted during a G/L transfer run. Transaction FJEPVC_IHB_ACC_DET allows you to define account determination rules that map specific in-house banking scenarios to the appropriate G/L accounts in your chart of accounts (see Figure 4.123).

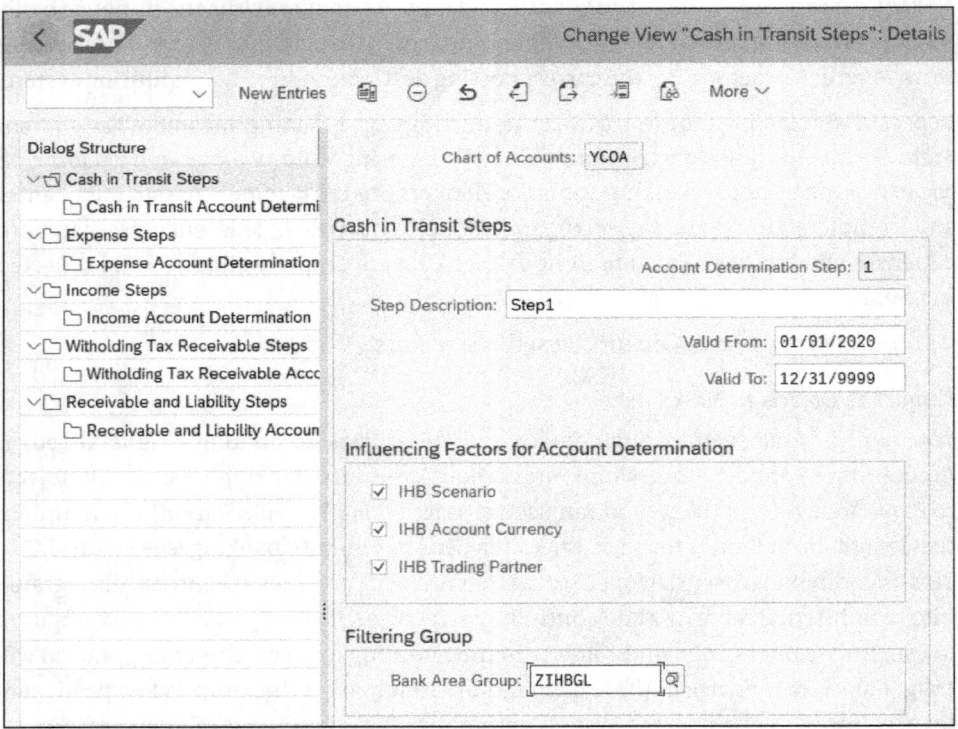

Figure 4.123 Define G/L Account for In-House Banking

When internal or external payments are processed between group entities, the system transfers the payment item data to accounting, generating a journal entry. One line item typically represents a receivable or payable item, while the other line item's category—such as cash in transit, expense, income, or withholding tax receivable—depends on the in-house banking scenario. By selecting each entry, you can provide a G/L account you would like to use.

First, for each category, you need to specify which chart of accounts should be used for the G/L account's determination. Then you enter your **Step Description** and then its validity by filling in the **Valid From** and **Valid To** fields. Next you determine which factors are affecting your account determination (**IHB Scenario**, **IHB Account Currency**, and/or **IHB Trading Partner**). Then you can provide your **Filtering Group**.

Each transaction type must be linked to the correct G/L account, and this is done by setting up rule steps in sequential order. These steps include key elements such as time dependency, filters by bank area groups, and influencing factors like the in-house banking scenario, account currency, or trading partner. The system evaluates these rules in the order of their step numbers, starting from the most specific. If no match is found, then the system prompts the user to complete the missing account determination.

It is also essential that activity types are already mapped to in-house banking scenarios beforehand as these mappings serve as a foundation for the G/L account determination. For simpler transaction categories such as receivables or liabilities, where only one account is used, influencing factors may not be necessary. However, for more complex postings, a precise and comprehensive set of rules ensures accurate financial integration between in-house banking operations and the G/L.

4.4 Integration of Intelligent Services

Payment transactions are among the most sensitive topics within a company and are therefore particularly protected with special authorization concepts, internal control systems, and encryption. Advanced payment management, in conjunction with SAP Multi-Bank Connectivity, can be expanded into a central gateway to the banking world, serving as the final instance for conducting fraud prevention, sanction checks, and embargo screening. As a central hub in payment processing, advanced payment management offers the flexibility to integrate additional intelligent services from third-party modules for the inspection and validation of transactions. SAP provides standard interfaces for connecting SAP Business Integrity Screening and SAP Watch List Screening.

> **[!] Fraud Detection Is Not Sufficient**
>
> We emphasize that fraud detection and sanction screening should not rely solely on advanced payment management; instead, they should be approached holistically. Ideally, this process should begin at the outset of a business relationship—starting with due diligence before any data is entered into the system—and extend throughout the entire accounting workflow. Because advanced payment management is limited to

4   Advanced Payment Management and In-House Banking

> analyzing data within the payment medium, such as structured address fields, recipient name, amount, and payment notes, it is advisable to assess both upstream and downstream processes. This ensures robust security in payment transactions and mitigates risks effectively.

In the following sections, we offer a brief overview of SAP Business Integrity Screening and SAP Watch List Screening and an introduction to their technical integration with advanced payment management.

4.4.1   Function and Processes

Figure 4.124 illustrates the screening of an example payment file that was created in a third-party SAP system and sent to advanced payment management for processing.

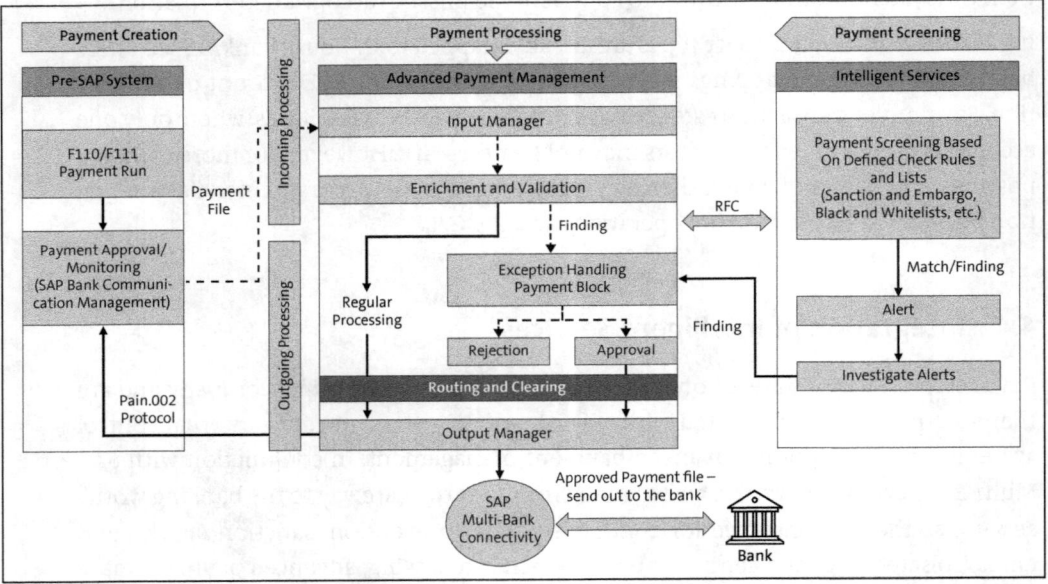

Figure 4.124 Payment Instrument

Let's walk through the main steps:

1. **Payment creation**
   In this example, the screening process begins with the upstream system by creating a payment run in the SAP system:
   – In SAP S/4HANA Finance, a payment is created using Transactions F110 or F111.
   – The payment is approved in the third SAP system via SAP Bank Communication Management and transmitted to the central advanced payment management system via SAP Multi-Bank Connectivity.

2. **Incoming payment processing**
   The payment instrument is transformed by the input manager into a proprietary internal format based on ISO 20022 (file handling database) in advanced payment management and checked through enrichment and validations. Intelligent services from SAP Business Integrity Screening or SAP Watch List Screening access the information of the payment instrument via enrichment and validation and a defined API interface.

3. **Payment screening**
   Payment items are examined for anomalies using the screening tool. This includes checking whether the beneficiary is listed on sanctions or embargo lists, or if the transaction appears suspicious based on defined criteria. If this is the case, an alert is triggered, and the payment instrument must be reviewed.

4. **Outgoing payment processing**
   After the payment has been reviewed, a decision must be made about whether the affected payment will be approved or rejected. If approved, the payment file can be generated in the next process step in advanced payment management via the output manager and sent to the bank through SAP Multi-Bank Connectivity.

If the payment is rejected, it will not be processed further. In the case of findings, it is always advisable to involve internal compliance and initiate downstream processes to prevent future fraud cases.

After successful payment processing, advanced payment management can send a status report in the form of a pain.002 message to the presystem, providing information on whether the payment was successfully processed or rejected. The pain.002 protocol can update the status of the SAP payment batch to either *accepted* or *rejected*.

4.4.2 Intelligent Service Integration

Now that you know how screening services can be connected to advanced payment management, let's examine the standard SAP screening models and introduce SAP Business Integrity Screening and SAP Watch List Screening.

SAP Business Integrity Screening

SAP Business Integrity Screening is a component of SAP assurance and compliance software for SAP S/4HANA, designed to monitor business processes and ensure compliance with regulatory requirements. The solution is designed to be generic and can be integrated into existing business operations.

The functionality of SAP Business Integrity Screening is based on predefined rules that examine both master and transaction data for potential fraud and compliance violations. These rules can be configured for SAP S/4HANA, SAP Process Control, and advanced payment management, among others.

4   Advanced Payment Management and In-House Banking

SAP Business Integrity Screening supports more complex analytical methods such as predictive analytics, network analysis, and machine learning to enable in-depth pattern recognition.

If the defined rules identify a suspicious payment, then an alert is generated that triggers an approval workflow. The suspicious payment must then be further investigated and either approved or rejected.

Figure 4.125 illustrates the following screening process with SAP Business Integrity Screening:

- Payment processing: In Figure 4.125, a payment request is created through the manual entry of a payment in advanced payment management. This payment request is processed as a payment order and payment item and is transferred via enrichment and validation to SAP Business Integrity Screening for screening.

- Detection: SAP Business Integrity Screening checks the payment medium based on predefined rules and screening lists. If the payment medium does not show any irregularities, then the payment is automatically approved, and the regular process in advanced payment management continues.

- Investigation: For suspicious payment items, an alert is generated, and the payment medium is stopped from further processing in advanced payment management through a payment block. If the alert turns out to be a false positive, the payment is approved in SAP Business Integrity Screening, and the payment block must be removed in advanced payment management.

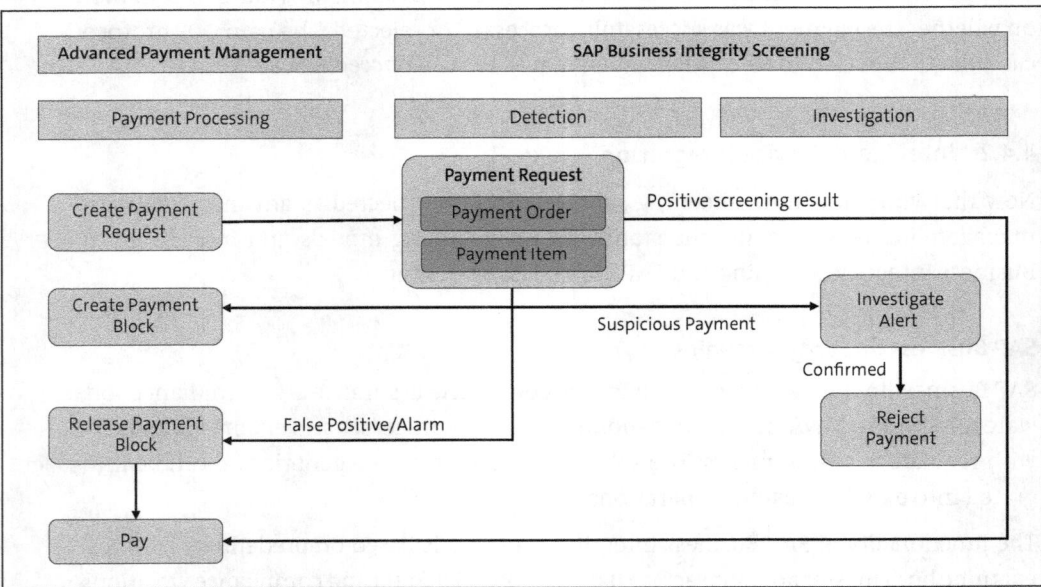

Figure 4.125 SAP Business Integrity Screening Process

SAP Watch List Screening

SAP Watch List Screening is a cloud-based application on SAP Business Technology Platform, designed for the automated screening of business partners and transactions against national and international sanction and watch lists. The solution supports both real-time and ad hoc screenings. Real-time screenings are conducted as part of business processes such as order-to-cash and procure-to-pay, particularly during the creation or modification of business partners. Ad hoc screenings allow manual checks of individual names or addresses—for example, when dealing with visitors or one-time transactions.

The application utilizes external data sources, such as sanction lists maintained by Mendel Verlag, which are regularly updated. SAP Watch List Screening integrates seamlessly with SAP S/4HANA and can also be connected to other systems via APIs. The analysis of hits is performed using a web-based user interface powered by SAP Fiori, where identified hits can be classified and documented for decision-making—as either true hits or false positives.

Why One Is Enough
As outlined in the descriptions of SAP Business Integrity Screening and SAP Watch List Screening, both are capable of performing sanction and embargo screening. In certain cases, it may be sufficient to use only one of these options for the sanction and embargo screening process.

4.4.3 Configuration

SAP Business Integrity Screening and SAP Watch List Screening must be configured before implementation. In the following sections, we explore how advanced payment management can be technically integrated with these solutions.

Screening Tool Integration of Third-Party Providers and Modules
The integration of third-party software is generally possible and can be managed via the application configuration in Transaction SPRO, under **Financial Supply Chain Management** • **Advanced Payment Management** • **Basic Configuration** • **Application Configuration**.
In addition to SAP Business Integrity Screening and SAP Watch List Screening, custom solutions or third-party solutions can also be connected for sanction and embargo screening as well as fraud detection.

SAP Business Integrity Screening

Since SAP S/4HANA version 2022 Initial Shipment Stack, advanced payment management has enabled the integration of SAP Business Integrity Screening. Payment media are processed centrally via advanced payment management, and the screening services are integrated into SAP Business Integrity Screening via enhancement spot /PF1/ES_BIS, which supports the screening process in both the online and batch process. Through the mentioned enhancement spot, SAP provides BAdI methods /PF1/BADI_BIS_DATA_MAPPING and /PF1/BADI_BIS_ONLINE_DETECTION, which can be extended and provide data from the file handler database.

Within SAP Business Integrity Screening, there is business content (FRA_S4_APM_BASIC_CONTENT) and BAdIs, such as FRA_BADI_APM_BLOCK_CHANGE, that can be activated and utilized. This significantly simplifies and accelerates the integration process.

If SAP Business Integrity Screening is not operated on the same machine as advanced payment management, then a basic technical connection via Transaction SM59 must be established between SAP Business Integrity Screening and the SAP system running advanced payment management.

SAP Business Integrity Screening is integrated into advanced payment management processes via the application management framework and the provided enhancement spot, /PF1/ES_BIS, as well as the **Define Local Application Management Areas** configuration activity.

Payment transactions are screened based on the parameters set in the local application area of SAP Business Integrity Screening. These parameters determine how and when compliance checks, such as embargo validations, are triggered. The configuration can be made in customizing for advanced payment management under **Basic Configuration** • **Application Configuration** • **Define Local Application Management Areas**.

The following parameters are available:

- Parameter **PI** enforces real-time screening for each individual transaction. This option is only effective if the item embargo check is configured in the system. It is especially useful in business scenarios that require immediate compliance validation for every single payment transaction.

- Parameter **PO** initiates a batch screening process for each payment order rather than for each transaction. To use this functionality, both the order embargo check and the item embargo check must be configured. This method is suitable for organizations processing a high volume of payments, where performance and system load need to be considered.

- Parameter **SIZE** implements a hybrid screening strategy based on the size of the incoming payment order. The system determines whether to perform online or batch screening by evaluating a configurable attribute related to order size, typically available in the payment file. Smaller payment orders are checked in online mode, while larger ones are handled through batch processing. The configuration requirements for this parameter are the same as for the **PO** parameter, meaning that both order and item embargo checks must be active. This approach is particularly beneficial when balancing system performance and compliance requirements in environments with varying payment volumes.

4.4 Integration of Intelligent Services

The parameters can be maintained within the **Define Local Application Management Areas** configuration activity. This activity is also used for the integration of additional modules, such as cash management. In Figure 4.126, we have added the application area for SAP Business Integrity Screening. Within this transaction, the clearing area can be selected, and an application ID (**App.ID**) is defined as a key field. SAP Business Integrity Screening can be selected via the local management system (**Lcl Mgt S**) field.

SAP Business Integrity Screening is predelivered by SAP and can be selected accordingly. The local application area description fields essentially exist twice and can be used for both short and long designations. The alternative local application area ID code field (**Alt.AppID**) is optional and only needs to be filled in if an alternative ID exists from the past.

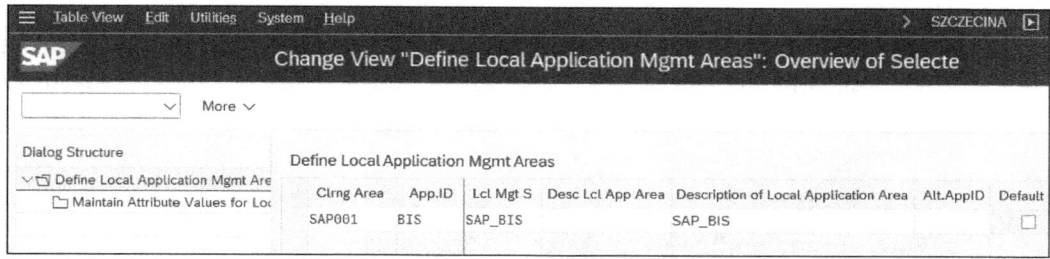

Figure 4.126 Application Area for SAP Business Integrity Screening

After the **Clrng Area** and application ID (**App.ID**) key fields have been defined and linked to the application, the parameters for the local application area can be maintained via the dialog structure.

In Figure 4.127, the **STRATEGY** attribute, which was predelivered, can be selected and maintained with the **PI** (post item) attribute value. In this case, we have set the parameter **PI** to enable individual transaction verification in online mode.

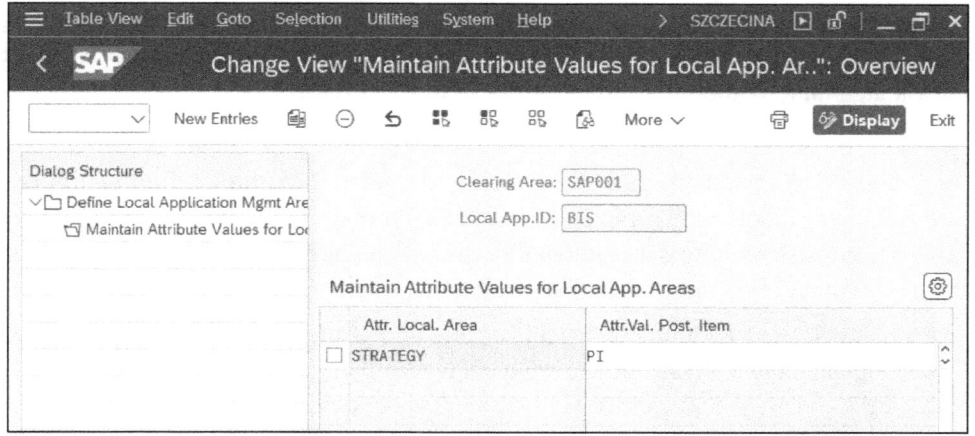

Figure 4.127 STRATEGY Attribute

345

Overall, the fact that SAP provides preconfigured content for SAP Business Integrity Screening, and specifically for the integration of advanced payment management, significantly simplifies the initial setup.

However, it is key to implement sanction, embargo, and fraud screening holistically within SAP Business Integrity Screening—ideally in the upstream system, long before the payment processing begins.

SAP Watch List Screening

With SAP Watch List Screening, recipient names and address details can be subjected to sanction and embargo screening. Sanction and embargo screening is triggered during enrichment and validation processing and performed through an external system which checks specified payment order (PO) and payment item (PI) types against predefined lists. The results of the check are returned to advanced payment management and indicate whether the payment can be further processed or, for a hit, if the payment is flagged and sent back to advanced payment management exception handling, where a response can be defined.

To integrate SAP Watch List Screening, there are prerequisites that need to be fulfilled:

- **Purchase and configure SAP Watch List Screening**
  SAP Watch List Screening must be licensed separately from SAP. Once licensing is complete, SAP Watch List Screening must be configured.

- **Connection between operating system and SAP Watch List Screening**
  SAP Watch List Screening must be connected to the advanced payment management operating system. For this, an HTTP(S) connection must be established between SAP Watch List Screening and the operating system on which advanced payment management operates. The communication is maintained via Transaction SM59. In Transaction SPRO activity **RFC Configuration for Watchlist Screening**, the integration between the Cloud Foundry application and the system operating advanced payment management can be maintained. **RFC Configuration for Watchlist Screening** can be found under **Financial Supply Chain Management • Advanced Payment Management • External Interfaces • SAP Watch List Screening**.

SAP Watch List Screening is linked to advanced payment management's enrichment and validation function. Within enrichment and validation, predefined configurations are already available, which can be customized. Through enrichment and validation, SAP Watch List Screening is integrated into the payment process. The respective validation rules must be configured in the system.

Within SAP S/4HANA Cloud Public Edition, predefined check set IDs for outbound validation are already provided, as shown in Table 4.7.

Business Object	Check	Check Set ID	Check Description
Payment order	Outgoing payment order cross item	14	External authorization check against SAP Watch List Screening on payment order level
Payment items	Payment Item in outgoing payment orders recipient	13	External authorization check against SAP Watch List Screening on payment item level

Table 4.7 Predefined Checkset IDs for Outbound Validation

4.5 Summary

With its multilayered functionalities and seamless integration into upstream and downstream processes—such as SAP S/4HANA Finance, cash management, and in-house banking—advanced payment management serves as a versatile central hub for both inbound and outbound payment transactions. Its adaptability makes it suitable for businesses of all sizes, from large corporations to smaller enterprises. We hope this chapter has helped convey a high-level understanding of advanced payment management's key processing steps and functionalities.

Chapter 5
SAP Multi-Bank Connectivity

At the heart of every payment lies a bank, the lifeblood of every transaction. In SAP, your bridge to this world of financial currents is SAP Multi-Bank Connectivity. It's more than just a tool: It's a portal that links your system with the pulse of global banks, ensuring that your payments move seamlessly and your business stays in harmony with the financial symphony of the world.

For years, SAP relied on third-party tools and providers to enable bank connectivity from the SAP system to the banks. These third-party solutions were essential for facilitating secure and seamless communication between the organization's financial systems and their banking partners, allowing for the execution of payments, the reception of bank statements, and the integration of various banking services.

To address this gap, SAP developed SAP Multi-Bank Connectivity, a solution that enables seamless communication between SAP systems and banks.

SAP Multi-Bank Connectivity provides a multibank digital channel between SAP systems and their banks, enabling organizations to manage their banking transactions more efficiently. To integrate an SAP system with SAP Multi-Bank Connectivity, a connector is utilized, which facilitates smooth communication between the two systems. SAP Multi-Bank Connectivity combines the benefits of SAP's robust financial management tools with seamless integration into the SWIFT network or the use of other banking channels. This integration allows businesses to execute payments, receive bank statements, and perform other banking functions securely and efficiently, all within a unified SAP environment. The solution ensures scalability and flexibility while offering secure, reliable, and real-time connectivity with various banking institutions.

In this chapter, we will first delve into an overview of SAP Multi-Bank Connectivity: what it is, why it was developed, and its core purpose in the landscape of financial integration (Section 5.1). You'll learn about the motivations behind this solution, how it has evolved, and why it is such a crucial tool for modern enterprises that deal with banking partners worldwide.

Next, we will explore the different connectivity options available through SAP Multi-Bank Connectivity (Section 5.2). You'll discover how you can connect your company's SAP system to banks using this framework and what options are available to suit your business needs.

Then, we'll move on to the functional and process-driven aspects of SAP Multi-Bank Connectivity (Section 5.3). We'll cover how messages are sent and received, how the system establishes and maintains connections, and what key applications you can use within the SAP ecosystem to manage these processes. We'll highlight the most important features and functionalities that enable secure, efficient, and standardized processing of bank communications.

We'll also briefly walk you through the onboarding process for SAP Multi-Bank Connectivity (Section 5.4). Finally, this chapter will walk you through the configuration landscape for SAP Multi-Bank Connectivity (Section 5.5). We'll outline the key steps and considerations needed to set up and configure this connectivity within your SAP system, ensuring a robust and reliable connection to your banking partners.

Let's get started and explore how SAP Multi-Bank Connectivity can become your company's lifeline to connect to global banking networks.

5.1 Introduction to SAP Multi-Bank Connectivity

After a payment file is created, SAP Multi-Bank Connectivity picks it up and sends it to the bank for further processing. SAP Multi-Bank Connectivity supports various channels for communication with the bank, including EBICS, SWIFT, host-to-host communication, and APIs, which are detailed further in Section 5.2. However, from the customer's perspective, there is no need to worry about the connectivity, as SAP handles all aspects of communication and integration with the bank. This ensures a seamless and secure transfer of payment files without requiring customers to manage the underlying technical complexities.

Once the payment file is received and processed by the bank, the bank generates a payment acknowledgment and sends it back to SAP S/4HANA via SAP Multi-Bank Connectivity. This acknowledgment updates the payment status monitor in SAP S/4HANA, providing the organization with real-time information about whether the payment was successfully processed or rejected by the bank. The payment status monitor ensures transparency by informing the organization of the outcome, allowing for timely follow-up actions in case of payment rejections or issues, thus streamlining the payment reconciliation process.

As a final step, banks send bank statements back to SAP S/4HANA via SAP Multi-Bank Connectivity to update financial accounting and cash management. These statements provide detailed information about all transactions processed through the bank account, including payments, deposits, and fees. Once received, the bank statements are automatically integrated into SAP, enabling organizations to reconcile their accounts and update cash positions in real time.

Currently, SAP Multi-Bank Connectivity is one of the easiest and most efficient ways to connect SAP systems to banks. Due to its growing popularity, SAP is continuously adding new functionalities to enhance its capabilities. One such feature is the ability to update real cash balances based on API connectivity, which is currently available for select banks. As SAP Multi-Bank Connectivity continues to evolve, we can expect more updates and improvements in the future, expanding its capabilities and supporting a wider range of banks and payment channels.

In this section, we will focus on the technical implementation and landscape, explaining how SAP Multi-Bank Connectivity is integrated with other systems and providing an overview of the technical architecture. In addition, we will discuss the recent migration of SAP Multi-Bank Connectivity to SAP Business Technology Platform (SAP BTP) and its deployment to Cloud Foundry.

5.1.1 System Landscape and Components

SAP Multi-Bank Connectivity fundamentally consists of two components that not only enable a bank-to-ERP connection but also facilitate an ERP-to-ERP linkage. These components are the connector for SAP Multi-Bank Connectivity and the cloud-based SAP Multi-Bank Connectivity solution itself, which operates as a service on SAP BTP or on Cloud Foundry.

Figure 5.1 illustrates how SAP Multi-Bank Connectivity is used to connect your SAP system to various banks.

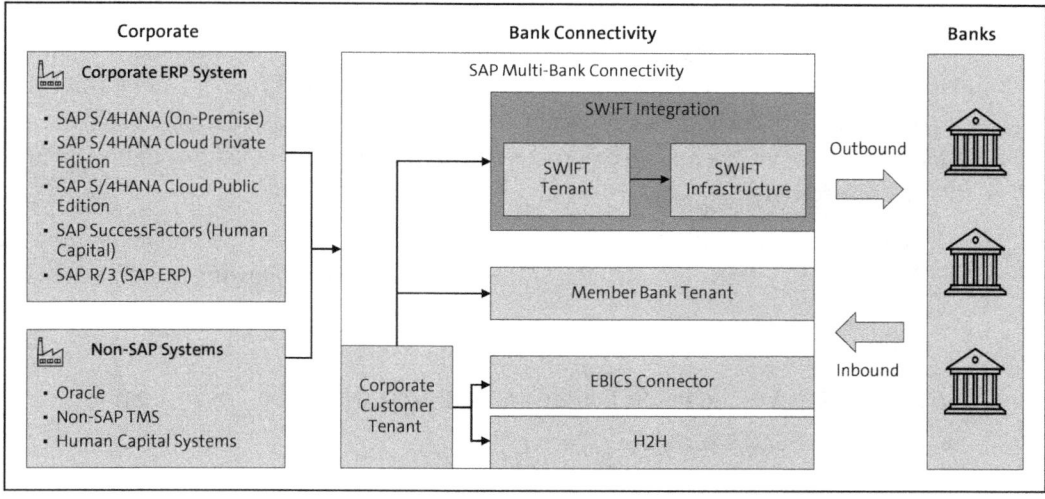

Figure 5.1 SAP Multi-Bank Connectivity Within an SAP Landscape

You can see how on the left side there are SAP systems and other, non-SAP systems, all of which are connected to SAP Multi-Bank Connectivity, which sits centrally. SAP Multi-Bank Connectivity acts as a single point of contact for all your banks, using various protocols such as SWIFT, EBICS, host-to-host communication, or other connectors. This setup supports both outbound communications, like sending payment files, and inbound communication, such as receiving bank statements and payment status updates. It's important to note that SAP Multi-Bank Connectivity works with all SAP versions and can also be connected to non-SAP systems, providing a unified and streamlined approach to bank integration.

In the following sections, we will dive into these two components.

Connector for SAP Multi-Bank Connectivity

The connector primarily serves to manage the secure connection between the ERP system and the solution hosted in the cloud. The connector for SAP Multi-Bank Connectivity establishes the link between a company's SAP system (applying to all of SAP's ERP solutions) and the SAP Multi-Bank Connectivity network, which in turn facilitates external connections to the banking world. It supports various communication protocols (e.g., SWIFT, EBICS, host-to-host) and ensures that message formats can be converted and interpreted correctly.

As an integrated component, it automates the exchange of financial messages, such as payment instructions (e.g., payment files in XML format according to ISO 20022, SWIFT message type, or local formats), account statements, and status reports. The connector handles tasks like secure transmission of messages, communication logging, and monitoring of message exchanges. Through integration with SAP S/4HANA or SAP ERP, the connector enables seamless end-to-end execution of payment processes, reduces manual interventions, and enhances transparency and security in payment transactions.

In addition to facilitating ERP-to-bank connections, the connector also enables connections between SAP S/4HANA Finance for advanced payment management and third-party systems. It includes a file picker, allowing payment instructions to be alternatively uploaded via file upload or written to a path via file download through SAP Multi-Bank Connectivity.

The connector was delivered with a downgrade by SAP, making it available not only for SAP S/4HANA systems but also for SAP ERP systems. The following systems are supported:

- SAP ERP (must be SAP ERP 6.0 EHP 0 or higher)
- SAP S/4HANA Cloud Public Edition and SAP S/4HANA Cloud Private Edition
- On-premise SAP S/4HANA

> **[»] How to Install the Connector**
>
> Starting with SAP S/4HANA 1809, the connector is automatically included. For older SAP S/4HANA or SAP ERP versions, if the connector is not directly supplied, it must be installed manually. The steps for installation are detailed in SAP Note 1781614 (*https://me.sap.com/notes/1781614uuup*).

5.1 Introduction to SAP Multi-Bank Connectivity

SAP Multi-Bank Connectivity

SAP Multi-Bank Connectivity is a cloud-based bank communication solution operated by SAP and offered as software as a service (SaaS). It runs on SAP BTP, Cloud Foundry environment. The solution supports common banking communication protocols such as EBICS, host-to-host communication, and SWIFT, enabling standardized and secure communication between corporations and banks.

Technically, SAP Multi-Bank Connectivity can be visualized like a high-rise building in which each corporate customer has its own apartment (tenant). These tenants are managed by SAP and ensure secure, compliant data segregation within the cloud infrastructure.

Access to SAP Multi-Bank Connectivity is provided to both corporate customers and banks. Banks can register as member banks and offer their clients a direct connection via SAP Multi-Bank Connectivity. Although banks do not receive a dedicated tenant like corporations, they are provided with a communication interface within the platform. This allows them to exchange messages with corporate tenants and offer value-added services, such as real-time cash management.

This architecture significantly streamlines bank connectivity, reduces implementation effort, and provides a centralized, standardized approach to connecting with banks worldwide.

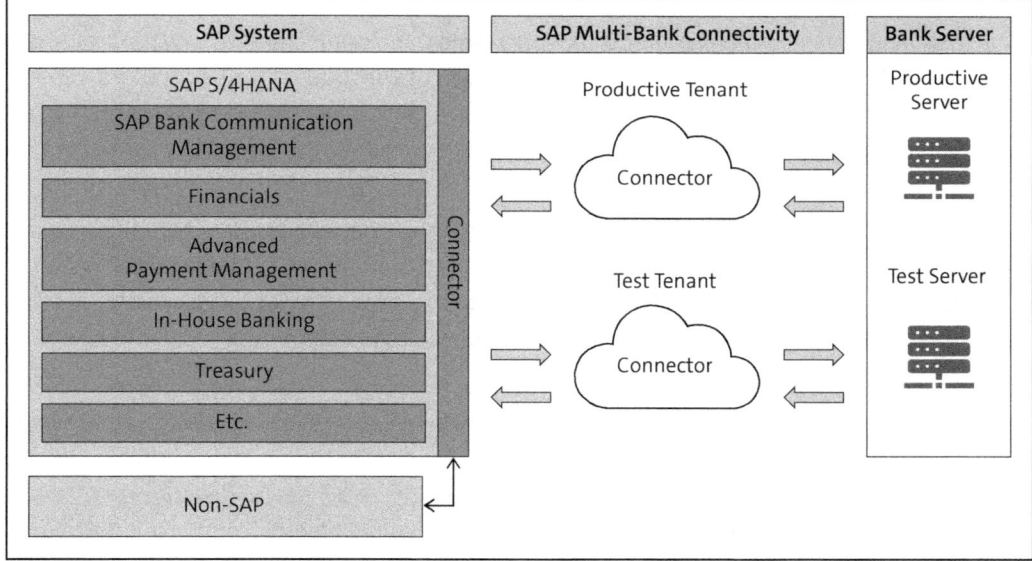

Figure 5.2 SAP Multi-Bank Connectivity: System Environment

As shown in Figure 5.2, the SAP system, incorporating all the solutions in SAP described in this book—including normal payment transformation finance, SAP Bank Communication Management, advanced payment management, in-house banking, treasury, and

353

other related areas—will communicate seamlessly with the banks via SAP Multi-Bank Connectivity. This integration ensures smooth, secure, and efficient data exchange between SAP and banking partners. The SAP Multi-Bank Connectivity setup consists of a productive tenant, which facilitates the real and live exchange of files with the banks for day-to-day operations, ensuring that all financial transactions are processed promptly and accurately. In addition, a test tenant environment is also established, enabling thorough testing, verification, and validation of configurations and processes before deploying them in the live environment.

5.1.2 Cloud Foundry Migration

The migration of SAP Multi-Bank Connectivity to SAP BTP, Cloud Foundry environment is part of SAP's broader strategy to modernize its cloud infrastructure. Previously, SAP Multi-Bank Connectivity operated on SAP BTP, Neo environment. However, SAP is phasing out Neo in favor of Cloud Foundry, which offers a more open, scalable, and cloud-native architecture.

For customers, this change means that new SAP Multi-Bank Connectivity tenants are now provisioned exclusively on Cloud Foundry. Existing customers with tenants on Neo are required to migrate to the new environment. The migration does not affect the customer's ERP system, but only the cloud-based SAP Multi-Bank Connectivity connector component.

The transition involves several technical steps, including the provisioning of a new Cloud Foundry tenant by SAP, reconfiguration of communication endpoints, reestablishment of certificates, and validation of payment flows. Integration points such as APIs, payment status tracking, and file exchange endpoints may change as part of the migration.

The migration itself is free of charge, but organizations are responsible for configuration efforts and thorough end-to-end testing with both banks and internal systems. Overall, the migration to Cloud Foundry brings improved scalability, security, and flexibility and ensures that SAP Multi-Bank Connectivity remains aligned with SAP's long-term cloud strategy.

> **Migration Deadline**
>
> SAP has announced that SAP BTP, Neo environment will be sunset by December 2028 for SAP Multi-Bank Connectivity. Therefore, all existing corporate tenants must be migrated to the Cloud Foundry environment by this date. For detailed guidance on the migration process, please refer to *Migration Guide for SAP Multi-Bank Connectivity from SAP BTP Neo to Cloud Foundry*, available at *https://help.sap.com/docs/SAP_MULTI_BANK_CONNECTIVITY/a24306513b4d4097848830dc1e3cc7c5/d37558e862584e1f998a3d dc73ec5553.html*.

To migrate from SAP Multi-Bank Connectivity on SAP BTP, Neo environment to SAP BTP, Cloud Foundry environment, you must follow a structured process, as shown in Figure 5.3.

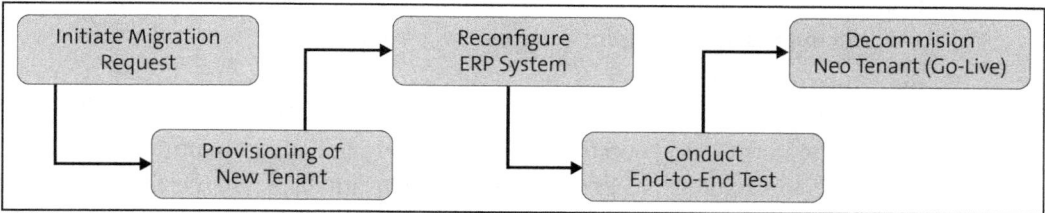

Figure 5.3 Cloud Foundry Migration Steps

Let's look at each of these steps in turn:

1. **Initiate migration request**
   The first step is to raise a customer case via SAP for Me or the SAP Support Portal, indicating the request to migrate to Cloud Foundry. This process must be initiated using a valid and authorized S-user account with the appropriate permissions to create support incidents. The case should be created with the following details:
   - **Component**: LOD-FSN-DEV-MGR
   - **Short Description**: Cloud Foundry migration [customer name]
   - **Environment Details**:
     - **Current Environment** (e.g., Neo)
     - **Target Environment**: Cloud Foundry
   - **Subaccount Details** (region, subaccount ID, and subaccount name)
   - **Migration Justification**: Reason for the migration
   - **Timeline**: Preferred and possible go-live date for the migration
   - **Additional Context**: Any relevant dependencies (e.g., active services, business-critical apps)

   In the **Comments** field of the customer case, you'll also need to include the following information:
   - Names of existing test and productive Neo tenants
   - Possible go live date
   - Contact persons at your organization whom SAP can collaborate with during the migration project

2. **Provisioning of new tenant**
   SAP will then provision a new SAP Multi-Bank Connectivity tenant in the requested Cloud Foundry region. As part of this process, SAP will provide the following:
   - Access credentials for the new SAP Multi-Bank Connectivity tenant
   - Endpoint URLs and region-specific details

- Initial configuration guidance (if applicable)
- Any necessary onboarding documentation

Once the new tenant is provisioned, you can begin with the migration of configurations and integrations from the existing Neo environment, in coordination with SAP support and based on the jointly defined migration timeline.

3. **Reconfigure ERP systems**
You must reconfigure your ERP systems (such as SAP S/4HANA or SAP ERP) to connect to the new Cloud Foundry tenant, including updating communication channels, RFC destinations, and certificates. All relevant bank partners must also be informed of the changes, as the communication endpoints and IP addresses will be different from those in the Neo environment.

4. **Conduct end-to-end tests**
A full set of end-to-end tests should be conducted to ensure that payment files, status messages, and bank statements are processed correctly in the new environment.

5. **Production switch and deactivation**
Once testing is successful, the production switch can take place, and the Neo tenant should be deactivated to prevent duplicate processing.

6. **Decommission Neo tenant**
Finally, the Neo tenant will be decommissioned by SAP no later than December 31, 2025. You must ensure that all operations have been fully transitioned by this date.

> **[+] Allocate Enough Time for the Migration**
> SAP recommends planning at least six months for the migration. SAP provides documentation, technical guidance, and support throughout the migration process. Starting early is strongly recommended to allow for sufficient testing and a smooth transition.

5.2 Connectivity Options

SAP offers more than just a simple connector with SAP Multi-Bank Connectivity. Because SAP Multi-Bank Connectivity can be used not only by companies but also by banks, it is increasingly evolving into a platform where messages can be exchanged.

As a prerequisite for establishing a bank connection, it is essential to determine in advance the type of connectivity that can and should be used for each bank. The contractual requirements also should be clearly defined before the commencement of such a project.

We have already described the standard connectivity types (EBICS, SWIFT, host-to-host) in detail in Chapter 2. In this section, we will explore these connectivity types as they relate to SAP Multi-Bank Connectivity, walking you through the questionnaire provided by SAP in the process. As Figure 5.4 shows, SAP Multi-Bank Connectivity can support multiple bank connectivity options. In SAP Multi-Bank Connectivity, *member banks* are the financial institutions that are connected to SAP Multi-Bank Connectivity. These banks participate in the SAP Multi-Bank Connectivity network and support the exchange of financial messages—like payment files, statements, and confirmations— via SAP Multi-Bank Connectivity.

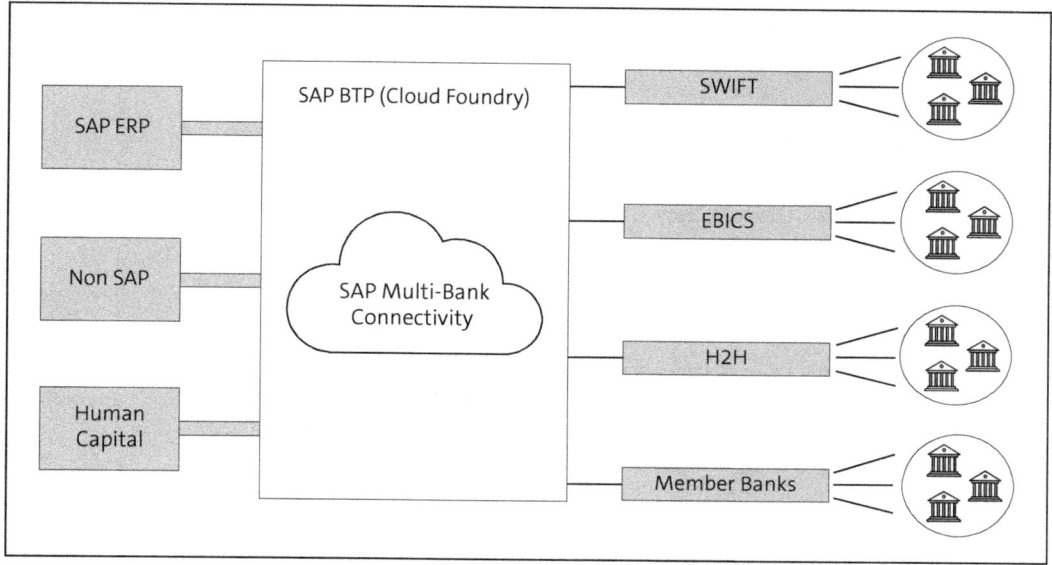

Figure 5.4 SAP Multi-Bank Connectivity: Connectivity Options

5.2.1 EBICS

The EBICS connectivity method is particularly widespread in the European region and is trending. As a cost-effective alternative to SWIFT and with a significantly higher standardization compared to host-to-host connections, EBICS is a popular method of integration both on the corporate side and the banking side. Because EBICS not only facilitates bank connections but also incorporates signature logic at the core of its solution, the following key information is relevant to the SAP Multi-Bank Connectivity team:

- **EBICS type**
  EBICS comes in various "flavors" and versions, with the most widely used being EBICS 2.5 and EBICS 3.0. These versions primarily differ in their order types and encryption methods. In addition to version differences, there are also country-specific and even bank-specific variations. SAP offers two general standards: EBICS in the format of the German Banking Industry Committee (DK) and the French variation, EBICS T.

> **German EBICS and EBICS T**
>
> The EBICS standard, according to Deutsche Kreditwirtschaft (DK), is the productive payment transaction standard in Germany. It supports the automatic processing of payments, including security checks and distributed electronic signatures (VEU). Common order types such as CCT, CCD, or C53 are standardized and strictly regulated. In contrast, the French variant EBICS T—where *T* stands for *transport*—is primarily used for pure file transfer without automatic processing by the bank. Signature checks are optional or may be omitted entirely.

- **EBICS server**
  Banks have the option to provide a testing environment alongside the EBICS production server. If a test environment is available, this should be indicated on the form provided by the SAP team during the scoping phase.

> **EBICS Test Server**
>
> Experience indicates that a test environment for EBICS is not guaranteed. Due to cost considerations, many banks opt not to have a test server. If no test server is available, this should be factored into the timeline for the go-live planning.

Figure 5.5 shows a portion of the questionnaire provided by SAP that must be completed and returned to the SAP team.

Section 2: Connectivity Protocol			
EBICS Protocol	TEST	PRODUCTION	NOTES (Eg. German Ebics, French Ebics "T")
EBICS v 2.5 (H004)	☐	☑	*German EBICS*
EBICS v 3.0 (H005)	☐	☐	*Please Select...*

Figure 5.5 EBICS Connectivity Protocol

- **EBICS connectivity parameters**
  The essence of the EBICS connection method, as detailed in Chapter 2, is rooted in the HTTP(S) communication protocol. To establish an EBICS connection, essential parameters provided by the bank are necessary, as follows:

  – Host ID

  The host ID identifies the bank's EBICS server. It is a unique identifier assigned by the bank and is used by the EBICS client (usually the corporate customer) to connect to the correct server.

  – EBICS server URL

  The EBICS server URL is the web address (endpoint) through which the EBICS client communicates with the bank's EBICS server. This is where the client sends EBICS messages and receives responses.

- **Partner ID**
  The partner ID represents the corporate customer (company) in the EBICS system. It is assigned by the bank and is used in combination with the user ID to identify and authenticate the client organization.
- **Bank BIC ID**
  This is the Bank Identifier Code (BIC), also known as the SWIFT code. It uniquely identifies the bank in international financial transactions and is also used within SAP Multi-Bank Connectivity to define the bank institution.
- **EBICS user IDs**
  These are the user-specific IDs assigned to individuals within the corporate customer's organization who are authorized to use EBICS. Each user ID is linked to a set of permissions and is used for authentication, signing, and transmission of orders.

> **Static and Dynamic User** [+]
>
> Using SAP Multi-Bank Connectivity, EBICS users can be determined dynamically through customization or assigned statically to the payment medium. For static assignment, an EBICS user with the necessary authorization is linked to the payment medium, independently of the SAP Bank Communication Management approver, authorizing the payment medium at the bank. The static EBICS user, commonly referred to as the *corporate seal* at financial institutions, should be coordinated with the bank prior to implementation.

Figure 5.6 shows the portion of the questionnaire provided by SAP that relates to environments.

Section 3: Environments			
Is there a dedicated Test environment available?		No	
		TEST	PRODUCTION
Host ID			TUBDDEDD
EBICS Server URL			https://ebics.hsbc.de/ebics
Partner ID			
Bank BIC ID			TUBDDEDDXXX
		USER ID	USERNAME
User ID(s)		01	
		01	
		01	
		01	
		01	
		01	
		01	
		01	Technischer Teilnehmer

Figure 5.6 EBICS Environments

5 SAP Multi-Bank Connectivity

- **Order types**

    Order types define whether a transaction is a pull or push order and specify the type of message being exchanged. It is essential for the SAP Multi-Bank Connectivity team to know the order types agreed upon with the bank to configure them in SAP Multi-Bank Connectivity effectively. They also must know which order types are applicable and which codes should be used, as shown in Figure 5.7.

MESSAGE TYPE	SELECT IF APPLICABLE	Order Type	File naming convention
Credit Transfer - PAIN.001	☑	CCT	
Direct Debit - PAIN.008	☑	CDD	
Other Payment format (Sepa Urgent)	☑	CCU	
Payment Status Reports - PAIN.002	☐	CRZ	
CAMT.053	☑		
CAMT.052	☑		
MT940	☑	STA	
MT942	☐		
BAI2	☐		
Other Statement format	☐		

EBICS v 2.5 (H004)

Figure 5.7 EBICS Order Types 2.5

5.2.2 Host-to-Host

The host-to-host connection (see Figure 5.8) through SAP Multi-Bank Connectivity offers direct integration with the bank's backend systems, eliminating the need for intermediary networks like SWIFT. This connection relies on a dedicated link between an enterprise's SAP Multi-Bank Connectivity tenant within SAP BTP and the respective banking system. This direct connectivity facilitates the exchange of various financial messages, such as outgoing payment instructions (e.g., pain.001), incoming electronic account statements (e.g., camt.053, MT940), and payment status updates (e.g., pain.002), alongside other possible message types. It's essential that only pre-agreed-upon messages with the bank are exchanged.

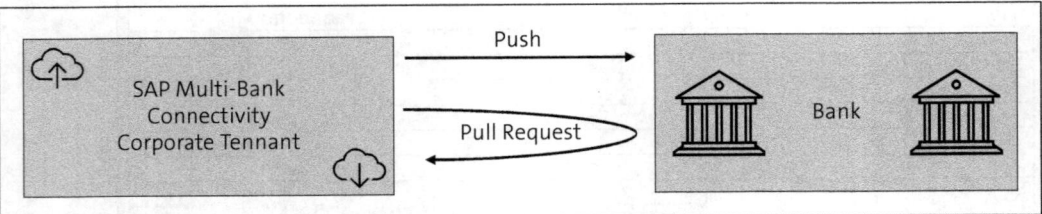

Figure 5.8 Host-to-Host Connectivity

Technically, host-to-host connections are typically established using secure protocols like SFTP or SOAP with mutual authentication through certificates. Within the SAP Multi-Bank Connectivity tenant, details of the host-to-host connections to banks are recorded (e.g., server addresses, ports, authentication credentials), and certificates are managed, with message mappings configured as necessary. During the preboarding phase, these specifics must be shared with the SAP team.

In this section, we delve into the host-to-host connectivity protocols that underpin direct corporate-to-bank integrations. Specifically, we will explore the three most common protocols supported within SAP Multi-Bank Connectivity: SFTP, AS2, and SOAP. These protocols represent advanced, secure methods for exchanging financial data and ensuring seamless communication between enterprise systems and banking platforms.

> **Host-to-Host Bank Integrations**
> A maximum of five host-to-host bank connections per corporate entity is recommended and supported via SAP Multi-Bank Connectivity. If more than five connections are needed, SAP recommends actively using SWIFT in combination with SAP Multi-Bank Connectivity.

Connectivity Protocol

SAP offers various types of host-to-host connectivity. The following connection types are supported by default and are the most common on the market for corporate-to-bank connection via host-to-host:

- **Secure File Transfer Protocol (SFTP): push/pull**
  SFTP is one of the most commonly used protocols for secure file exchange between companies and banks within the framework of a host-to-host connection. In a typical SAP Multi-Bank Connectivity connection, SFTP is used either in push mode (SAP Multi-Bank Connectivity actively sends data to the bank's system) or in pull mode (SAP Multi-Bank Connectivity actively retrieves data from the bank's server). Communication is encrypted and authenticated, typically through SSH key pairs (public/private key). Files are placed in predefined directories and are regularly retrieved or provided.

- **Applicability Statement 2 (AS2)**
  AS2 uses the HTTP or HTTPS protocol as a transport pathway, with the files to be transmitted additionally encrypted and digitally signed. The transmission is carried out synchronously or asynchronously, accompanied by an electronic acknowledgment of receipt (Message Disposition Notification [MDN]), ensuring that the data has been correctly received and decrypted.

- **Simple Object Access Protocol (SOAP)**
  SOAP is a standardized XML-based protocol for communication with web services. SOAP messages consist of strictly defined XML structures, which are transmitted via HTTP or HTTPS. Interaction follows a request-response model; SAP Multi-Bank

Connectivity submits a request to the bank's web service and immediately receives a response, such as the processing status or requested information.

In the questionnaire provided by SAP for host-to-host connections, you must indicate which type of protocol has been agreed upon with the bank, as well as whether it is a test or production environment, as shown in Figure 5.9.

Section 2: Connectivity Protocol		
TYPE	TEST / PRE-PRODUCTION	PRODUCTION
SFTP (Push/Pull)	☐	☐
AS2	☐	☐
SOAP	☐	☐

Figure 5.9 Section 2: Host-To-Host Connectivity Protocol

Environmental Details

Under the environmental details (shown in Figure 5.10), the address data of the bank server is recorded in detail. These details are provided by the bank. In this section, we will provide a high-level overview of the protocols supported by SAP Multi-Bank Connectivity, which are queried in the questionnaire by the SAP Multi-Bank Connectivity team.

Section 3: Environments Details		
	TEST / PRE-PRODUCTION	PRODUCTION
Mailbox URL or IP address (SFTP) or Endpoint (AS2/SOAP)		
Port Number		
Customer Username		
Inbound directory for payments		
Outbound directory for responses		

Figure 5.10 Section 3: Host-To-Host Environmental Details

Let's now take a closer look at each of the details:

- **Mailbox URL or IP address (SFTP) or endpoint (AS2/SOAP)**
  This is the address where the customer system connects to exchange data with the bank. It differs depending on the protocol:
  - **SFTP example: sftp.testbank.com or 192.168.200.55**
    SAP Multi-Bank Connectivity uses this IP or domain to upload or download files to/from the bank's SFTP server.
  - **AS2 example: https://as2.testbank.com/as2/receive**
    SAP Multi-Bank Connectivity sends AS2 messages (e.g., payment batches or acknowledgments) to this endpoint.
  - **SOAP example: https://api.testbank.com/soap/PaymentService**
    SAP Multi-Bank Connectivity sends SOAP requests (e.g., to initiate a payment or query a transaction status) to this web service endpoint.

The address usually differs between TEST (e.g., using testbank.com) and PRODUCTION (e.g., bank.com) to separate testing from live data processing.

- **Port number**
  This defines the network port used to establish the connection to the bank's system. Each protocol typically uses a standard port, though this can be customized:
  - SFTP
    - Default port: 22
    - Used for secure file transfers via SSH.
    - Example: sftp.testbank.com:22
  - AS2/SOAP (over HTTPS)
    - Default port: 443
    - Secure communication over SSL/TLS.
    - Example: https://api.bank.com:443/as2receive

  In rare cases (e.g., internal environments or legacy systems), custom ports like 2222, 8443, etc., might be used. Always confirm with the bank's technical team.

- **Customer user name**
  This is the authentication identifier used to access the bank's system. It also defines user-specific permissions, access directories, and security policies. It also differs depending on the protocol:
  - SFTP
    - User name: mbc_corp_xyz
    - Used to log into the bank's SFTP server and access dedicated folders for that customer.
  - AS2
    - AS2 identifier: XYZCompany_AS2
    - A globally unique string used to identify the sender and receiver in AS2 transactions.
  - SOAP
    - User name: apiuser_xyz
    - Used with basic authentication (username/password) or in combination with a certificate.

- **Inbound directory for payments**
  This directory is used to upload payment files to the bank for processing. These typically contain payment instructions like SEPA credit transfers, direct debits, or cross-border payments. Let's look at a few examples:
  - SFTP example: /in/payments/ or /upload/incoming
    SAP Multi-Bank Connectivity places files like pain.001.001.03.xml here for the bank to process.

- **AS2/SOAP example: No physical folder; data is sent via HTTP POST**
  The concept remains: This is the "incoming" stream of transactions to the bank.
- **Example file: SEPA_Payments_20250417_001.xml**
  A pain.001 file uploaded to the SFTP /in/payments/ folder.

> **Note**
> Some banks require specific file naming conventions or daily limits. Uploads may trigger immediate validation or a scheduled batch job on the bank's side.

- **Outbound directory for responses**
  This is the folder (or response channel) where the bank places feedback files or status reports, which SAP Multi-Bank Connectivity or a customer system retrieves for processing. Let's look at some examples:
  - **SFTP example: /out/responses/ or /download/acknowledgments**
    This contains files like the following:
    - pain.002 (status of submitted payments)
    - camt.054 (detailed transaction reports)
  - **AS2/SOAP: Responses often come as synchronous HTTP responses or via push callbacks**
    For example, SAP Multi-Bank Connectivity sends a payment request via SOAP and receives a status confirmation as part of the response XML.
  - **Example files in SFTP**
    - pain.002.001.03_Acknowledgement_20250417.xml
    - camt.054.001.04_CreditNotifications_20250417.xml

  SAP Multi-Bank Connectivity must regularly poll or monitor this directory to fetch and process the files. This step is crucial for reconciliation, monitoring errors, and compliance.

Transport Authentication

After specifying the exact address data and transport method, the authentication method agreed upon with the bank must be communicated to the SAP Multi-Bank Connectivity team, as shown in Figure 5.11. This can be through certified encryption or via user name and password, depending on the selected type of encryption.

The choice of authentication method depends on the protocol used, the security requirements of the bank, and regulatory obligations. Although user name and password combinations may be acceptable for testing, production environments generally require stronger methods: SSH key pairs for SFTP and X.509 certificates for AS2 and SOAP. Proper key and certificate life cycle management (creation, expiration, renewal, and revocation) is crucial for maintaining secure, reliable host-to-host banking interfaces.

For state-of-the-art encryption in bank connectivity, it is advisable to consult with the bank and cryptography experts.

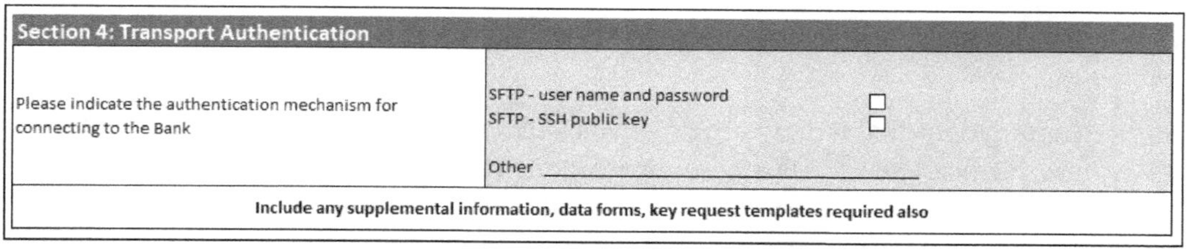

Section 4: Transport Authentication		
Please indicate the authentication mechanism for connecting to the Bank	SFTP - user name and password SFTP - SSH public key Other _____	☐ ☐
Include any supplemental information, data forms, key request templates required also		

Figure 5.11 Section 4: Host-To-Host Transport Authentication

5.2.3 SWIFT Connectivity

In March 2018, SWIFT partnered with SAP to enable a standardized and certified connection to the SWIFT network through SAP Multi-Bank Connectivity. This strategic collaboration has since evolved, incorporating additional SWIFT services—such as SWIFT GPI Tracker and SWIFTRef—into SAP's standard offering for corporate customers.

Chapter 2 outlined the functionality of SWIFT for corporations and highlighted key integration considerations. However, specific details related to SAP Multi-Bank Connectivity and how it interacts with SWIFT require further explanation. To establish connectivity to SWIFT, SAP itself operates key SWIFT infrastructure components on behalf of its customers. This includes running the following:

- SWIFT-certified software
- SWIFT-specific hardware (e.g., hardware security modules)
- Dedicated, secure internet links to the SWIFT network

This infrastructure is used by SAP Multi-Bank Connectivity in conjunction with the connector for SAP Multi-Bank Connectivity, which facilitates the end-to-end connection between the SAP S/4HANA system (corporate side) and the corporate's banks via SWIFT.

From a technical configuration perspective, SWIFT-specific parameters (e.g., BICs, message types, priorities) are included as part of the connector configuration and passed as additional attributes in the payload sent to SAP Multi-Bank Connectivity.

SAP supports communication over both the FIN and FileAct protocols:

- FIN for traditional MT-based message types (e.g., MT101, MT940)
- FileAct for bulk or large file transfers (e.g., XML payments, bank statements)

On the inbound side—from the SWIFT network back to the corporate customer—SAP Multi-Bank Connectivity performs signature validation on received messages and then forwards them to the customer's SAP S/4HANA system via the SAP Multi-Bank Connectivity connector. Notably, the message is stripped of its SWIFT-specific XML envelope, allowing the receiving ERP system to process the message natively.

Acknowledgments and delivery notifications from SWIFT are converted into the ISO 20022 pain.002 format. These are automatically delivered to SAP Bank Communication Management, enabling users to monitor payment batches and their corresponding status updates directly within their SAP S/4HANA system in a structured and auditable way.

Corporations that wish to use SWIFT via SAP Multi-Bank Connectivity must enter into an Alliance Lite2 for Business Applications (L2BA) agreement directly with SWIFT. During the onboarding process with SWIFT, it is critical to explicitly denote SAP Multi-Bank Connectivity as the intended business application. This ensures that the correct technical provisioning, such as setting up the proper connectors and certificate mappings, is done from the beginning.

5.2.4 Member Banks

Member banks are financial institutions that are already technically onboarded to the SAP Multi-Bank Connectivity platform with their own tenant. As previously mentioned, SAP Multi-Bank Connectivity provides services not only for corporations but also for financial institutions. Similar to corporations, these institutions can obtain their own tenants on the SAP Multi-Bank Connectivity platform to facilitate corporate connections. Because SAP member banks are integrated and technically preconfigured within the platform, SAP customers are relieved of the need for individual technical coordination to establish connectivity with these member banks.

Thanks to the pre-onboarding of member banks on SAP Multi-Bank Connectivity, the platform requires only the corporate entity to communicate with the bank's tenant, which is managed internally by SAP Multi-Bank Connectivity. Communication parameters, directory structures, protocols, and authentication methods are all predefined within the system.

5.3 Functions and Processes

Beyond simple bank connectivity, the SAP Multi-Bank Connectivity connector within the SAP S/4HANA system serves several key roles. The connector for SAP Multi-Bank Connectivity provides more than just connectivity; it includes essential elements such as a monitor and options for integrating other SAP functionalities like advanced payment management, in-house banking, and electronic bank statement processing. In addition, the connector features a file picker to read payment media or write account statements to a specified path.

This section explores the specific business functions, transactions, and relevant SAP Fiori applications for SAP Multi-Bank Connectivity. We'll cover the following topics:

- **Connector Monitor transaction**
  The transaction remains available for SAP ERP and SAP S/4HANA and includes some valuable reporting functions, which provide transparency regarding connectivity and message exchange between ERP-to-ERP systems, both external and internal banks.

- **Manage Bank Messages app**
  This app is available for SAP S/4HANA—on-premise, private cloud, and public cloud systems—and replicates the functions of the transaction. We provide an introduction here.

- **Pull messages**
  Messages can be pulled from your systems using this program. Here, we provide an overview of the transaction and its functionality.

- **Push messages**
  Messages can be actively sent using this program. Here, we introduce the transaction and its functionality.

- **Pickup files**
  The connector is equipped with a file picker. Here, we present its functionality and features.

- **Handling of sensitive data**
  Payment transactions are inherently sensitive. Some data not only must be protected from external access but also requires multiple layers of internal security—for example, HR payments. How can you handle such sensitive data, and what security measures does the connector offer? Here, we provide an overview of the solutions and functionalities available.

- **SWIFTRef integration**
  SWIFT has long provided SWIFTRef, an SAP-compatible master database for bank master data. Now, there is also a newly integrated SWIFTRef function within SAP Multi-Bank Connectivity. We are excited to introduce this functionality here.

5.3.1 Connector Monitor Transaction

One of the core components of the connector is the monitor, which tracks and clarifies both inbound and outbound communications using audit-proof log files, providing a comprehensive overview. The monitor helps users identify incoming messages such as bank statements, bank reports, incoming payment media for the in-house bank, and treasury deal confirmations. For outbound messages, the focus is typically on sent payment media (pain.001, pain.008, MT101, etc.) as well as forwarded account statements and reports.

The monitor is available via Transaction /BSNAGT/MONITOR for on-premise SAP S/4HANA and SAP S/4HANA Cloud Private Edition, as well as in the form of an SAP Fiori app. It is no longer available as a transaction for SAP S/4HANA Cloud Public Edition.

In addition to Transaction /BSNAGT/MONITOR, the connector monitor can be accessed through the **SAP Menu** under **Connector for Multi-Bank Connectivity • /BSNAGT/ MONITOR—Connector Monitor**, as shown in Figure 5.12.

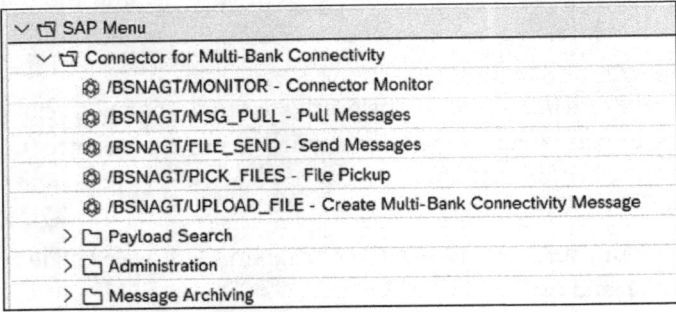

Figure 5.12 Connector Monitor: Customizing Menu

In the following sections, we'll examine the filters available for the connector monitor as well as its key components.

Filters

Upon opening the transaction, a filter appears, enabling you to sift through message categories and individual messages efficiently. Equipped with a range of functionalities and shortcuts, the filter enhances usability. The following features are accessible via the buttons on the header bar, as shown in Figure 5.13:

- ❶ **Execute**: Run the current selection.
- ❷ **Variant Management**: Manage selection variants.
- ❸ **Clearing Function**: Clears all individual fields in the selection.
- ❹ **Today**: Fills the current date in the **Create Date** field.
- ❺ **Current Week**: Selects the entire calendar week in the **Create Date** field.
- ❻ **My User**: Selects the user's own ID.
- ❼ **Manual Processing**: Selects messages with status **IBC95 Manual Processing Required** and message directory IN for manual processing requirements.
- ❽ **Error**: Displays messages with errors.
- ❾ **Log**: Provides direct access to the technical Transaction SLG1 log file in the connector monitor.

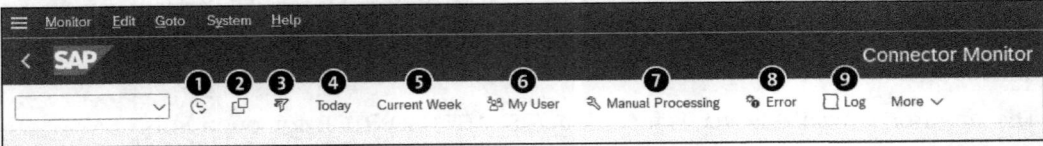

Figure 5.13 Connector Monitor: Header

Each message contains certain key fields, such as **Message Type**, **Sender ID**, and **Receiver ID**, as well as information indicating whether it is an incoming or outgoing message. Figure 5.14 shows an excerpt of the relevant fields and filters.

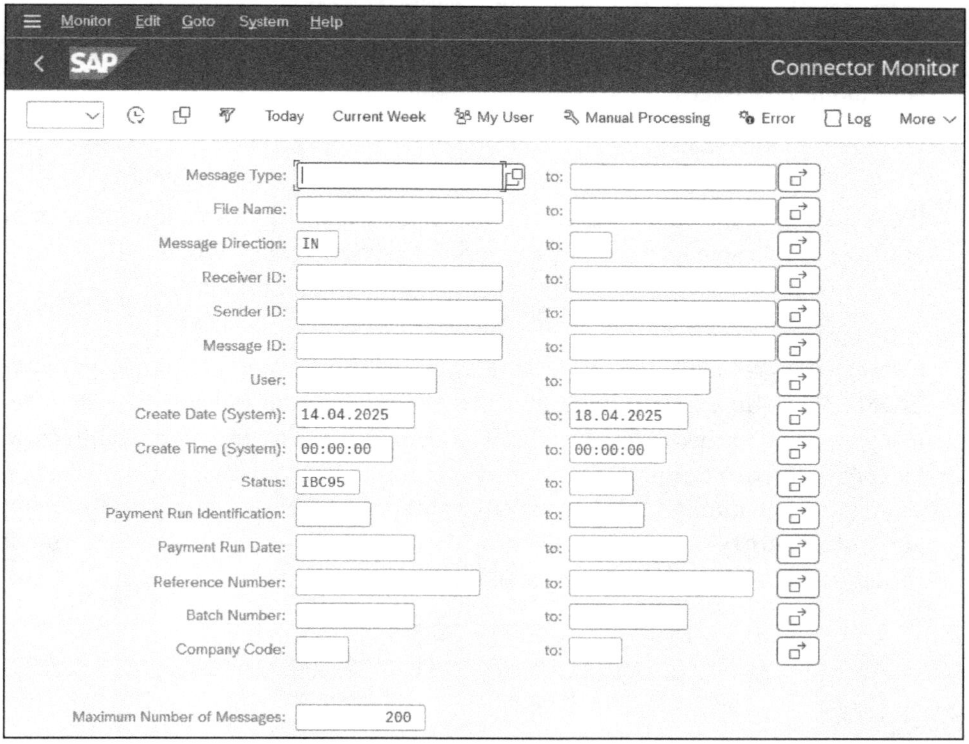

Figure 5.14 Connector Monitor: Filter

Let's now look at the most important fields you can use to filter your results:

- **Message Type**
  Message Type is used to categorize messages during the ERP-to-ERP or ERP-to-bank message exchange. Beyond categorization, it also involves customization that determines the next process steps. For example, a message type such as camt.053.001.02 can be linked with functionality like bank statement processing or advanced payment management for forwarding purposes. We explored this customization in detail in Chapter 4.

- **Message Direction**
  Message Direction determines whether a message is outbound (**OUT**) or inbound (**IN**).

- **Sender ID** and **Receiver ID**
  Sender ID and **Receiver ID** are used for identification and specify the sender and receiver of the message, playing a key role in managing the routing process within

SAP Multi-Bank Connectivity. A key role in this process is played by **Custom Field** within the bank master data. Depending on the perspective, it can represent either the sender ID or receiver ID and can be freely assigned. For outbound messages:

- **Sender ID**: Represents the customer ID within the bank master data.
- **Receiver ID**: Indicates the SWIFT/BIC code of the bank.

For inbound messages:

- **Sender ID**: Indicates the SWIFT/BIC code of the bank.
- **Receiver ID**: Represents the customer ID within the bank master data.

If the **Message Type** and/or **Receiver ID** fields are blank, the system interprets these entries as applicable to all message types and/or receiver IDs. When multiple valid entries exist, the system selects the most specific entry for processing.

- **Message ID**
  Message ID plays a vital role in tracking by ensuring each message is uniquely identifiable. When an external message ID is not provided, an internal ID is assigned using a preestablished number range. For example, messages generated internally by SAP Multi-Bank Connectivity are given an internal ID, whereas payment carriers use a reference number that is accessible both in the Manage Payment Media app and within the payment file itself. Figure 5.15 shows an excerpt from a filter result on the left and the output of a payment file on the right.

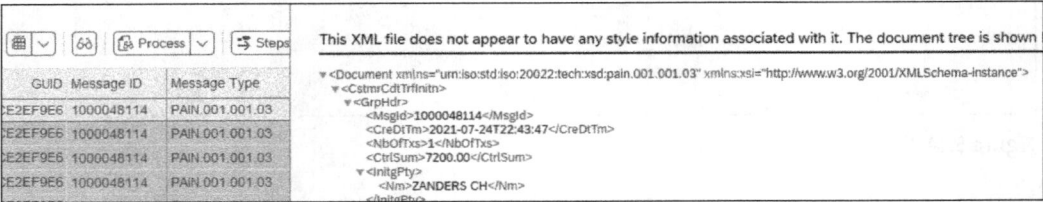

Figure 5.15 Message ID XML File Output

In addition, users can filter based on user creation date and time, along with other cross-area relevant fields like **Company Code**, **Payment Run Identification**, **Run Date**, and SAP Bank Communication Management's **Batch Number**. This functionality ensures precise data retrieval and enhanced cross-module navigation.

Components

After applying the filters, you can access the connector monitor, which features standard functionalities to customize and save the layout according to your preferences. Let's examine the monitor's key components.

As shown in Figure 5.16, adjacent to the transaction bar, the monitor enables a connection test between the connector monitor and SAP Multi-Bank Connectivity. The **Pull** button facilitates manual message retrieval from the SAP Multi-Bank Connectivity server, with a popup window to limit the number of messages. The **Alerts** button provides a summary of triggered alerts, requiring prior configuration through the standard alert functionality.

The connector monitor organizes message flows into **Outgoing**, **Incoming**, and **Error** tabs; the latter contains messages that are faulty or processed with errors, ensuring streamlined operations and efficient troubleshooting.

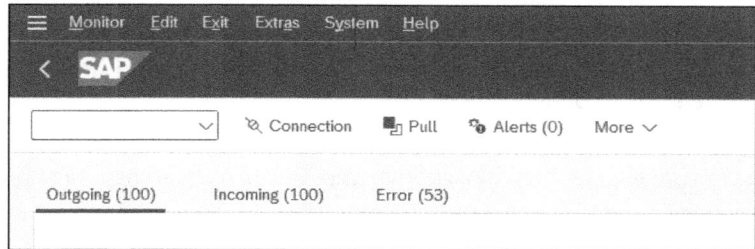

Figure 5.16 Connector Monitor: Header

The individual functions of the monitor are as follows (as shown in Figure 5.17):

- **AIF**
  The **AIF** button serves as a link to SAP Application Interface Framework, providing capabilities for centralized interface monitoring, analysis, and error management, thereby optimizing operational oversight and interface troubleshooting.

- **Send**
  Messages can be manually resent using the **Send** button.

- **Create Message**
  This button provides the ability to jump into the function to create an SAP Multi-Bank Connectivity message. This function is detailed in Section 5.3.3 and Section 5.3.4. Figure 5.18 shows the **Create Message** dropdown menu.

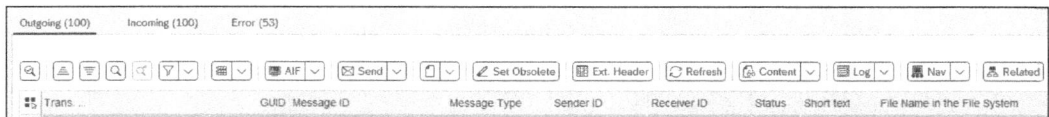

Figure 5.17 Connector Monitor: Navigation Panel

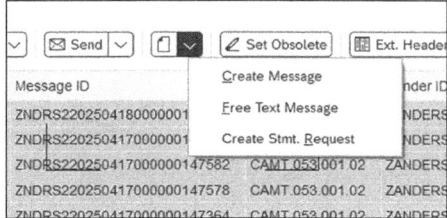

Figure 5.18 Connector Monitor: Create Message Dropdown Menu

- **Set Obsolete**
  Users can manually change a message's status to processed or obsolete, allowing for flexible management based on current needs.
- **Ext. Header**
  Every message includes header parameters, available through the **Ext. Header** button. By selecting a message and clicking the **Ext. Header** button, users can access a popup window that shows the technical header parameters, including the sending system, application type, and SWIFT parameters.
- **Refresh**
  The **Refresh** button lets you update the display.
- **View Content**
  The **View Content** button enables message viewing and, within its drilldown capabilities, offers a download function. You can manually download messages like payment files or account statements directly from the connector monitor as needed, enhancing accessibility and convenience.
- **View Log**
  Every message within the connector monitor is documented through an audit-proof log file. Selecting a message and clicking the **View Log** button to see the log file allows you to trace the processing steps, ensuring transparency and accountability.
- **Navigate to Related Functions**
  This button facilitates direct navigation from the monitor to the linked application or job, such as enabling access to advanced payment management, thus streamlining workflow transitions.
- **Show Related Messages**
  SAP Multi-Bank Connectivity generates separate entries for incoming and outgoing messages, regardless of whether they involve the same file. For instance, receiving a bank statement via SAP Multi-Bank Connectivity results in one message, while forwarding it to an external path creates another outgoing message. Using this button, you can easily review and track these entries for enhanced visibility.

Most of these navigation options and functions are also available in the Manage Bank Messages app (F4385), which we'll discuss in the following section.

> [+] **Connect the SAP Bank Communication Management Monitor**
>
> To enable navigation from SAP Bank Communication Management (Transaction BNK_MONI) to the monitor (see Figure 5.19), refer to SAP Note 3095350. After implementing the SAP Note, add the new status as detailed within. Once configured, you can seamlessly transition from Transaction BNK_MONI to SAP Multi-Bank Connectivity Transaction /BSNAGT/MONITOR for streamlined operations via **Navigate to MBC Monitor**.

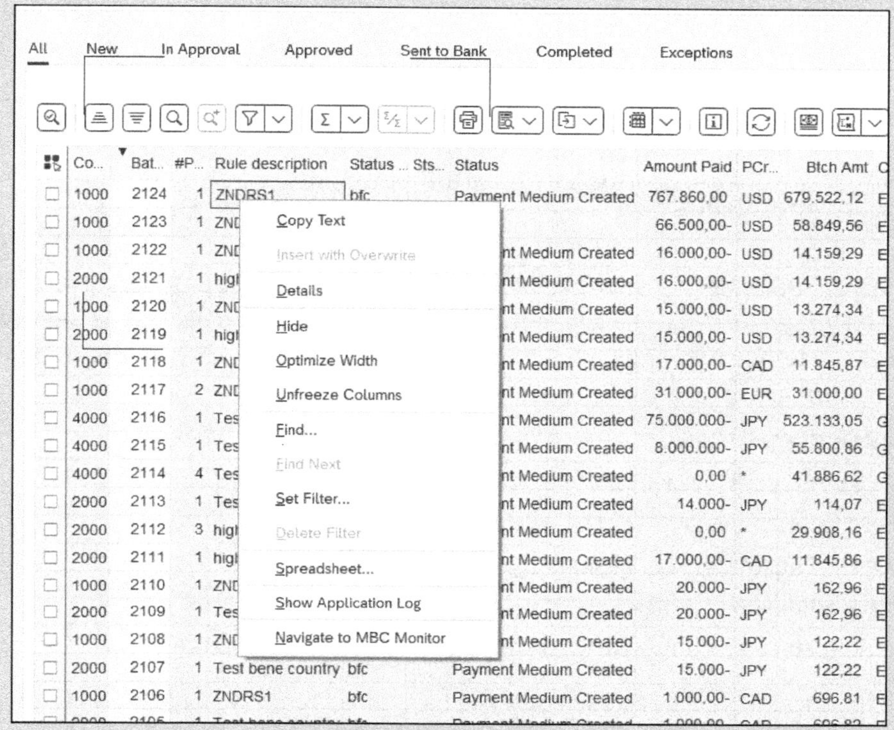

Figure 5.19 SAP Multi-Bank Connectivity Monitor: Navigate to SAP Multi-Bank Connectivity

5.3.2 Manage Bank Messages App

The Manage Bank Messages app is the SAP Fiori version of the connector monitor, available starting from SAP S/4HANA 1909 FPS 01. The Manage Bank Messages app essentially covers all the essential functions that are accessible in the transaction, with the added benefit of a significantly more modern interface and all the advantages that SAP Fiori offers. For each message, detailed information can be displayed, including administrative data, processing logs, and the message payload, which can also be downloaded if required. The app also offers visibility into the processing steps of inbound messages, along with their current processing status and any recorded log entries.

In addition, the application provides various functions for further message handling, such as viewing the sending and receiving details, modifying the status of inbound messages, displaying XML content, sending outbound messages, and creating free-text messages. You can also access the application log for error messages and view related messages.

Technical features such as performing a ping test and manually pulling messages via the **Connection Details** view are also supported.

5   SAP Multi-Bank Connectivity

> [»] **Transactions Are Not Available in the Public Cloud**
> The SAP Multi-Bank Connectivity connector monitor is only available as the Manage Bank Messages app for SAP S/4HANA Cloud Public Edition users. New developments are typically first implemented in SAP S/4HANA Cloud Public Edition before they become available in later releases for SAP S/4HANA Cloud Private Edition and eventually for on-premise SAP S/4HANA.

The Manage Bank Messages app has been delivered by SAP with an SAP Fiori catalog and an SAP Fiori group. It can be added to custom catalogs and groups via the SAP Fiori launchpad designer accordingly. For identification purposes, here are the technical details of the app:

- SAP Fiori app ID: F4385
- SAP Fiori type: Transactional app
- Technical catalog: SAP_TC_FIN_BSNAGT_COMMON
- SAP Fiori business catalog: SAP_FIN_BC_BSNAGT
- SAP Fiori business group: SAP_FIN_BCG_BSNAGT
- Standard business template role: SAP_BR_BANK_INT_SPECIALIST

Once the app is assigned to a user, they will have access to the **Manage Bank Messages** tile. Click the tile to enter the app (see Figure 5.20), which shows a header with an expandable filter option ❶. The filter can be expanded by clicking the **Adapt Filter** ❷ button.

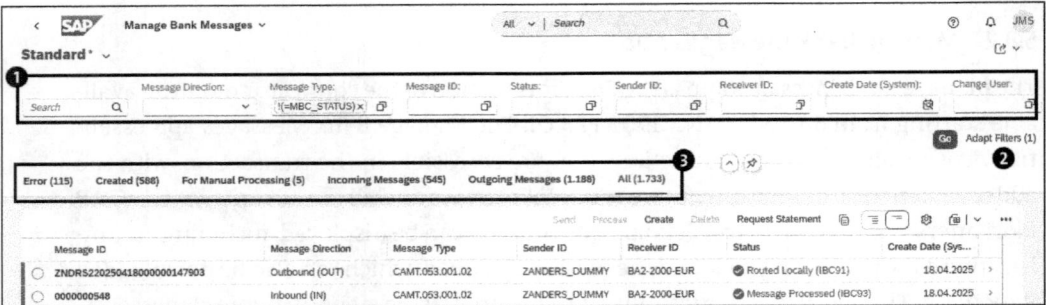

Figure 5.20 Manage Bank Messages App: Filters

Similar to the Transaction /BSNAGT/MONITOR, the filter results ❸ are divided into **Incoming Messages**, **Outgoing Messages**, and **Error** categories. The Manage Bank Messages app also includes tabs such as **Created**, for messages that have been created but not yet sent, and **For Manual Processing**, which includes everything that requires further manual processing. Under the **All** tab, you can view messages in any status, offering a comprehensive overview of all communications.

From within the app, you can initiate actions such as resending payment files to the bank, retrieving bank statements again, or requesting copies. The following functionalities are available within the app, as shown in Figure 5.21:

- **Request Statement**
  Upon clicking the **Request Statement** button, a popup window opens in which you can request bank statements. In this window, you can select the bank, specify the type of statement requested, and define a time span. Once you've confirmed your selections by clicking the **Send Request** button, a pull request is sent to the bank in the background, depending on the type of connection established.

- **Create**
  If you click the **Create** button, a new window opens in which a message can be sent to a financial institution. You can choose the type of message you wish to send. SAP Multi-Bank Connectivity offers the option between a free-form message in a custom format or a structured SWIFT message type. Available types include MT199 (Free Form Message), MT399, or MT999. After selecting the sender and receiver IDs, the message text can be composed in the free text field, as shown in Figure 5.22.

- **Process**
  The **Process** button can be used when incoming or outgoing messages cannot be automatically processed due to issues such as insufficient customization or if they have encountered errors. Once the error is resolved, the message can be reprocessed using the **Process** button. To do this, the message, which may be in **Error** status (IBC94), must be selected, and then you can click the **Process** button. After you click **Process**, a popup window will open in which the manual processing needs to be confirmed.

- **Send**
  Using this button, messages can be sent manually.

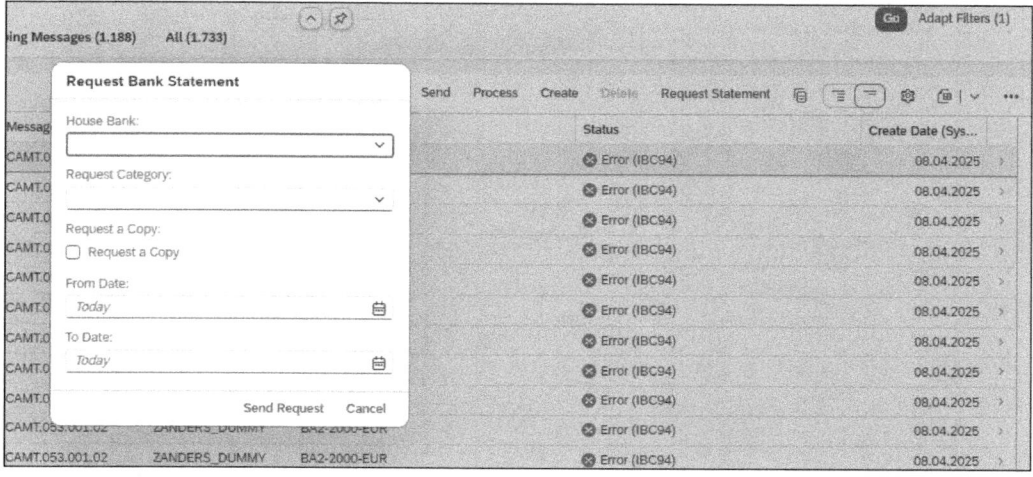

Figure 5.21 Manage Bank Messages: Request Bank Statement

5 SAP Multi-Bank Connectivity

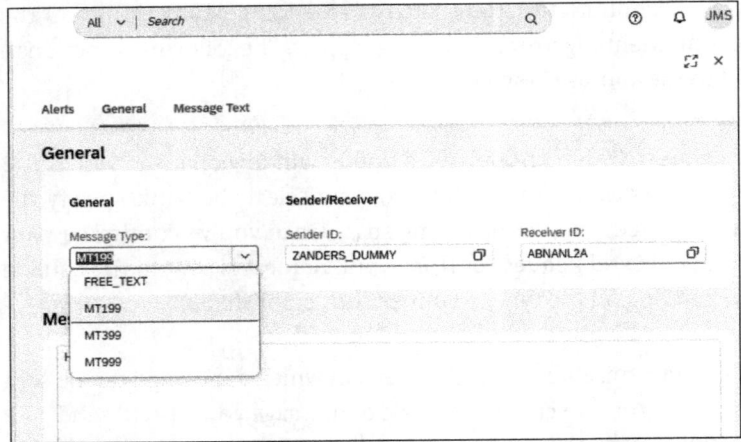

Figure 5.22 Manage Bank Messages: Create Bank Messages

When you select a message, a new popup window opens to the right, providing a detailed view of the message. Here you have the opportunity to examine the technical header information of the message and receive details about the sender, recipient, user, log files, and time.

As shown in Figure 5.23, when you click the **View Payload** button, a popup window featuring a preview of the message content opens. With the **Download Payload** button, you can manually download the message.

Figure 5.23 Manage Bank Messages: Drilldown

The app has the capability to transition into upstream or downstream process steps, as well as into the respective application. For instance, it can navigate to advanced payment management or to the bank statement processing area of the Manage Bank Statements app. Similar to a feature in the transaction, the app also allows you to view related messages by clicking the linked message ID, as demonstrated in Figure 5.24.

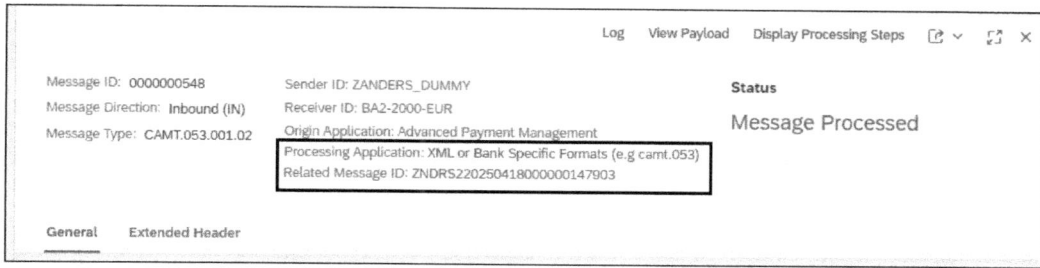

Figure 5.24 Manage Bank Messages: Application Links

Overall, the Manage Bank Messages app serves as a streamlined tool for centrally monitoring all incoming and outgoing messages. In conjunction with the alerting function and the ability to transition into upstream and downstream process steps and applications, the app offers nearly the same functionalities as the transaction. It is expected that all new features will be available first, or even exclusively, through the SAP Fiori app.

5.3.3 Pull Messages

In the context of SAP Multi-Bank Connectivity, the ERP system always acts as a client. To receive inbound bank messages such as status updates or bank statements, the ERP system must actively retrieve data from SAP Multi-Bank Connectivity using a pull mechanism. The pull functionality enables synchronous retrieval of such messages without requiring inbound connections to the corporate network—providing a secure and controlled integration scenario.

When the pull function is triggered, the system establishes a synchronous outbound connection to SAP Multi-Bank Connectivity to request new inbound messages. If new messages are available, SAP Multi-Bank Connectivity returns them in the response to the pull request. For each message received, the SAP Multi-Bank Connectivity connector then sends a receipt confirmation back to SAP Multi-Bank Connectivity. This process is repeated automatically until no further messages are provided.

For technical implementation in the background, report /BSNAGT/MESSAGES_PULL is used. This report can be scheduled as a background job to execute the pull process at regular intervals. Before using the functionality, the pull types must be properly maintained. Figure 5.25 shows the configuration setup of the **Maintain Pull Types** activity, which can be found within the IMG at **Multi-Bank Connectivity Connector • Routing and Connectivity • Maintain Pull Types**.

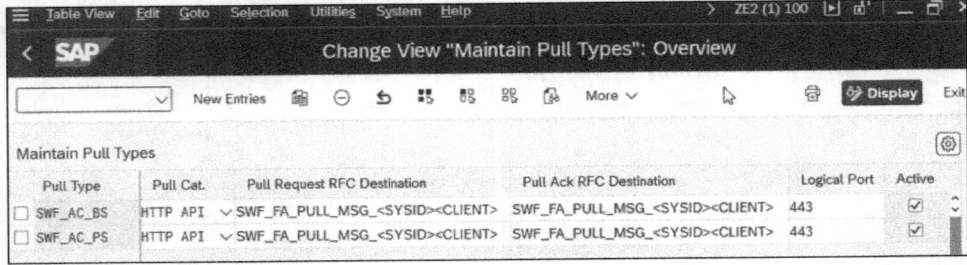

Figure 5.25 Pull Messages

In this configuration activity, you define the endpoints and protocols used to pull messages from SAP Multi-Bank Connectivity. Each pull type must be assigned a pull category, which determines the technical handling. The following pull categories are available:

- **Web Service Runtime**
  Messages in this category are pulled using the XI protocol through the web service runtime. If Web Services Reliable Messaging (WSRM) is used, an optional logical port can be specified for the pull operation. If no port is specified, the logical port marked as the default will be used.

- **HTTP API**
  Messages are pulled via a direct HTTPS connection to SAP Multi-Bank Connectivity. For this method, a corresponding RFC destination of type **G—HTTP Connections to External Server** must be created and assigned to the pull type definition.

The pull functionality offers an option for receiving inbound messages from SAP Multi-Bank Connectivity. By utilizing configurable endpoints, various protocol options, and automated background processing, this method enables seamless message transfer between SAP Multi-Bank Connectivity and the SAP S/4HANA system.

5.3.4 Push Messages

Messages created by the application that have gone through the approval process are directly handed over to SAP Multi-Bank Connectivity and sent to the receiving bank or recipient system. This functionality was described in Section 5.3.1 and Section 5.3.2. SAP S/4HANA also provides Transaction /BSNAGT/FILE_SEND, which offers the capability to select and resend files via a report and a selection screen. Figure 5.26 displays Transaction /BSNAGT/FILE_SEND and its selection fields.

The transaction is operated in the background by program /BSNAGT/SEND_FILES. This program identifies relevant messages by accessing the message status from table

/BSNAGT/FILE_INF. If the **Include Sent Msg.** checkbox is not selected, then the system omits any previously sent messages, even when they meet the selection criteria, ensuring that only pending messages are dispatched. By selecting the **Include Sent Msg.** checkbox, you can request to resend previously sent messages, in which case the program creates copies of these messages and initiates their sending; the original messages are never resent. When creating a copy for resending, the system assigns the original message as the parent message of the copy. You can access the original message by viewing the parent message in the connector monitor.

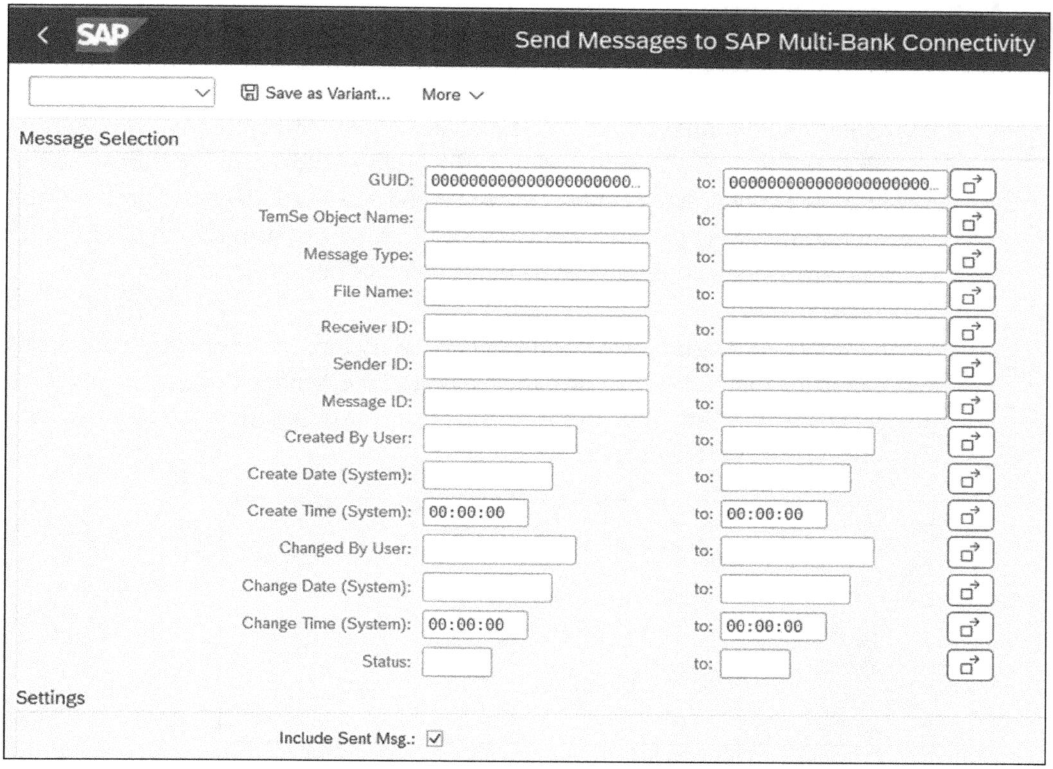

Figure 5.26 Send Messages

5.3.5 Pickup Files

The connector comes equipped with its own file picker, which is a crucial element for planning processes and technical infrastructure when integrating external, non-SAP systems. The file picker can be directly linked to advanced payment management through the process control in the SAP Multi-Bank Connectivity connector. This raises the question of whether files should be imported via advanced payment management or SAP Multi-Bank Connectivity when integrating non-SAP systems for which the file picker is the only connection method. There is some justification for using the connector; it allows for centralized monitoring of incoming and outgoing

5 SAP Multi-Bank Connectivity

payments via a central monitor, though it does have its limitations. In this section, we introduce the functionality of the file picker through the SAP Multi-Bank Connectivity connector and also highlight its limitations.

The file picker can be accessed via menu path **Connector for Multi-Bank Connectivity • Connector Monitor** or directly using Transaction /BSNAGT/MONITOR. The transaction provides the option to save settings and filters as a variant, allowing for job-controlled execution. For job execution, report /BSNAGT/REP_PICKUP_FILES is available.

When uploading files, there are essentially two different options available that you can use (see Figure 5.27):

- **Determine from File Names**
  In this option, all the necessary information is extracted from the filename itself. The filename must be formatted as a comma-separated list containing the following details: MessageType, SenderId, ReceiverId, NumberOfRecords, FileName, and an optional MessageId. When a file is retrieved, the SAP Multi-Bank Connectivity connector automatically parses this structured filename and uses the extracted values to populate the corresponding fields in the message header.

- **Use Fixed Values**
  With this option, the header information necessary for further processing of the file is defined through selection. After the **Message Type**, **Sender ID**, and **Receiver ID** are chosen in the selection mask (see Figure 5.28), the file is imported and assigned these parameters for subsequent processing.

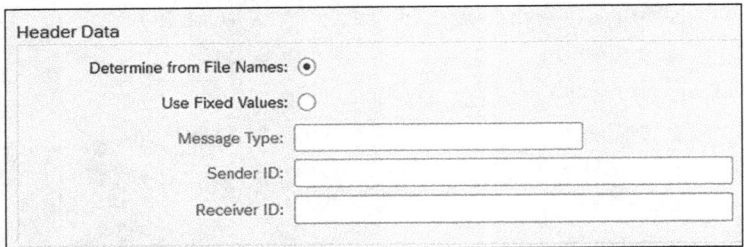

Figure 5.27 Pickup Files

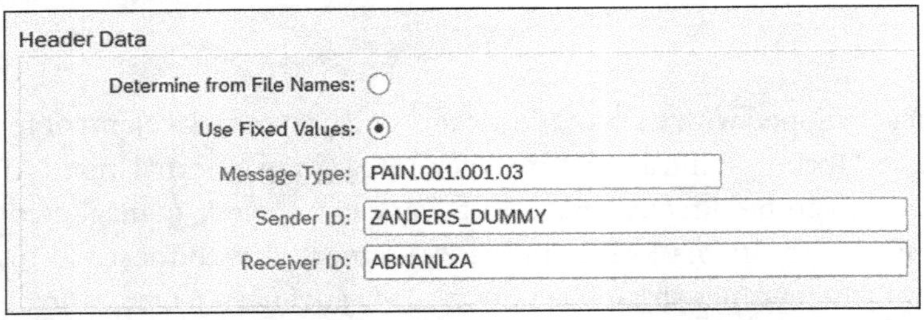

Header Data	
Determine from File Names: ○	
Use Fixed Values: ⦿	
Message Type:	PAIN.001.001.03
Sender ID:	ZANDERS_DUMMY
Receiver ID:	ABNANL2A

Figure 5.28 Fixed-Value Header Fields

The **Message ID** is automatically assigned by the SAP Multi-Bank Connectivity connector, using a number generated from the preconfigured number range object /BSNAGT/N2. The basic configuration is explained in Section 5.5.2.

If no specific file name is defined in the **File Name** field, then the name of the file being retrieved will be used without any modifications.

For file pickup, both the source and target directories need to be established and accessible. These directories can be configured using the standard Transaction FILE or Transaction SF01. SAP provides two logical directories for the SAP Multi-Bank Connectivity connector, which can be customized as needed. Alternatively, you can create and use your own logical directories within the system. The logical directories provided by SAP for Transaction FILE configuration are as follows:

- */BSNAGT/PICKUP_SRC_DIR*: Source directory
- */BSNAGT/PICKUP_TRG_DIR*: Target directory

Once the logical directories have been configured, they will appear in the selection options, as shown in Figure 5.29.

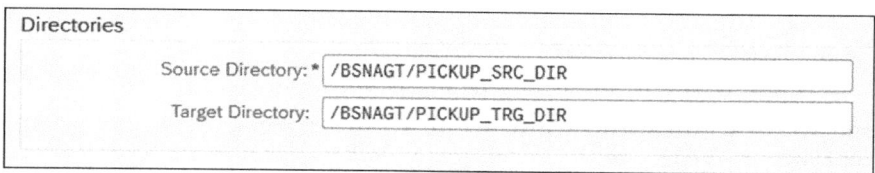

Figure 5.29 File Picker Directory

The **Technical Settings** (see Figure 5.30) are used for technical control. Here, you can configure EBICS settings, determine the code page for the file to be read, and also select a duplicate check:

- **Code Page**
  The code page is used during the file upload process, which is particularly important when importing files that contain different character sets, such as Cyrillic.
- **EBICS Order Type and BTF Type**
  With the file picker, you can directly associate EBICS order types with the payment medium for bank transmission. For EBICS 3.0, you can also select a **BTF Type**. The prerequisite is that EBICS communication has been configured in the system. This configuration is performed in outbound processing and is described in Section 5.5.3.
- **Duplicate Check**
  To determine if an incoming message is a duplicate, the system compares the following fields: **Message ID**, **Sender ID**, and **File Name**. If another message is identified with the exact same header information in these fields, the new message is marked

as a duplicate. During the pull job, only the duplicate flag is set and retained with the message. When messages are processed, the duplicate flag is checked, and any message with this flag will be ignored during processing.

Under **Send Options**, you can actively control whether the file should first be read and created as a message in the SAP Multi-Bank Connectivity connector or whether it should be directly processed up to the point of transmission. This is managed through the **Send Category** dropdown.

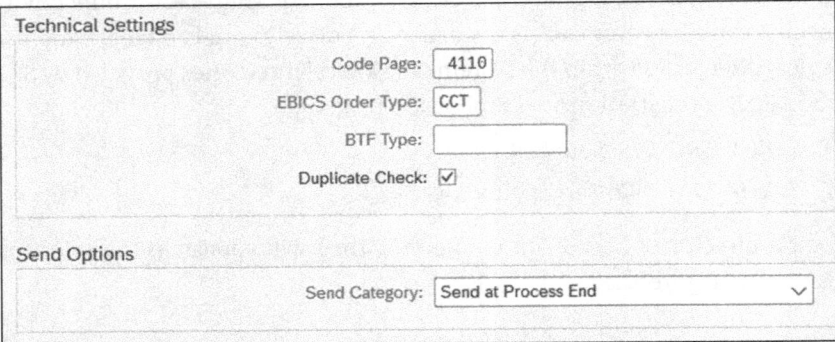

Figure 5.30 Pickup Files: Technical Settings

> **Handling Encrypted Files**
> Encrypted files or other storage options like SFTP or WebDAV are not supported by the file pickup function. The file picker is incapable of decrypting these files.

5.3.6 Handling of Sensitive Data

Within the SAP Multi-Bank Connectivity connector, certain messages can be marked as sensitive. Additional authorizations are required for marked messages, which must be assigned to a user in order for them to view or display sensitive messages in the Manage Bank Messages app or Transaction /BSNAGT/MONITOR of the connector monitor. The connector identifies sensitive messages based on the following criteria:

- Sender ID
- Receiver ID
- Message Type
- Origin

Upon receiving a message, the connector checks whether the combination of stored criteria matches. If it does, then the message is classified as sensitive.

> **Messages Classified via a Customized Header**
>
> SAP Multi-Bank Connectivity messages can also be designated as sensitive by adjusting the header parameters. To accomplish this, the *CONNECTOR.SensitivePayload* header parameter must be set to *true*. Messages with this parameter set will always be classified as sensitive.

To classify messages as sensitive, you can use customizing table /BSNAGT/V_SENS, which serves to define messages based on the criteria of **Sender ID**, **Receiver ID**, **Message Type**, and **Origin** (see Figure 5.31). If any of the fields are left unfilled, they are considered a wildcard. For example, if you only maintain the **Sender ID** and leave the other fields blank, all message types will be affected. If you define a message as sensitive data by selecting the **Sensitive** checkbox, then users will require additional authorizations to display or download the message.

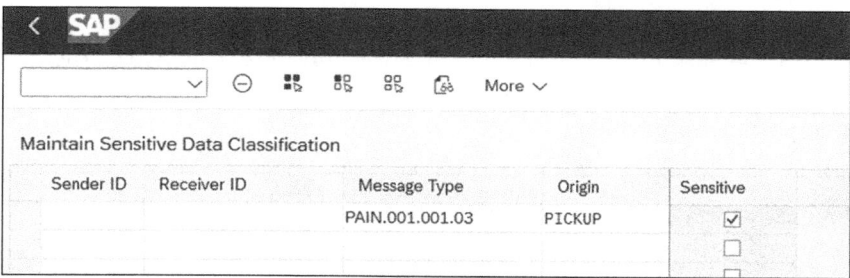

Figure 5.31 Maintain Sensitive Data Classification

These settings can be found in the IMG under the path **SPRO** • **Multi-Bank Connectivity Connector** • **Advanced Settings** • **Maintain Sensitive Data Classification for MBC Messages**.

5.3.7 SWIFTRef Integration

SWIFT provides a library of bank master data through its SWIFTRef product, which can be used not only by credit institutions but also by companies. This library contains nearly all global bank master data in a standardized, processable format. SWIFT maintains a dedicated team that centrally collects and tracks bank master data from central banks in the background. This master database is enriched with proprietary master data and information from the SWIFT network. The result is a comprehensive database that SWIFT makes available as a processable file through SAP Multi-Bank Connectivity.

At its core, SWIFTRef consists of two files. SWIFT provides an initial file with over 120,000 bank master records from 250 countries, which can be initially uploaded into the ERP system. SWIFT also offers an update file, known as a *delta file*, at regular monthly intervals for import. The delta file is intended to continuously update bank master data and mark bank master records as deleted due to closures or mergers.

To ensure this data is correctly imported into the SAP S/4HANA system, it is crucial to carefully analyze and adjust the relevant country settings and validation rules prior to import. Without this preparation, there is a risk of incorrect data imports.

SAP Multi-Bank Connectivity is directly connected to the SWIFTRef library. Global bank master data are automatically retrieved from the library and imported into the SAP S/4HANA system using standardized report RFBVBIC2 (BIC2). This automation significantly reduces the manual effort required to maintain bank master data. Updates occur regularly, ensuring that current information is always available.

The automatically maintained information includes address data, BICs, and national identifiers. Continuous updates contribute significantly to reducing errors in payment transactions. A central requirement for seamless operation is a properly configured country setting in Transaction OY17 and a functional technical connection to SAP Multi-Bank Connectivity. As depicted in Figure 5.32, after import, the master data is directly uploaded into table BNKA using the standard Transaction BIC2. This makes the data immediately available for BAM. See Chapter 3 for more information about why bank keys are so important for payments.

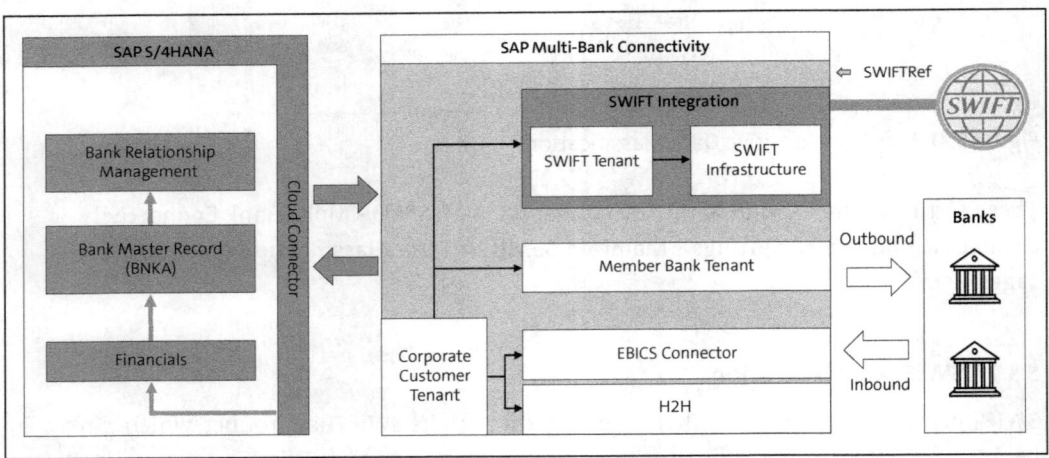

Figure 5.32 SWIFTRef Import

Particularly with increasing demands for fraud prevention and new format requirements from international clearing systems, complete banking master data in mass payments is a critical point to ensure stability in payment transactions. The ability to import a nearly complete database automatically via SWIFTRef is a benefit, especially in cross-border payment transactions.

5.4 Onboarding

Because SAP Multi-Bank Connectivity is a managed service solution, the bank connection is provided to the customer as a service. To gather the necessary information from both the corporate client and the bank, SAP has designed an onboarding process with six phases that guides you step by step through the SAP Multi-Bank Connectivity onboarding. Let's walk through them in detail:

1. **Scoping phase**

   During the scoping phase, you should already have a clear connection strategy and rollout plan for your bank integration. It is essential to have thoroughly planned your bank connection, discussing and defining critical aspects like the type of connection and formats with your banking partners. The scoping phase of SAP Multi-Bank Connectivity begins with a questionnaire sent via email, which includes key items such as the following:

   - Number of banks and bank details: Identify all banks you plan to connect with, including their respective addresses and contact information. This helps clarify the scope and scale of the integration needed.
   - Connectivity type (host-to-host, EBICS, member bank, or SWIFT): Determine the type of connectivity for each bank. Consider factors such as security, data volume, and transaction frequency to choose the most suitable protocol.
   - File formats: Specify the file formats required for both incoming and outgoing transactions (e.g., MT940, camt.053, pain.001). Ensure that these formats align with both your internal systems and the bank's capabilities.
   - Timeline for implementation: Develop a realistic timeline for the project, highlighting key milestones and delivery dates. Include buffer times for testing phases and any potential delays. The timeline should be tailored based on the type of connectivity and the specific bank involved, typically ranging from two to six months for the first bank.
   - BIC numbers: If you are already a SWIFT customer, SAP will request both your test and production BICs. These codes are essential to ensure that the test environment accurately simulates real-world scenarios for comprehensive SWIFT communication testing and to verify that the production BIC is configured correctly for seamless transaction processing and accurate routing in the live environment.
   - SAP system details: Prepare a summary of your system landscape to facilitate coordination with the SAP team. SAP Multi-Bank Connectivity provides both a test and a productive environment, both of which need to be integrated with your system landscape. Ensure you have key information ready, such as your system version, system ID, instance number, and host ID.
   - Project team details: List the project team members, including their roles and responsibilities. Ensure that team members from IT, finance, and other relevant departments are involved, and designate a project manager to oversee the process.

Upon submission of the required details, the SAP Multi-Bank Connectivity onboarding team will provide a welcome pack containing comprehensive information about the onboarding process.

2. **Preboarding phase**

   During the preboarding phase, there is a step-by-step checklist that is worked through in collaboration with the SAP onboarding team. Key components of this phase and the checklist include the following:

   – Detailed summary of bank connections: Utilizing a template, details are captured based on the type of connection. For example, with EBICS, information such as the EBICS link, EBICS host name, client name, and EBICS user must be collected, entered into the template, and sent to the team.

   – Technical integration: Both the test and productive SAP Multi-Bank Connectivity tenant need to be connected to the customer's environment. This requires a secure connection that includes a transport layer and message layer encryption. Details about the SAP system environment, including clients, are also gathered.

   SAP guides you through this process step by step and provides support as needed through regular catch-up calls.

3. **Technical phase**

   During the technical phase of the bank connectivity setup in the SAP Multi-Bank Connectivity process, the following steps are carried out:

   – Implementation of submitted details: The previously submitted details necessary for bank connectivity are implemented.

   – Integration of credentials: Input and configure the productive credentials for SWIFT, host-to-host, and EBICS within SAP Multi-Bank Connectivity.

   These steps are crucial to ensure that the system is correctly configured and operating properly before moving into full production mode.

4. **Validation phase**

   During this phase, user acceptance testing (UAT) is conducted. Banks typically offer test environments to facilitate end-to-end testing of bank connectivity. Approximately two weeks should be allocated per bank for this phase. We specifically recommend thoroughly testing the formats and obtaining clear feedback from the bank. In some countries, such as Germany, there may be instances in which the bank does not provide a test EBICS server. For these banks, it is advisable to plan a more intensive penny test phase.

5. **Promotion phase**

   During this phase, penny tests are conducted. Both incoming and outgoing payments should be tested. This includes sending payment media such as credit transfers and direct debits, as well as receiving and processing bank statements and account statements. Typically, up to two weeks of time is allocated for this phase.

Once the penny tests have been successfully completed, there is a handover to the global support team. This team provides 24/7 support for productive operations. From this point on, any issues that arise must be reported by opening tickets with SAP through SAP Help Portal. Figure 5.33 visualizes the steps of the SAP Multi-Bank Connectivity onboarding process.

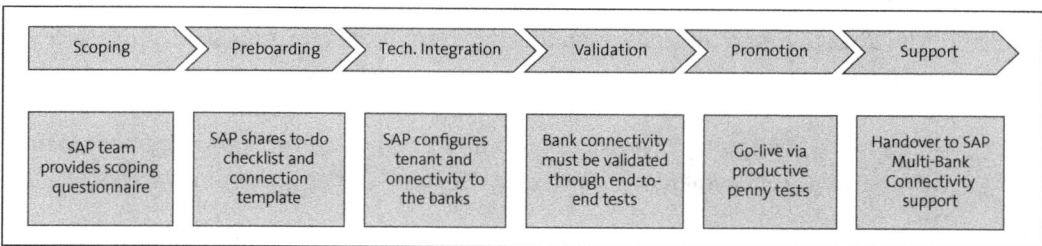

Figure 5.33 SAP Multi-Bank Connectivity Onboarding

5.5 Configuration

In this section, we outline the settings that need to be configured in the SAP Multi-Bank Connectivity connector to connect the SAP S/4HANA system with SAP Multi-Bank Connectivity. It is assumed that general settings in SAP S/4HANA Finance have already been completed to generate payment files or post bank statements within the SAP S/4HANA system. In addition, we address the prerequisites and specific settings required for EBICS and SWIFT integration.

In this section, we will focus on the configuration of SAP Multi-Bank Connectivity, starting with the creation of payment medium formats, maintaining bank master data, and the selection of variants for bank statements. We will then move on to basic configuration, including maintaining number ranges, building connections, and setting up the routing between systems to ensure proper file exchange. After that, we'll cover the inbound and outbound processing of files, as well as the configuration of the various connectivity options we discussed earlier. Finally, we'll discuss the specific authorizations required for SAP Multi-Bank Connectivity and the payment processes.

5.5.1 SAP S/4HANA Finance Configuration and Prerequisites

Payment files can be sent directly from SAP S/4HANA Finance to the SAP Multi-Bank Connectivity tenant. From there, they are transmitted to the bank through the predetermined channel (EBICS, SWIFT, host-to-host, or member bank). Essentially, there are two approaches in SAP S/4HANA Finance: either the payment file is uploaded to a path and imported via the connector's file picker, or it is sent directly from the system to the

SAP Multi-Bank Connectivity tenant through the connector. This setting must be configured for each type of file that is to be directly transferred from the SAP S/4HANA Finance system to the SAP Multi-Bank Connectivity connector. The following sections outline the key parameters needed for seamless transmission.

Create Payment Media Formats

In the IMG, there are settings related to payment media formats that can be accessed via Transaction OBPM1. Alternatively, this can also be done through IMG menu path **Financial Accounting (New) • Accounts Receivable and Accounts Payable • Business Transactions • Outgoing Payments • Automatic Outgoing Payments • Payment Media • Make Settings for Payment Medium Formats From Payment Medium Workbench • Create Payment Medium Formats.**

Upon opening the transaction, you will see an overview of all the payment medium formats configured in the system, as shown in Figure 5.34. If you double-click a specific format, you can access its settings.

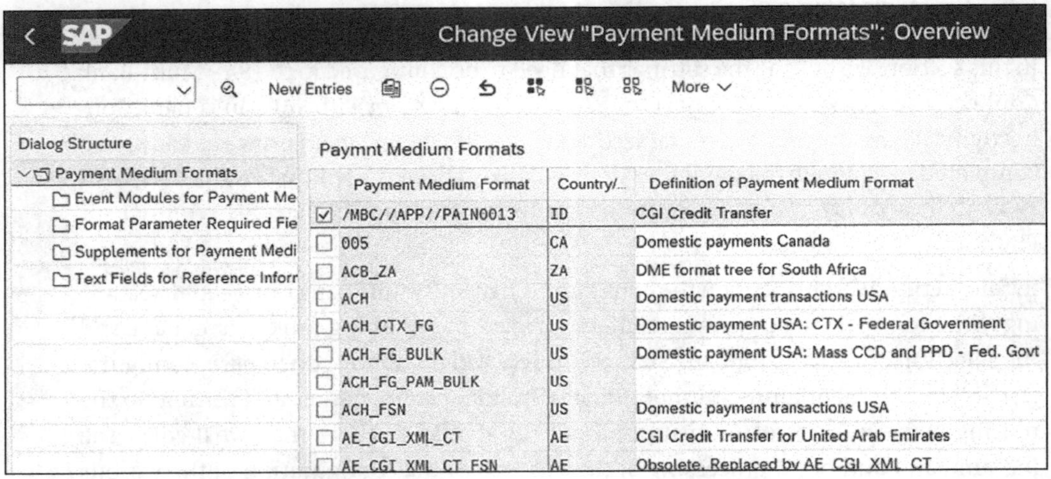
Change View "Payment Medium Formats": Overview
New Entries
More
Dialog Structure
Payment Medium Formats
Event Modules for Payment Me
Format Parameter Required Fie
Supplements for Payment Medi
Text Fields for Reference Inforr
Paymnt Medium Formats
Payment Medium Format
Country/...
Definition of Payment Medium Format
/MBC//APP//PAIN0013 ID CGI Credit Transfer
005 CA Domestic payments Canada
ACB_ZA ZA DME format tree for South Africa
ACH US Domestic payment transactions USA
ACH_CTX_FG US Domestic payment USA: CTX - Federal Government
ACH_FG_BULK US Domestic payment USA: Mass CCD and PPD - Fed. Govt
ACH_FG_PAM_BULK US
ACH_FSN US Domestic payment transactions USA
AE_CGI_XML_CT AE CGI Credit Transfer for United Arab Emirates
AE_CGI_XML_CT_FSN AE Obsolete. Replaced by AE_CGI_XML_CT

Figure 5.34 Create Payment Medium Formats

Within the transaction, as shown in Figure 5.35, the **Type** field can be set to **Connector for SAP Multi-Bank Connectivity** in the definition of the payment medium format. This setting triggers the automatic transfer of the file to the SAP Multi-Bank Connectivity connector for processing and dispatch upon successful generation.

After the setup, the program automatically assigns the file name. Once the setting is applied, it directly affects the payment medium selection variants, which are also responsible for assigning the file names. This field is hidden once the previously mentioned setting in Transaction OBPM1 is completed.

5.5 Configuration

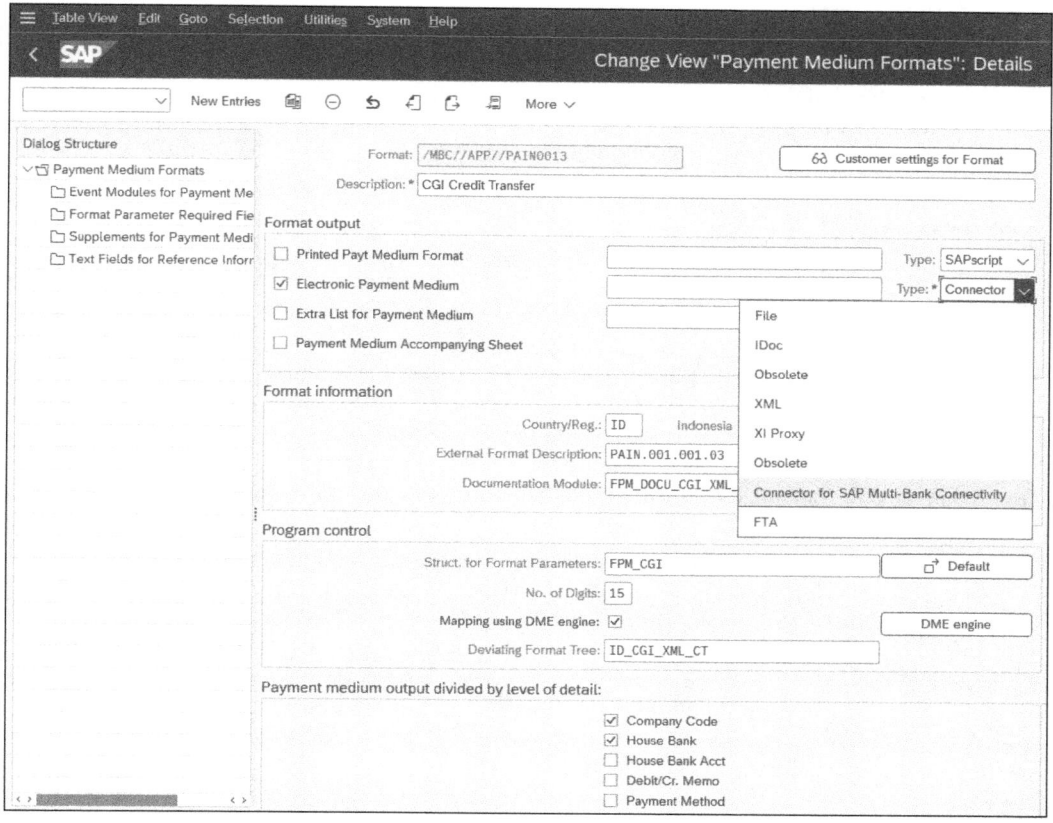

Figure 5.35 Configure Payment Medium Format

> **Individual File Name**
>
> In Transaction OBPM3, function module /BSNAGT/FI_PAYMEDIUM_DMEE_20 is available for EVENT 21. This function module is used to store global parameters during payment file creation. To influence the file name upon transfer to SAP Multi-Bank Connectivity, the function module can be copied to use as a base, and the C_FILENAME parameter can be customized.

Maintaining Bank Master Data

Sender and receiver IDs are used for identification, specifying who sends and receives a message, and they are crucial for managing routing within SAP Multi-Bank Connectivity. A key element in this process is the **Customer Number** field within the bank master data.

The **Customer Number** field can be adjusted via Transaction FI12 (Change House Banks) through the **House Banks** tab, or alternatively in SAP S/4HANA Cloud Private Edition and SAP S/4HANA Cloud Public Edition through the Manage Banks—Cash Management app (F1574A).

By navigating to the **House Banks** tab, you can access the house bank maintenance, as depicted in Figure 5.36.

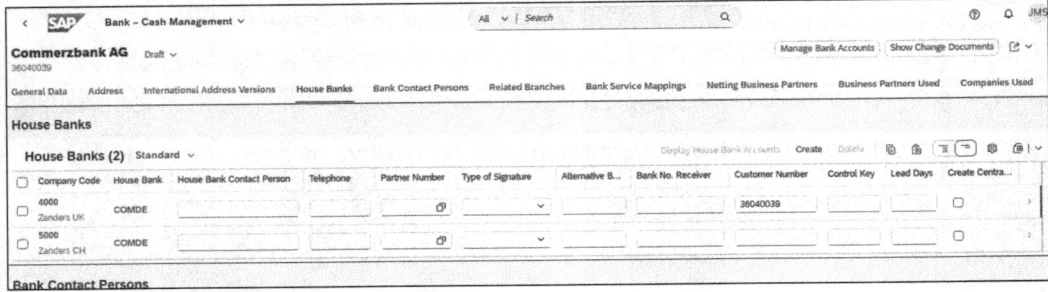

Figure 5.36 Manage Banks: Cash Management

By clicking a house bank, you enter the maintenance detail view of that house bank. In **Edit** mode, the sender ID can be updated using the **Customer Number** field, as shown in Figure 5.37.

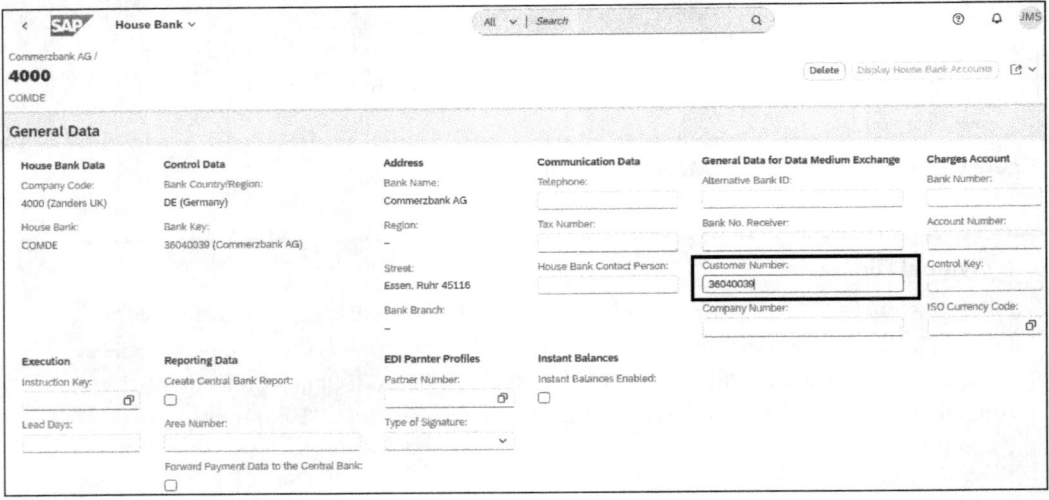

Figure 5.37 Manage House Banks

Maintain Selection Variants for Bank Statements

The account statement processing can be directly triggered as a process step from SAP Multi-Bank Connectivity, enabling a straight-through process for account statement handling. This requires configuring the account statement processing and maintaining a variant for program RFEBKA00. Further details on account statement processing can be found in Chapter 9. Once the variant for account statement processing is configured, it can be linked as a process step with the SAP Multi-Bank Connectivity connector in customizing and is automatically invoked after pulling the account statements. This process is illustrated in Figure 5.38.

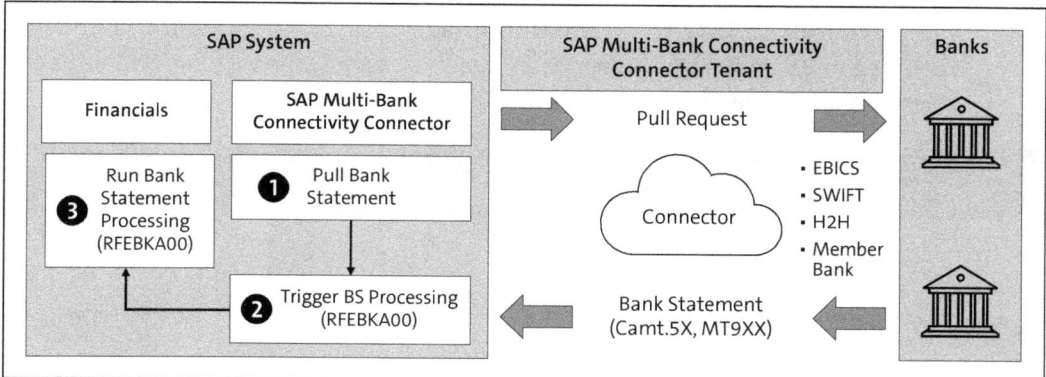

Figure 5.38 Bank Statement Processing

To configure the settings, customizing table /BSNAGT/V_VARI must be updated. This can also be accessed via the IMG through the path **Multi-Bank Connectivity Connector** • **Inbound Processing** • **Assign Variants for Bank Statement Processing**.

To configure this step (as shown in Figure 5.39), click **New Entries**, select the **Sender ID**, **Message Type**, and **Variant**, and click **Save**.

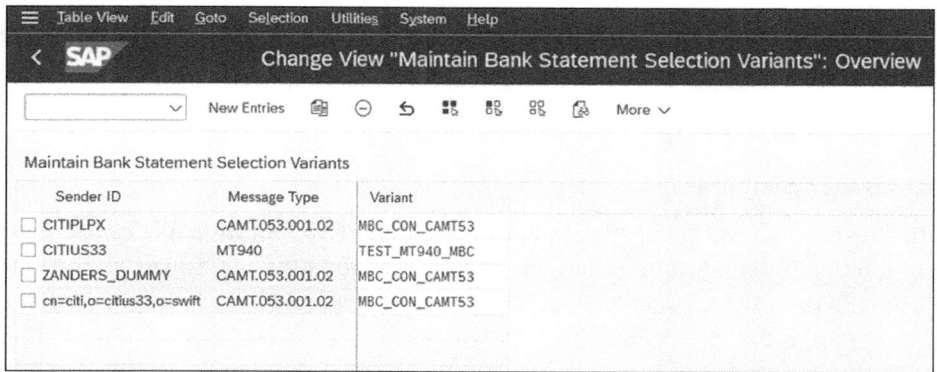

Figure 5.39 Maintain Bank Statement Variants

If parameter sets for account statement processing have been defined, they can be linked via customizing as well. This is done through table /BSNAGT/V_PSETID, which is located in the IMG under **Multi-Bank Connectivity Connector** • **Inbound Processing** • **Assign Bank Statement Parameter Sets**. Through this transaction, the file format can also be linked for importing account statements and for SAP Multi-Bank Connectivity, as illustrated in Figure 5.40. This step is particularly important for SAP S/4HANA Cloud Public Edition customers.

5 SAP Multi-Bank Connectivity

To create new entries, click **New Entries**, then select the **Sender ID** and the **Message Type** you're receiving. Specify the **Parameter Set ID** from the available help documentation and select the format for the statement you're expecting to receive. You'll also need to indicate whether the format is in XML or some other bank-specific format defined in the system. Next, select whether this is an **Intraday** format, which means you anticipate receiving multiple statements per day. Finally, assign a name to your format for easy reference, click **Save**, and the new entries will be created.

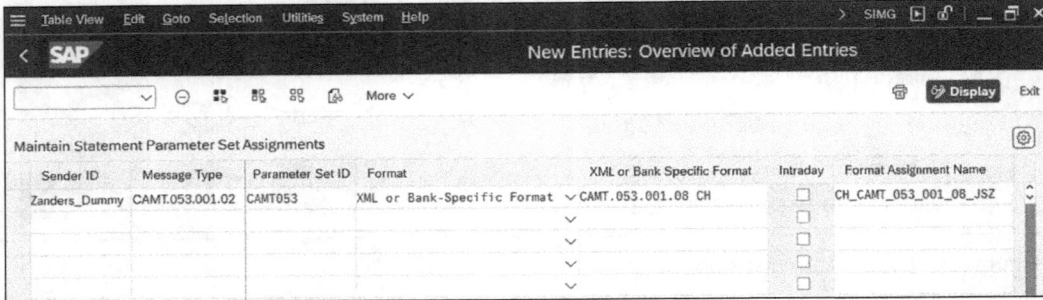

Figure 5.40 Maintain Bank Statement Parameter Sets

> **Bank Statement Parameter Sets**
> Bank statement parameter sets control the basic processing of account statements. Defined parameter sets determine whether account statements should be read without posting, read and fully posted, or posted only up to the first posting step in bank accounting. This is a standard SAP S/4HANA Finance setting and can be configured in SAP S/4HANA Cloud Public Edition through configuration step ID 103635.

5.5.2 Basic Configuration

In this section, we will cover the basic configuration of SAP Multi-Bank Connectivity, including maintaining number ranges, integrating the SAP Multi-Bank Connectivity tenant with the SAP S/4HANA system, and routing.

Maintaining Number Ranges

For all messages that do not receive a message ID through the preprocess application, a message ID will be assigned by the SAP Multi-Bank Connectivity connector. To facilitate this, number range object /BSNAGT/N2 must be maintained. This can be done via Transaction SNRO or through IMG path **Multi-Bank Connectivity Connector • Basic Settings • Maintain Number Range for Message ID**.

Figure 5.41 shows an example of number range maintenance for the SAP Multi-Bank Connectivity connector. To maintain a number range, click **Edit Interval**, then enter your **Number Range No.** and specify the range from **From No.** to **To Number**. Ensure this interval is not marked **External** as otherwise it won't have any effect. After specifying the range, click **Save**.

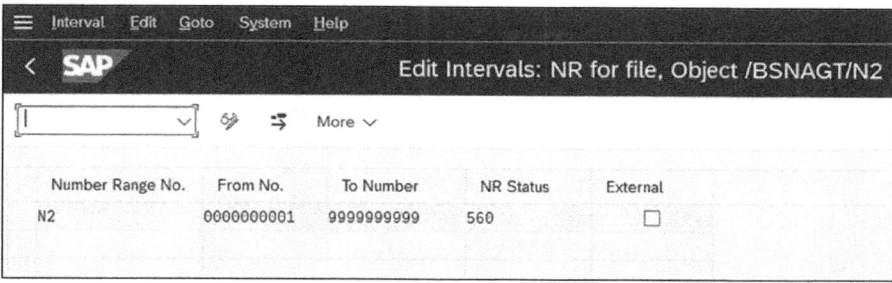

Figure 5.41 Maintain Number Ranges

Building a Secure Connection to SAP Multi-Bank Connectivity

For establishing a secure connection between the SAP S/4HANA system and the SAP Multi-Bank Connectivity tenant, the connection is secured through multiple layers of encryption. Simply put, there is encryption that secures the transport channel itself and encryption that secures the message itself. Those encryption types are as follows:

- Transport Layer Security (TLS): Channel encryption
- Message Layer Security (MLS): Message encryption

This section aims to provide a high-level overview of how this process works and what configuration settings are required for the connector. We'll walk through certificate exchange, maintaining Secure Store and Forward (SSF) application parameters, and maintaining SSF profile data.

Certification Exchange

To enable secure message exchange between the SAP S/4HANA system and the SAP Multi-Bank Connectivity tenant, it is mandatory to obtain a certificate authority (CA)-signed Secure Sockets Layer (SSL) client standard Personal Security Environment (PSE) certificate (your own certificate) for transport-level security prior to the SAP Multi-Bank Connectivity onboarding phase. The procedure begins with logging into the SAP S/4HANA system and executing Transaction STRUST. The existing SSL client standard certificate must be exported and submitted to an SAP-trusted certificate authority for signing. After receiving the signed certificate, it is imported back into the system. Once the certificate is successfully signed, it must be exported again and forwarded to the SAP Multi-Bank Connectivity onboarding team, who will upload it to the corresponding SAP Multi-Bank Connectivity tenant.

To streamline certificate maintenance across nonproduction environments, a single CA-signed certificate can be used by including Subject Alternative Names (SAN) for all relevant systems during the certificate signing request (CSR) generation.

In addition, the SSL client standard PSE must be configured to trust SAP Multi-Bank Connectivity load balancers. This involves downloading the root and intermediate certificates from SAP Multi-Bank Connectivity, logging into the SAP S/4HANA system via Transaction STRUST, importing the certificates into the SSL client standard PSE, and saving the configuration.

Settings in Transaction STRUST are typically managed by the basis team, who should be involved in the configuration process. The certificate exchange must take place in both the test environment and the production environment with the respective assigned SAP Multi-Bank Connectivity tenant.

Within Transaction STRUST, there is a folder for the **SSF BSNAGT** application. The first step here is exporting the system's own key and securely providing it to the SAP Multi-Bank Connectivity team through a predetermined method.

To access the certificate, double-click the **SSF BSNAGT** folder. This will open a settings window on the right side of the screen. You can export the certificate using the **Export Certificate Request** button, as shown in Figure 5.42.

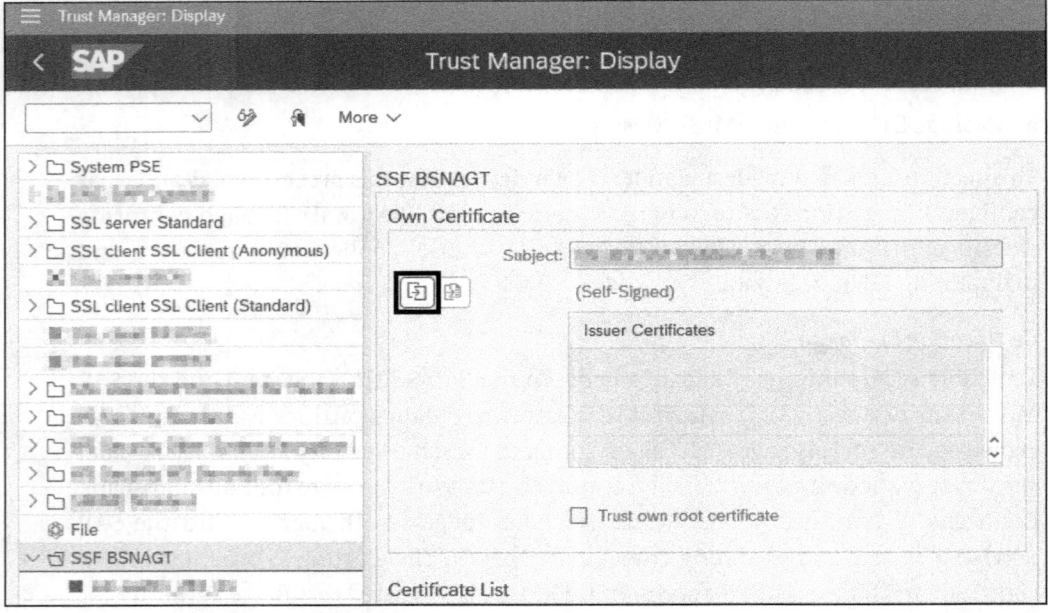

Figure 5.42 Transaction STRUST Configuration

This certificate must be transmitted to the SAP Multi-Bank Connectivity team. The certificate exchange must occur for both the test and production systems.

Finally, to establish a secure connection, the SAP Multi-Bank Connectivity team will provide a Message Layer Security (MLS) certificate. A separate certificate will be provided for each system to be connected (test system and production system). This certificate must also be imported into the **SSF BSNAGT** folder. The import is done by clicking the **Import Certificate** button, as shown in Figure 5.43.

5.5 Configuration

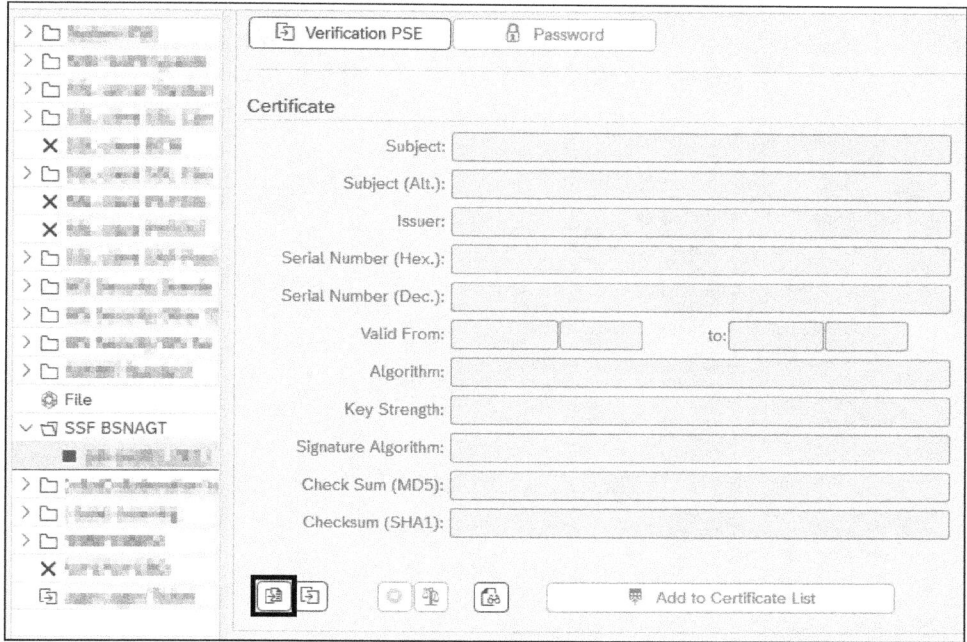

Figure 5.43 Import Certificate

Maintain SSF Application Parameters

Next, parameters are set for a secure and encrypted connection between the SAP S/4HANA system and the SAP Multi-Bank Connectivity tenant hosted in the cloud. This configuration is particularly relevant for MLS. SAP provides two standard applications here: BSNAGT (Payload Security) and FSNCFP (Tamper Protection), which can be tailored individually.

For a standard secure connection, the BSNAGT application is used with the following parameters:

- **Hash Algorithm**: SHA256
- **Encryption Algorithm**: AES128-CBC
- **Include Certificate:** Select this checkbox
- **Digital Signature with Data**: Select this checkbox

Configuration is done via Transaction SSFA, which can also be accessed through the menu path **Multi-Bank Connectivity Connector** • **Basic Settings** • **Maintain SSF Application Parameters**. Click the **New Entries** button to set up the SSF application. If the entry already exists, you can adjust the parameters by double-clicking the **BSNAGT** application. Figure 5.44 provides an illustration of the SSF application.

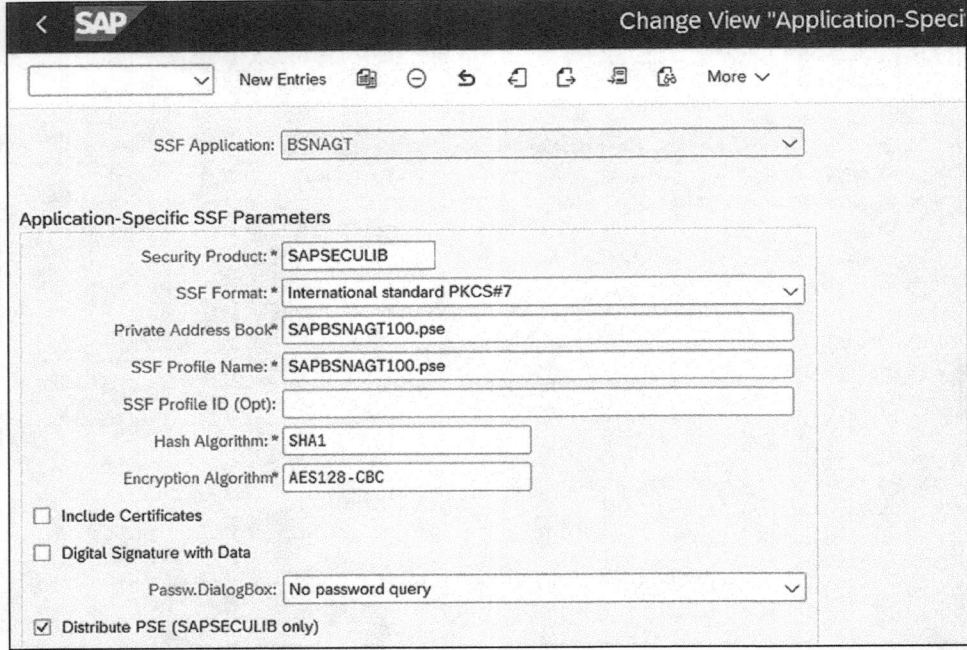

Figure 5.44 Store and Forward Application Setup

Maintaining SSF Profile Data

This transaction manages the encryption settings for the BSNAGT application. You can specify whether outbound messages should be encrypted and verified and whether inbound messages should be decrypted. Encryption can optionally be controlled at the message type and sender/receiver ID level. It is sufficient to maintain the application and set the encryption flags.

The transaction is located in the IMG and can be accessed via the following path: **Multi-Bank Connectivity Connector • Basic Settings • Maintain Secure and Forward (SSF) Profile Data**.

A new profile for the application (**Appl.**) can be created using the **New Entries** button or alternatively by pressing F5. After that, the profile can be restricted to the **Message Type** field as well as the sender or receiver ID (**SenderID/ReceiverID**) field. Figure 5.45 illustrates the configuration example.

The field for **Signatory/Recipient Name** is automatically populated with the certificate name provided by the SAP Multi-Bank Connectivity team. This field is populated in the **Maintain Secure Store and Forward (SSF) Profile Data** activity after the profile has been saved.

5.5 Configuration

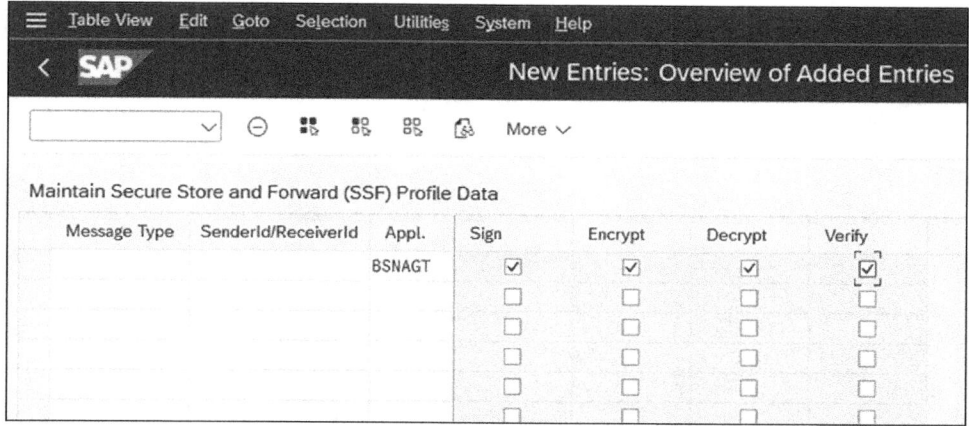

Figure 5.45 Maintain SSF Profile Data

Routing

SAP Multi-Bank Connectivity offers various integration options and includes a routing function that enables the redirection of payment processes from external systems to specific areas of SAP S/4HANA, such as advanced payment management or in-house banking. Alternatively, payment files provided by an external non-SAP system can be imported and sent directly to the bank. In this section, we will introduce you to the options available within SAP Multi-Bank Connectivity. Specifically, we will discuss local routing, ERP-to-ERP routing, and connectivity from the connector to the SAP Multi-Bank Connectivity tenant.

Before we begin, we should note that payment files and account statements that are processed using standard message types in SAP Multi-Bank Connectivity are sent directly to the bank by default, or they are forwarded to account statement processing. In Section 5.5.3, Table 5.3 lists message types and standard processing methods.

Local Routing

The SAP Multi-Bank Connectivity connector is primarily used for local routing within SAP S/4HANA by the advanced payment management system and its associated in-house banking solution. Payment files for internal transactions, for example, are handed over from SAP S/4HANA Finance to in-house banking via the SAP Multi-Bank Connectivity connector after executing Transaction F110. For this, local routing must be configured in the system. The configuration is done in the customizing transaction, which can be found in the IMG menu under **Multi-Bank Connectivity Connector • Routing and Connectivity • Maintain Routing Settings for Outbound Messages**.

Figure 5.46 illustrates this transaction with configurations for local routing. In this transaction, you have the **Sender ID**, **Receiver ID**, **Message Type**, and **Origin** fields; the latter represents the originating application. These selection criteria are used in the

397

mapping to forward messages to local advanced payment management. Within this setup, you expect messages from the Payment Medium Workbench program in SAP S/4HANA Finance (**Origin: FIPMW**), which you transfer to advanced payment management via local routing. To add new entries, use the **New Entries** button or press F5.

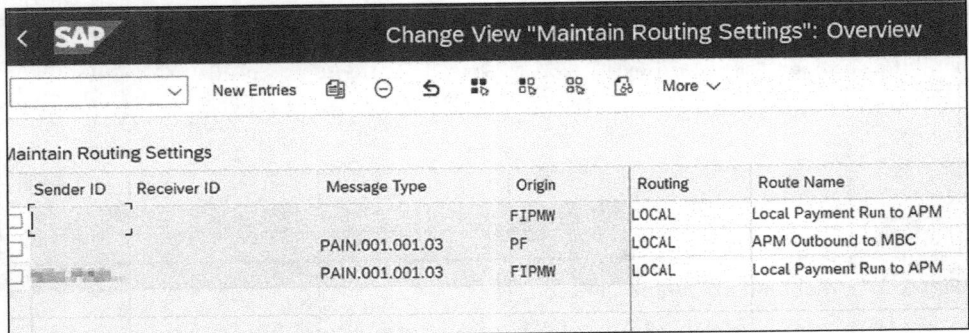

Figure 5.46 Maintain Local Routing

ERP-to-ERP Routing

The SAP Multi-Bank Connectivity connector enables the exchange of incoming and outgoing messages between SAP S/4HANA systems. A prerequisite for this is that the SAP Multi-Bank Connectivity connector is available on both the sender and the receiver side. If this condition is met, an RFC connection of type 3 (RFC connection to an ABAP system using TCP/IP) can be established between the systems, granting the necessary permissions to read and write data, as shown in Figure 5.47.

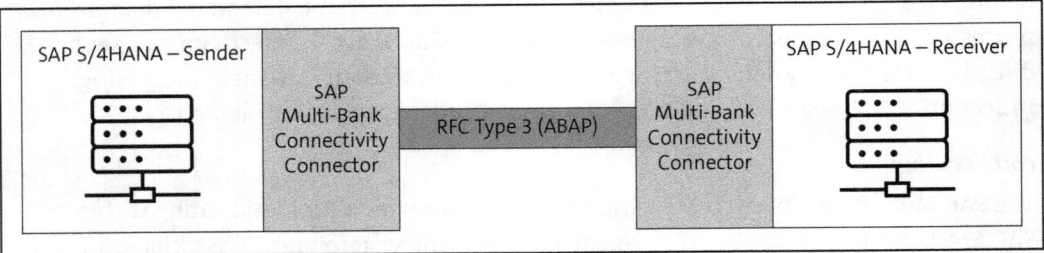

Figure 5.47 ERP-to-ERP Routing

Once the RFC ABAP connection between the systems is established, the customization can be performed via menu path **Multi-Bank Connectivity Connector • Routing and Connectivity • Maintain Routing Settings for Outbound Messages**. It must be configured in parallel on both systems.

Figure 5.48 shows an example of how to create a connection between two SAP S/4HANA systems. To add the new connection, use the **New Entries** button or press F5. In this example, we select messages for which the **Sender ID** is Zanders_Dummy and route them via RFC connectivity (**Routing: RFCAPI**) between those two systems.

5.5 Configuration

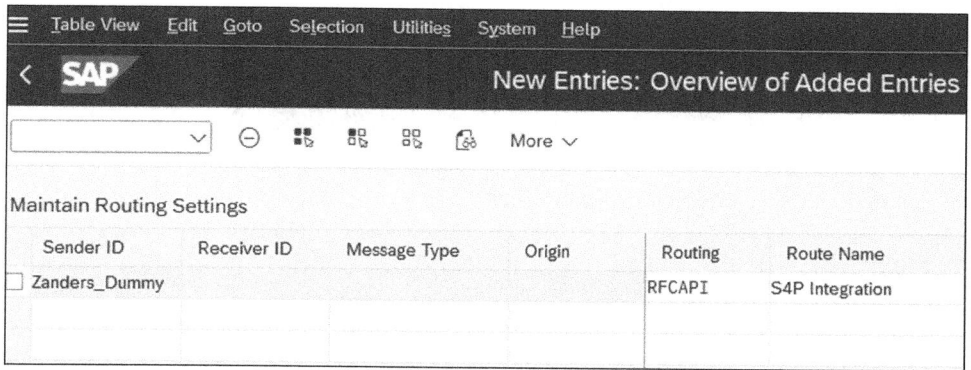

Figure 5.48 ERP-to-ERP Routing Configuration

Connectivity from Connector to SAP Multi-Bank Connectivity Tenant

In general, there are two possible connectivity channels for integrating your SAP S/4HANA system with SAP Multi-Bank Connectivity. The first (and SAP-recommended) method is HTTP connectivity, which establishes a synchronous communication channel between you SAP S/4HANA system and the SAP Multi-Bank Connectivity tenant. This method ensures a direct and real-time message exchange.

> **XI Protocol Replaced with HTTP(s) in Version 2202**
>
> Since February 28, 2022, the connection method using the XI protocol has been replaced by an HTTP(s)-based connection method. The following SAP Notes are relevant for this:
>
> - 3091189: Objects for Direct HTTP Communication (HTTP API)
> - 3103718: Mention RFC Destination for HTTP API
> - 3133824: Maintenance View for HTTP API Settings
>
> SAP Multi-Bank Connectivity is being continuously developed by SAP. You can learn about the latest developments at *https://help.sap.com/whats-new/a63104d9ec664e7d be5f03750f110e2f?locale=en-US*.

Let's now look at the settings and configurations specifically based on the use of the HTTP connectivity channel. In this context, an RFC destination of type G (HTTP connection to external server) is required because communication with the SAP Multi-Bank Connectivity tenant takes place over an HTTP-based interface. Within Transaction SM59, a new HTTP connection is created, in which parameters such as the target host (the URL of the SAP Multi-Bank Connectivity tenant), service path, and login credentials and proxy settings, if applicable, are configured. After saving, the connection can be validated using the integrated test function to ensure error-free communication. Figure 5.49 shows an example of how the connection can be set up using Transaction SM59. To create a new RFC connection, you have to click the **Create** button. After clicking the

5 SAP Multi-Bank Connectivity

button, a popup window will show up in which you can choose the **Connection Type** and **Destination Description**.

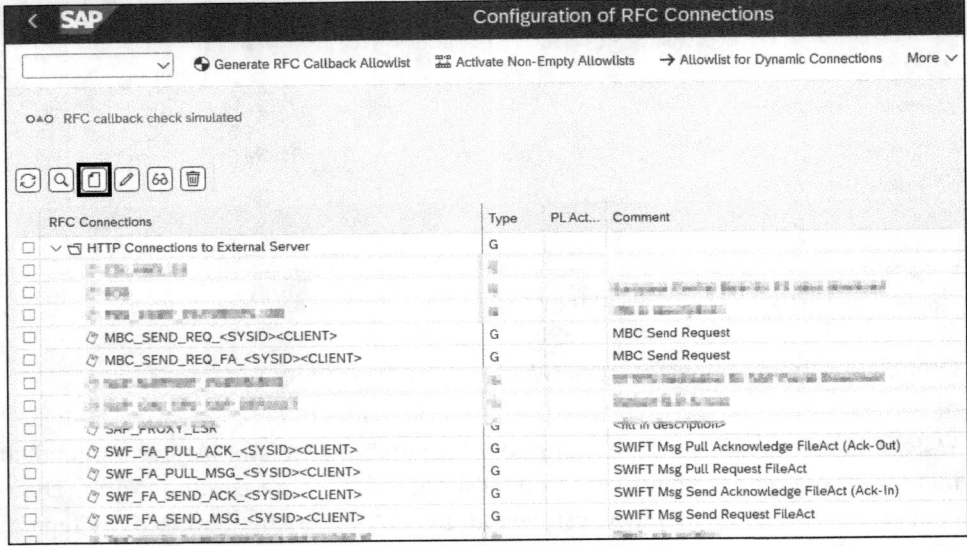

Figure 5.49 RFC Configuration

The parameters provided ahead are examples for an HTTP connection and based on the tabs shown in Figure 5.50.

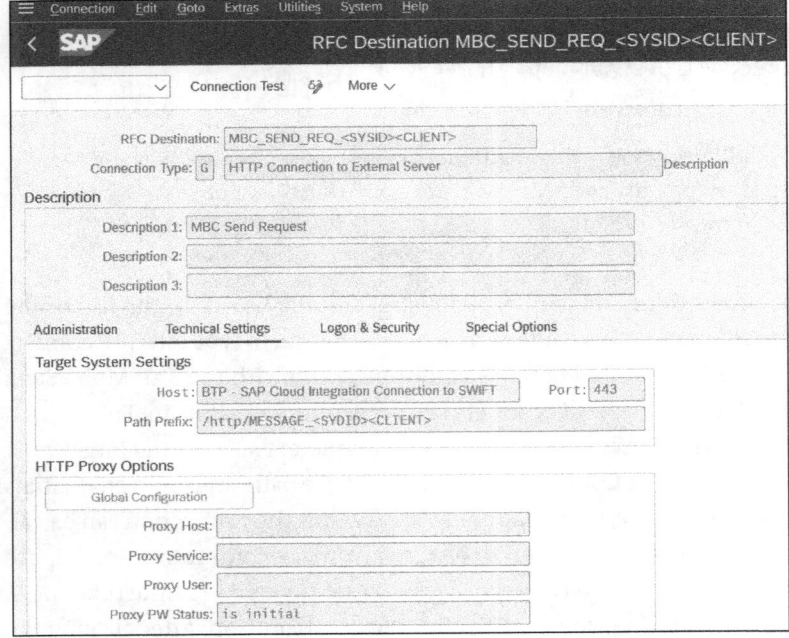

Figure 5.50 RFC Destination MBC

Table 5.1 shows details needed for an example setup of an RFC connection between an SAP system and the SAP Multi-Bank Connectivity client to send data. The <SYSID> and <CLIENT> values represent the SAP system ID and your own client; enter this info in place of these placeholder values.

Field	User Action and Value to Be Entered
RFC Destination	MBC_SEND_REQ_<SYSID><CLIENT>
Connection Type	G (HTTP Connection to External Server)
Technical Settings Tab	
Host	Corporate SAP Multi-Bank Connectivity tenant host
Path Prefix	/http/MESSAGE_<SYDID><CLIENT>
Port	443
Logon & Security Tab	
SSL	Active (radio button)
SSL Certificate	DFAULT SSL Client (Standard)

Table 5.1 SAP Multi-Bank Connectivity Send Acknowledge (Receipt)

Table 5.2 shows details needed for an example setup of an RFC connection between an SAP system and the SAP Multi-Bank Connectivity client to pull messages. The <SYSID> and <CLIENT> values represent the SAP system ID and your own client; enter this info in place of these placeholder values.

Field	User Action and Value to Be Entered
RFC Destination	MBC_PULL_REQ_<SYSID><CLIENT>
Connection Type	G (HTTP Connection to External Server)
Technical Settings Tab	
Host	Corporate SAP Multi-Bank Connectivity tenant host
Path Prefix	/http/PULL_<SYDID><CLIENT>
Port	443
Logon & Security Tab	
SSL	Active (radio button)
SSL Certificate	DFAULT SSL Client (Standard)

Table 5.2 SAP Multi-Bank Connectivity Pull Acknowledgement (Receipt)

5.5.3 Inbound and Outbound Processing

In this section, we will discuss the incoming and outgoing payment transactions and provide an overview of how not only locally integrated SAP S/4HANA solutions but also third-party systems and non-SAP systems can be integrated with and connected to the SAP Multi-Bank Connectivity connector.

We'll cover the following topics in this section:

- **Sender/receiver ID mapping**
  Here, we take a detailed look at the sender and receiver IDs that are crucial for SAP Multi-Bank Connectivity, including how and where to maintain them in the system.
- **Maintaining inbound processing steps**
  The processing steps can be sequentially controlled via the connector. Here, we present how you can configure the processing of incoming messages.
- **Configuring EBICS**
  EBICS is a local standard in Europe, particularly prevalent in Germany, Austria, Switzerland, and France. Here, we describe how you can successfully implement the EBICS protocol with SAP Multi-Bank Connectivity in your next project or rollout in Europe.
- **Configuring SWIFT**
  SWIFT, a widely used international communication channel, has long been the standard. Here, we show you how to configure the SAP Multi-Bank Connectivity connector to ensure that the companion files are correctly generated in the background.

Maintaining Sender/Receiver ID Mapping

Maintaining sender/receiver ID mapping is relevant for all payments for which assignment of a sender or receiver ID from the master data of the house bank is not possible. This situation may arise when payment carriers from third-party systems are imported or when the sender ID is simply not maintained in the master data. In such cases, a fallback can be set up in customizing table /BSNAGT/V_SENDID. The setting can be accessed through the following IMG menu path: **Multi-Bank Connectivity Connector** • **Advanced Settings** • **Maintain Sender/Receiver ID Mappings**.

The table can be maintained with considerable flexibility. The settings can be configured in combination based on **Message Type**, **Receiver ID**, and **Origin**. If, as shown in Figure 5.51, only the **Sender ID** field is maintained, then all payments lacking an assigned sender ID will receive the sender ID specified in the customizing table.

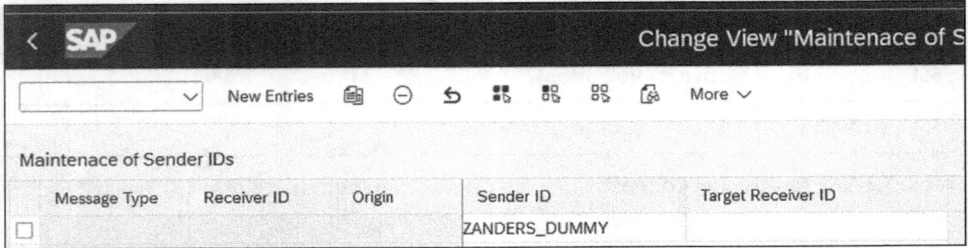

Figure 5.51 Maintain Sender/Receiver ID

5.5 Configuration

Maintaining Inbound Processing Steps

Using the **Maintain Processing Steps for Inbound Messages** customizing activity, processing steps can be actively overridden (see Figure 5.52). SAP Multi-Bank Connectivity has default processing steps for each message type that handle the further processing of incoming and outgoing messages; for example, bank statements are forwarded directly to bank statement processing upon receipt and posted according to the settings. If a forwarding action is required alongside the posting of the bank statement, this is the appropriate customizing transaction.

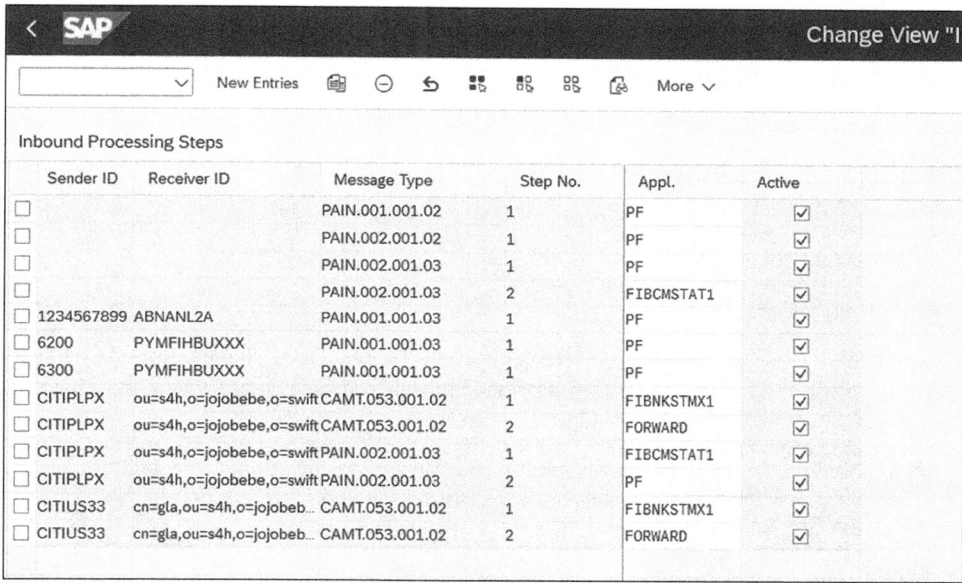

Figure 5.52 Inbound Processing Steps

The maintenance of process steps is performed via the IMG and can be accessed through the following menu path: **Multi-Bank Connectivity Connector** • **Inbound Processing** • **Maintain Processing Steps for Inbound Messages**.

New processing steps can be added via the **New Entries** button. A selection can be made based on **Sender ID**, **Receiver ID**, and **Message Type**, with all fields treated as wildcards. The **Step No.** field controls the processing frequency. The **Appl.** field links the application through which the message will be processed. The **Active** checkbox determines whether the processing step is enabled and in use or deactivated.

During maintenance, it is crucial to ensure that different step numbers are used in sequence for further processing. This sequential assignment ensures that each process is properly executed according to its designed sequence and logical order.

Table 5.3 provides an overview of message types that have default processing embedded within the system.

Message Type	Triggered Processing
pain.002*	For incoming bank protocols, the standard payment status report processing is directly initiated.
camt.052*	Camt bank statements are directly read into the SAP system. In the background, the system checks which variants for processing have been maintained in the customizing settings. This applies to end-of-day bank statements as well as intraday and credit/debit notifications.
camt.053*	
camt.054*	
MT940*	
MT942*	
CFONB_STATEMENT	For these message types, bank statement processing will be triggered.
CSB43_STATEMENT	
BAI_STATEMENT	
BAI2_STATEMENT	
ITAU_STATEMENT	
camt.086*	Local bank statements in the CFONB, CSD43, and BAI formats are also automatically processed further, provided no processing steps have been overridden.
BAI_LOCKBOX	All message types that have incoming BAI_LOCKBOX or BAI2_LOCKBOX in the header are automatically processed further.
BAI2_LOCKBOX	

Table 5.3 Inbound Message Types

Configuring EBICS

For an introduction to EBICS, refer back to Chapter 2. In this section, we provide an overview of the configuration options in SAP Multi-Bank Connectivity related to EBICS. The EBICS configuration in SAP Multi-Bank Connectivity supports both the classic EBICS 2.5 with traditional order types and EBICS 3.0 with integrated business transaction formats (BTFs). All EBICS signature types are supported by SAP Multi-Bank Connectivity. Users who have authorized payments in SAP Bank Communication Management can be assigned directly as dynamic EBICS users. The corporate seal is also supported via the static user. In the following sections, we offer a step-by-step look at the configuration of both EBICS 3.0 and EBICS 2.5.

EBICS 3.0 Configuration

EBICS 3.0 is a combination of the French-influenced EBICS standard and the German Credit Industry (DK) standard. For configuration, the bank's parameters are first required and must be provided to the SAP team during the scoping phase. Figure 5.53 shows an excerpt from the SAP Multi-Bank Connectivity questionnaire for EBICS 3.0.

5.5 Configuration

Administrative Order Type (Eg. BTD-download- and BTU-upload- are the most commonly used.)	Service Name (Eg. "SCT" for the SEPA services, "EOP" for End Of Period statements, etc.)	Scope (Eg:. FR, DE, CH, GLB etc.)	Service Option (Eg. COR, MCT, SDD, B2B)	Container (optional attributes: Type. eg. XML, ZIP)	MESSAGE NAME (Eg. mt940, pain.001, cfonb160, etc) Optional attributes: - variant (eg. 001) - version (eg. 03) - format (eg. XML)	File Naming convention
BTU	SCT	DE		XML	pain.001	
BTU	SDD	DE	B2B	XML	pain.008	

Figure 5.53 EBICS Business Transaction Format Parameters

Once these parameters have been communicated to the SAP Multi-Bank Connectivity team, the SAP Multi-Bank Connectivity connector must be configured. For this, SAP provides customization options via the following menu path in the IMG: **Multi-Bank Connectivity Connector • Outbound Processing • EBICS BTF**:

- **Maintain EBICS BTF Types**
  For the connector, a distinctly identifiable BTF type must be defined as a key field. The field is freely definable, and in our example, we opted for a key combination of BTF and service name—for example, business transaction upload (BTU) and SEPA credit transfer (SCT) in the **BTF Type** field, as shown in Figure 5.54. You can also assign an optional description in the **Description** field.

EBICS BTF Type Definition	
BTF Type	Description
☐ BTU_SCT	Upload SEPACredit Transfer
☐ BTU_SDD	Upload SEPA Direct Debit

Figure 5.54 BTF Type Determination

- **Maintain EBICS BTF Parameter Assignments**
  With the order type, the individual EBICS parameters must also be assigned. The parameters are provided by the bank and typically consist of the following components (which appear in the **Parameter Name** field with their entries in the **Parameter Value** field, as shown in Figure 5.55):

  - SERVICE_NAME
    The service name is a three-letter abbreviation that indicates the type of payment involved. For instance, *SCT* stands for *SEPA credit transfer*.

  - SCOPE
    The scope specifies the set of rules (e.g., a particular payment or billing system) used for processing the message. For example, a scope set to **DE** might indicate that the message has been created in accordance with German law and its corresponding regulations.

- **Service Option**
  Service option is one of the fields of the selection not shown in Figure 5.55 that allows for indicating a more specific variant within an existing service, such as a standard or urgent transfer. In Germany, three-digit standardized codes like EXP for an express payment are used for this purpose. If you specify service option EXP during a transfer, the bank recognizes that the payment should be given priority treatment. Without a service option, the payment is processed using the standard procedure.

- **MSG_NAME**
  The message name denotes the type of message transmitted within a service, such as a transfer, direct debit, or account report. Typical message names are standardized formats like pain.001 for payment instructions or camt.053 for account statements. The message name helps accurately identify the specific type of data being transmitted. Without specifying a message name, the receiving system may not be able to correctly interpret the message's content.

- **CONTAINER_TYPE**
  The container type (not shown in Figure 5.55) defines the technical packaging format via which the message is transmitted, such as a single file or a ZIP archive containing multiple files. ZIP is a common container format for compressed content. The container ensures that data can be correctly bundled and processed, especially when dealing with multiple or large files. Without a defined container, transmission errors or misunderstandings during processing could occur.

Figure 5.55 shows a configuration of the parameters for a pain.001 SEPA credit transfer and for a pain.008 SEPA direct debit.

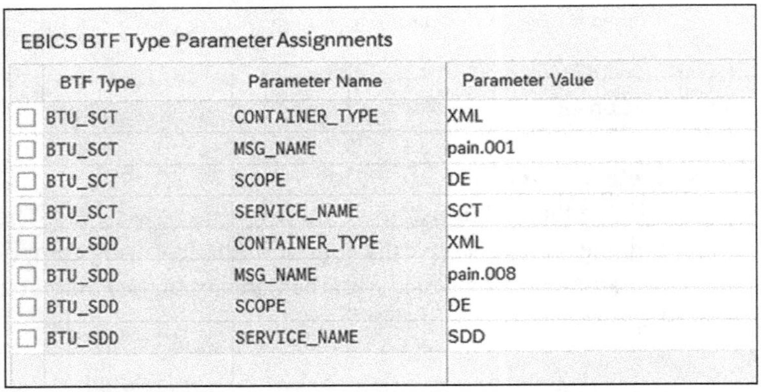

EBICS BTF Type Parameter Assignments		
BTF Type	Parameter Name	Parameter Value
☐ BTU_SCT	CONTAINER_TYPE	XML
☐ BTU_SCT	MSG_NAME	pain.001
☐ BTU_SCT	SCOPE	DE
☐ BTU_SCT	SERVICE_NAME	SCT
☐ BTU_SDD	CONTAINER_TYPE	XML
☐ BTU_SDD	MSG_NAME	pain.008
☐ BTU_SDD	SCOPE	DE
☐ BTU_SDD	SERVICE_NAME	SDD

Figure 5.55 EBICS BTF Parameter Assignment

- **Maintain EBICS BTF Type Determination Rules**
  Not every bank currently prefers or is capable of supporting the EBICS 3.0 standard. To allow separation within the connector and to determine which bank, as well as which payment method, should be assigned to a specific BTF type, the connector

includes a determination table. This table enables the configuration for such assignments.

Figure 5.56 illustrates the determination table. In this transaction, selection criteria such as company code (**CdCd**), house bank **(Hou...)**, house bank account (**Acct...**), payment method (**PM**), and instruction keys (**IK**) can be linked to the respective BTF order type (**BTF Type**). All fields function as wildcards, allowing them to be maintained independently or in combination with one another. You can create a new entry by clicking the **New Entries** button or pressing F5. Selecting the **Urgt** checkbox marks a payment request as urgent.

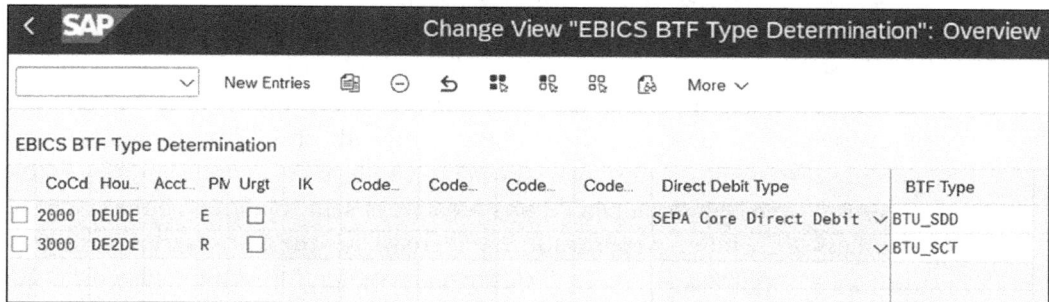

Figure 5.56 EBICS BTF Type Determination

EBICS 2.5 Configuration

For the assignment of order types in EBICS 2.5, the SAP Multi-Bank Connectivity connector provides a direct determination table. Unlike EBICS 3.0, a direct assignment of order types to the respective payment process is sufficient.

Figure 5.57 shows an example of maintaining **SDD** (SEPA direct debit) and **CCT** (SEPA credit transfer) for **Order Type**. New entries can be added via the **New Entries** button or the F5 shortcut.

Maintain EBICS Order Type Determination											
CoCd	Hou..	Acct...	PM	Urgt	IK	Code...	Code...	Code...	Code...	Direct De...	Order Type
4000	COMDE		E	☐							SDD
4000	COMDE		R	☐							CCT
				☐							

Figure 5.57 EBICS Order Type Determination

In this transaction, selection criteria such as company code (**CdCd**), house bank **(Hou...)**, house bank account (**Acct...**), payment method (**PM**), and instruction keys (**IK**) can be directly linked to the respective order type. All fields function as wildcards, allowing them to be maintained independently or in combination with one another. Selecting the **Urgt** checkbox marks a payment request as urgent.

For the authentication of payment files, EBICS users with sufficient signature authority are required. As mentioned in Chapter 2, there are different EBICS users that can be assigned to authenticate payment files. Essentially, the system offers two ways to identify the payment file, as follows:

- **Authorization by a dynamic user**
  In the SAP system, payment files can be approved through an approval process in SAP Bank Communication Management. Approvers involved in the SAP Bank Communication Management approval process can be linked to the respective EBICS users in the SAP Multi-Bank Connectivity connector. This ensures transparency within the company and for the bank regarding who approved the payment file.

- **Corporate seal—authorization by a static user**
  A corporate seal, also known as a *CFO user*, consists of one or two EBICS users that are statically assigned to all payment files during transmission. In this procedure, two EBICS users are used as permanent approvers to authorize the payment file for the bank. It is irrelevant who approved the payment file in the approval process within the SAP system; the same EBICS users are always sent to authorize the payment, regardless of the internal approvals. This approach has the advantage of making the management of EBICS users significantly more straightforward for all banks using the corporate seal, thereby drastically reducing the administrative effort.

The following are the relevant customization points, which can be accessed via the IMG using the following path: **Multi-Bank Connectivity Connector • Outbound Processing**:

- **Maintain Approval User Determination Rules**
  In the **Maintain Approval User Determination Rules** customization transaction, EBICS users are identified for the approval of the payment files from the approval process. For complete EBICS authorization, EBICS participants with sufficient authority need to be included. Specifically, either two EBICS users with an A (first signature) or two EBICS users with one A (first signature) and one B (second signature authorization) are required.

  Let's look at a process example. If payment files are approved using the four-eyes principle, then the customization table can be set to always select the last two approver pairs as the EBICS approvers. The following combinations are possible:
  - USER_A 1: First approval user
  - USER_B 2: Second approval user
  - USER_C-2: Last but two
  - USER_D-1: Last but one
  - USER_E LAST: Last approval user

  For example, if you want to send the first two approver signatures as EBICS users within the process, this can be set through the following parameters:
  - 1 - First Approval User • Parameter 1
  - 2 - Second Approval User • Parameter 2

In the **Maintain Approval User Determination Rules** activity, shown in Figure 5.58, determination can be made per **Sender ID**, **Receiver ID**, **Message Type**, or **Origin**. Fields that are not used are considered as wildcards. Once the key fields are determined, the parameters will be assigned to designate the approval user.

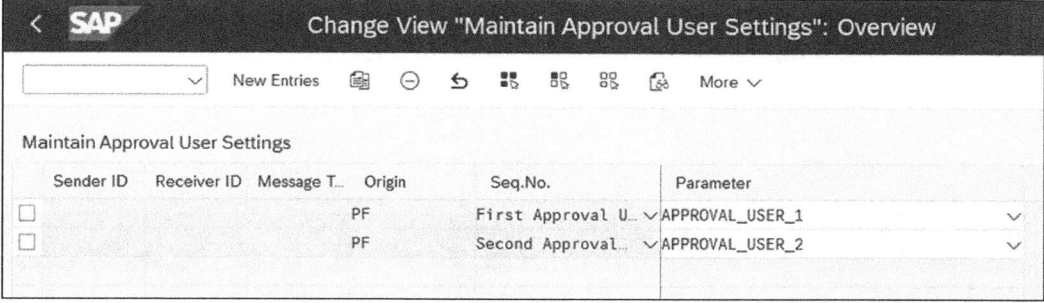

Figure 5.58 Maintain EBICS Approval User

- **Map Approval Users to EBICS Users**

    In the **Map Approval Users to EBICS Users** transaction, a direct mapping occurs between the bank, the SAP user, and the EBICS user. **Sender ID**, **Receiver ID**, and SAP **Approval User** are available as key fields and can be mapped to the respective **EBICS User**, as shown in Figure 5.59. Once this configuration is done, during the payment approval process via SAP Bank Communication Management, the respective **EBICS User** is identified and transmitted in the header parameters to the SAP Multi-Bank Connectivity tenant and, therefore, to the bank. Fields that are not filled in are treated as wildcards.

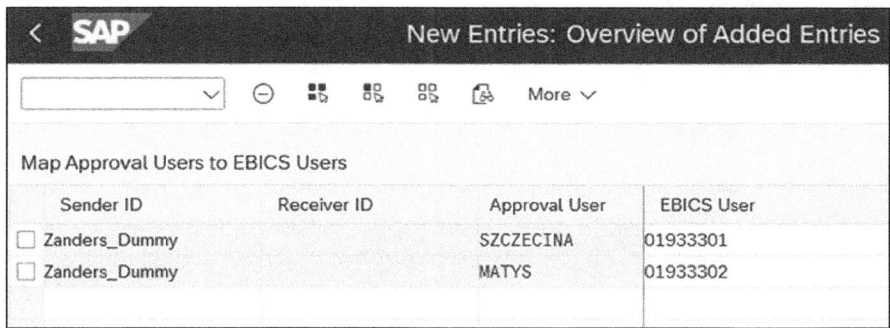

Figure 5.59 Map Approval User to EBICS User

- **Maintain Static EBICS Users**

    If the corporate seal is to be set up via EBICS, this is the transaction you need. In this transaction, either one EBICS E (single signature) user or two EBICS users, each with an A (first signature) and EBICS B (second signature), can be configured as static users per bank or even per message type. The following key fields are available for this purpose:

5   SAP Multi-Bank Connectivity

- Sender ID
- Receiver ID
- Message Type
- Origin

Using these fields, the static EBICS user is identified and, regardless of the approval process within the SAP S/4HANA system, communicated via the extended header parameters to the SAP Multi-Bank Connectivity tenant and finally to the bank.

Figure 5.60 shows the configuration by storing two EBICS A users (first signature) as static users.

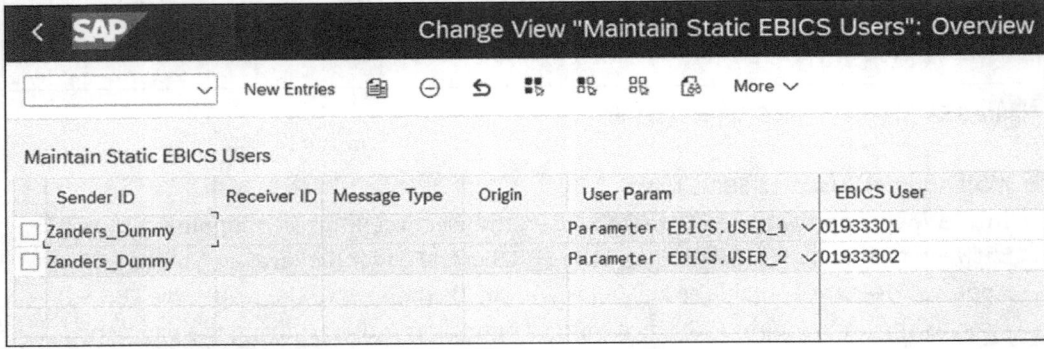

Figure 5.60 Maintain Static EBICS Users

> [!] **EBICS Distributed Electronic Signature Not Supported**
> The EBICS distributed electronic signature (DES) is not fully supported as of SAP S/4HANA 2303. SAP generally follows a centralized and protocol-independent approach. Because DES is a process based purely on an EBICS functionality specification, we assume it will not be fully supported by SAP Multi-Bank Connectivity in the future either.

Overall, SAP Multi-Bank Connectivity provides everything required for state-of-the-art communication via EBICS, given proper configuration. Further developments regarding EBICS are planned; not only is the communication standard evolving, but so too is SAP Multi-Bank Connectivity. For example, in the future, the initialization of EBICS users will no longer be conducted by the SAP Multi-Bank Connectivity team. Instead, an SAP Fiori app is being developed to enable the key user to carry out the initialization process. Unfortunately, we don't know when or if this development will be announced and released by SAP.

Configuring SWIFT

SWIFT communication is conducted via the SAP Multi-Bank Connectivity tenant using SWIFT's Alliance Lite2 for business applications, making it completely cloud based. For the SWIFT network, companion files are needed during the transmission of payment files, containing parameters that the SWIFT network reads for routing purposes. Companion files include information on whether the communication is inbound or outbound, as well as the sender and recipient of the message. This information is required for both SWIFT's FIN and FileAct channels. These parameters are typically provided by the bank and can be configured in the IMG menu under the following path: **Multi-Bank Connectivity Connector • Outbound Processing • Assign SWIFT Parameters for Outbound Messages • Maintenance of SWIFT Parameters**.

The maintenance of parameters can be determined based on the **Receiver ID**, **Message Type**, and **SystemRole**. The system role specifies whether the parameters are for test connectivity (**Non-Productive only**), production connectivity (**Productive only**), or if they are applicable to both connections (**All**). The values in the **Parameter Value** field are provided by the bank. Figure 5.61 shows an example.

Maintenance of SWIFT Parameters				
Receiver ID	Message T...	Parameter Name	SystemRole	Parameter Value
CITIGB2L	PAIN.001.0...	FileDescription	Non-Productive o...	5678
CITIGB2L	PAIN.001.0...	FileDescription	Productive only	1234
CITIGB2L	PAIN.001.0...	IsNotificationRequested	All	true
CITIGB2L	PAIN.001.0...	NonRepudiation	All	true
CITIGB2L	PAIN.001.0...	Priority	All	Normal
CITIGB2L	PAIN.001.0...	RequestType	All	pain.001.001.03
CITIGB2L	PAIN.001.0...	Responder	All	cn=citiconnect,ou=eu,o=citius33,o=swift
CITIGB2L	PAIN.001.0...	SenderDN	All	ou=s4h,o=jojobebe,o=swift
CITIGB2L	PAIN.001.0...	Service	Non-Productive o...	swift.corp.fa!p
CITIGB2L	PAIN.001.0...	Service	Productive only	swift.corp.fa
CITIPLPX	PAIN.001.0...	FileDescription	Non-Productive o...	5678
CITIPLPX	PAIN.001.0...	FileDescription	Productive only	1234
CITIPLPX	PAIN.001.0...	IsNotificationRequested	All	true
CITIPLPX	PAIN.001.0...	NonRepudiation	All	true
CITIPLPX	PAIN.001.0...	Priority	All	Normal
CITIPLPX	PAIN.001.0...	RequestType	All	pain.001.001.03
CITIPLPX	PAIN.001.0...	Responder	All	cn=citiconnect,ou=eu,o=citius33,o=swift
CITIPLPX	PAIN.001.0...	SenderDN	All	ou=s4h,o=jojobebe,o=swift
CITIPLPX	PAIN.001.0...	Service	Non-Productive o...	swift.corp.fa!p
CITIPLPX	PAIN.001.0...	Service	Productive only	swift.corp.fa
CITIUS33	PAIN.001.0...	RequestType	All	pain.001.001.03

Figure 5.61 SWIFT Parameters

SWIFTRef Configuration

To import the SWIFTRef directory, Transaction BIC2 (Bank Master Data Cockpit) must first be properly configured. This setup requires two variants. The first variant is intended for importing the *full* file of the SWIFTRef directory, which typically only needs to be imported once, depending on the configuration. The second variant is designed for the delta file, and for this variant, the **Delta Upload** flag (see Figure 5.62) must be selected.

For processing, there are essentially two options:

- **File import using the two-step approach**
  The file is first imported through SAP Multi-Bank Connectivity, then written to a directory, and then separately imported using report RFBVBIC2. The processing occurs in two steps:
  - First, the SWIFTRef directory is imported via SAP Multi-Bank Connectivity and written to a designated path.
  - The SWIFTRef file then is read into the SAP S/4HANA system through the standard import report (Transaction BIC2/report RFBVBIC2), which populates table BNKA according to the settings.

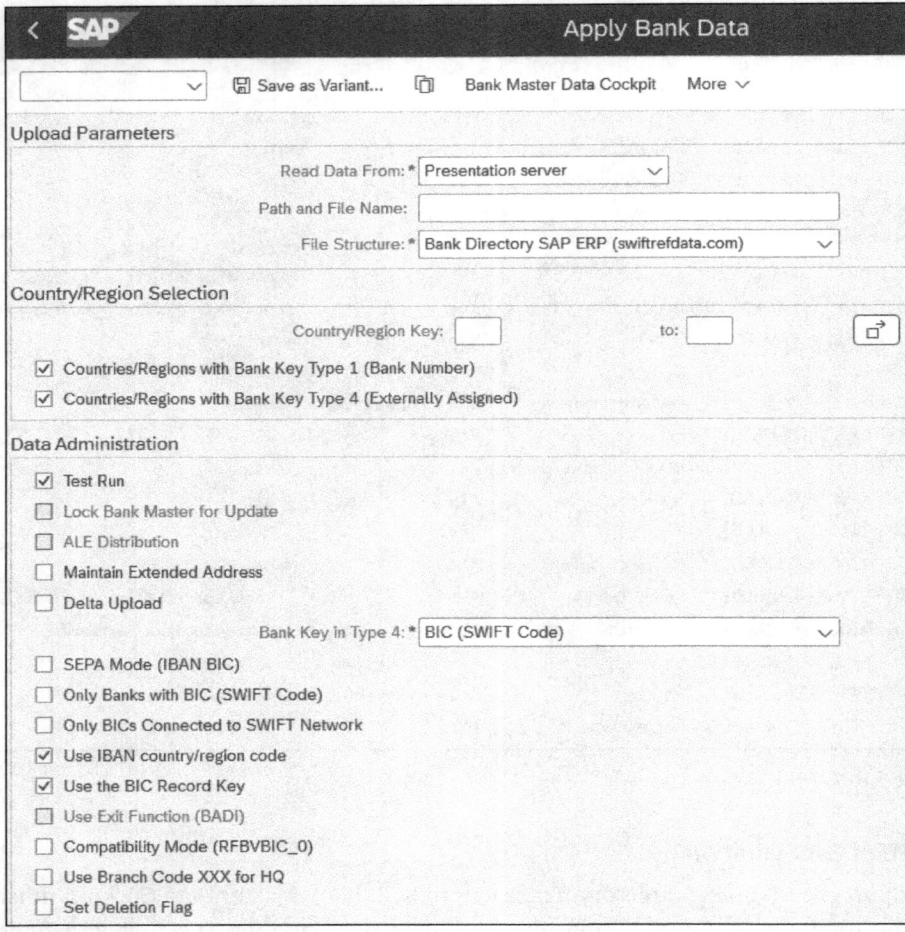

Figure 5.62 Delta Upload Flag

- **Direct process linking via SAP Multi-Bank Connectivity**
  The variant created in Transaction BIC2 is stored in customizing table /BSNAGT/V_SREF. You can access this customizing table via IMG path **Multi-Bank Connectivity • Inbound Processing • Assign Variants for SwiftRef Processing**.

Within this transaction, you can link the respective variant, which has been maintained in the BIC transaction, with SAP Multi-Bank Connectivity. The transaction offers four wildcard fields that can be used independently for selection: **Sender ID, Receiver ID, Message Type**, and **Origin**.

These fields can be assigned to a sequence via the **Seq.No.** field, mapping them to a BIC2 variant. To add a new entry, use the **New Entries** button or press F5.

> **Review and Verify the Country Settings**
>
> SAP provides a wide range of preconfigured country settings and checks as part of the standard features. These can be configured using Transaction OY17. Experience has shown that these settings may be partially incomplete for processing the SWIFTRef file, which can result in information being truncated during import or the import failing altogether due to errors. It is strongly recommended to review these settings thoroughly before conducting the first import.

5.5.4 Business Add-Ins

For on-premise SAP S/4HANA and SAP S/4HANA Cloud Private Edition, business add-ins (BAdIs) are available to extend the solution. We have compiled a list of existing extension points in this section based on SAP S/4HANA 2023 FPS 02 (which was released in October 2024).

BAdIs can be found in the IMG under the menu item **Multi-Bank Connectivity • Business Add-Ins**. BAdIs are exit points in the coding that enable customization through individual extensions; they can be accessed and modified via Transaction SE18.

The BAdIs in Table 5.4 serve to extend the outbound and inbound message processing. If you are looking for ways to incorporate additional conditions into process control or, for example, to develop a logic for assigning sender or receiver IDs when integrating external systems, you will find them here.

BAdI	Description
/BSNAGT/BADI_MESSAGE	Central BAdI for processing incoming messages; supports message enrichment, transformation, and error handling.
/BSNAGT/BADI_MESSAGE_DEFAULT	Standard implementation for generic incoming message types.

Table 5.4 Message Processing BAdIs

BAdI	Description
/BSNAGT/BADI_IDOC_PAYMENT	Specialized implementation for processing incoming IDoc payment messages.
/BSNAGT/BADI_MESSAGE_STATUS	Handles processing of incoming status messages.
/BSNAGT/MSG_OUT_BADI	Used for modifying outgoing messages, adding custom data, or supporting new message formats.
/BSNAGT/BADI_MESSAGE_OUT_DEF	Standard implementation for outgoing messages and handler for specific formats like pain.001.
/BSNAGT/BADI_MESSAGE_FACTORY	This BAdI is used to customize the sender ID of a message. This is particularly necessary when the **Customer Number** field is already occupied for other purposes and must not be changed, or when the same bank is used for both accounts payable (AP) and human resources (HR) payments. By implementing this BAdI, the sender ID can be set flexibly based on specific criteria. The details of the required programming are described in a separate document.

Table 5.4 Message Processing BAdIs (Cont.)

The BAdIs in Table 5.5 allow for additional security settings to be configured.

BAdI	Description
/BSNAGT/BADI_SECURITY	Enables custom authentication or authorization checks—for example, validating security tokens or implementing multifactor authentication.
EPIC_SECURITY_TOKEN_CONTROL	Controls and integrates security tokens for advanced authentication beyond standard certificate-based security.

Table 5.5 Security-Related BAdIs

The BAdIs in Table 5.6 are designed to extend the SAP Multi-Bank Connectivity monitor in Transaction /BSNAGT/MONITOR and the Manage Bank Messages app with custom fields and views.

BAdI	Description
/BSNAGT/BADI_MON_FIORI	Extends the Manage Bank Messages app by adding custom views, fields, or functionalities for more specific insights.

Table 5.6 Monitoring BAdIs

BAdI	Description
/BSNAGT/MONITOR	Standard monitoring tool to track message flows and identify potential issues.
/BSNAGT/BADI_MON_DISP_CONT	Allows customization of the display of monitoring data within the SAP Multi-Bank Connectivity monitor.
/BSNAGT/BADI_MON_NAV_2_APP	Enables navigation from monitored messages to related applications based on processing status or message content.

Table 5.6 Monitoring BAdIs (Cont.)

5.5.5 Authorizations

In a sensitive field like payment transactions, a well-thought-out authorization concept plays a key role. In this section, we have compiled the relevant authorization objects that are important not only for Transaction /BSNAGT/MONITOR but also for the Manage Bank Messages app as backend authorizations.

Backend Authorizations

The authorization objects in Table 5.7 are particularly important for the backend of the Manage Bank Messages app (F4385) as well as for Transaction /BSNAGT/MONITOR.

Object	Description
/BSNAGT/AD	SAP Multi-Bank Connectivity Connector Administration
/BSNAGT/AP	SAP Multi-Bank Connectivity Connector API
/BSNAGT/BL	Connector Block
/BSNAGT/CR	Creation of SAP Multi-Bank Connectivity Messages
/BSNAGT/FM	SAP Multi-Bank Connectivity Connector Monitor
/BSNAGT/ID	Payload Search Index Maintenance
/BSNAGT/MS	SAP Multi-Bank Connectivity Message Authorization
/BSNAGT/PI	File/Message Pickup
/BSNAGT/PL	Access to Message Payload
/BSNAGT/PS	Payload Search
/BSNAGT/PU	Pull Messages
/BSNAGT/RP	Reset/Reprocess Already Processed Steps

Table 5.7 List of Standard Authorization Objects for SAP Multi-Bank Connectivity

Object	Description
/BSNAGT/SA	Send Message Again
/BSNAGT/SE	Access to Sensitive Data
/BSNAGT/SN	Send Message to SAP Multi-Bank Connectivity
/BSNAGT/ST	Update Status of Message
/BSNAGT/TR	Trace Messages (SAP Multi-Bank Connectivity Connector)
/BSNAGT/AD	SAP Multi-Bank Connectivity Connector Administration

Table 5.7 List of Standard Authorization Objects for SAP Multi-Bank Connectivity (Cont.)

The permissions for technical users involved in the automated execution of jobs, as well as the permissions granted to users for daily use, should be carefully analyzed and specified in an implementation project. This includes, among others, permission object /BSNAGT/CR (Creation of SAP Multi-Bank Connectivity Messages), in conjunction with the permission for sending messages via /BSNAGT/SA. In addition, permission object /BSNAGT/SE, which we discussed in Section 5.3.6, is considered another sensitive authorization object.

Frontend Authorization

For SAP Multi-Bank Connectivity, there is a standard frontend role that includes the Manage Bank Messages app (F4385), which has been available for SAP S/4HANA and SAP S/4HANA Cloud Private Edition since SAP S/4HANA 1909 FPS 01: SAP_BR_BANK_INT_SPE-CIALIST. This role includes a template featuring an SAP Fiori catalog and a group in which the app is located on the SAP Fiori launchpad upon assignment of permissions:

- SAP Fiori catalog: SAP_FIN_BC_BSNAGT (Business Integration—Bank Integration)
- SAP Fiori group: SAP_FIN_BCG_BSNAGT (Bank Integration)

Figure 5.63 shows the standard SAP Fiori user catalogs for SAP Multi-Bank Connectivity.

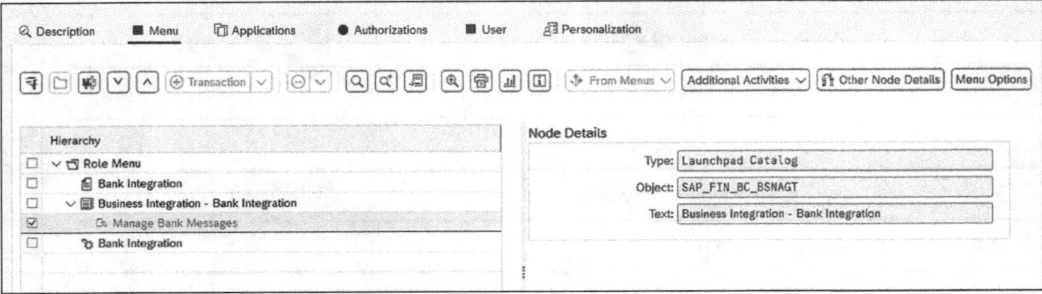

Figure 5.63 Frontend Authorizations

5.6 Summary

SAP Multi-Bank Connectivity is steadily evolving into more than just a simple bank connector. It has long established itself as a platform for financial institutions and companies, as well as a routing platform for message orchestration between systems with an "as a service" approach. The integration of SAP S/4HANA systems with SAP Multi-Bank Connectivity is under continuous development and is becoming increasingly simplified, with configuration steps being streamlined, enhancing the value of the "as a service" approach. We are eager to see how SAP Multi-Bank Connectivity is progressively expanded in the future and remain curious about its development, especially in the current dynamic payment environment with its alternative connection approaches and payment methods.

Chapter 6
SAP S/4HANA Finance for Cash Management

Cash management is an essential process because it ensures that payments can be made. By maintaining an up-to-date view of cash positions and accurately forecasting future cash needs, organizations can confidently execute payments and meet their financial obligations. Cash management helps answer the fundamental question of how much money you have available and how much you will need in the future. This crucial visibility enables better decision-making, supports payment execution, and ensures financial stability.

By this point, you know how Bank Account Management (BAM) in SAP S/4HANA can serve as your central bank account repository and how to set up bank accounts, and you have also learned how payments can be executed using SAP S/4HANA Finance advanced payment management and in-house banking for SAP S/4HANA Finance for advanced payment management. You have also explored different bank connectivity options to ensure secure and efficient payment flows. Now, it's time to dive into the world of SAP S/4HANA Finance for cash management, which plays a vital role in supporting our payment processes.

In this chapter, we will take a closer look at how cash management helps track your balances, assess short-term liquidity, and forecast cash positions with precision. We will also explore how treasury transfers can be used to fund your internal accounts, ensuring optimal use of available cash. Although cash management encompasses a wide range of functionalities—such as long-term liquidity management, detailed actuals analysis, snapshot comparisons, and forecast accuracy testing—this chapter will focus on the core features most relevant to payment processes. It's important to note that cash management is a broad and complex topic that could easily fill an entire book on its own. However, here we will focus on the key aspects that support and complement the payment processes we've discussed so far.

6.1 Functions and Processes

As described in Chapter 1, cash management is a critical prerequisite for our payment processes. To execute payments effectively, it's essential to know your available liquidity and how much cash you have on hand. If you have insufficient funds, payments may be rejected by banks. SAP S/4HANA provides various functionalities to help you manage liquidity and monitor cash balances accurately. These tools enable you to assess your cash position, forecast liquidity needs, and ensure that funds are available for the execution of payments. In this section, we will explain how to use these functionalities and outline the most important applications that support effective cash management and help ensure that payment transactions are processed smoothly and without issues.

In this chapter, you will learn about the key functions and processes that are most relevant for cash management from a payment perspective. We will cover how to monitor and review your bank account balances using essential SAP Fiori apps like Bank Account Balances and Cash Position. You will also see how to analyze short-term liquidity through apps such as Short-Term Positioning and Cash Flow Analyzer, which provide valuable insights into your cash flow needs and available resources. Furthermore, we will explore the critical role of the One Exposure from Operations table, which serves as a central data source for all cash-related applications. This chapter will also explain how to update your cash management data with memo records and how to manage the funding of your internal accounts through bank-to-bank transfers, free-form payments, and cash concentration processes. In other words, we will focus on the most important applications that support you in executing payments efficiently and maintaining a solid understanding of your short-term liquidity position.

6.1.1 Bank Account Balances

Bank balances are essential for managing liquidity and tracking account activity across your in-house bank accounts. There are several applications available to review and manage these balances, but the primary one is the Manage Bank Account Balances app (F5175). This app not only allows you to monitor current and historical balances from uploaded bank statements but also provides the capability to manually enter balances when needed. This is especially useful in cases in which automated statements are not yet available or when validating data during testing phases.

The Manage Bank Account Balances app enables you to view and manage bank account balances that are updated through imported end-of-day bank statements. These balances are stored based on the bank statement dates provided in the statements. If a balance for a specific date has already been updated via an electronic bank statement, it cannot be modified through manual entry or spreadsheet import. The app also allows for manual balance entry and importing balances from spreadsheets when electronic statements are not available. In addition, you can review the full balance history for each bank account, providing greater transparency and control over your financial data.

The default and recommended method for updating bank account balances is through automatic uploads from bank statements, ensuring real-time accuracy and minimal manual effort. However, in certain rare cases—such as when a connection to a bank cannot be established, or the volume of transactions is too low to justify an automated setup—it may be necessary to update balances manually or via spreadsheet import. In such scenarios, the relevant bank accounts must be marked as eligible for manual or spreadsheet-based balance updates by selecting the **Enable Balance from Manual Entry** or **Enable Balance from Spreadsheet Import** checkboxes. This setting can be configured using the Define Bank Account Settings—Bank Statements app (F5488), as shown in Figure 6.1, ensuring proper alignment with the balance management process.

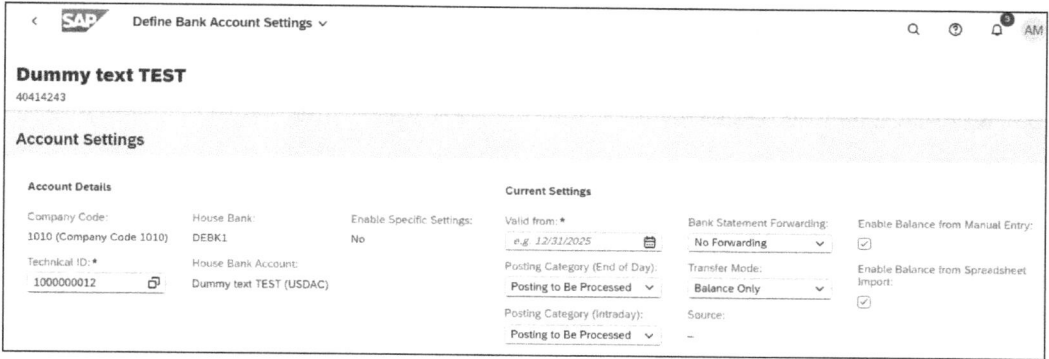

Figure 6.1 Define Bank Account Settings

However, the app still provides valuable configuration options. You can specify the **Technical ID** (BAM account), define whether postings should be generated for end-of-day (**Posting Category (End of Day)**) or intraday (**Posting Category (Intraday)**) statements, enable bank statement forwarding from an advanced payment management perspective (by selecting a value under **Bank Statement Forwarding**), and choose the transfer mode to cash management (**Transfer Mode**)—whether you are updating only balances, line items and postings, or not updating cash management at all. In addition, the app allows you to indicate whether manual entry or spreadsheet uploads are permitted for specific bank accounts, offering flexibility for hybrid or manual balance management setups.

Once the relevant entries are selected, the system will update the bank account accordingly. This update influences several important behaviors, including how bank statements are processed, how balances are created, and how updates are reflected in cash management. It determines whether the account will allow manual or automated balance updates, whether end-of-day or intraday postings are generated, and how bank statement data flows through the system. These settings are crucial for ensuring accurate and timely reflection of financial data across your treasury and cash management processes. If you selected **Enable Balance from Manual Entry** or **Enable Balance from**

Spreadsheet Import, you can update balances using the Manage Bank Account Balances app (F5175).

With the Manage Bank Account Balances app, you can monitor and maintain your bank account balances. This app allows you to view balances that are automatically updated by imported end-of-day bank statements. It also supports manual entry and spreadsheet imports for scenarios in which automatic updates are not feasible. Balances are stored based on the bank statement date, and if a balance has already been updated through an electronic bank statement for a given date, it cannot be modified manually or via spreadsheet. In addition to entering and importing balances, the app also enables you to review the balance history for each bank account, providing full visibility into past financial data. Once you enter the app, you can click the **Create** button to manually enter a bank balance (see Figure 6.2).

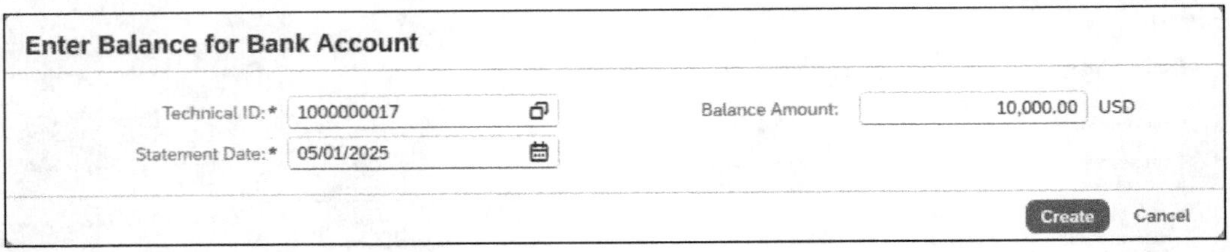

Enter Balance for Bank Account			
Technical ID: *	1000000017	Balance Amount:	10,000.00 USD
Statement Date: *	05/01/2025		
			Create Cancel

Figure 6.2 Enter Balance for Bank Account

Here, you need to enter the **Technical ID**, which represents the ID for the BAM account; the **Statement Date**, representing the date on which the statement was received (so, the balance for a particular day); and the **Balance Amount**, corresponding to the balance for that day. Note that you cannot select the currency here as the system derives it from the bank account currency. Once saved, the system will create and update the balance in both the app and the system itself, ensuring that other cash management applications are also updated accordingly. Alternatively, within the Manage Bank Account Balances app, you can import balances directly from a spreadsheet if needed. You also have the option to navigate to related applications such as Monitor Bank Account Balances or Manage Bank Statements, where you can view the balances that have been uploaded. These functionalities will be discussed in more detail in Chapter 9. The key purpose of the Manage Bank Account Balances app (see Figure 6.3) is to display and verify the closing balances for your selected bank accounts (based on either manual entry or balances coming from the bank statements).

The easiest way to check your bank balances across all bank accounts that are updated in the system is by using the Monitor Bank Account Balances app (F5176), which is shown in Figure 6.4. This app provides a centralized and real-time overview of your balances, allowing you to quickly assess the financial position of your accounts without having to check each one individually. This app allows you to monitor whether bank account balances have been updated on time for your accounts (and how they were uploaded: manually, by spreadsheet, or from the bank statement). It provides key functionality such as filtering bank accounts by status and other attributes to quickly identify any issues. In addition, you can export the monitoring results to a spreadsheet for further analysis or reporting.

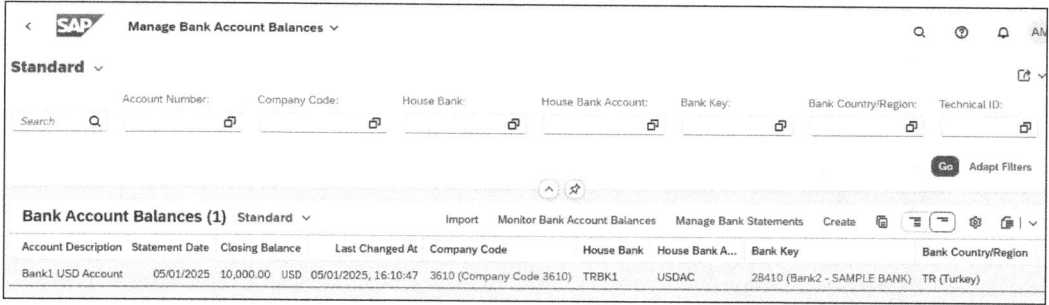

Figure 6.3 Manage Bank Account Balances

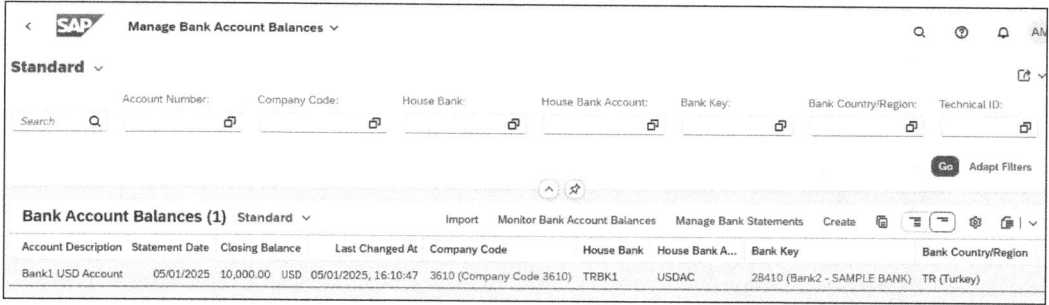

Figure 6.4 Monitor Bank Account Balances

It is worth mentioning here the Define Bank Account Settings—Instant Balances app (F7805), which will become increasingly important as we move toward API-based integration and real-time banking connections. In the future, when SAP Multi-Bank Connectivity is fully set up and the system is enabled to receive instant balance updates via an API, this app will be used to activate those real-time connections for eligible bank accounts. If SAP Multi-Bank Connectivity is not configured or not available in your system, then the instant balance option will remain disabled and the app won't be available. Also, instant balances will work only for some bank accounts—but the number of banks providing this option will grow.

The functionalities described in this section are available starting from SAP S/4HANA 2023. In earlier versions, a different application, called Bank Account Balance (F3940), was used for monitoring balances. This app served as an analytical tool for displaying bank account balances but offered fewer features and more limited functionality compared to the newer Monitor Bank Account Balances app. Starting with SAP S/4HANA 2023, the Bank Account Balance app has been deprecated.

6.1.2 Cash Position

The Cash Position app (F1737) is one of the classic tools for displaying cash positions and cash balances, and it has been available since the earliest release of SAP S/4HANA. Although its interface and features have evolved over time, it remains a staple in every version, providing a reliable way to view your organization's current and projected liquidity across all accounts.

You can use this app to review forecasted cash positions for the current date, segmented by location, company, and currency. When you open the app (see Figure 6.5), the system will automatically display predefined values by default. However, you can review and adjust the selection parameters as needed. Based on your chosen criteria, the system will then display the available cash accordingly.

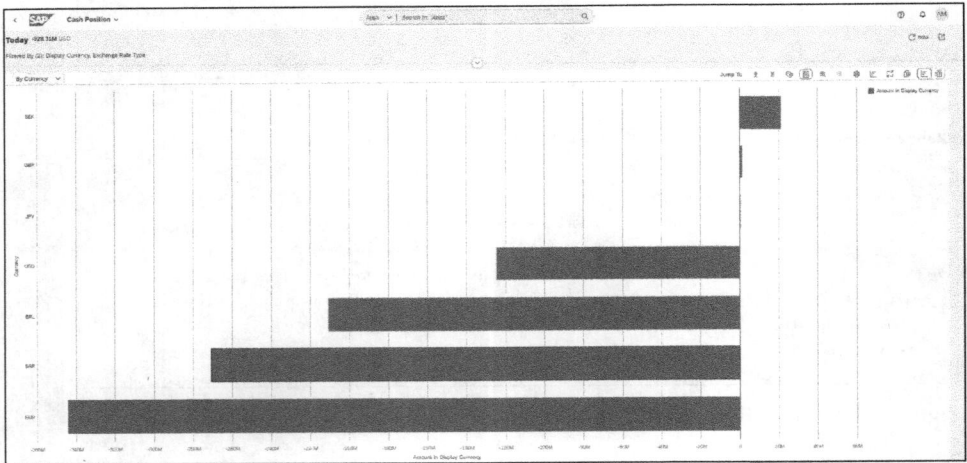

Figure 6.5 Cash Position Graph

You can use this app to review forecasted cash positions for the current date, segmented by location, company, and currency. The system calculates cash positions by combining memo records with data from the One Exposure from Operations hub. Depending on your assigned role, you can display cash positions across various analytical dimensions—such as bank country or region, bank, bank group, company, and currency combinations (for bank and bank account). You can switch between chart and table views (see Figure 6.6) to suit your needs and export the results to a spreadsheet for further analysis.

The Cash Position app also includes all cash flows for the selected bank accounts and currencies—both those derived from imported bank statements and those forecasted by the system. If you prefer to view only actual flows or wish to exclude specific items (such as intercompany balances), you can apply additional filters. For example, you might filter by planning level to include only **Actual** data sources (e.g., bank balances) or select specific planning levels that correspond to the types of flows you want to see. This flexibility ensures that you can tailor the display to show exactly the information you need.

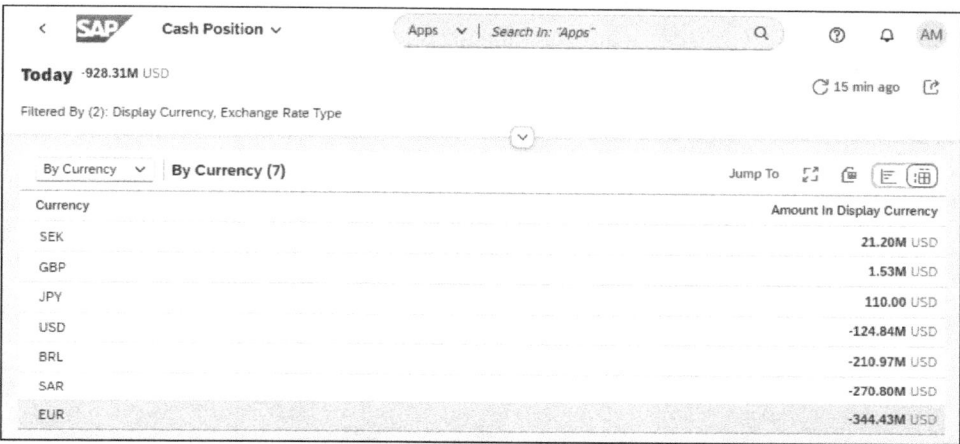

Figure 6.6 Cash Position Tabular View

The **Cash Position** tile on the SAP Fiori launchpad provides a consolidated view of your full available cash by aggregating all cash flows from cash management and translating them into the selected display currency. You can tailor exactly what this tile shows—such as which planning levels, currencies, or account groups are included—by adjusting the tile settings in the Cash Position app's configuration. These changes allow you to customize the SAP Fiori launchpad title to reflect the most relevant snapshot of your organization's liquidity immediately.

However, the Cash Position app primarily aggregates all cash flows into a single bulk amount in the selected display currency. It doesn't provide direct visibility into individual flows, making it less effective for detailed liquidity analysis. As a result, this app has limited usage for in-depth cash management. To address these limitations, SAP S/4HANA offers additional applications that are more suitable for a detailed and comprehensive analysis of liquidity and cash position, offering enhanced functionality for tracking and managing cash balances, flows, and forecasts in a more granular way.

6.1.3 Cash Flow Analyzer

Another application that will help you perform a short-term or long-term payment review and liquidity forecast is Cash Flow Analyzer (F2332). The Cash Flow Analyzer app is the primary SAP Fiori application used for liquidity forecasting in SAP S/4HANA. It is designed to replace the classic SAP ERP Transactions FF7A (Cash Position) and FF7B (Liquidity Forecast). Although these classic transactions can still be used in SAP S/4HANA through Transaction FF7AN and Transaction FF7BN, SAP strongly recommends transitioning to the Cash Flow Analyzer app due to its modern interface, enhanced features, and integration with other SAP Fiori apps and analytics tools. With the Cash Flow Analyzer app (see Figure 6.7), you can gain a comprehensive overview of aggregated amounts and line-item details across your cash position, medium- and long-term liquidity forecasts, and actual cash flows. It supports flexible analysis over multiple time intervals—daily, weekly, monthly, quarterly, or yearly—across all bank accounts and liquidity items. This app is particularly useful for providing clear liquidity insights to management.

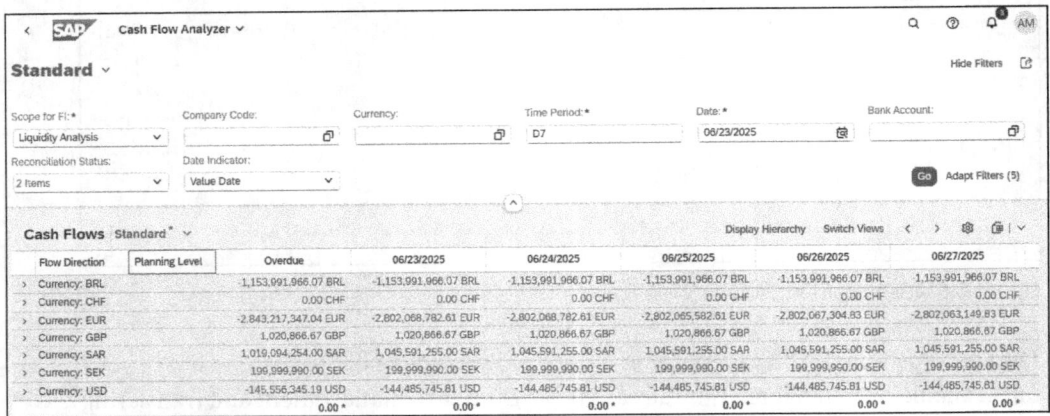

Figure 6.7 Cash Flow Analyzer

The Cash Flow Analyzer app is packed with powerful functionalities designed to give you deep insight into your liquidity position. One of the first things you'll notice upon opening the application is the abundance of blue text throughout the interface. In SAP Fiori, blue text is a key indicator: it means the item is interactive. Every blue field or figure can be clicked, leading you to detailed breakdowns, additional information, or even navigation to other related applications. This interactive design makes it easy to drill down into specific cash flows, analyze data from different angles, and take quick action based on real-time insights.

Throughout this section, we will focus on the core functionalities of the application and explore the key actions you can perform directly within it. This will help you understand how to navigate, analyze, and leverage the tool for effective cash and liquidity management.

One important feature in the Cash Flow Analyzer app is the **User Settings** area (see Figure 6.8), which can be accessed by clicking your user name and selecting **User Settings** from the dropdown menu. This allows you to customize how information is displayed in the app.

6.1 Functions and Processes

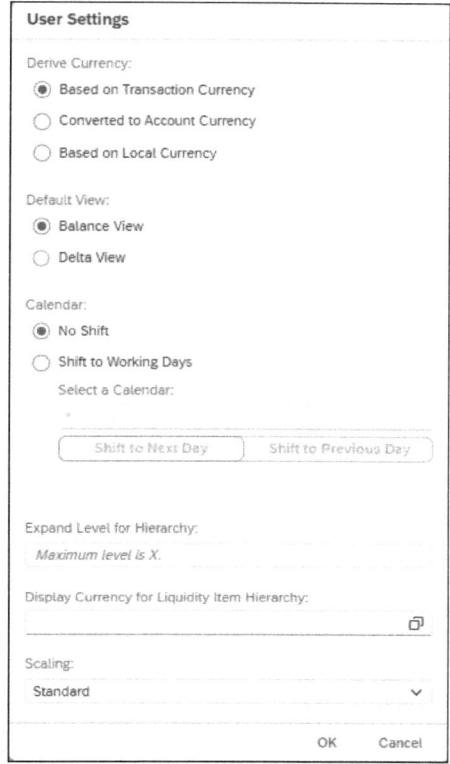

Figure 6.8 Cash Flow Analyzer User Settings

For example, you can choose to display data **Based on Transaction Currency**, which means that if your bank account is in USD but contains flows in GBP or EUR, those flows will appear in their original currencies. Alternatively, if you select **Converted to Account Currency**, all flows will be shown in USD, regardless of their original currency. You can also opt to display data **Based on Local Currency**, depending on your reporting needs. These settings help ensure that your cash flow overview aligns with your preferred financial view and analysis structure.

Next, within the **User Settings** of the Cash Flow Analyzer app, you can define how balances are displayed:

- For **Balance View**, the system aggregates previous balances for the specific account category or planning level and incrementally adds new values, giving you a cumulative perspective.
- For **Delta View**, only the transactions occurring on the selected day are shown, which helps you focus on daily cash movements.

You can also select whether to apply a calendar shift, which ensures that any flows due on nonworking days are shifted to the next or previous working day. This is especially useful for accurate liquidity planning.

Additional options allow you to do the following:

- Select how many hierarchy levels to expand automatically.
- Choose the display currency (especially relevant when using the liquidity item hierarchy, which will be discussed later).
- Set a scaling factor—for example, to display values in full, thousands, or millions—depending on the level of detail you want.

It's important to note that when converting currencies, the system always uses the latest available exchange rate for the selected display date, not the historical rate from the original transaction date.

After setting your user settings, the next step is to define your selection parameters, which determine what data you want to see and how it will be displayed. This is where the real flexibility of the Cash Flow Analyzer app comes into play. At first glance, you'll already see several powerful filters to narrow down your data—like **Company Code**, **Bank Account**, **Currency**, and **Liquidity Items**. But that's just the beginning!

By clicking **Adapt Filters** (see Figure 6.9), you unlock even more advanced options that let you fine-tune the data view to your exact needs. Whether you're focusing on a specific planning level, date range, or cash flow type, these filters allow you to tailor your analysis with incredible precision—ensuring you only see what's relevant for your treasury review.

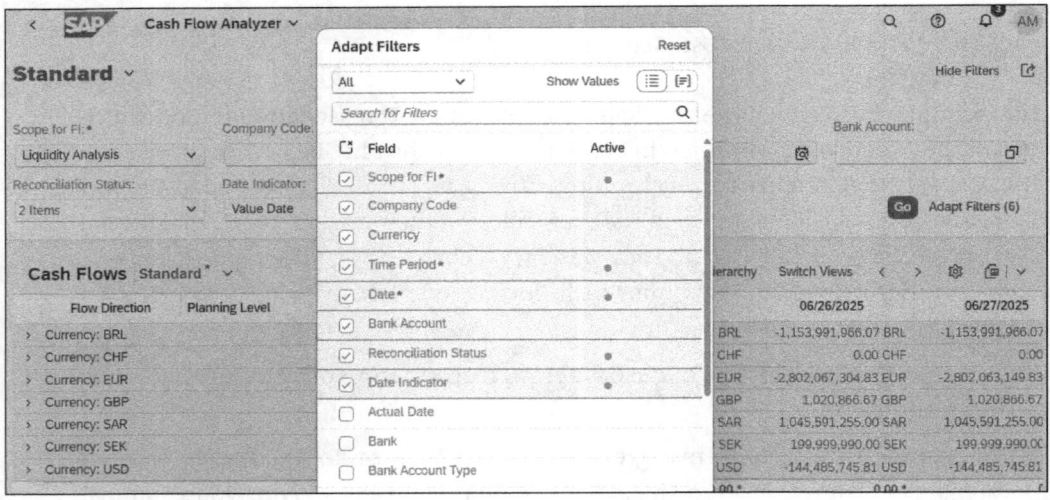

Figure 6.9 Cash Flow Analyzer: Adapt Filters

You can begin by selecting multiple company codes that are relevant for your analysis. Then, you have the flexibility to filter by currencies—either by displaying flows in their original transaction currency or translating all values into a single display currency for consistency across accounts.

Next, you define the **Time Period** for your analysis. You can work with specific days (e.g., enter "D7" for a time period covering the next seven working days), or aggregate data by months (e.g., enter "M12" to analyze the last 12 months), which is especially useful when reviewing liquidity trends, liquidity item hierarchies, or bank account hierarchies.

You also can specify a starting date for your analysis—whether you're reviewing historical performance or forecasting into the future. You can narrow down your scope by selecting specific bank accounts or, more importantly, by filtering based on certainty levels. For instance, choosing **Actual** will restrict the view to real, posted data such as bank statements and confirmed balances. With this flexibility, you can tailor your selections to show only the most relevant data and exclude less useful data sources to keep your analysis focused and meaningful.

Once you've selected your preferred parameters and determined how you want to display the report, you can save these settings as your personal variants for quick and consistent access in the future. By default, the system displays the standard variant, but you can easily switch between predefined system variants—such as **Cash Position**, **Liquidity Forecast**, and **Actual Cash Flows**—or any custom variants you've created. This allows you to tailor views for different analysis purposes without having to reconfigure settings each time. Once you have your desired selection parameters and view selected, you can start analyzing the report.

The analysis in the Cash Flow Analyzer app runs based on the selected hierarchy structure, allowing you to navigate from a high-level overview down to specific bank account details (see Figure 6.10). In this example, we begin the analysis at an aggregated level, first filtering by currency, then by company code, and finally down to individual bank accounts. For each bank account, you can view the opening balance for the selected day, followed by categorized flows based on planning levels, and then the closing balance. This closing balance automatically becomes the opening balance for the next day, continuing the analysis forward. You can examine the flows occurring each day, helping you understand how your liquidity position evolves over time. This rolling analysis offers a clear picture of your predicted cash availability across the chosen timeframes, such as the next 14 days—helping you identify whether a bank account is expected to face a shortfall requiring funding, or if it will have a surplus that can be used elsewhere.

The Cash Flow Analyzer app offers flexible options to change how your data is displayed, making it easier to tailor the view to your specific needs. You can do this by clicking the **Settings** icon and selecting the **Group** option (see Figure 6.11). This allows you to modify the structure of your report by choosing how the data is grouped. For example, instead of displaying data by company code, you can switch the view to show bank keys, general ledger (G/L) accounts, or other dimensions relevant to your analysis. This customization helps you align the display with your internal reporting needs and gain better insights into your liquidity from different perspectives.

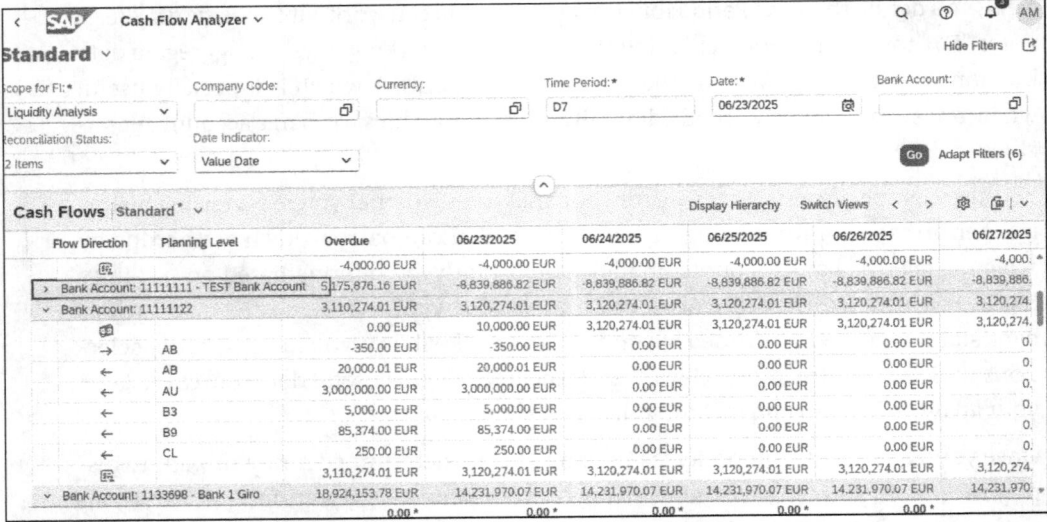

Figure 6.10 Cash Flow Analyzer: Display Amounts

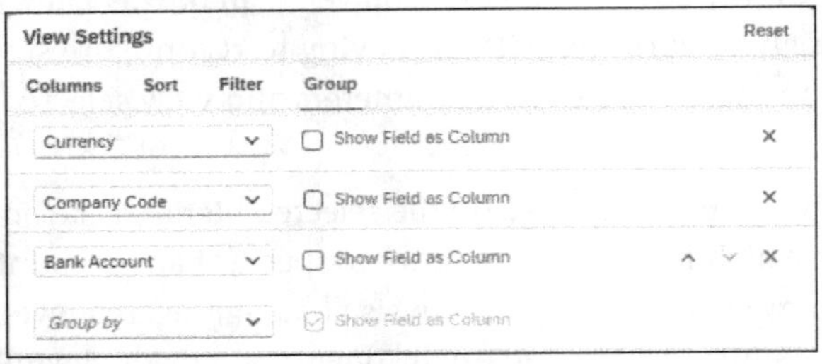

Figure 6.11 View Settings: Display Hierarchy

Every time you display a bank account in the Cash Flow Analyzer app, the system shows the opening balance, the closing balance, and the inflows and outflows in between, represented by green and red arrows respectively. Each flow is accompanied by a planning level, which categorizes the source of the transaction. There may be multiple lines for different flows as each planning level—based on master data and configuration—represents a different data source. For example, outgoing payments are typically shown under planning level B2, incoming flows under B1, bank statement postings under F0, accounts payable under F1, and accounts receivable under F2. The exact setup may vary depending on your configuration. Planning levels are crucial for the Cash Flow Analyzer app, Short-Term Cash Positioning app, and other related applications, as they define how and from where the data is pulled and interpreted.

> [!] **Not Assigned Flows**
>
> It's important to pay attention to **Not Assigned** positions in the Cash Flow Analyzer app (see Figure 6.12). These represent flows that could not be mapped to any specific house

bank or bank account—whether they are payments or incoming items. This typically occurs when transactions, such as invoices, are not linked at the accounting level to a house bank or bank account ID (including the technical account ID). For these flows to appear under the correct bank account in the analyzer, such mappings must be done in accounting. If this mapping is missing, the flows appear as **Not Assigned**, which reduces the accuracy and reliability of your liquidity forecast as these amounts are not visible where you're monitoring actual cash availability. To improve data quality and ensure full visibility, you can implement enhancements or enrichment logic during data loading into cash management to supplement missing bank assignment details.

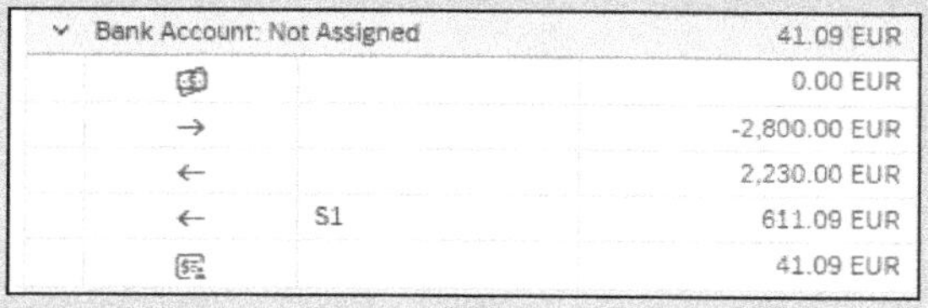

Bank Account: Not Assigned		41.09 EUR
💰		0.00 EUR
→		-2,800.00 EUR
←		2,230.00 EUR
←	S1	611.09 EUR
📊		41.09 EUR

Figure 6.12 Not Assigned Flows

When analyzing cash data and available liquidity in the Cash Flow Analyzer app, the key insight you're aiming to obtain is the expected balance of a particular bank account within the selected time. Once you identify the forecasted balance, you may need to take further action based on this insight. For instance, if one account has surplus cash while another requires funding, you might want to initiate a bank-to-bank transfer. If you notice missing data, you can manually add a memo record to reflect expected flows. Similarly, if you have excess cash in USD but need EUR, you can create a foreign exchange (FX) spot trade request to exchange the currencies. By clicking any blue amount in the analyzer, you can see a menu of available actions (see Figure 6.13) and related applications that you can access directly from the Cash Flow Analyzer app for further processing or decision-making.

When you open the list of available actions from the Cash Flow Analyzer app, you'll see a wide range of applications that you can launch directly. By default, you'll typically have access to key functionalities such as bank-to-bank transfers, memo record creation, and cash trade requests—which is particularly useful if your system is integrated with a trading platform and SAP Treasury and Risk Management. This enables you to initiate FX or money market transactions right from the Cash Flow Analyzer app when, for example, you want to invest surplus cash. In addition to the default options, there are other applications that may be available depending on your system setup. These are often country- or company-specific and can be restricted or customized based on user roles and SAP Fiori configuration.

Another important functionality of the Cash Flow Analyzer app is its powerful drill-down capabilities. This feature allows you to break down aggregated data into detailed components. For example, from a high-level planning level overview, you can click any amount to view the individual cash flow items that make up that total. This includes drilling down into customer (AR) or vendor (AP) invoices, giving you full visibility into the transactions driving your liquidity forecast. This level of detail helps ensure transparency, traceability, and better decision-making when analyzing your company's cash position.

Figure 6.13 Links Shown After Clicking Blue Amount

When you click a blue amount in the Cash Flow Analyzer app, one of the available options is **Display Cash Flow Items**. If you select this option, the system opens a new view (see Figure 6.14) in which you can see a detailed list of all cash flow items stored on the One Exposure from Operations table that make up the selected amount. The Cash Flow Analyzer app shows aggregated figures, but this view breaks them down into individual line items, giving you complete insight into the underlying transactions. This helps you understand exactly what is contributing to your cash position or forecast at a granular level.

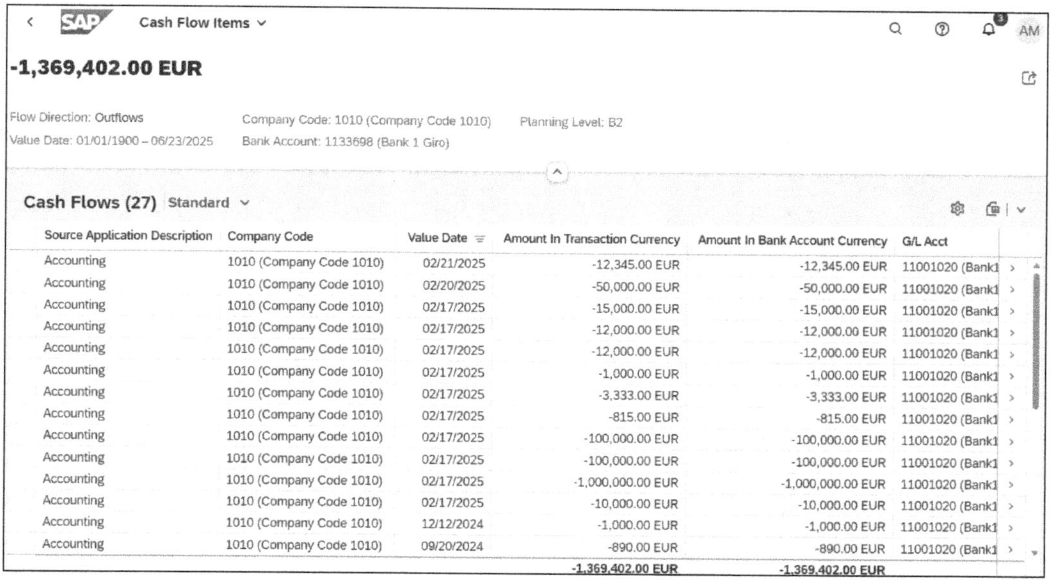

6.1 Functions and Processes

Figure 6.14 Display Cash Flow Items

You can also click any line item in the detailed view (see Figure 6.15), and the system will display additional information about that specific flow. From this view, you also have the option to edit the flow; for example, you can change the value date or liquidity item. In theory, it's also possible to adjust the amounts and assignments, but this should be done with caution. Any changes must still match the original document; otherwise, you may encounter errors or create data discrepancies in your liquidity forecast. Always ensure consistency and validate changes before saving.

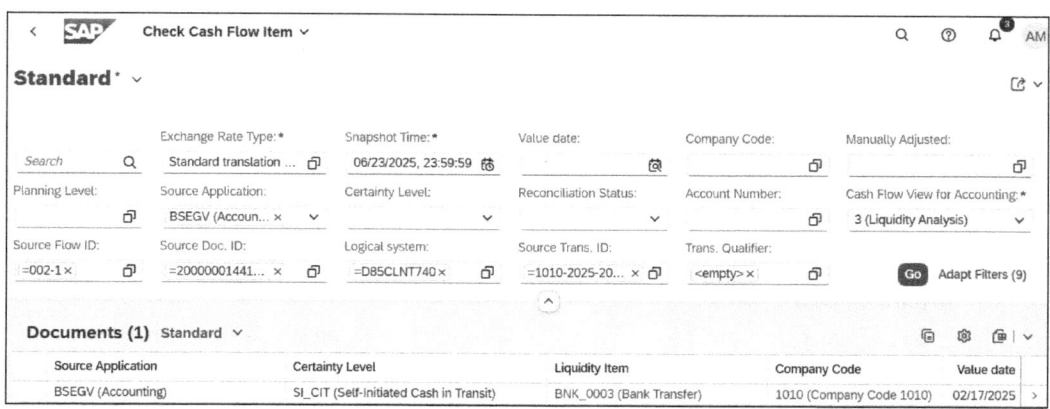

Figure 6.15 Check Cash Flow Items

The way cash flow items are displayed and the options for editing them depend on the release of SAP S/4HANA you are using. Newer releases offer more advanced functionality

433

6 SAP S/4HANA Finance for Cash Management

and additional SAP Fiori apps that provide enhanced capabilities for viewing, checking, and analyzing cash flow items.

You can also review your cash flow items directly—without going through the Cash Flow Analyzer app—by using the dedicated Check Cash Flow Items app (F0735). This app allows you to display all cash flow items stored in the One Exposure from Operations table, offering a centralized view of your financial flows. It's particularly useful for users who want to perform more detailed analysis or validations independently from the aggregated views in the Cash Flow Analyzer app.

In addition to the classic view of the liquidity forecast for bank accounts, which we just analyzed, the Cash Flow Analyzer app also offers the flexibility to display data using different views and hierarchies. To switch the view and modify how data is analyzed, simply click **Select Hierarchy** in the application. This option allows you to choose from three distinct hierarchies (see Figure 6.16): **Bank Account Hierarchy**, **Cash Concentration Simulation**, and **Liquidity Item Hierarchy**. Each hierarchy provides a different perspective and level of detail for cash flow analysis, helping you to tailor the reporting to your organization's structure and business needs.

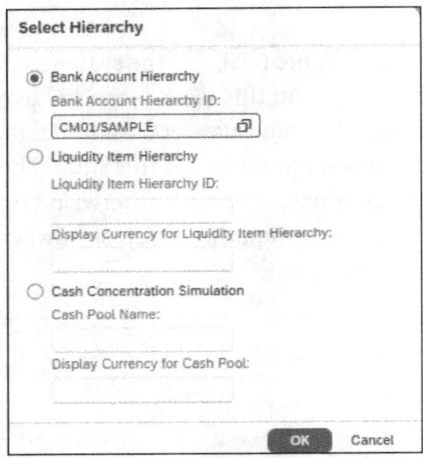

Figure 6.16 Select Hierarchy

Let's look at each option in a little more detail:

- **Bank Account Hierarchy**
  In the bank account hierarchy view (see Figure 6.17), you use the hierarchies previously created in the Bank Account Hierarchy app, as described earlier. This perspective allows you to analyze cash flows and forecasted liquidity based on the structure and grouping you define—whether by currency, legal entity, region, bank, or any other relevant classification. This enables you to perform liquidity analysis in a way that aligns with your organizational structure and reporting needs. This view also allows you to build and monitor cash pools, showing consolidated forecasted flows

on the header account level. You can easily evaluate the available balance for a selected period and determine whether a cash pool needs funding or if it has a surplus that could be invested. This functionality supports efficient cash management and strategic liquidity planning.

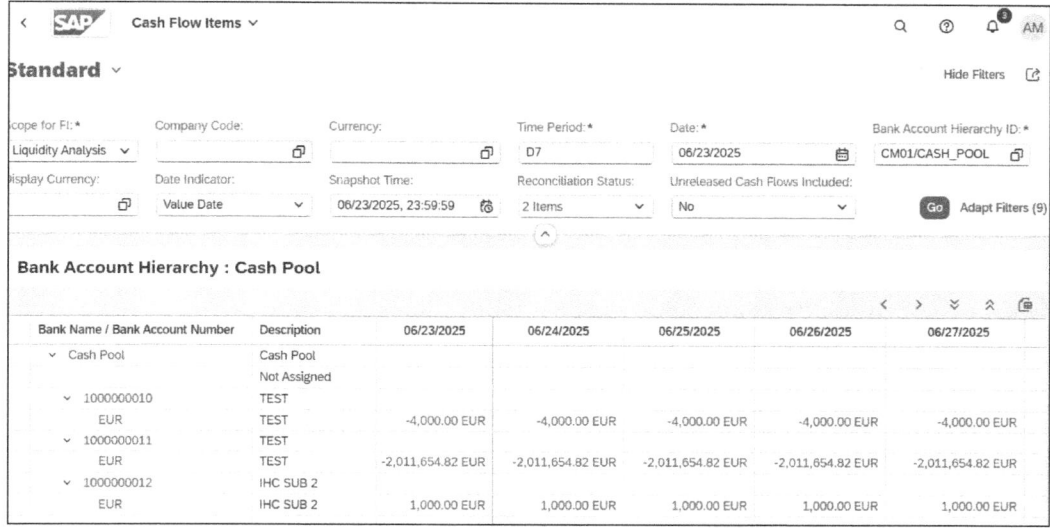

Figure 6.17 Bank Account Hierarchy

> [!]
> **Payments on Behalf of Hierarchy**
>
> As seen in the Cash Flow Analyzer app, you can view cash flows, forecasted flows, and payments associated with specific bank accounts. This always involves the connection between the company code, in-house bank, account ID, and technical ID. However, it can become problematic when payments are not easily linked from one entity to another bank account. This is particularly challenging when working with payments on behalf of (POBO) and receivables on behalf of (ROBO) scenarios, as in these cases it's difficult to predict the expected balance on your head bank account for all POBO or ROBO transactions because the corresponding payments are in different accounts. A workaround for this issue is to create a hierarchy with the POBO header account at the top; beneath it, include all the in-house bank accounts from which payments will be made for POBO transactions or those that will be used for ROBO transactions. By working with the bank account hierarchy, you should be able to see the forecasted balances on your POBO or ROBO header account.

- **Liquidity Item Hierarchy**

    In the Cash Flow Analyzer app, the liquidity item hierarchy (see Figure 6.18) does not focus on the transactions happening in specific bank accounts, where you want to see the forecasted balances and flows for those accounts. Instead, it focuses on the nature of the transaction itself, categorizing flows based on specific liquidity items. The liquidity item hierarchy allows you to group and analyze cash flows according to their type, such as operating cash, investment cash, or financing cash flows, which helps you understand the cash movement from a broader perspective.

This type of hierarchy is useful when you want to track liquidity movements tied not directly to specific bank accounts but to the transaction type or liquidity category. For example, you can separate flows coming from operational expenses, loan repayments, or capital investments. By using a liquidity item hierarchy, you can gain insights into your liquidity forecast based on transaction categories, giving you a better understanding of where cash is coming from and how it is being utilized across your operations.

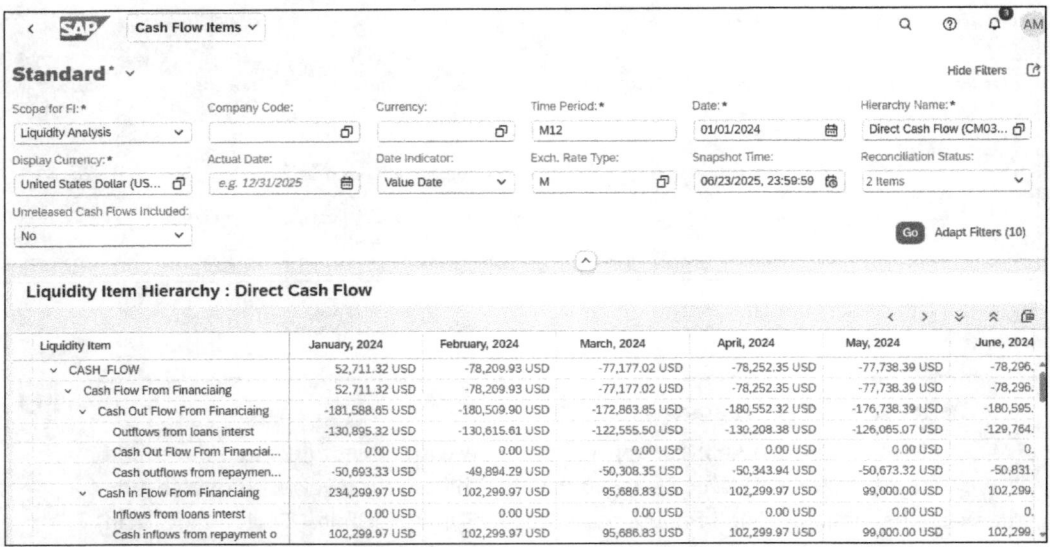

Figure 6.18 Liquidity Item Hierarchy

There are many ways to design your liquidity item hierarchy, but the guiding principle is that every cash flow in One Exposure from Operations must be assigned to a specific category. These categories typically mirror the sections of your cash flow statements, such as operating activities, investing activities, and financing activities—or whatever classifications best suit your analysis and reporting needs. For example, you might break down inflows and outflows into subcategories like customer receipts, vendor payments, loan proceeds, or interest expenses. In the example, we've built a direct cash flow hierarchy to analyze actual cash movements over the past 12 months, with each flow mapped to its appropriate category. This structured approach not only streamlines the assignment of flows but also enables you to generate clear, actionable reports that can be shared with management.

Similarly, you can build an indirect cash flow analysis within your Cash Flow Analyzer app, depending on how you choose to assign flows to your cash flow items and the categories you want to include in your reporting. This approach gives you the flexibility to analyze cash flows that are indirectly related to your core operations, such as adjustments for changes in working capital or noncash items. However, one key issue to be aware of is that certain cash flow items may not appear in your Cash Flow Analyzer app if they aren't properly assigned to a category. As seen in the example, the **Not Assigned** category represents flows that haven't been mapped correctly. This results in missing or incomplete data, which can compromise the accuracy of your liquidity forecast and reporting. Proper assignment of cash flow items to liquidity item categories is crucial to ensure that your analysis is comprehensive and reliable.

Assigning liquidity items to cash flow items is crucial and can be challenging, so it must be carefully planned. You can map payment transactions to liquidity categories automatically—by linking them to specific G/L accounts, using queries, or developing custom logic routines that assign flows based on your chart of accounts. Alternatively, you can handle the assignment manually, selecting the appropriate liquidity item for each cash flow line. Whichever method you choose, ensuring that every transaction is correctly categorized is essential for accurate cash flow analysis and reliable liquidity forecasting.

- **Cash Concentration Simulation**

    You can also create cash pools within your system, which will be triggered automatically by the system. These cash pools allow you to group various bank accounts into a hierarchy, where you can link a header account and define how money flows from this header account to its subaccounts. You can set target balances for each subaccount, and the system will trigger payments to transfer any remaining cash to the appropriate accounts. The forecasted and predicted balances for the cash pool can be monitored directly from the Cash Flow Analyzer app (see Figure 6.19). However, there are certain prerequisites that must be met before you can set up and utilize cash pools, which we will explain in Section 6.1.10. Creation of cash pools or cash concentrations depends on your system version.

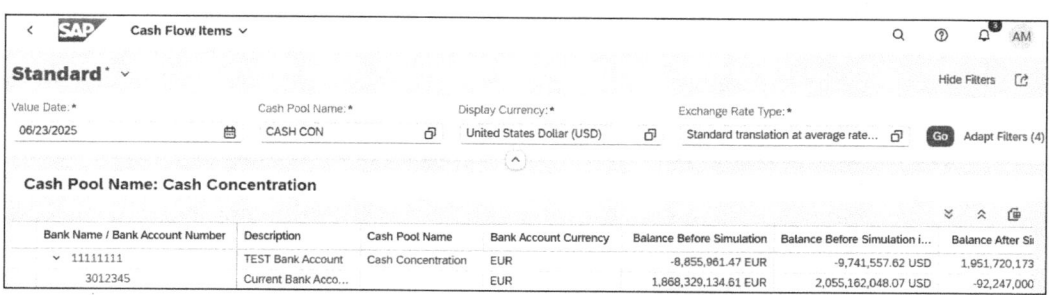

Figure 6.19 Cash Concentration Simulation

The system will analyze the forecasted balance and compare it with the target balance, then predict the necessary payments to ensure the target balance is reached. It will display the payments triggered, the amounts involved, and the balances after the payments are made.

6.1.4 Short-Term Positioning

The Short-Term Cash Positioning app (F5380) complements the Cash Flow Analyzer app by providing a focused, day-by-day view of your immediate liquidity needs. Whereas the Cash Flow Analyzer app helps you model and forecast cash flows over a broader horizon, the Short-Term Cash Positioning app zeroes in on each bank account's opening balance, incoming and outgoing flows for the selected days, and the predicted closing balances for today and the next few days. Together, these apps give cash managers the detailed insights they need to ensure sufficient liquidity is available to execute upcoming payments from the chosen bank accounts.

With this app, you can view cash positions by location, company, and currency. The cash position data is calculated based on the cash position profiles that you define in the Define Cash Position Profiles app. Key features for this app (specified and configured in Transaction FCLM_CP_PROFILE) are as follows:

- You can display cash positions based on various analytical dimensions, such as country/region, company code, bank account, and currency.
- You can view cash positions according to the settings defined in the selected cash position profile using selected data sources.
- You can simulate cash concentration for cash pools defined in the Manage Cash Pools (Version 2) app.
- When the hierarchy source is set to **03 (Derived from Cash Pool and Bank Account Master Data)** in the cash position profile, bank accounts assigned to cash pools of the **Bank—Time Dependent** service provider are displayed within the cash pool hierarchies.
- You can automatically aggregate the balances of subaccounts into the header account of the cash pools to simulate the cash concentration performed by banks.
- You can enable custom scaling to adjust the value amounts displayed in the app. You can specify the scaling factor (e.g., enter "3" for showing values in thousands, or "6" for millions) and define the number of decimal places to display.
- You can export your search results to a spreadsheet for further analysis.

As mentioned, before you can use the Short-Term Cash Positioning app, you must first define cash position profiles. These profiles determine which bank accounts, currencies, and forecast horizons are included in the analysis. You can create and maintain these profiles either in the IMG configuration or directly through the Define Cash Position Profiles app, which provides an SAP Fiori–optimized interface for the underlying SAP GUI transaction.

When you open the Define Cash Position Profiles app, click **New Entries** and enter a name (in the **Cash Pos. Profile** field) and a **Description** for your profile. You may create multiple profiles. In this example (see Figure 6.20), we'll set one up to display balances for a specific bank account over selected days.

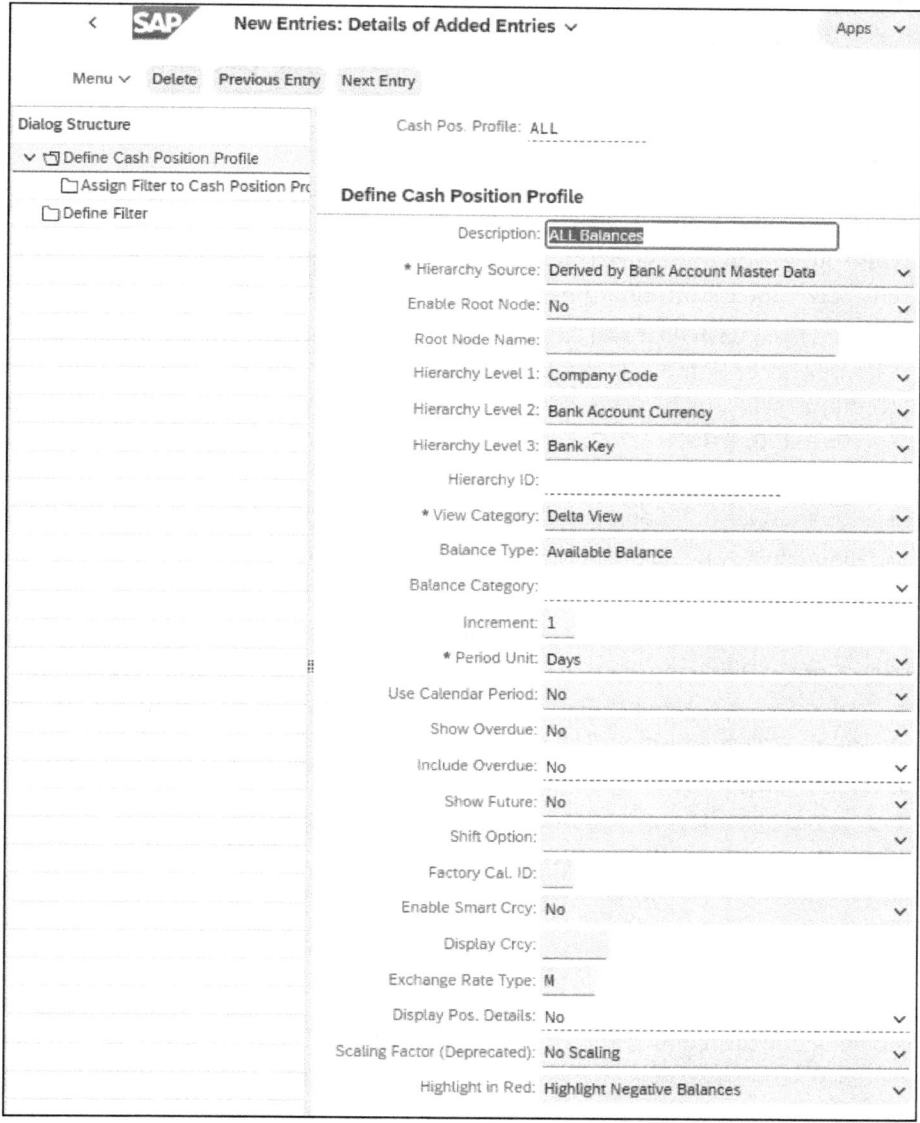

Figure 6.20 Define Cash Position Profiles

The Define Cash Position Profiles app helps you to build reporting for your short-term positioning based on the following:

- **Hierarchy Source of cash position profile**
  The hierarchy source determines how bank accounts are organized in a hierarchical structure within the Short-Term Cash Positioning app. There are three available hierarchy sources:
  - **Derived by Bank Account Master Data**
    Choose this option if you want the hierarchy to be based on the attributes defined in the bank account master data, such as company code or account currency. When you select this source, you must define an attribute for each hierarchy level to establish the structure. You can also set a root node if needed.
  - **Derived from Bank Account Hierarchy**
    Choose this option to use an existing bank account hierarchy as defined in the Manage Bank Account Hierarchies app, described in Chapter 3. When you select this source, you must specify a hierarchy ID you created in this app (e.g., a hierarchy created to monitor funding for a POBO hierarchy).
  - **Derived from Cash Pool and Bank Account Master Data**
    Choose this option if you want the hierarchy to be derived from the cash pool combined with the bank account master data. A cash pool structure is created in Section 6.1.10. In this case, the root node is not supported, and the hierarchy levels are predefined as follows:
    - Hierarchy Level 1: Company Code
    - Hierarchy Level 2: Bank Account Currency
- **Enable Root Node**
  You can define a root node when **Hierarchy Source** is set to **Derived by Bank Account Master Data**. If **Hierarchy Source** is set to **Derived from Bank Account Hierarchy**, then the root node is automatically taken from the hierarchy specified in the bank account hierarchy definition. The root node and hierarchy levels are disabled.
- **Root Node Name**
  When you enable the root node for a hierarchy with **Hierarchy Source** set to **Derived by Bank Account Master Data**, you have the option to specify a custom root node name. For example, you might choose a name like "Cash Pool Header Deutsche Bank EUR." If you enable the root node but do not specify a name, then the system will use a default root node name for the cash position profile.
- **Hierarchy Level 1, 2, 3**
  You can define the hierarchy levels when you set **Hierarchy Source** to **Derived by Bank Account Master Data**. At each hierarchy level, you select an attribute from the bank account master data—such as company code or currency—that groups the bank accounts accordingly. These hierarchy levels ultimately determine how bank accounts are displayed in a structured, hierarchical format within the Short-Term

Cash Positioning app and how data is displayed when you drill down from the top (in the example, currency to the individual bank account).

By default, you can set up your hierarchy following the default setup from the Cash Flow Analyzer app:

- Hierarchy Level 1: Currency
- Hierarchy Level 2: Company Code
- Hierarchy Level 3: Bank Account

- **Hierarchy ID**
The hierarchy ID represents the name of a tree structure, and the root node of the hierarchy carries the same name. This identifier can have several versions, each corresponding to a different timeframe.

- **View Category**
The view category determines which amounts and flows are displayed on the Cash Position Overview page in the Short-Term Cash Positioning app. The possible values for this view category are as follows:
  - **Balance View**: Sums all flows together with the balance from the previous day; the balance increments every day.
  - **Delta View**: Displays only delta flows—flows happening each particular day—and their aggregated amounts.

- **Balance Type**
  - **Available Balance**
  The available balance indicates the funds available for immediate use. This balance should come from the bank statement, or you can enter it manually.
  - **Value Date Balance**
  The value date balance is the closing balance that bears interest for a particular day, coming from the G/L balance.

- **Balance Category**
When the **View Category** is set to **Balance View**, you must specify a balance category. The balance category determines whether the opening balance or the closing balance is displayed on the Cash Position Overview page of the Short-Term Cash Positioning app.

- **Increment**
Shows the number which will be used to increment selected period units. If you select 1 and **Month** as period units, then each column in the report will represent a single month.

- **Period Unit**
This can be days, weeks, months, or years, and is set in conjunction with an increment value.

- **Use Calendar Period**
  This option defines whether periods in the Short-Term Cash Positioning app are displayed as intervals or based on calendar periods, such as calendar weeks, months, or years, starting from the current date. If the option is set to **Yes**, the periods are displayed according to calendar weeks, months, or years. When the period unit is weeks, then seven-day intervals are shown. For months, periods start on the current date and end on the day before the same date in the following month. For years, periods start on the current date and end on the day before the same date in the following year.

- **Show Overdue**
  If the **Show Overdue** option is selected, the overdue amount will be displayed in a separate column in the Short-Term Cash Positioning app. This overdue amount includes all cash flows with a value date prior to today.

- **Include Overdue Amount in Closing**
  If you select the **Include Overdue** checkbox, the sum of the overdue amount will be included in the calculation of the first period's closing balance when displaying the cash position. However, it is common for accounting sources to have many unreconciled items that can distort cash management reports. To avoid these inaccuracies and prevent such items from showing up, you can select **No** here. In this case, the system will exclude overdue amounts and items from the cash position overview.

- **Show Future**
  Defines whether to show the future amount (all flows with a value date after the analyzed period) in the Short-Term Cash Positioning app. If you set the option to **Yes**, a column named **Future Amount** will be displayed after your specified reporting period. The opening balance comes from the closing balance of the last day of the reporting period. Cash flows that have a value date later than the reporting period are aggregated and displayed in this column.

- **Shift Option**
  Determines whether you shift flows to the next working date—Saturday/Sunday to Monday, nonworking day to next working date, and so on—or not.

- **Factory Cal. ID**
  You specify a factory calendar to determine nonworking days—for example, Fridays and Saturdays for countries in the Middle East.

- **Enable Smart Crcy**
  The enable smart currency feature allows you to switch between different currency types for an aggregated balance view. The available currency types are company code currency, bank account currency, and display currency. There are three types of hierarchy sources:
  – Bank account master data

6.1 Functions and Processes

- Bank account hierarchy
- Cash pool and bank account master data

When the hierarchy source is cash pool and bank account master data, and smart currency is enabled, the **Display Crcy** field is optional. However, if the hierarchy source is either bank account master data or bank account hierarchy and smart currency is enabled, specifying the **Display Crcy** field is mandatory.

- **Display Crcy**
  Once selected, the system will translate all flows and amounts into the selected currency using the specified exchange rate type.

- **Exchange Rate Type**
  This is the exchange rate type used for translation to the display currency—for example, **M**.

- **Display Pos. Details**
  The display cash position details function allows you to display cash position details across all node levels. This function can be enabled only when smart currency is active or a display currency has been specified. If this function is not enabled, cash position details can only be displayed for the cash pool header and bank account levels.

- **Scaling Factor (deprecated)**
  Define the number of places you want to show before the decimal point of the amount. If you enter "3", the system will cut the last three digits and show numbers as thousands (e.g., 500.000 will be shown as 500).

- **Highlight in Red**
  The **Highlight in Red** field determines whether negative balances are highlighted in red in the cash position overview and cash pool structure of the Short-Term Cash Positioning app. When selected, items with negative balances will be displayed in red and a highlight indicator will appear in front of the row within the cash position overview and cash pool structure. This feature can be used as an alert for the treasury team to quickly identify and address accounts with negative balances.

You can also assign filters to profile you just created (see Figure 6.21) by selecting **Define Filter** in the **Dialog Structure**. On this screen, after clicking **New Entries**, you enter the name for your filter and its description.

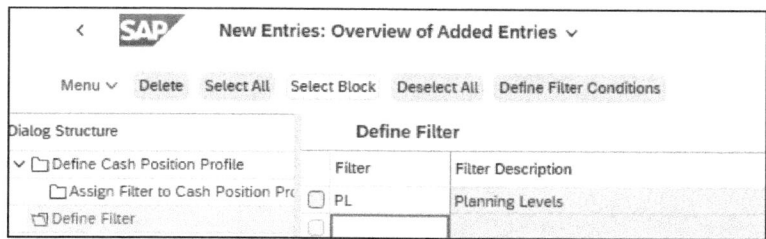

Figure 6.21 Define Filter

443

When you launch the Short-Term Cash Positioning app, simply select the profile you want to display—such as the one you just created—and click **Go**. The screen shown in Figure 6.22 will appear.

You can then use the **Collapse All** feature to neatly organize the view, and you'll see your bank accounts arranged according to the hierarchy defined in the profile setup. In this example, the accounts are grouped first by company code, then by currency, and finally by bank. You can easily switch between different cash position profiles to view the data in the format you prefer. After selecting a profile, click any level in the hierarchy—such as a specific company code, currency, or bank—to expand it. You'll then see the cash position details (see Figure 6.23), detailed balances, and cash flows for that selection, broken down by currency, giving you a clear, drilldown view of your short-term liquidity position.

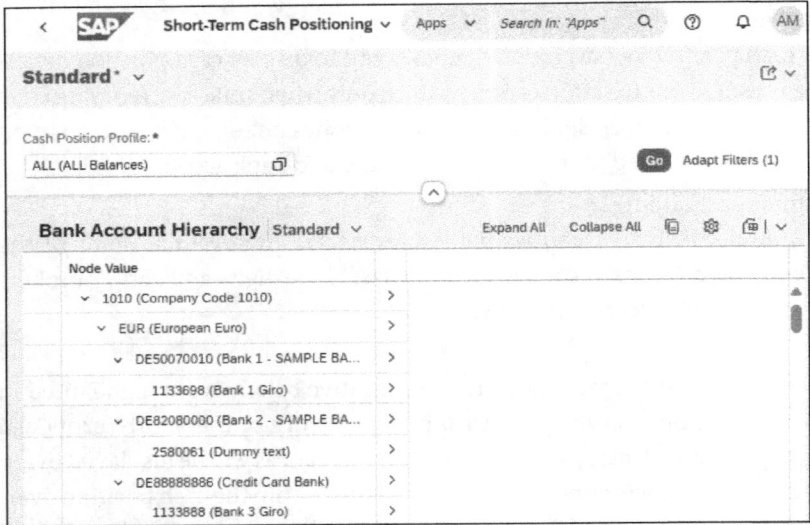

Figure 6.22 Short-Term Cash Positioning

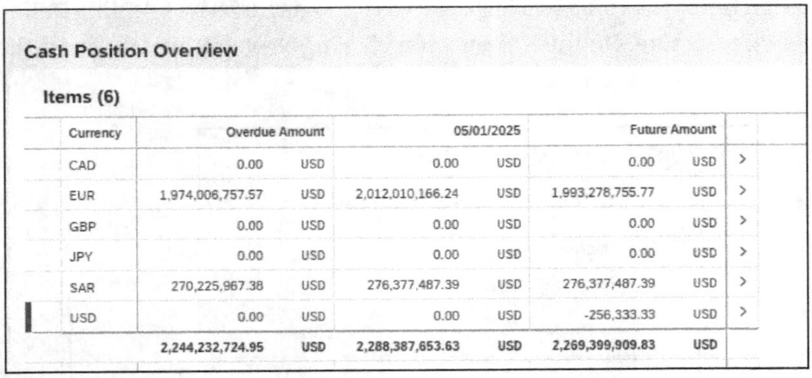

Currency	Overdue Amount		05/01/2025		Future Amount		
CAD	0.00	USD	0.00	USD	0.00	USD	>
EUR	1,974,006,757.57	USD	2,012,010,166.24	USD	1,993,278,755.77	USD	>
GBP	0.00	USD	0.00	USD	0.00	USD	>
JPY	0.00	USD	0.00	USD	0.00	USD	>
SAR	270,225,967.38	USD	276,377,487.39	USD	276,377,487.39	USD	>
USD	0.00	USD	0.00	USD	-256,333.33	USD	>
	2,244,232,724.95	USD	2,288,387,653.63	USD	2,269,399,909.83	USD	

Figure 6.23 Cash Position Details

6.1 Functions and Processes

You can view the opening balances for your selected bank accounts and currencies, the predicted transactions for each day, and the resulting future balances. Any future balance that falls below zero is highlighted in red, signaling a potential liquidity shortfall that requires your attention. This visual cue helps you quickly identify where funding adjustments are needed and takes proactive measures to ensure all accounts maintain sufficient coverage. Once you click further, you'll be able to view the detailed cash position. This includes all relevant data sources (see Figure 6.24), along with categorized inflows and outflows, each assigned to their respective planning levels for better financial analysis and tracking.

If you continue clicking and checking more details, the system will take you to the lowest level of the hierarchy and open the Check Cash Flow Items app (see Figure 6.25). Here, you can view the specific flows that contribute to the balance. In this case, these will be the outgoing flows—such as payments—that occurred in the past or are responsible for building the current balance. This detailed view allows you to analyze the individual transactions affecting your cash position.

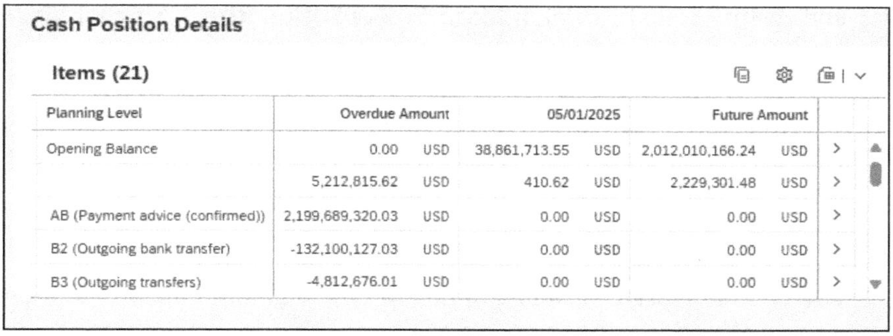

Figure 6.24 Cash Position Details Categories

Figure 6.25 Cash Flow Items Debit

6   SAP S/4HANA Finance for Cash Management

> **Cash Management Data Sources**
>
> All cash management applications in SAP S/4HANA rely heavily on the underlying accounting data, which means that any transactions that aren't properly cleared will continue to appear in your cash management views unless you explicitly filter them out. If you're seeing outdated or irrelevant items in the Cash Flow Analyzer—or in any other cash app—consider creating filters to exclude those entries and focus only on the current, treasury-relevant data. It's common to encounter certain AP/AR line items that were posted for accounting reasons but aren't meaningful for treasury operations; filtering them out keeps your cash management dashboards clean and actionable. Remember, data quality is critical for effective cash management, so taking the time to maintain accurate, well-filtered data will greatly improve your visibility and decision-making.

The Short-Term Cash Positioning app is a vital tool for managing your day-to-day liquidity. It displays each bank account's opening balance, all incoming and outgoing cash flows, and the forecasted closing balance. You can also model cash pool movements by defining hierarchies—linking subaccounts or entities to a header account used for POBO or ROBO transactions—to predict the header account's balance. This functionality gives you clear, real-time insights into your cash position and helps you plan funding needs proactively.

6.1.5   One Exposure from Operations

The One Exposure from Operations table in the system functions similarly to traditional SAP Business Warehouse (SAP BW) applications, where all relevant data is gathered from underlying sources. These sources can include data from within SAP S/4HANA, such as accounting, or from external systems, depending on the configuration. The process involves extracting data from these sources, transforming it to meet specific requirements (such as assigning it to a house bank and bank account IDs, or liquidity items), and performing any necessary adjustments, like changing value dates. Once this data is processed, it is loaded into the One Exposure from Operations table. All cash management applications within the system rely on this table, using the data stored within it to perform various cash management functions and generate accurate forecasts.

You don't need to worry about loading or deleting data in the One Exposure from Operations table; it functions like any other database, where you can freely import new datasets and remove old ones as needed. Data is typically loaded into One Exposure from Operations via scheduled jobs or manual triggers that extract information from your configured sources (e.g., accounting tables, bank statement files, or external systems). During the load process, the system transforms the raw data by assigning it to house banks, account IDs, and liquidity items and adjusting value dates—before writing it into the table. When you need to refresh the data (e.g., to correct errors or update for a new reporting period), you can simply delete the existing records and reload the new dataset. This flexibility ensures that your cash management applications always work with the most accurate and up-to-date information.

In SAP S/4HANA 2021 and earlier, data can be loaded into cash management using Transaction FCLM_FLOW_BUILDER. This tool, which can be scheduled as a background job, allows users to manually load and process data flows from various sources into the system, making it easier to manage cash and liquidity information. In SAP S/4HANA 2022 and up, you use Transaction FCLM_FLOW_BUILDER_2 (see Figure 6.26) for loading data into cash management.

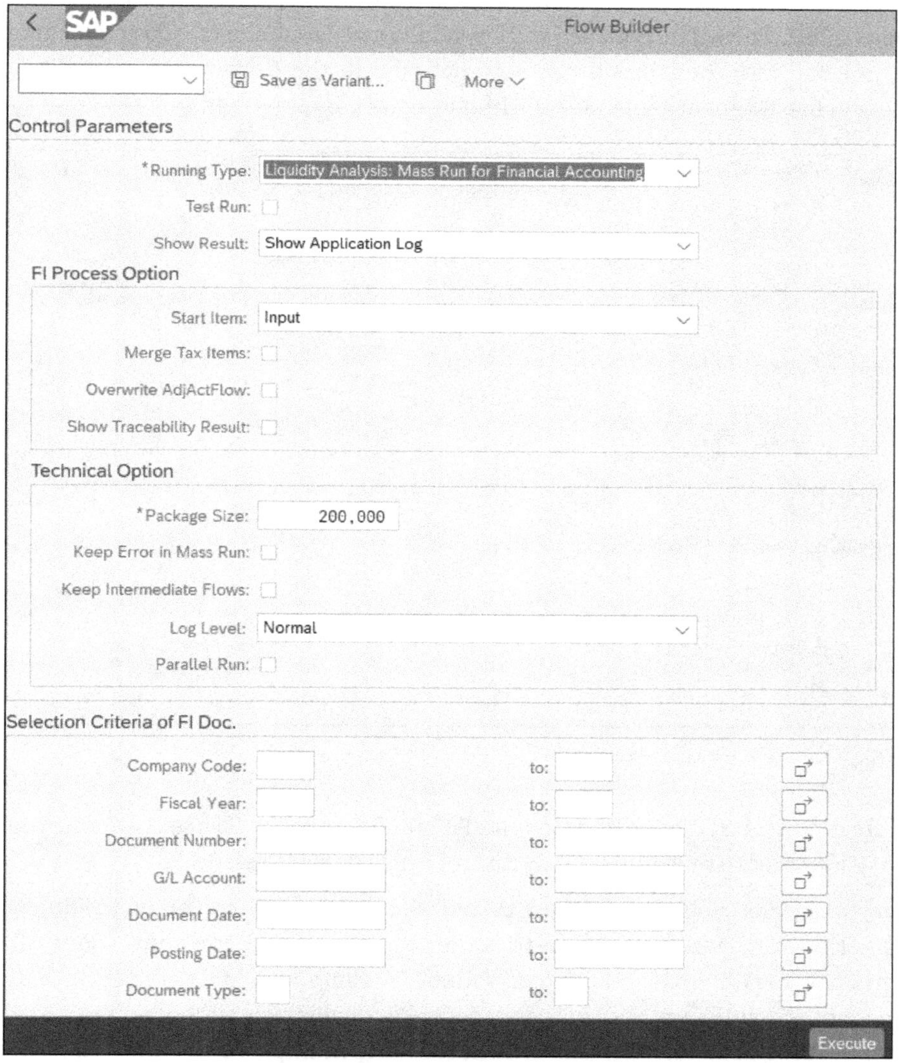

Figure 6.26 Transaction FCLM_FLOW_BUILDER_2

6 SAP S/4HANA Finance for Cash Management

Starting with SAP S/4HANA 2023, the activation and scheduling of cash management data loads has been fully transitioned to SAP Fiori. The new Schedule Jobs for Flow Builder app (F3804) enables you to configure and launch background jobs that automatically extract, transform, and upload data into the One Exposure from Operations table at defined intervals. This SAP Fiori–based scheduling tool provides an intuitive interface for setting up recurring or one-off data loads, ensuring your cash management applications always work with the latest information without manual intervention.

When you open the Schedule Jobs for Flow Builder app (see Figure 6.27), simply click **Create** to define a new job. Enter a descriptive job name under **Job Template** and **Job Name**, select the start date and start time (in the **Start** field), and choose the **Recurrence Pattern** (such as **Daily**, **Weekly**, or **Monthly**). You'll also specify the **Flow Builder Variant** that determines which data is loaded and any additional parameters required. Once you save these settings, the job will run automatically in the background, periodically updating your cash management data without manual intervention.

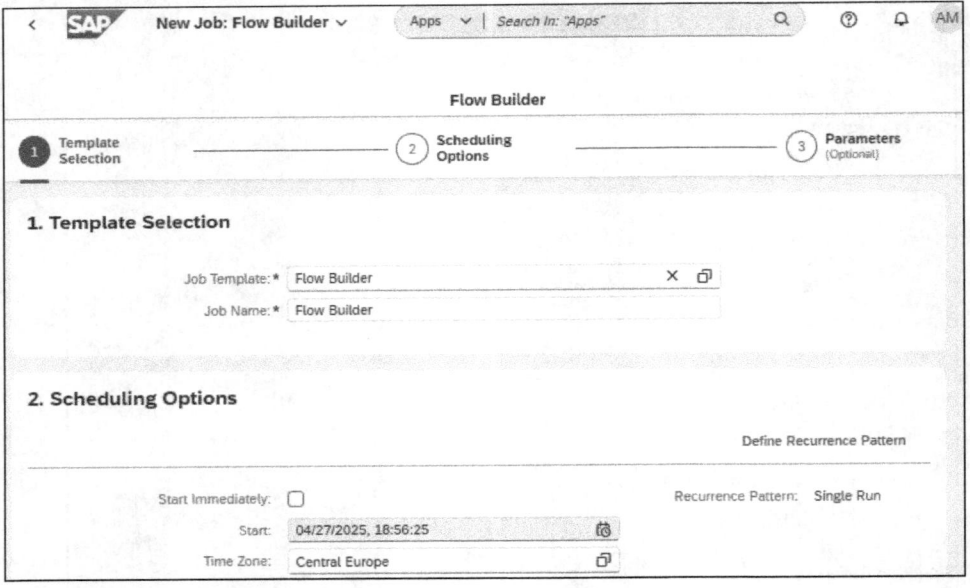

Figure 6.27 New Job for Flow Builder

When creating a job in the Schedule Jobs for Flow Builder app, you'll need to define your **Control Parameters** (see Figure 6.28) to control which data is loaded.

In simple scenarios, you might choose to load all documents for a complete liquidity analysis. However, you can also restrict the load to specific company codes, G/L accounts, or other criteria to reduce data volume. Be cautious: Loading the entire dataset can amount to millions of entries and may take considerable time. If you're reloading or appending data, consider splitting the load into multiple runs—for example, by date range or company code—to optimize performance and manage processing time.

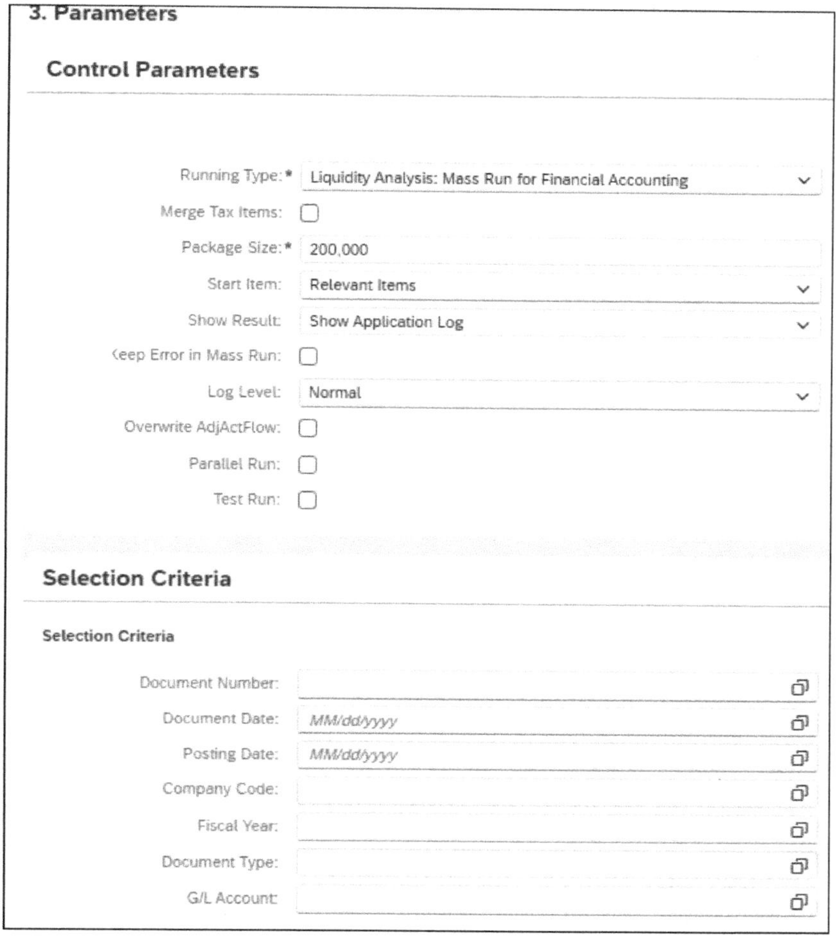

3. Parameters

Control Parameters

Running Type:*	Liquidity Analysis: Mass Run for Financial Accounting
Merge Tax Items:	☐
Package Size:*	200,000
Start Item:	Relevant Items
Show Result:	Show Application Log
Keep Error in Mass Run:	☐
Log Level:	Normal
Overwrite AdjActFlow:	☐
Parallel Run:	☐
Test Run:	☐

Selection Criteria

Selection Criteria

Document Number:	
Document Date:	MM/dd/yyyy
Posting Date:	MM/dd/yyyy
Company Code:	
Fiscal Year:	
Document Type:	
G/L Account:	

Figure 6.28 Flow Builder Selection Parameters

When scheduling a job with the flow builder (Transaction FCLM_FLOW_BUILDER_2), you are setting up how the system should extract and process data to generate cash flows in cash management. This job pulls information from both financial accounting (tables BKPF and BSEG) and material management (e.g., EKKO, EKEP, EKKN) and loads it into the One Exposure from Operations hub (table FQM_FLOW), which is the central database for all cash-related analytics in SAP S/4HANA. Key selections in the job parameters are as follows:

- **Running Type (determines which items you can select on the screen)**
  You'll choose among the following:
  - **Cash Position: Delta Run from Staging**—delta flows, so newly created from the previous run
  - **Cash Position: Mass Run**—for loading large historical data or rebuilding flows after config changes for cash position

- **Delta Run from Staging for Material Management**—loads newly created data from purchase orders and requisitions
- **Liquidity Analysis: Delta Run from Staging**—delta flows, so newly created from the previous run
- **Liquidity Analysis: Mass Run**—for loading large historical data or rebuilding flows after config changes for liquidity analysis
- **Mass Run for Purchase Ord./Sch. Agreement**
- **Mass Runf for Purchase Requisition**

- **Merge Tax Items**
  If not selected, value-added taxes (VAT) will create separate flows in cash management; if selected, the full invoice amount will flow into items created in cash management.

- **Package Size (number of data objects in package)**
  The smaller the number, the better the performance of the load, but will increase time needed for a load.

- **Start Item**
  Defines how the program selects the starting items for generating flows:
  - **Relevant Items:** Finds all related documents and identifies start items.
  - **Input:** Treats your selected documents as start items.
  - **Actual Input:** Considers only start items with a certainty level of **Actual**.

  Note: The delta run from staging only works with the default related items value.

- **Show Result**
  Determine whether you want to see application log or results of the load (used for manual loads).

- **Keep Error in Mass Run**
  Even if there is an error, the system will try to load the rest of the data.

- **Log Level**
  Select what granularity you would like to see.

- **Overwrite AdjActFlow**
  If you select this checkbox, the system will overwrite flows that were already loaded, but that you changed manually. If not checked, then your changes will not be overwritten.

- **Parallel Run**
  If selected, the system will create parallel runs to load data faster, but will use more memory to process them.

- **Test Run**
  This lets you simulate the results of the load.

> **Selection Criteria**
>
> You can filter what data is processed by company code, G/L accounts, document dates, posting date, fiscal year, and document type.
>
> Be careful: Loading everything at once can be very heavy and time-consuming. For large volumes, it's recommended to split the job by criteria like company code or date range.

If you are scheduling a job to run in the background, it's recommended to use the **Delta Run** option. This ensures that the system loads only the newly created or changed data into cash management, keeping your One Exposure from Operations table up to date without reprocessing all historical entries. Once you've completed all your selection parameters—such as company codes, G/L accounts, or parallel execution settings—simply click **Schedule**, and the system will start the job in the background according to your defined time and frequency. This helps automate and streamline data updates for liquidity analysis without manual intervention.

We've described how to load data into cash management, but it's also important to know that you can delete data from the One Exposure from Operations table when needed. This is especially useful if you notice incorrect or obsolete data, or if you want to refresh and reload the information after changes to configuration or source data. To do this, you can use the Delete Data from One Exposure app (FQM_DELETE; see Figure 6.29), which corresponds to Transaction FQM_DELETE in the SAP GUI.

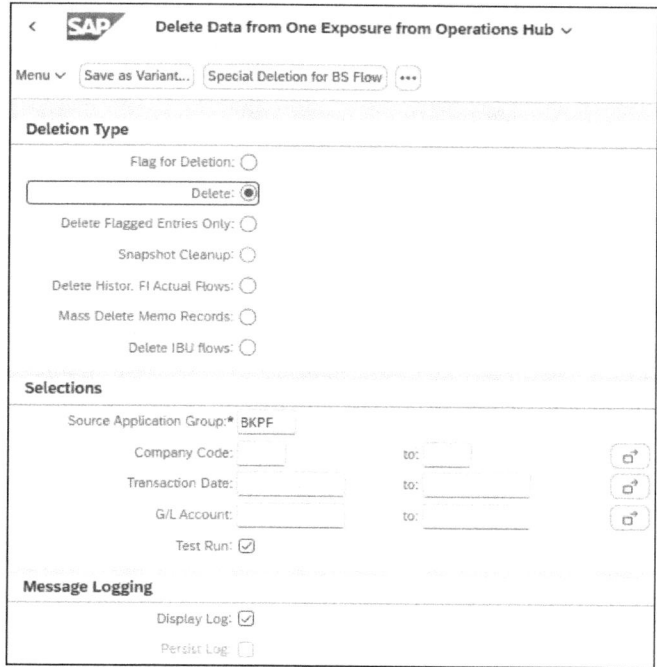

Figure 6.29 Delete Data from One Exposure from Operations

This tool allows you to selectively delete records based on criteria such as company code, document type, or posting date, ensuring you retain full control over the data stored in cash management.

In this app, you simply need to select the data you want to delete by specifying parameters such as **G/L Account**, **Transaction Date**, **Company Code**, and **Source Application Group**. The source refers to the origin of the data and is determined by the configuration settings; each application that feeds data into One Exposure from Operations has its own source identifier. You can view these sources directly in the One Exposure from Operations table (FQM_FLOW) to understand which entries belong to which application. After setting your selection criteria, executing the transaction will delete the matching records from One Exposure from Operations. This app also offers options to clean up snapshots, delete memory cards, and perform other maintenance tasks, ensuring your cash management data stays optimized and up to date.

6.1.6 Manage Memo Records

Memo records in cash management are used to represent expected cash flows that are not captured in the SAP S/4HANA system through standard transactional data. These are manually created entries that allow treasury teams to plan for upcoming cash needs or inflows that are not otherwise visible in accounting, purchasing, or other operational modules. For example, salary payments, merchant acquisition costs, or planned financing activities might not yet exist in SAP S/4HANA through invoices or payment runs but are known to the treasury and must be considered in liquidity planning. Memo records help ensure these flows are included in the liquidity forecast, so the organization can proactively secure funding or manage available cash to cover them. Their manual creation fills the gap between system-derived flows and real-world financial expectations.

Traditionally, memo records were entered and managed using the Manage Memo Records app (F2986), which allowed you to create, edit, delete, and monitor memo entries in cash management. These records served as placeholders for expected cash flows that were not yet reflected in system transactions. However, starting from SAP S/4HANA 2023, a new application called Manage Memo Records 2.0 (F2986A) was introduced. This enhanced version brings improved functionality, with a stronger focus on differentiating between actual and forecasted data sources and offering a more user-friendly interface for easier data input and classification. The upgrade supports more efficient and accurate cash planning by enabling better tracking and categorization of manually maintained cash flow data.

When creating a memo record, the first step is to specify the **Company Code** and the **Memo Record Type** (see Figure 6.30), which are defined in the system configuration. In the newer Manage Memo Records 2.0 app, the data entry process is significantly simplified. Much of the required information—such as the planning level and other key attributes—is automatically derived from the customizing settings of the selected memo record type. In contrast, the earlier version of the app required you to input many of these details manually.

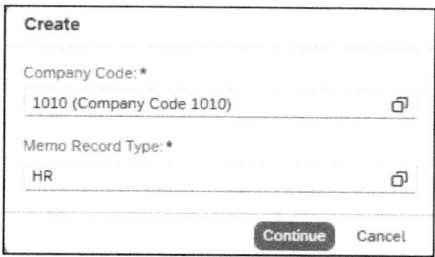

Figure 6.30 Create Memo Record

The fields you are required to fill in (see Figure 6.31) when creating a memo record depend on the configuration and field selection settings defined in your system. This configuration determines which fields are mandatory and which are optional, ensuring that the data entry process aligns with your organization's requirements.

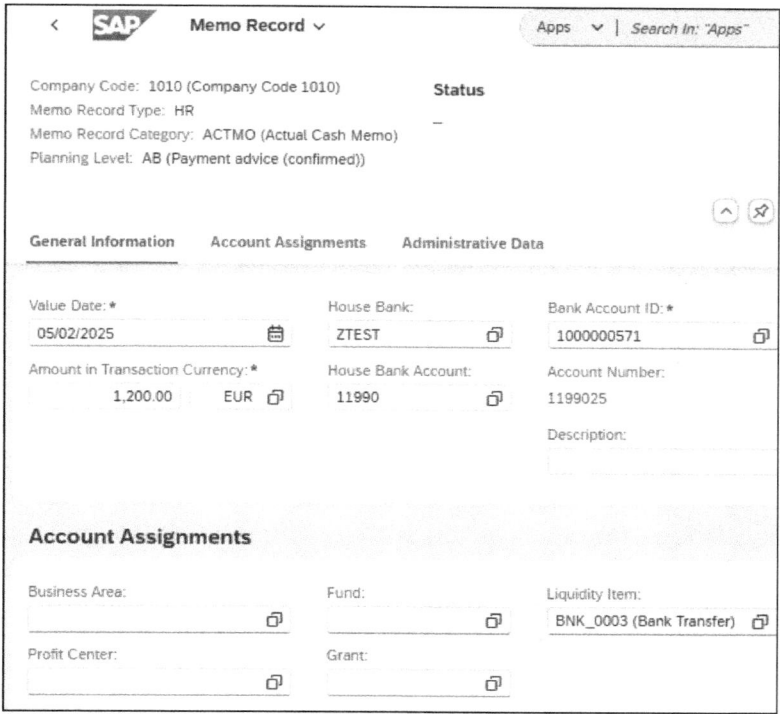

Figure 6.31 Create Memo Record: Detailed Information

It's also important to note that memo records should be autoarchived to maintain the accuracy of your liquidity forecasts. In the example, we are working with an actual memo record, so it will remain visible in the system. This ensures the system automatically removes or deactivates the memo record when the date passes. If no expiration is set, then outdated forecasted records will appear as overdue and can distort your liquidity analysis, leading to misleading data in your cash management reports. Alternatively, you can run a background job to delete all older memo records with a date value that falls before today's date. Once you click **Save**, the system will automatically create and activate the memo record (see Figure 6.32).

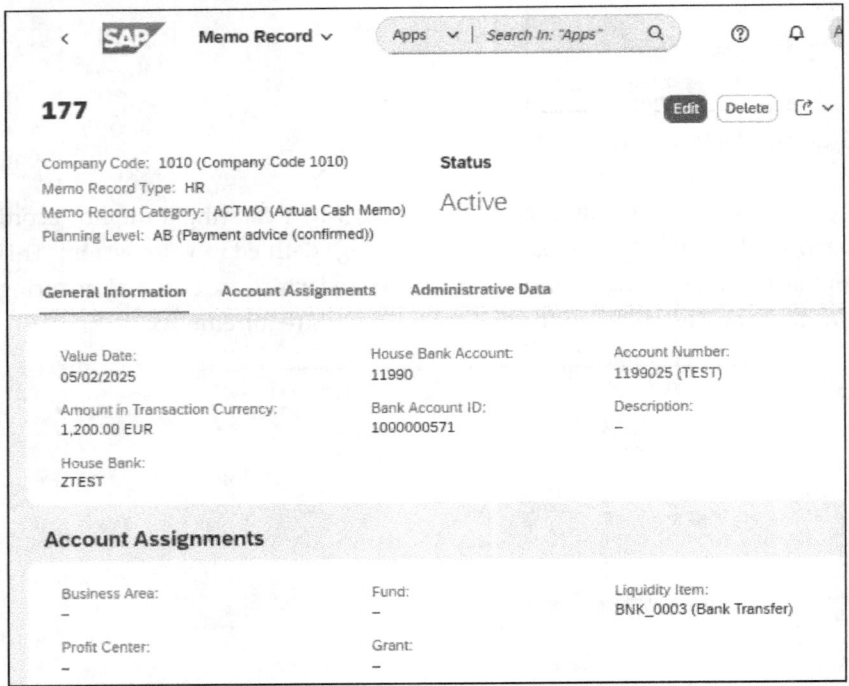

Figure 6.32 Memo Record Created

When you create a memo record, it becomes immediately available across all cash flow applications—including Cash Flow Analyzer, Short-Term Cash Positioning, and Monitor Bank Account Balances. This unified visibility allows you to verify that the memo entries are correctly factored into your liquidity analyses. From any of these apps, you can drill down to view memo details, make edits, or delete records that are no longer required. This ensures that your forecasts remain accurate and that only relevant manual entries impact your cash management processes.

6.1.7 Import Memo Records

Managing memo records manually can be both time-consuming and prone to human error, especially when you have a large volume of anticipated cash flows to enter. To streamline this process, SAP S/4HANA provides the option to upload memo records in bulk from an Excel file. By preparing your memo data in a spreadsheet—complete with amounts, value dates, expiration dates, account assignments, and liquidity item codes—you can import multiple records at once into the Manage Memo Records app. This capability not only saves significant data-entry effort but also ensures consistency and accuracy across your liquidity forecasts.

In SAP ERP and releases prior to SAP S/4HANA 2023, memo records had to be uploaded using ABAP program RFTS6510. You can run this program from Transaction SE38 (or SA38) or by entering Transaction RFTS6510 directly. The program prompts you for the file path and format—typically a CSV or text file exported from Excel—and then maps each column (such as amount, value date, account, and liquidity item) to the corresponding memo record fields. Once executed, RFTS6510 processes the file and creates or updates all specified memo records in bulk, streamlining what would otherwise be a labor-intensive manual entry task.

The old program was not very user-friendly, especially when it came to assigning flows to memo records and linking them to the proper bank accounts. Users were required to select either the G/L account or the technical ID, which made the process cumbersome and prone to errors. This complexity often led to difficulties in ensuring the correct assignment of flows to the right accounts, adding to the challenges of managing memo records. To address these limitations and improve the user experience, SAP introduced the Import Memo Records 2.0 app (F6124). This newer approach offers a more streamlined process for importing memo records, providing enhanced features, better usability, and improved integration with other cash management applications.

When you open the Import Memo Records 2.0 app, simply click **Download Template** to retrieve a preformatted spreadsheet. Open the downloaded file (see Figure 6.33), populate it with your memo record details—such as amounts, value dates, account assignments, and liquidity items—and save your changes. Back in the app, click **Upload File**, select your filled-in template, and click **Import**.

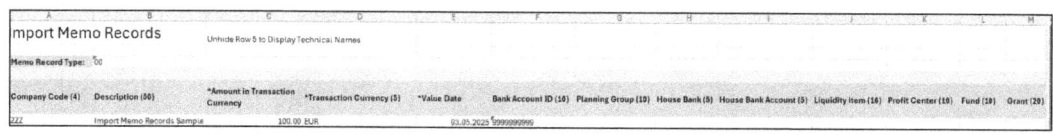

Figure 6.33 Import Memo Records Template

Once you upload the completed spreadsheet in the Import Memo Records 2.0 app, the records will initially appear with the status **In Process** (see Figure 6.34). After a short moment or a manual refresh, the system completes the processing, and the memo records will be fully created and available through your cash management and Cash Flow Analyzer applications.

6 SAP S/4HANA Finance for Cash Management

Figure 6.34 Import Memo Records Results

6.1.8 Bank-to-Bank Transfers

Once you have a clear understanding of our liquidity situation and know how much money is available in each bank account and what the forecasted balances look like, you can start making informed decisions about how to manage your liquidity. One of the simplest scenarios is identifying accounts that require funding due to expected outgoing payments and other accounts that may have surplus cash. In such cases, the most straightforward option is to transfer money between these accounts. This can be done using the Make Bank-to-Bank Transfer app (F0691), which allows you to initiate internal transfers quickly. As shown in Section 6.1.3, these transfers can also be triggered directly from the Cash Flow Analyzer app or, alternatively, you can access the dedicated app and execute the transfer based on predefined templates.

Making a bank-to-bank transfer (see Figure 6.35) in SAP S/4HANA is a very straightforward process. All you need to do is select the paying bank account, the receiving bank account, the amount, the date of the transfer, and the currency. The system will automatically populate the remaining fields based on your configuration and defaults. If you choose the **Release and Pay** option, the transfer will be executed and paid immediately; there's no need to run Transaction F111. You also can enter other optional information, such as central bank indicators, instruction keys (e.g., purpose of the payment), who will bear the bank charges, or any payment method supplements. You also have the option to fill in a reference field for internal tracking or reconciliation purposes.

When you click **Create**, the system will generate a bank-to-bank transfer and automatically create a payment request. If you select **Release and Pay**, the payment run will be triggered immediately. The next step typically involves batching and approval, depending on your payment setup, and then sending the payment file to the bank for execution. Bank-to-bank transfers are generally straightforward and require minimal approvals because users can only initiate payments between bank accounts that are already approved and maintained in BAM. This ensures strong control as no unauthorized payments can be made to external parties. In most cases, you also can select the **Pay and Reverse** option, but in the example scenario, the payment run has already been executed with identification UAR-R, which can be verified in Transaction F111.

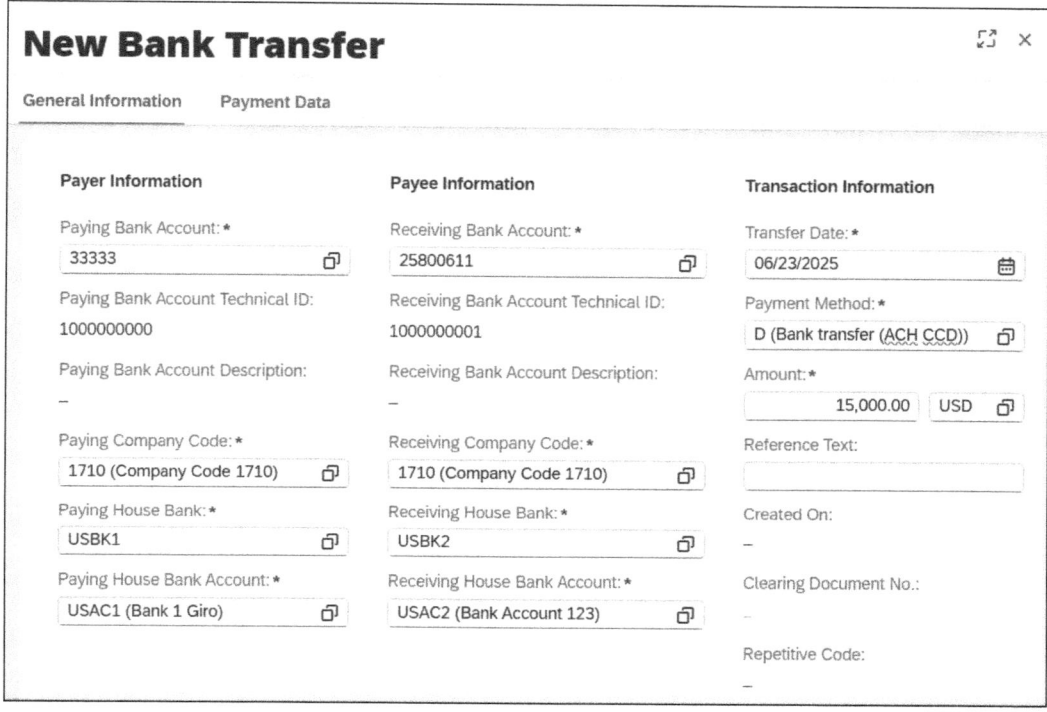

Figure 6.35 Make Bank Transfer

Once the bank-to-bank transfer is executed, you can display it in the Make Bank-to-Bank Transfer app (see Figure 6.36).

> **Payment Request**
>
> As described in Chapter 1, there are situations where you need to create payments even though no invoice exists. Bank-to-bank transfers and the payments we're currently discussing are good examples of this. These types of transactions generate payment requests directly in the system. You can view and monitor them using the Display Payment Request app or Transaction F8BT, and the data is stored in table PAYRQ. In the example, the payment was released immediately, but the system also supports defining approval patterns and workflows for payment requests based on your configuration. This becomes especially relevant for freeform payments or any other payment types that require stronger controls. In many scenarios, payment requests are generated from external systems and sent to a central SAP S/4HANA system, where having proper workflows in place ensures the data can be reviewed, enriched, and approved before payment execution.

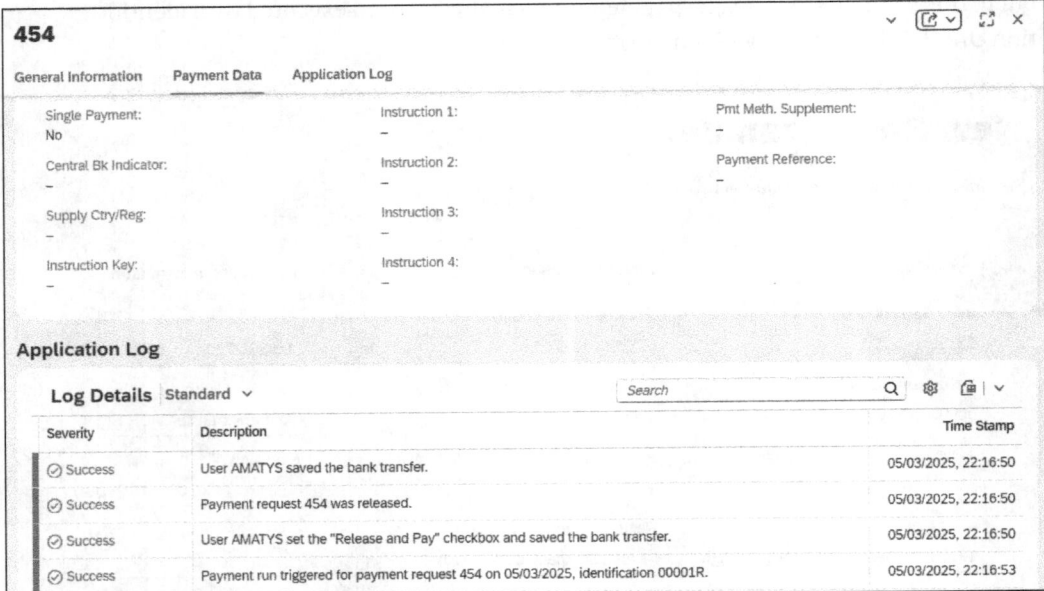

Figure 6.36 Bank Transfer Executed

To simplify and speed up the processes—especially because treasury teams often execute many bank-to-bank transfers—SAP S/4HANA allows you to create bank transfer templates. These templates help automate repetitive transfer tasks. You can define them using the Define Bank Transfer Templates app (F3759; see Figure 6.37).

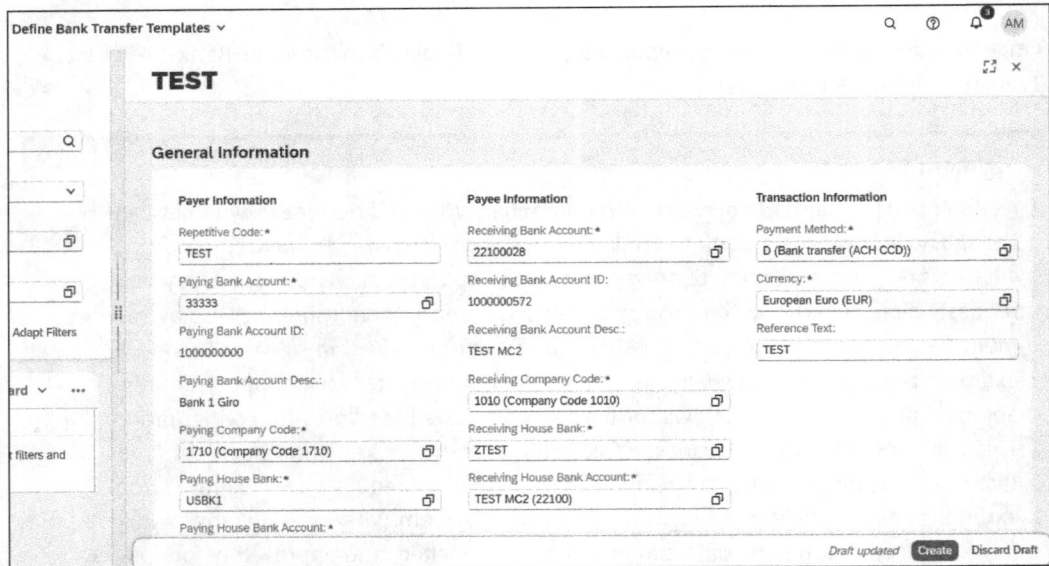

Figure 6.37 Define Bank Transfer Template

When creating a bank transfer template, the process is nearly identical to creating a regular bank-to-bank transfer. You simply select the paying bank account, the target (receiving) bank account, and the payment method, and enter any relevant reference information. Once all necessary details are entered, you save your template. After saving, the template becomes available for reuse in the Make Bank Transfers—Create with Templates app (F3760), allowing you to quickly execute bank-to-bank transfers directly from the template without reentering data each time (see Figure 6.38), thus streamlining and standardizing treasury operations.

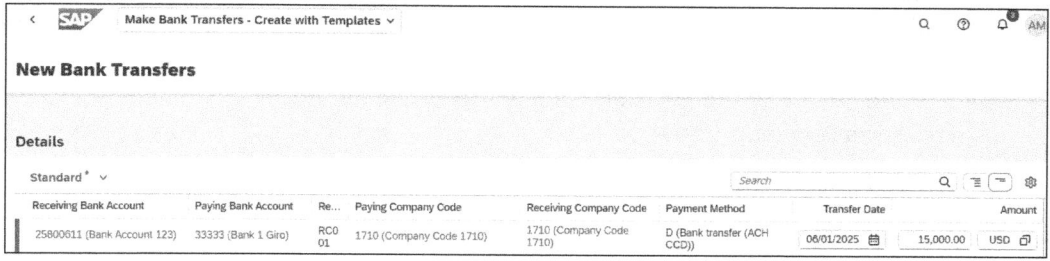

Figure 6.38 Create Bank-to-Bank Transfers Using Template

Using the Define Bank Transfer Templates app is very straightforward. All you need to do is select the appropriate amount, optionally modify an existing template, and then click **Create**. The system will automatically generate the corresponding payment request(s) based on the template data. These payment requests can then be processed in the next steps, typically by using Transaction F111 or the Automatic Payment Transactions for Payment Request app. From there, the payment will follow your configured approval workflow, go through batching, and finally be sent to the bank for execution, ensuring compliance and control over the process.

In SAP ERP and SAP S/4HANA systems, there's also the option to execute bank-to-bank transfers using repetitive codes, which are predefined templates storing payment details for frequent transfers. This can be done using Transaction FRFT_B, which allows you to select a repetitive code, fill in necessary payment details like the amount, date, and currency, and then create the transfer. There's also an analogous SAP Fiori app available that offers the same core functionality via a modern user interface. This method is particularly useful when working with pre-agreed-upon payment formats and partners, and it serves as an alternative to using payment templates or free-form entries; however, it is recommended to use the previously mentioned apps in the SAP S/4HANA environment.

6.1.9 Free-Form Payments

An important point to emphasize is that SAP S/4HANA provides capabilities to execute payments without an invoice directly from the system, as an alternative to executing payments manually via online banking platforms, which should be avoided whenever possible. Relying on external platforms fragments the payment process and reduces visibility, traceability, and control. The strategic goal is to maintain one streamlined system—SAP S/4HANA—as the single source for initiating, monitoring, and approving all payments. That's why SAP S/4HANA offers solutions like free-form payments, which are specifically designed for scenarios in which a payment must be made but there is no corresponding invoice. These payments are often needed by the treasury—for example, to pay a third party, to handle intercompany settlements, or in any other situation in which an invoice document doesn't exist. Free-form payments ensure such cases are handled within SAP S/4HANA, maintaining compliance, control, and auditability across the full payments landscape. Free-form payments can be made using the Process Free Form Payments app (F2564; see Figure 6.39). As announced, starting with SAP S/4HANA 2025 there will be a new app available for the public cloud only: Process Free Form Payments (F8654).

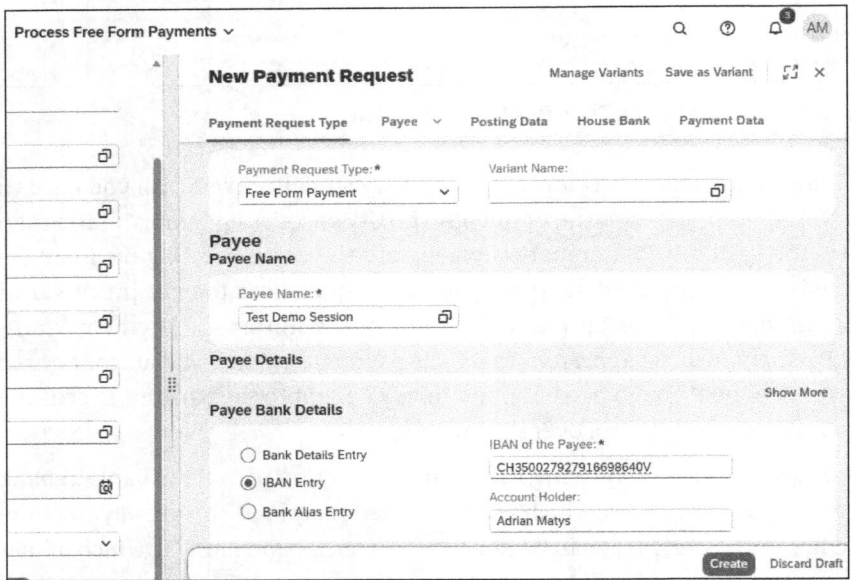

Figure 6.39 Process Free-Form Payments

After you open the Process Free Form Payments app, click **Create** to initiate a new payment. Now you can choose the **Payment Request Type**, which determines how the recipient data will be entered. You can select **Vendor**, **Customer**, or **Free Form Payment**. If you choose **Vendor** or **Customer**, the system will allow you to select an existing business partner in SAP S/4HANA, and it will automatically derive the payment details—such as bank account information—from the supplier or customer master data. However, if you select **Free Form Payment**, you will need to manually enter all bank details for the recipient. In a basic case, you might enter just an **IBAN Entry**, but if you opt to use **Bank Details Entry** or **Bank Alias Entry**, then you must provide additional fields like **Bank Key**, **Bank Country**, and possibly **SWIFT Code**. This distinction ensures flexibility while preserving data integrity based on the source of the payment. You can also select a variant that was created earlier to quickly derive and prefill the payment details for the transaction you want to execute. After you enter the **Payee Bank Details**, you also need to enter **Posting Data** (see Figure 6.40).

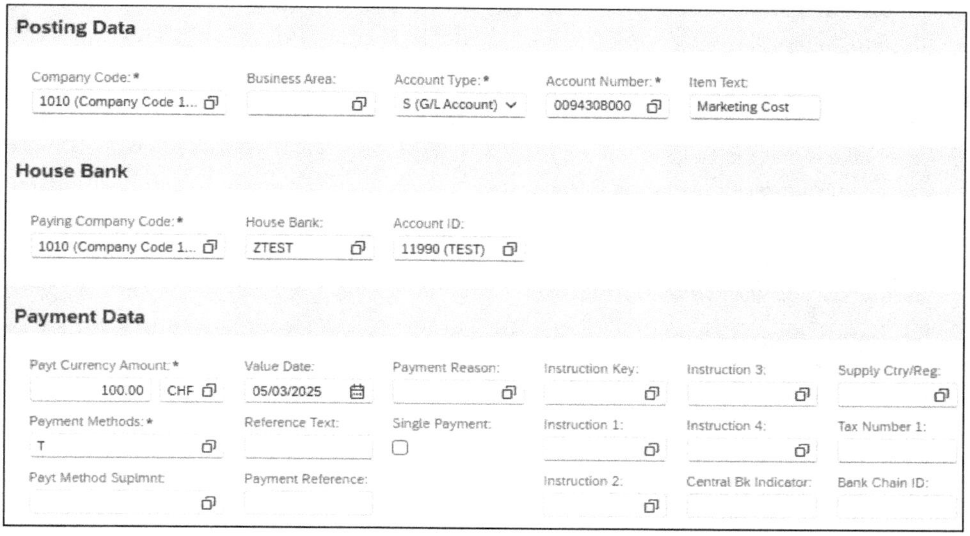

Figure 6.40 Free-Form Payment: Posting Data

Then you need to select the **Company Code** where the free-form payment will be posted. Next, you define the **Account Type**; this could be a vendor, customer, or G/L account, depending on how the payment should be recorded. You'll also need to provide the **Account Number** (customer, vendor, or G/L) and **Item Text** that describes the transaction. This account becomes the offsetting account, while the main payment account is automatically derived from the payment request account configured in your system setup.

After that, you must select the **Paying Company Code** and the **House Bank** account from which the payment will be executed. You also input the amount, payment method, value date, and any additional information required to complete the payment (e.g., purpose, reference, or central bank reporting fields).

Once all data is entered, click **Create**. The system will generate a payment request, which can then be processed through batching and approval. From there, the payment file can be created and sent to the bank for execution.

The payment you executed, along with all the details entered, can be saved as a variant, making it easier to reuse and execute similar payments in the future without reentering all the information. This is particularly useful for recurring standardized payments. It's also important to note that SAP S/4HANA is actively developing the Process Free Form Payments app to support digital payments, including those made via digital currencies like Bitcoin. These enhancements are expected to integrate with the bank alias functionality, enabling more flexible and modern payment methods. This development is part of SAP's broader digital payments roadmap, and further details can be found in Chapter 11.

> **[!] Free-Form Payments**
>
> Free-form payments, although flexible and useful in specific scenarios, carry a significant risk from both operational and fraud perspectives. Because users can manually enter all payment details—including recipient bank accounts—there is a high potential for mistakes or even intentional fraud, such as transferring money to unauthorized or personal accounts. For this reason, such payments should ideally be minimized or avoided entirely, and companies should instead rely on payments made to authorized counterparties with validated standing instructions maintained in the system. If free-form payments must be used, it is critical to have strong internal controls in place, including multistep payment approval workflows and rigorous validation processes, to mitigate these risks and ensure the integrity of treasury operations.

6.1.10 Cash Concentration

Cash concentration and cash pooling in SAP S/4HANA differ somewhat from the traditional cash pooling services offered directly by banks. In a bank-managed cash pool, the bank itself performs the sweeps or intercompany transfers automatically on your behalf based on predefined rules. In contrast, SAP S/4HANA provides similar functionality, but with more control and flexibility, as the sweeps are initiated and triggered by the company itself through apps like Manage Cash Concentration. This means you are not limited to accounts within a single bank; SAP S/4HANA cash pools can include bank accounts from different banks, giving treasury teams broader scope and centralized control over their liquidity.

Starting from SAP S/4HANA 2023, SAP S/4HANA streamlined the creation and maintenance of cash pools by moving the full setup process to dedicated SAP Fiori applications. This change significantly simplifies the configuration process and provides a more intuitive interface for treasury users.

Once you create a cash pool by assigning a header account and corresponding subaccounts in the Cash Pools app, the **Cash Pool** tab in the Manage Bank Accounts app is automatically updated for the related bank accounts. This synchronization eliminates the need for manual maintenance. On each bank account level, you can then specify key parameters such as the target balance (the ideal end-of-day balance for the account), the minimum balance transfer amount (the threshold that must be reached before a transfer is triggered), and the payment method to be used for executing the cash concentration transactions and internal bank-to-bank transfers. These settings ensure that cash movements within the pool are automated and align with treasury policies.

There are multiple SAP Fiori applications available for managing cash pooling in SAP S/4HANA, but the preferred and most up-to-date version is Cash Pools (Version 2) (F3266A). This app offers a modern interface and enhanced usability for defining and managing your cash pool structures. The earlier version, Manage Cash Pools (F3266), provides similar core functionalities but lacks some of the refinements and improvements found in version 2. For consistency and to align with SAP's latest innovations, it is recommended to use Cash Pools (Version 2) when setting up or maintaining cash pools. If you open this app and click **Create**, the system will open the popup shown in Figure 6.41.

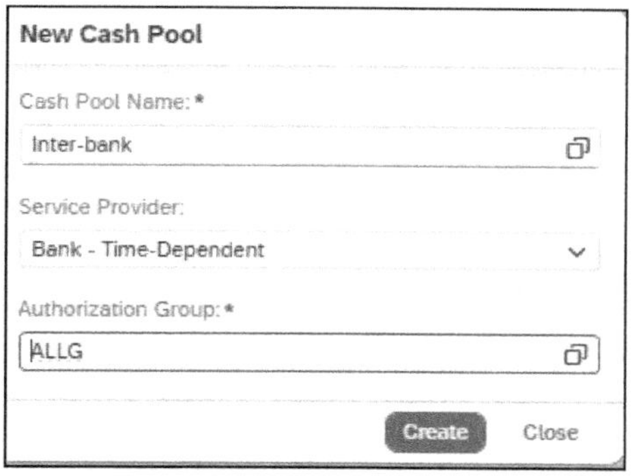

Figure 6.41 New Cash Pool

To create a cash pool, you first need to provide a **Cash Pool Name**, which will serve as the unique identifier for your cash pool structure. Next, specify the **Service Provider**, indicating whether the cash pool or cash concentration functionality is managed by an in-house bank or by external banks. You will also need to assign an **Authorization Group**, which defines access control and ensures that only authorized users can view or modify the cash pool setup. Once all required information is entered, simply click the **Create** button to initiate the setup process.

After creating the cash pool, the next step (see Figure 6.42) is to provide a description that clearly defines the purpose or scope of the cash pool for easier identification. You then need to assign the head bank account, which will act as the central account in which excess cash from subaccounts is concentrated. Following that, you must add the relevant subaccounts that belong to this cash pool structure. Optionally, you can assign custom names or labels to each bank account to enhance clarity. It is also important to define the validity period for the cash pool setup—by specifying the start and end dates during which the cash pooling rules and structure will be applied.

Once you assign bank accounts to a cash pool, the system automatically populates the cash pool details at the managed bank accounts level (see Figure 6.43), making the setup much more efficient. All you need to do is go into each relevant bank account and specify the payment method that will be used for executing the cash concentration transfers. In addition, you can define the target balance, the minimum transfer amount (to

6   SAP S/4HANA Finance for Cash Management

avoid triggering small, unnecessary transfers), and the maximum transfer amount (to control large movements of cash). You also have the option to enter a reference text, which will be automatically included in each generated transfer, making it easier to track and audit cash movements.

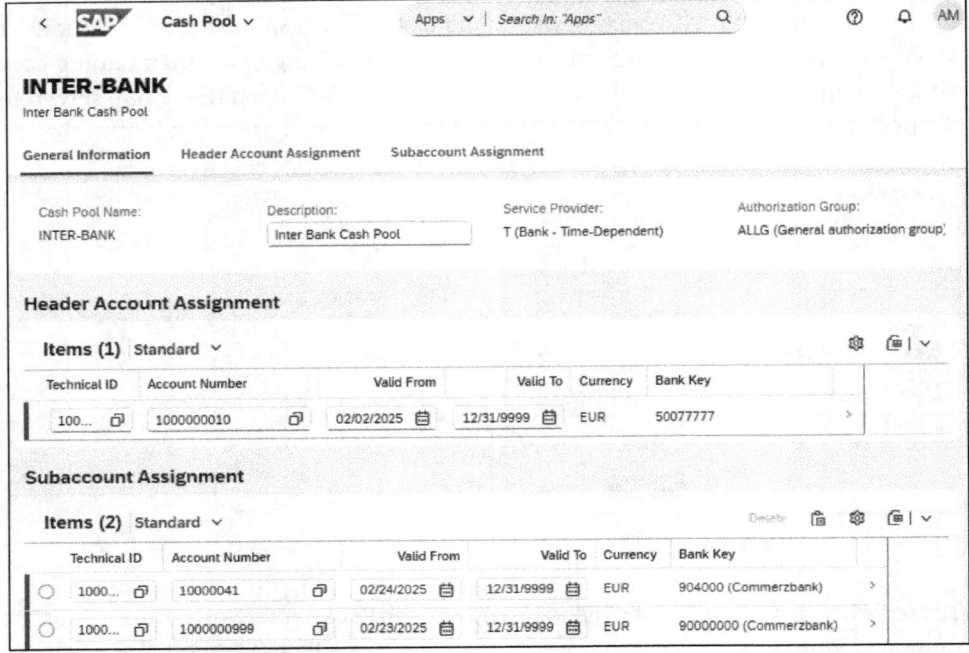

Figure 6.42 Cash Pool Details

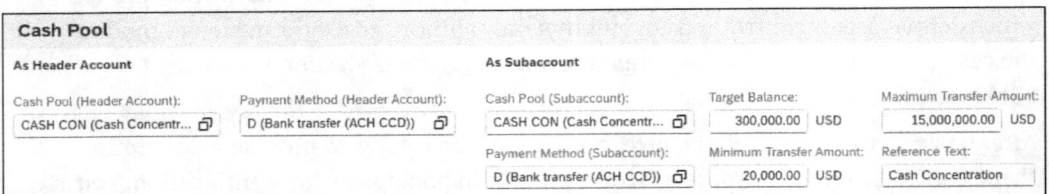

Figure 6.43 Cash Pool Details in Manage Bank Accounts App

SAP S/4HANA provides a dedicated report called Display Cash Pool Hierarchies (F6123), which allows you to view a comprehensive list of all created cash pool hierarchies in the system (see Figure 6.44). This report presents the hierarchies in a tabular format, making it easy to review their structure and attributes. It also offers convenient features such as downloading the list to a spreadsheet for offline analysis or documentation purposes. From this report, you can also navigate directly to the Manage Cash Pool app to view or modify the details of a specific cash pool hierarchy.

6.1 Functions and Processes

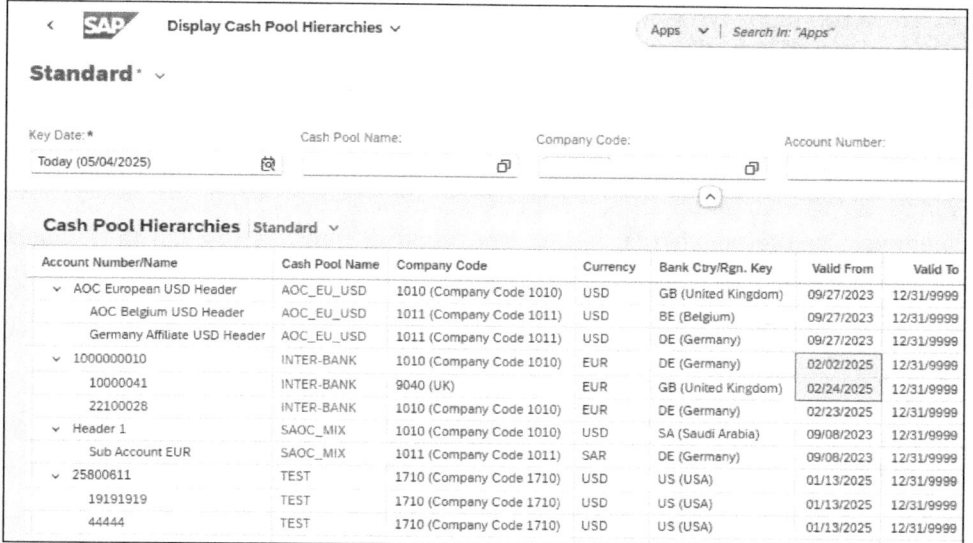

Figure 6.44 Display Cash Pool Hierarchies

Once you have maintained your cash pool and defined the necessary execution details, you are ready to run the cash concentration using the Manage Cash Concentration app (F3265). When you open this application, you will see a list of all previously created cash concentration transfers, along with their statuses. To create a new set of transfers, simply click the **Create** button. A popup will appear (Figure 6.45) in which you need to specify the **Cash Pool Name** and the **Plan Date** on which the concentration should be executed.

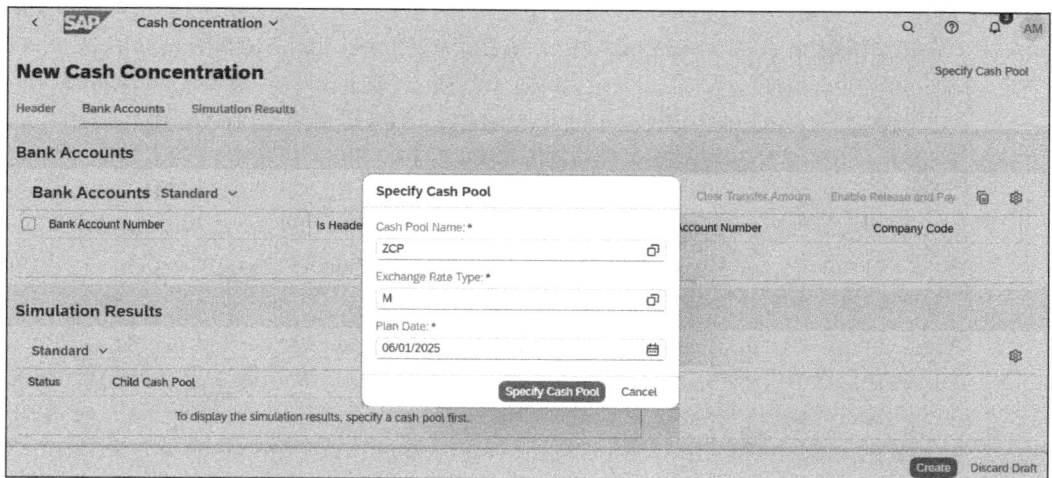

Figure 6.45 New Cash Concentration

Based on this input, the system will evaluate the balances of the accounts in the pool and propose transfers needed to bring them in line with the defined target balances. Click **Create** again, and the system will show you the proposed transfers for a particular date.

Based on the setup defined at the bank account level and within the cash pool configuration, the system automatically evaluates current account balances and compares them against the target balances. It then proposes transfers to align the balances accordingly (see Figure 6.46).

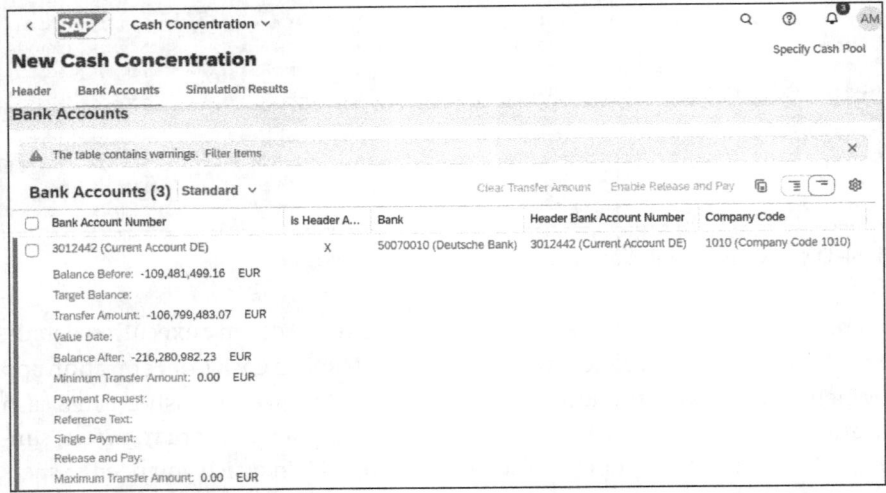

Figure 6.46 Cash Concentration Simulation

In the example, the system identifies that funds need to be transferred from the header account to the subaccounts to meet their target balances. However, as the header account does not have sufficient funds, the system issued a warning indicating that additional funding is required for the cash pool. You have the flexibility to adjust the proposed transfer amounts and simulate the results to review the projected balances after execution. Once you're satisfied with the setup, you can click **Create**, and the system will generate a payment request. This request will follow the configured workflow for batching and approvals, and ultimately generate the payment file and instructions to send the transfer to the bank.

The Manage Cash Concentration app allows you to trigger cash concentration processes manually by selecting the relevant cash pool and planning date, which is particularly useful when testing or fine-tuning your setup. However, once you have established confidence in your bank balances and properly configured your rules for cash concentration, SAP S/4HANA offers the ability to automate this process using the Schedule Cash Concentration Jobs app (F3688). This app enables daily scheduling of cash concentration runs, automatically generating payment requests based on the current cash positions and configured target balances. Despite the automation, it is strongly recommended to have a layer of payment approvals in place for these automatically created transfers. This ensures that all payments align with internal treasury policies and that liquidity is maintained as required across all participating accounts.

6.2 Configuration

There are relatively few customizing activities required for cash management in SAP S/4HANA today, as much of the setup now comes from the SAP Fiori applications and master data configuration. However, there are still a few essential elements that need to be created in the system configuration. Key objects such as planning levels, liquidity items, and planning groups must be defined through traditional configuration steps, along with the activation of data sources. Despite this, most cash management functionalities—particularly related to data assignment and behavior—are now managed directly through master data setup in SAP Fiori, reducing the need for extensive backend customization.

Especially as most of the traditional configuration is still tied to the old cash management setup, such as Transaction FF7AN, many of the customizing activities are no longer relevant when using the new SAP Fiori–based applications. With the shift to modern tools and master data–driven processes, much of what used to be handled through backend configuration is now managed through SAP Fiori apps.

SAP is increasingly shifting more functionalities and configurations directly to SAP Fiori applications, streamlining the user experience. However, there are still key configurations that need to be set up to determine how information is displayed, loaded, and updated within cash management. The most crucial configurations involve defining how data is categorized, including planning levels, planning groups, liquidity items, and the sources for memo record determination and behavior. It's essential to specify these elements carefully to ensure accurate and meaningful reporting. Note that here we will describe only the most important configuration steps. Although there might be additional steps required for other reporting scenarios, as cash management includes a broader range of functionalities, these activities should be sufficient for you to effectively use cash management for payment purposes.

6.2.1 Planning Levels

Planning levels in cash management represent the different data sources that feed into your cash position and liquidity forecasts. These levels are essential for structuring the information displayed in your reports and applications. You need to create dedicated planning levels for key sources such as accounts receivable (AR) items, accounts payable (AP) items, bank balances, intraday bank statements, and financial transactions from the treasury, along with any other relevant data you want to include. When designing planning levels, the key question to answer is: What kind of information do you want to see as a separate line in your cash planning and reporting? The planning level acts as a label that organizes and distinguishes different types of financial data for accurate and meaningful analysis.

The next step in the configuration involves assigning planning levels to specific types of items to further refine your cash visibility. For example, you should assign a dedicated planning level for blocked items, allowing you to separate invoices that are blocked for payment from regular AR and AP invoices. It is also important to define a planning level for logistics-related data, such as purchase orders and sales orders, so that this information can be reflected in your cash forecasts. You may also need to set planning levels for other relevant activities or specific items in the AR and AP subledgers, ensuring that all expected inflows and outflows are properly categorized and reported in your cash management setup.

You can find planning level creation (see Figure 6.47) under **Financial Supply Chain Management** • **Cash and Liquidity Management** • **Planning Levels and Planning Groups** • **Define Planning Levels**.

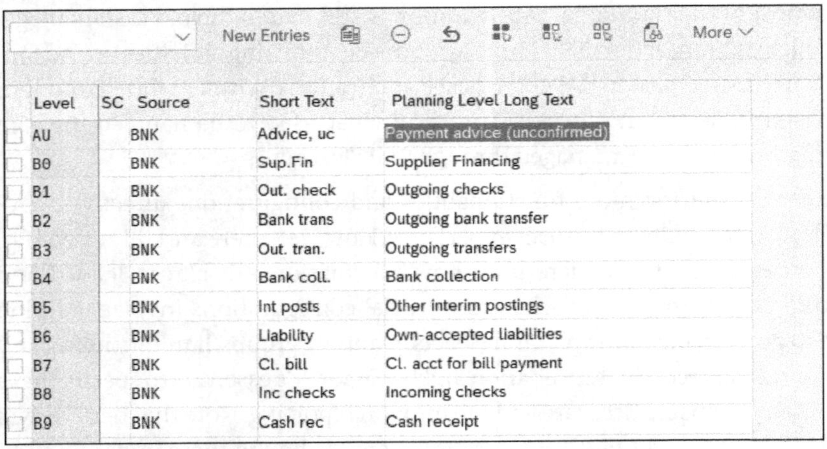

Level	SC Source	Short Text	Planning Level Long Text
AU	BNK	Advice, uc	Payment advice (unconfirmed)
B0	BNK	Sup.Fin	Supplier Financing
B1	BNK	Out. check	Outgoing checks
B2	BNK	Bank trans	Outgoing bank transfer
B3	BNK	Out. tran.	Outgoing transfers
B4	BNK	Bank coll.	Bank collection
B5	BNK	Int posts	Other interim postings
B6	BNK	Liability	Own-accepted liabilities
B7	BNK	Cl. bill	Cl. acct for bill payment
B8	BNK	Inc checks	Incoming checks
B9	BNK	Cash rec	Cash receipt

Figure 6.47 Define Planning Levels

To create new planning levels in cash management, simply click **New Entries**. You will need to define a two-character planning level identifier and specify the source, whether it's from subledgers (such as AP/AR) or from the bank. Be sure to enter a short-text and a long-text description for clarity. Once you have provided all the necessary information, click **Save**, and your new planning level will be created and ready for use.

Typically, the configuration of cash management in SAP S/4HANA comes with a set of predefined planning levels that are already aligned with the most common financial processes, such as AR, AP, and bank data. In most cases, you don't need to make significant changes to these settings, as they are designed to support standard reporting and cash visibility requirements. However, if your organization has specific reporting needs or wants to track certain information—like blocked invoices or custom financial flows—separately, then you can create or adjust planning levels accordingly. Otherwise, the default configuration provided by SAP S/4HANA is usually sufficient. In the following sections, we'll look at assigning planning levels to payment requests and payment blocks.

Assign Planning Levels to Payment Requests

Planning levels for invoices are automatically derived based on the planning group assigned to the vendor for which the invoice is posted. However, for payment requests, as you don't have this vendor-specific information readily available, you need to map the payment requests and their associated payment methods to the appropriate planning levels manually. This mapping ensures that even payment requests, which lack direct vendor planning group information, are correctly categorized and included in your cash planning and forecasting. You can perform this configuration (see Figure 6.48) by navigating to the following path in the Transaction SPRO configuration: **Financial Supply Chain Management** • **Cash and Liquidity Management** • **Cash Management** • **Planning Levels in Payment Request**.

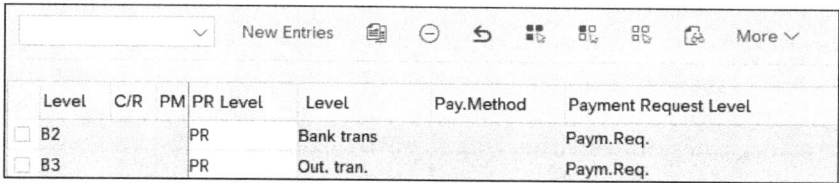

Figure 6.48 Define Planning Levels for Payment Requests

Here, you simply click **New Entries**, specify the planning level that should be automatically assigned based on the payment run, and then map the payment method to the corresponding planning level for a payment request. You can create and map separate planning levels specifically for payment requests tied to payment methods, ensuring that whenever you create a payment request using a specific payment method, it is linked with the correct planning level in cash management. However, our recommendation would be to keep the setup as simple as possible by using just one planning level for all payment requests (in this example, PR). This is because payment requests are typically short-lived in the Cash Flow Analyzer app; the expectation is that they are quickly followed by an automatic payment run, and the associated cash flows are then updated and transferred to the planning level dedicated to actual payments.

Define Planning Levels for Payment Blocks

Normally, when you analyze payments, you would only want to see invoices that you actually intend to pay in your next payment run. Invoices that are blocked for payment should therefore appear in cash management as a separate line and category, clearly showing that they're not expected to be paid right away. SAP supports this by allowing you to maintain separate planning levels for all invoices that have payment blocks (see Figure 6.49). You can set this up in the configuration under the following Transaction SPRO path: **Financial Supply Chain Management** • **Cash and Liquidity Management** • **Cash Management** • **Maintain Blocked Levels**.

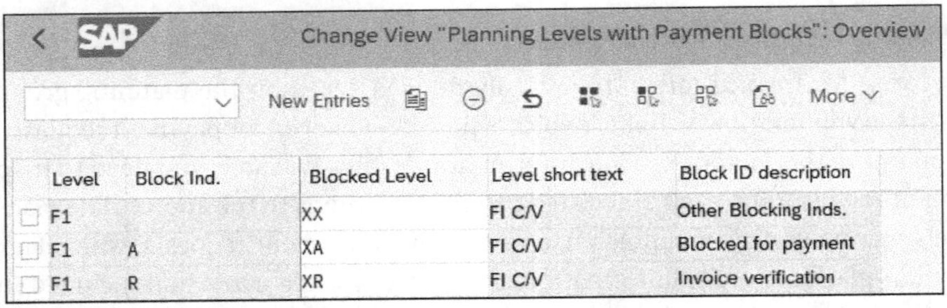

Level	Block Ind.	Blocked Level	Level short text	Block ID description
F1		XX	FI C/V	Other Blocking Inds.
F1	A	XA	FI C/V	Blocked for payment
F1	R	XR	FI C/V	Invoice verification

Figure 6.49 Define Planning Level for Payment Blocks

In this activity, to create and map payment blocks to planning levels, click **New Entries**, specify the original planning level that should be used, and then map the payment block to the blocked item. Here you can see the default values from SAP that appear automatically when you create these entries, but it's recommended to map all the payment blocks to a single planning level. This ensures that all the activities related to blocked items are displayed as a separate line in cash management, making it easier to identify and manage these items within your financial planning.

6.2.2 Planning Groups

Planning groups are typically used only for AR and AP items. They are maintained directly in the customer and vendor master data and serve as an additional way to categorize and analyze financial transactions within cash management. In the Cash Flow Analyzer app and other related applications, AR and AP data is aggregated based on the planning levels assigned to those items. However, by using planning groups, you can further segment and analyze your inflows and outflows according to the specific groupings you've defined—such as by customer category, region, or vendor type. When designing planning groups, it is important to consider how you want to classify your customers and vendors. For example, you can group them based on risk, strategic relationship, importance to your business, or the likelihood that they will pay on time. These groupings provide valuable insight for cash forecasting and liquidity analysis.

You can find planning groups (see Figure 6.50) under **Financial Supply Chain Management** • **Cash and Liquidity Management** • **Planning Levels and Planning Groups** • **Define Planning Groups**.

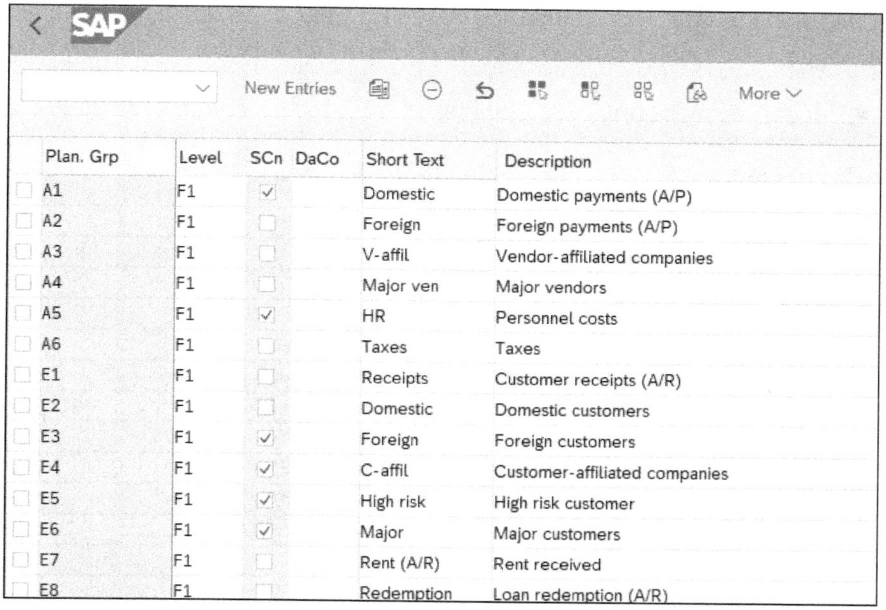

Plan. Grp	Level	SCn	DaCo	Short Text	Description
A1	F1	✓		Domestic	Domestic payments (A/P)
A2	F1			Foreign	Foreign payments (A/P)
A3	F1			V-affil	Vendor-affiliated companies
A4	F1			Major ven	Major vendors
A5	F1	✓		HR	Personnel costs
A6	F1			Taxes	Taxes
E1	F1			Receipts	Customer receipts (A/R)
E2	F1			Domestic	Domestic customers
E3	F1	✓		Foreign	Foreign customers
E4	F1	✓		C-affil	Customer-affiliated companies
E5	F1	✓		High risk	High risk customer
E6	F1	✓		Major	Major customers
E7	F1			Rent (A/R)	Rent received
E8	F1			Redemption	Loan redemption (A/R)

Figure 6.50 Define Planning Groups

SAP S/4HANA comes with many predefined values, but most people want to define their own. To do this, you need to define your planning by creating planning groups. First, click **New Entries** and enter a two-digit code for your planning group. Assign this planning group to the appropriate planning level, then provide both a short text and a detailed description to make it clear and understandable. In addition, you can determine whether you want the planning date on the invoice level to be modifiable. If you'd like to allow moving planning dates for the invoices, simply select the select the **SCn** checkbox.

6.2.3 Liquidity Items

If you want to use liquidity item assignment for analyzing cash flows through liquidity item hierarchies, you first need to create liquidity items in the configuration. These items serve as categories that classify cash flow data as either inflows or outflows. Once created, liquidity items need to be maintained through assignment rules—such as linking them to specific G/L accounts, queries, or other derivation logic. This setup determines how incoming data is categorized. The actual derivation of liquidity items takes place when data is loaded into the system using Transaction FCLM_FLOW_BUILDER_2 (Flow Builder), which interprets the predefined rules and assigns the correct liquidity items accordingly. This enables structured, detailed analysis of cash movements in liquidity reporting.

You can find the liquidity item configuration (see Figure 6.51) under **Financial Supply Chain Management • Cash and Liquidity Management • Liquidity Items • Edit Liquidity Items**.

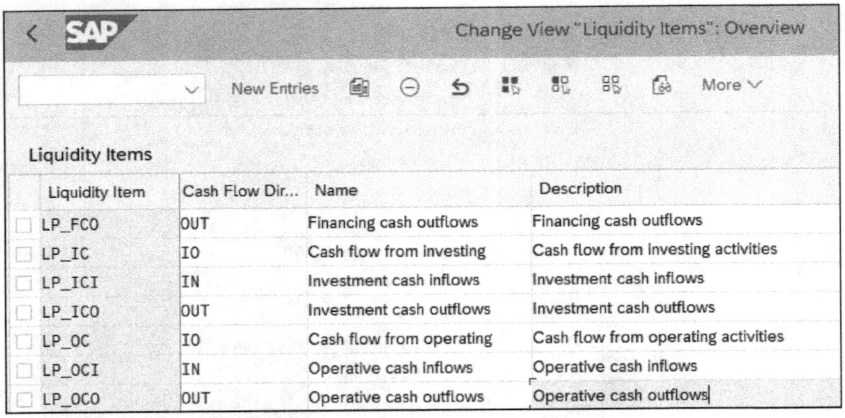
Change View "Liquidity Items": Overview
New Entries
More
Liquidity Items
Liquidity Item | Cash Flow Dir... | Name | Description
LP_FCO | OUT | Financing cash outflows | Financing cash outflows
LP_IC | IO | Cash flow from investing | Cash flow from investing activities
LP_ICI | IN | Investment cash inflows | Investment cash inflows
LP_ICO | OUT | Investment cash outflows | Investment cash outflows
LP_OC | IO | Cash flow from operating | Cash flow from operating activities
LP_OCI | IN | Operative cash inflows | Operative cash inflows
LP_OCO | OUT | Operative cash outflows | Operative cash outflows

Figure 6.51 Define Liquidity Items

Building liquidity items is a straightforward process: Click **New Entries**, provide your liquidity item, and then determine the cash flow direction—incoming or outgoing. After that, enter the name and description for the liquidity item. Once you're done, click **Save** to finalize your new liquidity item.

6.2.4 Memo Records

Before you can use Manage Memo Records 2.0 app, there are a couple of configuration activities that need to be performed in advance. These configurations are essential to ensure that memo records can be properly created and updated within cash management. They include setting up the relevant authorization roles, defining the cash management planning levels, and ensuring that the memo record types and categories are correctly configured. Only after these preliminary configurations are in place will you be able to fully utilize the functionality of Manage Memo Records 2.0 in your cash management processes.

Define Memo Records

In this step, you define the memo record types that are used for manual planning (see Figure 6.52). Each memo record type determines how planned memo records are manually entered and managed.

For each memo record type, you can specify the planning level it is linked to, the archiving category for expired records, whether memo records should expire automatically, which fields are displayed, and whether they are mandatory or optional. In addition, you can assign a mnemonic name that appears when memo records are created.

You can find the configuration for this step in **Financial Supply Chain Management • Cash and Liquidity Management • Cash Management • Memo Records 2.0 • Define Memo Record Types**.

Memo Record Types					
Memo Tp	Plan. Lvl	Arch.cat.	Auto Expir	Memo Record Type Des	Memo Record Category
AB	AB	A	✓	Confirmed advices	Payment Advice
AU	AU	A		Unconfirmed advices	
CL	CL	A	✓	Cash concentration	
DE	DE	A		Loan revenue	Planned Item
DI	DI	A		General planning	Planned Item
HR	AB	A	✓	Salary Transfer	Actual Cash Memo
ID	ID	A	✓	Intraday Statement	Actual Cash Memo

Figure 6.52 Define Memo Records Type

To create new memo record types, click **New Entries**. Specify the memo record type, which must be two characters long; this is the value you will select from the dropdown list when creating a new memo record in the app. Then specify the planning level, which determines how the memo records will be displayed in cash management. Define the archiving category to determine where expired memo records will be stored, and set whether the memo record will autoexpire. In addition, provide a description and select the memo record category. (The memo record categories are predefined by SAP. To use the new cash flow reconciliation model and the Reconcile Cash Flows—Intraday app, you must define at least one planning type with the memo record category **Residual Forecast**. This planning type is dedicated to residual flows. During cash flow reconciliation, residual flows help track unreconciled forecasted flows that you intend to include in the next reconciliation round. These residual flows are created as memo records using the planning type assigned to the **Residual Forecast** memo record category.) Once all the required information is entered, save your entries.

Define Number Range for Memo Records

Before you can create memo records using the Memo Records 2.0 app, you need to specify the number ranges (see Figure 6.53) that will be used in the background to generate these memo records. This configuration can be done by navigating to the following path: **Financial Supply Chain Management • Cash and Liquidity Management • Cash Management • Memo Records 2.0 • Define Number Range for Memo Records**.

To define a number range for memo records, proceed as follows. First, in this activity, click the **Change Intervals** button. In the **Number Range No.** field, enter "01"; this is the only number that can be used for memo records as no other numbers are permitted. Next, specify a start number and an end number for the range. Ensure that the **External** indicator remains unchecked. Once you've entered all the necessary information, save your changes.

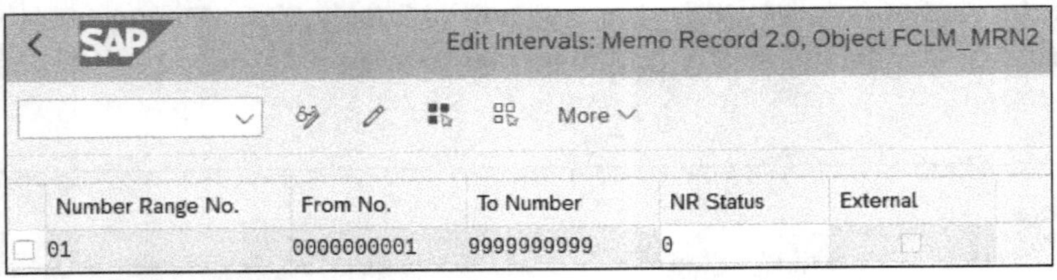

6 SAP S/4HANA Finance for Cash Management

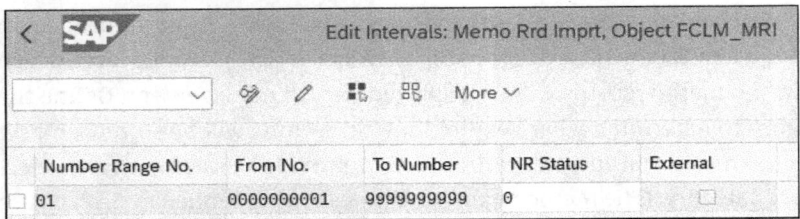

Figure 6.53 Define Number Range for Memo Records

Define Number Range for Memo Records Imports

There is a dedicated number range that is used for all memo records imported from a file. You must also specify this number range to successfully upload memo records (see Figure 6.54). You can find this configuration under Transaction SPRO following menu path **Financial Supply Chain Management** • **Cash and Liquidity Management** • **Cash Management** • **Memo Records 2.0** • **Define Number Range for Memo Records Imports**.

Figure 6.54 Define Number Range for Memo Records Import

You use this activity to define a number range for memo record imports, which is required before using the Import Memo Records 2.0 app. To define a number range for memo record imports, start by selecting the **Change Intervals** button. In the **Number Range No.** field, enter "01"; this is the only number that can be used for memo record imports as no other numbers are allowed. Next, specify a start number and an end number for the range. Leave the **External** indicator unchecked. Once all entries have been made, save your changes.

Manage Field Status Groups for Memo Records

In this configuration activity, you determine which fields will be available during the creation of a memo record. In addition, you can decide the status of each field, specifying whether it should be mandatory, hidden, or optional. Depending on the selected manage record type, you can customize the available fields to ensure that only the relevant information is displayed during the memo record creation process. This flexibility allows you to tailor the memo record entry to your organization's specific requirements and workflows.

6.2 Configuration

You can find this configuration activity under the following menu path: **Financial Supply Chain Management** • **Cash and Liquidity Management** • **Cash Management** • **Memo Records 2.0** • **Manage Field Status Groups for Memo Records**.

Once you open this activity, you need to create a field status first (see Figure 6.55). Click **New Entries** and provide text values for **Field Status Group** and **Field Status Group Description**.

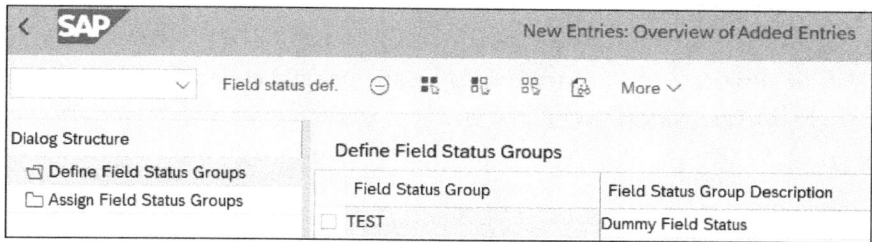

Figure 6.55 Define Field Status Groups

Once the field status is created, you can select it and click **Field status def**. The system will display the types of fields available. Click each type, and you will see a list of fields and set their statuses (see Figure 6.56). You can choose **Suppress** when a field should not be available, **Req. Entry** when a field is mandatory, or **Opt. entry** when it is optional.

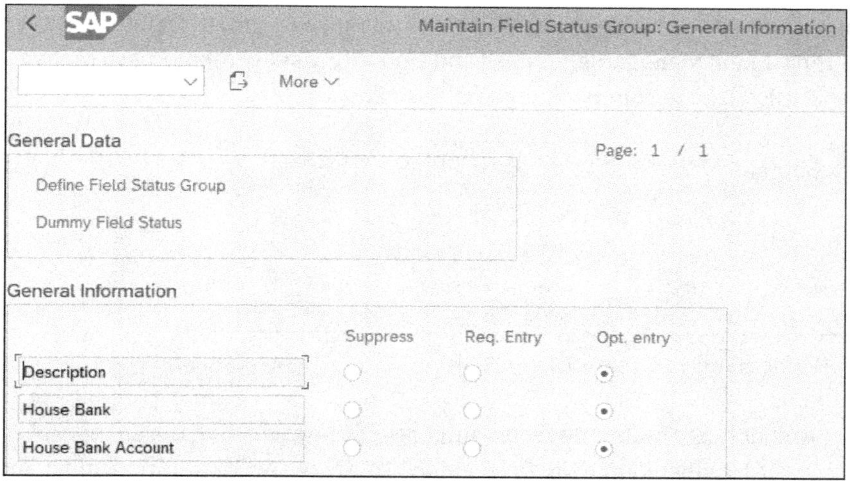

Figure 6.56 Maintain Field Status Group

Once you have maintained the field statuses, you can assign them to memo record types (see Figure 6.57).

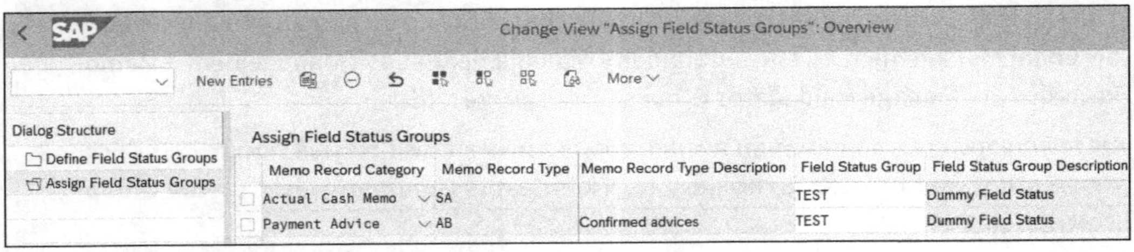

6 SAP S/4HANA Finance for Cash Management

![Assign Field Status Groups screen]

Figure 6.57 Assign Field Status Groups

Here, click **New Entries**, select the **Memo Record Category** and **Memo Record Type**, and assign them to the **Field Status Group** that you created earlier.

6.2.5 Cash Pools

Before you can use the Manage Cash Pools 2.0 app, there are certain configuration activities that need to be completed. These configurations are essential for setting up the cash pool structure and defining cash pool hierarchies.

Define Number Range for Cash Pools

First, you need to create a number range for the creation of specific cash pools (see Figure 6.58). This ensures that each cash pool has a unique identifier for tracking and management purposes. You can define this number range in the configuration under **Financial Supply Chain Management • Cash and Liquidity Management • Cash Management • Cash Pools • Define Number Range for Cash Pools**.

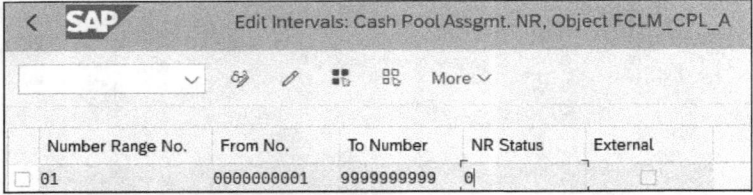

Figure 6.58 Define Number Range for Cash Pools

To define a number range for memo record imports, start by selecting the **Change Intervals** button. In the **Number Range No.** field, enter "01". Next, specify a start number and an end number for the range. This number is not super important. Leave the **External** indicator unchecked. Once all entries have been made, save your changes.

Define Authorization Groups for Cash Pools

In this activity, you define authorization groups for time-dependent cash pools. These authorization groups control access to cash pools in the Manage Cash Pools (Version 2) app. It's important to set this up before you start working with this type of cash pool as assigning an authorization group restricts access to specific users. Only users who have the corresponding authorization can create, edit, or view these cash pools.

Creation of authorization groups is available under **Financial Chain Management** • **Cash and Liquidity Management** • **Cash Management** • **Cash Pools** • **Define Authorization Groups for Cash Pools**.

Define Authorization Group for Cash Pool	
Auth Group	Description
DB	Authorization Group for Deutsche Bank

Figure 6.59 Define Authorization Group for Cash Pool

To create a new authorization group, click **New Entries** and then provide a value for **Auth Group** and a **Description**, as shown in Figure 6.59. Click **Save**. Once created, the authorization group will be available in the Manage Cash Pools (2.0) app.

6.2.6 Define Source Application Accounting

In the **Define Source Application Accounting** activity, you can perform additional configuration for SAP S/4HANA Finance in One Exposure from Operations. For company code entries, if the **Company Code** field is empty, these entries are treated as default configurations. However, entries with specific company codes take higher priority than the default ones. For **Accounting Scope**, there are several options to choose from: **Basic Cash**, which helps you manage daily cash operations and provides information for your cash position and liquidity forecast; **Advanced Cash: Cash Position Only**, which generates only cash position cash flows from accounting documents without document chain tracing; **Advanced Cash: Liquidity Analysis Only**, which generates only liquidity analysis cash flows using document chain tracing; and **Advanced Cash: Cash Position & Liquidity Analysis**, which generates both types of cash flows.

For **Liquidity Analysis Update Mode**, you can select different modes based on your requirements. The **Not Relevant** mode is reserved for the cash position only scope. The **Deferred—Delta Table Update** mode allows the system to monitor FI postings and insert newly posted or changed documents into the staging table without triggering the Flow Builder Plus technical job automatically, requiring you to set up your own job to process the entries. Keep in mind that in busy systems, the data volume in the staging table can grow quickly. The **Close to Real Time** mode also monitors FI postings and inserts new or changed documents into the staging table, but it automatically triggers the Flow Builder Plus technical job to process these entries. Finally, the **Deferred—No Delta Table Update** mode does not monitor FI postings and does not insert documents into the staging table. In this case, the *find missing documents* function in the Flow Builder: Utility Toolset program does not support company code set in this mode.

This configuration activity (see Figure 6.60) can be found at **Financial Chain Management • Cash and Liquidity Management • Cash Management • Data Setup • Define Liquidity Item Derivation for Flow Builder**.

Source Application Accounting

CoCd	Accounting Scope	Liquidity Analysis Update Mode
1010	Cash Position & Liquidity Analysis	Close to Real Time

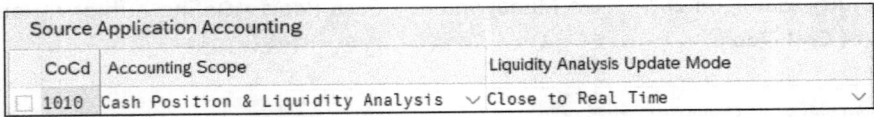

Figure 6.60 Define Source Application Accounting

To create an entry, click **New Entries**, enter the **Company Code**, and select the appropriate options from the dropdown lists for **Accounting Scope** and **Liquidity Analysis Update Mode**.

6.2.7 Activate Source Application

To populate data into cash management, you need to define and activate data sources (see Figure 6.61). If you want to derive data from various sources, such as bank statements, accounting, logistics, and others, then you must activate these data sources individually or in bulk per company code. Once activated, the data will be automatically loaded as part of the Flow Builder process. This ensures that relevant data from the selected sources is incorporated into your cash management system, allowing for accurate and up-to-date liquidity analysis and reporting.

Activate Source Applications

Src. App Grp	CoCd	Company Name	Active
Accounting	1010	Company Code 1010	Active
Bank Statement	1010	Company Code 1010	Active
One Exposure	1010	Company Code 1010	Active
Materials Manage...	1010	Company Code 1010	Active
Lease and Contra...	1010	Company Code 1010	Active
Sales And Distri...	1010	Company Code 1010	Active
Treasury and Ris...	1010	Company Code 1010	Active

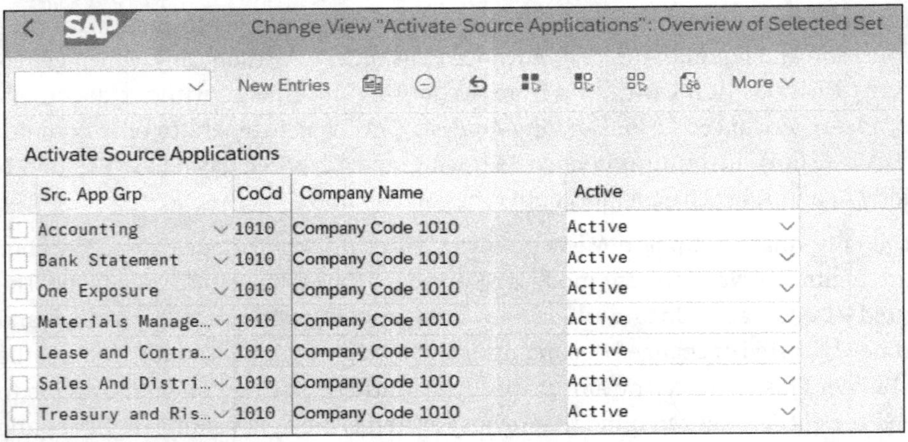

Figure 6.61 Activate Individual Sources for Application

Under **Financial Supply Chain Management • Cash and Liquidity Management • Data Setup • Activate Individual Source Applications**, you can specify which data sources you would like to load into cash management.

Activation of source applications is rather simple. You just need to select the source application, choose the company code, and then specify whether the source should be active or inactive. Once you set the source as **Active**, you will be able to start loading.

6.3 Summary

In this chapter, we took a deep dive into the most important functionalities of cash management in SAP, particularly focusing on the processes and tools that support payments and liquidity management. Cash management is vast, with many additional capabilities that go far beyond what we presented here. We have focused on key applications and configurations that give you a comprehensive and practical foundation for managing your cash balances and forecasting your short-term liquidity.

We discussed the concept of One Exposure from Operations, how to ensure data is properly loaded and updated within cash management, and the most important configuration activities you'll need to perform to make these applications work effectively.

We also covered the critical role of bank statements in providing accurate and timely data for your cash management processes. You learned how to load balances for bank accounts when statements aren't available, ensuring that you always have up-to-date information for your decision-making. In addition, we explored the key applications in cash management, such as the Cash Position, Cash Flow Analyzer, and Short-Term Cash Position apps, which provide you with powerful insights and tools to manage your cash and liquidity more efficiently.

Remember, the activities and applications we discussed in this chapter are more than enough to support cash management for payments-related purposes. However, it's important to note that cash management has a much broader scope and offers many more functionalities supporting your historic data analysis, actuals, snapshots and long-term liquidity planning. After reading this chapter, you should have a solid understanding of how to use cash management to support your payments processes and keep your cash flow under control.

Chapter 7
SAP Bank Communication Management

If you want to centralize payments, you need a single place where you can group and manage all payment activities effectively. In SAP, this role is served by SAP Bank Communication Management, which consolidates payment processes, ensuring seamless tracking, approval, and execution of payments from one central hub.

This chapter focuses on SAP Bank Communication Management and its key functionalities. We will explore how SAP Bank Communication Management can be used for monitoring payments, tracking payment statuses, managing payment approvals, and handling batching processes. You will also learn how to set up SAP Bank Communication Management from the configuration level to ensure that it supports your specific payment processing needs. As mentioned in Chapter 3, there are two ways to set up and manage payment approvals: one option involves using the setup and payment approvers specified in Bank Account Management (BAM) in SAP S/4HANA or in the SAP Bank Communication Management configuration, while the other option involves additional configuration covered here in this chapter. Regardless of the option chosen, batching rules and payment approvals are handled using the same applications, ensuring a streamlined and consistent process.

The purpose of SAP Bank Communication Management is to streamline payment processes and provide a structured framework for managing outgoing payments by batching payments and giving payment approvals in line with corporate requirements. SAP Bank Communication Management ensures that approvals are granted only after a liquidity check confirms that there are sufficient funds to execute the payment. This approach ensures both compliance with internal controls and efficient management of payment workflows, supporting a secure and reliable payment execution process.

Once you understand how the generation of payment files and the approval of payments are handled by SAP Bank Communication Management, the next step is to see how all of this comes together in practice. In this chapter, we will explain how you can use SAP Bank Communication Management to manage these processes effectively.

7 SAP Bank Communication Management

7.1 Functions and Processes

As explained in Chapter 1, SAP Bank Communication Management was originally developed and integrated within the SAP ERP system, which is why many of its functionalities are still accessible through traditional SAP GUI transactions. However, in SAP S/4HANA, SAP Bank Communication Management is also supported by a set of dedicated SAP Fiori applications that enhance usability and streamline payment workflows. As we discussed in Chapter 1 and Chapter 4, SAP Bank Communication Management serves as a complementary system for SAP S/4HANA Finance for advanced payment management. Whereas advanced payment management is primarily focused on the incoming side of payments—handling the receipt, transformation, and processing of inbound transactions— SAP Bank Communication Management is responsible for the outgoing side. This includes managing payment approvals, tracking payment statuses, and handling the overall execution of outgoing payment flows. This section will explain how SAP Bank Communication Management can be utilized within SAP S/4HANA to efficiently manage and control outbound and inbound payment processes.

We will focus on how to manage payments from a batching perspective using SAP Bank Communication Management. Specifically, we will cover batching rules, payment approvals, and the processes for uploading and updating payment statuses. You will also learn how to monitor the status of your payments and stay on top of any updates. Importantly, these steps are the same whether you use the configuration options available in SAP Bank Communication Management or derive payment approval rules from the BAM setup. Because SAP Bank Communication Management can also be used in SAP ERP systems, we will include examples and references to both traditional SAP GUI transactions and the new SAP Fiori applications where relevant. By following these steps, you ensure a consistent and streamlined payment processing experience, regardless of your chosen configuration.

> [!]
> **SAP Bank Communication Management in SAP GUI**
>
> Note that SAP Bank Communication Management primarily runs in the background, so we will present SAP GUI transactions in this section. Many of these transactions have already been "Fiori-fied," meaning they are available through SAP Fiori apps or in SAP S/4HANA Cloud Public Edition. Although some functionality is still accessed via SAP GUI screens, even when launched from within SAP Fiori, SAP continues to transition these features into fully native SAP Fiori apps. As a result, this landscape is likely to evolve further with each new release.

In SAP S/4HANA, a significant volume of payments are executed across various areas, as outlined in Chapter 1. These include payments for supplier invoices, treasury-related payments such as cash concentration and bank-to-bank transfers, and other types of financial transactions. The key objective of managing all these payments through the system in a standardized and automated way is to establish a single point of control and visibility. By doing so, you ensure consistency, reduce manual errors, and improve process efficiency. The Monitor Payments app (F2388) in SAP S/4HANA plays a crucial role in this as it allows users to track the status of all payments across different sources and processes in real time, ensuring transparency and better control over cash flows.

Typically, you can simply select the relevant parameters—such as **Batch Number**, **Paying Company Code**, payment **Run ID**, or **Run Date** (when the payment was executed)—and run the report. Once executed (see Figure 7.1), the system will display the statuses of all payments matching those criteria. This provides a clear and centralized overview of payment processing, helping users quickly identify which payments have been approved, are still pending, or encountered errors.

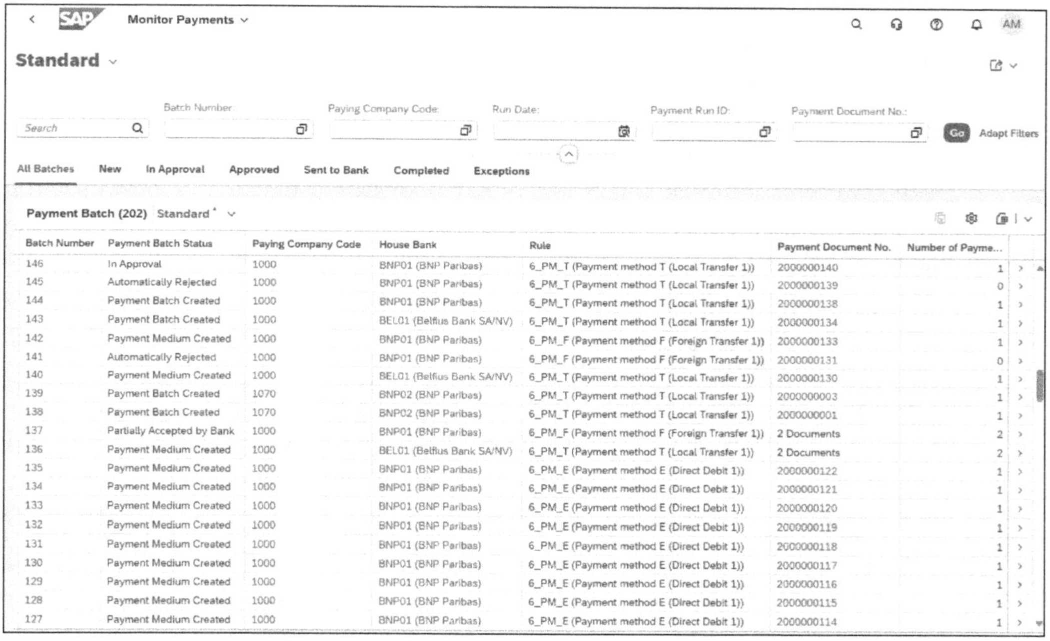

Figure 7.1 Monitor Payments

In the Monitor Payments app, you can see the **Batch Number** along with the approval status (**Payment Batch Status**). All batches are shown together: those that are newly created, those for which the payment medium has already been generated (indicating approval), and those for which the payment batch has just been created. You can view the status of each batch by clicking each of the tabs, including **In Approval**, **Approved**, **Sent to Bank**, and **Completed**, the latter of which usually implies that a corresponding bank statement has also been received. The report also has a tab for **Exceptions**—payments that were rejected either by the system or during the user approval process for various reasons.

As is typical for many SAP Fiori applications, you will notice blue-colored amounts or indicators throughout the interface. These elements act as interactive links and can be clicked to navigate to related details or other applications. For instance, you can click links to access additional information about the payment batch, view the associated company code or house bank, or directly display the related payment document. This interactive design enhances usability and enables seamless navigation across payment-related data within the system.

7.1.1 Payment Approvals

As explained in Chapter 1, in today's complex payments environment, the risks of fraud and human error are persistent challenges that can have serious financial and reputational consequences. Within SAP S/4HANA, the implementation of a robust payment approval process is a critical control mechanism to mitigate these risks. Without proper approval procedures, organizations become vulnerable to fraudulent activities such as unauthorized payments or internal embezzlement. In addition, manual errors—such as incorrect payment amounts or transactions directed to the wrong recipients—can result in significant financial losses and strained relationships with vendors or business partners. Therefore, enforcing structured and consistent payment approvals within SAP S/4HANA is essential to safeguard an organization's financial integrity and ensure that each payment is authorized, accurate, and aligned with internal controls.

Designing a payment approval matrix requires thoughtful planning and consideration of multiple factors. These include the type of payment (e.g., vendor invoices, employee reimbursements, or intercompany transfers), the urgency of the transaction, the bank involved, and especially the amount of the payment. Higher-value transactions should be subject to stricter oversight and approval by more senior personnel to minimize risk. Time-sensitive payments may require an expedited approval path, while internal payments between bank accounts might be preapproved based on predefined templates.

Although SAP Bank Communication Management offers the technical infrastructure to manage a variety of approval workflows, its foundation lies in clearly defining functional requirements. This ensures that the system configuration reflects the organization's policies and provides a tailored approach to payment approvals. By aligning functional needs with SAP Bank Communication Management capabilities, businesses can automate and streamline payment approvals without compromising on control or compliance.

SAP Bank Communication Management supports a wide range of configuration options to accommodate various approval scenarios. Some payments—such as internal bank transfers or those initiated via approved templates—may qualify for auto-approval. Others may require single-level approval or adhere to stricter principles such as dual control (following the four-eyes principle). In more complex environments, approvals can involve up to four separate approvers. Approval rules in SAP Bank Communication Management can be defined based on multiple attributes, including payment method, urgency, and transaction amount. For example, it is common for high-value payments to require authorization from senior finance or treasury stakeholders.

7.1 Functions and Processes

In this section, we will explore how payment approvals are structured and processed within SAP Bank Communication Management. We will demonstrate the approval process using both classic SAP GUI Transaction BNK_APP and the Approve Bank Payments (Version 2) app (F0673A; the Approve Bank Payments app [F0673] was used in releases lower than SAP S/4HANA 2023), showcasing how each tool supports effective payment governance within the system.

Once you open the Approve Bank Payments (Version 2) app, you can see all payments awaiting your approval (see Figure 7.2).

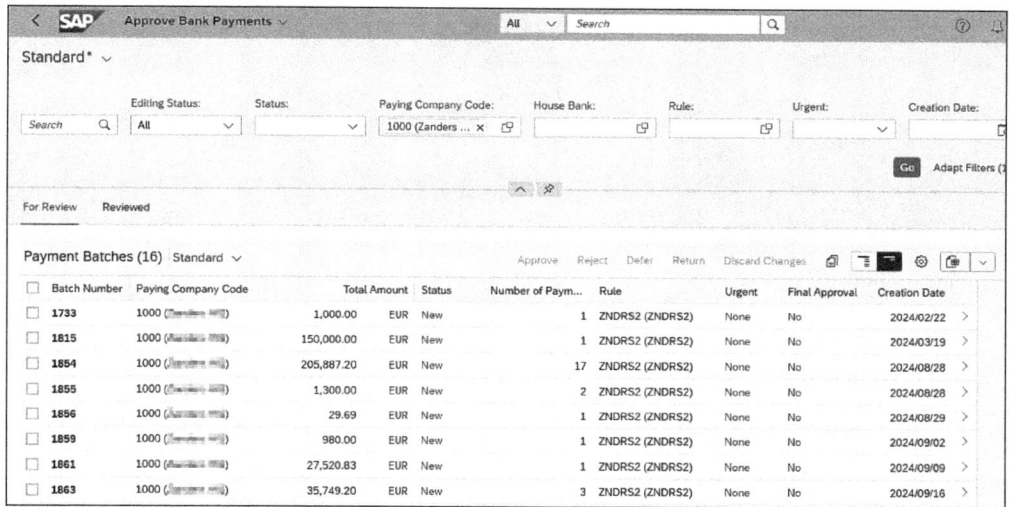

Figure 7.2 Approve Bank Payments App

Similarly, you can approve bank payments using Transaction BNK_APP (see Figure 7.3). This transaction has the same functionalities as the SAP Fiori app, and you should only use it when you are not using an SAP S/4HANA system. Otherwise, we recommend using the newest SAP Fiori apps.

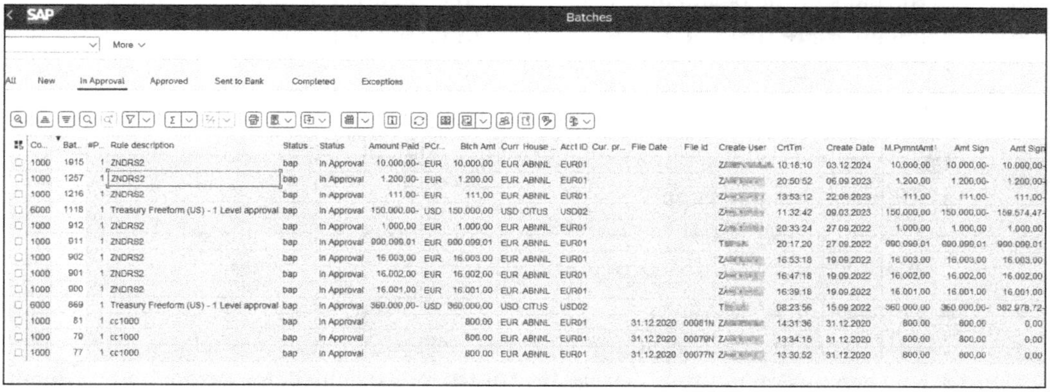

Figure 7.3 Transaction BNK_APP

485

7 SAP Bank Communication Management

When approving payments, you simply need to select the relevant payment to review its details. Depending on the system version, configuration, and application in use, you may also have the option to defer the payment; this means adjusting the payment terms or modifying the value date.

To approve payment in the Approve Bank Payments (Version 2) app (see Figure 7.4), simply review the relevant data and click the appropriate action button based on your checks (**Approve and Submit**, **Approve**, **Reject**, or **Defer**).

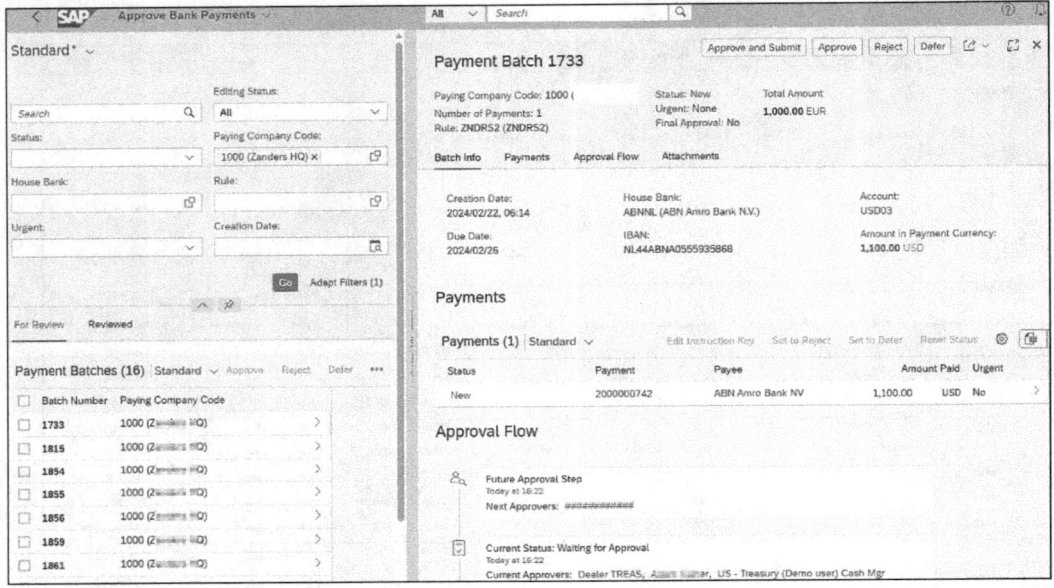

Figure 7.4 Approve Bank Payment

If you approve a payment, depending on your setup and the number of required approvers defined for that payment, it will either be forwarded to the next approver or marked as fully approved, triggering the generation of the payment file and its transfer to the bank. If the payment is rejected, the related document will need to be reversed manually. Note that SAP S/4HANA does not reverse a payment automatically unless BAdI `BNK_BADI_PAYM_ALRT` is implemented. After rejection, the data must be corrected and the payment recreated, depending on the specific reason for the rejection.

To check the status of batches and payments, you can always use the following:

- **Monitor Payments app**
  You can navigate to the app and check the various statuses (like whether payments are processed, have failed, or are pending).

- **Transaction BNK_MONI**
  This transaction allows you to view detailed logs and statuses related to payments and banking operations. You can drill down into specific transactions and payments here.

For setup purposes, payment approvals can be set up either through BAM or through the setup and configuration of SAP Bank Communication Management. The BAM master data is easier to maintain (you assign users directly to the bank accounts), but it only allows payment approvals based solely on the payment amount; it has limited flexibility as it only supports approvals directly tied to the amount itself. On the other hand, the SAP Bank Communication Management configuration provides more complex and flexible approval rules, allowing for criteria such as payment methods, payment types, or other specific conditions. Although SAP Bank Communication Management offers greater flexibility, it requires more effort for the maintenance of users and approvers, making the setup process more complicated. In summary, BAM is the preferred option for simpler approval processes, while SAP Bank Communication Management is better suited for more intricate approval rules, though it demands more configuration effort.

> **Two-Factor Authentication**
>
> Two-factor authentication (2FA) for approving payments in cloud-based SAP solutions is supported and can be enabled easily with minimal setup effort. This additional security layer enhances protection by requiring a second authentication factor during approval processes. For SAP's cloud solutions, enabling 2FA—specifically, a time-based, one-time password (TOTP)—is straightforward and user-friendly. In contrast, for on-premise solutions, enabling 2FA may require additional custom developments or third-party integration to achieve similar functionality.

7.1.2 Payment Run

Before any payment file can be generated or before any processing can happen in SAP Bank Communication Management, the first essential step is executing a payment run. In fact, executing the payment run is the starting point of every payment process in the system. In SAP S/4HANA, you typically do this using the Manage Automatic Payments app (F0770). This app allows you to create and process payment proposals and to execute payment runs that initiate the entire workflow of payments, including the steps that follow in SAP Bank Communication Management.

When you open the Manage Automatic Payments app, you will see an overview of all the payments made in your system, displayed based on the selection parameters you choose—most commonly, by date. To execute a new payment, click **Create Parameter**. The system will open a popup (see Figure 7.5), in which you provide an **Identification** and specify the **Run Date** on which you would like to execute the payment. This step is the entry point for managing and processing payments within SAP S/4HANA, setting the stage for the subsequent steps in SAP Bank Communication Management.

Click **Create**. Next, you will need to enter parameters for your payment (see Figure 7.6). Step 1 involves providing the date on which you would like to post the payment documents. This date determines the cutoff for document validation: Invoices posted before this date will be revalidated and included in the payment process. The system will automatically move to the next scheduled payment date according to your payment plan if applicable. During this step, you also need to specify the payment method you are executing, and you have the option to provide additional selection criteria, such as vendor number, supplier number, customer number for direct payments, and even document number if needed. In most cases, however, you'll want to execute the payment run for all documents with a value for the date up to the selected date.

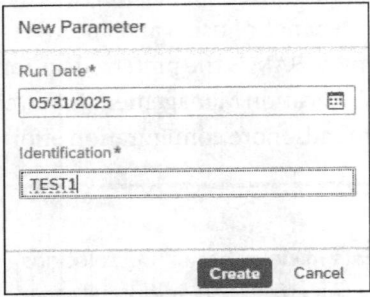

Figure 7.5 Initiate Payment

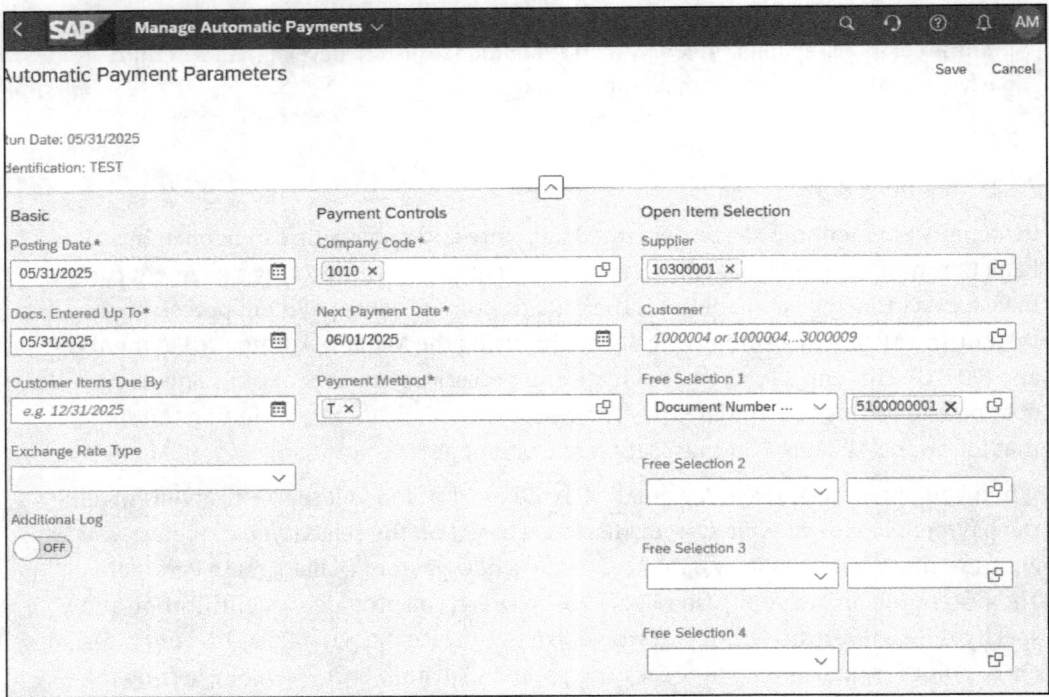

Figure 7.6 Create Parameters for Payment Run

Once you're done, click **Save**. The system will save the parameters (see Figure 7.7), and you can click **Schedule** to schedule the proposal or payment run.

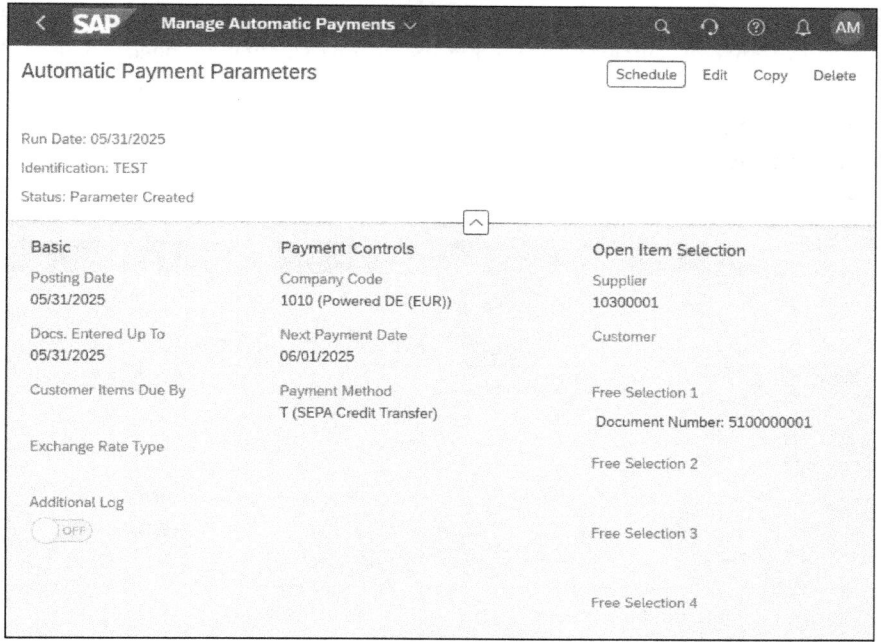

Figure 7.7 Schedule Payment

You have two options when it comes to executing payments. First, you can schedule a proposal, which is essentially a test run to see exactly what the system is going to do with the payments—what will be paid and how. Alternatively, you can schedule an actual payment run, which processes the payments directly. It's worth noting that creating proposals is typically done when executing payments manually. However, the main idea is that in most organizations, payments are executed automatically and independently, ensuring a streamlined and efficient payment process.

Select **Proposal** and a new popup will open (see Figure 7.8). Here you can select a date and time for your run. To perform the run manually, select **Start Immediately** and then click **Schedule**.

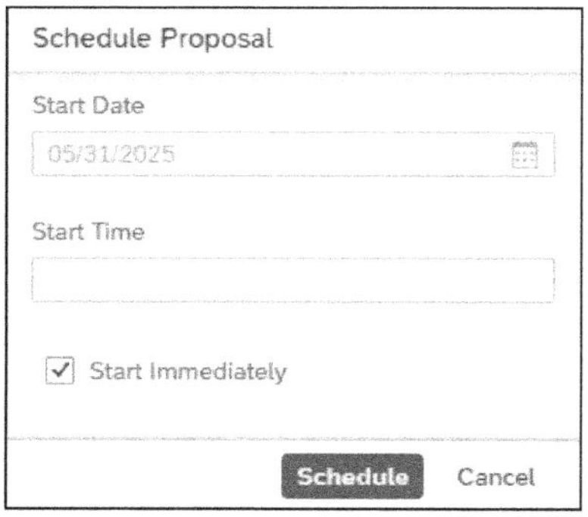

Figure 7.8 Schedule Proposal

Once you schedule a proposal, you'll be able to see it under the **Proposals** section (see Figure 7.9). Here, you can review all the details, including exactly what needs to be paid and how it will be processed. If everything looks correct and you're ready to proceed, simply select the line item and click **Schedule Payment** to initiate the actual payment run. This step transitions your proposal into a real payment run, ensuring that the payments are executed according to the settings you reviewed.

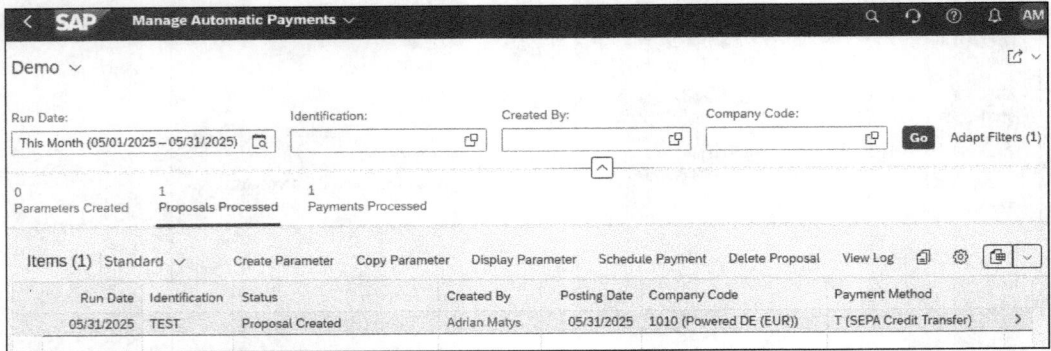

Figure 7.9 Proposal Scheduled

Once you click **Schedule Payment**, you'll see the same popup window that appeared in the earlier step (see Figure 7.8). Here, you'll need to select the time at which you want to run your payment. If you're executing the payment run manually, simply choose the **Start Immediately** option and then click **Schedule** to proceed. This step finalizes the scheduling of your payment run and initiates the actual processing of your payments. The payment run will be visible under **Payments Processed** (see Figure 7.10).

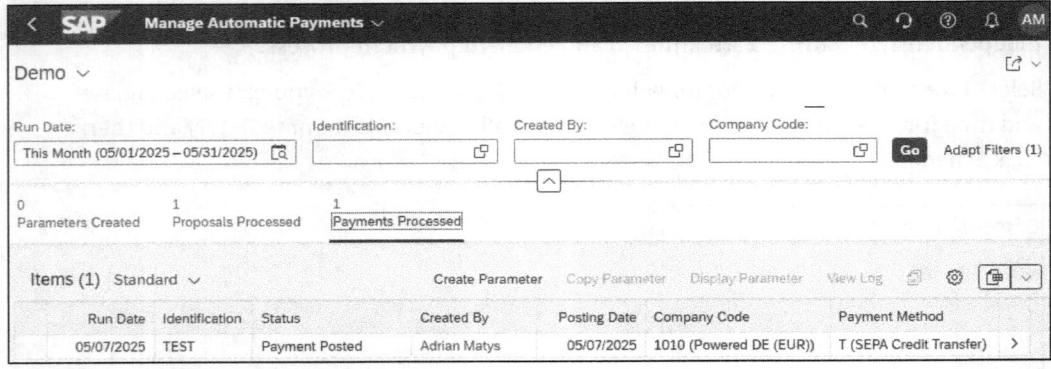

Figure 7.10 Payment Executed

The payment run is now executed. You can click the arrow to see details of the payment (Figure 7.11).

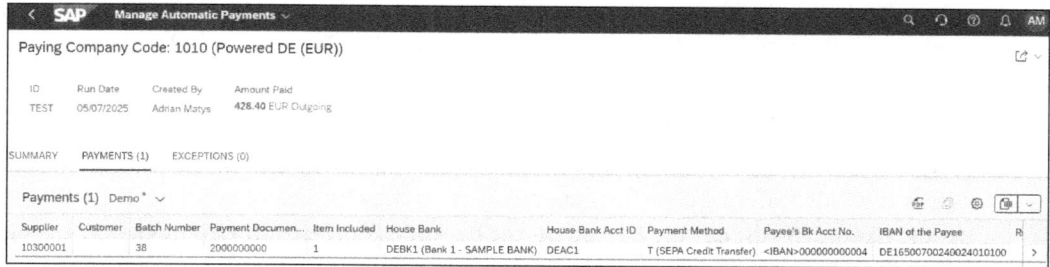

Figure 7.11 Check Payment

Here you can see all the important details of your executed payment run, including how much was paid, who executed it, which supplier, what the batch number is, and the payment document associated with it. You can also see from which account the payment was made. If you click the batch number, the system will take you to a different application to check the details of the payment batch (see Figure 7.12). There's also a further drilldown available that lets you see the individual invoices that were paid in this run, giving you a clear and detailed view of the entire payment process.

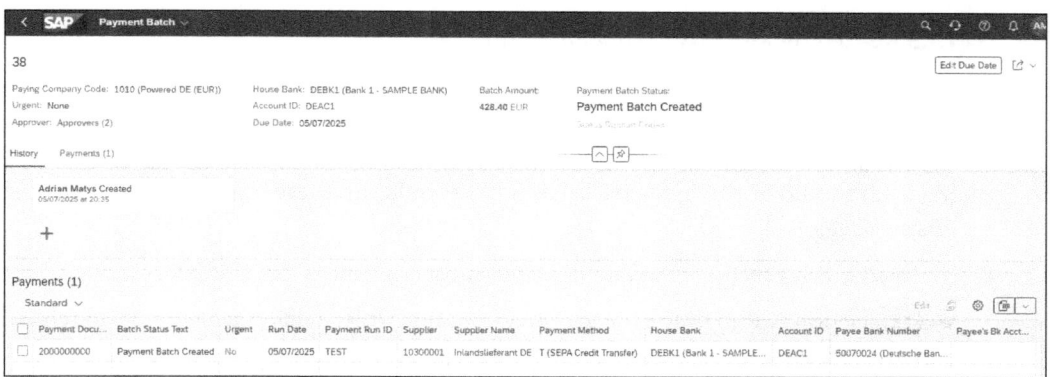

Figure 7.12 Batch Details

Here you can see the batched payment and how many payment runs were executed and included in this batch. In this example, there is only one payment run, but the key point of batching is that you can combine multiple payment runs together to generate a single payment file. You can also see who has the authority to approve this payment file or batch. Later, we will show you exactly how to batch payments and multiple payment runs together, how the approval process works, and how to configure everything to fit your organization's needs.

One of the most important functionalities of SAP Bank Communication Management is to ensure that no one can modify the payment files. SAP Bank Communication Management only generates the payment file once payment approval has been completed. After that, the file is created and remains available in the system for only a very short period of time, as it should be sent to the bank immediately. This approach ensures that no one has the opportunity to alter the file, safeguarding the integrity and security of your payments.

A payment file is generated in the format specified in SAP S/4HANA customizing for the payment method chosen. Payment formats such as pain.001 can be generated directly from the SAP system. These formats are used to structure payment data according to standardized XML schemas, often required by banks for electronic payment processing. For example, the pain.001 format, which is commonly used for customer credit transfers, can be generated using Transaction DMEX in SAP S/4HANA. SAP S/4HANA has lots of tools supporting you with the creation of various formats. There are also a lot of standard formats that you can use during the configuration of your payments.

The payment file can be displayed using Transaction BNK_MONI or the Monitor Payments app. Once you open Transaction BNK_MONI or the Monitor Payments app, a list of payment batches will be shown. Payment files can be checked only for batches in with the status set to **Payment Medium Created**. A file might be available for only a selected period depending on the setup in the system.

7.1.3 Batching and Processing Payments

In this section, you will learn how batching and processing payments works in SAP Bank Communication Management. You will also learn how you can troubleshoot problems that happen during the batching process. Note that this is an IT-heavy topic; and batching itself should be run in the background. IT should run these transactions only in case of any issues.

In the following sections, we'll review the transactions used for this process:

- Transaction FBPM1 or the Merge Payments app, which creates a payment "batch" for a given payment run.
- Transaction BNK_MONIP (Status of Payments), where payment runs and payment documents can be displayed.
- Transaction BNK_MERGE_RESET or the Reset Payment Media Batch Runs app; this transaction or app can be used to reset a created payment batch.

Then we'll discuss payment status management and how payment status reports are uploaded.

Create and Display Payment Batch

Transaction FBPM1 (see Figure 7.13) should be run as a background job, grouping all relevant payments into one batch (payment file) per bank.

As shown in Figure 7.13, enter the **Run Date** and **Identification** that was used in the Transaction F110 payment program. Then click the **Execute** button. The system will execute the batch and will follow the workflow set up for the approval of the payments. Now when you check Transaction BNK_MONI (Batch and Payment Monitor) or the status of the batch, you will see that a new batch appeared in the report and is ready for further processing depending on the configuration setup.

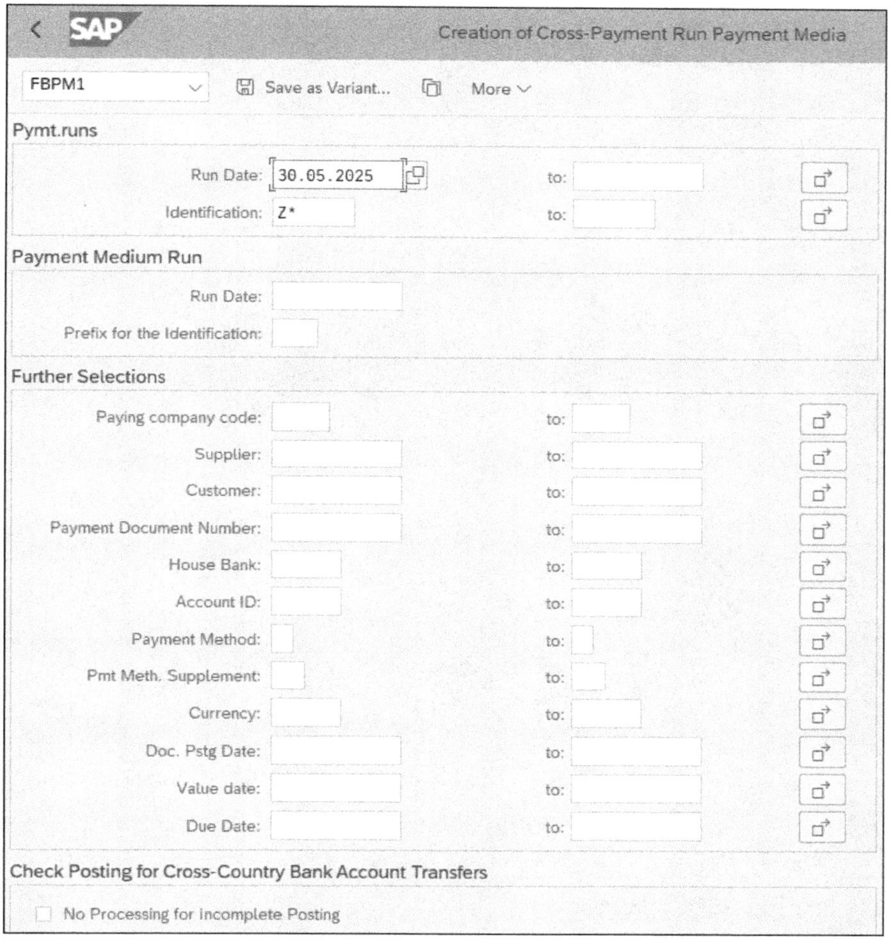

Creation of Cross-Payment Run Payment Media

FBPM1

Pymt.runs
- Run Date: 30.05.2025 to:
- Identification: Z* to:

Payment Medium Run
- Run Date:
- Prefix for the Identification:

Further Selections
- Paying company code: to:
- Supplier: to:
- Customer: to:
- Payment Document Number: to:
- House Bank: to:
- Account ID: to:
- Payment Method: to:
- Pmt Meth. Supplement: to:
- Currency: to:
- Doc. Pstg Date: to:
- Value date: to:
- Due Date: to:

Check Posting for Cross-Country Bank Account Transfers
☐ No Processing for Incomplete Posting

Figure 7.13 Batching Run

Display Status of Payment Documents

Using Transaction BNK_MONIP (see Figure 7.14), you can check the status of all batches and see what potentially caused errors during the processing. As shown, enter the **Run Date**(s) of the Transaction F110 payment programs, and/or the **Payment Document Number** that was generated by payment program F110. You can also use any other selection criteria.

Then click **Execute**, and the system will show you the status of all batches (see Figure 7.15).

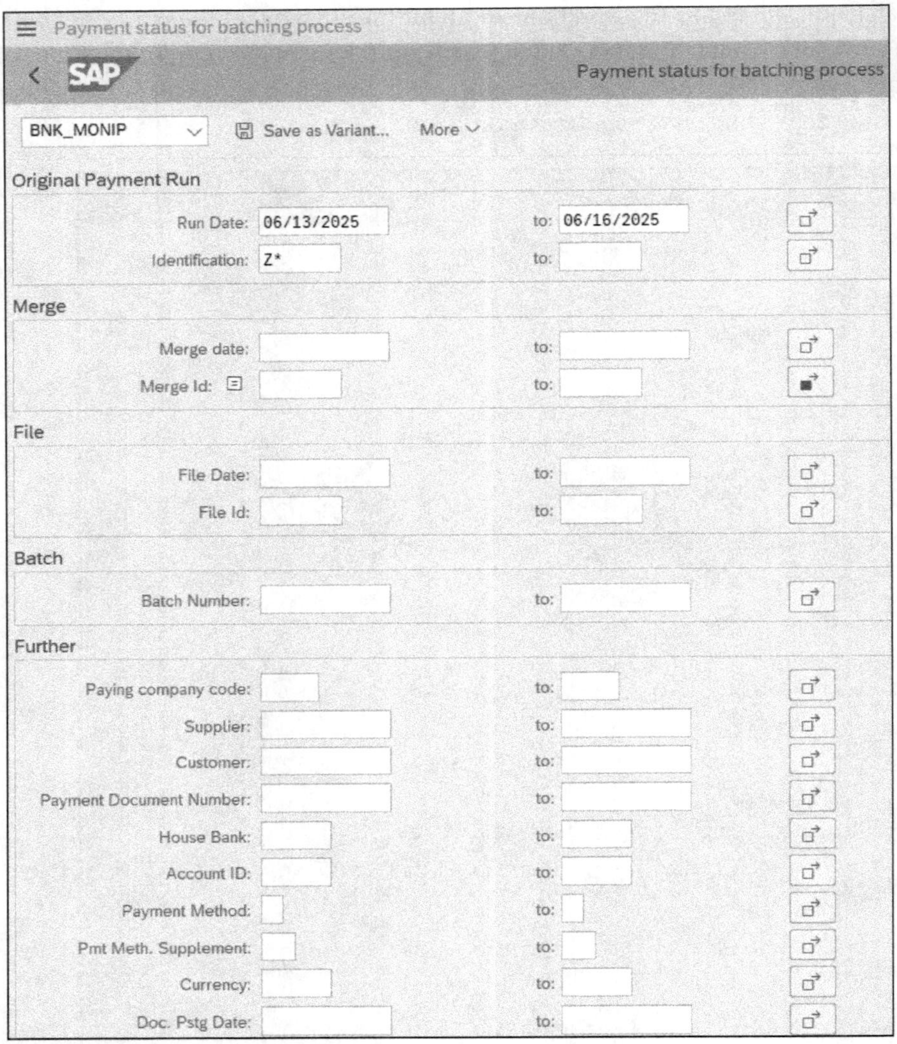

7   SAP Bank Communication Management

Payment status for batching process			
Original Payment Run			
Run Date:	06/13/2025	to:	06/16/2025
Identification:	Z*	to:	
Merge			
Merge date:		to:	
Merge Id:		to:	
File			
File Date:		to:	
File Id:		to:	
Batch			
Batch Number:		to:	
Further			
Paying company code:		to:	
Supplier:		to:	
Customer:		to:	
Payment Document Number:		to:	
House Bank:		to:	
Account ID:		to:	
Payment Method:		to:	
Pmt Meth. Supplement:		to:	
Currency:		to:	
Doc. Pstg Date:		to:	

Figure 7.14 Transaction BNK_MONIP

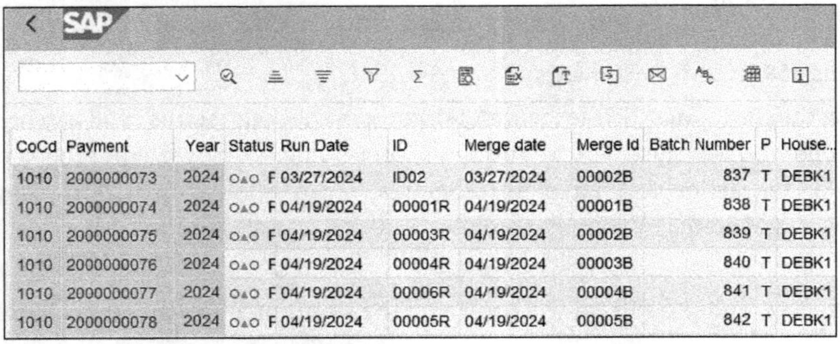

CoCd	Payment	Year	Status	Run Date	ID	Merge date	Merge Id	Batch Number	P	House...
1010	2000000073	2024	O▲O	F 03/27/2024	ID02	03/27/2024	00002B	837	T	DEBK1
1010	2000000074	2024	O▲O	F 04/19/2024	00001R	04/19/2024	00001B	838	T	DEBK1
1010	2000000075	2024	O▲O	F 04/19/2024	00003R	04/19/2024	00002B	839	T	DEBK1
1010	2000000076	2024	O▲O	F 04/19/2024	00004R	04/19/2024	00003B	840	T	DEBK1
1010	2000000077	2024	O▲O	F 04/19/2024	00006R	04/19/2024	00004B	841	T	DEBK1
1010	2000000078	2024	O▲O	F 04/19/2024	00005R	04/19/2024	00005B	842	T	DEBK1

Figure 7.15 Status of Payment Batches

Reset/Delete a Payment Media Batch Run

As demonstrated in the previous section, Transaction FBPM1 is used to generate a payment batch in SAP Bank Communication Management based on a particular Transaction F110 payment run. It may sometimes be necessary to reset (delete) the payment batch. This can sometimes be necessary if there were technical issues at the time of file creation (second approval of the batch) or if the list of approvers was not configured correctly.

To start this process, use Transaction BNK_MERGE_RESET. By resetting (deleting) the batch, a new batch can be generated using Transaction FBPM1, and the approval process can start again. As shown in Figure 7.16, you first enter the **Batch Number** that needs to be reset/deleted.

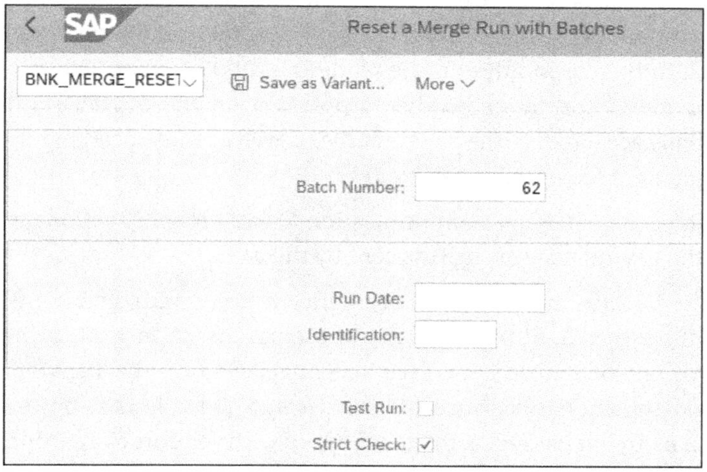

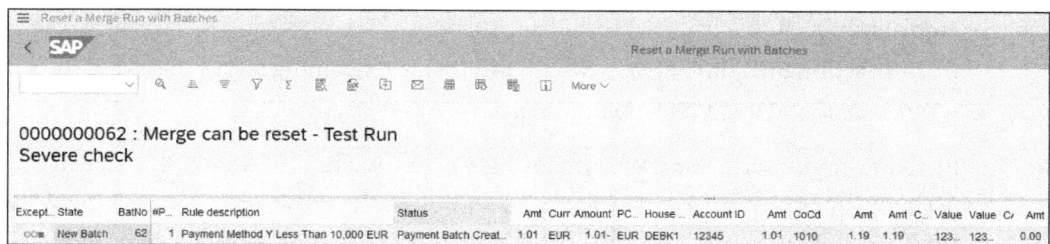

Figure 7.16 Reset Batch

Initially, you should execute the reset with the **Test Run** and **Strict Check** flags activated.

Click the **Execute** button, and the system will reset the payment status (see Figure 7.17). You can see whether the batch can be reset (deleted).

Figure 7.17 Reset Payment Status

If the system says OK, then you can go back to the initial screen, remove the check from the **Test Run** checkbox, and re-execute. The batch will be reset/deleted. Once deleted, you can recreate it using Transaction FBPM1 and following the necessary and predefined approval process again.

Payment Status Management

A payment status report in ISO 20022 pain.002.001.03 XML format is issued by the bank in response to each payment file that is sent by the corporation to the bank. We recommend you check out ISO 20022 for the latest information, as new payment formats and payments statuses based on this standard are expected to be adopted and widely used by the public in the near future.

As described in Chapter 1 and Chapter 2, corporations send payment files to banks in the pain.001.001.03 format. When a bank receives these payment files, it checks the files to see if the structure and content of the files are OK. The bank automatically sends a payment status report back to the corporation in the pain.002.001.03 format. This payment status report provides information to the corporation on the overall status (accepted, rejected, partially accepted) of the payment file, as well as the processing status of each individual payment (accepted or rejected).

In this section, we describe the SAP Bank Communication Management functionality to receive and process the payment status reports sent by the bank.

Thanks to these payment status reports, the corporation's accounts payable (AP) department can see within the Monitor Payments app or Transaction BNK_MONI whether individual payment files have been accepted or rejected by the bank and whether each individual payment has been accepted or rejected. This allows a corporation to quickly identify if payment has failed and to follow up with vendors as appropriate. For example, if the payment failed due to an incorrect bank account number, the corporation could request the correct bank account number from its supplier so that the payment can be made.

Within SAP Bank Communication Management, there are three transactions that are used specifically for payment status reports, as follows:

- Transaction S_EBJ_98000208: Upload payment status report (to be run in the background)
- Transaction BNK_INCMNG_MSG_MONI: Display incoming status messages
- Transaction ALRTCATDEF: Define alerts

In addition, the status of the payment files in Transaction BNK_MONI is updated to **Accepted by Bank** or **Rejected by Bank**, depending on the status sent in the payment status report file.

In the next section, we will look at the functionality within Transaction S_EBJ_98000208, which is used to upload payment status reports into SAP S/4HANA, and Transaction BNK_MONI, where the updated status (payments file accepted or rejected by bank) can be displayed.

Upload Payment Status Report

The bank communication tool used (as discussed in Chapter 2, there can be many, but SAP recommends SAP Multi-Bank Connectivity) will place the payment status reports in an SAP folder location. Transaction S_EBJ_98000208 or the program behind it, RBNK_IMPORT_PAYM_STATUS_REPORT, can be scheduled to run automatically every 10 minutes or so. This will update the payment file status in Transaction BNK_MONI.

As shown in Figure 7.18, the following technical settings are made and saved in a variant:

- Logical path from which to import the PSR files.
- Logical path where the imported PSR files should be archived.
- Logical path where PSR files with errors should be placed.
- Name of the XSL transformation used to transform the incoming XML file. Normally this is the standard SAP XSLT PAIN002_V3_TO_CPON transformation, or a Z copy of the same.

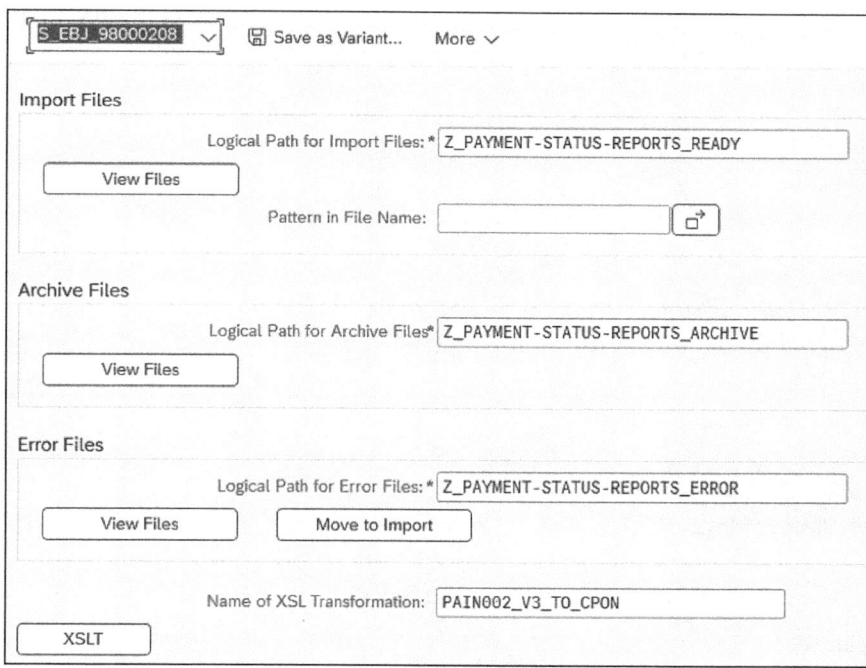

Figure 7.18 Upload Payment Status

Program RBNK_IMPORT_PAYM_STATUS_REPORT is set to run every 10 minutes or so in order to pick up and process the payment status report.

Transaction S_EBJ_98000208 is a technical transaction that runs automatically in the background to pick up and process payment status reports. Most users are generally not even aware that this transaction exists.

7 SAP Bank Communication Management

7.2 Configuration

The configuration activities for the subprocesses in SAP Bank Communication Management are essential for ensuring a smooth and efficient payment workflow. As described earlier, if you are not using the setup for payment approvals from the BAM solution, the system will automatically rely on the batching configuration within SAP Bank Communication Management and the corresponding workflow for managing payment approvals. This approach ensures that payments are processed consistently and that all approvals and batching rules are managed directly within SAP Bank Communication Management, providing a streamlined and centralized framework for your payment processing activities.

7.2.1 Basic Configuration

The first step is to enter the basic settings that will be used for approvals (see Figure 7.19). You can find this configuration via menu path **SAP IMG • Financial Supply Chain Management • Bank Communication Management • Basic Settings • Basic Settings for Approval**.

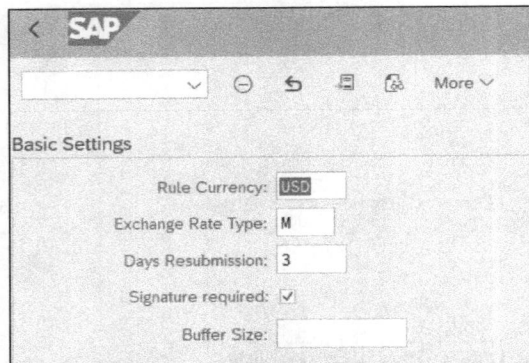

Figure 7.19 Basic Settings

The fields you will fill out on this screen are as follows:

- **Rule Currency**
  This is the default currency. When a payment batch is evaluated for release (such as via dual control, triple control, automatic release, and so on), the total batch amount can be evaluated based on the default currency if the payments in the batch consist of different currencies or different company codes (and hence different local currency).

- **Days Resubmission**
  This is the default resubmission date. For resubmission of payments, the resubmission date from the current date can be set.

7.2 Configuration

- **Signature Required**
  If this checkbox is marked, the signature popup appears before the payment batch is approved.

- **Exchange Rate Type**
  Enter the currency in which most of the rules are supposed to occur, the exchange rate type for foreign currencies, the days for resubmission, and whether or not an electronic signature will be required.

When you're done, save your entries.

The next step is to assess the payment media configuration (see Figure 7.20). To do this, follow menu path **SAP IMG • Financial Supply Chain Management • Bank Communication Management • Basic Settings • Payment Medium Create/Assign Selection Variants** or use Transaction OBPM4.

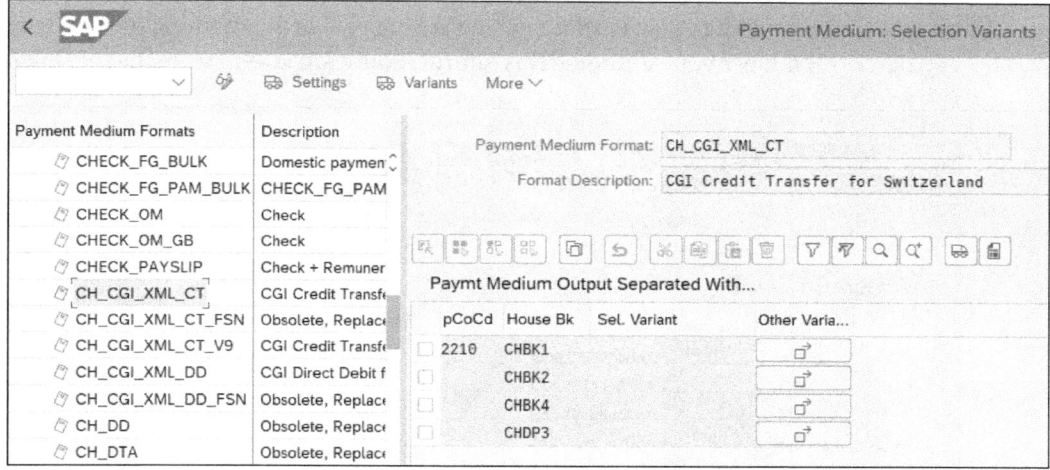

Figure 7.20 Payment Medium in Transaction OBPM4

In this example, for **Payment Medium Formats**, we have selected the pain.001 CGI XML format for Switzerland. You can also see here the company code and bank that have been selected and the payment medium assigned.

After the required format is created in the payment media workbench, select it, then click **Variants** on the left. After you access that screen, you link your format to a paying company code. After you complete this configuration, it will be added to a transport request.

This configuration step is necessary because if you are not using SAP Bank Communication Management when you select your payment method, the payment method is configured to use a certain DMEEX format, and you enter the variant for the program linked to that DMEEX format when you enter the parameters for the payment run. In SAP Bank Communication Management, however, you don't necessarily need to create

7 SAP Bank Communication Management

the payment media for each payment run; instead, you can group them by previously configured criteria, like merge ID, amount, and so on. The batches are then created according to those grouping rules. As a result, the system doesn't determine the payment media format based on the payment run parameters and instead uses this configuration to determine the proper format.

7.2.2 Batching Rules Configuration

In this configuration step, you will specify rules for the creation of batches (see Figure 7.21) using the following menu path: **SAP IMG** • **Financial Supply Chain Management** • **Bank Communication Management- Payment Grouping** • **Rule Maintenance**.

On this screen, you can see the list of the rules configured so far (**Rule ID**), the **Priority** of each, and a **Rule description**. In this example, the first rule, **6_PM_&**, is about merging batches with a certain ID that fall within a certain value range. When you open this transaction, simply click **New Entries**, then create or provide the batching rule, priority, and description. Once you've entered this information, click **Save** to store the rule in the system.

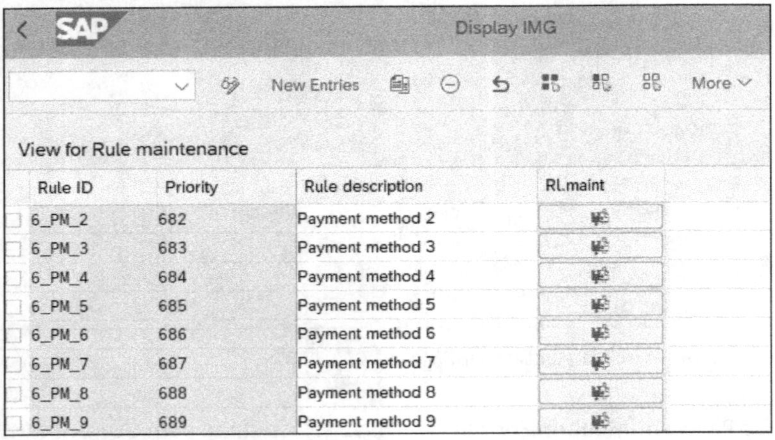

Rule ID	Priority	Rule description	Rl.maint
6_PM_2	682	Payment method 2	
6_PM_3	683	Payment method 3	
6_PM_4	684	Payment method 4	
6_PM_5	685	Payment method 5	
6_PM_6	686	Payment method 6	
6_PM_7	687	Payment method 7	
6_PM_8	688	Payment method 8	
6_PM_9	689	Payment method 9	

Figure 7.21 SAP Bank Communication Management Rules

If you double-click a **Rule ID** entry or click the **Rl.maint** button for an entry, you will go to the detail screen, as shown in Figure 7.22.

If you need to create a new rule, simply click the **New** button, which is represented by a white card icon on your screen. From there, select the value you want to base the rule on; for instance, in this example, we're looking for the payment method (you can add multiple selection criteria). Once you've selected a value, you can define the details of the rule and proceed with your configuration.

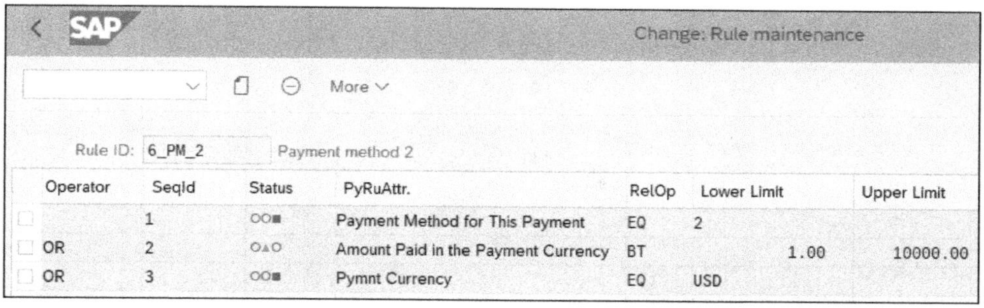

7.2 Configuration

Operator	SeqId	Status	PyRuAttr.	RelOp	Lower Limit	Upper Limit
	1	○○■	Payment Method for This Payment	EQ	2	
OR	2	○▲○	Amount Paid in the Payment Currency	BT	1.00	10000.00
OR	3	○○■	Pymnt Currency	EQ	USD	

Rule ID: 6_PM_2   Payment method 2

Figure 7.22 Rules Maintenance

Now, enter any **Operator** that applies (AND, OR, etc.) on the left. Then enter the sequence for each step in the **SeqId** column. In this example, the rule says to merge payments based on specific payment methods. You can also set up multiple other rules; the most common rules center on the origin of the payment or amounts from, for example, 000 to ZZZ whose value is between 2,000,000 and 99,999,999.9 (or the same negative values).

> **Batching Rules**
>
> It's quite common to build rules in SAP Bank Communication Management based on the origin of the payment, tailoring the process to specific scenarios. For example, if the payment is an internal bank-to-bank transfer within your organization, then you might choose to skip the approval process entirely, creating a rule that allows automatic out-of-approval processing. This makes sense because users typically wouldn't be sending payments to external accounts in these cases, so no additional application or approval step is needed. Conversely, for domestic payments, you might create rules that allow users to initiate payments throughout the day, then at the end of the day, group and batch all domestic payments into one file for the bank. This approach can reduce costs as banks often charge based on the number of payment files transmitted. On the other hand, for urgent payments, you may set up immediate batching rules to ensure funds reach your vendors as quickly as possible, balancing cost efficiency with business-critical priorities.

Now access the following path: **SAP IMG • Financial Supply Chain Management • Bank Communication Management • Payment Grouping • Additional Criteria for Payment Grouping**. On this screen (see Figure 7.23), you can select two additional fields to be used as criteria for grouping your existing rules. In this example, we are also grouping all the payments based on the paying bank account, to group all the payments going to the same bank and potentially reduce the number of payment files. In many banks' connectivity options, you pay based on the number of payment files sent.

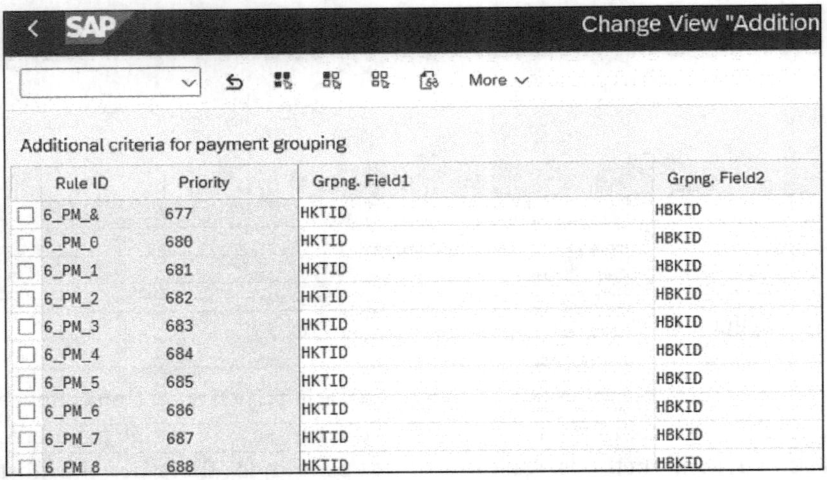

7 SAP Bank Communication Management

Rule ID	Priority	Grpng. Field1	Grpng. Field2
6_PM_&	677	HKTID	HBKID
6_PM_0	680	HKTID	HBKID
6_PM_1	681	HKTID	HBKID
6_PM_2	682	HKTID	HBKID
6_PM_3	683	HKTID	HBKID
6_PM_4	684	HKTID	HBKID
6_PM_5	685	HKTID	HBKID
6_PM_6	686	HKTID	HBKID
6_PM_7	687	HKTID	HBKID
6_PM_8	688	HKTID	HBKID

Figure 7.23 Additional Grouping Criteria

7.2.3 Workflow Activation and Configuration

In this configuration, you will set up workflow rules (see Figure 7.24). To do so, follow menu path **SAP IMG · Financial Supply Chain Management · Bank Communication Management · Release Strategy · Mark Rules for Automatic Payments**. The configuration already contains predefined templates for the approval process that can be adapted to suit the processes in your company. The workflow tool distributes the approval tasks to the corresponding staff members defined as approvers in SAP Bank Communication Management.

SAP Bank Communication Management works with two release objects, BNK_INI (Edit Workflow) and BNK_COM (Main Approval Workflow). Because the first approver has the possibility of changing the batch by, for example, resubmitting certain payments or rejecting some others, release object BNK_INI is used to enable the editing possibilities. The subsequent approvers can only approve or return at a batch level; for this scenario, release object BNK_COM is used. A combination of the two objects enables dual, triple, and quadruple control.

On the screen shown in Figure 7.24, enter a **Rule ID** and the rule **Priority**. After this is complete, select whether the rule results in automatic approval (**Auto**) and whether drill-down is required (**Dri…**).

The next step is to assign rules to release strategies. Like release strategies in purchasing, whenever cash is sent outside of the company, several parties will likely have to approve it. At the same time, however, you don't want to burden a very high-level executive with the job of reviewing hundreds of payments that are repetitive in nature (e.g., local tax payments), that were formulated by an expert department (e.g., the tax department), and for which it is extremely unlikely that high-level management will have a comment or a reason to stop it. Thus, to differentiate those payments that require little oversight from those that require one, two, or more levels of approval, you create release strategies.

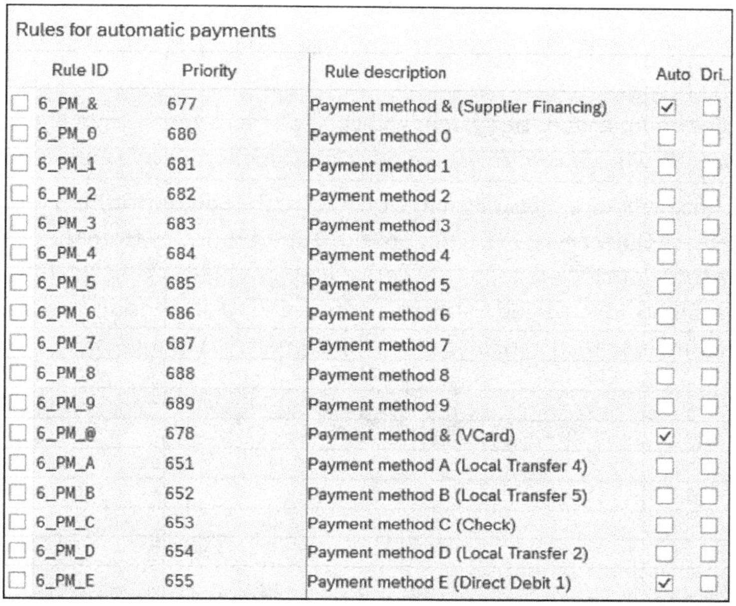

Rules for automatic payments				
Rule ID	Priority	Rule description	Auto	Dri.
☐ 6_PM_&	677	Payment method & (Supplier Financing)	☑	☐
☐ 6_PM_0	680	Payment method 0	☐	☐
☐ 6_PM_1	681	Payment method 1	☐	☐
☐ 6_PM_2	682	Payment method 2	☐	☐
☐ 6_PM_3	683	Payment method 3	☐	☐
☐ 6_PM_4	684	Payment method 4	☐	☐
☐ 6_PM_5	685	Payment method 5	☐	☐
☐ 6_PM_6	686	Payment method 6	☐	☐
☐ 6_PM_7	687	Payment method 7	☐	☐
☐ 6_PM_8	688	Payment method 8	☐	☐
☐ 6_PM_9	689	Payment method 9	☐	☐
☐ 6_PM_@	678	Payment method & (VCard)	☑	☐
☐ 6_PM_A	651	Payment method A (Local Transfer 4)	☐	☐
☐ 6_PM_B	652	Payment method B (Local Transfer 5)	☐	☐
☐ 6_PM_C	653	Payment method C (Check)	☐	☐
☐ 6_PM_D	654	Payment method D (Local Transfer 2)	☐	☐
☐ 6_PM_E	655	Payment method E (Direct Debit 1)	☑	☐

Figure 7.24 Rules for Automatic Processing

To do so, follow menu path **SAP IMG • Financial Supply Chain Management • Bank Communication Management • Release Strategy • Change and Release • Assign Rule to Release Steps**. On the screen that you see, enter the workflow **Release Object**, **WF Release Step** (representing the number of approval steps required), and the **Rule** ID (see Figure 7.25).

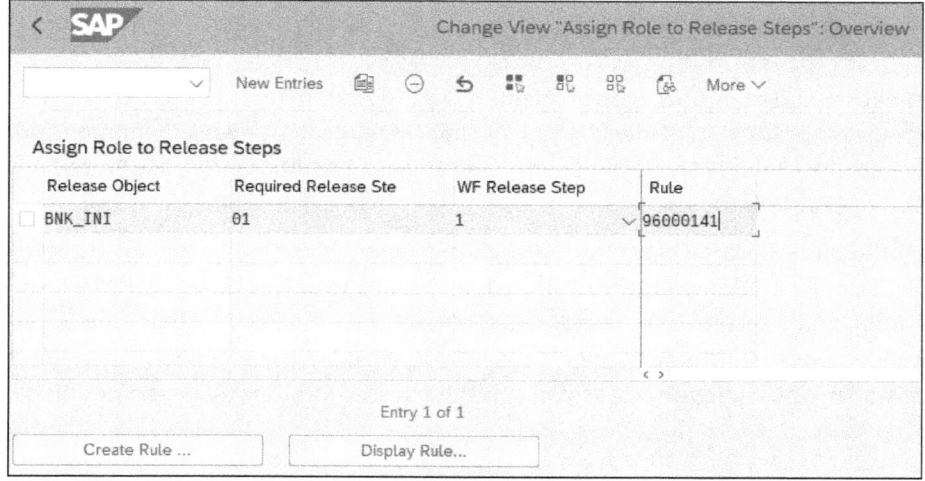

Figure 7.25 Assign Role to Release Step BNK_INI

7 SAP Bank Communication Management

The rule ID identifies the workflow rules assigned to the release procedure. Then, click the **Create Rule** button to define the rule.

BNK_INI always represents the first initializing workflow step, while BNK_COM represents any subsequent approval processes (if needed and specified in the rule), which are configured in the next step.

On this screen, click **Create Rule**, and the system will go to the screen shown in Figure 7.26, where you can specify who can approve payments for the selected rule and step.

Here, click the **Create** button (white sheet of paper icon next to the **Edit** button) and then enter the workflow **Release Object** (BNK_INI, if not provided automatically) from the previous screen. Add the individuals that need to approve it by double-clicking the rules. You can also give a certain validity period to the rule; it is useful to do periodic updates to make sure the structure is kept current and to ensure segregation of duties.

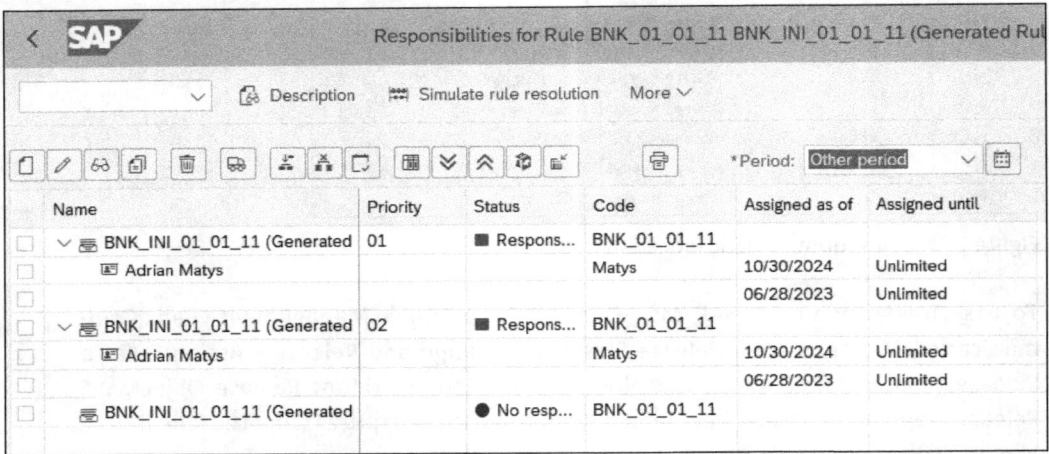

Figure 7.26 Define Responsibilities

Finally, test your configuration with the **Simulate Rule Resolution** button.

The next step is to define release procedures (Figure 7.27). To do this, access the following menu path: **SAP IMG • Financial Supply Chain Management • Bank Communication Management • Release Strategy • Change and Release • Additional Release Procedure • Define Release Procedures**.

Select the same **Release Object** BNK_COM and then enter a number for the release procedure (in this example, we entered "01", so there will be two release steps: one from initialization and a second one from BNK_COM, but you can enter more steps. The release procedure determines how many approvers you need for the payment.) Then determine under which circumstances you run the release workflow (**Always**, **Conditional**, etc.).

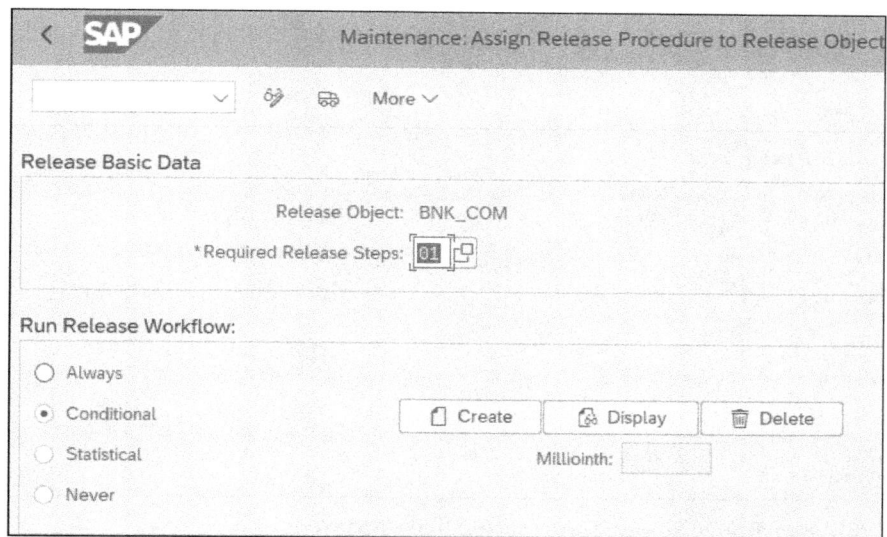

Figure 7.27 Assign Release Procedure to Release Object

Normally, you should save this configuration available in SAP and you do not need to change it here; you only need to do so in specific scenarios. If you do not need more than one payment approval, you do not need to specify anything here.

The next step is to configure additional release steps (see Figure 7.29). To do this, access the following menu path: **SPRO • SAP IMG • Financial Supply Chain Management • Bank Communication Management • Release Strategy • Additional Release Steps • Assign Role to Release Steps**.

As shown in Figure 7.28, for release object BNK_COM, you need to assign both the steps and the workflow rule. You do this by clicking **New Entries**, and the process is exactly the same as for the first initializing workflow step. The key difference here is that with BNK_COM, you are specifying the second or any subsequent steps in your payment approval workflow.

This configuration ensures that if additional approval steps are needed, they are clearly defined and linked to the appropriate workflow rules. However, it's important to note that this setup is only required if you need multilevel payment approvals. If your process only requires a single level of approval, you won't need to configure these additional steps.

The next step requires you to assign workflow templates to a release procedure. Go to **SPRO • SAP IMG • Financial Supply Chain Management • Bank Communication Management • Release Strategy • Additional Release Steps • Assign Workflow Template to Release Procedure**.

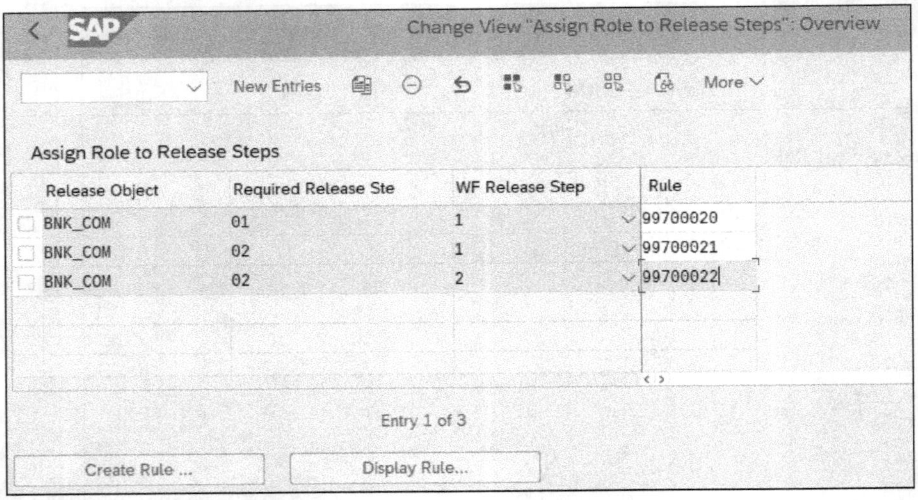

7 SAP Bank Communication Management

Figure 7.28 Assign Role to Release Steps for Next-Level Approvals

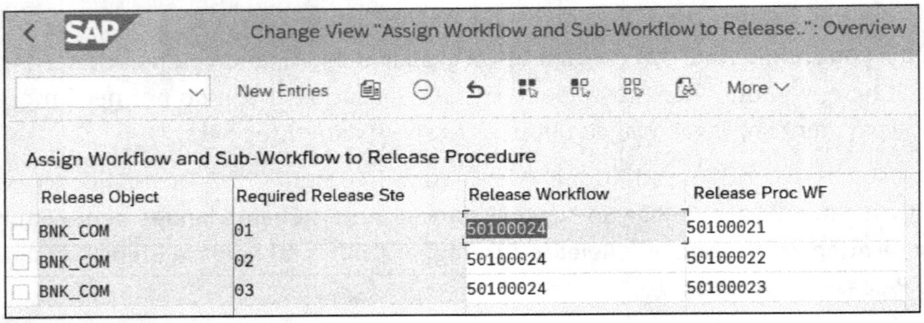

Figure 7.29 Assign Workflow to Release Object

If you need to specify these details, simply click **New Entries** and enter the required values for your release object. You'll need to specify release object BNK_COM, define how many release steps are required, assign the relevant workflow, and choose the release procedure. In most cases, you'll find that standard workflows are already defined in your system, and you can use these standard values to configure your release process. However, if you need to make adjustments to better fit your organization's needs, you may need to configure a new workflow. In general, though, it's recommended to use the standard procedures already available in the system for simplicity and consistency.

7.2.4 Alerts

The next configuration activity defines alerts, and you access it via menu path **Financial Supply Chain Management** • **Bank Communication Management** • **ALRTCATDEF** • **Define Alerts** or Transaction ALRTCATDEF.

7.2 Configuration

Transaction ALRTCATDEF can be used to automatically send alert messages to managers and employees in the AP team if, for example, the following situations occur:

- A payment file is sent to a bank, and no payment status report is received back within, say, 30 minutes. This would indicate that there is a problem with the interface between the corporation and the bank.
- A payment status report is received with a "reject" notification. This means that either an entire payment file, or individual payments within the file, have a problem.

Figure 7.30 shows an example of the rule setup for rejected files (click **Create**, the white sheet of paper icon on the screen, when you want to create a new rule).

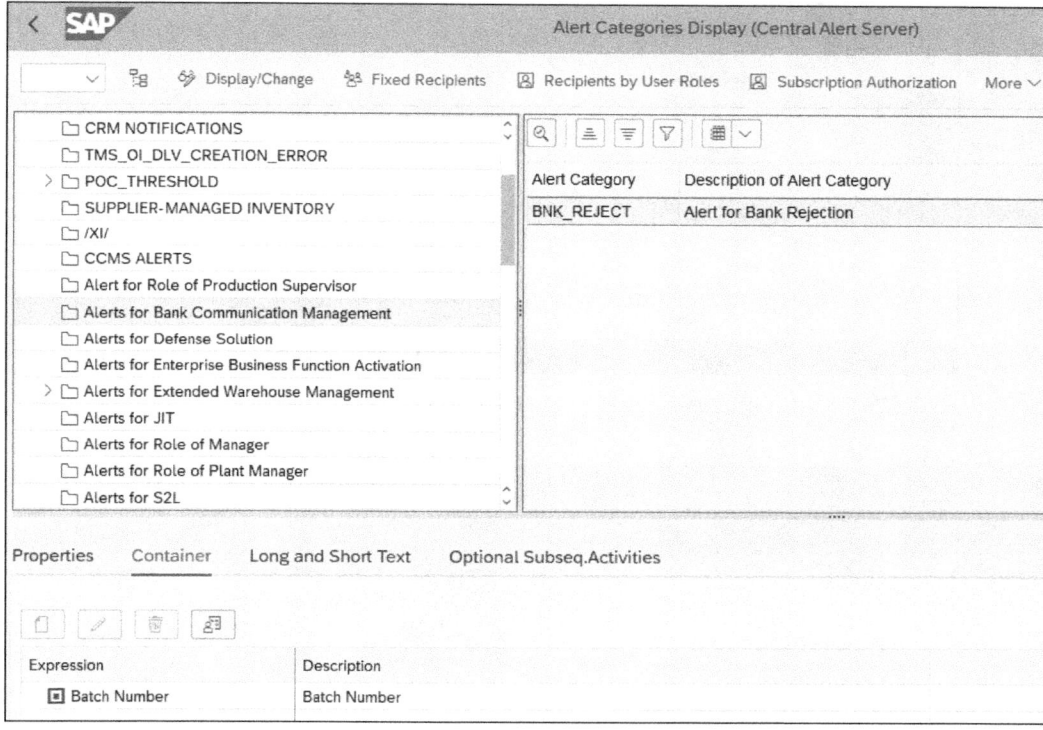

Figure 7.30 Alerts for Bank Rejections

Then go to **Container**, click the **Create** button, and enter "BNK_STR_PAY_STAT" in **Element**, enter your **Name** and **Short Descript.**, and enter the dictionary types in the **Type Name** field on the **D. Type** tab (see Figure 7.31).

Scroll down to where you can create your own text, which will be sent when an alert is generated, in the **Short Text** and **Long Text** tabs (see Figure 7.32).

507

7   SAP Bank Communication Management

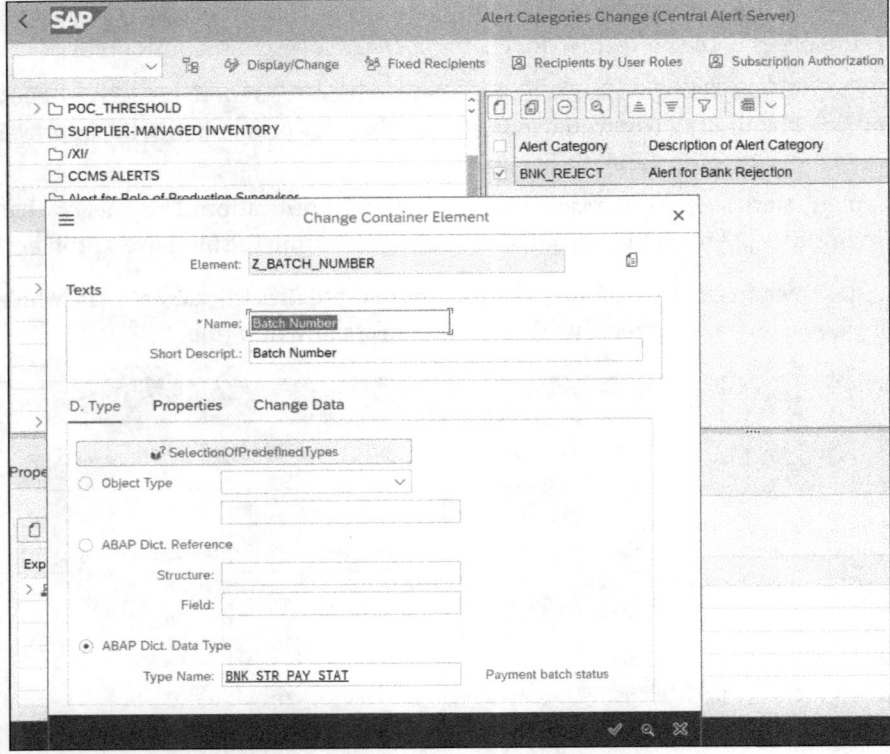

Figure 7.31 Alerts Container

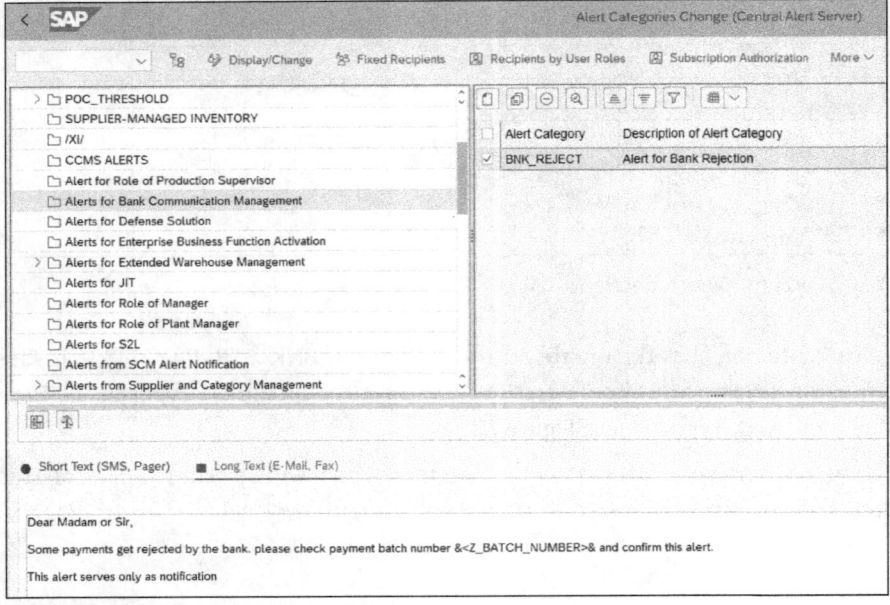

Figure 7.32 Alerts Text

7.2.5 Useful BAdIs

By default, SAP Bank Communication Management does not automatically handle the original payment in the event of rejection, whether by the user or the bank. To automate the reversal of the payment document, the implementation of BAdI BNK_BADI_PAYM_ALRT is required. Within this BAdI, you can, for example, establish a rule that will automatically reverse the payment document as soon as the payment batch is reversed. This enables business users to regenerate the payment run and file. The triggering point for this BAdI could be an alert created in the preceding step. Utilizing this BAdI allows for the implementation of any business requirements tied to actions to be taken after a payment batch is rejected.

7.3 Summary

In this chapter, you learned how to execute a payment run using SAP and learned how SAP Bank Communication Management can help you to merge payments into batches to streamline your processes. We also explored how to approve payments and manage updates to ensure payments are processed correctly and efficiently. In addition, you discovered the purpose of updating payments, which helps you maintain accurate and up-to-date information in your system. Finally, you saw how to configure SAP Bank Communication Management as an alternative approach to managing payments, instead of relying solely on the setup provided by BAM.

You also learned that the final step in the payment process is to provide payment approval. This step can also be used in conjunction with advanced payment management to ensure proper authorization of all outgoing payments. In general, everything that should be sent to the bank must first be approved in SAP Bank Communication Management. Only after obtaining this approval can the payment file be generated and sent to the bank. Once you've selected the payment and sent it to the bank, the bank statement will be generated to confirm the transaction. These payment processes and the handling of bank statements will be further described in the next chapter. With these skills, you're now well-equipped to handle payment processes in SAP in a streamlined and controlled way.

Chapter 8
Mapping Format Data

This chapter outlines the functions and processes of the Map Format Data apps for processing payments and bank statements, giving click-by-click instructions for performing the corresponding configuration activities.

The Map Format Data apps provide advanced mapping functionalities for data fields and structures across nearly all file formats pertinent to structured bank communication under ISO 20022, including the camt.053/54 format. This capability ensures the systematic transformation of these formats into structures that are seamlessly usable within the SAP S/4HANA environment. The configurable mapping rules and transformations enable precise and accurate data transfer, accommodating specific business requirements for enhanced flexibility.

Built on DMEEX technology, the Map Format Data apps integrate with various systems and software applications, ensuring compatibility across diverse IT environments. This foundation supports smooth operations without encountering compatibility issues. The tool's customization and configuration capabilities are designed to preserve critical information during data transformation, ensuring that essential data remains intact and correctly mapped. Flexible mapping and transformation rules help retain significant data elements or transaction details, which is crucial for accurate financial reporting and analysis.

To ensure reliable and precise data processing, the Map Format Data apps include comprehensive testing and validation tools. These tools verify that mapped data meets expected standards and formats, with a thorough validation process against anticipated bank statement structures. This proactive approach helps identify and rectify discrepancies early on.

In this chapter, we aim to not only explore the functionality of the tool but also establish a connection to payment transactions and the business requirements it addresses. Therefore, we will not only focus on the Map Format Data apps but also highlight the importance of ISO 20022 and how it can support SAP journeys during the migration process. We also delve into the details of the apps, examining the variants in depth—from operation and usage to an introduction to configuration. We explore how each variant functions, providing practical insights to optimize its application within financial processes.

8.1 Transition from SWIFT MT to ISO 20022 MX

The banking world is undergoing a transformation. Both local and global formats, such as the flat file standard SWIFT message type (MT), are on the verge of being replaced by the XML-based ISO 20022 format. This transformation presents companies with challenges in the areas mentioned ahead. Using the Map Format Data apps, format transformations in account statement processing, payments, and treasury banking communication can be simplified and orchestrated across systems, in collaboration with, for example, SAP S/4HANA Finance for advanced payment management. Let's examine the challenges this transition brings:

- **Data mapping and transformation**
  Mapping data fields from existing MT formats to the new message XML-based (MX) structure is a complex and time-consuming process. Achieving accurate data transformation without the loss of crucial information is essential to maintaining the integrity and reliability of the data. This involves careful planning, thorough testing, and meticulous execution to ensure that all data elements are correctly mapped and that no critical information is lost or misinterpreted during the conversion.

  Integrating MX message formats into existing systems and applications requires substantial coordination and intensive testing. Compatibility issues, data validation, and ensuring seamless connectivity are common integration hurdles that need to be addressed. This process often involves updating or modifying existing software, ensuring that different systems can communicate effectively, and maintaining data consistency across various platforms. Close collaboration between IT teams, system architects, and business users is necessary to overcome these challenges and ensure a smooth integration process.

  Adapting existing business processes to accommodate new MX message formats presents its own set of challenges. Ensuring smooth workflow transitions while maintaining operational efficiency is crucial. This requires a comprehensive review of current business processes to identify areas that need adjustment and implement changes without disrupting daily operations. It's essential to provide adequate training and support to staff to help them adapt to the new formats and workflows. Change management strategies should be deployed to facilitate a smooth transition and mitigate any resistance or confusion among the staff. In addition, monitoring and continuous improvement are key to ensuring that the new processes are optimized for efficiency and effectiveness in the long term.

- **System integration**
  Integrating MX message formats into existing systems and applications requires significant coordination and rigorous testing. This process involves ensuring that all systems can interpret and process the new message formats correctly, which often means updating software and making necessary adjustments to ensure compatibility. It's essential to address compatibility issues promptly to prevent data mismatches

or processing errors that could disrupt business operations. Data validation is another critical aspect as it ensures that the data exchanged between systems meets necessary quality standards and complies with regulatory requirements. Achieving seamless connectivity between different system components is also a common hurdle. This can involve implementing new communication protocols, establishing secure data transfer methods, and ensuring that all system interfaces work harmoniously together. Effective project management, stakeholder collaboration, and continuous monitoring are key to overcoming these integration challenges and ensuring a successful implementation.

- **Business process alignment**
  Adapting existing business processes to accommodate the new MX message formats is a challenging task that requires careful consideration and strategic planning. Current processes need to be reviewed and potentially reengineered to support the new format's requirements. This adaptation might include updating internal workflows, retraining staff, and adjusting compliance checks. Ensuring that these transitions are smooth while maintaining operational efficiency is crucial. It is essential to minimize disruptions during the changeover by implementing changes in phased stages if possible and providing continuous support to employees throughout the transition period. In addition, maintaining clear communication with all stakeholders about the changes and their implications is vital to foster understanding and buy-in. Continuous improvement practices and feedback loops can help in fine-tuning the new processes and ensuring they align well with organizational goals and provide the intended benefits.

8.2 Functions and Configuration

The Map Format Data apps are based on Extended Data Medium Exchange Engine (DMEEX) technology, delivered with SAP S/4HANA Finance in the SAP S/4HANA 2023 release (in both the on-premise version and SAP S/4HANA Cloud Private Edition). The Map Format Data apps are linked with the account statement processing (report RFEBKA00 [Bank Statement Processing]) and advanced payment management. In SAP S/4HANA Cloud Public Edition, these apps have now fully replaced DMEEX and have become the main tools for payment format configuration. As the Map Format Data apps were natively developed on SAP S/4HANA, they are configured and operated as SAP Fiori apps. The Map Format Data apps fundamentally consist of the same core, but multiple apps with different filters are delivered.

Incoming File Mapping Engine

Prior to SAP S/4HANA 1909, program fragments of the Map Format Data apps can be found under the name Incoming File Mapping Engine (IFME). This is due to a rebranding

8   Mapping Format Data

> of the tool in 2023. This leads to the outdated designation Incoming File Mapping Engine or IFME still being found occasionally in the customizing settings.

Here is an overview of the next sections:

- We'll first provide a holistic overview of the Map Format Data apps, detailing the available app variants and offering insights into their functionality.
- Then we'll focus on the apps themselves and explore their configuration and customization options in greater depth.
- The Map Format Data apps are not only deeply integrated into SAP S/4HANA Finance functionalities but also serve as core components for the inbound converter of advanced payment management. To conclude, we'll provide an insight into their role and capabilities in this area.

8.2.1   Map Format Data Apps Overview

Figure 8.1 shows how the Map Format Data apps fit into your banking and payments processes. Basically, the Map Format Data apps consist of preprocessing, in which the file is read and temporarily stored; the mapping engine, in which rule-based mapping rules are executed; and postprocessing, in which, similar to DMEEX, the format is created.

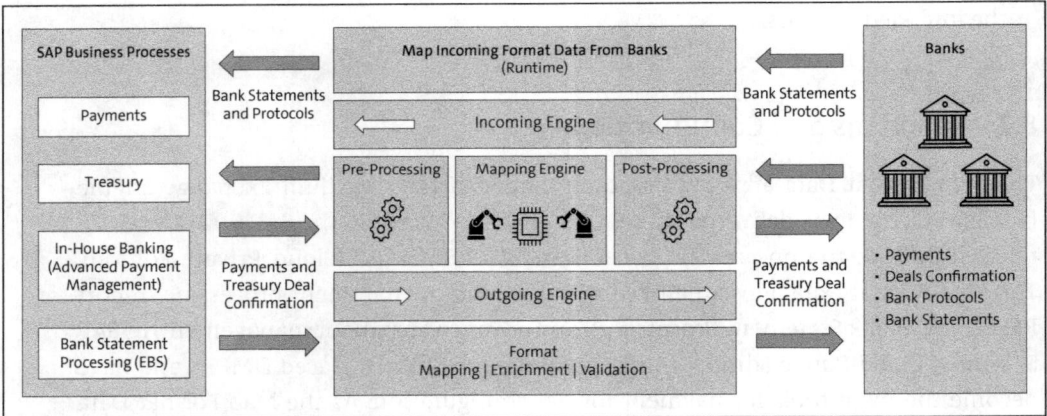

Figure 8.1 Map Format Data: Process Diagram

There are four Map Format Data apps delivered by SAP, as shown in Figure 8.2. In the following sections, we'll walk through features that are shared by all four apps. Then, in Section 8.2.2 and Section 8.2.3, we'll discuss the two apps most relevant to the process discussed in this book.

8.2 Functions and Configuration

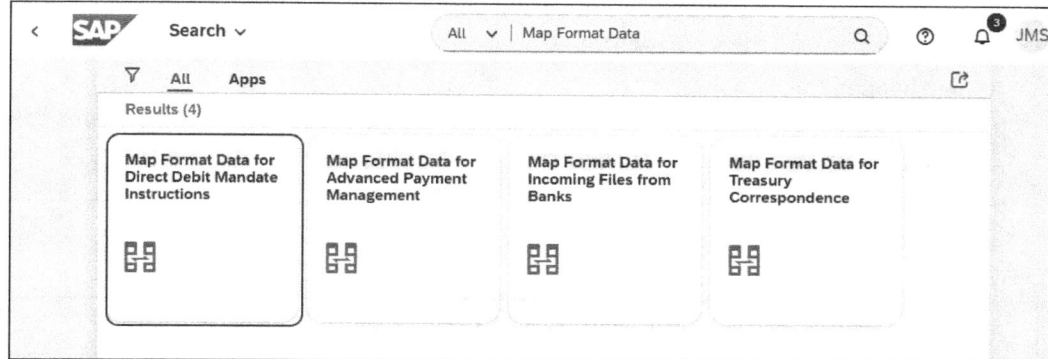

Figure 8.2 Map Format Data Apps

Filters

The Map Format Data apps offer a selection of filters specifically designed for them, allowing you to select your desired formats. In addition to drilldown fields, you can also use free-text searches with wildcards (*) for selection. Figure 8.3 illustrates the most important filters, which are as follows:

- **Format Mapping Name**

  The format mapping name is used as an identifier and is subsequently displayed in drilldown fields in apps such as Manage Incoming Files from Banks (F5608) or in account statement processing (Transaction FF.5). Almost all characters can be used for the description. A common practice is to use the format abbreviation along with the version, country code, and/or house bank—for example, camt.053.001.08_PL_PK001.

- **Status**

  There are generally three statuses available, as follows:
  - **Maintenance (M)**: This status indicates that the object is still under development and is therefore not available for use or selection.
  - **Active (A)**: This status indicates that the item has been saved.
  - **Released (R)**: This status indicates that the object has been developed, tested, and approved. No critical syntax errors have been detected by the system, and the user has saved and enabled the object for use. With this status, the object can be used in other areas, such as account statement processing or advanced payment management.

- **Data Format**

  In principle, the Map Format Data apps can process both flat files and XML files. Flat files typically include traditional SWIFT Message Types or MT formats, such as MT940 end-of-day account statements, as well as treasury correspondences like MT300 or MT320 messages. XML formats encompass ISO 20022 formats like

515

8 Mapping Format Data

camt.053.001.08, as well as bank protocols like pain.002.001.10. A comprehensive deep dive into the world of formats is provided in Chapter 2 and Appendix A.

- **Changed On**
  It is possible to filter by recent changes. The respective changes on the specified date or within the date range will be displayed accordingly.

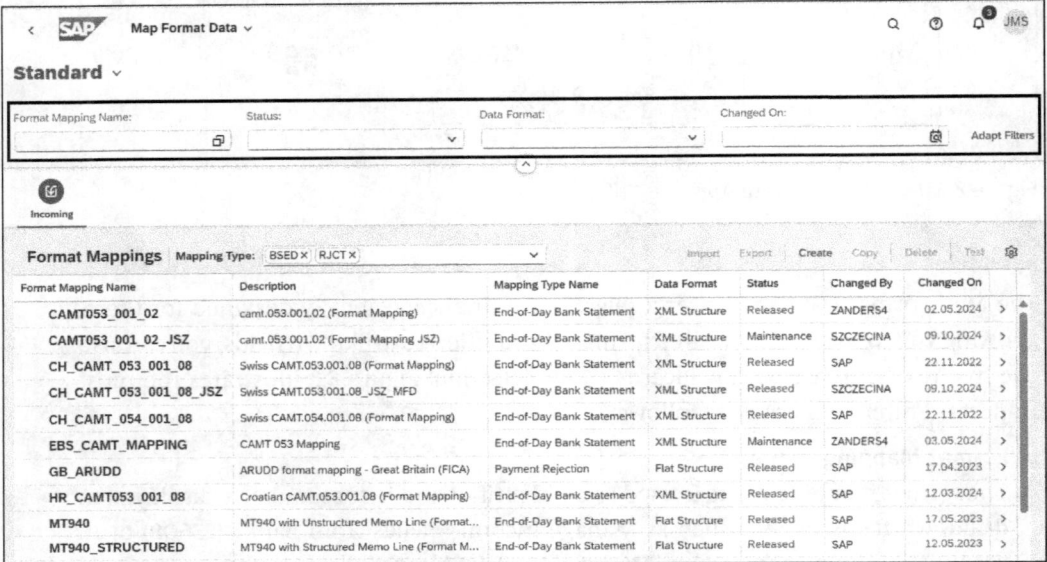

Figure 8.3 Map Format Data Filter

Mapping Types

A mapping type defines the format and structure of sent or received data. It is often tailored to a specific business object or business process data structure. In relation to the Map Format Data apps, the selection of a mapping type influences which areas of SAP S/4HANA and for which functions the mapping is available and can be utilized. The following are the most important mapping types:

- **APM customer credit transfer (APCT)/APM direct debit (APDD)**
  This category specifically deals with format mappings particularly configured for in-house banking and advanced payment management. These configurations are accessible in the import converter and can be utilized for importing external data into advanced payment management and mapping them to the meta-structure of advanced payment management. The affected formats include the following:

  - **Payment advices (pain.001)**
    These are payment instructions sent to or from a bank. These are used for managing payment processes, transferring funds, and ensuring smooth interactions between the business and banking institutions.

- **Direct debits (pain.008)**

    These are formats for initiating direct debits from customer or vendor accounts. These formats facilitate the automatic collection of payments directly from customer or vendor accounts, thereby streamlining billing and accounts receivable processes. The proper configuration and mapping of these formats in advanced payment management ensures seamless integration and efficient processing of financial transactions within the advanced payment management system. This helps in maintaining accurate financial records, easing reconciliation, and improving overall financial management.

- **End-of-day bank statement (BSED)**

    This mapping type is used for processing account statements. This applies to both flat files and XML-based account statements. The affected formats include the following:

    - MT940: SWIFT end-of-day bank statement
    - MT942: SWIFT intraday bank statement
    - camt.052: Intraday bank statement
    - camt.053: End-of-day bank statement
    - camt.054: Debit and credit notifications

- **Direct debit mandate: Notification of change (DMI1)**

    These mapping types are used for direct debit notifications in SAP S/4HANA. The affected format is ADDACS (for *automated direct debit amendment and cancellation service*). This format is used for notifications related to changes in direct debit mandates, such as amendments or cancellations. It helps in managing and updating the direct debit instructions efficiently.

    The proper processing of these mapping types ensures that any changes to direct debit mandates are accurately reflected in the system, thereby maintaining the integrity and accuracy of direct debit transactions. This is crucial for effective financial management and compliance with direct debit regulations.

- **Payment rejection (RJCT)**

    Mapping type RJCT is used for bank reports in the format pain.002, an XML format for providing detailed information on the status report of payment instructions, including rejections.

- **Treasury correspondence (TRMC)**

    Mapping type TRMC is primarily used for the transaction manager within SAP Treasury and Risk Management. The affected formats include the following:

    - MT300: A SWIFT message format used for foreign exchange confirmations. Provides details of foreign exchange and currency option transactions, including the terms and conditions agreed upon between parties.

8 Mapping Format Data

- MT320: A SWIFT message format for fixed loan / deposit confirmations. Used to confirm details of time deposits and fixed loans, including the principal amount, interest rate, and maturity date.
- MT535: A SWIFT message format for settlement of securities transactions. Provides statements of holdings, detailing the securities positions held in a custodian's account, which is crucial for portfolio management and reconciliation.

Using the TRMC mapping type ensures that the transaction manager within SAP Treasury and Risk Management accurately processes these formats, aiding in the management of financial transactions, securities, and risk. This enhances the capability to effectively track, confirm, and settle transactions, thereby supporting robust treasury and risk management operations.

In Figure 8.4, the mapping types relevant to payment transactions and bank correspondence are marked and highlighted.

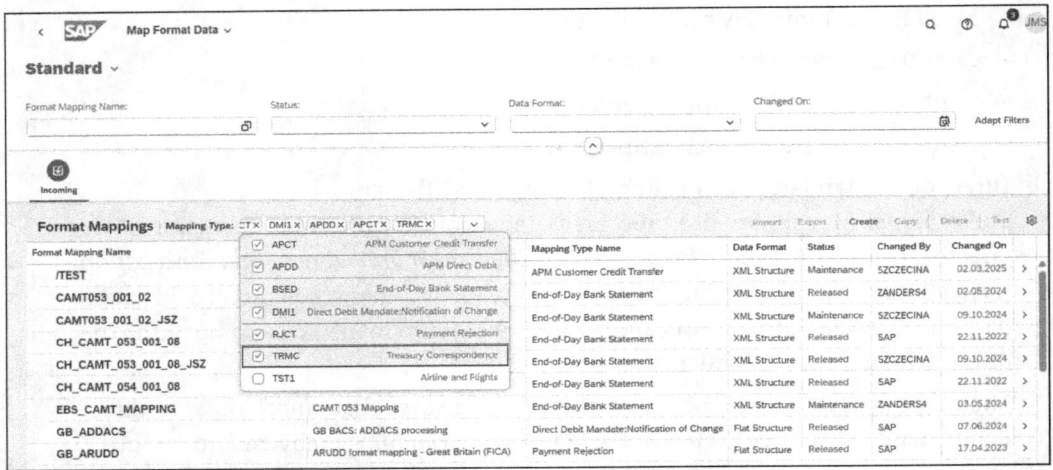

Figure 8.4 Mapping Types

Transport

Settings and mappings created for the Map Format Data apps are subject to transportation requirements. A workbench transport is required for this purpose, which is assigned to a user.

Workbench Transport

Workbench transports are a type of transport in the SAP S/4HANA system used to move development objects, such as programs, screens, and development modules from one SAP S/4HANA system to another, typically from a development system to a quality assurance system or a production system. These objects are usually related to the technical components of the system and require accurate version control and secure transport mechanisms to maintain system integrity and ensure consistent functionality across different environments. Workbench transports enable developers to package and transfer these objects efficiently while preserving the dependencies and relationships between them.

When you save your settings, the workbench transport window opens, in which the desired transport can be selected, as shown in Figure 8.5.

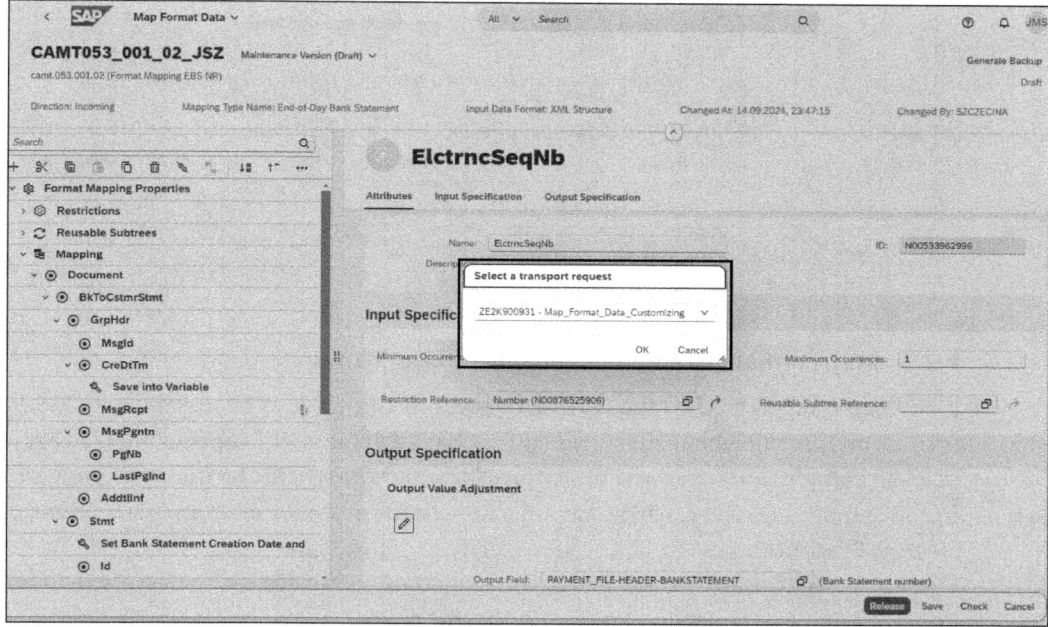

Figure 8.5 Transport Map Format Data

Now that you are familiar with the basic structure of the apps and the transport functionality, let's take a closer look at two of the specific apps: Map Format Data for Incoming Files from Bank and Map Format Data for Advanced Payment Management.

Incoming File Mapping Engine

As shown in Figure 8.6, the term *incoming file mapping* is used here instead of *map format data*. With each release, SAP strives to standardize its naming; however, almost all backend tables are also labeled with the prefix *IFME* for *incoming file mapping engine*.

8 Mapping Format Data

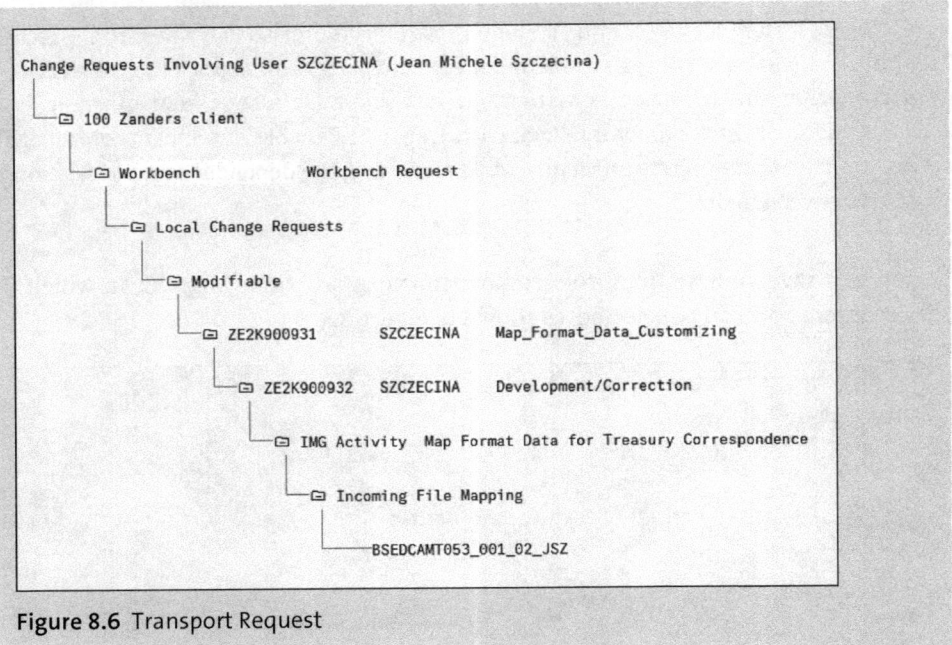

Figure 8.6 Transport Request

8.2.2 Map Format Data for Incoming Files from Banks

As its name suggests, the Map Format Data for Incoming Files from Bank app (F5608) focuses on incoming bank messages such as account statements or bank logs. To avoid starting completely from scratch, SAP provides country-specific format configurations in the standard delivery, which are aligned with local standards from associations or clearing institutions, such as the German Banking Industry Committee (DK) or the Swiss SIX. SAP continually delivers new format templates with each release, which can be copied and customized to meet your specific needs.

In this section, we first explore the mapping formats and guide you through the process of creating your own mapping. Then we explore the configuration options of the Map Format Data apps and present the available settings to help optimize their functionality.

Different node types may feature various editable properties. Next, we examine their impact on system functionality. Particularly within sensitive customizing steps, like setting up payment transactions, caution is essential. With version management, settings can be tracked and, in some cases, restored if necessary.

Created mappings must be approved. Therefore, we also show you how to perform the approval process within the system and outline important considerations. The Map Format Data apps are used, among other things, for importing bank statements. We will provide a general overview of the system prerequisites required for this functionality.

8.2 Functions and Configuration

Finally, artificial intelligence is increasingly being integrated into financial processes, including customizing. Therefore, we present a use case and beta functionality within Map Format Data.

Create Mapping

To open a mapping, click the **Create** button. A popup window will open in which **New Format Mapping Name**, **Description**, **Data Format**, and **Mapping Type Name** need to be entered as essential data, as shown in Figure 8.7.

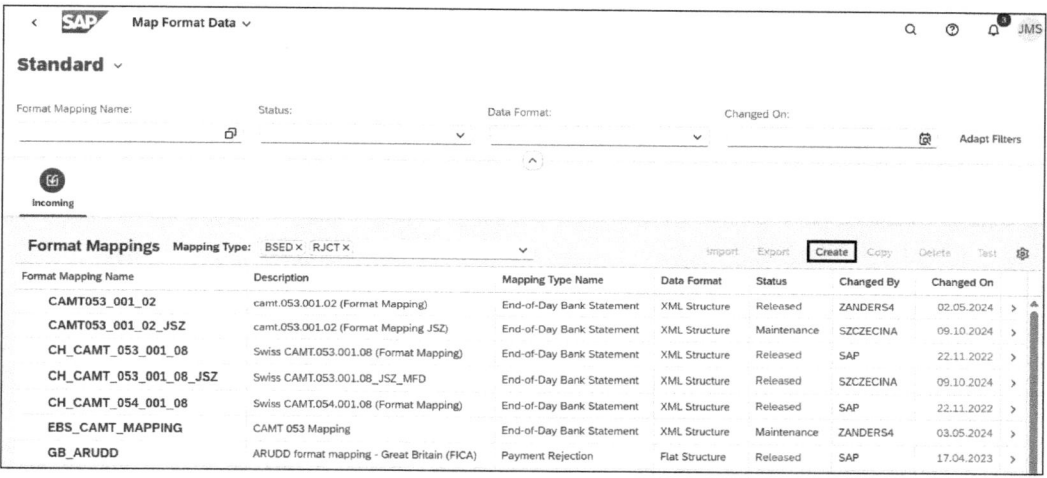

Figure 8.7 Create Format Mapping

In particular, the **Data Format** and **Mapping Type Name** fields are crucial for the use of the mapping. The **Data Format** determines whether this is a mapping for a flat file (**Flat Structure**) or an XML file (**XML Structure**). The flat file data format is used for processing classic SWIFT message files like MT940 end-of-day account statements or MT300/MT320 treasury deal confirmations. XML is used for classic camt.053 account statements and for all other XML-based formats. The **Mapping Type Name** determines the application. For account statement processing, **BSED—End-of-Day Bank Statement** should be selected for **Mapping Type**, as shown in Figure 8.8.

SAP offers a selection of common formats in the ISO 20022 standard. For efficient handling of mappings, it's worthwhile to make use of these offering by copying an existing mapping to use as a template for your individual processing. To do so, select the delivered mapping and click the **Copy** button, as shown in Figure 8.9.

8 Mapping Format Data

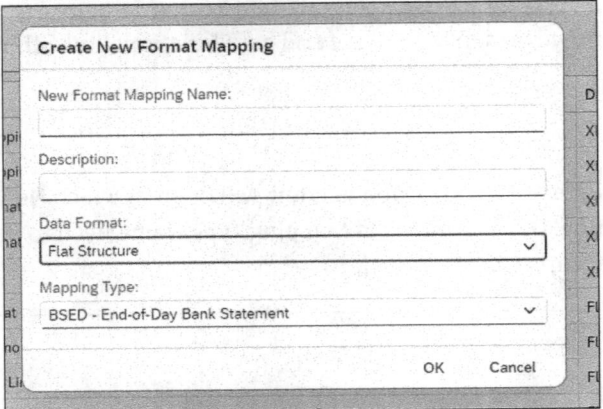

Figure 8.8 Create New Format Mapping

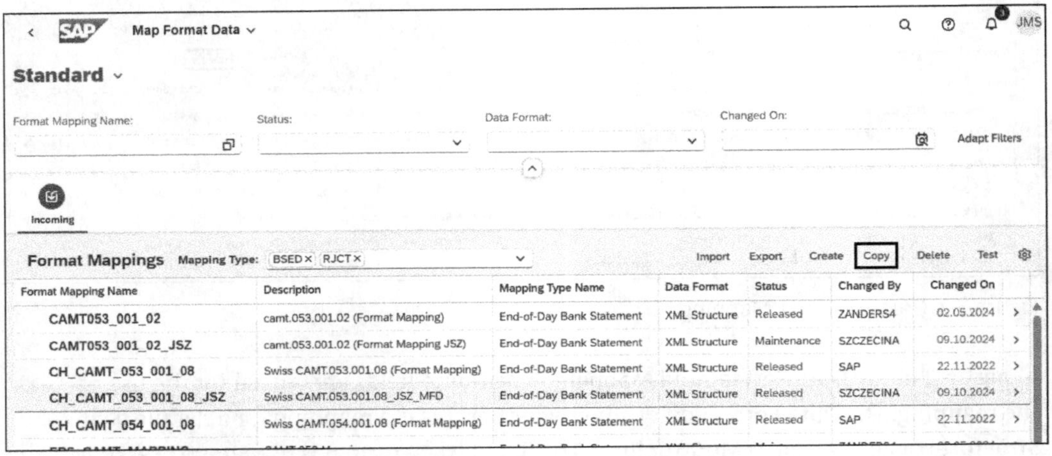

![Figure 8.9]

Figure 8.9 Copy Format Mapping

After the tree has been copied, it is displayed in the Map Dormant Data app for selection. After you click the newly created mapping, a window with a tree structure similar to the DMEEX tool appears; this is called the Map Format Data Workspace. On the right side is the tree structure—in this case, for a camt.053 account statement. In the menu bar, individual tags can be edited, as shown in Figure 8.10. With the **+ Create Node** icon ❶, you can add individual nodes. Use the **Cut** icon ❷ to cut individual nodes and add them in the desired locations via the **Paste** icon ❹. The **Copy** icon ❸ serves to copy individual nodes. With the **Delete** icon ❻, you can permanently delete tags. Alternatively, the **Deactivate Node** icon ❼ is available to deactivate individual nodes. Icons like **Inline Reusable Subtree** ❽, **Expand All** ❾, and **Collapse All** ❿ are used to sort the view.

After a node has been selected, the detail view for the respective node opens on the right side of the window. Here too, similarly to how the DMEEX tool works, individual nodes can be edited.

8.2 Functions and Configuration

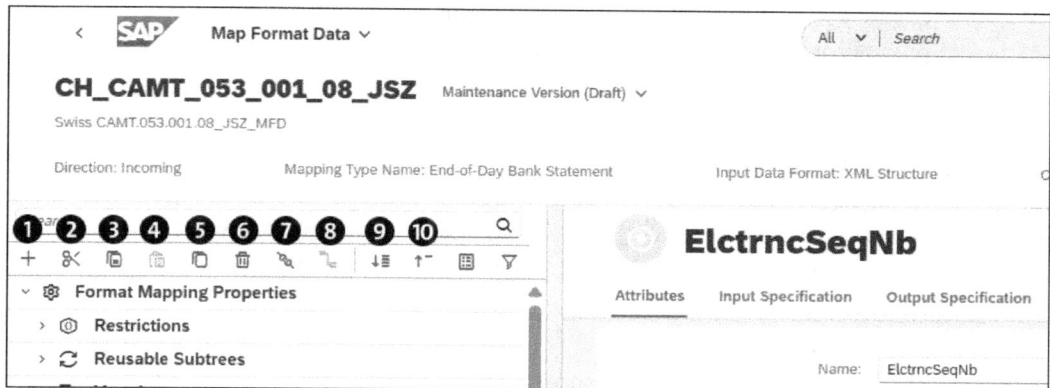

Figure 8.10 Map Format Data Configuration

Node Types

Upon clicking the **Add Node** button (see Figure 8.11), a selection of elements and attributes will open from which you can choose.

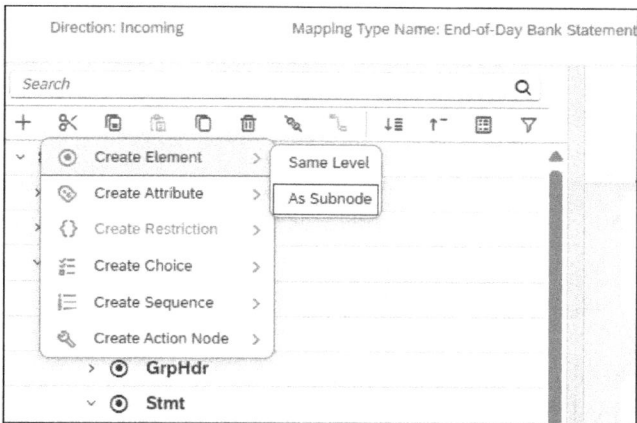

Figure 8.11 Node Types

The following are some of the important note types:

- **Element**
  An element, in the context of the Map Format Data app, represents an XML element in the input file and its corresponding mapping to the output structure. This element is crucial for translating and transferring data accurately between different systems. It defines how the original data in the XML input file should be transformed and placed into the target format to ensure compatibility with the receiving system. Proper mapping ensures data integrity and seamless communication across different SAP S/4HANA functionalities or external applications. The customized mapping logic can also accommodate specific business requirements, facilitating tailored data

523

transformation processes. Ultimately, elements in the Map Format Data app play a pivotal role in maintaining data consistency and enabling effective system integration.

- **Attributes**
Attributes represent specific properties or characteristics of an XML element present in the input file. They facilitate the mapping of these elements to the corresponding structure in the output file. By selecting appropriate attributes, you can ensure accurate data representation and transformation between the input and output XML schemas.

- **Action nodes**
These nodes represent one or more actions that are executed each time the node is encountered during the processing of the input file. Unlike elements and attributes, action nodes do not correspond to any actual data within the input file. Instead, they serve a functional role in the transformation process.

For instance, if your input file consists of multiple items, you can insert an action node to specify where to append a new row in the output structure. This is particularly useful for tasks such as the following:

– Data aggregation: Combining data from multiple input elements into a single output element.

– Row insertion: Indicating the precise location for adding new rows in a table or list in the output structure.

– Trigger actions: Initiating specific actions like counter increments, data formatting, or conditional executions based on the structure or content of the input file.

Using action nodes effectively ensures that the output structure accurately reflects the desired format and includes any necessary transformations or operations.

- **Choice nodes**
These nodes represent a selection between multiple subnodes. They provide flexibility by allowing you to specify multiple potential subnodes, but typically, only one subnode should be present by default. However, you can further customize this behavior by defining the minimum and maximum occurrences, which dictate how many subnodes are expected to be processed from the input file. Some examples of this are as follows:

– Single choice (default): By default, a choice node will allow exactly one subnode to be present. This is useful when your input file contains mutually exclusive elements and you only expect one of them to appear in each instance.

– Multiple choices: By adjusting the minimum and maximum occurrences, you can allow or require multiple subnodes to be processed. This is useful for scenarios in which an input file might contain a variable number of related elements and you need to handle each one appropriately in the output structure.

By specifying these parameters, you can fine-tune how choice nodes should behave, ensuring that the output structure accurately represents the variability and complexity of the input data. For instance, in a purchase order XML file, if items are listed under different categories like *electronics*, *furniture*, or *clothing*, then a choice node can be used to process each type of item accordingly, ensuring that the output structure adapts to whichever category is present while maintaining correct formatting and data integrity.

- **Sequence**

  These nodes dictate the order in which the specified subnodes are expected to appear within the input file. By default, the system processes nodes in the order you define within the mapping hierarchy. This means that if the nodes only occur once, you may not need a sequence node. However, by utilizing a sequence node, you gain the ability to define multiple occurrences of the same sequence of nodes, ensuring that the input data is processed correctly in repeated patterns. Some examples of this are as follows:

  – Default order processing: In scenarios in which nodes occur only once and in a fixed order, the system will naturally follow the sequence you have defined in the mapping hierarchy. This approach is straightforward and effective for simple, linear data structures.

  – Multiple occurrences: When your input file contains repeated sequences of nodes, a sequence node becomes essential. It allows you to explicitly define that the same sequence of nodes can appear multiple times. This is useful for handling complex data structures or records that naturally occur in repeated series.

  By using sequence nodes, you can do the following:

  – Maintain order: Ensure that the nodes are processed in the exact order they appear, preserving the hierarchy and structure of the input data.

  – Handle repetitions: Accurately process multiple instances of the same node series, which is crucial for datasets like transaction logs, repeatable forms, or grouped data entries.

  – Data integrity: Maintain the logical and contextual integrity of the data by ensuring that related nodes are processed together and in the correct order.

  For instance, in an XML file representing a series of customer orders, each order might include a sequence of `Item` nodes. Using a sequence node in this context ensures that each item is processed correctly, and any repeated sequences of `Item` nodes are handled accurately, maintaining the integrity and structure of the order information.

Node Properties

When in edit mode, the properties of existing nodes can be modified. The specific properties available for editing or specification depend on the type of node. When a node is

8 Mapping Format Data

selected on the **Format Mapping Details** screen, various sections dedicated to maintaining the node's parameters are displayed. These sections, as outlined in the following documentation, allow for detailed customization and adjustment of node attributes.

Different node types may include diverse editable properties, such as the following:

- **Attributes**
  Properties that define specific characteristics of a node
- **Child nodes**
  Configuration of subnodes or nested elements within the main node
- **Occurrence constraints**
  Rules specifying how often the node should appear
- **Data mappings**
  Adjustments of how data from the input file maps to the node

In addition to editing the tag attributes, both input and output specifications can be adjusted. You can adjust the output values and add further conditions within the **Output Value Adjustment** editing options, as shown in Figure 8.12. A wide range of functions and operators are available for use here.

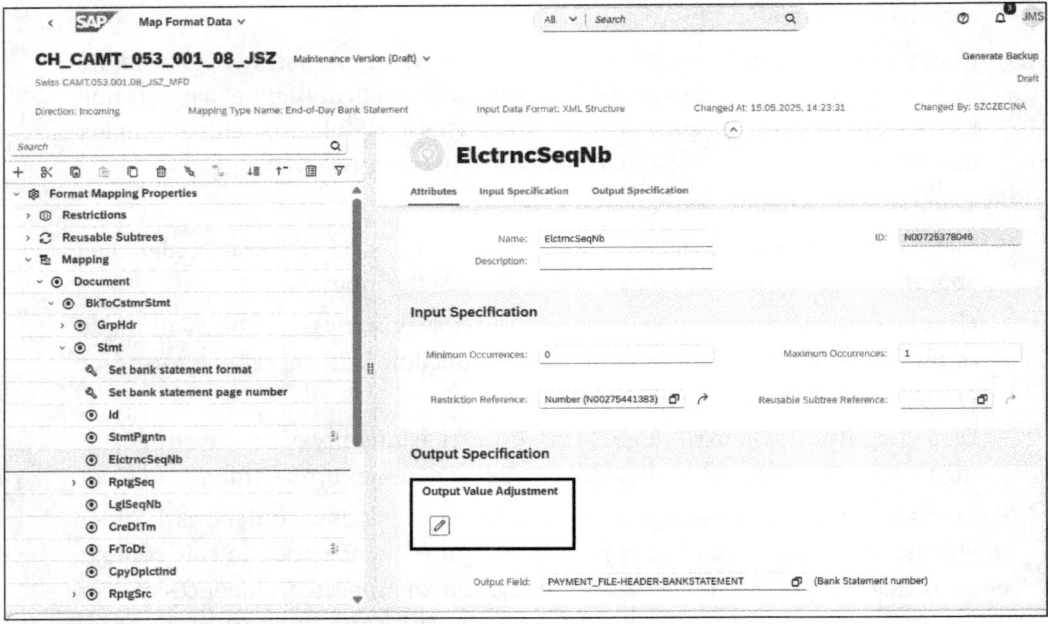

Figure 8.12 Output Specification

If you double-click a node, as we did in Figure 8.13 for the example node called **ElctrnSeqNb**, where the bank account statement number is expected, then you can edit the node and adjust output values.

8.2 Functions and Configuration

A detailed list of functions can be found on the SAP Help page at *http://s-prs.co/v605300*.

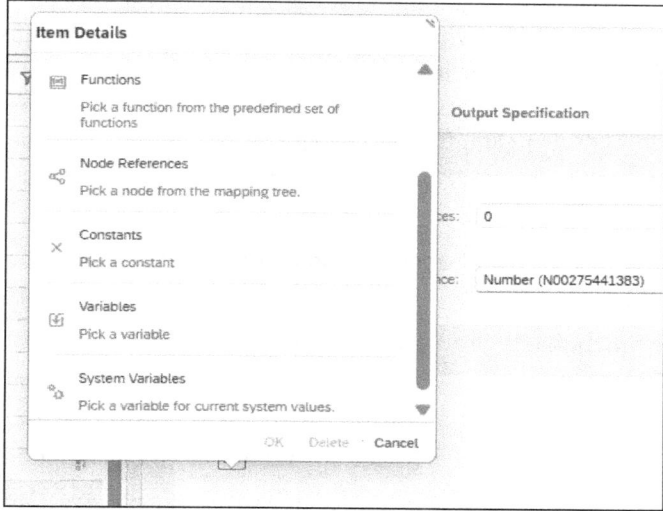

Figure 8.13 Item Details

Managing Versions

The similarity to DMEEX is also evident in version control. You can save intermediate results and manage them in versions here. If you click the **Generate Backup** button (see Figure 8.14), a backup of the current draft maintenance version is created.

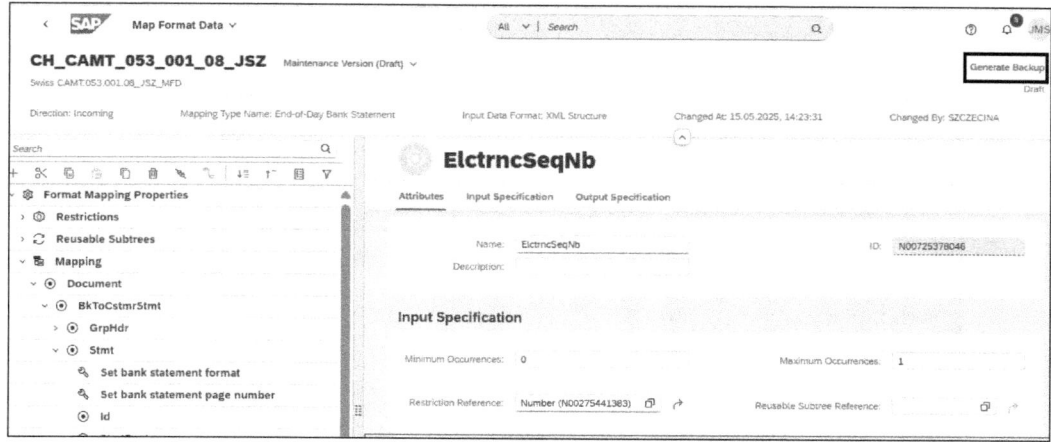

Figure 8.14 Generate Backup Button

In the Map Format Data apps, version management is a crucial feature that ensures data integrity and consistency. This function allows you to maintain multiple versions of the

8 Mapping Format Data

dataset, thereby providing a safety net in case of data corruption, accidental deletions, or other unforeseen issues.

With version management, you can create backups at various stages of the data mapping process, which can be restored to revert the dataset to a stable state. These backups can then be activated to replace the current working version. This feature is particularly useful in complex data transformation and migration projects, where changes are frequent and the risk of errors is high. By having a robust version management process in place, organizations can ensure the continuity and reliability of their critical data workflows.

To get into version management, click the **Released Version** button next to the format description in the upper-left corner of the app.

As shown in Figure 8.15, the functionality provided by the buttons for managing backups is straightforward, yet powerful:

- **Delete** (🗑)
  This button allows you to permanently delete selected backups. It is crucial to note that once a backup is deleted, it cannot be recovered. Therefore, it is advised to use this feature with caution, ensuring that the backups being deleted are no longer needed.

- **Restore Backups Version** (↺)
  This button enables you to restore a selected backup. When a backup is applied, the current settings and data mappings are overwritten by the contents of the backup. This feature is particularly useful when you need to revert to a previous stable state due to issues or errors in the current configuration.

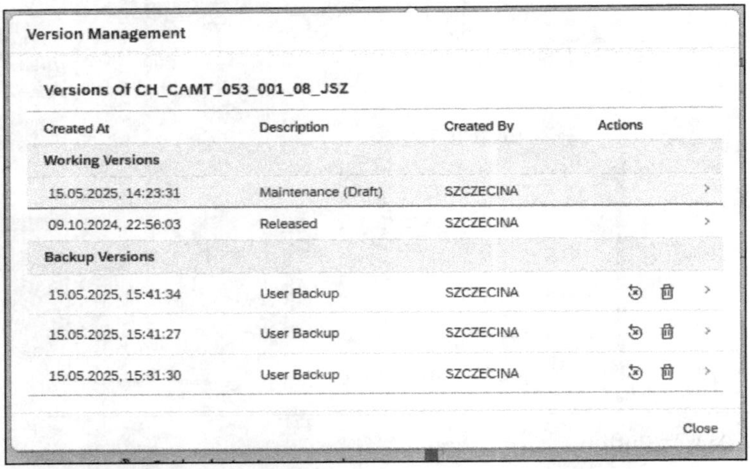

Version Management			
Versions Of CH_CAMT_053_001_08_JSZ			
Created At	Description	Created By	Actions
Working Versions			
15.05.2025, 14:23:31	Maintenance (Draft)	SZCZECINA	>
09.10.2024, 22:56:03	Released	SZCZECINA	>
Backup Versions			
15.05.2025, 15:41:34	User Backup	SZCZECINA	↺ 🗑 >
15.05.2025, 15:41:27	User Backup	SZCZECINA	↺ 🗑 >
15.05.2025, 15:31:30	User Backup	SZCZECINA	↺ 🗑 >

Figure 8.15 Version Management

Releasing the Mapping

After all settings have been configured, you can perform a check before saving and releasing the mapping. In editing mode, clicking the **Check** button (see Figure 8.16) will verify your format mapping for any issues. Any problems identified will be presented in a list of individual error or warning messages.

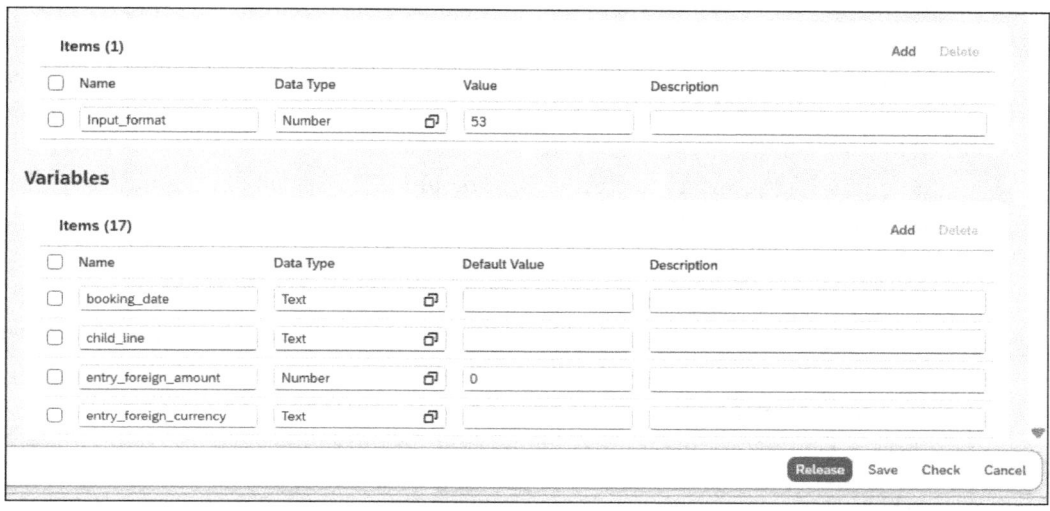

Figure 8.16 Check Button

If no critical errors are present, the mapping can be released with the **Release** button. It will then be approved for operation and ready for deployment, and the status will be set to **Released**.

Account Statements

For importing account statements, the general basic settings of account statement processing must be established. For customization, the BAM master data (banks, house banks, and house bank accounts) is used. The basic settings for account statement processing can be found under the path **Financial Accounting • Bank Accounting • Business Transactions • Payment Transactions • Electronic Bank Statement • Make Global Settings for Electronic Bank Statement**.

Processing bank statements generally occurs in two steps. Step one, for posting area 1, involves posting against a reconciliation account; step two, for posting area 2, involves posting against the open item. The following design considerations mainly describe the requirements for posting area 1. For posting area 1, reconciliation accounts from the general ledger (G/L) team are needed, as well as business transaction codes from external and internal in-house banks.

When you open the **Make Global Settings for Electronic Bank Statement** activity, you have to choose the chart of accounts based on which you will perform the configuration. The following configuration points are used for the general settings:

- **Account Symbols**
  Account symbols are used to represent bank accounts within the SAP S/4HANA system. They act as placeholders that link external bank account numbers to corresponding accounts in SAP S/4HANA. The account symbol is defined by the user during customizing. It specifies which G/L account is posted to. The following must be considered when setting up house banks: In this customizing activity, you maintain the bank details as well as the accounts at the house bank. For each of these accounts, it is necessary to create a G/L account in the system. In the master records of each of these G/L accounts, you enter a currency key. This currency key must match the currency in which you operate the respective account at the house bank.

- **Assign Accounts to Account Symbol**
  Assigning accounts to the account symbols defined in the previous step ensures that each symbol is linked to a specific G/L account.

- **Keys for Posting Rules**
  Posting keys determine the specific accounting entry type (debit or credit) for each line item in the bank statement.

- **Posting Rules**
  Posting rules dictate how the system processes bank statement transactions and posts them to the corresponding accounts.

- **Transaction Types**
  Transaction types categorize different kinds of bank transactions into groups that can be processed in a similar manner.

- **Assign External Transaction Types**
  External transaction types are the codes that the bank uses to identify different types of transactions. These need to be mapped to internal transaction types in SAP S/4HANA.

- **Assign Bank Accounts to Transaction Types**
  This step links specific bank accounts to the predefined transaction types, ensuring that transactions are routed correctly.

Using the report accessed through Transaction FF.5 (Import Electronic Bank Statements), account statements can be imported in both flat and XML formats. The previously described basic settings are prerequisites for using Transaction FF.5. The Map Format Data app mappings can be selected via the format selection. There is a new option in the drilldown menu under the **File Specifications** section. **Format with Format Assignment** must be selected, and only upon this selection will the mappings configured in the Map Format Data app with type BESD appear, as shown in Figure 8.17.

8.2 Functions and Configuration

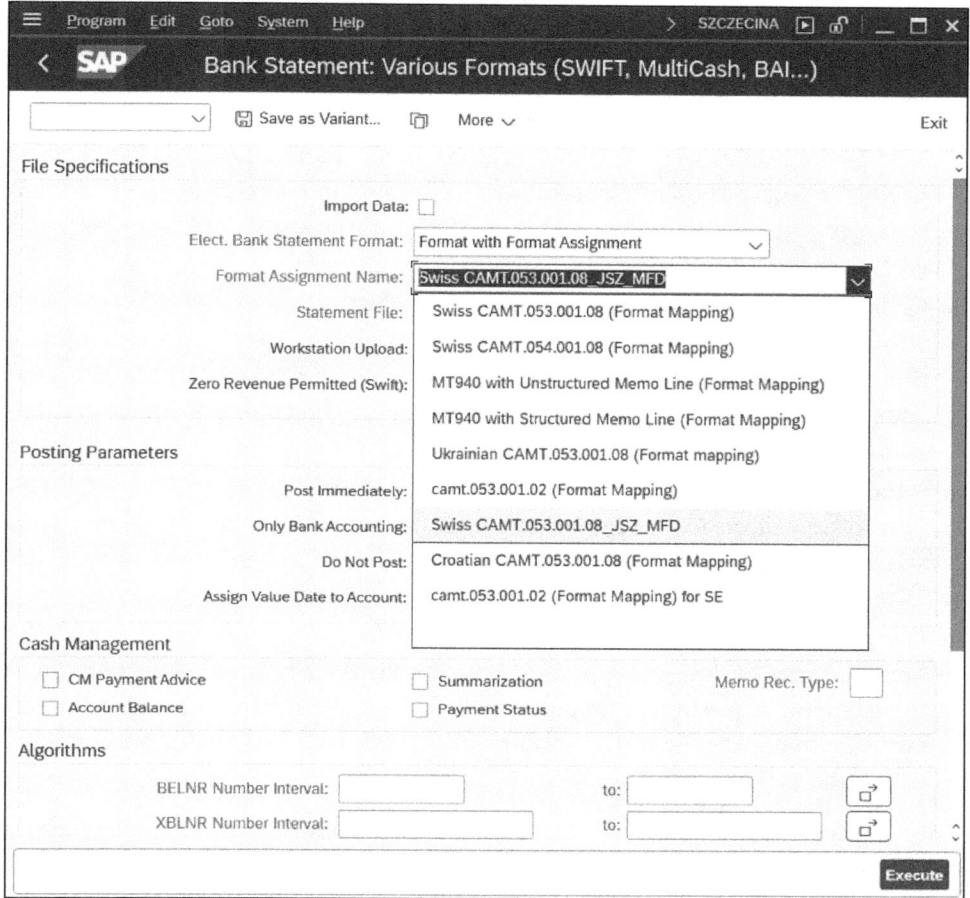

Figure 8.17 Transaction FF.5: Upload Bank Statement via Map Format Data

In general, according to SAP, Transaction FF.5 will be replaced by the new Manage Incoming Payment Files app (F1680). This app offers not only the import of account statements but also the import of bank protocols. In addition to the Map Format Data app mappings, it has a direct connection with advanced payment management, allowing it to be used for the import of external payment carriers. Furthermore, the app comes with all the advantages and disadvantages of SAP Fiori, such as an improved user interface, better integration with analytical apps, and other transactions and applications like account statement management.

Once you're in the app, click the **Import** button (see Figure 8.18). A popup window will open, with the following sections:

- **Bank Statement**
  This section is intended for the import of account statements in flat or XML format. These statements typically provide a detailed overview of transactions for a specific

8   Mapping Format Data

period and are essential for reconciling bank accounts with the company's internal records.

- **Lockbox Batch**
  This section is for importing lockbox account statements. Lockbox services are often used by companies to process customer payments more efficiently. Banks collect payments sent to a post office box and process them, and the corresponding account statements can be imported here to update the company's accounts receivable records promptly.

- **Payment Rejection**
  This section handles the import of pain.002 bank protocols, which provide information on rejected or returned payments. This enables users to quickly identify issues with payments, such as incorrect account details or insufficient funds, and take appropriate corrective actions.

- **Intraday Statement**
  This section allows for the import of MT942 or camt.54 intraday statements. These statements provide real-time updates on transactions throughout the business day, enabling more accurate and timely cash flow management.

- **Payment File for Advanced Payment Management**
  This section is for importing payment carriers via the advanced payment management import converter. It supports a wide range of payment types and formats, ensuring that all payment transactions can be efficiently integrated and managed within the SAP system.

For the import of end-of-day account statements, as shown in the example, click the **Bank Statement** button.

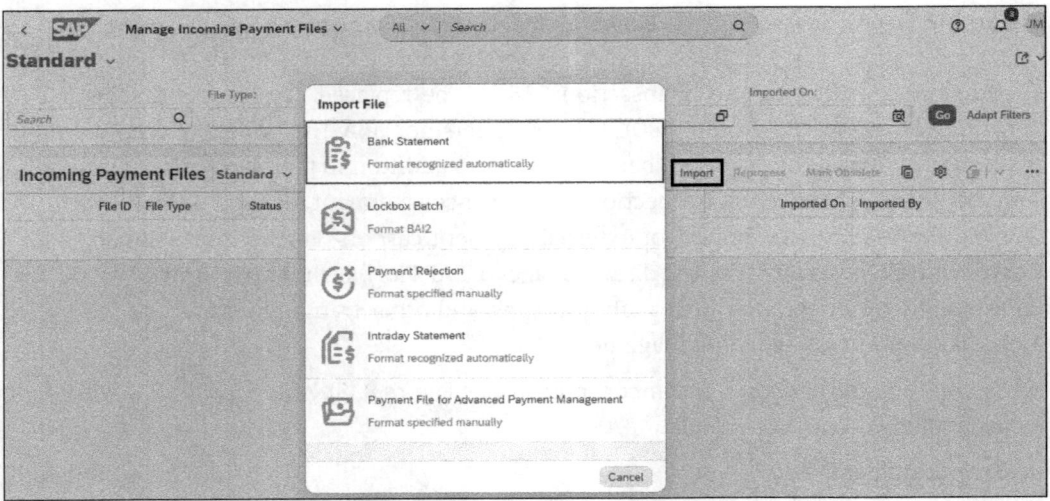

Figure 8.18 Import Bank Statement via Incoming Payment Files from Bank App

The Manage Incoming Payment Files app is directly linked with the Map Format Data apps. All mappings are indicated in parentheses, such as **(Format Mapping)**, as shown in Figure 8.19. In addition to format mapping, there is an option to automate the determination of the XSLT transformation by selecting the camt.53 format. Your custom format should be displayed and selectable here after release. Once the parameters have been determined, the account statement can be imported via drag and drop.

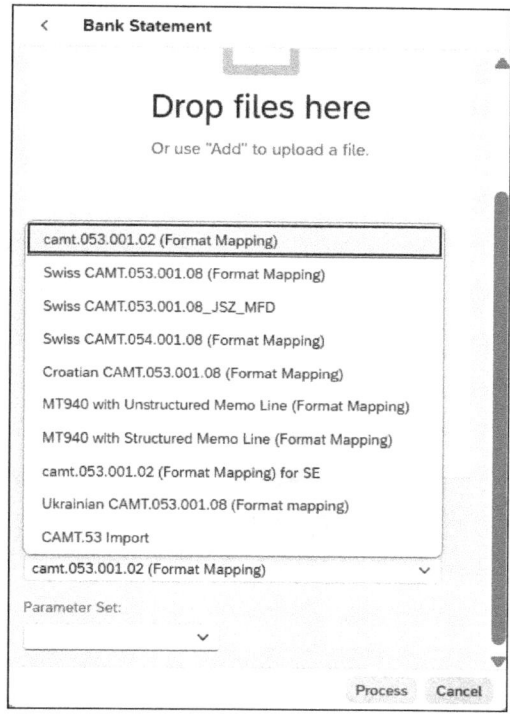

Figure 8.19 Manage Incoming Payment Files: Import Bank Statement

AI Assistant

At regular intervals, XML-based formats undergo updates. Global standard setters such as Common Global Implementation (CGI) regularly update these formats. We have described this in detail, particularly in Appendix A on the ISO 20022 transformation. SAP also uses artificial intelligence to assist with formats and has integrated AI intelligence into the Map Format Data apps.

> **AI-Assisted Merge of Format Mappings Is Only Available for the Public Cloud**
>
> The functionality is available across industries but is currently only accessible in SAP S/4HANA Cloud Public Edition. As a prerequisite, SAP must activate the function, and the configuration between SAP S/4HANA Cloud and the large language model (LLM) must be completed. In addition, the user must be assigned the SAP_BR_BPC_EXPERT role. At the

8 Mapping Format Data

> time of writing (June 2025), the functionality is still in beta status and may therefore not yet be available. For more about SAP's AI functionalities, visit *http://s-prs.co/v605301*.

SAP regularly delivers format templates with version updates aligned with global and local standards. These templates can be copied and adjusted to meet the specific needs of banks. During an upgrade, SAP S/4HANA provides the merge function within the Map Format Data apps. With the help of AI, format upgrades—for example, from version 03 to 09—are significantly simplified.

The AI compares the individually customized version with the template delivered by SAP S/4HANA and suggests adjustments to meet the specific requirements of the bank. To do this, the version to be adjusted must be selected, and the target version must be chosen via the **Merge** button. The AI functionality can then be activated to display comparison results (see Figure 8.20).

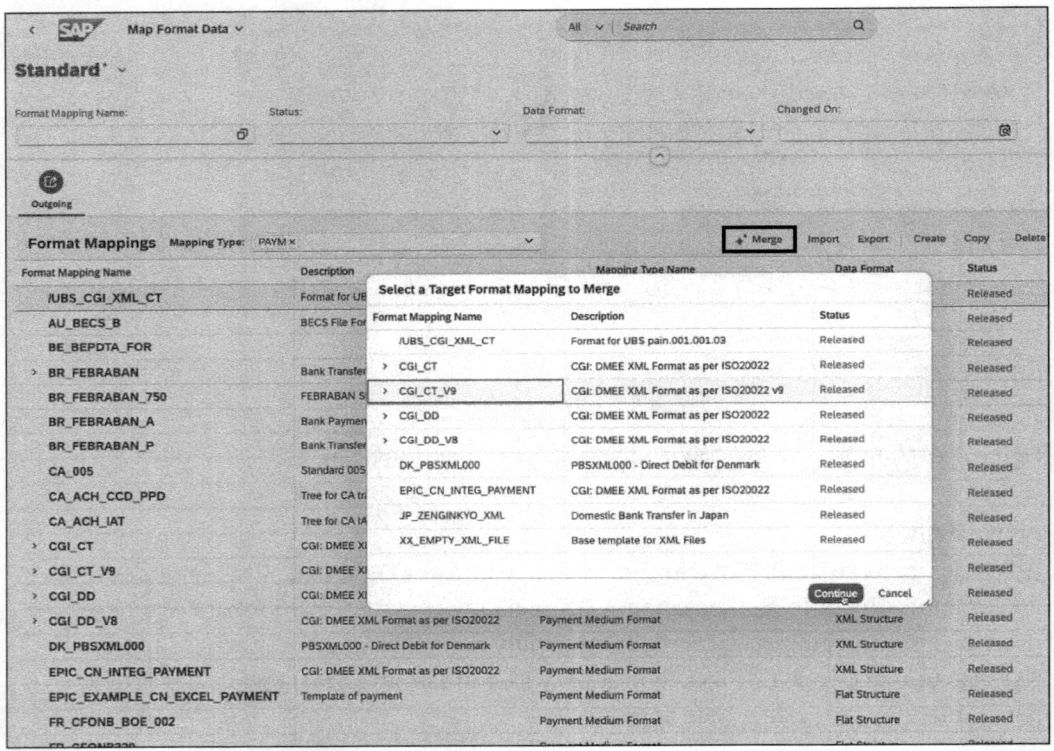

Figure 8.20 Merge Format Mapping with AI

After you click the **Merge** button, you enter the AI-supported merge management, which provides field suggestions to help transfer the old format into the new version. Before doing so, you need to define a new format name for your target format (see Figure 8.21).

8.2 Functions and Configuration

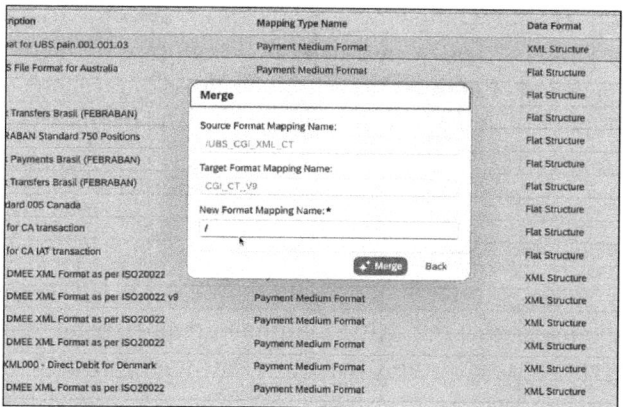

Figure 8.21 New Format Mapping Name

After you click the **Merge** button here, the new target format is created with AI-generated adjustment suggestions to finalize it. If you click the **Merge Log** button (see Figure 8.22), you can review which fields have been modified by the AI. These suggestions are not mandatory and can be individually adjusted as needed.

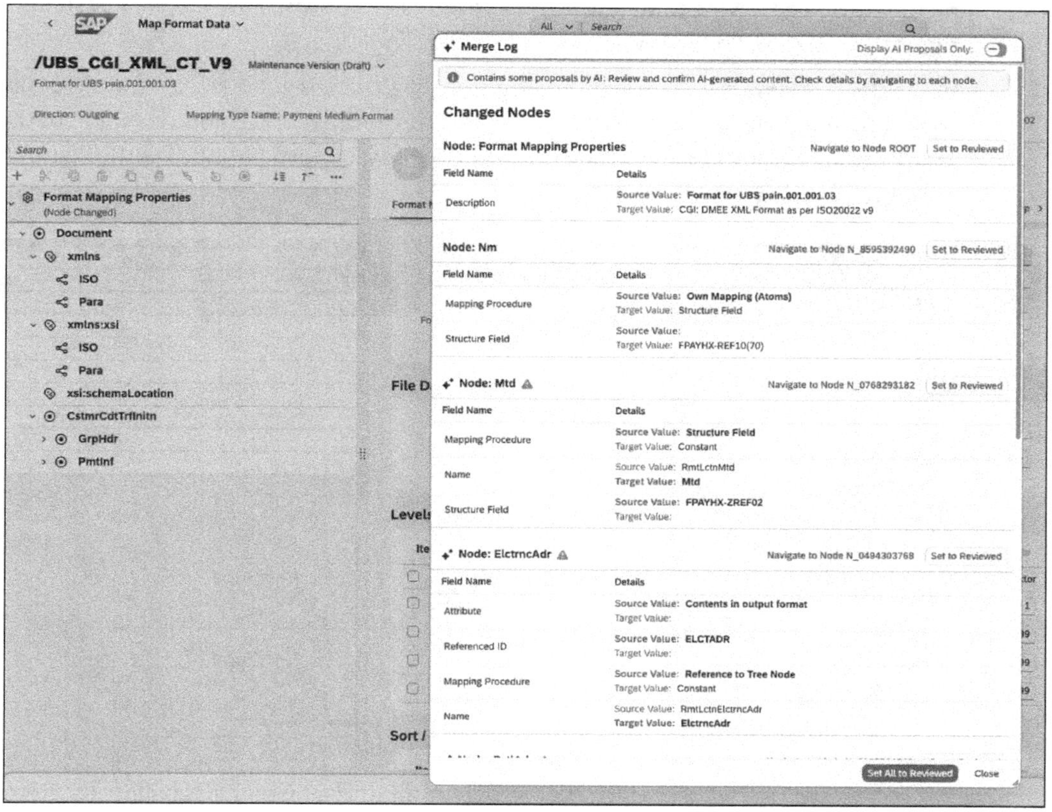

Figure 8.22 Merge Log AI

Overall, the Map Format Data apps not only complement the classic DMEEX options but can also, as already implemented in SAP S/4HANA Cloud Public Edition, completely replace it. In particular, the integration of AI into the technical process of format customization will significantly simplify configuration and drastically reduce implementation times.

8.2.3 Map Format Data for Advanced Payment Management

For advanced payment management, SAP provides a dedicated variant that, when accessed, applies filters to mapping types APCT (APM customer credit transfer) and APDD (APM direct debit). This section discusses the Map Format Data for Advanced Payment Management app.

The configuration options within the Map Format Data for Advanced Payment Management app are largely similar to those introduced in Section 8.2.1 and Section 8.2.2. Therefore, this section focuses solely on the integration within advanced payment management. We covered advanced payment management in detail in Chapter 4. Here, we will explain how to use the Map Format Data for Advanced Payment Management app with the input manager, within the customizing tree under **Financial Supply Chain Management** • **Advanced Payment Management** • **External Interfaces** • **File Handler** • **Basic Configuration** • **Define Converter (New)**.

In the **Define Converter (New)** customizing activity, there is a folder with the old designation **Define IFME Settings**. Within this converter setting, the format mapping of the Map Format Data for Advanced Payment Management app can be linked to the inbound converter, as shown in Figure 8.23.

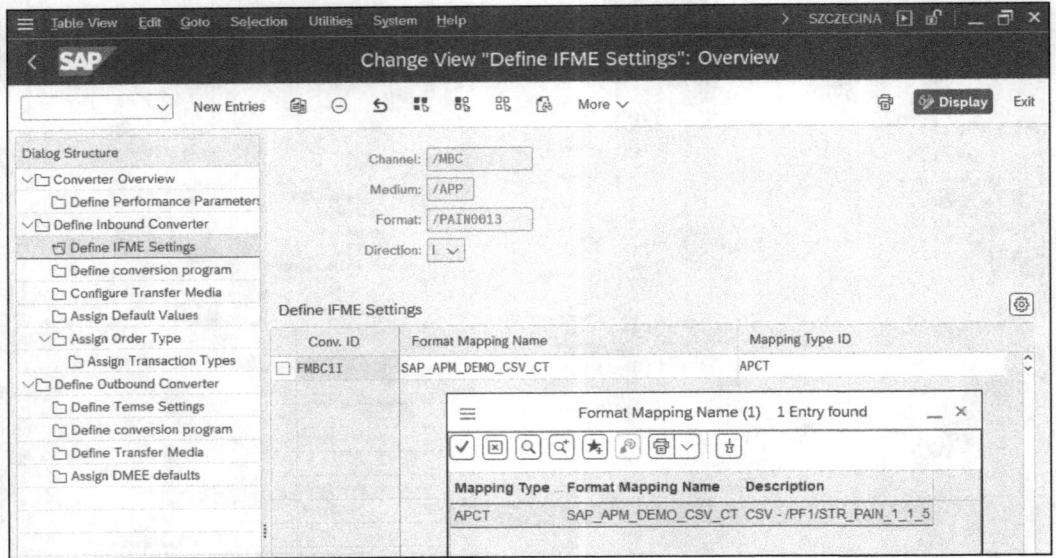

Figure 8.23 Advanced Payment Management Input Converter Configuration

8.2 Functions and Configuration

Via the **New Entries** button, you can add a line in the configuration for **Define IFME Settings** and link the converter with the respective Map Format Data app mapping.

After the configuration, the converter can be selected in the Manage Incoming Payment Files from Banks app (F1680), allowing the file to be imported into the system via drag and drop with the **Import** button, as shown in Figure 8.24.

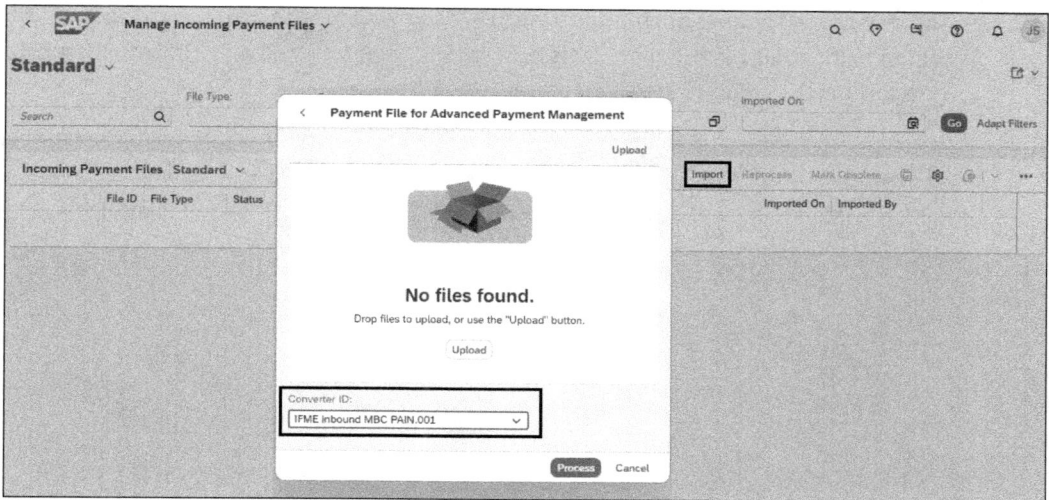

Figure 8.24 Manage Incoming Payment Files: Advanced Payment Management Import

After a successful import, the files will be displayed in the overview within the Manage Payment Files from Bank app (see Figure 8.25).

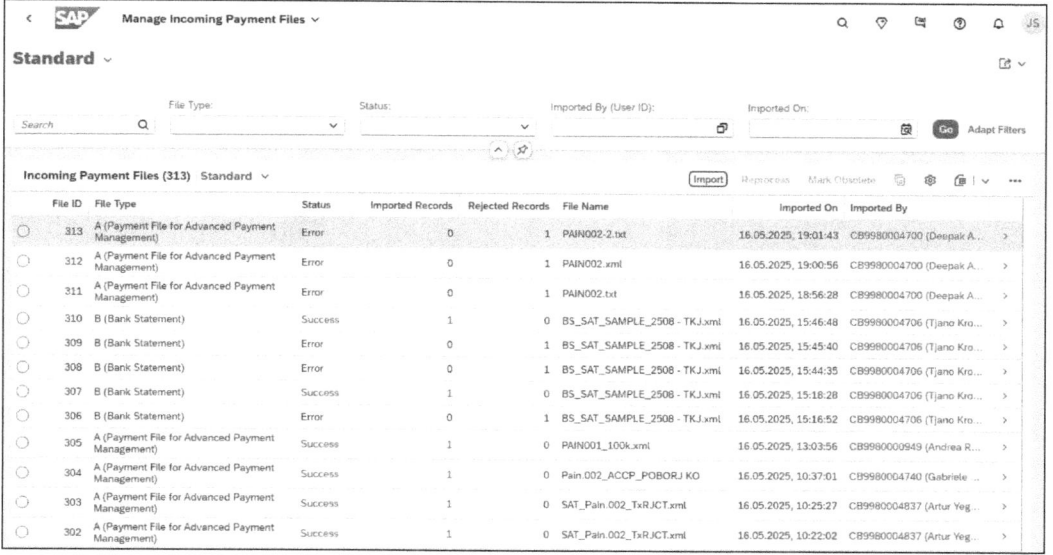

Figure 8.25 Manage Incoming Payment Files: Overview

537

8.3 Summary

The Map Format Data apps have evolved from a simple successor to DMEEX in SAP S/4HANA Cloud Public Edition into an all-round tool with AI features, offering the flexibility needed to tackle challenges in payment formats. In addition to advanced payment management, the Map Format Data apps can also be used in treasury correspondence for trade confirmations and FX deals (MT300/MT320). This expands the flexibility beyond payment formats and makes the apps valuable tools for financial transactions that require structured messaging.

Understanding the Basics of Map Format Data for Payments

For the latest info about the Map Format Data apps, visit the following helpful link for the SAP Learning Journey for Map Format Data: *http://s-prs.co/v605302*.

Chapter 9
Bank Statements

Bank statements serve as the primary means of communication between a company and its banks, making them a crucial element in maintaining accurate and up-to-date financial records. This communication ensures alignment with the external banking environment, providing transparency and confirmation of all executed transactions.

The bank statement acts as a window into the financial relationship with a bank, offering the final and most reliable validation of payments, collections, and overall account activity. It is through this daily or periodic exchange that organizations can verify the integrity of their internal records and maintain trust in their financial operations.

In this chapter, we will focus on the most important aspects of bank statement processing in SAP, providing you with a clear starting point for understanding how this critical function supports payment operations. Bank statement processing can be a complex area, involving numerous variations, formats, and configuration options that must be carefully tailored to your organization's needs. Our goal here is to highlight the key elements and foundational setup required to help you understand where and how to begin with a bank statement implementation in SAP S/4HANA. Because this book is primarily centered around payments, we have intentionally limited the scope of this chapter to cover only the most essential topics. This approach offers a solid introduction to the bank statement process without delving too deeply into specific scenarios or exceptions. Processing bank statements in SAP S/4HANA is a vast topic that could easily fill its own book. In this chapter, we just want to give you a high-level overview of the most important functionalities. We'll cover what formats can be used and how they're loaded, the purpose of the bank statements themselves, and how to monitor them. We also briefly touch on reconciliation and postings, and the key configurations needed to post, process, and monitor statements. Of course, there are many other functionalities, but the main goal of this chapter is to provide a comprehensive overview of what bank statements are, why they matter for payments, and how they connect to treasury applications.

9.1 Bank Statement Processing

To run payments and successfully run a company, you need to have sufficient funds available. One of the key sources of information about your financial position is the *bank statement*, which provides a detailed record of all incoming and outgoing transactions. Bank statement processing in SAP S/4HANA is critical for effective cash management, as described in Chapter 1, which emphasized that cash visibility and accurate liquidity tracking are fundamental to a company's financial operations. Each cash and treasury process starts with bank statements: They provide available cash balances, which are essential data needed for subsequent financial activities. Bank statements play a crucial role in financial accounting, along with treasury functions, as they provide a comprehensive record of all financial transactions made through the company's bank accounts. The treasury focuses on managing cash flow, liquidity, and financial risk, whereas bank statements support accurate financial accounting by offering vital data for recording and monitoring cash movements.

In this section, we'll dive into the critical topic of bank statements. We'll explain what bank statements are, their significance in your organization's financial processes, and how to load them into SAP. You'll also learn about the most commonly used formats for bank statements worldwide and in the US and how these formats can be integrated into SAP S/4HANA. In addition, we'll cover the vital role that bank statement information plays in confirming and reconciling transactions—including both outgoing payments and incoming collections. We'll also explore the most important applications in SAP to monitor end-of-day and intraday bank statements, along with the reconciliation activities tied to them. As we discussed earlier, bank statements form the foundation of cash management activities, allowing you to confirm and update the transactions your organization executes every day.

9.1.1 Overview

During reconciliation, bank statements provide essential insights into payments executed by the company, returned payments that could not be processed by banks, exchange rates applied, and bank charges incurred. This information helps identify discrepancies and ensures that the accounting records accurately reflect your company's financial position. In addition, it is only while processing bank statements that you learn which customers have paid you and what the payments are for, allowing for proper allocation and recording of incoming funds.

Moreover, bank statements are indispensable for updating both the bank ledger and the subledger, maintaining consistency between the general ledger (G/L) and subsidiary accounts. This process supports accurate financial reporting, subledger position management, compliance with accounting standards, and effective cash management. By processing bank statements, SAP S/4HANA automatically updates cash balances and provides insights into available funds, helping the treasury team make informed decisions about liquidity.

Due to all this, it is essential to actively monitor the processing of bank statements and promptly raise alerts when statements are missing. The timely processing of bank statements in SAP S/4HANA is crucial because it not only supports liquidity tracking but also ensures the smooth functioning of reconciliation processes and accurate cash position reporting. Without this foundation, other treasury processes, such as cash forecasting, payment execution, and credit limit management, are significantly impacted, leading to potential inefficiencies and financial risks.

Accurate and timely bank statement processing also is vital for clearing subledger customer and vendor invoices. This is crucial for promptly identifying which payments have been made, allowing the business to stop unnecessary dunning and other follow-up activities. Without clearing these invoices efficiently, there is a risk of sending payment reminders or initiating collection activities for already settled accounts, which can damage customer relationships and create administrative inefficiencies.

Monitoring the timely receipt and processing of bank statements helps maintain accurate financial records and allows for quick resolution of discrepancies. Delays or gaps in processing can hinder the company's ability to track available funds, forecast cash flow, and execute payments efficiently. Consequently, ensuring that bank statements are processed without delay is vital to maintaining the financial stability and operational efficiency of your organization.

Bank statements contain a wide range of important information, including the following most important elements:

- **Account balances**
  Opening and closing balances for the statement period
- **Statement number**
  Sequential number for tracking and archiving
- **Bank account number**
  The bank account number
- **Transaction dates**
  Bank execution date, posting, and value dates for each transaction
- **Payment references**
  Information on payments executed and their unique identifiers
- **Currency information**
  Details about the currency in which each transaction occurred

There are multiple bank statement formats used around the world, each providing different types of information and varying levels of detail. The goal is to have the camt formats, based on the ISO 20022 standard, eventually become the global norm (see Appendix A), but this transition may take several more years to be fully implemented.

9 Bank Statements

In the meantime, companies continue to work with a variety of formats depending on the region, bank, and specific requirements. Throughout this chapter, we will focus on some of the most important formats commonly used worldwide (although there are more available):

- **MT940**
  A SWIFT message format (TXT format) commonly used around the world for end-of-day account statements. It provides detailed information about account transactions and balances, making it suitable for reconciliation and cash management.

- **Camt.053**
  Part of the ISO 20022 standard, designed to replace MT940 and other legacy formats. Camt.053 offers a richer and more structured data format, allowing for more detailed reporting and improved interoperability between financial systems. It is gradually becoming the global standard for electronic bank statement reporting.

- **Zengin**
  A format used predominantly in Japan, managed by the Zengin system for domestic fund transfers and bank statement reporting. Recently, it has moved to XML formats as part of modernization and standardization efforts.

- **Lockbox**
  A format used primarily in the United States and Canada, designed to support the processing of incoming payments—particularly checks—via a bank's lockbox service. In this setup, customers send their payments directly to a special post office box managed by the bank, which then collects, processes, and deposits the payments on behalf of the company. Lockbox services help accelerate cash application, reduce manual effort, and improve overall efficiency in accounts receivable processing.

  Lockbox data is typically transmitted using the BAI or BAI2 format. These standardized flat file formats contain detailed transaction and remittance information, including check numbers, deposit dates, payment amounts, and lockbox location details. BAI2, the more modern version, offers greater flexibility and more structured data, making it the preferred format for many organizations.

- **Multicash**
  Once a popular format and software solution mainly used in Europe for centralized cash management and reporting. Although it has become somewhat obsolete, some banks—particularly in Australia and New Zealand—still provide statements in this format.

Understanding these formats and their specific uses is essential for efficient bank statement processing and reconciliation, especially for multinational companies dealing with diverse banking requirements across different regions. The existence of various bank statement formats around the world presents technical challenges for corporations, as each format has its own structure, data fields, and processing requirements. This diversity requires proper planning and management within financial systems to ensure seamless integration and accurate reconciliation. Implementing support for multiple formats in SAP S/4HANA can be complex as it involves configuring the system to handle different data structures and ensuring compatibility with both local and global banking practices. Throughout this chapter, we will highlight how to effectively implement and manage diverse bank statement formats within SAP S/4HANA, focusing on best practices and configuration tips to optimize processing and minimize errors.

9.1.2 Bank Statement Reconciliation

Bank statement reconciliation in SAP is the process of matching and verifying the transactions recorded in the company's bank accounts against the actual entries provided by the bank in the electronic bank statement. This reconciliation is typically done using the electronic bank statement functionality in SAP, where incoming bank files are automatically uploaded and processed. The system attempts to clear open items in the subledger—such as customer and vendor invoices—based on predefined rules and algorithms. Reconciliation is important because it ensures the accuracy of financial records, confirms that all payments and receipts have been correctly posted, and provides a real-time view of the company's cash position. It also supports compliance, audit readiness, and reliable financial reporting.

However, issues may arise due to inconsistent or incomplete data in bank statements, incorrect configuration, poorly maintained master data, or missing references in payments. These issues can prevent automatic clearing, resulting in large volumes of open items that need to be reviewed and cleared manually—a process that is both cumbersome and time-consuming. All transactions that cannot be reconciled automatically require manual intervention, which significantly increases the workload for finance teams and delays financial closing processes. Therefore, implementing proper tools and configurations that support high levels of automation in reconciliation should be a top priority for the treasury and the teams responsible for cash reconciliation. Automating this process not only improves efficiency and accuracy but also frees up resources to focus on more strategic tasks.

In SAP S/4HANA, bank statement processing typically involves two levels of postings to ensure accurate tracking and reconciliation of cash movements. These two levels are reflected in different subledgers and serve distinct purposes in the reconciliation process:

- **Bank subledger posting (level 1)**
  This is the initial posting that occurs when the bank statement is processed. It records the movement of cash between the bank main account and the bank clearing account. This entry reflects the actual cash flow in and out of the bank as reported by the bank statement. For example, when a payment is received, the system debits the bank clearing account and credits the bank main account, indicating that funds have been received by the bank but not yet applied to a specific subledger item.

- **Subledger clearing (level 2)**
  This is the second level, where the transaction is matched against open items in the customer or vendor subledger. At this stage, the system clears the bank clearing account by posting against the final G/L account or settling an open item such as a

customer invoice, vendor payment, or payment run. This step ensures that the transaction is fully accounted for and correctly reflected in the financial statements.

This two-step process allows SAP S/4HANA to clearly separate the actual cash movement from the settlement of specific subledger items, providing greater control, traceability, and accuracy in financial accounting and reconciliation processes. Using a two-level posting approach for all bank statement transactions in SAP S/4HANA is highly recommended as it provides better control and transparency in financial accounting. Although it is technically possible to configure the system to post directly from the bank main account to the cleared subledger items (e.g., customer or vendor accounts), this one-step method has several disadvantages. If an incorrect posting is made and subsequently reversed, then the reversal impacts the main bank G/L account directly. This can lead to discrepancies between the G/L account balance and the actual bank statement balance.

By using two-level postings—where the first level posts between the main bank account and the clearing bank account, and the second level clears the clearing bank against the final G/L or subledger—only the bank statement transactions affect the main bank account. This approach ensures that the G/L balance in the main bank account always aligns with the bank statement balance, making reconciliation more accurate and reducing the risk of errors or mismatches. It also simplifies audit trails and improves the overall reliability of cash reporting and financial close processes.

Bank clearing accounts in SAP S/4HANA play a central role in separating the movement of funds from the final settlement of transactions. In the classic approach, it was common practice to maintain two separate clearing G/L bank accounts per bank—one for incoming payments and one for outgoing payments. This separation helped simplify reconciliation by clearly distinguishing between receipts and disbursements, making it easier to track and clear items. In some cases—especially in the past, when system capabilities were more limited—companies would even create multiple clearing accounts per bank or payment type to support manual reconciliation processes.

However, with advancements in SAP S/4HANA solutions like automation, improved reconciliation tools, and better reporting capabilities, it has become increasingly common to use a single clearing G/L bank account per bank for both incoming and outgoing flows. This streamlined setup reduces the number of accounts to manage, simplifies configuration, and can still support efficient reconciliation when properly structured. Ultimately, the decision depends on a company's specific operational needs and reconciliation practices, but the trend is clearly moving toward a more consolidated and simplified bank clearing account structure.

9.1.3 Bank Statement Monitor

As described in this chapter, it is expected that all bank statements are loaded automatically into the system, minimizing the need for manual intervention. You are primarily

9.1 Bank Statement Processing

responsible for monitoring the bank statements and their statuses to ensure smooth processing. This monitoring can be carried out in several places within SAP S/4HANA; however, the most important and user-friendly tool for this purpose is the Bank Statement Monitor app (F6388; see Figure 9.1). This app provides a centralized view of all incoming bank statements, their processing status, and any potential issues that may require attention.

If your organization is not yet on SAP S/4HANA, the only viable option to access advanced bank statement monitoring functionalities is SAP Bank Communication Management, via Transaction FTE_BSM. This transaction provides similar capabilities for tracking and analyzing bank statement processing in SAP ERP systems.

House Ban...	Compan...	House ...	Latest Statement...	Processing Status	Difference Status	Serial Number Status	Reconciliation Status	Closing Balance		Created On	
EUR01	1000	ABNNL	02.04.2025	⊗	⊗	⊘	⊗	50,00	EUR	31.05.2025, 10:50:20	>
EUR02	1000	ABNNL	20.05.2025	⊗	⊗	⊘	⊘	0,00	EUR	31.05.2025, 10:50:20	>
EURM1	1000	ABNNL	20.05.2025	⊗	⊗	⊘	⊘	-9.850,00	EUR	31.05.2025, 10:50:20	>
JPY01	1000	ABNNL	10.03.2025	⊗	⊗	⊗	⊘	500	JPY	31.05.2025, 10:50:20	>
JPY02	1000	ABNNL		⊗	⊗	⊗	⊗	0	JPY	31.05.2025, 10:50:20	>
GBP01	1000	CITGB	23.01.2024	⊗	Not Applicable	⊗	⊘	0,00	GBP	31.05.2025, 10:50:20	>
GBP04	1000	CITGB		⊗	⊗	Not Activated	⊗	0,00	GBP	31.05.2025, 10:50:20	>
EUR01	1000	DHBNL	08.03.2023	⊗	⊗	⊗	⊗	5,00	EUR	31.05.2025, 10:50:20	>
EURM1	1000	DHBNL	14.03.2023	⊗	⊗	⊘	⊗	7,00	EUR	31.05.2025, 10:50:20	>
EUR01	2000	ABNNL	15.12.2023	⊗	⊗	⊗	⊗	-7.123,00	EUR	31.05.2025, 10:50:20	>
EUR15	2000	ABNNL	08.03.2023	⊗	⊗	⊗	⊗	5,00	EUR	31.05.2025, 10:50:20	>
01	2000	BARC		⊗	⊗	⊗	⊗	0,00	GBP	31.05.2025, 10:50:20	>
EUR01	2000	DHBNL	21.02.2024	⊗	⊗	⊗	⊗	5,00	EUR	31.05.2025, 10:50:20	>

Figure 9.1 Bank Statement Monitor

The main purpose of the Bank Statement Monitor app is to provide a clear and efficient way to track the status of each bank statement. It allows you to quickly identify which statements have been successfully received, which are still missing, and the current processing status of each. Statements are displayed based on a selected **Key Date** and **Company Code**, giving flexibility in how the data is reviewed. The app offers two viewing options: **Single Day**, which shows the status of statements for a specific date, and **Last 14 Days** (see Figure 9.2), which is particularly useful for identifying any statements that have not been received over the past two weeks. The monitor works based on the Bank Account Management (BAM) in SAP S/4HANA setup described in Chapter 5.

When bank statements are received electronically, you should expect all statements to appear with a green status, indicating they have been successfully loaded into the system. A fully automated and consistent electronic statement process is essential for ensuring accurate and timely reconciliations. If any statements are missing, it can lead to a range of downstream issues, including incomplete reconciliations, mismatches in G/L balances, and delays in financial reporting. Exceptions to this expectation typically apply only to manually loaded bank statements or in scenarios where certain banks do not provide statements daily. In such cases, monitoring must account for these known gaps.

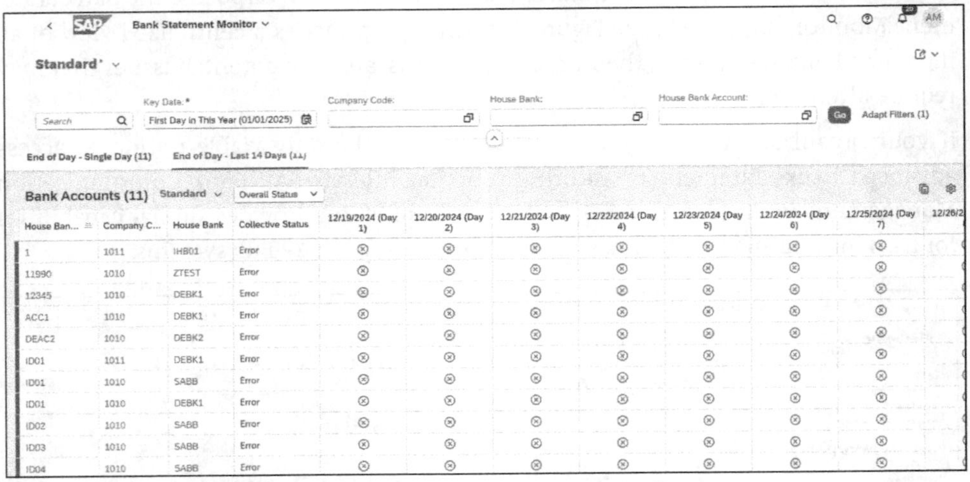

Figure 9.2 Bank Statement Monitor: Last 14 Days

The main purpose of the Bank Statement Monitor app is to verify and compile a clear overview of the status and content of each bank statement. This includes checking whether the statement was received, reviewing the opening and closing balances, confirming whether the statement was successfully reconciled, ensuring the serial number sequence is correct, and identifying any discrepancies between the G/L balance and the bank statement balance. These checks are crucial for maintaining data consistency and financial accuracy. You also can customize the view by clicking on the **Settings** button, where you can select which fields to display in the report (see Figure 9.3), allowing for a tailored and efficient monitoring experience based on individual or organizational needs.

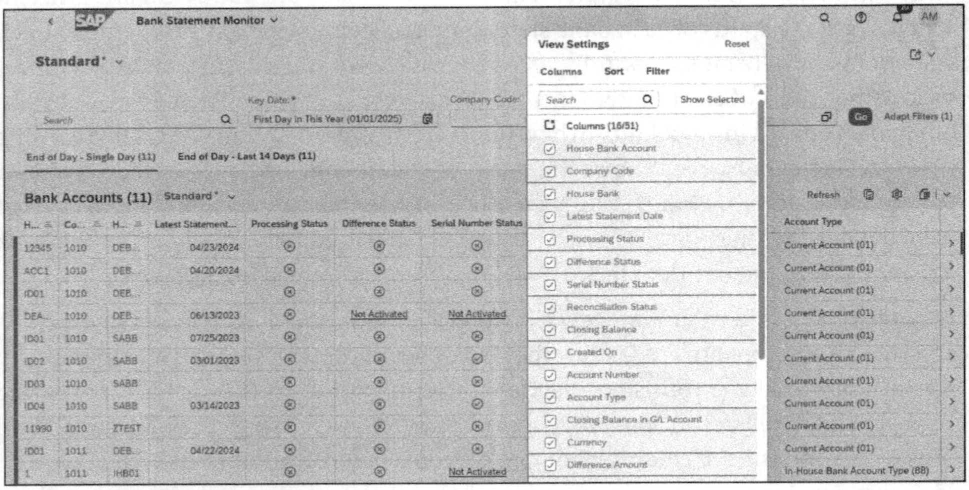

Figure 9.3 View Settings: Bank Statement Monitor

If you click the > arrow at the far right of the line item, you can display statement details (see Figure 9.4) and, most importantly, the **Reconciliation Status** tab, showing the reconciliation status.

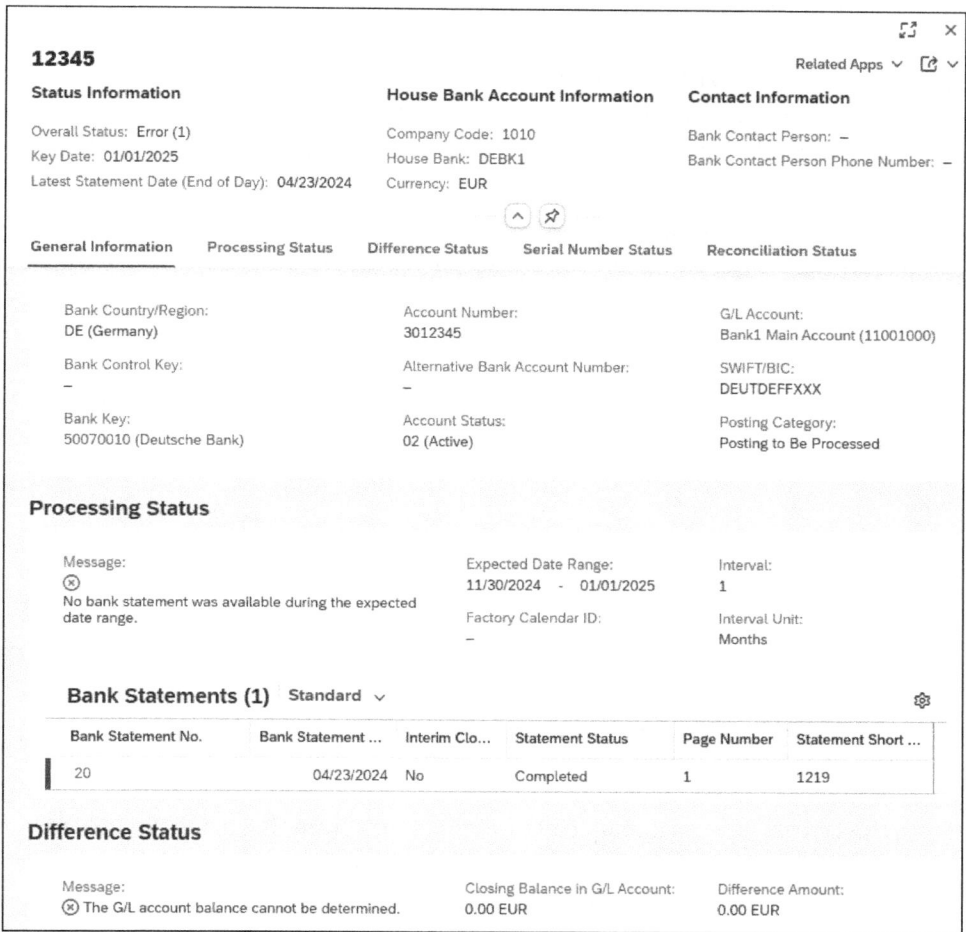

Figure 9.4 Display Bank Statement Details

The most important activity related to all bank statements is ensuring that reconciliation is performed accurately so that the balance sheet and all relevant subledgers are properly aligned and up to date. A successful reconciliation process ensures that all incoming and outgoing transactions recorded in the bank statement are correctly reflected in the company's financial records, maintaining the integrity of financial accounting. Any discrepancies can lead to reporting errors, missed transactions, or unbalanced accounts. In Section 9.1.9 and Section 9.2.3, we will explain in more detail how to perform reconciliation using the Reprocess Bank Statement Items app (F1520). We'll provide step-by-step guidance to support a smooth and reliable reconciliation process.

9.1.4 Manage Bank Statements

The Bank Statement Monitor app focuses on checking which bank statements have been received for each specific bank account on a particular day, helping users identify missing or unprocessed statements. Meanwhile, the Manage Bank Statements app (F1564; see Figure 9.5) serves a different but complementary purpose. This app provides a comprehensive view of all bank statements, regardless of their processing status, and allows you to monitor, review, and process them directly. In addition to tracking existing statements, the app also supports the manual upload of new bank statements, which is particularly useful in cases in which statements are not received electronically.

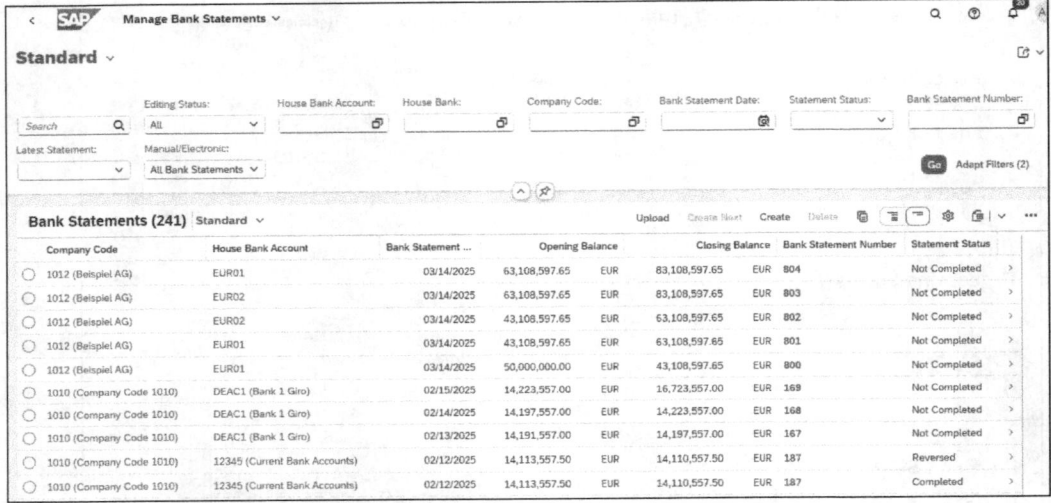

Figure 9.5 Manage Bank Statements

From this screen, you can also upload new statements from files using the **Upload** button, or you can manually create new ones by clicking **Create New**.

If you click any line within the Bank Statement Monitor app, the system will open a detailed screen that shows the selected bank statement (see Figure 9.6). Here, you can check the overall status and review the statement details, including the header information, all line items, and their posting statuses. This view gives you full insight into how the statement was processed and whether any follow-up is needed.

From this screen, you can also post the statement if it's not yet been done and perform postprocessing tasks, such as reconciling bank line items; if necessary, you also have the option to reverse the entire statement. However, it's important to note that reversal should only be done in exceptional cases: Once a statement is reversed, it must be manually reuploaded into the system. This can create extra work and increase the risk of inconsistencies if not handled carefully.

9.1 Bank Statement Processing

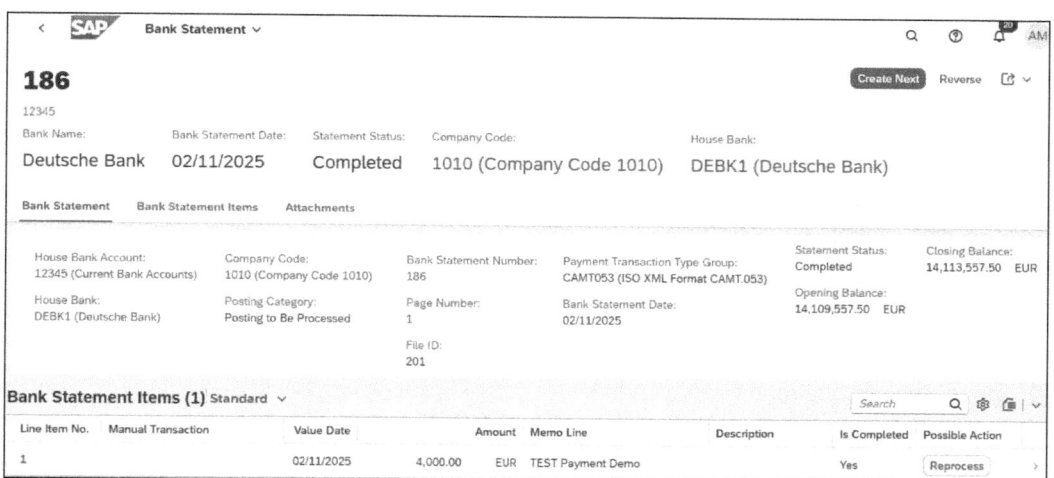

Figure 9.6 Display Bank Statement

When you click the **Reverse** button, the system will immediately reverse the bank statement, and its status will be updated to **Reversed** (see Figure 9.7). Once a statement is reversed, it can no longer be processed or corrected in the system. The only available action at that point is to recreate or reupload the statement manually. This action should be taken with caution as it effectively resets the processing of that statement and may require coordination with the bank or your internal team to obtain and upload the correct version again.

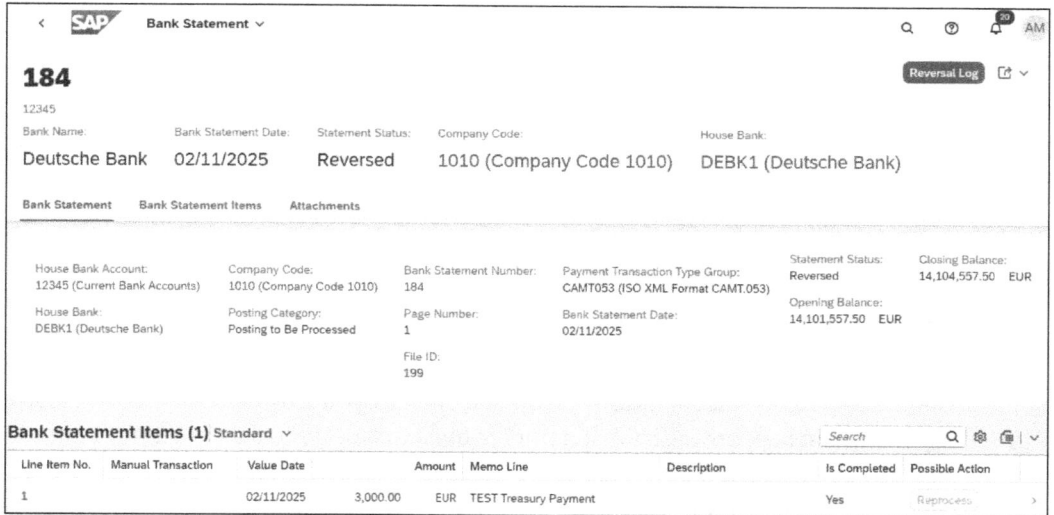

Figure 9.7 Statement Reversed

Reversing a bank statement directly within the Bank Statement Monitor app is a feature available after SAP S/4HANA 2022. In earlier versions of SAP S/4HANA or in SAP ERP, this functionality was not available through the user interface. Instead, users had to execute program RFEBKA96 via Transaction SE38 to delete an already uploaded statement. This process was cumbersome, typically required involvement from superusers, and often demanded additional approvals and oversight from the IT department to ensure that statements were correctly removed without compromising financial data. To address these challenges, SAP introduced more user-friendly and accessible functionalities in recent releases. However, despite the convenience, reversing statements should remain a tightly controlled and restricted action as it can impact financial accuracy and requires subsequent reprocessing.

When you click the highlighted **Bank Statement Number**, the system will allow you to navigate to several related apps (see Figure 9.8), including the Bank Statement Monitor app, the Manage Bank Statements app (which you are already in; selecting this will take you back to the overall view), the Reprocess Bank Statement Items app, and the Upload Bank Statements app. This provides quick access to key functionalities for reviewing, correcting, and managing bank statement data efficiently.

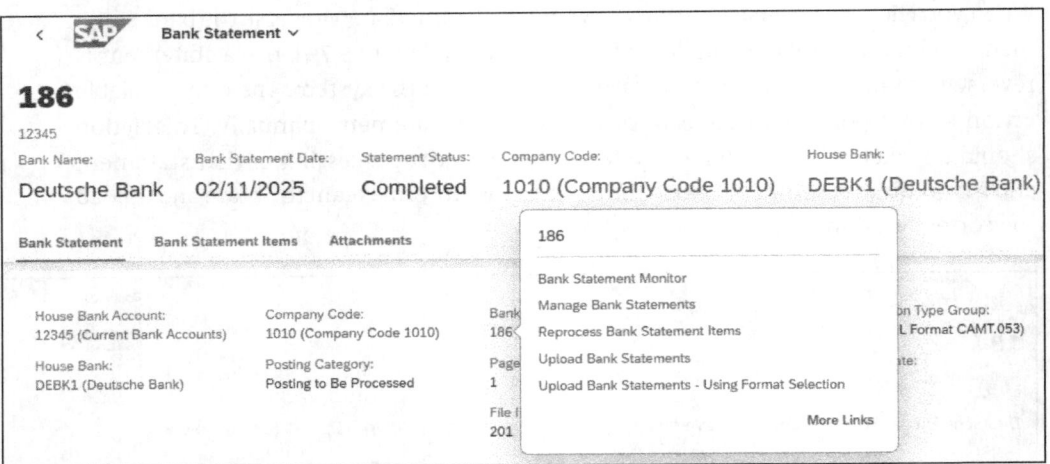

Figure 9.8 Bank Statement Options

9.1.5 Upload Bank Statement

So far, we have covered how to check, monitor, and manage bank statements within the system, including what actions can be taken once they are available. However, before any of this can happen, the bank statements must first be uploaded into the system. Ideally, this should be done automatically, to ensure consistency, reduce manual effort, and minimize the risk of errors. In this section, we will explore the different options available for uploading bank statements, including how to do it manually, when necessary, and how to configure and run automatic uploads to streamline the process and support daily operations.

9.1 Bank Statement Processing

Management Bank Statements App

When you click the **Upload** button within the Manage Bank Statements app, the system will open the Manage Incoming Payment Files app (F1680; see Figure 9.9), which allows you to upload a bank statement directly. This interface simplifies the manual upload process, making it more intuitive and accessible. Alternatively, you can use the Upload Bank Statement app (FF.5), which serves as a modernized version of traditional SAP GUI Transaction FF.5. Both options offer the same core functionality, enabling you to upload bank statements into the system, but the SAP Fiori app provides a more streamlined and visually enhanced user experience.

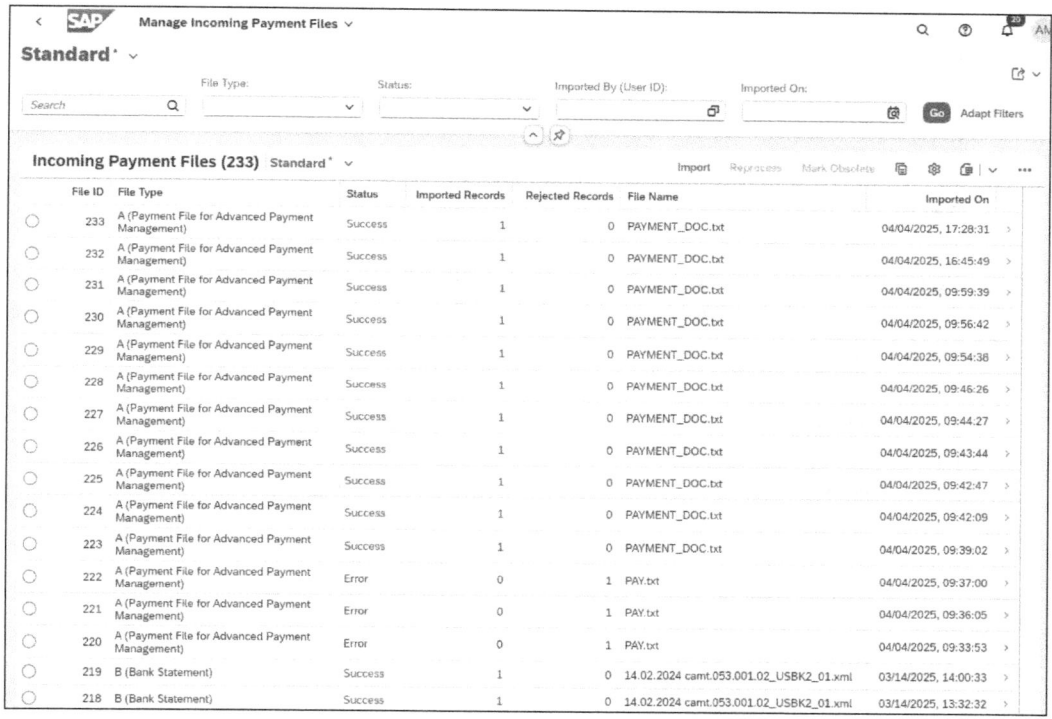

Figure 9.9 Manage Incoming Payment Files

You can use this app to import electronic payment files, such as bank statements, payment rejections, intraday bank statements, lockbox batches, and payment files for advanced payment management. After the bank statements and lockbox batches are imported and posted, they can be processed further using the Manage Bank Statements, Manage Lockbox Batches, or Manage Payments apps, if needed.

If you want to upload a new bank statement, click the **Import** button and select the format (see Figure 9.10) of the statement. Each of the formats, and their usage and functions, were described in Chapter 2, Section 2.2.4.

9   Bank Statements

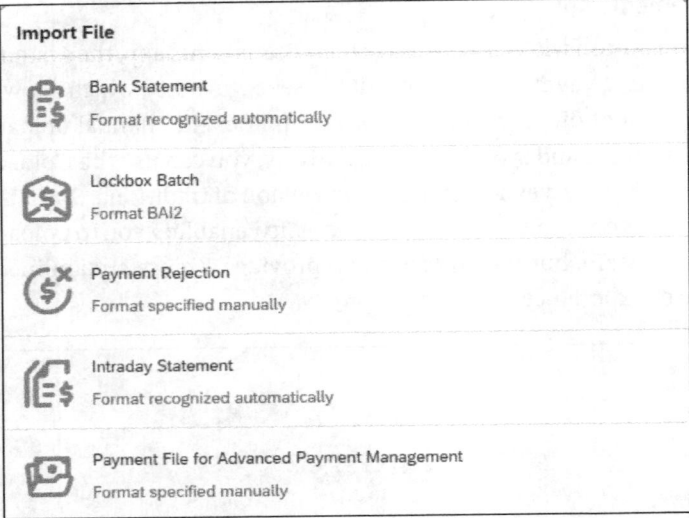

Figure 9.10 Import New File

To select the file, just drag it from your desktop or workstation and drop it in the **Drop files here** area—or click the **Add** button and navigate to where the file is saved (see Figure 9.11).

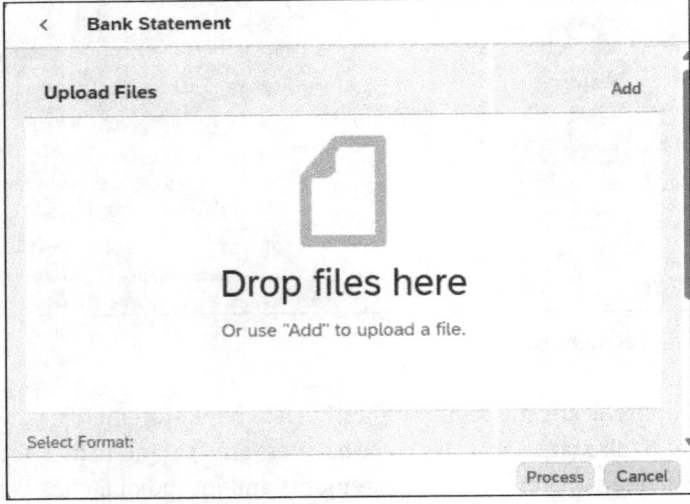

Figure 9.11 Drop Files Here

Select the proper format (e.g., **Bank Administration Institute (USA)**, **MT940 (Germany)**, **camt.053.001.02 (Format Mapping)**, or any other relevant one; see Figure 9.12), then click the **Process** button. The system will now upload the statement.

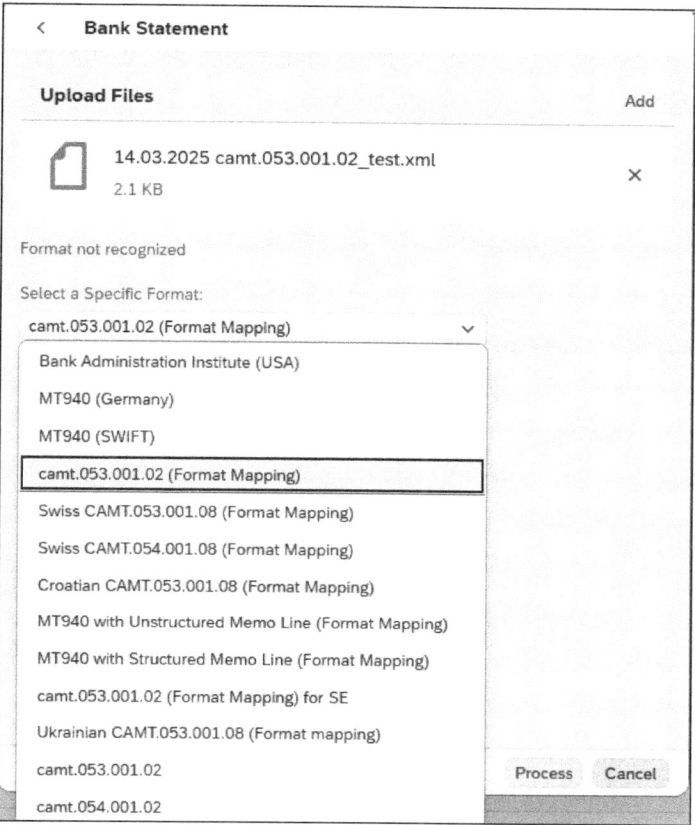

Figure 9.12 Select Statement Format

Navigate back to the initial screen. Finally, you can check whether the statement was uploaded (see Figure 9.13) and created correctly.

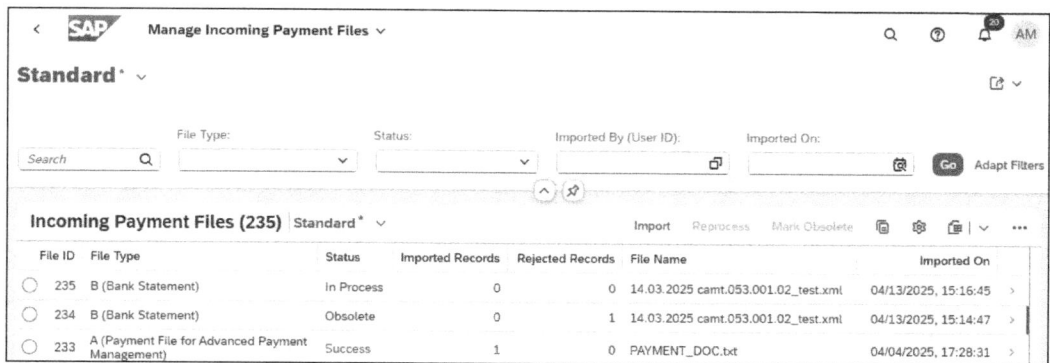

Figure 9.13 Statement Uploaded

9 Bank Statements

Transaction FF.5

An alternative to using the SAP Fiori app for uploading bank statements is the classic Transaction FF.5 (Upload Bank Statement; see Figure 9.14), which is available in SAP GUI. This traditional method is commonly used in SAP ERP systems and remains fully supported in SAP S/4HANA for users who prefer or require the classic interface. Although it lacks the modern look and user experience of the SAP Fiori app, Transaction FF.5 provides the same core functionality for manually uploading bank statements into the system.

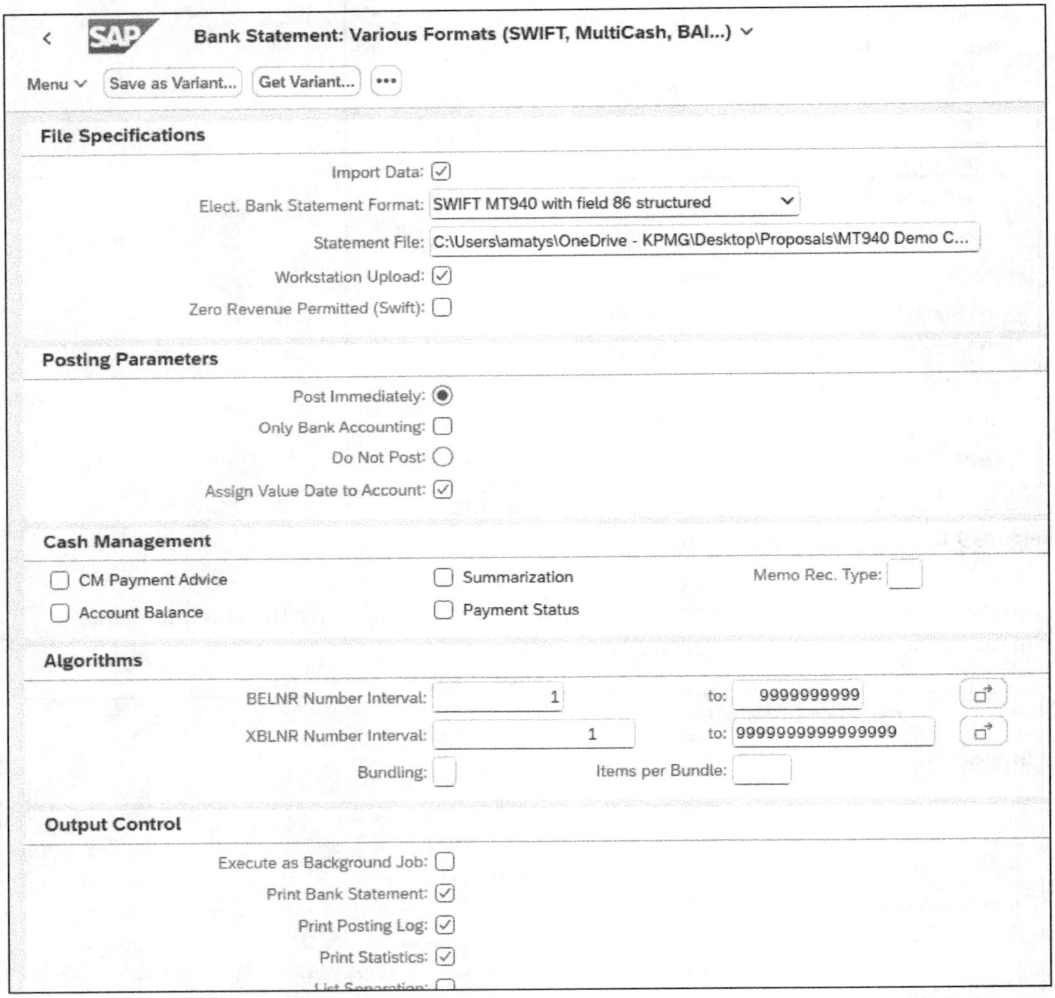

Figure 9.14 Upload Statement with Transaction FF.5

In addition, Transaction FF.5 can be scheduled to run as a background job, enabling the automatic upload of bank statements without user interaction. This setup allows for regular and timely processing of incoming statements, reducing manual effort and supporting a more automated and efficient reconciliation process.

When using Transaction FF.5 to upload a bank statement, the first step is to select the format of the statement you want to upload in the **Elect. Bank Statement Format** field. Transaction FF.5 primarily supports older formats such as MT940 or Multicash, but it is possible to extend the configuration to support newer formats like camt.053 or other XML-based standards, depending on your system landscape.

You then need to select the file to upload in the **Statement File** field. If the file is located on your local machine, choose the **Workstation Upload** option to browse to and select the file directly from your desktop.

Next, you'll need to define the posting parameters, as follows:

- **Do Not Post**
  This option uploads the bank statement without posting any entries to the G/L bank account. You will need to manually post the entries later using Transaction FEBP if you want to post. This is also useful for updating bank balances based on the statements without posting in cash management, as outlined in Chapter 5.
- **Post Immediately**
  With this option, entries in the bank statement are automatically posted to the G/L bank account at the time of upload, allowing for real-time processing.

Another critical part of the upload process is the use of interpretation algorithms, which are applied during the upload to help the system identify and match incoming payments to the correct open items or reference documents. These algorithms typically look for document numbers such as invoice numbers, customer or vendor references, payment references, or clearing documents in the "note to payee" fields of the bank statement. Proper configuration of interpretation algorithms is essential for automated clearing and accurate posting. Once all required fields are selected and the setup is complete, click **Execute**; the system then will process and upload the bank statement accordingly.

For background processing using Transaction FF.5, it is crucial to ensure that the bank statement file is stored daily in the same folder and path as specified in the **Statement File Path** field during the job configuration. The system relies on this fixed path to locate and process the file automatically. In addition, the file name must either match exactly what is defined in the job parameters or be managed dynamically using file name extensions that update based on the date or other variables.

To support this automation, it is common to use STVARV parameters, which allow you to define dynamic variables (e.g., current date) and pass them into the background job to correctly identify and select the file for each day. For background processing, you need to save the selection parameters as a variant and run it together with program RFBASM00—just as you do for all other jobs.

9 Bank Statements

Once Transaction FF.5 is run and the bank statement is successfully uploaded, it's important to ensure that the file is moved to another folder to avoid duplicate processing. This can be done using Transaction CG3Z or through any other automated extension or custom script. However, due to the limitations mentioned earlier—such as strict file paths, naming requirements, and manual handling—SAP began exploring and introducing better alternative solutions, like program FEB_FILE_HANDLING, to streamline and automate the upload and processing of bank statements more efficiently. We'll discuss that next.

Program FEB_FILE_HANDLING

Due to the limitations and rigidity associated with managing fixed file names and paths in background processing using Transaction FF.5, SAP introduced program FEB_FILE_HANDLING to simplify and automate the upload of bank statements. With this program, there is no need to manage or predefine specific file names. Instead, you simply need to prepare a designated folder, and any file placed in this folder will be automatically picked up and uploaded by the system when FEB_FILE_HANDLING (see Figure 9.15) is executed. This significantly reduces manual effort, eliminates dependency on strict naming conventions, and allows for a more flexible and user-friendly upload process. All the required configurations are handled in the background, making it an efficient solution for automated bank statement processing.

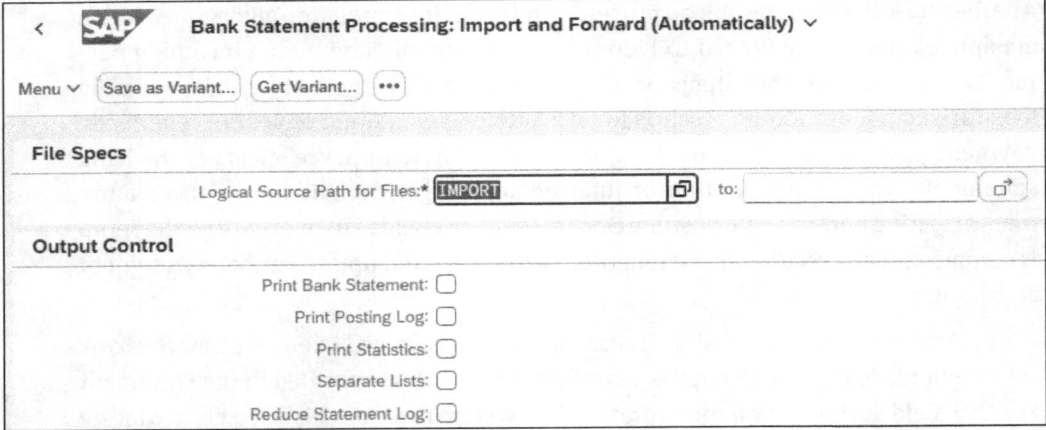

Figure 9.15 FEB_FILE_HANDLING

In Transaction FEB_FILE_HANDLING, the user does not need to manually select any files or parameters. The only input required is to fill the **Logical Path for Files** field, which is derived from the configuration settings maintained in the background. More details on how this configuration is set up will be explained later in Section 9.2.1. The transaction is specifically designed to be executed automatically in the background, making it ideal for scheduled jobs. Once configured, it enables fully automated processing of all files placed in the specified directory, with minimal user involvement, significantly improving efficiency and reliability in bank statement uploads.

There are several reasons that a bank statement upload might fail in the system. One of the main limitations is that only one statement per date can be uploaded. In the background, the system will create a unique entry based on the combination of the date and statement ID. Once this entry is created, it will not be possible to upload the same statement again for that specific date (see Figure 9.16). In addition, the system will automatically verify that the opening and closing balances in the bank statement match correctly. If there is any discrepancy or error, such as mismatched balances or an attempt to upload the same statement more than once, then the upload will fail, and the statement will not be processed.

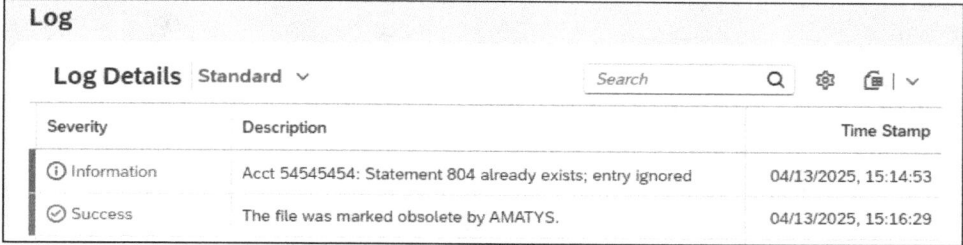

Figure 9.16 Bank Statement Exists

To maintain data integrity, the system stores the uploaded bank statement information in the following tables:

- FEBKO (Electronic Bank Statement Header Records)
- FEBEP (Electronic Bank Statement Line Items)
- FEBRE (Reference Records for Electronic Bank Statement Line Items)

These tables hold the relevant details for each bank statement, allowing for proper reconciliation and ensuring that only valid statements are processed and stored in the system.

9.1.6 Japanese Bank Statements

Due to the specific requirements and structure of Japanese bank statements, particularly the Zengin format, SAP provides two dedicated standard transactions for uploading these files: IBS_JP and IBSX_JP. Transaction IBS_JP is used for uploading TXT format bank statements, while Transaction IBSX_JP is designed for handling XML format statements. Both transactions are available in SAP GUI as well as through the SAP Fiori launchpad, offering flexibility in how users interact with them. These tools are tailored to meet local banking requirements in Japan and ensure proper handling of Zengin-compliant files during the upload and reconciliation processes.

9   Bank Statements

The screens for Transactions IBS_JP and IBSX_JP (see Figure 9.17) are very similar to the Transaction FF.5 upload interface, but with additional features tailored to Japanese banking specifics. One of the key differences is the ability to select local calendars in the **Wareki** field, as Japanese banks often operate on a calendar system different from the Gregorian calendar. In addition, these transactions include a crucial field for the **File ID**, which is essential for handling the fact that Japanese banks send multiple bank statements per day. Each of these statements requires separate processing and posting, and the **File ID** ensures that all statements are properly identified and posted, even when they originate on the same day. This functionality addresses the unique challenges posed by Japanese banking practices, allowing for accurate and efficient processing of multiple daily bank statements.

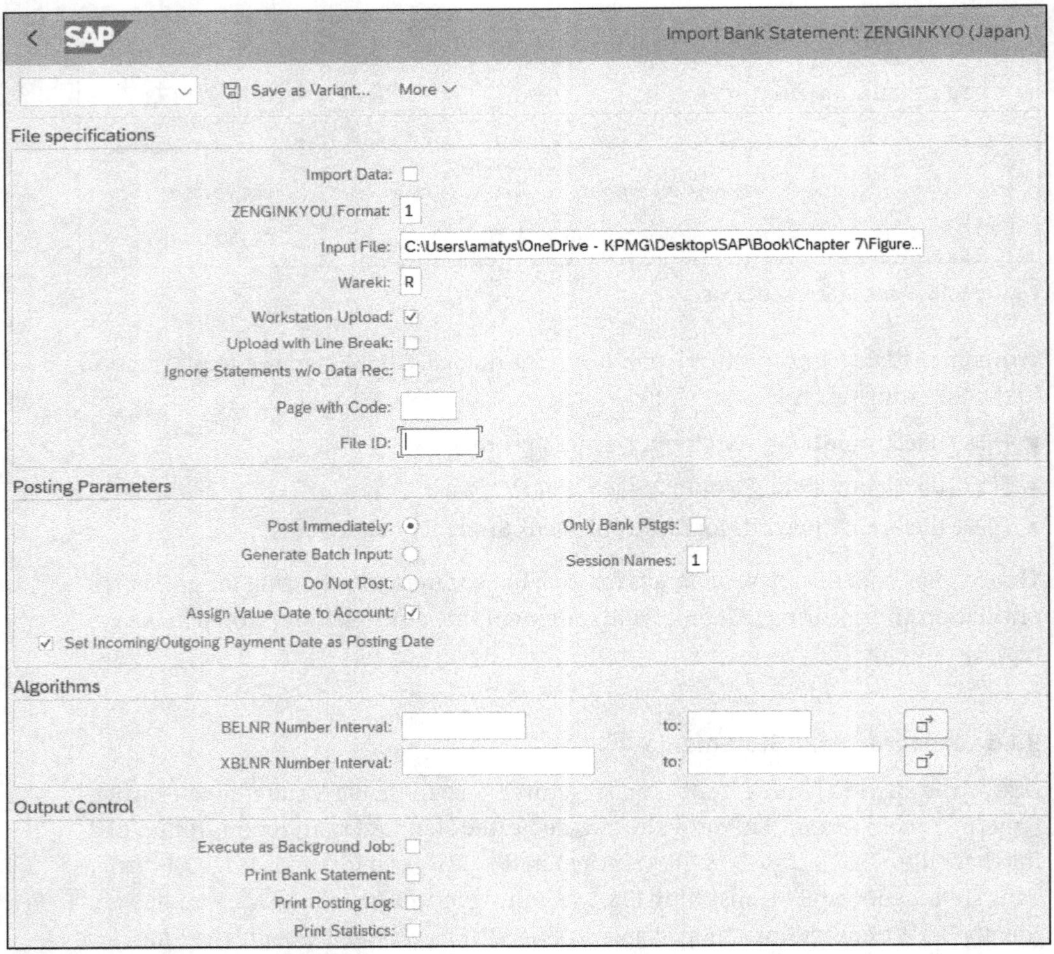

Figure 9.17 Zengin Upload

9.1.7 Bank Statement Forwarding Using Advanced Payment Management

Bank statement forwarding, which was described in detail in Chapter 4, demonstrates how advanced payment management can serve as a centralized bank statement repository, enabling the routing and forwarding of both end-of-day and intraday bank statements to various systems within a company's internal landscape.

With advanced payment management, bank communication can be fully centralized. This includes routing bank statements received from external banks (via SAP Multi-Bank Connectivity or the Manage Incoming Payment Files app) to the appropriate internal systems, such as subsidiary ERP systems, based on external bank account information.

The following is a high-level overview of how bank statement file forwarding works in advanced payment management:

1. The process begins when a bank statement is received from an external house bank through SAP Multi-Bank Connectivity or imported manually using the Manage Incoming Payment Files app.
2. If bank statement forwarding is activated, the import of the bank statement automatically creates a corresponding payment order in advanced payment management.
3. This payment order is then processed and passed to output management, which manages the outbound communication.
4. During outbound processing, the original bank statement file is handed over to SAP Multi-Bank Connectivity, in which the message is routed to the designated target system using the receiver configuration.

This centralized approach not only improves efficiency and traceability but also reduces manual intervention and potential errors. By leveraging advanced payment management as a central point for bank statement reception and distribution, companies can streamline treasury operations across multiple systems and entities.

9.1.8 Intraday Bank Statements

Analogously to end-of-day bank statements, you can also receive intraday bank statements from banks. The most popular formats for these are MT942 and camt.052. Unlike end-of-day statements, which provide a snapshot of transactions and balances from the previous day, intraday statements are received multiple times per day, depending on your bank. Loading these into the system gives you much faster visibility into your cash position and the transactions happening in real-time. From a processing and configuration perspective, working with intraday bank statements is quite similar to end-of-day statements: They use the same transactions to load, and although monitoring and processing differ slightly, the approach is largely the same. In this section, we'll briefly

9   Bank Statements

explain these differences and how to leverage intraday statements to get a better handle on your cash visibility throughout the day.

Before you can process intraday bank statements, you need to create intraday bank statement rules using the Define Monitoring Rules—Intraday Statements app (FCLM_BRM_RULE). Once you open this app (see Figure 9.18), click the **Create Rule** button and provide a **Rule ID** (a code you will assign in the BAM master data to an account) and a **Description** for your rule. Then simply click the **Continue** button, and the system will move you to the next screen, where you can specify the details of your rule.

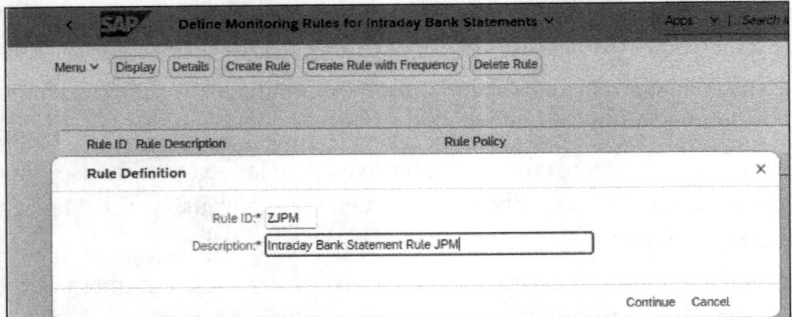

Figure 9.18 Define Monitoring Rules for Intraday Bank Statements

On the next screen (see Figure 9.19), click **More** to create rules for how often you expect to receive the intraday bank statement (**All**) and between which times (**Fr.** and **To**), including the time zone.

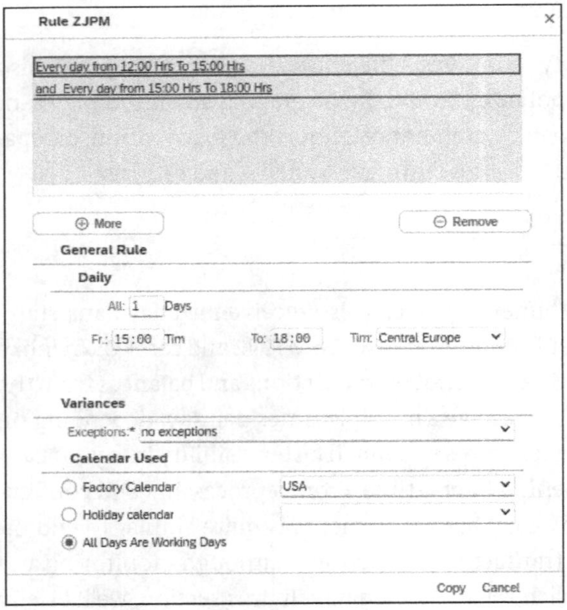

Figure 9.19 Create Rule

9.1 Bank Statement Processing

For instance, here we're setting up a rule stating that the system expects to receive a statement every three hours starting from 12:00. You'll need to align this with your bank to confirm how often they plan to send statements; most banks will typically send intraday statements three or four times per day. You can also specify any exceptions to these rules for receiving statements, and you can set the factory calendar or simply decide if all days are considered working days (in some cases, statements are sent daily regardless). Once you've configured all these settings to match your needs, click the **Copy** button to save the rule.

Once you create your rule, you can add it to the accounts for which you are expecting to receive intraday bank statements using the Manage Bank Accounts app. You also need to schedule appointments for the account by scheduling report SAP_FIN_FCLM_BAM_GEN_APPTS. For more information about this job, see SAP Note 2190119. Once you've created your rule, assign it to a specific bank account (by doing this, you're effectively telling the system which account the rule applies to and when the system should expect to receive intraday bank statements). Now you can start working with intraday bank statements.

When the statements are loaded into the system, they're typically loaded through the same channel and in the same way as end-of-day bank statements, which we already covered. After they're loaded, you can start using the Bank Statement Monitor—Intraday app (F3671) to see which statements have been received (see Figure 9.20). This monitor will show you how many intraday bank statements you've received, indicate how many you were expecting, and highlight any that are missing or haven't come through for a particular bank account. This helps you keep a close eye on your cash visibility throughout the day.

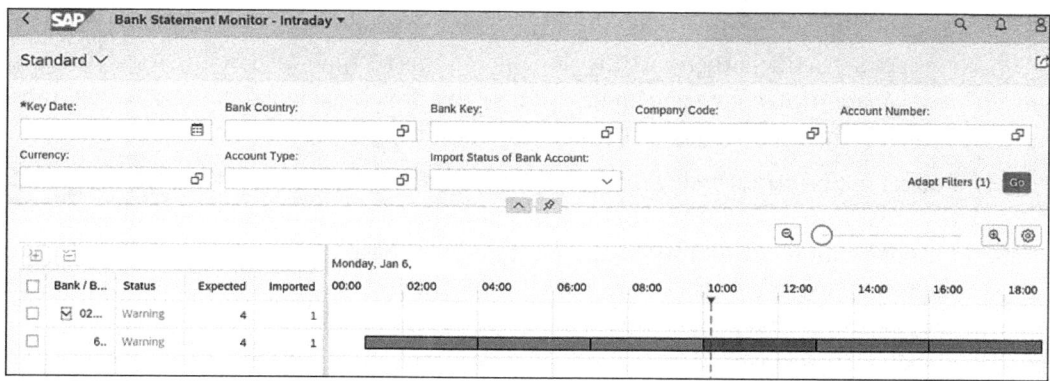

Figure 9.20 Intraday Bank Statement Monitor

In addition, line items from intraday bank statements are designed to update your cash management applications, significantly enhancing your cash visibility throughout the day. These updates are especially important for the Cash Flow Analyzer app, where the real-time or near-real-time updates from intraday statements allow you to see an accurate and up-to-date picture of your company's liquidity. By integrating these statements directly into cash management applications, you can better anticipate your cash needs, monitor movements, and make informed financial decisions much faster.

Starting from SAP S/4HANA 2023, intraday bank statements can also generate postings in the system. This has been an important improvement for certain companies, particularly ones in retail, that deal with a high volume of transactions. For these businesses, it's crucial to clear customer receipts as quickly as possible as doing so frees up credit lines and enables customers to make additional purchases sooner. However, it's worth noting that this feature is still in its early stages and is not yet fully mature. Although it offers a promising solution for faster cash flow visibility and reconciliation, it may require some refinement in future releases. For more information about posting intraday bank statements, refer to SAP Note 3502319.

9.1.9 Processing Rules

Traditionally, all the posting rules and setup for processing bank statements in SAP were handled through detailed configuration work, making it quite cumbersome and often difficult for business users to maintain. Because these configurations typically had to be managed by IT and released in a controlled way, it was challenging and time-consuming to create rules and automate bank statement processes. Due to these complexities, many third-party solutions were developed over the years to support bank statement reconciliation and automate this work. Starting from SAP S/4HANA 2023, SAP has begun to move some of this functionality out of the traditional configuration and into SAP Fiori apps, making it more accessible to business users. However, this is still a bit limited for private cloud and on-premise versions, whereas public cloud users are seeing significantly more functionality. In this section, we'll briefly touch on these new capabilities to give you an idea of how this area is evolving and to offer a glimpse of the future, when much of the current customizing setup may no longer be needed.

You can manage processing rules using the Manage Bank Statement Reprocessing Rules app (F3555; see Figure 9.21). This app allows you to create and manage processing rules for performing G/L and AP/AR account postings based on the processed bank statement line items. You can create and edit rules, define the specific conditions in which they should be applied, and set the actions they should take with bank statement items. The app also lets you share manually applicable processing rules with your colleagues, deprecate outdated or no longer useful rules, and use these rules to automatically process payment advice. In addition, you can download and upload processing rules from one system to another, automate the rules to run in the background, and manage templates for processing rules, making it easier to maintain consistency and ensure efficient processing across your organization.

9.1 Bank Statement Processing

When you open this app, you'll start by selecting the key date or another selection parameter to upload and check data. The app will show you all the unprocessed line items from your bank statements so that you can see exactly what still needs to be reconciled. When you highlight a rule, you'll be able to see the details of what kind of posting it's set up to perform. You also have the option to manually assign an existing posting rule to a specific line item, or you can even ask the system to automatically process items for you. However, keep in mind that automatic processing typically requires additional licenses or solutions, so this option might not be available by default; you may need to process postings manually instead. Click **Create Rule** to see how to create a new rule and configure it as needed.

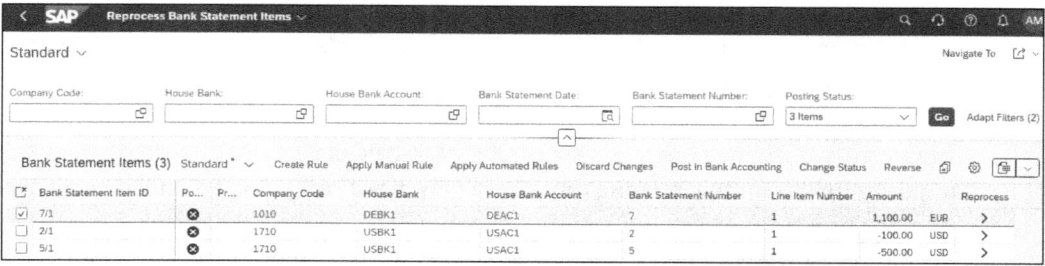

Figure 9.21 Reprocess Bank Statement Items

On the next screen (see Figure 9.22), you'll need to provide a description for your new rule, which will help you and others understand its purpose later on.

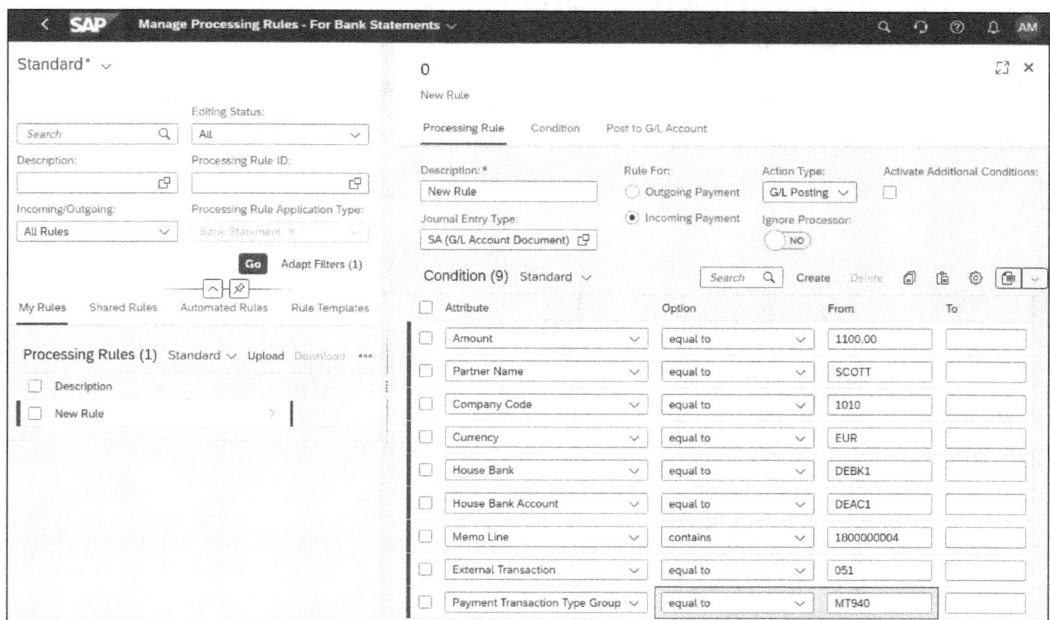

Figure 9.22 Create Processing Rule

9   Bank Statements

You'll also choose whether it's meant for G/L postings or for clearing AR or AP items (via the **Action Type** dropdown) and select the **Journal Entry Type**. The system will automatically derive certain criteria from the bank statement posting, but you can review these and remove or adjust them to make the rule more universal if you want. Alternatively, you can keep them as they are for future use. This step allows you to tailor your rule to your specific needs and make sure it aligns with your reconciliation processes.

Scroll down to the **Post to G/L Account** section (see Figure 9.23) and select which G/L accounts should be used for postings and all other corresponding items, like **Tax Code**, **Cost Center**, and so on. Once you're ready, click **Create**. Your rule now will be ready to be used.

This app is a step into the future as it's designed to reduce reliance on the traditional configuration described in the next section and to enable more automation. However, as you can see, posting line items here is still quite a manual process, especially in older versions of SAP or when additional licenses are not in place. In those scenarios, you might still need to rely on the configuration described in Section 9.2. Nevertheless, this application is definitely moving things in a new direction, giving you full control over how you process and post items. From this app, you have full flexibility to reprocess and post as needed. We wanted to mention this alternative to using the configuration, but it's up to you to decide how to proceed. Over time, the system is expected to rely more on this application and on master data rather than on configuration. There are also upcoming innovations like cash applications powered by AI learning that will play a role, but these are larger topics that could easily fill their own dedicated book. For now, we simply wanted to introduce you to the trends and show you some of the options available in the system.

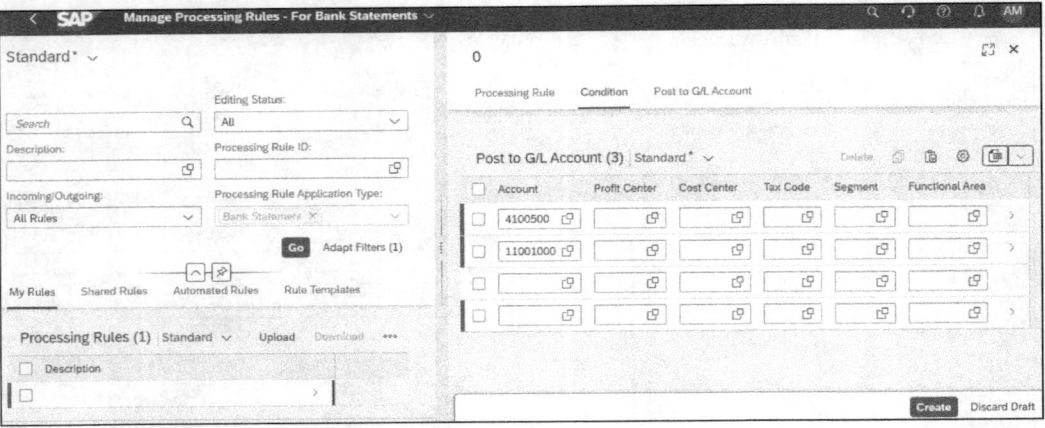

Figure 9.23 Processing Rules for G/L Accounts

9.2 Bank Statement Configuration

In this section, we'll focus specifically on the configuration of bank statements and maintaining the IT activities that go along with it. In general, SAP S/4HANA bank statement processing involves master data maintenance, but there will always be technical tasks that require IT involvement and transport activities to ensure stability. This is because, for instance, posting rules—if maintained directly in the master data—can cause significant issues if changed incorrectly, potentially leading to incorrect or invalid postings. It's better to avoid posting altogether than to risk incorrect postings that can disrupt your financial processes. Here, we'll look closely at the main configurations for bank statement processing, which are essential for running the upload programs in the background and ensuring that everything functions smoothly.

9.2.1 New Bank Statement Import Program: FEB_FILE_HANDLING

The prerequisite for program FEB_FILE_HANDLING is configuration of the logical paths via Transaction FILE, which you can view in Transaction AL11. A bank connectivity tool will retrieve electronic bank statement files from the bank's SFTP folder and place it in the directory specified by us in this step. It is considered best practice to create three distinct logical paths for processing bank statements using FEB_FILE_HANDLING:

- **Inbound**
  The folder where new bank statements are initially placed for processing.
- **Archive**
  The folder where successfully processed statements are moved after upload.
- **Error**
  The folder where failed or incorrectly formatted statements are moved.

Once a bank statement is processed by FEB_FILE_HANDLING, the system automatically moves the file to either the archive or error folder, depending on the outcome of the upload. This structure allows for clear segregation and easy monitoring. If a file lands in the error folder, it signals that the upload failed, enabling you to quickly identify and troubleshoot issues. By maintaining these dedicated folders, organizations can ensure better transparency, control, and traceability in the bank statement upload process, as well as efficiently rerun and recreate any failed uploads.

The configuration related to program FEB_FILE_HANDLING can be found in the SAP S/4HANA system by navigating to the following path: **Financial Accounting** • **Bank Accounting** • **Payment Transactions** • **Electronic Bank Statement** • **Settings for the Data Import**.

In this section, we'll focus on the configuration of FEB_FILE_HANDLING, which runs in the background to upload statements from dedicated folders. The first step is to define

9 Bank Statements

the logical paths for the directories and locations in the system where you expect to receive bank statements. Once you have the paths set up, you can specify the import parameters that are used to import and upload these statements, ensuring the system knows how to process them correctly. Finally, you can configure whether the statements should be posted automatically and define the key parameters needed for posting and processing. This setup helps ensure that bank statements are loaded efficiently, minimizing manual effort and errors.

Define Logical Paths

In **Financial Accounting** • **Bank Accounting** • **Business Transactions** • **Payment Transactions** • **Electronic Bank Statement** • **Settings for the Data Import** • **Define Logical Paths** (or via Transaction FEB_IMP_FILEPATH), you define the key details necessary for automated bank statement processing using FEB_FILE_HANDLING. First, click **New Entries** to begin creating the logical paths for your bank statement upload process. In this step, you'll enter the logical path name and give a clear, descriptive label that helps you and others identify the purpose of this path. Next, specify how the path is going to be used—whether it's meant to store the bank statement files or to import them for processing. Then, define the directory, which refers to the exact server file path. This depends on how your server is set up; you should usually get this information from your technical team. Finally, provide the file name or file pattern that will be stored in this logical path. This ensures that the system knows exactly which files to look for in the designated directory.

Figure 9.24 shows logical paths for import, archive, and error. Figure 9.25 shows how to assign proper directories used on the server to these paths (defined by the technical team depending on the server you use).

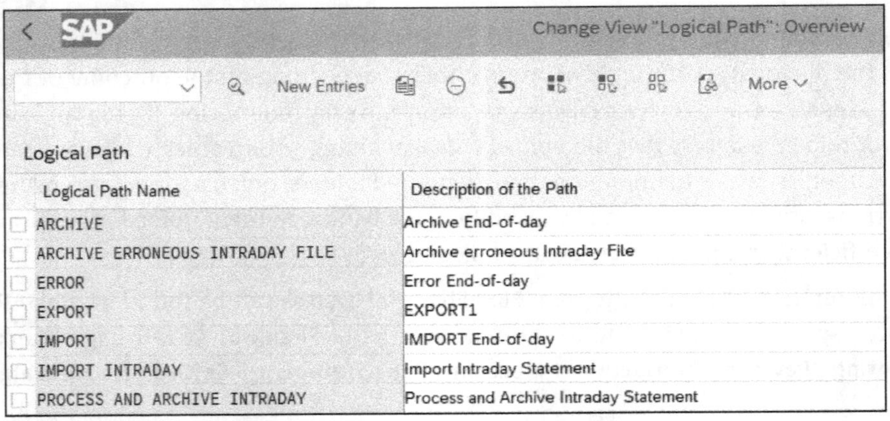

Logical Path Name	Description of the Path
ARCHIVE	Archive End-of-day
ARCHIVE ERRONEOUS INTRADAY FILE	Archive erroneous Intraday File
ERROR	Error End-of-day
EXPORT	EXPORT1
IMPORT	IMPORT End-of-day
IMPORT INTRADAY	Import Intraday Statement
PROCESS AND ARCHIVE INTRADAY	Process and Archive Intraday Statement

Figure 9.24 Logical Paths

9.2 Bank Statement Configuration

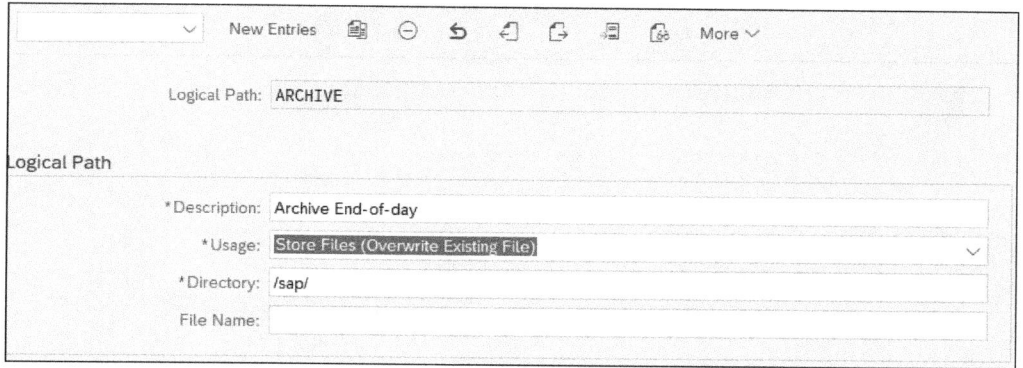

Figure 9.25 Directories on Server

Define Import Parameters

In **Financial Accounting** • **Bank Accounting** • **Business Transactions** • **Payment Transactions** • **Electronic Bank Statement** • **Settings for the Data Import** • **Define Import Parameters** (or via Transaction FEB_IMP_SOURCE), you define the directories from which the system should import bank statement files (see Figure 9.26). As part of the setup, you also specify the target directories where the files should be moved after processing—either to an archive folder following successful upload or to an error folder if the processing fails. This ensures a clear and automated file management process throughout the bank statement import workflow. Here, you can define source paths by clicking **New Entries** and entering a meaningful description for your source path so that it's clear to everyone what this particular path is for. Next, you specify the format of the incoming file you want to process for the bank statement. For example, you might choose camt.052 for intraday bank statements or camt.053 for end-of-day statements. You can select this format from the dropdown list under the **Format** field. Then, select the storage path from the dropdown list; you created this path in the previous step. This is where you expect to receive the statements from the bank. Next, specify the path for errors, again by choosing from the dropdown list of entries created earlier. If the system's file handling is unable to process the file and there's an error, the file is automatically stored in this path. If the file is processed successfully, it's moved from the storage path to the archive path you specified previously. That's why you create logical paths for import (where you expect to receive the statements), archive (where processed statements are stored), and error (where files that couldn't be processed will be stored). Click **Save** to save your entries.

9   Bank Statements

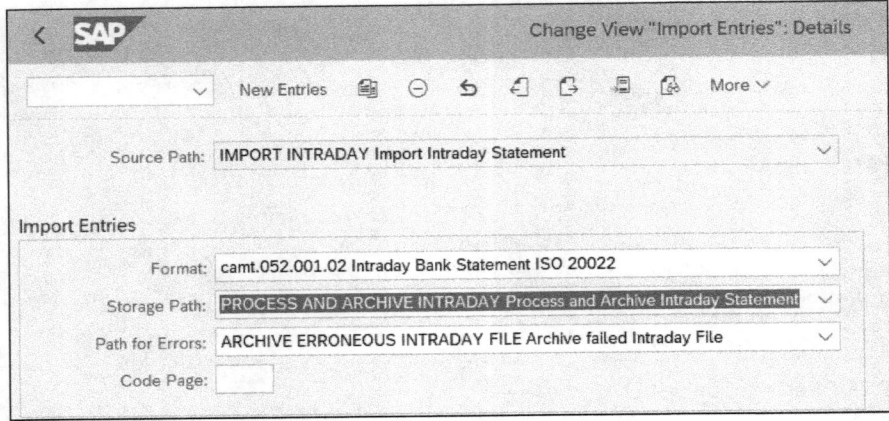

Figure 9.26 Add Import Parameters

Define Posting Parameters

Via **Financial Accounting** • **Bank Accounting** •**Business Transactions** • **Payment Transactions** • **Electronic Bank Statement** • **Settings for the Data Import** • **Define Posting Parameters**, you define how bank statements should be handled after they are imported into the system (see Figure 9.27). This includes specifying which bank accounts are expected to receive end-of-day statements that should be uploaded and posted, as well as which accounts are expected to receive intraday bank statements. To configure these settings, click **New Entries**, then select the company code that you want to use. You can also select the house bank account and account ID if needed. Next, specify if the statements processed from this activity are to be imported and posted, then choose the mode for posting—whether you want it to be posted automatically or not. Then, specify if this setup is for an account balance statement, such as intraday bank statements. In the past, SAP wasn't generating postings for these types of statements, but, as you might have seen in earlier examples, SAP has recently started to include functionalities for posting even these statements—so this activity might feel a bit outdated, but it's still available in the configuration. Once you're done with these entries, just click **Save** to confirm your settings.

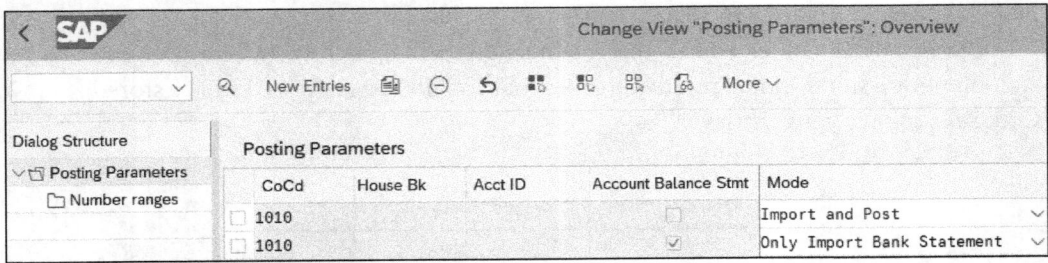

Figure 9.27 Define Posting Parameters

If you want to apply the same processing rules to all bank accounts within a company code, you can simply leave the **House Bk** field blank. This enables a more streamlined and consistent approach to statement handling across the organization.

If your goal is to upload bank statements solely for creating cash management balances or to import intraday bank statements without posting any accounting entries, then you should select the **Account Balance Stmt** option and choose **Import Only** for **Mode**. This setting ensures that the statements are uploaded into the system purely for informational and reporting purposes, such as updating liquidity positions in cash management, without triggering any postings to the G/L. This is particularly useful for real-time cash visibility and for accounts that receive multiple updates throughout the day.

9.2.2 Lockboxes

The configuration related to lockboxes can be found in the SAP S/4HANA system by navigating to the following path: **Financial Accounting • Bank Accounting • Payment Transactions • Lockbox**. In the following sections, we'll walk you through defining control parameters, posting data, and lockboxes for house banks.

Define Control Parameters

In this activity, you configure control data for the lockbox procedure. Currently, only the "LOCKBOX" **Procedure** is supported, and this data is required for importing lockbox files sent by banks. SAP currently supports the BAI and BAI2 file formats. In the BAI format, you must specify the document number length (which is 10 in the standard SAP system) and the number of document numbers in record types 6 and 4 of the BAI file. Your bank must agree to use this format. For the BAI2 format, it is not necessary to specify the document number length or the number of document numbers in record types 6 and 4. The BAI2 file is designed so that each document number is stored in a separate record type 4, with its corresponding payment and deduction amounts. Again, your bank must agree to this format.

The BAI and BAI2 formats are standard, but they can vary by bank. Many SAP clients have opted for customized formats. The format you receive must be mapped to reconcile with the SAP-delivered data dictionary layout (tables such as FLB01, FLB05, FLB06, etc.). If the format does not reconcile and you do not want to have the bank change it, SAP recommends using a custom-written ABAP program to reformat the file. Alternatively, you can modify the SAP data dictionary directly, though this requires a repair that must be reapplied in future releases.

Under **Batch Input Sessions (Posting Functions)**, you specify which postings the system should create—G/L cash postings and/or customer cash applications. It is recommended to select both options if you use the G/L and AR modules. For the G/L, you can decide whether to post an aggregate amount for all incoming cash or post one line per check, depending on your reconciliation with the bank.

9 Bank Statements

The lockbox program first tries to identify the customer using the unique bank information stored in the customer master record, matching it to the MICR information on the check. Maintaining accurate, unique MICR data for customers is critical for successful processing. If more than one customer has the same MICR bank and account details, then the program cannot apply the check. This may indicate duplicate customer records or that what should be only ship-to or sold-to customers have been set up as separate payers. In such cases, you may need to establish a relationship in which one customer is designated as the main payer (with MICR info), and related customers do not have MICR info but instead have the main payer's customer number in the **Alternate Payer** field of the customer master record. The payment information in the bank data file will generate a payment advice for each check. This payment advice is then used by the lockbox procedure to clear subledger open items. In **Financial Accounting • Bank Accounting • Payment Transactions • Lockbox • Define Control Parameters**, you store control data for the lockbox procedure (alternatively, you can use Transaction OBAY). Note that there might be various procedures available depending on your system version (see Figure 9.28).

Figure 9.28 Lockbox Procedure

When you select a procedure by double-clicking it, the system will allow you to set up more rules and define how the lockbox can be processed. As shown in Figure 9.29, you'll have the following fields:

- **Document Number Length**
  This field is applicable only for BAI format records.
- **Num. of Doc. Numbers in Type 6**
  This setting applies only to BAI format records.
- **Num. of Doc. Numbers in Type 4**
  This setting also applies only to BAI format records.
- **G/L Account Postings**
  Enable this option to allow postings to the cash account in the G/L for deposits. We recommend activating this field.
- **Incoming Customer Payments**
  Enable this setting to post to the AR subledger. This allows for the clearing of customer accounts and creation of residual items. Activation is recommended.

- **Insert Bank Details**
  Used for batch input sessions that update customer master records with bank details—when bank data either has changed or was previously missing.
- **Type of G/L Account Posting**
  - Entering "1" creates a separate G/L account posting for each check in the file.
  - Entering "2" creates a single G/L account posting for the entire lockbox file.
  - Entering "3" creates a single G/L account posting for each batch.

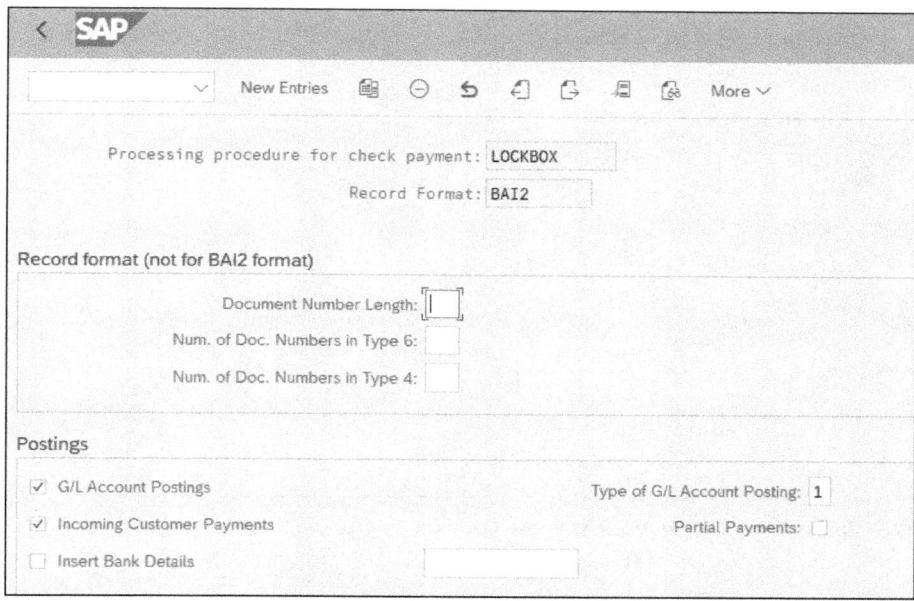

Figure 9.29 Control Parameters

Define Posting Data

In **Financial Accounting • Bank Accounting • Payment Transactions • Lockbox • Define Posting Data** (or via Transaction OBAX), you define the necessary information to process lockbox data and generate the appropriate accounting entries. You can select or create a new destination here (see Figure 9.30).

The **Destination** and **Source** fields represent routing information provided by your bank. For each unique combination of destination and origin, you must define the posting details required to create the following entries in general:

- **Bank G/L posting**
  Debit the bank account (for incoming checks) and credit the payment clearing account.
- **A/R posting—second step, clearing**
  Debit the payment clearing account and credit the customer account.

9   Bank Statements

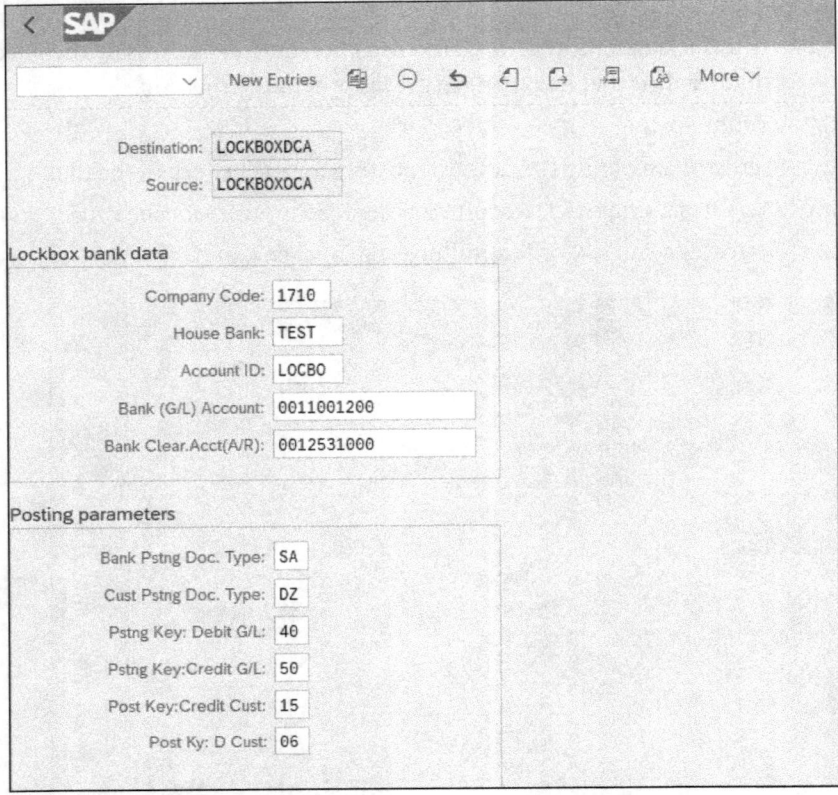

Figure 9.30 Posting Data for Autocash with Lockbox

To configure the posting data, enter the following:

- Specify the **Company Code** that owns the lockbox.
- In the **Bank (G/L) Account** field, enter the G/L account number corresponding to the lockbox (clearing) bank account.
- In the **Bank Clear. Acct (A/R)** field, enter the G/L account number representing the A/R clearing account. This account will hold the balance of all unapplied lockbox payments.
- Also define the document types and posting keys to be used for these postings. It is quite typical for bank statements and lockbox processing to involve two levels of postings. In the case of a lockbox, the first level of posting generally uses debit posting key 40 with the bank document type SA, while the credit side is usually posted with posting key 50 to the G/L account, typically involving the main GL bank account and a clearing bank account (debit and credit depending on the incoming or outgoing transaction). The second level of posting, associated with document type DZ, typically involves clearing the customer or vendor's open items. This involves credit postings to the customer account using posting key 15 and debit postings using posting key 06 to clear the vendor's respective items. In this structure, the first level

reflects the movement between the main bank and clearing bank accounts, while the second level reconciles the clearing bank account with the customer or vendor account, depending on whether the payment is incoming or outgoing. Usually, posting rules will be predefined and you do not need to change these values.

Define Lockboxes for House Banks

In this step, you define the lockbox accounts associated with your house banks. This information can then be included on outgoing invoices to inform customers where their payments should be made. By clearly specifying the lockbox details, you can streamline and optimize the incoming payment process. It's important to note that the lockbox procedure is primarily used in the United States.

You can find this configuration activity under **Financial Accounting · Bank Accounting · Bank Accounts · Define Lockboxes for House Banks**.

Define your lockbox assignments (see Figure 9.31) by specifying the company code (**CoCd**), the lockbox key (**Lockbox**) designated for customer payments, the house bank ID (**House Bk**), and the lockbox number (**LBox No**) provided by your house bank. Within the customer master data, you can assign the appropriate lockbox number that the customer should use for payments. This can be done in the application menu under **Payment Transactions · Company Code Data**.

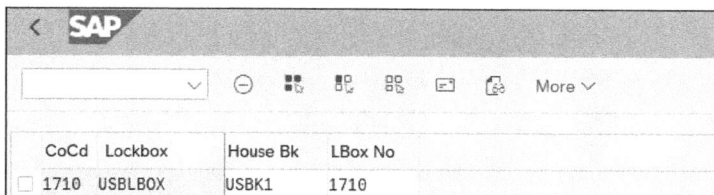

Figure 9.31 Define Lockboxes for House Banks

9.2.3 Bank Statement Posting

In this customizing activity, you will find all the necessary configuration required to enable proper posting of bank statements in the system. This includes defining posting rules, assigning external transaction types, setting up posting specifications for different scenarios such as incoming and outgoing payments, and mapping bank statement details to the correct G/L accounts and subledger postings. This setup ensures that once bank statements are imported, the system can automatically interpret the data and create the corresponding financial entries accurately and efficiently. Proper configuration here is essential for achieving seamless bank reconciliation and financial reporting.

It's important to note that all these activities are part of system configuration and therefore require transport requests, which makes it quite challenging for business users to update or adjust them easily. This limitation has historically hindered achieving a high

9   Bank Statements

level of automation as each change needed involvement from IT or functional consultants. To overcome this, various third-party providers developed tools that allow business users to define rules and logic directly in the production environment. In response, SAP introduced solutions like SAP Cash Application and postprocessing rules, which empower users to handle matching logic and exception processing directly in production without the need for system transports, significantly improving agility and automation in the bank statement processing landscape.

You can find the bank statement configuration under **Financial Accounting • Business Transactions • Payment Transactions • Electronic Bank Statement • Make Global Settings for Electronic Bank Statement**. You can also use Transaction OT83. Once you select this configuration option, the system will ask you for a **Chart of Accounts**; select it from the list, and it will open a new screen in which you can define your setup.

Define the **Account Symbol** for G/L accounts (see Figure 9.32) and transactions (e.g., main bank account, clearing bank account, cash receipt, outgoing checks, bank fees) that should be used for postings from the bank statement. The **Account Symbol** will be used for determining posting rules.

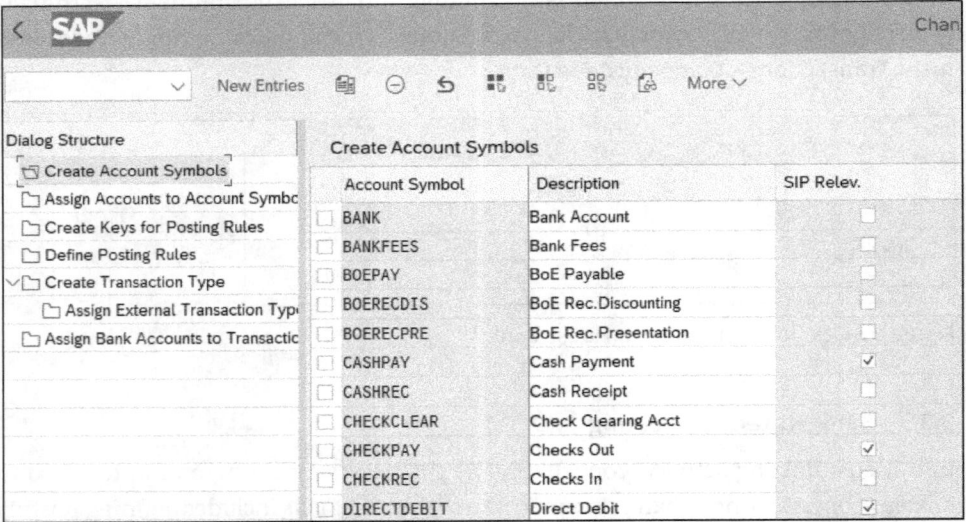

Figure 9.32 Account Symbol

The self-initiated payments checkbox (**SIP Relev.**) indicates that the payment is initiated by the bank account holder rather than by an external source, such as an electronic bank statement. This feature applies to bank transfer orders, direct debit orders, and payments by check created through a company's automatic payment program.

When this checkbox is selected, SAP simplifies and standardizes how it determines the appropriate bank clearing account. It does this by using account symbols, which apply

9.2 Bank Statement Configuration

to both the bank statement process and the automatic payment process. This checkbox is specifically used for payment processes that reduce the G/L account.

You assign account symbols to the G/L account numbers in the next step. First, click **Assign Accounts to Account Symbols** in the **Dialog Structure**, and then define the account determination procedure for each account symbol (see Figure 9.33). You can either assign specific **G/L Accounts** directly to the symbols or use a masked output to simplify the setup. This avoids the need to create a separate account symbol for every individual house bank account with a different G/L account. In addition, you can further refine the account determination by using the **Acct Mod.** and **Currency** fields to control how accounts are selected during posting. Masked values will be determined based on the G/L account assigned to a house bank and bank account ID.

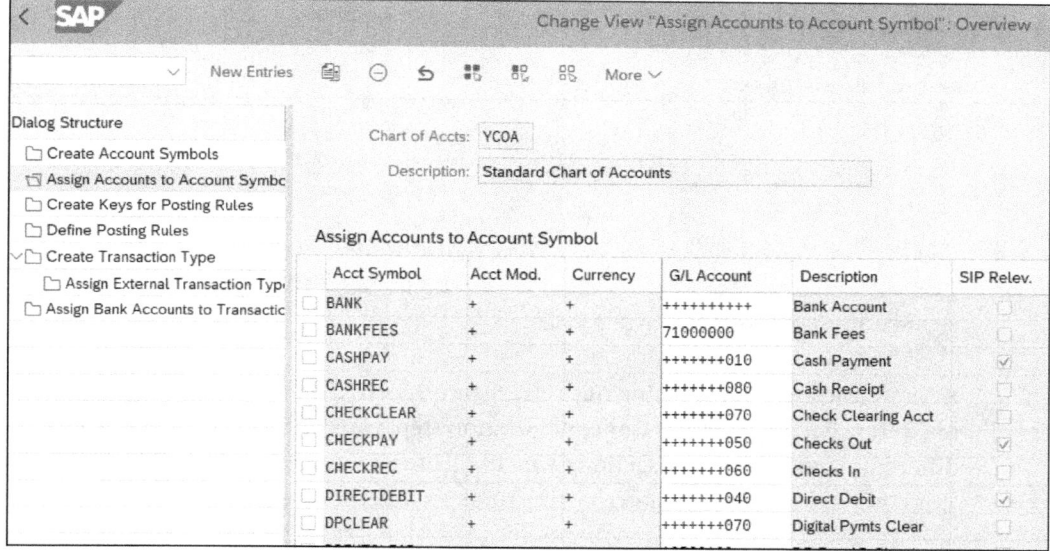

Figure 9.33 Assign Accounts to Account Symbols

In SAP, the masked output feature (with the **+++** symbols) is used in bank account determination to match flexible patterns in bank account numbers. The **+** sign here does not ignore characters; instead, it represents a placeholder for any digit (0–9). Literal characters (like specific digits) must be present exactly as defined in the mask.

In the configuration for house bank accounts, you create a mask using a combination of literal characters (like 1100) and placeholders (+). For example, 1100+++ means that the first four digits must be exactly 1100, while the next three digits can be any number (+ placeholders).

This pattern matching helps SAP find the right G/L account or house bank subaccount for the bank account number provided in the statement. It ensures that the system can handle cases in which banks send account numbers with variable endings or extensions, providing the needed flexibility for accurate account determination.

575

In the next step, accessed by clicking **Create Keys for Posting Rules**, you can create keys for posting rules (see Figure 9.34). To create a new key for posting rules, click **New Entries**, then enter a value for the posting rule and a description for the rule. Click **Save** to store your entry.

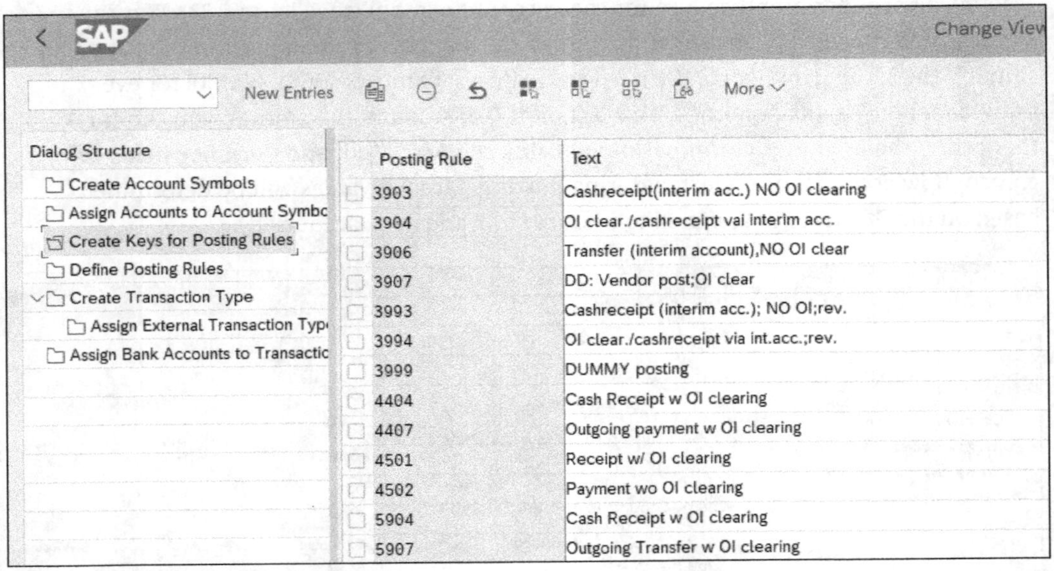
SAP
Change View
New Entries
More
Dialog Structure
Create Account Symbols
Assign Accounts to Account Symbo
Create Keys for Posting Rules
Define Posting Rules
Create Transaction Type
Assign External Transaction Typ
Assign Bank Accounts to Transactic
Posting Rule
Text
3903 Cashreceipt(interim acc.) NO OI clearing
3904 OI clear./cashreceipt vai interim acc.
3906 Transfer (interim account),NO OI clear
3907 DD: Vendor post;OI clear
3993 Cashreceipt (interim acc.); NO OI;rev.
3994 OI clear./cashreceipt via int.acc.;rev.
3999 DUMMY posting
4404 Cash Receipt w OI clearing
4407 Outgoing payment w OI clearing
4501 Receipt w/ OI clearing
4502 Payment wo OI clearing
5904 Cash Receipt w OI clearing
5907 Outgoing Transfer w OI clearing

Figure 9.34 Create Keys for Posting Rules

Keys are used to create posting rules (see Figure 9.35) in the next step, accessed by clicking **Define Posting Rules**. In this configuration step, you define the rules for processing different types of bank transactions, which will be executed based on the incoming data from the bank statements. Each posting rule specifies the relevant G/L accounts for various transaction types (e.g., payments, fees, transfers), as follows:

- Posting keys are used to define whether the transaction will be a debit or a credit.
- For each posting rule, you assign a posting key to the corresponding account symbol.

To create a posting key for your posting rule, click **New Entries**. Next, select your posting rule from the values created in the previous steps and specify the posting area. The posting area indicates whether this rule applies to the first level of posting (for the bank level). If a second-level posting is required, this step needs to be repeated. After that, specify the posting key for the debit entry, along with any special G/L indicator used and the account symbol assigned for posting. Then, define the posting key, special G/L indicator, and account symbol for the credit entry. Specify the document type to be used, as well as the posting type, which indicates whether it's posting to a G/L account, a clearing account, or another activity. Also enter the posting for the account key if you're posting at the account level and, optionally, include a reversal reason. Once all entries are complete, click **Save** to store your posting rule.

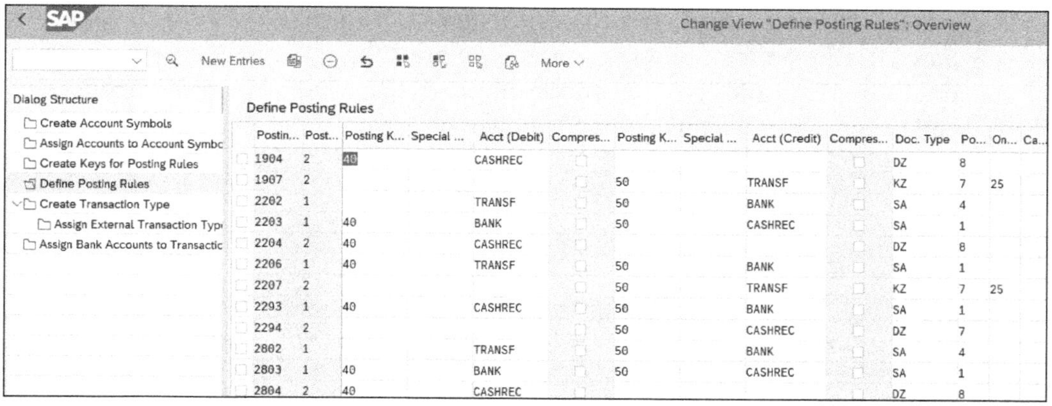

9.2 Bank Statement Configuration

Define Posting Rules

Postin...	Post...	Posting K...	Special ...	Acct (Debit)	Compres...	Posting K...	Special ...	Acct (Credit)	Compres...	Doc. Type	Po...	On...	Ca...
1904	2	40		CASHREC						DZ	8		
1907	2					50		TRANSF		KZ	7	25	
2202	1			TRANSF		50		BANK		SA	4		
2203	1	40		BANK		50		CASHREC		SA	1		
2204	2	40		CASHREC						DZ	8		
2206	1	40		TRANSF		50		BANK		SA	1		
2207	2					50		TRANSF		KZ	7	25	
2293	1	40		CASHREC		50		BANK		SA	1		
2294	2					50		CASHREC		DZ	7		
2802	1			TRANSF		50		BANK		SA	4		
2803	1	40		BANK		50		CASHREC		SA	1		
2804	2	40		CASHREC						DZ	8		

Figure 9.35 Define Posting Rules

For example, when posting a customer payment, the posting key for the bank account (debit) and for the customer account (credit) would be assigned in the posting rule to ensure the system knows how to record the transaction (see Figure 9.36).

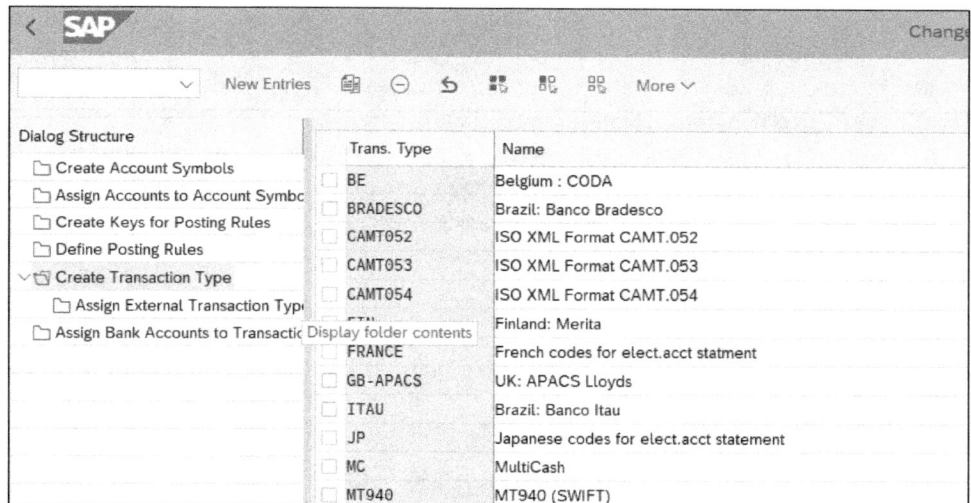

Figure 9.36 Create Transaction Type

In the next activity, accessed by clicking **Create Transaction Type** (see Figure 9.36), you define the names and descriptions for the different transaction types needed—usually representing statements or banks providing statements.

In the next step, accessed by clicking **Assigned External Transaction Type**, you need to assign each business transaction code (BTC) and the type of transaction provided by the bank to a corresponding posting rule, as shown in Figure 9.37. This assignment defines how the transaction will be posted in the system. You also must assign an interpretation algorithm, which supports automatic reconciliation by helping the system identify and match relevant information such as open items, document numbers, or invoice references. It is important to select the correct interpretation algorithm; choosing the wrong one can led to incorrect postings or clearings, which may result in reconciliation issues and require manual intervention to resolve.

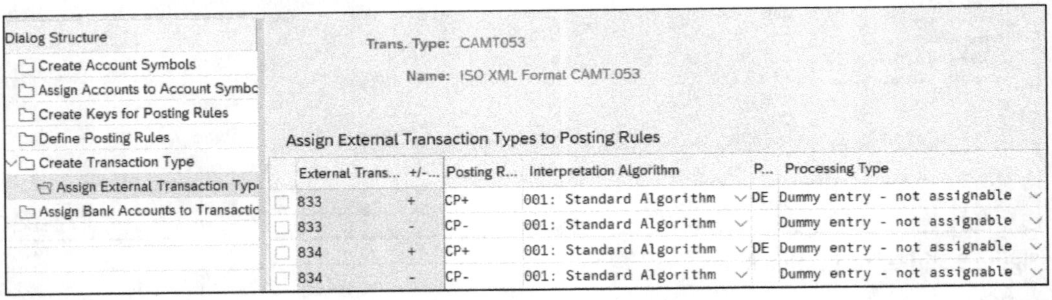

Figure 9.37 Assign External Transaction Type

Business Transaction Codes
BTCs are key identifiers used in bank statements to classify different types of transactions, and they can often be negotiated with your bank to fit your company's specific needs. Although many banks use predefined BTC portfolios, it is usually possible to request that certain transactions be sent using specific BTCs. This flexibility can be very useful in SAP, as it allows you to better control how transactions are interpreted, matched to posting rules, and automatically processed within the system. In some cases, banks even offer the option for you to select which BTCs to use, giving you greater influence over how data is structured and how automation rules are applied in your bank statement processing.

In the final activity (see Figure 9.38), you map each bank key and bank account to a specific transaction type.

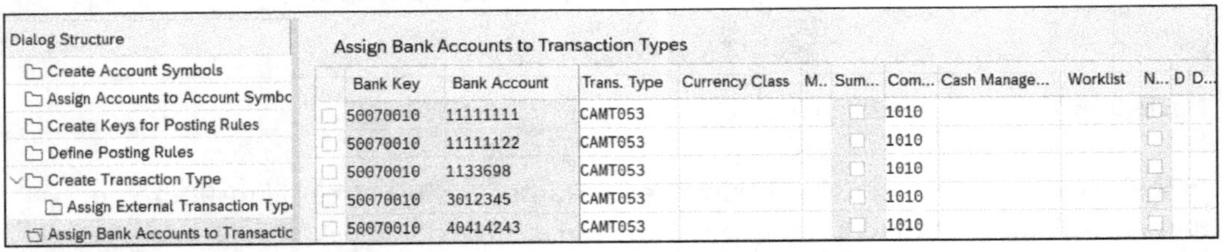

Figure 9.38 Assign Bank Accounts to Transaction Types

In other words, you define which set of posting rules should be applied to process the bank statements received for a particular bank account. This ensures that when a statement is uploaded, the system knows exactly how to interpret the external transaction codes and which posting logic to follow based on the predefined transaction type.

9.2 Bank Statement Configuration

To assign bank data to a transaction type, first click **New Entries**. Enter the bank key in the **Bank Key** field, which represents the bank account details. Then, enter the bank account number under **Bank Account**. Next, map the transaction types; for example, in this case, the type is camt.053 (the value created in the previous step). Add the **Company Code** value to complete the entry. The other fields are typically no longer used in SAP S/4HANA.

9.2.4 Search String Configuration

When importing an electronic bank statement, SAP S/4HANA uses interpretation algorithms and predefined settings to identify business transactions and determine how they should be recorded. Typically, the system searches for document numbers in the "note to payee" field to identify and clear open items—such as matching a cash receipt to a customer invoice.

However, the information in the note to payee may be incomplete or altered—characters might be missing, added, or in a different format—making it difficult for the system to correctly identify and clear items. In such cases, postprocessing is required.

To improve automatic clearing and reduce manual effort, you can define search strings. These allow the system to look for additional reference information in the note to payee—such as invoice reference numbers (generated and sent to the customer), customer names, or internal entity numbers. Any of these details can help identify the correct invoice or customer account for clearing.

Search strings work by targeting a specific field—for example, enriching the note to payee during interpretation—and directing the system where to write the result (the "target field"). Although document number search is the most common use, search strings can also populate other fields, like cost centers or posting rules.

You can find search string configuration under **Financial Accounting • Business Transactions • Payment Transactions • Electronic Bank Statement • Make Global Settings for Electronic Bank Statement • Define Search String for Electronic Bank Statements**. As shown in Figure 9.39, the first step here is to define the search string.

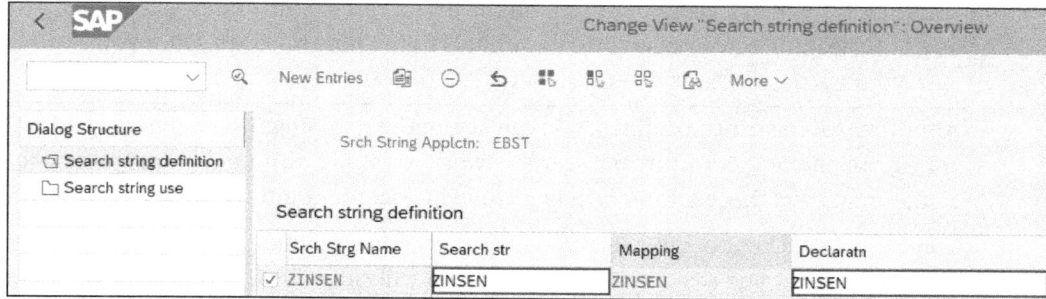

Figure 9.39 Search String Definition

9 Bank Statements

The search string identifies patterns in the note to payee field of a bank statement. Special characters can be used within the search string, definitions for which are provided at the end of this section.

To create a new search string, first click **New Entries**. Under **Srch Str Name**, enter the name of the search string, followed by a description to help identify its purpose. In the **Search str** field, enter the pattern or keyword you're looking for within the note to payee information in the bank statement. Then, create a mapping. In the mapping, you specify which fields you want to populate and how you want to map the information extracted from the note to payee during interpretation. In the example in Figure 9.40 we've used Zinsen (translation: interest) information.

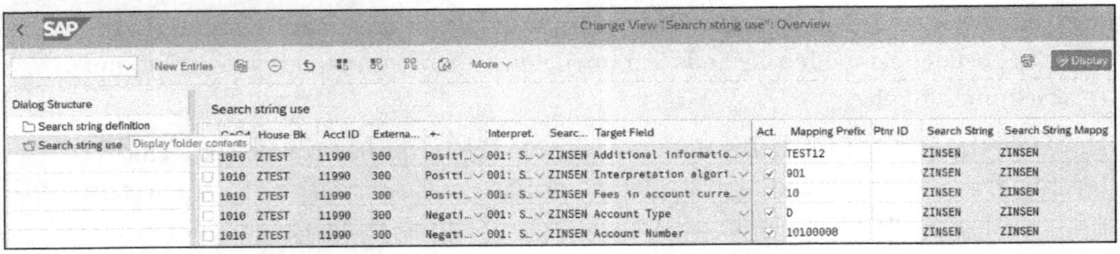

Figure 9.40 Search String Definitions

Let's look at another example, this one with an IBAN number (see Figure 9.41). For the cash pooling determination or ROBO payments mentioned previously, banks send virtual IBANs to help you determine who money really belongs to. Once you have defined the search string, you need to assign it and specify how it will be used. Click **New Entries** and enter the company code in the corresponding **CoCd** field. Next, specify the house bank (**House Bk**) and account ID (**Acct ID**). If needed, you can map external transactions from the bank statement using the **External Transaction** field. Then, indicate whether the search string applies to positive or negative entries (select this from the dropdown list available under **+-**). Select the interpretation algorithm you are using and provide the search string name; this is where the previously created search string is assigned. Specify the target field you want to populate, such as customer or vendor master data. Mark the **Act.** checkbox to activate the search string. Finally, enter any mapping prefix information, which allows you to extend the found values with additional details, such as adding leading zeros or document numbers to the note to payee. Once you're done, save your entries.

After the search string is defined, you can set up the mapping. The mapping is used to transform characters within the search string—typically, to remove or replace additional characters that customers may include in document numbers.

You can test the logic of your search string by entering a sample note to payee entry in the detail screen and selecting **Test**. The system will show the result of the search and mapping.

9.2 Bank Statement Configuration

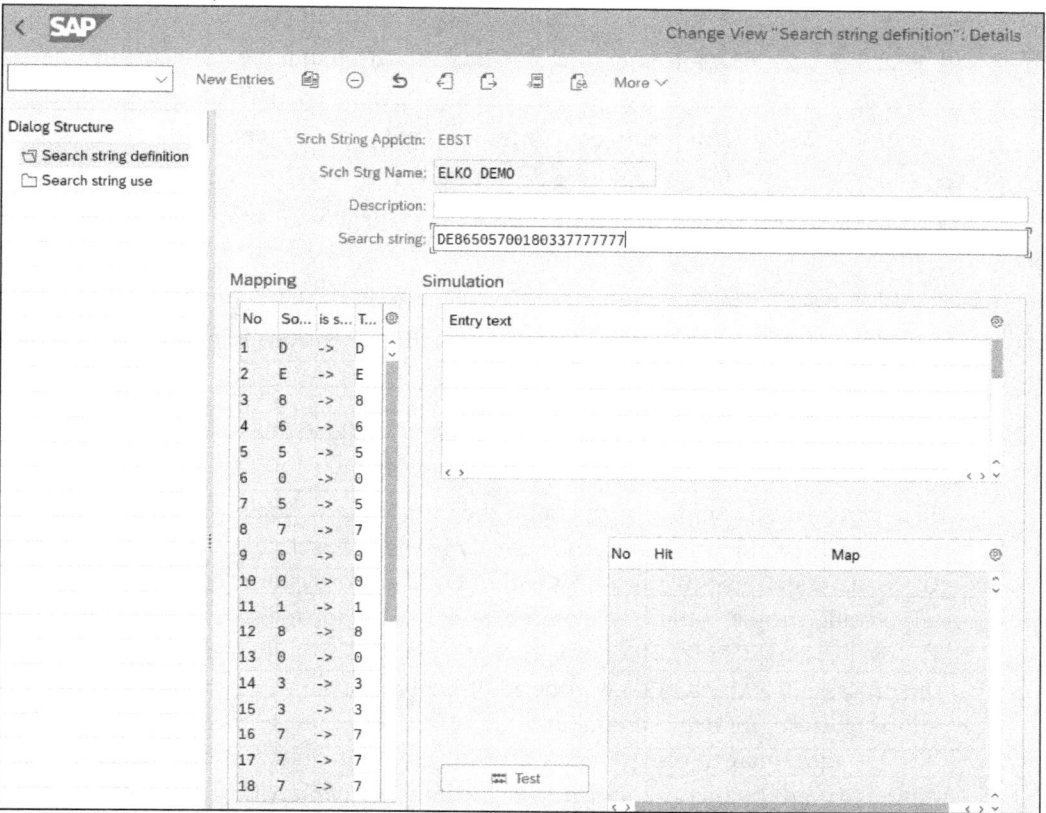

Figure 9.41 Virtual IBAN Determination

Once your search string and mapping are defined, you must assign it to the **Note to Payee** field, which is where the system will look for relevant information (e.g., document numbers or references).

You can define multiple search strings for each bank account and interpretation algorithm. In addition, you may specify a mapping basis and enable the **ID** flag, depending on your needs.

9.2.5 BAdIs

It has probably quickly become clear that there could be tens of thousands of possible combinations when building search strings—especially if you aim to cover every IBAN, customer name, reference, or other detail that may assist in identifying open items. The configuration of search strings in SAP is not straightforward, and this activity requires transport requests, making the deployment process even more complex and time-consuming. To support more efficient and flexible solutions, SAP offers a set of standard BAdIs, such as the following:

- `BADI_FEB_BANK_STATEMENT`, for enhancing bank statement processing in general.
- `BADI_FEB_DOC_POST`, for extending document posting and creating custom posting logic.
- `BADI_FEB_BS_POST`, for posting enhancements of bank statements.
- `BADI_FEB_BS_DTI`, for determining the interpretation algorithm.
- `BADI_FEB_CASH_RECON`, for customizing cash management and cash reconciliation processes.
- `BADI_FEB_ADD_FIELDS`, for adding additional fields to the bank statement data.
- `BADI_FEB_CUST_EXT`, for overall customer-specific extensions in bank statement processing.

These BAdIs provide extensive customization and allow you to tailor the SAP bank statement processing and reconciliation to your organization's specific needs

These can be used to enhance and tailor the interpretation logic. A common approach is to create a custom mapping table directly in the productive environment, enabling you to maintain reconciliation rules without waiting for system deployments. This not only avoids cumbersome transport processes but also significantly accelerates and improves the quality of the reconciliation process. It's also quite common to write a custom program to extend or determine additional rules, or even to use it as an interpretation algorithm for bank statement processing. There are many possibilities available to tailor and enhance this process according to your organization's needs. We just wanted to mention this as a starting point to help you explore how you can extend and enrich your information later on.

9.3 Summary

In this chapter, we explored the vital role of bank statements in the broader context of payments and cash management processes. We examined the most popular bank statement formats and their importance in ensuring accurate cash reconciliation and streamlined processes. You learned how to load and monitor your bank statements in SAP, as well as how to configure the necessary activities to support this integration.

We also touched on the specific characteristics of intraday bank statements and highlighted the trend in SAP to introduce more control and query applications for bank statement processing. Throughout this chapter, we presented an overview of different bank statement formats and the various options for loading and managing them within your SAP system. Although we kept the discussion at a high level, it's clear that there are many variations and considerations in bank statement processing. Our aim was to provide you with a solid starting point from which to explore and expand on these essential processes in your payment landscape.

Chapter 10
The Payment Factory

Managing payments in organizations that operate across multiple systems presents a significant challenge. With payment data originating from various sources, ensuring consistency, control, and efficiency becomes increasingly complex. In such environments, a centralized approach—commonly referred to as a payment factory—can bring clarity and order to the chaos.

In today's complex corporate environments, managing payments efficiently across multiple systems, entities, and geographies is a significant challenge. This chapter explores how implementing a payment factory within an SAP landscape can address these challenges by offering a centralized, streamlined approach to payment processing. By consolidating payments into a single system, organizations can realize substantial quantitative benefits, such as lower transaction costs, reduced bank fees, and improved liquidity through centralized cash visibility. At the same time, qualitative advantages—including enhanced compliance, improved internal controls, and greater transparency—support a more robust and agile financial structure. Key concepts such as in-house banking and a reconciliation factory will also be examined, demonstrating how they complement the payment factory by reducing intercompany complexity and automating transaction matching. Together, these components form a scalable and intelligent framework that empowers finance and treasury teams to manage payments with precision, control, and strategic insight within the SAP environment.

10.1 Designing a Payment Factory

In globally operating companies, payment transactions are often organized in a decentralized way, resulting in a variety of different processes, bank connections, and systems. This fragmentation complicates control over payment flows, increases complexity in cash management, and causes high operational costs.

A payment factory provides a central solution by acting as a *payment processing center* for all corporate entities. The goal is to manage all payments—whether supplier, salary, or intercompany payments—through a unified platform and with standardized processes. The payment factory takes care of the operative handling of payments and acts as the interface between the individual business units and external banks. In other words, the concept of a payment factory connects multiple systems into one centralized payment platform.

This centralization not only significantly reduces transaction costs and bank fees but also enhances the efficiency of the payment process. Furthermore, a payment factory increases transparency over company-wide payment flows and establishes uniform governance in areas of compliance, security, and reporting.

In an increasingly regulated and digital environment, the payment factory thus becomes a strategic component to harmonize company-wide payment processes and prepare them for future requirements.

The main idea behind a payment Factory is to connect multiple systems into one centralized payment platform. This central system serves as the hub for payment orchestration (see Figure 10.1), where all payment service providers and funding sources are integrated.

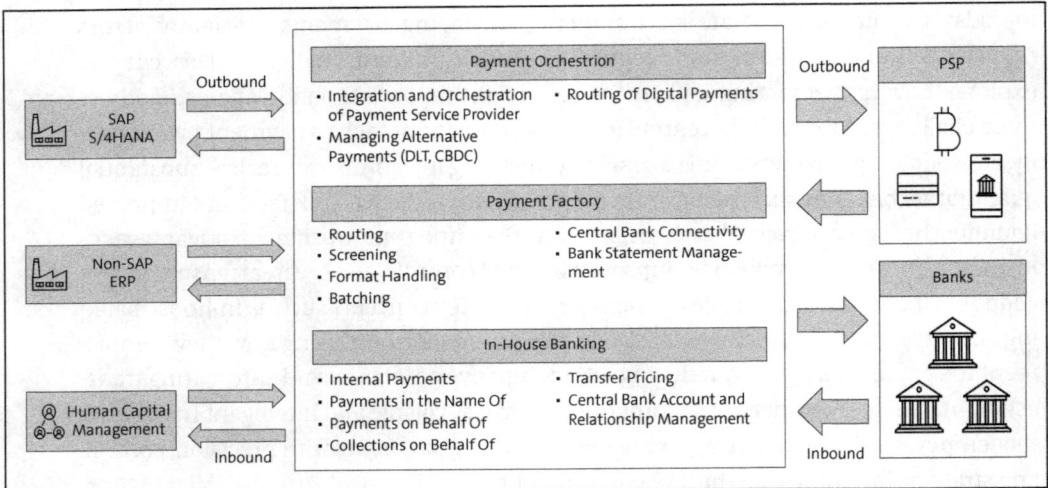

Figure 10.1 Payment Orchestration

Instead of establishing separate connections from each underlying system to banks or payment providers, only the payment factory needs to maintain those external connections—significantly simplifying the architecture and reducing operational complexity. When outbound payment files are received, the payment factory acts as a smart processing layer: It identifies the nature of each payment—whether it's a PINO, POBO, or COBO payment, an internal transfer, or another type defined in this book—and applies appropriate checks and treatment for this payment. These include validations, fraud detection, sanctions screening, and even payment optimization, such as identifying opportunities to convert cross-border payments into more efficient local transfers. Once verified, the system processes the payment and routes it to the appropriate bank or payment service provider. In return, the payment factory receives bank statements and payment acknowledgements, which it also processes and integrates into the system. By consolidating these tasks into a single platform, the payment factory enables a unified, transparent, and efficient payment landscape—providing a single point of control and complete visibility for reporting and reconciliation.

A payment factory is fundamentally always as individual as the company itself and can be modularly structured. Just like every company is unique, not all services of a comprehensive payment factory are always necessary. Let's look at the key components of a payment factory:

- **In-house bank**
  In a group with numerous subsidiaries and cost centers, extensive intragroup services arise, which culminate in intercompany invoicing processes and intragroup payments. In the worst-case scenario, these intragroup payments are processed through external banks, leading to fees and potentially also to foreign exchange (FX) conversion costs. This is where an in-house bank can provide support. An in-house bank, in its broadest definition, functions as a centralized financial management vehicle in which the group's treasury acts as the primary provider of diverse financial services to its operating units, maintaining an arm's-length distance in transactions. The services provided by an in-house bank include central cash and liquidity management, which ensures optimal use of financial resources across the corporation. It also centralizes the management of bank relationships, thereby streamlining interactions and potentially enhancing negotiation leverage with banking partners.

- **Payment factory function**
  A payment factory acts as a centralized payment processing unit within a corporation, designed to enhance the efficiency and effectiveness of handling financial transactions. It is primarily responsible for the execution of external payments on behalf of (POBO) subsidiaries, thereby relieving them from the complexities involved in managing individual payment processes. This centralized approach allows for the standardization and automation of payment flows, streamlining operations and reducing manual intervention. By controlling external bank connections, the payment factory ensures secure, efficient, and consistent interactions with financial institutions, maintaining a standardized protocol across the organization's banking activities.

  The focus of a payment factory is predominantly on cost efficiency and process standardization, aiming to achieve significant reductions in transaction costs through economies of scale while minimizing risks and errors associated with decentralized payment processes. This centralization not only improves the financial control and oversight within the corporation but also enhances liquidity management and provides greater strategic support for treasury operations.

- **Payment orchestration**
  Payment orchestration refers to the centralized management and control of various payment providers, payment systems, and alternative digital currencies within a

unified platform. By integrating multiple payment service providers (PSPs), a payment orchestrator enables centralized and flexible control of these PSPs during transaction processing. Transactions can be dynamically orchestrated based on criteria such as cost, execution method, or geographical requirements. In addition to supporting PSP orchestration, a modern payment orchestration solution also includes the integration of digital currencies such as Bitcoin, stablecoins, or central bank digital currencies. Cryptocurrencies can be processed either directly through blockchain transactions or via specialized payment gateways like BitPay or Lightning Network. Furthermore, payment orchestration offers centralized reporting and analytics functions to monitor transaction flows in real time and optimize payment strategy.

Table 10.1 provides a comprehensive overview of three key concepts in corporate financial management: in-house banking, the payment factory, and payment orchestration. Each of these financial strategies offers unique advantages in terms of efficiency, cost management, and strategic control over the financial operations within a company. By centralizing various functions, they streamline processes and enhance the effectiveness of treasury activities across complex organizational structures.

In-House Banking: Internal Corporate Bank	Payment Factory: Centralized Payment Processing Unit	Payment Orchestration: Payment Service Providers and Payment Method Orchestration
Management of internal (virtual) accounts	Execution of external payments for subsidiaries	Optimization of customer payments in e-commerce
Intercompany payments and financing	Standardization and automation of payment flows	Smart routing across multiple payment service providers
Internal interest calculation and cash pooling	Control of external bank connections	Flexible integration of payment methods and channels
Focus: Liquidity and financing within the group	Focus: Cost efficiency and process standardization	Focus: Conversion optimization and customer experience

Table 10.1 Concepts for Payment Management

The implementation of individual solutions, such as in-house banking or an entire payment factory, can be driven by a variety of factors. Prior to undertaking such initiatives, it is advisable to conduct a thorough analysis of your current business processes. This provides an overview of areas where a payment factory could deliver significant value. It is rare for transaction costs alone to serve as the primary motivator for establishing a payment factory, especially as these costs—particularly for cross-border payments—have markedly declined in recent years.

10.1 Designing a Payment Factory

Today, the primary drivers for implementing a payment factory are the enhanced security of payment transactions and the ability to orchestrate payment flows across the entire group. This also encompasses the group-wide management of liquidity, offering a unified and strategic approach. A fully developed payment factory serves as a central organizational model, delivering greater control over global bank relationship management, payment processes, and banking connectivity. This centralized approach creates a powerful lever for optimizing efficiency and ensuring transparency across financial operations.

The following key design questions should be considered when designing a comprehensive payment factory for an in-house bank:

- **Number of in-house bank areas**
  Define the organizational scope of the in-house banking divisions.
- **Geographical orientation of the in-house bank**
  Determine the regions where the in-house bank is active and their financial strategies.
- **Participating entities in the in-house banking model**
  Identify which subsidiaries or companies are integrated into the in-house banking framework.
- **Centralized services offered**
  Outline the key financial services provided centrally, such as liquidity management and treasury operations.
- **Regulatory restrictions in each country**
  Assess any local regulatory limitations or compliance requirements that may impact operations.
- **Method of executing intercompany payments**
  Specify the approach for intercompany transactions, whether credit- or debit-based.
- **Core currencies in use**
  Identify the major currencies utilized within the group that the in-house bank should manage.
- **Cash pools deployed group-wide**
  Detail the group's cash pooling arrangements and their structures.
- **Transfer pricing requirements**
  Address any specific guidelines or standards concerning transfer pricing.

Cash Management Restrictions

Some countries impose restrictions on capital export, impacting cash pooling and payment processes like POBO. The affected countries and payment types vary globally. For precise details on restrictions in each country, consulting the local central bank and a bank advisor is recommended. Some banks, like BNP and its Cash Management Atlas,

> provide public information as a helpful guide. For more information, visit *https://cash-management.bnpparibas.com/atlas-countries*.

Similarly, for a payment factory, you must consider the following questions:

- **Upstream systems within scope**
  Identify existing systems integrated into the payment factory infrastructure.
- **Payment types in scope**
  Determine the types of payments managed, such as credit transfers, direct debits, and so on.
- **Special requirements (e.g., encryption of HR payments)**
  Address unique security or data protection needs.
- **Number of banks to connect**
  Specify the number of banks that need integration with the payment factory.
- **Payment formats managed (inbound and outbound)**
  Define the specific payment formats and standards handled within the scope.
- **Need for format enrichment**
  Assess whether additional information needs to be appended to certain formats.
- **Need for format conversion**
  Evaluate whether formats require conversion to meet operational or regional requirements.
- **Centralized screening services offered**
  Detail the compliance checks and validation services provided by the payment factory.

And finally, for payment orchestration:

- **E-commerce payment requirements**
  Outline the specific payment capabilities needed to support e-commerce activities.
- **Direct integration of payment service providers or use of gateways**
  Determine whether payment service providers should be connected directly or if using intermediaries is preferable.
- **Information provided by payment service providers on sales transactions**
  Specify what transaction data is made available for analysis and reconciliation.

10.2 Processes and Functions

Through automation, standardized workflows, and unified interfaces, a payment factory reduces operational complexity, lowers costs, and enhances transaction transparency. This section outlines the essential processes within a payment factory, including payment initiation, routing, authorization, accounting, and reporting. Both technical and organizational aspects are covered to provide a comprehensive understanding of the benefits of centralized payment management.

Implementing a payment factory offers strategic advantages and directly affects internal company processes, especially in payment transactions. The connection between upstream systems and the payment factory becomes evident here. Payment execution and transaction security start in preliminary systems and continue with the secure transfer of payment files to the bank via a connector.

Key processes are those that establish essential prerequisites for the core functions of the payment factory, as well as subsequent processes that benefit from its improved infrastructure and efficiency. These interdependencies demonstrate that a payment factory significantly impacts both operational and strategic business processes, as follows:

- **Upstream processes**
  Security and compliance in payment transactions begin within a payment factory with the centralized collection and validation of master data. This data is systematically checked against current sanctions and embargo lists to ensure compliance with international regulations. Managing this data centrally provides a consistent and verified foundation for all subsequent processes, enhancing accuracy, reliability, and security in payment processing.

  A robust master data management system minimizes risks such as erroneous payments, fraud attempts, or regulatory violations, ensuring smooth initiation of payment workflows. Modern technologies like machine learning and advanced analytics further improve efficiency by automating verification processes, identifying potential risks in real time, and adapting continuously to regulatory requirements.

- **Core processes**
  The core functions of a payment factory include managing outbound payments, central validation and processing of incoming account statements, and operating a unified platform for both automated and manual payment processing. These processes standardize and provide transparency within payment workflows. In addition, payment hubs facilitate structured orchestration of diverse payment streams across multiple banks.

  Key features like role-based payment approvals, limit controls, and standardized communication channels (e.g., SWIFT, EBICS, or APIs) enhance governance in payment transactions. Real-time monitoring and traceability of each transaction are vital for audit purposes and internal control systems. Advanced functions, such as intercompany netting and POBO, consolidate group-wide payment flows and unlock internal efficiency potential, contributing to transparency and auditability and laying the foundation for strategic financial management.

- **Downstream processes**
  The payment factory also oversees downstream processes like reporting, daily cash

management, and liquidity planning. By leveraging centralized data and real-time analytics, it generates insightful reports that enable effective liquidity management. Central objectives include optimizing liquidity allocation, precise cash flow forecasting, and reducing idle balances or funding gaps.

- **Regulatory reporting**
  Moreover, the payment factory fulfills regulatory reporting obligations, both locally and internationally, through standardized reporting frameworks, ensuring compliance and operational transparency across jurisdictions. In addition, having all payment and banking transaction data centralized in a single system greatly simplifies audit and compliance processes. When all relevant information is stored and processed within one platform, it becomes much easier to trace payment flows, verify approvals, track user actions, and generate audit reports. This level of transparency not only supports internal controls and governance but also ensures that regulatory and external audit requirements can be met more efficiently and with greater accuracy. A centralized payment factory eliminates the need to gather fragmented data from multiple systems, reducing the risk of inconsistencies and missing information during audits.

In the following sections, we'll examine the overarching in-house banking processes, the functions of the payment factory, and payment orchestration. Every company is unique; therefore, the design of a comprehensive payment factory must also be tailored to each company's specific needs. The subsequent sections are intended to serve as inspiration for how a payment factory can be structured and which processes can be integrated under this organizational model.

10.2.1 In-House Banking Processes

An in-house bank is a strategic framework within a company that provides internal banking services and maintains proprietary bank accounts. This setup allows the organization to centralize the execution of POBO subsidiaries, intercompany transactions, and other financial operations.

Before a company decides to implement an in-house bank, it is crucial to address several key questions to ascertain its feasibility and benefits. Companies must evaluate whether the strategic advantages outweigh the investment required.

In addition, assessing whether the organization possesses sufficient expertise and resources internally to manage an in-house bank is vital. Alignment with the organizational setup, the degree of centralization, and the treasury department's mandate is equally important to ensure seamless integration and function.

Several considerations must be addressed to ensure successful implementation and operation. An in-house bank may not cater effectively to local in-country nuances such as tax obligations and specific local payment formats. Furthermore, affiliates located in jurisdictions with strict legal, fiscal, and monetary regulations may face participation limitations. It is essential to monitor internal pricing vigilantly to align with arm's-length principles and avoid compliance issues. The concentration of operational risk within an in-house bank also requires robust risk management strategies to safeguard against potential challenges.

Overall, while an in-house bank can offer significant efficiency and strategic advantages, companies must weigh these benefits against potential limitations and risks associated with its implementation and operation. As corporation increasingly seek centralized solutions for financial management in a globalized economy, in-house banks represent a proactive approach to optimizing financial operations and enhancing corporate agility.

Key functions of an in-house bank include the following:

- **Internal account management**
  The in-house bank manages dedicated bank accounts for each member of the corporate group, facilitating efficient handling of cash and liabilities across the organization without reliance on external banking services.

- **On-behalf-of payments**
  This model allows the in-house bank to process POBO subsidiaries, reducing the need for individual entities to maintain separate external bank relationships and thereby streamlining overall payment processes.

- **Intercompany transactions**
  By using internal bank accounts, the in-house bank simplifies the management of intercompany payments. Transactions are directly recorded between accounts within in-house banking, lowering transaction costs and enhancing transparency and control over internal cash flows.

  For more in-depth information about the payment processes within in-house banking, refer to Chapter 1, Section 1.3. We discussed payments processing and provided a detailed overview there.

- **Intercompany clearing**
  Intercompany clearing is the process in which receivables and payables between affiliated companies within a corporate group are reconciled and settled. For instance, when one group company provides services to another group company, intercompany receivables and payables arise. As part of intercompany clearing, these mutual claims are recorded, offset, and balanced to ensure that the financial relationships within the corporate group are accurately reflected.

  The goal of intercompany clearing is to make the financial transactions between group companies efficient and transparent and to enable proper consolidation of the group's financial statements.

- **Investment and funding**
  In the context of an in-house bank, the central unit within the corporate group

assumes both the funding and investment functions for the affiliated companies. The in-house bank acts as an internal financial service provider that manages and optimizes liquidity across the corporate group. In the area of funding, it provides group companies with resources as needed—whether through internal loans, cash pooling, or other financing instruments—and thus largely replaces external financing through banks.

On the investment side, the in-house bank consolidates surplus liquidity from the group, centralizes it, and invests it either within the internal cycle—for example, by forwarding it to companies with financing needs—or externally in the financial market according to the group-wide treasury strategy.

By doing so, the in-house bank reduces external bank transactions across the group, optimizes the internal use of funds, and contributes to efficiency and cost optimization in liquidity management.

- **Central cash and liquidity management**
Central cash and liquidity management within an in-house bank is the centralized control and optimization of company-wide cash flows and liquidity reserves by the internal banking unit. The in-house bank is responsible for aggregating and centrally managing all cash flows of the group companies. This is often achieved through cash pooling structures (physical or virtual), where the liquidity of operational units is consolidated into central accounts. This provides group-wide transparency over the current liquidity position and the short- to medium-term liquidity needs of the entire corporate group.

Within this central cash management framework, the in-house bank ensures that excess liquidity is efficiently distributed or invested and that deficits within the group are internally compensated before resorting to external financing sources. At the same time, the in-house bank supports liquidity planning, forecasting, and ensuring the solvency of the corporate group. Through the central cash and liquidity management of the in-house bank, liquidity costs are reduced, interest benefits are realized, and risks, such as currency or refinancing risks, are better managed across the group.

- **Central management of bank relationships**
Instead of having each subsidiary maintain its own banking relationships, the in-house bank functions as a centralized interface with external financial institutions. By consolidating payment volumes, the company can negotiate better terms with banks, such as lower transaction fees, improved interest rates, and optimized credit lines. This also reduces the number of external bank accounts, cutting administrative costs and minimizing the complexity of banking relationships. In some cases, it might still be necessary to maintain local bank accounts for specific subsidiaries. The in-house bank can implement centralized bank relationship management; functionalities like BAM are ideal for establishing workflows across subsidiaries, centrally

orchestrating the opening, modification, and closing of bank accounts. This centralization allows the treasury to maintain oversight and orchestrate bank relationship management organization-wide, saving costs and ensuring that compliance requirements are centrally met.

For instance, centrally tracking the approved signers with authority over each of the company's bank accounts across various banks, countries, and entities is crucial. This role is vital for reducing fraud risk, particularly concerning duplicate signers on accounts or failing to remove signers after they leave the company. Maintaining a clean signer list is also important for compliance with banking regulations impacting corporations, such as FBAR in the US.

- **Bank fee management**
  Although it can be tedious, the process of analyzing bank fees and service charges to identify cost-saving opportunities can become a significant part of the treasury's bank account management responsibilities. These projects are often necessary for ensuring the company's account structure is streamlined and optimized. In addition, reconciling bank account statements with internal records to identify discrepancies, ensure accurate accounting, and detect potential fraud is an area in which the treasury will have a vested interest, typically working in alignment with accounting.

- **Central foreign exchange and exposure management**
  Depending on the global setup, in-house bank accounts can be opened in both the local currency and the balance sheet currency of the in-house bank. Based on the business model and activities of the company operating the in-house bank, it may be necessary to maintain in-house bank accounts in different currencies. This can result in currency risks affecting either the in-house bank or the subsidiaries.

> **Note**
> As a general rule, all transactions for which the transaction currency differs from the account currency should be converted. The system can store various exchange rates and use them for these conversions.

To explain the currency risk of the in-house bank through an example, we have prepared a brief case study. The company used in our example is purely fictional and serves illustrative purposes only. It does not represent any real organization.

> **Case Study: GlobalTech LLG**
> GlobalTech LLG is an American industrial group based in Chicago that produces high-precision machinery for the automotive and aerospace industries. The group operates production facilities in several locations in the US and in other countries, including Poland, to benefit from lower production costs and a strategically advantageous location within the EU.

The corporate structure and currency risk details are as follows:

- **Holding company (GlobalTech LLG, USA)**
  The holding company in Chicago serves as the central management unit and reports in USD. It sets the financial strategy, coordinates investments, and bears the central currency risk.
- **In-house bank**
  The holding company operates an in-house bank, which also reports in USD. It manages cross-group financial flows, consolidates liquidity, and handles currency management for the subsidiaries.
- **Subsidiary (GlobalTech Polska Sp. z o.o.)**
  The Polish production facility manufactures machine parts and reports in PLN. It sells its products to both external customers and other corporate entities.

Let's take a look at two scenarios based on this case study:

- **Scenario 1: In-house bank account in central currency (USD)**
  In this scenario, all in-house bank accounts are maintained in the reporting currency of the in-house bank, which is USD. This eliminates currency risk for both the in-house bank and the holding company as all financial flows are centrally managed in the same currency.

  However, the Polish subsidiary (GlobalTech Polska Sp. z o.o.), which reports in PLN, bears all the currency risk. Because its In-House Bank account is maintained in USD, all deposits and withdrawals need to be converted from PLN to USD. A depreciation of the Polish złoty against the US dollar would increase the costs of USD payments for the subsidiary, whereas an appreciation of PLN could reduce these costs. Therefore, the subsidiary must independently implement hedging measures or rely on the in-house bank for hedging in order to minimize exchange rate risks.

- **Scenario 2: In-house bank account in local currency (PLN)**
  In this scenario, the in-house bank account of the Polish subsidiary is directly maintained in PLN. This setup eliminates immediate currency risk for the subsidiary, as its transactions occur directly in its balance sheet currency.

However, the currency risk shifts to the in-house bank, which reports in USD. Because the in-house bank holds a foreign currency account in PLN, it faces an exchange rate risk when PLN balances need to be converted to USD. A depreciation of the Polish złoty against the US dollar would result in losses, while an appreciation could lead to gains. Therefore, the in-house bank would need to implement appropriate hedging strategies, such as FX swaps or forward contracts, to protect against exchange rate fluctuations.

Table 10.2 shows a comparison of the effect of having the in-house banking account in the group versus the functional currency.

Scenario	Currency Risk for Subsidiary	Currency Risk for In-House Bank
1: In-house bank account in USD	High: PLN must be converted to USD	No risk: Everything is accounted for in USD
2: In-house bank account in PLN	No risk: PLN is used directly	High: Foreign currency balances must be converted to USD

Table 10.2 Comparison of Currency Used in In-House Bank Accounts

When setting up an in-house bank, there is no obligation to adhere to a specific scenario. Local subsidiaries can maintain accounts in both their respective balance sheet currencies and foreign currencies. However, from a strategic exposure management perspective, this choice is crucial. Before establishing an in-house bank, it is essential to determine whether the FX risk should be distributed across the subsidiaries or centralized within the in-house bank.

Another important factor to consider when selecting the currency for in-house bank accounts is the impact of interest calculations and related charges. Interest income, interest payables, and any applicable bank charges will be calculated and settled in the currency of the in-house bank account. This means that the chosen currency will directly affect how these financial elements are handled within the system. Some organizations will open an in-house bank account for each currency they have transactions in.

It is quite common for organizations to maintain separate bank areas for different regions, such as the US and Europe. Typically, all European entities are grouped under a European header entity, while US-based and other American entities are grouped under a US header. This structure is primarily driven by operational reasons, such as time zone differences, as well as tax and legal requirements. From a treasury perspective, it's much more efficient to manage a single intercompany payable relationship between the US and Europe—specifically, between their respective holding entities—than to maintain numerous intercompany payables and receivables between individual entities across regions. This simplifies reconciliation, accounting, and compliance. However, different companies may adopt different approaches. In some cases, entities from one region may participate directly under a header entity in another region. The same logic applies to Asia, where it is also common to have a dedicated header entity located in financial hubs such as Singapore or Hong Kong, depending on the organization's presence and business structure in the region.

10.2.2 Payment Factory Functions

A payment factory serves as a central platform for standardizing, automating, and optimizing payment processes within a company. It acts as an interface between internal financial systems such as ERP or treasury management systems and external banking interfaces. Through its various functions, a payment factory contributes to making payment transactions more efficient, secure, and cost-effective. The following are the key processes and functions of payment factory operations:

- **Format conversions as a payment factory function**
  Different banks and countries use various payment formats for instructions and confirmations. A payment factory automatically converts and harmonizes these formats so that a company can operate internally with a consistent standard. This reduces manual adjustments, minimizes errors, and improves process quality. Companies benefit from easier integration with banks and increased automation of payment processes. The payment factory will provide a significant advantage during upcoming format changes, like MT to MX (ISO 20022), centralizing this task as a service unit within the company. Format adjustments do not necessarily have to be performed locally but can be handled centrally by the payment factory.

- **Bank connectivity provided by the payment factory**
  A payment factory enables centralized bank connectivity, eliminating the need for companies to manage individual bank access points. With a unified interface to all banks, payment transactions are simplified, more transparent, and secure. This function allows companies to centrally manage their banking relationships, consolidate payment processes, and negotiate better terms. In addition, it enhances security in payment transactions and reduces maintenance efforts associated with maintaining different online banking systems and bank channels.

- **Batching payments as a payment factory function**
  The payment factory can collect individual payments and process them in larger batches. This function leads to greater efficiency in payment processing and can reduce transaction costs. Banks often offer more favorable conditions for larger volumes of payments, enabling companies to benefit from economies of scale. Simultaneously, batching reduces administrative efforts and improves control over payment flows.

- **Central verification and routing by the payment factory**
  The payment factory provides a platform for verifying and managing payments. Through this function, risks, irregularities, or errors in payments can be detected early. The payment factory can automatically route payments through the most cost-effective path to minimize fees and ensure secure processing. This function also enhances liquidity management and allows optimized control of currency risks.

- **Cost optimization as a payment factory function**
  Centralized payment management by the payment factory allows for strategic selection of favorable payment methods, exchange rates, or banks. By consolidating the entire payment volume, the treasury department can negotiate better terms and use automated processes to reduce the overall costs of payment transactions.

In addition, companies can strategically control payment timing to optimize the use of discounts and rebates.

- **Automation and process optimization by the payment factory**
  Implementing a payment factory automates recurring payment processes, reducing manual effort and minimizing error rates. Automated workflows ensure faster processing and allow employees to focus on value-added activities. In addition, the payment factory supports compliance requirements with complete audit trails and automatic verification routines.

- **Security and fraud prevention as a payment factory function**
  The centralized management of all payments enables better control and reduces the risk of fraud. Automatic verification mechanisms allow the payment factory to detect and block suspicious transactions early. Clear access rights and detailed user management ensure that only authorized personnel can initiate or approve payments.

- **Scalability and flexibility of payment factory functions**
  A payment factory is designed to grow with the company and adapt to changing business requirements. New bank connections, subsidiaries, or business units can be integrated easily. Thus, the payment factory remains a future-proof solution for both medium-sized companies and large corporations, remaining efficient even with increasing transaction volumes or expansion into new markets.

Another topic that needs to be discussed when it comes to a payment factory is compliance. New technologies, evolving banking products, and increasing transaction speeds are contributing to greater complexity in payment processes. Concurrently, regulatory requirements are becoming more stringent, resulting in increased compliance efforts both at the group level and within individual subsidiaries.

A payment factory provides a centralized solution to effectively manage and meet these expanding compliance obligations across the entire organization. Let's examine some of the key compliance-related requirements that can be centrally managed through a payment factory:

- **IT security/secure data transfer**
  It's important to ensure compliance with current IT security standards (e.g., ISO/IEC 27001, PCI DSS) for end-to-end encryption and data integrity among payment factory participants, the payment factory, and banks. This includes TLS-encrypted communication channels, two-factor authentication, and system- and process-level access controls.

- **Inbound and outbound validation**
  Payment runs must be approved before processing by the payment factory (e.g., four-eyes principle, role-based workflows). Incoming payment files are validated for technical completeness (e.g., format checks according to ISO 20022, SEPA, or MT standards) and content, including fraud detection (e.g., anomaly recognition) and

regulatory compliance (e.g., automated embargo and sanctions list screening per EU/OFAC regulations).

- **Bank authorizations**
  Payment files are authorized by the payment factory before being submitted to the bank—depending on the integration method (EBICS, SWIFT, APIs, or host-to-host communication). Authorization follows defined corporate governance rules, such as limit checks and user-based permissions.

- **Service-level agreements and internal pricing**
  Responsibilities for payment services are governed by bilateral service-level agreements between the payment factory and subsidiaries. These define key performance indicators (KPIs) such as processing speeds, error rates, and response times. Internal service pricing is based on transparent, market-comparable principles (e.g., a cost-plus or market-price approach).

- **Policies**
  Mandatory group-wide policies apply to all payment-related activities, covering master data handling, approval procedures, and compliance with regulatory requirements (e.g., GDPR, MaRisk, GoBD). These policies are centrally managed by the payment factory and cascaded down to each group entity.

- **Audit-proof logfiles**
  All relevant process steps (e.g., payment creation, approval, transmission, error handling) are recorded in an audit-proof, traceable, and tamper-resistant manner—including timestamps, user IDs, and system data. These logs support internal control systems and serve as documentation for external audits (e.g., by auditors or compliance units).

- **Reporting obligations**
  The payment factory supports regulatory reporting requirements, such as foreign trade reports (e.g., Z4/Z5 under Germany's AWV), §24/§25 KWG reporting, and country-specific obligations (e.g., FBRA in France or DFI in the US). Reporting data is automatically generated from processed transactions and exported via defined interfaces (e.g., balance of payments reporting).

Through its diverse functions, a payment factory can significantly optimize a company's payment processes. In addition to automation and standardization, a payment factory also contributes to cost reduction, security, and efficiency improvements in the financial domain. Centralized control of bank connections, currency risks, and payment flows enables transparent and strategic management of the entire payment operation.

10.2.3 Payment Orchestration

Payment systems are evolving rapidly, driven by new technologies and innovations in e-commerce. Mobile payments, blockchain technologies, and artificial intelligence are revolutionizing the market, leading to the emergence of numerous payment service providers and alternative payment methods. Independent payment service providers such as Stripe, Adyen, and PayPal are setting new standards, while traditional credit institutions like Giropay and Wero are adapting to the digital transformation and expanding their product portfolios. Businesses face challenges in this arena, such as the following:

- **Lack of standardization**
  Businesses face challenges regarding the standardization of payment processes. There are no uniform formats or connectivity standards, like EBICS and SWIFT, for payment service providers, complicating the seamless integration of new providers into existing payment systems. For example, consider the diverse handling of transaction formats: Some payment service providers use proprietary APIs, whereas others rely on standardized protocols like ISO 20022.

- **Capital ties and counterparty risk**
  Capital ties vary depending on the payment service provider and country. An online retailer operating in multiple markets often has to work with several payment service providers, which increases counterparty risk and can lead to liquidity shortages if settlement cycles vary significantly. Therefore, centralized management of these risks is essential to ensure financial stability.

- **Management of alternative currencies**
  Digital currencies that use distributed ledger technology are becoming increasingly significant. Governments and central banks are advancing the development of central bank digital currencies globally, while businesses are progressively exploring cryptocurrencies like Bitcoin and stablecoins as viable payment options. Prominent companies such as Tesla and Microsoft already accept cryptocurrency, underscoring the importance of effective treasury management.

Modern ERP systems like SAP S/4HANA provide robust solutions for integrating and managing diverse payment options. These systems can either complement existing payment factory structures or serve as foundational building blocks for future payment infrastructures. By leveraging the SAP digital payments add-on, businesses gain access to standardized connectivity with payment service providers, streamlining the automation of transaction data processing. This not only enhances operational efficiency but also reduces the complexity of managing multiple payment channels.

In addition, SAP Digital Currency Hub offers advanced capabilities for handling digital currencies, including cryptocurrencies and stablecoins. It supports real-time analytics to optimize liquidity and mitigate risks, making it particularly valuable for cross-border transactions. SAP Digital Currency Hub enables 24/7 instant payments with reduced transaction costs, eliminating the need for intermediaries. This aligns with the growing trend of blockchain-based payment solutions, which are reshaping global financial operations by offering faster settlement times and enhanced transparency.

The adoption of diverse payment service providers and acquirers facilitates adaptable payment handling, enabling businesses to seamlessly integrate varied payment methods into their platforms. Beyond conventional options, innovative payment forms—including central bank digital currencies, Bitcoin, and stablecoins—are gaining traction, especially in the international B2B arena. These innovations pave the way for faster transactions and lower costs.

Effective counterparty management is vital to payment orchestration. Centralized oversight of balances across multiple providers and continuous evaluation of transaction expenses are crucial to streamlining payment operations. For example, a global online marketplace collaborates with numerous providers, leveraging real-time monitoring of payment flows to enhance liquidity and promptly adjust to market fluctuations. By orchestrating these processes, companies can establish scalable and transparent payment frameworks that address evolving demands for agility and efficiency.

Overall, structured management and integration of new payment technologies enable efficient adaptation to digital transformation. Businesses benefit from optimized cost structures, reduced risk, and increased flexibility in payment systems. The ongoing digitization in this area presents numerous opportunities while also requiring companies to integrate innovative and secure payment solutions into their business models.

10.3 Quantitative and Qualitative Factors

By conducting a detailed analysis, a company can evaluate whether adopting a strategic organizational framework, such as a payment factory, delivers substantial benefits, both quantitatively and qualitatively. As highlighted in earlier sections, a payment factory rarely proves financially advantageous based solely on savings from cross-border payment transactions, as costs continue to decline. However, the enhanced security and improved control over liquidity and risk management that it offers often justify implementing specific functionalities or a comprehensive payment factory. Consider the following curated list of key factors to examine during the implementation process.

First, let's start with a look at a qualitative factory:

- **Compliance**
  A payment factory provides a controlled and standardized environment for processing payments, which is a critical enabler for regulatory and internal compliance. With centralized payment release, companies can implement consistent approval workflows and enforce segregation of duties across all entities and geographies. All changes, actions, and approvals are logged in an auditable format, ensuring transparency and full traceability throughout the payment lifecycle. This reduces the risk of unauthorized payments and supports regulatory audits, internal control assessments, and Sarbanes-Oxley (SOX) compliance, particularly in complex multinational organizations.

- **Fraud prevention**
  Fraud risk is significantly reduced within a payment factory due to the high degree of automation and control available. Integrated fraud-detection tools can analyze transaction patterns to identify anomalies or suspicious behavior in real time. In addition, automated sanction list screening and embargo checks ensure that payments are compliant with international regulations, reducing legal and reputational risks. Manual payments—often a major vulnerability—are minimized or eliminated entirely, lowering the likelihood of errors, fraud, or unauthorized transactions.

- **Liquidity management and transparency**
  The payment factory acts as a central visibility and control point for all outgoing payments across the group. This allows treasury and finance teams to gain real-time insight into global cash flows and make more informed liquidity decisions. With centralized control over payment timing and execution, companies can manage working capital more effectively, extending payment terms where possible or optimizing funding across entities. This also enables pooling or central funding structures (e.g., in-house banks), which helps to reduce idle cash and optimize interest income or expense.

- **IT security and simplified system landscape**
  From an IT and infrastructure perspective, the payment factory consolidates payment interfaces between ERP systems and banks into a single secure platform. This not only reduces complexity but also significantly improves cybersecurity. Fewer interfaces mean fewer points of failure or vulnerability, while centralizing bank connectivity and format management reduces maintenance effort and dependency on local systems. Standardized file formats (e.g., ISO 20022) and centralized updates ensure consistency and compliance with evolving banking standards, improving system resilience and reducing operational risk.

Next, let's take a look at a quantitative factory:

- **Initial investment**
  The setup of a payment factory typically requires an up-front investment, including the purchase of product licenses for treasury or payment hub software. In addition, companies must account for internal costs (e.g., dedicated project teams, IT resources) and external costs (e.g., consultants, system integrators). Although this represents a financial entry barrier, it lays the foundation for long-term operational and financial efficiencies.

- **Transactional costs**
  Many payment factory solutions use a licensing model that includes volume-based pricing. This means that a portion of the software cost is tied to the number or value of transactions processed. As payment volumes grow, this can lead to scalable pricing, but it also requires careful monitoring to avoid unexpected cost increases.

- **Reduced transactional costs**
 By consolidating payment activities and enabling netting or intercompany settlement, a payment factory significantly reduces the number of external (i.e., bank-processed) transactions. Fewer external transactions translate directly into lower transaction fees charged by banks, especially for high-volume, high-frequency payment environments.

- **Optimal payment routing**
 This helps with reduction of cross border payments and elimination of bank converted payments. With centralized payment control, the payment factory allows for intelligent routing of payments through the most cost-effective channels. This reduces reliance on expensive cross-border transactions and avoids costly currency conversions performed by banks. Instead, internal netting or centralized FX execution strategies can be applied, increasing efficiency and reducing costs.

- **Reduced bank fees**
 This happens through bank relationship rationalization. Centralizing payments within a payment factory enables a company to reduce the number of banking partners and accounts needed globally. By consolidating transaction volumes with fewer banks, companies gain negotiating power and can secure better terms and lower service fees. This also simplifies account management and treasury operations.

- **Reduced foreign exchange spreads**
 A payment factory can support internal FX management strategies, such as intercompany hedging and internal netting, which reduce the need for external currency conversions. This leads to significantly lower FX spreads compared to relying on banks' conversion rates, thereby optimizing FX costs.

- **Process efficiency**
 The automation of payment workflows, approvals, compliance checks, and reporting in a payment factory eliminates manual intervention and reduces operational errors. Furthermore, centralizing payment expertise in a single team or center of excellence allows for better standardization, faster issue resolution, and overall increased operational efficiency.

Integrating these qualitative and quantitative benefits emphasizes the strategic value of implementing a payment factory, leading to more streamlined and cost-effective payment processes.

10.4 The Reconciliation Factory

A reconciliation factory is used for the centralized reconciliation of open items. The principle is theoretically simple: Instead of each unit, each company code, and even each individual ERP or mass contract system performing its bank statement reconciliation in its own system, open items instead are technically transferred to the central system. In this central system, all the corporation's bank statements are also centrally imported for reconciliation.

The advantages are clear: By centralizing reconciliation activities into a shared service or platform, companies can reduce redundancy, eliminate siloed approaches, and enforce consistent rules and controls. This leads to improved data quality, faster issue resolution, and greater transparency across the entire reconciliation lifecycle. Automation capabilities within a reconciliation factory significantly reduce manual effort, allowing teams to focus on exceptions rather than routine matching. Posting rules for automating bank statement processing achieve significantly higher effectiveness in a centralized setup and streamline the process massively. This also has positive effects on cash visibility, as it provides an immediate group-wide overview of liabilities, thus optimizing cash management.

The reconciliation factory in SAP S/4HANA is typically built using standard SAP S/4HANA Finance capabilities, with a strong reliance on Central Finance (see Figure 10.2. The core idea is to move all open items from the underlying local ERP systems into a central system, where reconciliation is performed centrally. This approach eliminates the need to maintain complex reconciliation rules across multiple systems, greatly simplifying the process and improving consistency. Centralized reconciliation also enables better reporting and more efficient cash management as all relevant data is consolidated in one place. The process involves pulling open items from local ERP systems, receiving bank statements centrally, executing the reconciliation in the central system, and then sending the reconciliation results back to the local ERP systems. In addition, the central system updates cash management and in-house banking based on the statement information. When reconciliation is handled centrally, there's no longer a need to distribute bank statements and related information back to each local ERP system, further streamlining operations. Although the reconciliation factory is less widely adopted than the payment factory, it is a powerful model for organizations seeking centralized control, efficiency, and visibility in their treasury operations.

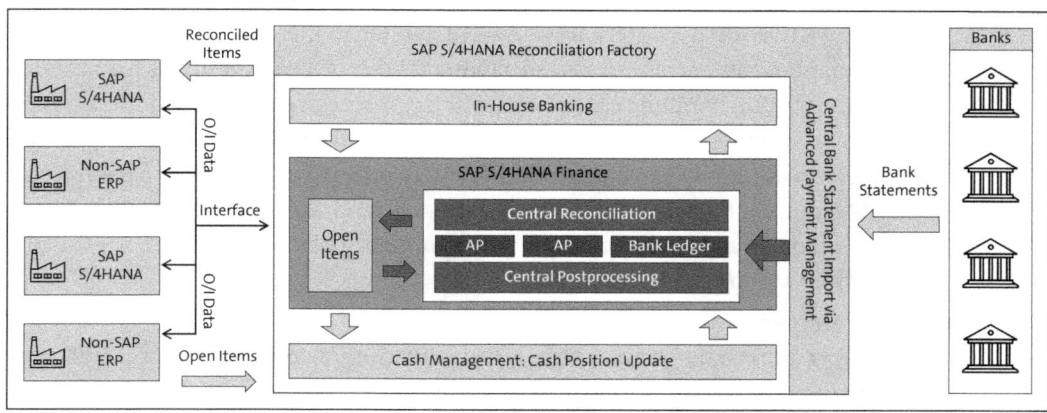

Figure 10.2 Reconciliation Factory

However, the implementation of a reconciliation factory does not come without challenges. To achieve the highest possible level of automation, well-coordinated interfaces between the systems are necessary, allowing the receipt of open items as well as the dispatch of reconciled items via, for example, an RFC.

The challenges are not only limited to systems within the corporation. Consider the following:

- **System interfaces**
  One of the primary challenges in integrating upstream systems with a reconciliation factory lies in the heterogeneity of source systems, each potentially using different data formats, transmission protocols, and update frequencies. Ensuring consistent and accurate data ingestion from these systems requires robust data mapping, transformation, and normalization processes. In addition, discrepancies in data quality, completeness, and timeliness can significantly impact the effectiveness of the reconciliation process. Another key issue is the alignment of data semantics across systems; differing definitions or interpretations of key data fields can lead to mismatches and false positives. Furthermore, managing the volume and velocity of data from multiple sources requires scalable infrastructure and efficient processing pipelines. Security, data lineage, and auditability are also critical concerns. Ultimately, seamless integration demands a combination of flexible architecture, strong governance, and ongoing collaboration between system owners and the reconciliation team.

- **Processing payment advises centrally**
  Processing remittance advices within a reconciliation factory presents several key challenges. First, advices often arrive in a wide variety of formats—from structured EDI messages and PDFs to unstructured emails—which makes automated ingestion and parsing complex. Many advices lack standardization, requiring sophisticated logic to extract relevant data such as references, amounts, and counterparties. Furthermore, timing mismatches between the arrival of the advice and the corresponding transactions can hinder real-time reconciliation. There are also challenges in matching advice data to internal system entries due to missing or inconsistent identifiers, manual entry errors, or partial payments. This often necessitates fuzzy matching and exception handling capabilities within the reconciliation engine. In addition, ensuring auditability and traceability for compliance purposes adds another layer of complexity, especially when advices are received from a multitude of external partners. Overall, high variability, lack of standards, and the need for intelligent data extraction make advice processing a critical but difficult component in a reconciliation factory setup.

- **Bank statement processing**
  Processing bank statements within a reconciliation factory presents several cross-industry challenges, primarily due to the variety and complexity of bank statement

formats. Depending on the financial institutions involved, statements can come in formats such as MT940, camt.053, BAI2, CSV, or even unstructured PDFs—each with its own structure, field definitions, and data granularity. This lack of standardization demands adaptable parsing capabilities and format-specific mapping logic to ensure accurate data ingestion. In addition, banks often populate fields differently or inconsistently, which can lead to discrepancies in transaction references, value dates, or counterparty details. Another key challenge is the timing of bank statement availability, which may not align with internal postings, causing temporary mismatches in the reconciliation process. Some statements are delivered daily, others multiple times a day (intraday statements), and some only upon request, which further complicates automation. Moreover, organizations must deal with multibank, multicurrency environments, increasing the need for robust normalization and currency handling. Finally, for regulatory and audit purposes, the reconciliation process must ensure transparency, traceability, and secure handling of financial data across all statement sources.

Corporations that address these challenges greatly benefit from the scaling possibilities in bank statement processing, as well as increased visibility in cash management, which significantly enhances the effectiveness in both cash management and working capital while reducing opportunity costs. However, the effort required, especially in an SAP environment using SAP standard processes, should not be underestimated. Custom programming and extensions are almost inevitable in such an endeavor and should be planned from the start. In conjunction with a payment factory, where bank connectivity is centrally managed and formats are centrally standardized, the payment factory becomes an intriguing construct for companies with high reconciliation effort and a heterogeneous system landscape.

10.5 Summary

In this chapter, we described the different expansion options and functions that companies can bundle under the concept of a payment factory. In the context of the increasingly rapid developments in payment transactions and the resulting increased need for companies to react, centralization with an extendable platform is almost indispensable. Particularly when adapted to a company's business model, a payment factory and reconciliation factory are increasingly valuable today due to their ability to streamline financial operations and enhance efficiency.

The payment factory centralizes the management of payment processes and standardizes formats, providing companies with better cash flow visibility, reducing operational costs and minimizing risks associated with decentralized systems. It simplifies bank connectivity, offering greater control and oversight of transactions. Similarly, a reconciliation factory centralizes the reconciliation of open items, allowing for improved automation and accuracy in processing. By consolidating bank statements and reconciling transactions in a single system, companies can achieve higher efficiency and reduce manual efforts, thus optimizing cash management and enhancing financial reporting. Together, these factories facilitate more strategic financial decision-making and contribute to improved resource allocation and operational effectiveness.

Chapter 11
Outlook on Payments and Bank Communication with SAP

This chapter provides an overview of current developments and SAP's responses to alternative payment methods and currencies. Here, we focus purely on business processes and offer food for thought for the design of your system and process architecture.

This chapter summarizes ongoing and previews future development in the payments and bank communication space, covering both functional development within SAP solutions and legislation-driven changes. We will provide an introduction and suggestions for dealing with alternative payment methods and currencies based on distributed ledger technologies (Section 11.1). After that, we will discuss the outlook for SAP solutions moving forward (Section 11.2).

11.1 SAP Solutions

In recent years, consumer behavior and payment habits have drastically changed. Along with bank-owned alternatives for payments such as GIRO or Wero in the Eurozone, payment service providers like PayPal, Alipay, or BLIK have become standard in the payment landscape, posing serious competition to established payment methods like credit cards. Since March 2020, with the onset of the COVID-19 pandemic, e-commerce and digital payment methods have experienced a new bloom and massive growth. Companies handling digital payments in an international environment face a range of complex challenges. One of the central issues is the multitude of different payment service providers and payment methods required for various markets. Each of these solutions comes with its own interfaces, security requirements, and integration efforts. This heterogeneity makes it difficult to establish a unified payment process and often leads to isolated solutions within the system landscape.

In addition, regulatory requirements such as PCI DSS, local compliance mandates, and changing legal frameworks in different countries must be continuously considered and technically implemented. The issue of security is also growing in importance as sensitive payment data must be protected against cyberattacks and fraud attempts.

Furthermore, customers today expect a seamless and secure payment experience, whether they are shopping via an online store, a mobile app, or traditional channels. Delays or technical issues in payment processing can directly lead to abandoned purchases and negatively impact customer satisfaction.

Finally, companies are confronted with the effort of continuously maintaining and updating multiple payment providers and their interfaces. This binds IT resources and makes it challenging to quickly respond to new trends in payment transactions—such as new payment methods or changing customer requirements.

In this section, we present opportunities for expanding and offering alternatives in payment processing with SAP. Alternative payment service providers and distributed ledger technology (DLT)–based currencies are expected to gain significant momentum, especially in the wake of the US election in 2024. But how can companies integrate these payment methods into their SAP architecture? Let's examine the expansion possibilities to find inspiration.

11.1.1 SAP Digital Payments Add-On

Similar to SAP Multi-Bank Connectivity, the SAP digital payments add-on provides a gateway between payment service providers, covering alternatives in payment processing. The solution is based on the cloud-based SAP Business Technology Platform (SAP BTP) and can be seamlessly integrated with the SAP system. It is designed to connect credit card providers and payment service providers and integrate them into ERP processes such as accounts payable (AP) and accounts receivable (AR).

The SAP digital payments add-on thus offers a central platform for managing various payment providers. Businesses can integrate different payment service providers such as PayPal, Stripe, Adyen, or Worldpay to centrally orchestrate their payment processes.

One of the key challenges is the heterogeneity of payment service providers. This includes the services themselves as well as formats and types of integration. Although bank integrations now commonly use standardized account statements in SWIFT message type (MT940/MT942) and ISO 20022 camt.053/052 formats, account statements are widespread; there is a jumble of formats among payment service providers, creating the challenge of orchestrating different formats and integration methods. A market has formed for orchestrating the integration of payment service providers—like Adyen, Worldline, and Stripe—with multiple payment service providers and credit card providers from a single source. Nevertheless, companies with a strong e-commerce focus face the challenge of managing this heterogeneous system landscape.

The system architecture is fundamentally divided into three application areas, as shown in Figure 11.1.

In the following sections, we'll introduce the consumer applications, the SAP digital payments add-on connector, and payment service providers in more detail.

11.1 SAP Solutions

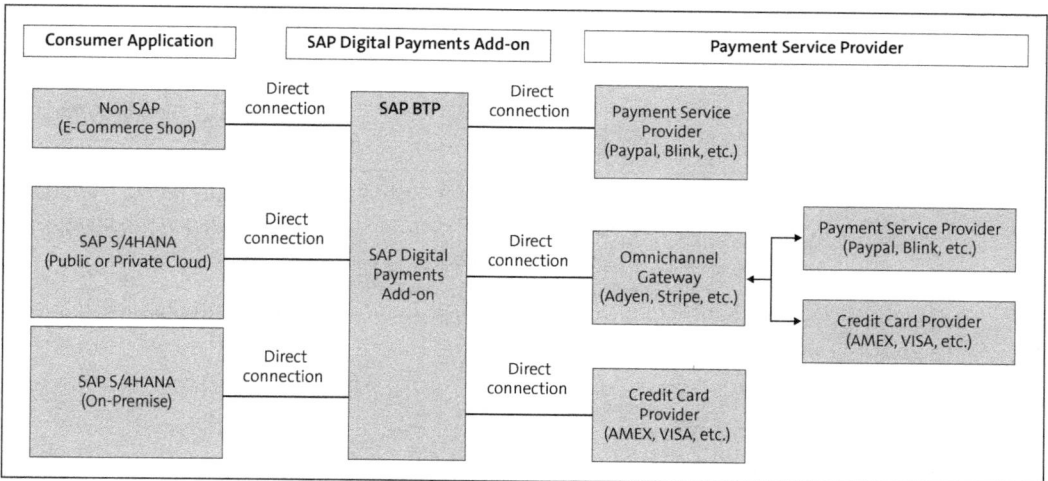

Figure 11.1 Example System Landscape

Consumer Application

The initiation of payments and the reconciliation of advice and statement files generally occur in consumer applications. SAP offers partial direct integration into the processes of various consumer applications, as indicated in Table 11.1.

System	Consumer Application	Integration Type
SAP S/4HANA Cloud Private Edition	<ul><li>Customer management</li><li>Sales (sales order)</li><li>Sales (billing)</li><li>Finance (accounts receivable)</li><li>Contract accounting</li><li>Business partner</li><li>SAP Billing and Revenue Innovation Management (convergent invoicing)</li></ul>	Out-of-the-box API provided by the SAP digital payments add-on
SAP S/4HANA Cloud Public Edition	<ul><li>Customer management</li><li>Sales (sales order)</li><li>Sales (billing)</li><li>Finance (accounts receivable)</li><li>Contract accounting</li><li>Business partner</li><li>SAP Billing and Revenue Innovation Management (convergent invoicing)</li></ul>	Out-of-the-box API provided by the SAP digital payments add-on

Table 11.1 Consumer Applications

System	Consumer Application	Integration Type
SAP ERP	- SAP ERP	Out-of-the-box API provided by the SAP digital payments add-on
Other SAP solutions	- SAP S/4HANA Cloud for Customer Payments - SAP Commerce Cloud - SAP Subscription Billing	Depending on the backend system, requires an individual API

Table 11.1 Consumer Applications (Cont.)

SAP Digital Payments Add-on

As previously mentioned, the SAP digital payments add-on is designed, simply put, to connect financial processes between online shops, ERP finance systems, payment service providers, and payment gateways. The SAP digital payments add-on can receive information from multiple systems simultaneously and process it centrally.

Here are the key processes that the add-on covers:

- **Payment card registration (payment card handling)**
    - Storage and management of credit card data via tokenization.
    - Passing card data to a payment service provider for validation and token creation.
    - Management of card tokens for recurring payments (e.g., in e-commerce or subscription business).
- **Payment authorization**
    - Execution of authorization requests for payments through the connected payment service provider.
    - Feedback of authorization results (e.g., approval or rejection) to the SAP system.
    - Management of authorization references for later billing or cancellation.
- **Payment capture**
    - Capture of the authorized amount (full or partial capture).
    - Transmission of the capture request to the payment service provider and feedback to the SAP system.
- **Cancellations and refunds**
    - Cancellation (void) of open authorizations.
    - Processing of refunds through the payment service provider.
- **Token management**
    - Management and maintenance of the transferred payment tokens.
    - Automatic transfer of tokens to various SAP applications like SAP Commerce Cloud, SAP S/4HANA Sales, or SAP Subscription Billing.

- **Reporting and monitoring**
  - Provision of status messages and logs for payments and transactions.
  - Integration into SAP standard tools for monitoring and error handling.
- **Omnichannel processes**
  - Support for payment processes both in e-commerce (B2C/B2B) and in traditional ERP scenarios (e.g., order processing in SAP S/4HANA).
  - The SAP digital payments add-on serves as a central bridge between SAP applications and external payment service providers, decoupling SAP systems from the direct processing of sensitive payment data.

Payment Service Provider: Credit Card

Figure 11.2 clarifies the scope and functions of the SAP digital payments add-on for credit card payments.

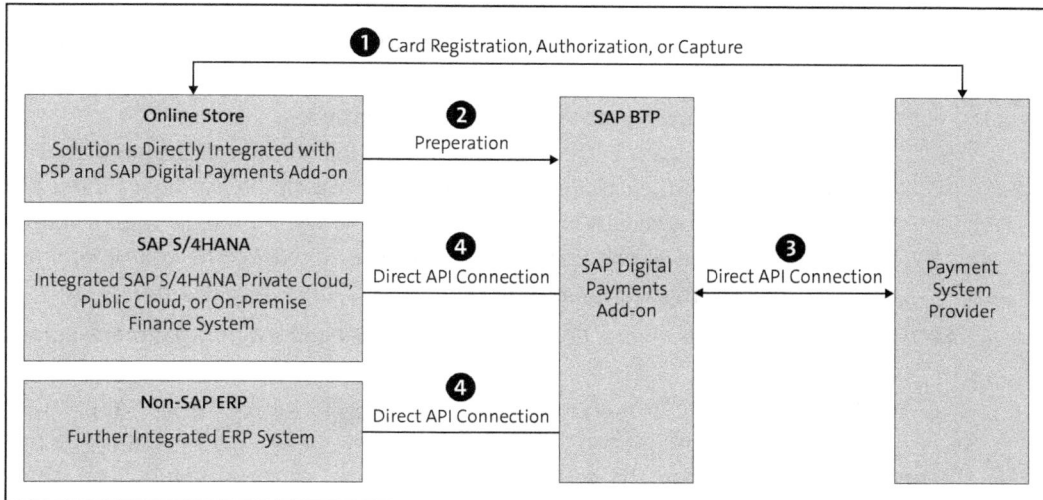

Figure 11.2 Digital Payments Process

Let's look at each step in a little more detail:

❶ **Card registration, authorization, and capture**
One of the key use cases of digital payments is when you want to centrally orchestrate payment processes triggered by sales orders through the online store and also mirror sales orders with the ERP system. For an online shop, there is usually a direct connection between it and the payment service provider. Customers can directly record a payment in the online shop, enter credit card details for the payment, and authorize the payment.

❷ **Preparation**
Using the SAP digital payments add-on, there are several scenarios in which payment card data and authorizations are transferred from an external consumer application to the add-on:

- **Registration of the payment card**
  When a user registers their credit card directly with the payment service provider, a card token is generated in the background and transmitted to the consumer application. The consumer application sends this card token to the SAP digital payments add-on, where the card is validated by the payment service provider and stored within the add-on. The add-on creates its own token as a reference and returns the token to the consumer application, where the card can then be used.

- **Transmission of a payment authorization**
  If the consumer application has performed a payment authorization directly with the payment service provider, it can pass this authorization to the SAP digital payments add-on. The authorization data is stored within the add-on, and a reference is returned to the consumer application, ensuring the authorization is available in SAP systems.

- **Combination of card registration and authorization**
  The consumer application can also carry out the registration of the payment card and the payment authorization in one step directly with the payment service provider. The combined results are transmitted to the SAP digital payments add-on, which stores both the card data and the authorization data. The add-on provides a token and an authorization reference to the consumer application.

❸ **Direct connectivity between SAP digital payments add-on and payment service provider**
The SAP digital payments add-on offers a wide range of out-of-the-box API connections to various payment service providers and gateways, such as Stripe. The payment gateway is directly connected with credit card providers and sends notifications or advices and statements to the SAP digital payments add-on or receives payment requests from the add-on, which then can be executed. Furthermore, the SAP digital payments add-on checks that the transaction really exists at the payment service provider and is consistent with the data provided.

❹ **Connection between the SAP digital payments add-on and SAP S/4HANA**
In this step, the add-on acts as a transformer for messages and formats between the payment service provider and the ERP system. With respect to the process, the SAP digital payments add-on provides notifications with revenue from transactions conducted via the payment service provider or the payment gateway. These transactions can be reconciled in SAP against open items.

> **SAP Digital Payments Add-On Is PCI DSS Certified**
>
> Payment Card Industry Data Security Standard (PCI DSS) is a global security standard developed by major credit card organizations to ensure the secure handling of credit card data and prevent fraud. Companies that process, store, or transmit credit card data must comply with this standard.
>
> For SAP digital payments, PCI DSS means that SAP outsources the handling of sensitive payment data to certified payment service providers. SAP digital payments itself does not store credit card data but transmits it directly and encrypted to the payment service providers. This significantly reduces effort for SAP customers: They only need to meet minimal PCI DSS requirements as no direct credit card data is processed within their own SAP system.
>
> In addition, by utilizing tokenization and secure interfaces, SAP digital payments ensures that real card data does not enter customer systems. This allows SAP customers to benefit from simplified compliance and reduced security and liability risks in the payment process.

11.1.2 SAP Digital Currency Hub

Cross-border payments are critical to the global economy but come with challenges such as high costs and long processing times due to multiple intermediaries. Digital currencies—specifically, stablecoins and central bank digital currencies—offer solutions by reducing intermediary involvement and leveraging blockchain for direct transactions, minimizing fees and improving transparency.

SAP Digital Currency Hub integrates digital currencies for transactions, enabling efficient cross-border payments and real-time tracking. It processes invoices and payment requests directly in ERP systems, interacting with crypto exchanges for seamless fund transfers. Although still in development, the hub aims to streamline and enhance cross-border transaction transparency, potentially integrating with traditional treasury management functions. SAP Digital Currency Hub holds promise in addressing cross-border payment inefficiencies, even as further ERP integrations are anticipated.

SAP Digital Currency Hub is a cloud solution from SAP that helps companies integrate digital currencies—particularly central bank digital currencies and stablecoins—into their existing business processes. The goal of SAP Digital Currency Hub is to enable the use of digital currencies in a corporate context while meeting regulatory and technical requirements. The solution acts as a link between SAP systems and the blockchain-based infrastructures on which digital currencies are issued and managed.

Figure 11.3 illustrates the high-level approach in SAP Digital Currency Hub, outlining the key steps and functionalities involved:

❶ Payment request
An incoming invoice is recorded in SAP S/4HANA and settled via the payment run. This payment run generates a payment request, which is then sent directly to SAP Digital Currency Hub for further processing.

❷ Crypto exchange
SAP Digital Currency Hub converts fiat currency into a tradeable cryptocurrency, enabling seamless digital transactions in the evolving financial landscape.

❸ Outgoing supplier payment
After the currency exchange is completed, the payment is executed in digital currency and sent to the supplier. A confirmation is then transmitted to the SAP S/4HANA ERP system, ensuring seamless transaction tracking.

❹ Incoming payment from customer
The customer pays an invoice in cryptocurrency, and the payment is credited to a digital wallet, ensuring seamless transaction flow.

❺ Crypto-fiat exchange
If needed, the digital currency can be converted into fiat currency such as USD, offering flexibility in financial transactions.

❻ Bank statement
SAP Digital Currency Hub generates a bank statement in camt.53 format and makes it available for reconciliation, ensuring transparent financial tracking.

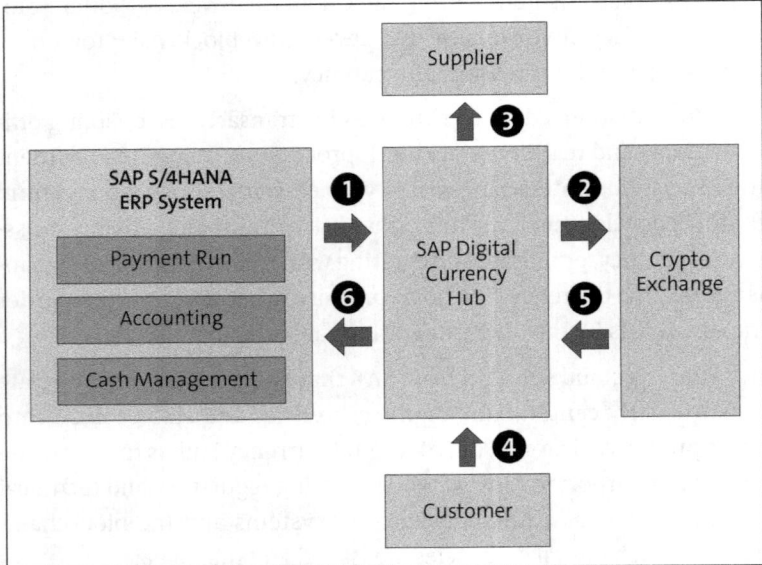

Figure 11.3 SAP Digital Currency Hub

Overall, what are digital currencies, tokens, and distributed ledgers? How are these supported? This section provides an overview of digital networks and clarifies the key terminology before we delve into the processes within the SAP system and how SAP Digital Currency Hub can help integrate alternative currencies into established financial processes.

Payment transactions using crypto-based digital currencies are carried out via distributed ledger networks and tokens. But what does that mean? Entire academic programs now focus on these concepts. We'll aim to provide a high-level overview of these terms to offer insight into the foundation on which SAP Digital Currency Hub operates and functions.

In the world of cryptocurrencies, a *network* is the underlying blockchain infrastructure through which transactions are processed, smart contracts are executed, and data is stored. Each network is composed of a decentralized group of computers (*nodes*) that collectively achieve consensus on the state of the ledger. This ledger—known as a *distributed ledger*—is a digital register that transparently and securely documents all transactions and states of a network. Unlike traditional, centrally managed databases, a distributed ledger is synchronized across many independent participants, ensuring data integrity without the need for a central authority.

Different networks can have varying characteristics, fees, speeds, and use cases. A primary application of these networks is the trading and transfer of cryptocurrencies and tokens, including well-known assets like Ethereum (ETH) or MATIC (Polygon), as well as tokens like USD Coin (USDC). These tokens are stored on their respective networks and transferred between users—either directly or via decentralized applications, exchanges, and wallets.

To ensure that developers and users can securely work with blockchain applications, networks are classified into two types—live networks (main nets) and test networks (test nets):

- **Main net**
  The "live" network where real transactions with actual monetary value occur.
- **Test net**
  A testing environment that simulates the main net but uses free and valueless test money. It allows developers to safely test their applications.

The networks supported by SAP Digital Currency Hub are as follows:

- **Ethereum**
  Ethereum is one of the most well-known and oldest smart contract platforms. It serves as the foundation for many decentralized applications, decentralized finance projects, and NFTs. Ethereum is characterized by high security and decentralization.
- **Polygon**
  Polygon is a Layer-2 scaling network for Ethereum that enables cheaper and faster transactions. It leverages Ethereum's security while offering greater efficiency through its sidechain architecture.

11 Outlook on Payments and Bank Communication with SAP

Tokens are digital assets that can be issued and transferred via existing blockchain networks like Ethereum or Polygon. Unlike native cryptocurrencies such as ETH or MATIC, which are directly used to secure and operate the network, tokens are based on specific standards—such as ERC-20 for Ethereum—and serve a variety of purposes. These can include functioning as a medium of exchange, representing real-world assets, or acting as access keys to specific applications.

> **What Are ERC-20 Tokens?**
>
> Ethereum Request for Comment 20 (ERC-20) is a technical standard on the Ethereum blockchain for so-called fungible tokens—meaning tokens that are fully interchangeable with one another, such as digital currencies or stablecoins. These tokens are designed to be compatible with one another, making it easier for developers to create new tokens and integrate them into existing applications, such as wallets and exchanges. The standard defines a set of rules that a token must follow in order to be compatible with wallets, exchanges, and smart contracts within the Ethereum ecosystem.

Stablecoins are a special type of token. They peg their value to a stable reference currency like the US dollar.

The following tokens are supported by SAP Digital Currency Hub:

- USDC: Issued by the company Circle, USDC is a stablecoin that is fully backed by US dollar reserves and undergoes regular audits. It is supported on multiple blockchains, including Ethereum and Polygon, and is commonly used in payment transactions and the decentralized finance sector.
- Paypal USD (PYUSD): A newer stablecoin initiated by Paypal, also pegged 1:1 to the US dollar. It is based on Ethereum (ERC-20) and is designed for integration within the existing Paypal ecosystem as well as blockchain applications.
- ETH: The native cryptocurrency of the Ethereum network, ETH is used, among other things, to pay transaction fees (gas fees) for executing smart contracts or sending tokens.
- MATIC: The native cryptocurrency of the Polygon network, MATIC is used to pay transaction fees on the network and plays a central role in the security and scalability of Polygon.

Tokens such as USDC, PYUSD, ETH, and MATIC enable fast, transparent, and cost-effective transactions, serving as a bridge between the traditional financial world and decentralized digital networks.

SAP Digital Currency Hub translates traditional financial processes into payment transactions using alternative methods like digital currencies. Existing SAP S/4HANA Finance settings can be used to seamlessly integrate the SAP Digital Currency Hub into established processes—for example:

- **Payment processing**
  Companies can use digital currencies for payment transactions, such as settling invoices, managing supplier or customer claims, and making cross-border payments. To facilitate this, new payment methods are integrated directly into the payment process and can be used through the traditional Transaction F110 payment program to settle invoices.

- **Treasury management**
  SAP Digital Currency Hub enables the management of digital currencies within liquidity and cash management, including conversions between fiat and digital currencies. SAP Digital Currency Hub converts credit and debit notifications into processable account statements in the camt.053 format. These transactions can be recorded and directly imported into One Exposure from Operations and thus be integrated into cash management.

- **Order-to-cash processes**
  SAP Digital Currency Hub converts transaction information and wallet balances into processable XML-based account statements in the camt.053 format. This allows outstanding items to be automatically reconciled using the familiar functions of standard account statement processing in SAP.

11.2 SAP Outlook by Kolja Ewering

The overall strategic direction of SAP aims to provide easy-to-use payment services embedded in SAP S/4HANA and ensure a seamless user experience for the surrounding SAP BTP, applications like SAP Multi-Bank Connectivity or the SAP digital payments add-on. To achieve this, SAP is working on a multitude of initiatives within payments, as follows:

- **Unified payment processing: Expanding flexibility and insight across SAP systems**
  One significant focus area aims to unify the processing of payments regardless of their sources. Advanced payment management, an integral part of SAP S/4HANA, now accepts payment files from any system and imports them into the payment factory. In the future, open APIs are planned to allow customers to integrate payment sources more seamlessly. SAP sources such as account payables, account receivables, contract accounting, treasury and risk management, or SAP SuccessFactors Payroll intend to utilize these APIs natively. On top of this, SAP also plans to enable new payment types natively out of the SAP S/4HANA payment process, including wallets or digital currencies, leveraging integration with SAP digital payments add-on for outbound payments and SAP Multi-Bank Connectivity. With more detailed data, the SAP system will provide even better insights into all of a company's payments, improving accuracy and compliance in areas like monitoring, approvals, and sanction party list screening.

- **Payment processing enhancements**
Existing functionalities of advanced payment management, including enriching and validating payment data, flexible bank account determination, payment scheduling, and (re)batching of individual payments, can be used more efficiently for these sources. Moreover, SAP plans to continually enhance these capabilities. One area slated for broader coverage is advanced support for centralized bank statement handling. Currently, the solution can dispatch statements into the operational system and update cash management centrally. In the future, it aims to harmonize bank statements before forwarding them to downstream systems. Harmonization examples include standardizing business transaction type codes or adjusting bank statement formats, allowing local systems to remain unaffected by regulatory or bank-specific changes.

- **Expand the scope of in-house banking**
The embedded in-house banking functionality currently offers robust features for running an internal bank that provides internal bank accounts to affiliates. These accounts can be used for intercompany payments, payments on behalf of, or incoming payments, reducing external payments and the number of external bank accounts. SAP plans to expand the functional scope over time to offer more out-of-the-box support for customers with advanced requirements. For instance, the enablement of simplified cross-bank area intercompany payments is expected to be extended to cross-bank area on-behalf-of payments.

- **Payment factory sidecar based on SAP S/4HANA Public Cloud**
Regarding deployment options, SAP focuses on enabling customers to run cash and payments, sometimes referred to as a payment factory, as a sidecar based on SAP S/4HANA Cloud Public Edition. This setup can be optionally enriched by SAP Treasury and Risk Management. The sidecar allows customers to operate their systems at their own pace and make changes when convenient while adopting innovations faster in the public cloud sidecar. Customers will continually receive new functionality seamlessly without the need to perform upgrades through their IT departments as SAP takes care of this in a public cloud setup. The journey to get such a public cloud sidecar running is also much more simplified due to best practice content delivered with the solution, which only needs to be adjusted to specific business needs. In such a setup, the ultimate goal is to decouple payments from the underlying general ledger (G/L) postings and, in general, a financial accounting setup in the public cloud sidecar if it's used only for cash and payment purposes.

- **Bank APIs**
In the realm of bank integration today, SAP Multi-Bank Connectivity facilitates payment initiations, status reporting, account statements, and various treasury and risk management processes. Recently, instant scenarios have been incorporated into the scope, enabling real-time balances and transaction reporting through the consumption of bank-specific APIs via the SAP cloud services. The current scope is scheduled

for continuous expansion into areas that generate additional value for customers and reduce reliance on manual data collection processes from various sources. The integration of APIs provided by banks contributes to the development of new operational models, as data will be available more rapidly, thus allowing decisions to be made with greater confidence. SAP Multi-Bank Connectivity allows customers to utilize these proprietary APIs through a single generic channel without encountering bank-specific complexities. In this context, SAP plans to offer opportunities for integration with partners to enhance global coverage. Therefore, SAP is actively assessing initiatives such as SWIFT instant cash reporting, given the lack of standardization in this domain.

- **SAP and non-SAP data**
  In addition to data obtained from bank integrations, SAP is creating ready-to-use data products that provide information on entities such as open items, payments, and liquidity in a standardized format. This is facilitated by SAP Business Data Cloud, a fully managed software-as-a-service (SaaS) solution that unifies and governs all SAP data and integrates with third-party data seamlessly. One intelligent application utilizing these data products is the working capital dashboard, which consolidates data from various sources into one platform, offering insights to prompt actionable outcomes

- **AI-infused payment processes**
  In the future, the full end-to-end payment process is planned to be powered by AI. Today, SAP provides the *Joule* digital assistant. Joule scenarios can be clustered into four patterns:
  - *Informational*, for retrieving knowledge-based content
  - *Navigational*, for guiding users to the right apps or screens
  - *Transactional*, for executing business processes via natural language
  - *Analytical*, for generating insights and visualizations from business data

  These patterns are designed to streamline workflows and enhance productivity across SAP systems. SAP is working on implementing those patterns also within payments. Applying these patterns to the area of payments the informational pattern provides insights into how certain functionalities within, for example, in-house banking works. The navigational pattern guides you to an object, such as a specific account or payment. In the transactional pattern, you will be able to ask about the status of a certain payment or to start the creation of a payment, whereas the analytical pattern is planned to create tailored insights into payment history or account balances in a certain region.

On top of this, SAP plans to provide Joule agents that run autonomously and provide recommendations for actions. One example could be a cash management agent that could detect a liquidity gap and recommend a specific bank transfer or even financing options through SAP Taulia.

11.3 Summary

The pressure for innovation in digital payment solutions and payment service providers continues to grow, remaining dynamic and volatile. In this ever-changing landscape of rapidly evolving trends, it is key to have a future-oriented and modularly controllable platform—one that allows for adaptability in response to internal business demands, such as those from e-commerce, as well as external trends and market developments.

SAP Digital Currency Hub serves as a bridge between cryptocurrency transactions and established accounts payable and accounts receivable processes, enabling seamless integration within financial operations. We are beginning to see early projects and use cases emerge in the market, shaping the future of digital payments. It will be exciting to observe how this sector evolves alongside established fiat currencies, and how payment behaviors adapt to new financial technologies.

Appendices

A ISO 20022 Transformation 623
B Ongoing Regulations 639
C The Authors 645

Appendices

Appendix A
ISO 20022 Transformation

As mentioned in Chapter 2, the ISO 20022 banking standard was established to harmonize and standardize communication in the financial industry. As an XML-based format, it is continuously developed to meet both global and local requirements. The adoption of the ISO 20022 standard brings significant changes for companies. This standard provides a unified language for payment data worldwide, leading to faster processing times and improved reconciliation processes. For companies, transitioning to ISO 20022 requires adapting their internal systems to support the new XML-based message formats, particularly impacting payment processing, liquidity management, and accounting—key areas within SAP systems.

One major advantage of ISO 20022 is the ability to transmit more comprehensive and structured payment information. This enhances process automation and increases transparency in payment transactions. However, companies must ensure their systems can process and store this enhanced data.

Migrating to ISO 20022 is not just a technical shift; it also offers the opportunity to rethink and optimize existing business processes. Companies should seize this chance to modernize their payment strategies and prepare for future developments in the financial sector.

A.1 Opportunities and Risks

Overall, the implementation of ISO 20022 represents a significant step forward in the evolution of global payment systems, from which companies can benefit through more efficient processes and improved data quality. The ISO 20022 transformation affects several core areas of a company. The following are some areas that are affected by the ISO 20022 transformation within a company:

- **Treasury**
  Within treasury and cash management, the introduction of ISO 20022 leads to improved transparency in payment processing. Companies benefit from more detailed payment information, allowing for more accurate monitoring and control of liquidity. Real-time information on incoming and outgoing payments significantly eases liquidity management. In addition, reconciliation processes are automated and accelerated due to the structured data formats.

A ISO 20022 Transformation

- **Financial accounting and controlling**
  Financial accounting and controlling are also directly impacted. Standardized and more comprehensive payment information simplifies the reconciliation between incoming payments and invoices. This reduces error sources and speeds up accounting processes. Furthermore, the quality of financial data improves, enabling more precise analyses and reports. However, accounting systems must be adapted to the new message formats to fully leverage the benefits of ISO 20022.

- **ERP and online banking**
  ERP and payment systems play a crucial role in the implementation of the ISO 20022 standard. Companies need to update their existing SAP ERP or SAP S/4HANA systems to process the new ISO 20022–compliant formats. This involves both internal IT infrastructures and interfaces with banks and SAP, as well as non SAP ERP systems. Particularly, the processing and storage of structured payment information pose a challenge as many existing systems are designed for the older MT format.

- **Compliance**
  The compliance and regulatory sector also benefits from the migration to the new standard. More detailed transaction data aids in fraud prevention and enhances anti-money-laundering efforts. The compliance with regulatory requirements is simplified through standardized and more comprehensive payment information. In addition, sanctions and embargo checks can be conducted more efficiently thanks to the availability of more structured data for analysis.

- **Human capital**
  The impact of the ISO 20022 transformation on human capital or human resources (HR) is indirect but significant, especially in payroll and compliance. Within payroll payments, ISO 20022 enables more efficient processes by providing standardized formats with detailed and structured information. This improves the reliability and speed of salary transfers, particularly for companies with cross-border payments, reducing errors and decreasing reversals.

  Overall, the ISO 20022 migration results in more efficient and transparent payroll operations but necessitates process adjustments, system updates, and employee training to meet new requirements.

- **Customer and supplier relationship management**
  Finally, the transition impacts customer and supplier management. Companies can optimize payment processes and reduce error rates in payment processing. Automated and more precise reconciliation with business partners contributes to improved business relationships. Invoicing also benefits, as payments and corresponding invoices can be matched more easily.

The ISO 20022 transformation offers significant opportunities and some risks for companies. One major benefit is improved data quality and transparency, allowing for detailed and structured transmission of payment information, which enhances transaction traceability and simplifies reconciliation with invoices, reducing errors and speeding up processes.

Another key advantage is increased efficiency and automation. Standardized data formats optimize payment processing, minimizing manual interventions. This allows for better monitoring of payment flows and refining of liquidity management strategies. ISO 20022 also aids regulatory compliance, as detailed payment information supports anti-money-laundering efforts and sanction checks, reducing the risk of regulatory breaches.

Global harmonization and enhanced interoperability are also crucial benefits. As ISO 20022 is adopted worldwide, companies can collaborate more seamlessly with banks and partners across countries, reducing international payment barriers and minimizing error rates. Long term, this results in cost savings through automated processes and standardized operations, lowering administrative workload and the need for error corrections.

Despite these benefits, challenges exist. One major hurdle is the implementation effort, requiring IT system adjustments, new software, and employee training. The transition can be time-consuming and demands additional resources. Until November 2025, companies must manage two standards, MT and ISO 20022, which can lead to complexity and processing errors if systems are not compatible.

Increased data management requirements pose another risk as the more extensive payment information of ISO 20022 demands efficient processing and storage capabilities. Regulatory uncertainties also play a role, as different countries or banks might implement the standard slightly differently. Companies need flexible systems to meet various requirements.

Organizational challenges include training employees in relevant departments like accounting, payments, and compliance. Resistance to change or lack of expertise could slow the implementation process, necessitating early investment in change management.

A.2 From SWIFT Message Type to ISO 20022 XML

The decision to migrate from SWIFT message type (MT) to ISO 20022 (MX) was made by the global financial community, coordinated by SWIFT.

The migration to ISO 20022 primarily affects cross-border, financial institution to financial institution (FI-to-FI) payment instructions. In this process, the old MT message formats are being replaced by the new XML-based ISO 20022 format, enabling the transfer of more comprehensive and structured data in payment messages. The main reason

A ISO 20022 Transformation

for this migration is to harmonize and modernize global payment processes. ISO 20022 offers a unified language for financial messages, enhancing efficiency, facilitating process automation, and increasing transparency. In addition, the standard supports compliance requirements by providing more detailed information.

The migration began in March 2023 with a coexistence phase, during which both MT and ISO 20022 message formats are supported. This phase will end on November 22, 2025, giving a total coexistence period of approximately two and a half years. After this date, MT messages for cross-border payment instructions will no longer be accepted, and ISO 20022 will become the sole standard.

It is important to note that these deadlines specifically apply to cross-border payments between financial institutions. Other financial message types and market infrastructures may have different timelines for migration.

Although this involves a transformation of global financial infrastructure and settlement systems, it still has a massive impact on the processes and functions of ERP, HR, and treasury management system (TMS) setups if not addressed. Many banks have already informed their customers that the transformation will be passed directly to them, as traditional MT formats like the MT940 end-of-day account statement or DTAZV will no longer be supported starting in November 2025—and at some financial institutions, even earlier.

In Table A.1 and Table A.2, we provide an overview of which MT formats are being replaced by which MX formats, as well as an overview of the MX formats that already have an end-of-life date in the SEPA area and the formats by which they are expected to be replaced. For more information, visit *http://s-prs.co/v605303*.

MT Format	Description	MX Format Description	Format Version	End of Lifecycle	Replacement Format
MT101	Request for Transfer	Foreign Payments (Local format)	DTAZV	11/2025	pain.001.001.09
MT101	Request for Transfer	SEPA Payments	pain.001.001.03	11/2025	pain.001.001.09
MT104	Direct Debit and Request for Debit Transfer Message	Direct Debit	pain.008.001.03	11/2025	pain.008.001.08
MT199	Free Format Message	Bank Protocol	pain.002.001.02	11/2025	pain.002.001.10

Table A.1 SWIFT Message Type Category 1: Customer Payments and Checks

MT Format	Description	Format Version	Format Description	End of Lifecycle	Replacement Format
MT942	Intraday Bank Statement (SWIFT Message Type)	camt.052.001.02	Intraday Bank Statement	11/2025	camt.052.001.08
MT940	End of Day Bank Statement (SWIFT Message Type)	camt.053.001.02	End of Day Bank Statement (ISO20022)	11/2025	camt.053.001.08
■ MT900 ■ MT910	■ Confirmation of Debit ■ Confirmation of Credit	camt.054.001.02	Credit/Debit Notification (ISO20022)	11/2025	camt.054.001.08

Table A.2 SWIFT Message Type Category 9: Cash Management and Customer Status

A.3 Format Transformation Within Payments and Protocols

Starting in November 2025, significant changes will come into effect for the pain.001, pain.008, and pain.002 payment formats, which are used for initiating SEPA credit transfers, SEPA direct debits, and payment status reporting, respectively. These adjustments aim to further harmonize and enhance efficiency in payment processing.

The key changes in the version (release 2019) used starting in November 2025 are to the following formats:

- For pain.001 (SEPA Credit Transfers): version pain.001.001.09
- For pain.008 (SEPA Direct Debits): version pain.008.001.08
- For pain.002 (Payment Status Reports): version pain.002.001.10

These new versions add structural modifications, including changes to field structures and the introduction of new data elements.

A.3.1 Changes to Specific Fields

There is a change to the remittance information: The maximum character length for remittance information remains 140 characters. However, it is recommended to use structured remittance details to improve automation and payment processing efficiency.

A ISO 20022 Transformation

In the future, structured or hybrid addresses will be mandatory for new payment formats. Address data must be structured or hybrid if provided, with foreign payments requiring addresses as before, and domestic ones being optional. Since March 2024, the ISO 2019 version requires structured addresses, but most data isn't currently stored this way. To address this, a hybrid model was introduced, combining structured fields with unstructured lines. From October and November 2025, the hybrid format will be allowed for SEPA, express, and international payments. Cities and countries are mandatory, but additional info can be added via unstructured lines—though using structured fields is recommended.

Listing A.1 shows an unstructured address under the old ISO version, while Listing A.2 shows a structured address under the new version. Listing A.3 shows the hybrid model.

```
...
<Nm>ABC Handels GmbH</Nm>
<PstlAdr>
 <Ctry>DE</Ctry>
 <AdrLine>Zentrale1, Dorfstrasse 23/2</AdrLine>
 <AdrLine>80995 Muenchen / Bogenhausen</AdrLine>
</PstlAdr>
...
```

Listing A.1 Unstructured Address: Old ISO Version

```
<Nm>ABC Handels GmbH</Nm>
<PstlAdr>
 <Debt>Zentrale1</Debt>
 <StrtNm>Dorfstrasse</StrtNm>
 <BldgNb>23</BldgNb>
 <Flr>2</Flr>
 <PstCd>80995</PstCd>
 <TwnNm>Muenchen</TwnNm>
 <TwnLctnNm>Bogenhause</TwnLctnNm>
 <Ctry>DE</Ctry>
</PstlAdr>
...
```

Listing A.2 Structured Address: New ISO Version

```
...
<Nm>ABC Handels GmbH</Nm>
<PstlAdr>
 <PstlCd>80995</PstlCd>
 <TwnNm>Muenchen</TwnNm>
```

```
<Ctry>DE</Ctry>
<AdrLine>Dorfstrasse 23, 2. Stock</AdrLine>
</PstlAdr>
```

Listing A.3 Hybrid Address

Finally, a new optional field for the legal entity identifier (LEI) has been introduced, enabling the unique identification of legal entities. Providing the LEI can improve transparency and security in payments.

> **Changes to XML Tag Names**
> With the new versions, some XML tag names are updated to align with current standards. For example, the BIC tag changes from <BIC> to <BICFI>, enabling a more precise identification of financial institutions.

A.3.2 Changes for pain.002

The Payment Status Report (pain.002) is used to provide status updates on submitted payment transactions. With the introduction of version pain.002.001.10 in November 2025, the following adjustments will be made:

- **Extended status codes**
  New status codes provide more detailed information about the processing status of payments, improving transparency.
- **Structured error descriptions**
  Error messages are now provided in a structured format, making it easier to identify and correct issues.
- **Additional response fields**
  New fields are introduced to provide more specific feedback on individual transactions, such as instructions for required corrections.

The updated pain.002 format introduces additional status codes for greater granularity in payment processing responses—for example:

- **RJCT (Rejected)**
  More specific rejection reasons, such as incorrect IBAN or regulatory compliance failure.
- **PDNG (Pending)**
  Indicates transactions awaiting processing due to checks or manual intervention.
- **ACSC (Accepted, Settlement Completed)**
  Confirms that the transaction has been successfully settled.
- **ACSP (Accepted for Processing)**
  Confirms that the payment has been accepted but is not yet settled.

A ISO 20022 Transformation

Table A.3 highlights the key tags that have changed in the new version (release 2019). For more information, visit *http://s-prs.co/v605304*.

Format	Changed Field/Tag	New Version	Change
pain.001 (SEPA Credit Transfer)	<BIC> → <BICFI>	pain.001.001.09	Change of tag name for more precise identification of financial institutions
	<UltmtDbtr>	pain.001.001.09	New optional tag for the ultimate debtor for better traceability
	<PstlAdr> (Postal Address)	pain.001.001.09	Introduction of structured address fields (street, postal code, city, country separately)
	<RmtInf> (Remittance Information)	pain.001.001.09	Recommendation to use structured remittance information
	<InstrPrty> (Priority)	pain.001.001.09	Adjustment of permissible values to align with ISO 20022 standards
pain.008 (SEPA Direct Debit)	<CdtrSchmeId> (Creditor Scheme ID)	pain.008.001.08	Refinement of field for creditor identification number
	<Dbtr> (Debtor)	pain.008.001.08	Expansion with additional optional fields for more detailed debtor data
	<PmtTpInf> (Payment Type Information)	pain.008.001.08	Introduction of new code options for direct debit types
	<PstlAdr> (Postal Address)	pain.008.001.08	Change to structured address fields (instead of unstructured input)
pain.002 (Payment Status Report)	<StsRsnInf> (Status Reason Information)	pain.002.001.10	Introduction of new status codes for more detailed status reporting
	<AddtlInf> (Additional Information)	pain.002.001.10	Expansion of error descriptions in structured format

Table A.3 ISO 20022 Changed Tags within Version 2019

Format	Changed Field/Tag	New Version	Change
	<TxSts> (Transaction Status)	pain.002.001.10	Introduction of new status codes like RJCT, PDNG, ACSC, ACSP
	<UETR>	pain.002.001.10	New field for the SWIFT gpi UETR. The pain.002 will support structured GPI with the new version

Table A.3 ISO 20022 Changed Tags within Version 2019 (Cont.)

Overall, planned upcoming changes in pain.001, pain.008, and pain.002 starting in November 2025 represent a significant step in the evolution of payment processing. Companies are advised to adapt their systems and workflows accordingly to fully leverage the benefits of the new standards and to ensure efficient and error-free transactions.

A.4 Format Transformation Within Bank Statements

Starting in November 2025, not only will the MT940 and MT942 formats be replaced by the ISO 20022 equivalents camt.053 and camt.052, but older versions of account statements, such as the common versions camt.053.001.02, camt.052.001.02, and camt.054.001.02, also will be replaced by release 2019 version 08. This affects current clearing systems and banks worldwide and thus also impacts corporations. Some banks offer an indefinite parallel operation of MT940 and camt or even may continue to offer older versions beyond November 2025. However, corporate entities should be prepared for the possibility that their main bank may not maintain parallel operation indefinitely or may not offer it at all. It is recommended to contact your bank regarding this matter.

However, a new format does not only bring challenges and disadvantages during system transition. The updated version 08 of the new camt format will be more detailed and structured. This significantly increases matching rates in a well-configured account statement processing system and thereby immensely reduces the manual processing effort.

Nevertheless, XML account statements are not entirely without problems. More details also mean a greater data volume that needs to be processed. The German banking industry recommends that banks cut account statements over 20 MB to ease processing for corporations. When transitioning, both data volume and performance should definitely be considered.

A ISO 20022 Transformation

Listing A.4 shows an example of how an incoming payment is represented in an MT format. The same payment is shown structured in an XML format in Listing A.5.

```
:20:STARTUMS
:25:DE12345678901234567890
:28C:00001/001
:60F:C200123EUR1234,56
:61:2001240124DR567,89NMSCNONREF
:86:Überweisung von Max Mustermann
:62F:C200124EUR1802,45
```

Listing A.4 MT Format

```
<Stmt>
  <Id>STATEMENT_ID</Id>
  <Acct>
    <Id>
      <IBAN>DE12345678901234567890</IBAN>
    </Id>
  </Acct>
  <Ntry>
    <Amt Ccy="EUR">567.89</Amt>
    <CdtDbtInd>DBIT</CdtDbtInd>
    <BookgDt>
      <Dt>2025-01-24</Dt>
    </BookgDt>
    <ValDt>
      <Dt>2025-01-24</Dt>
    </ValDt>
    <NtryDtls>
      <TxDtls>
        <Refs>
          <EndToEndId>UETR1234567890</EndToEndId>
        </Refs>
        <RltdPties>
          <Dbtr>
            <Nm>Max Mustermann</Nm>
            <PstlAdr>
              <StrtNm>Musterstraße</StrtNm>
              <BldgNb>1</BldgNb>
              <PstCd>12345</PstCd>
              <TwnNm>Musterstadt</TwnNm>
              <Ctry>DE</Ctry>
            </PstlAdr>
          </Dbtr>
```

```
        </RltdPties>
        <RmtInf>
          <Ustrd>Überweisung von Max Mustermann</Ustrd>
        </RmtInf>
      </TxDtls>
    </NtryDtls>
  </Ntry>
</Stmt>
```

Listing A.5 ISO 20022 XML Example

Table A.4 provides an overview of the changes between camt version 02 and the new version 08.

Area	Old Format/Tag	New Format/Tag	Change	Format
Bank Identification	<BIC>	<BICFI>	More precise identification of financial institutions with the new <BICFI> tag.	camt.052, camt.053
Postal Address	<PstlAdr>	<PstlAdr> (structured)	Introduction of structured fields for street, postal code, city, and country.	camt.052, camt.053
Transaction Amount	<Amt>	<Amt>	Detailed specification of transaction amounts including taxes and fees.	camt.052, camt.053
Purpose of Payment	<RmtInf>	<RmtInf> (structured)	Introduction of structured payment purpose fields, including <Ustrd> and <Strd>.	camt.052, camt.053
Ultimate Debtor	-	<UltmtDbtr>	New tag to identify the ultimate debtor (end customer) if the payer is a company.	camt.052, camt.053
Status Codes	<TxSts>	<TxSts> (with new codes)	Expansion to include new status codes like RJCT, PDNG, ACSC, ACSP.	camt.052, camt.053

Table A.4 Changes Between camt Versions 02 and 08

A   ISO 20022 Transformation

Area	Old Format/Tag	New Format/Tag	Change	Format
Error Handling	`<AddtlInf>`	`<AddtlInf>`	Expanded error handling with more detailed and structured error messages.	camt.052, camt.053
Transaction Reference	`<Ref>`	`<Ref>`	Expanded transaction reference for better matching of payments.	camt.052, camt.053
Chargeback Details	-	`<RtrRsnInf>`	New tag for specifying chargeback reasons and details.	camt.052, camt.053
Payment Types	`<PmtTpInf>`	`<PmtTpInf>` (with new codes)	Introduction of new payment types such as CCT (Credit Transfer) and DD (Direct Debit).	camt.052, camt.053

Table A.4 Changes Between camt Versions 02 and 08 (Cont.)

A.5   Changes Within Transaction Banking

The transition to ISO 20022 primarily affects payment messages, particularly MT categories 1, 2, and 9, which must be migrated to ISO 20022 messages by November 2025. However, the MT300 and MT320 message formats, used for trade confirmations in transaction banking, fall under category 3 (Foreign Exchange, Money Market & Derivatives) and category 5 (Securities Transactions).

Currently, there is no official announcement from SWIFT regarding the mandatory migration of MT300 and MT320 to ISO 20022 messages. The focus of the migration is on payment-related messages, while trade confirmations and securities transactions continue to use the existing MT formats.

However, although there is no immediate requirement to replace MT300/MT320 with ISO 20022, SWIFT and financial institutions may consider migrating these message types in the future.

A.6   Electronic Bank Account Management

Electronic bank account management (eBAM) aims to digitize and streamline the communication and administration of banking relationships. Global companies, especially those with dynamic business developments, face the challenge of continuously opening new bank accounts, making adjustments (such as limits and authorizations), and closing accounts.

eBAM seeks to digitize the entire lifecycle of an account, from opening and adjustments to fee management and closure. The challenges include country-specific regulatory requirements and the varied services offered by banks. There is also only limited market support for eBAM. The fundamental idea is quite straightforward: to provide an electronic upgrade to current banking solutions, digitizing everything from account maintenance, authorization management, and legal compliance to bank fee overviews.

To support this, ISO 20022 has established a standardization based on XML to facilitate digital workflows. Since 2013, the CGI-MP working group has been dedicated to digitizing this process. The messages intended for this purpose are managed and continuously updated under the **Payments** section in **Account Management (acmt)**.

> **Latest ISO 20022 Format Updates**
>
> You can always check the latest developments on format updates on the official ISO 20022 homepage at *http://s-prs.co/v605305*.
>
> The latest formats are published there. However, please note that when the CGI releases a new format, it does not necessarily mean that banks immediately adopt the latest version. Format updates are typically published via banking associations and often lag behind by several versions. For example, many banks are currently working on implementing formats from 2019 for use in 2025.

Figure A.1 illustrates an example of an account opening process handled via eBAM. A prerequisite for this process is an eBAM-capable system that can process XML-based acmt messages and transmit them to the bank via a channel such as SWIFT, host-to-host communication, or EBICS.

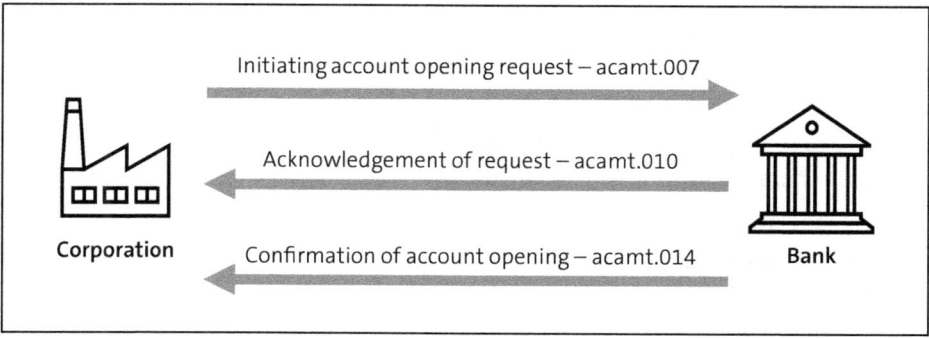

Figure A.1 BAM Process

A ISO 20022 Transformation

This process has the following steps:

1. **Initiating the account opening request**
   The company sends an account opening request message (acmt.007) to the financial institution. This message includes all relevant details about the new account, such as account type, currency, authorized signatories, and any other necessary information.

2. **Verification and authentication by the bank**
   Upon receiving the request, the financial institution verifies the authenticity and authorization of the request. This includes identity verification of the requester, compliance with regulatory requirements (e.g., KYC and AML checks), and validation of the provided information.

3. **Acknowledgment of the request**
   If the initial validation is successful, the financial institution sends an account request acknowledgement message (acmt.010) back to the company. This message confirms the receipt of the account opening request and may include a reserved account number, serving as a reference for further communication.

4. **Setting up the account in the bank's system**
   The financial institution processes the account setup based on the details provided in the initial request. This includes configuring account parameters, linking the account to relevant services (e.g., online banking, payment channels), and completing internal approval processes.

5. **Confirmation of account opening**
   Once the account is fully established, the financial institution sends an account report message (acmt.014) to the company. This message contains all final details of the newly opened account, including the assigned account number, account status, and any specific attributes. This allows the company to verify the account characteristics against the original request and start using the account for financial transactions.

In addition to the account opening process, as mentioned initially, eBAM also supports modification and closure processes. For digitalized bank communication, a whole series of standardized acmt messages has been developed under the umbrella of ISO 20022. The following list describes XML-based acamt messages sent from the corporation to the bank, as well as messages that are received from the bank by the corporation:

- **Messages from the corporation to the financial institution**
  These messages are sent by the company to the financial institution to manage bank accounts:
  - **acmt.007, Account Opening Request**
    Request to open an account. The customer requests the opening of a new bank account with specific details.

A.6 Electronic Bank Account Management

- acmt.008, Account Opening Amendment Request
  Modification of an account opening request. The customer submits an adjustment or correction to an already submitted account opening request.
- acmt.013, Account Report Request
  Request for an account report. The customer asks the bank for detailed information or a report on an existing account.
- acmt.015, Account Excluded Mandate Maintenance Request
  Management of excluded mandates. The customer updates or manages exclusions for certain individuals or organizations that are not authorized to operate the account.
- acmt.016, Account Excluded Mandate Maintenance Amendment Request
  Amendment to an excluded mandate management request. The customer modifies a previously submitted request regarding excluded mandates.
- acmt.017, Account Mandate Maintenance Request
  Management of signatory rights. The customer requests changes to the individuals or institutions authorized to conduct transactions on behalf of the company.
- acmt.018, Account Mandate Maintenance Amendment Request
  Amendment to a signatory rights management request. The customer submits corrections or modifications to a previously submitted request.
- acmt.019, Account Closing Request
  Request to close an account. The customer asks the bank to close an existing account.
- acmt.020, Account Closing Amendment Request
  Modification of an account closing request. The customer makes changes or corrections to an already submitted account closure request.

- **Messages from the financial institution to the corporation**
  These messages are sent by the bank to the customer to respond to requests or request additional information:
  - acmt.009, Account Opening Additional Information Request
    Request for additional information for an account opening. The bank requires more details to complete the account opening process.
  - acmt.010, Account Request Acknowledgement
    Acknowledgment of an account request. The bank confirms the receipt of a request, such as an account opening, modification, or closure.
  - acmt.011, Account Request Rejection
    Rejection of an account request. The bank rejects an application for account opening, modification, or closure and may provide the reasons for the rejection.
  - acmt.012, Account Additional Information Request
    Request for additional information for an account request. The bank asks for

supplementary data or documents related to an existing request in order to proceed with processing.

- **acmt.014, Account Report**
  Provision of an account report. The bank provides the customer with the requested account information.
- **acmt.021, Account Closing Additional Information Request**
  Request for additional information for an account closure. The bank requires further data or confirmations to process the closure of an account.

A.7 Summary

ISO 20022, as an XML-based standard, offers significantly more opportunities than MT and is considerably more flexible due to its XML structure. This flexibility is apparent in regulatory requirements and payment schemes, which actively integrate interactive APIs alongside traditional connection protocols like EBICS and host-to-host communication. Although companies must adapt to new formats, the optimization potential cannot be fully realized until banks have completely transitioned to the format for interbank transactions. Unfortunately, many banks still convert old MT formats into camt or pain formats without fully leveraging the new tags. Those banks that have successfully transitioned can provide corporations with significant advantages, such as increased information availability, higher matching rates, and additional services that can markedly improve accounts payable, accounts receivable, and treasury processes. However, the transition poses challenges, as evidenced by ongoing delays in meeting deadlines.

By November 2025, ISO 20022 will become the sole globally recognized standard for interbank cross-border payments, with the new MX messaging format set to fully replace the old MT format.

Appendix B
Ongoing Regulations

This appendix will provide you with an overview of regulations in payment transactions that you should be familiar with before embarking on your next payment transaction project. These are essential legislative frameworks that aim to unify and enhance the security of payment transactions, or at least influence their execution. This appendix does not aspire to be a legal reference guide, but rather aims to provide an overview and raise awareness.

B.1 Sarbanes-Oxley Act

The Sarbanes-Oxley (SOX) Act of 2002 is a United States federal law that was enacted in response to several major corporate and accounting scandals. It establishes requirements for financial practice and corporate governance. One of its most significant sections is Section 404, which mandates that companies report on the effectiveness of their internal controls over financial reporting. The company's outside auditor must also attest to and provide an independent assessment of the company's processes. Sarbanes-Oxley is also known for fostering transparency and accuracy in corporate disclosures. Compliance with the act is monitored and enforced by the Securities and Exchange Commission (SEC). SOX is intended to prevent companies from manipulating their books and ensures that managers are held accountable if they deceive investors.

Overall, SOX has three general requirements:

- **Submission of accurate financial reports certified by the company's management**
  According to Section 302 of SOX, "Corporate Responsibility for Financial Reports," the CEO, CFO, or equivalent officers of a company must sign off on every annual and quarterly financial report filed with the SEC. When signing the reports, the CEO and CFO must certify that the financial statements are completely accurate. They must also affirm that the appropriate internal controls are in place and have been validated within the last 90 days. According to SOX Section 404, "Management Assessment of Internal Controls," every annual financial report filed with the SEC must include a detailed report on internal controls. The internal controls report states that management is responsible for the internal controls and assesses the effectiveness of the company's internal controls at the end of the most recent fiscal year. Although SOX does not specifically mandate cybersecurity incident reporting, companies

must integrate cybersecurity risks into their internal control assessments under Section 404. In addition, the SEC introduced separate cybersecurity disclosure requirements in 2023, which mandate that companies report significant cybersecurity incidents within four days after determining their material impact. These SEC rules complement but do not replace SOX requirements.

- **Implementation of appropriate internal controls and passing regular audits**
  Companies implement internal SOX controls to prevent internal and external actors from fraudulently altering financial data or using it for illegal purposes. SOX does not explicitly list all the controls that companies must implement. Instead, organizations often rely on established corporate governance frameworks, including the following:
    - **Committee of Sponsoring Organizations of the Treadway Commission (COSO)**
      The most widely adopted framework for designing and evaluating internal controls, updated in 2013.
    - **Control Objectives for Information and Related Technologies (COBIT)**
      A governance framework that helps manage IT and financial reporting risks.
    - **NIST Cybersecurity Framework and ISO 27001**
      Increasingly adopted for SOX compliance in IT security-related internal controls.

  Business process controls include aspects such as training employees on SOX requirements and establishing secure reporting channels for whistleblowers. Many companies also apply the principle of segregation of duties, where workflows are divided into multiple parts and different employees are responsible for each step. The idea is that no single employee controls the entire workflow; instead, each participant checks the work of others. A typical example would be ensuring that the person who approves payments is not the same individual who issues checks from the company account. Companies can also establish processes for storing and retaining records to meet SOX document retention requirements. For instance, auditors are required to keep all work papers associated with an audit for seven years.

- **Passing regular audits**
  As mentioned earlier, the CEO and CFO are responsible for the accuracy of each financial report and the effectiveness of internal controls. Regular audits provide executives with the evidence they need for these statements. By conducting regular internal audits of financial reporting practices and data controls, companies can monitor compliance over time, identify gaps, and address vulnerabilities. The results of internal audits can also assist external auditors who conduct the annual SOX compliance audits. During the annual audit, an independent accounting firm makes its own assessment of the internal controls and financial reporting. The results of this audit are often included in the company's annual SEC report.

Companies implement controls at both the business process level and the IT infrastructure level. Overarching SOX compliance initially sounds bureaucratic, but it is an important protection for investors and corporations. To ensure SOX compliance within the SAP landscape, several critical configurations and process enhancements are implemented, particularly in the areas of payment management, authorization control, and auditability. Within SAP Bank Communication Management, the release of payment batches and individual payments can be not only monitored but also configured with workflow-based approvals tailored to specific business processes. This enables organizations to enforce dual control mechanisms and ensure that no single individual can initiate and approve a payment, in line with SOX principles.

Similarly, advanced payment management supports SOX compliance by allowing approval workflows to be integrated into the handling of sensitive master data, such as payment routing rules or clearing agreements. In addition to master data, approval controls can also be applied to payment orders and even at the individual payment item level. For example, in scenarios where manual payments are directly entered through advanced payment management, the system can be configured to require an explicit approval step before execution, thereby mitigating risks associated with unauthorized transactions.

Furthermore, SAP systems generate log files during key activities such as file ingestion, payment processing, and exception handling. These logs serve as audit trails that record all relevant events and decisions, supporting transparency and providing verifiable documentation for internal and external audits, which is a cornerstone of SOX compliance.

Finally, the authorization concept within SAP must be aligned with the SOX requirement of segregation of duties. This involves designing role-based access controls so that no user has conflicting permissions—for instance, being able to create, approve, and execute a payment. By enforcing strict role assignments and utilizing SAP's capabilities for access governance (e.g., SAP Access Control), the system ensures the integrity of financial processes through clear responsibilities and oversight.

B.2 Payment Services Directive

The original idea of the Payment Services Directive (PSD) was to create a unified, secure, and competitive payment market in the EU. Before its introduction, the European payment system was fragmented, with different rules and infrastructures in each country. This led to high costs, slow transactions, and limited competition.

The primary objectives of the original Payment Services Directive (PSD1) introduced in 2007 aimed to create a Single Euro Payments Area (SEPA), facilitating faster and more cost-effective cross-border payments. This initiative sought to make international transactions as straightforward as domestic ones, thereby enhancing the efficiency of the payment landscape. Another key objective was to promote competition in the sector by allowing nonbank institutions, such as fintech companies, to provide payment

B   Ongoing Regulations

services, thus preventing traditional banks from maintaining a monopoly over the market. In addition, PSD1 placed a strong emphasis on enhancing consumer protection by introducing explicit rules regarding fees, liability, and transparency, ensuring quicker refunds in cases of fraud or transaction errors. Furthermore, the directive encouraged innovation in payment systems by supporting the development of new digital payment services and fostering improvements in efficiency through technological advances.

Before PSD1, there were no standardized rules for payments across the EU. Each country had its own regulations, making payments slow, expensive, and inefficient. The directive aimed to remove these barriers and modernize the European payment landscape.

Figure B.1 illustrates the development of the Payment Service Directory from 2007 to 2025.

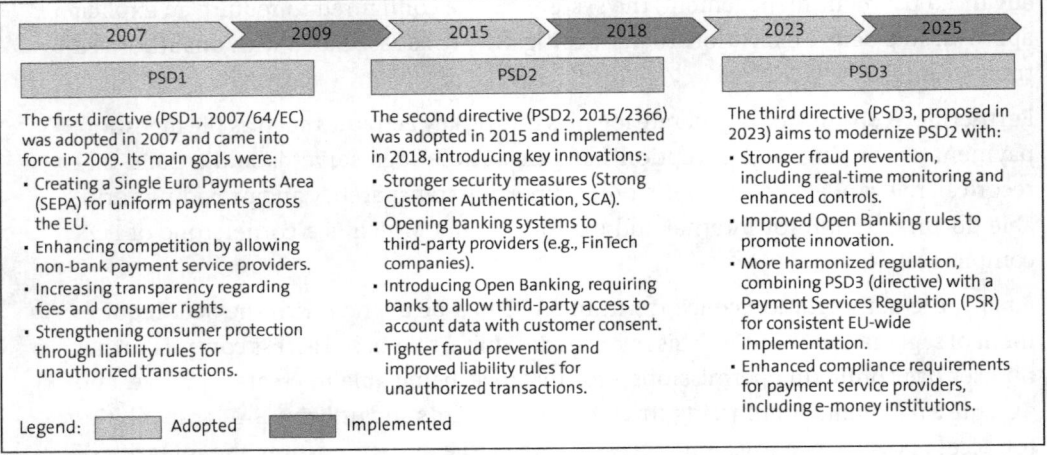

Figure B.1 Payment Service Directory

In the following sections, we provide an overview of PSD2 and PSD3 and offer examples of their impact on the SAP setup.

B.2.1   PSD2

The second Payment Services Directive (PSD2) is an EU directive that modernizes and harmonizes payment transactions within the European Union. It came into effect on January 13, 2018, aiming to foster competition, support innovation, and enhance the security of payments. The key points of PSD2 are as follows:

- **Promotion of new payment services**
  PSD2 allows third-party providers, such as payment initiation services and account information services, to access customers' bank accounts with their explicit consent. This encourages innovative financial services and increases competition.

- **Strengthening consumer protection**
  The directive improves consumer rights by establishing transparent information obligations for payment service providers and regulating liability in the case of unauthorized payments.

- **Enhancing payment security**
  PSD2 introduces Strong Customer Authentication (SCA), requiring authentication for electronic payments using at least two independent factors (e.g., knowledge, possession, inherence).

- **Expanding the scope**
  The directive now includes payments in all currencies within the European Economic Area (EEA) and certain transactions with a third-country link, expanding its scope compared to the previous directive.

- **Implementation in Germany**
  In Germany, PSD2 was implemented through adjustments to the Payment Services Supervision Act (ZAG). The Federal Financial Supervisory Authority (BaFin) and the Deutsche Bundesbank are responsible for monitoring compliance with the directive.

Thus, PSD2 represents an important step in modernizing payment transactions in the EU by promoting competition, strengthening consumer protection, and enhancing payment security.

Particularly in terms of payment authentication and bank connectivity, PSD2 has left a significant footprint, as follows:

- **Payment approval**
  Payments should be executed with two-factor authentication according to PSD2. This can be implemented, for example, through an extension or a specific setup via SAP Bank Communication Management, where the authorization is done through a token generated or transmitted via another device, such as a smartphone.

- **Bank connectivity**
  Through PSD2, payment service providers are required to provide secure interfaces for third-party providers. This has led to an increase in the offering of API interfaces by banks. Banks are mandated by PSD2 to offer these interfaces.

In practice, however, it turns out that this has been implemented differently by banks. There have been variations in standards and functionalities concerning bank API interfaces. Initiatives like the Berlin Group are working on developing cross-bank standards to address this issue. However, overall, a positive development can be observed, leading to further advancements in bank connectivity. Alongside EBICS and SWIFT as connectivity standards, PSD2 has given the API method of connectivity new significance.

B.2.2 PSD3

The Payment Services Directive 3 (PSD3), along with the new Payment Services Regulation (PSR), aims to modernize the EU's payment landscape by strengthening security, enhancing fraud prevention, improving Open Banking, and ensuring regulatory consistency across member states. One of its key changes is the introduction of stricter fraud prevention measures, including real-time fraud monitoring and improved SCA for online transactions. In addition, banks and payment service providers will be required to report fraud cases more efficiently and provide better refund rights for victims of authorized push payment (APP) fraud, protecting consumers from scams and unauthorized transactions.

A major update in PSD3 is the enhancement of open banking. Banks will be required to provide standardized and more reliable API access for third-party providers, ensuring smoother and more secure data sharing. Consumers will also have greater control over their consent management, making it easier to authorize and revoke third-party access to their financial data. Furthermore, PSD3 will completely phase out screen scraping, replacing it with more secure and regulated access methods.

To improve overall security and transparency in digital payments, PSD3 introduces stricter licensing and supervision requirements for payment and e-money institutions, ensuring they meet higher compliance standards. In addition, fee transparency will be enhanced, making it easier for consumers to understand currency conversion costs and transaction charges.

Another significant change is the merging of PSD2 with the Electronic Money Directive (EMD2). This integration will eliminate inconsistencies between payment institutions and e-money providers, subjecting both to the same regulatory framework and ensuring fair competition. At the same time, the introduction of the PSR will harmonize enforcement across the EU, reducing regulatory fragmentation and ensuring a level playing field for all payment service providers.

Finally, national and EU regulators will receive increased supervisory powers, enabling stronger oversight and better coordination at the European level. Overall, PSD3 is designed to make digital payments safer, faster, and more efficient, while fostering innovation and maintaining a secure and competitive EU payment market.

Appendix C
The Authors

Adrian Matys is an experienced SAP finance professional specializing in treasury and risk management, payments, advanced payment management, in-house banking, cash management, collections and dispute management, liquidity planning, and AI-driven solutions for treasury management.

He has worked with numerous companies across Europe, Asia, and the United States, spanning a wide range of industries and organizational models. Adrian has successfully completed more than 20 end-to-end SAP implementations, supporting multinational enterprises in transforming and optimizing their treasury operations.

Adrian began his career at BPX, then moved to Lodestone Management Consulting (now part of Infosys Consulting), and continued at Hanse Orga (now Serrala) and PwC. He is currently at KPMG, where he leads the SAP Treasury Consulting Practice, focusing on delivering innovative, scalable solutions based on SAP S/4HANA, including the latest capabilities around in-house banking and advanced payment processing.

Jean-Michele Szczecina is an SAP finance and treasury professional and subject matter expert in the areas of payments and cash management at Zanders. Previously, he led the Cash and Liquidity Management focus group at PwC Germany, driving developments in this field.

Jean-Michele began his career in the financial industry in 2012, and since 2016 he has held various senior consultant and manager positions within SAP projects. Among these roles, he worked at Serrala, a software provider specializing in SAP integrated payment and cash management solutions, where he analyzed customer requirements and implemented SAP-integrated cash management and payment solutions. He also served as a treasury consultant in Vienna at Schwabe, Ley, and Greiner. Jean-Michele has participated in numerous payment and cash management projects throughout the DACH region with many global rollouts, executing several full-cycle projects within the SAP domain.

Index

A

Account balancing	271
Account management lifecycle	262
Account symbols	575
Advanced payment management	35, 215, 641
configuration	295
internal payments	41
key components	216
master data	245
number ranges	300
payment order	301
PINO	39
POBO	40
process flow	38
transaction types	307
Alerts	506
Alliance Cloud	99
Alliance Lite2	97
Applicability Statement 2 (AS2)	361
Application programming interfaces (APIs)	108
Approve Bank Account Applications app	155, 185
Approve Bank Account Changes—Two-Person Verification app	183
Approve Bank Payments (Version 2) app	485, 486
Audits	640
Authorized representatives	195
AutoClient	97

B

BAI and BAI2 file formats	130, 569
Bank	
account approvals	183
account contracts	33
account hierarchies	31
account replication via IDoc	182
address	161
APIs	618
authorizations	598
balances	290
chains	190
charges	47
connectivity	54, 93, 356, 596

Bank (Cont.)	
create	153
create account	167
end-of-day statement	169
fee analysis	148
fee analyzer	32
fee management	593
fees	197, 279
intraday statement	171
key	152, 264
limits	278
master data	389
migration	164
relationship	168, 592
requirements	25
statement BAdIs	581
statement configuration	565
statement forwarding	559
statement frequency	272
statement group	169
statement information	541
statement management	32
statement posting	573
statement processing	540
statement reconciliation	543
statement upload	550
statements	51, 539, 604
Bank Account Change Requests app	184
Bank Account Hierarchy app	434
Bank Account Management (BAM)	30, 151
approvals	208
basic settings	201
close bank account	189
configuration	201
contract types	209
enable payment approval	204
field status groups	212
master data	206
number ranges	205
workflows	186
Bank Relationship Overview app	192
Bank Statement Monitor app	545, 546
Bank Statement Monitor—Intraday app	172, 561
Bank subledger posting	543
Batching rules	501
Business partners	269
Business transaction code (BTC)	577

Index

C

Camt format 631
Camt.053 542
Capital ties 599
Cash concentration 33, 462
Cash Flow Analyzer app 425, 470
   cash concentration 437
   display amounts 429
   display cash flow 432
   display hierarchy 429
   filters 428
   liquidity item hierarchy 435
   not assigned flows 430
   select hierarchy 434
   user settings 427
Cash management 56, 419
   bank balances 420
   cash position 424
   configuration 467
   data sources 446
   integration with advanced payment
      management 298
   memo records 452, 454
   process flow 57
   short-term positioning 438
Cash pooling 47, 80, 178
Cash Pools (Version 2) app 463
Cash position analysis 31
Cash Position app 424
Cashless settlements 69
Central cash and liquidity management 592
Central Finance 42, 603
Centralized netting 65
CHAPS 139
Check Cash Flow Items app 434, 445
Check payments 128
Check sets 224, 326
CHIPS 139
Clearing agreements 252
Clearing area 246, 296
   assign internal bank keys 298
   SLA 247
Collections on behalf of (COBO) 28, 76, 216
Compliance 600
Connectivity standards 93
Connector monitor 367
   components 370
   filters 368
Contact management 32
Corporate seal 359
Cost optimization 596
Create Payments app 230, 235, 326
Credit transfer 121
Cross-border credit transfer 122
Cross-border payments 72
Cryptocurrencies 615
Currency management 23
Customer SLA 247

D

Decentralized netting 64
Define Bank Account Settings—Bank
   Statements app 172, 421
Define Bank Account Settings—Instant
   Balances app 423
Define Bank Transfer Templates app 458
Define Cash Position Profiles app 439
Define Monitoring Rules—Intraday
   Statements app 560
Delete Data from One Exposure app 451
Direct debit 61, 125
Display Cash Pool Hierarchies app 464
Distributed ledger 615
DMEE defaults 325
Domestic credit transfer 121

E

EBICS 109, 357
   3.0 119
   configuration 404
   corporate seal 118
   data transfer 117
   distributed electronic signature (DES) 117
   entities 110
   onboarding process 115
   order types 111
   signature types 110
   users 359
Electronic bank account management
   (eBAM) 152, 634
Electronic Money Directive (EMD2) 644
Enrichment and validations 222, 248, 326
   check sets 328
   payment items 329
   payment order type group 327
Error codes 255
Ethereum 615
EURO1 140
Exception control 255
Exception Control app 255
Exception handling 226

Index

Exposure management 593
Extended Data Medium Exchange
    Engine (DMEEX) 513
External financial transactions 87

F

FedNow 123
File handler 311
    define channels 313
    define formats 312
    define media 312
Financial Action Task Force (FATF) 99
Financial standards 121
Foreign Bank Accounts app 191
Foreign exchange (FX) 593
Format conversions 596
Format transformation 627, 631
Four eyes approach 25
Fraud prevention 25, 597, 601
Free-form payments 86, 460

G

General ledger account 175
Group SLA 247

H

Host-to-host connection 108, 360
    connectivity protocol 361
    environment details 362
    transport authentication 364
House bank 160
    account connectivity 173
    bank account ID 175

I

Implement Powers of Attorney for
    Banking Transactions app 196
Import and Export Bank Accounts app 181
Import Bank Services Billing Files app 198
Import Memo Records 2.0 app 455
Import parameters 567
Incoming file mapping engine (IFME) 519
India UPI 142
In-house banking 36, 37, 43, 260, 585
    account management 334
    accounting integration 335
    assign G/L objects 337
    cash pooling 82

In-house banking (Cont.)
    COBO 78
    configuration 331
    create accounts 265
    define activity types 333
    define bank areas 331
    define G/L accounts 338
    end-of-day process 292
    functions 46
    intercompany payment 69
    number ranges 334
    outlook 618
    payment factory processes 590
    POBO 74
    posting expenses 336
    prerequisites 261
    ROBO 80
    scenario 48
Input conversion program 317
Input manager 218
Instant payments 123
Intelligent services 339
Intercompany agreements 261
Intercompany clearing 591
Intercompany netting 46
Intercompany payments 63
Intercompany transactions 591
Interest 276
Intermediary banks 26
Internal account management 591
Internal payments 41, 288
Intraday bank statements 559
Investment and funding 591
Invoice verification 24
Invoices 28
ISO 20022 .... 521, 542, 623
    CGI 137, 146
    migration from SWIFT MT 625
    MX 512

J

Joule 619

K

Know your customer process 99

L

Legal entity identifier (LEI) 629
Liquidity items 471

Index

Liquidity management 601
Liquidity planning 56
Lockboxes 542, 569
   control parameters 569
   house banks 573
   posting data 571
Logical paths 566

M

Maintain Payment Blocks app 258
Maintain Route and Clearing Agreement
   app 250
Make Bank Transfers—Create with
   Templates app 459
Make Bank-to-Bank Transfer app 456
Manage Automatic Payments app 54, 282, 487
Manage Bank Account Balances app 420, 422
Manage Bank Account Hierarchies app 187
Manage Bank Account Reviews app 188
Manage Bank Accounts app 165, 176, 266
Manage Bank Chains app 190
Manage Bank Fee Conditions app 198, 279
Manage Bank Messages app 373
Manage Bank Statement Reprocessing
   Rules app 562
Manage Bank Statements app 548
Manage Banks app 158, 162
Manage Banks—Master Data app 157
Manage Business Partner app 288
Manage Cash Concentration app 465
Manage Cash Pool app 464
Manage Cash Pools (Version 2) app 179
Manage Cash Pools 2.0 app 476
   define authorization group 476
   define number range 476
Manage Incoming Payment Files
   app 219, 220, 284, 531, 533, 551
Manage Incoming Payment Files from
   Banks app 537
Manage In-House Bank Account Balances
   app 290
Manage In-House Bank Account Templates
   app 273
Manage In-House Bank Accounts
   app 268, 274
Manage In-House Bank Application Jobs
   app 293
Manage In-House Bank Conditions app 275
Manage In-House Bank End of Day app 294
Manage In-House Bank Limits app 278

Manage In-House Bank Payment Items
   app 291
Manage Memo Records 2.0 app 452, 472
   define number range 473
   field status group 474
   memo record types 472
Manage Payment Agreements app 252
Manage Payment Batches app 240, 286
Manage Payment Items app 240, 241
Manage Payments app 239, 289
Manage Powers of Attorney for Banking
   Transactions app 194, 196
Manage Workflows—For Bank Accounts
   app 186
Map Dormant Data app 522
Map Format Data app 36, 511
   filters 515
   mapping types 516
   overview 514
   settings 317
   transports 518
Map Format Data for Advanced Payment
   Management app 536
Map Format Data for Incoming Files
   from Bank app 520
   account statements 529
   action nodes 524
   AI assistant 533
   attributes 524
   choice nodes 524
   create mapping 521
   element 523
   merge formatting 535
   node properties 525
   node types 523
   release mapping 529
   sequence nodes 525
   versions 527
Master data 88, 152
Member banks 366
MICR data 570
Monitor Bank Account Balances app 422
Monitor Bank Fees app 199
Monitor Payments app 483, 486, 496
Multicash 542
My Inbox—All Items app 189

N

Net settlement system 138
Netting 64
   currencies 66

Index

Notional cash pooling 81
Number ranges 270

O

On-behalf-of payments 591
One Exposure from Operations table 424, 446, 451
Order types 320
Output enrichment 228
Output manager 229
Overdraft limits 32, 177

P

pain.002 629
Payment
   approval 52, 176, 484
   batching 55, 491, 596
   business imperative 22
   create and display batch 492
   decentralization 25
   display document status 493
   duplicate 29
   in-house operations 282
   instruction keys 123
   instruments 121
   items 223, 305
   media batch run 495
   media configuration 499
   media formats 388
   methods 24
   orchestration 585, 598
   order types 302, 304
   orders 222
   originator 231
   process flow 38
   processing 22, 59
   processing center 583
   recipient item 233
   requests 28, 457
   routing 602
   run 54, 487
   schemes 130
   service providers 608
   Services Directive (PSD) 641
   signatories 31
   status 496
   systems 137
   upload status report 497
   with SAP 26

Payment Card Industry Data Security Standard (PCI DSS) 613
Payment factory 583
   design 583
   design considerations 587
   functions 595
Payment formats 142
   CSV 143
   fixed-length 143
   SWIFT MT 144
Payment Services Directive (PSD)
   PSD2 642
   PSD3 644
Payments Analyzer apps 243
Payments in the name of (PINO) 39, 59, 216
Payments on behalf of (POBO) 28, 40, 47, 71, 216, 282
Penny tests 387
Physical bank transfers 63
Physical cash pooling 81
Planning groups 470
Planning levels 467
   assign 469
   define 469
Polygon 615
Posting keys 576
Posting parameters 568
Power of attorney 33, 193
Process Free Form Payments app 460
Program
   FEB_FILE_HANDLING 556, 565
   RBNK_IMPORT_PAYM_STATUS_REPORT 497
   RFEBKA00 390
   RFEBKA96 550

Q

Qualitative factory 600
Quantitative factory 601

R

Real-time gross settlement system 138
Real-Time Payments (RTP) 123
Receivables on behalf of (ROBO) 78
Reconciliation 540
Reconciliation factory 602
Regulations 639
Repair Payments app 226, 236
Report
   /BSNAGT/REP_PICKUP_FILES 380

651

Index

Report (Cont.)
SAP_FIN_FCLM_BAM_GEN_APPTS 561
Request for forwarding 120
Routes 251
Routing control 227

S

SAP Bank Communication Management 50, 481, 641
    advanced payment management 310
    alerts 506
    assign roles to release steps 505
    BAdIs 509
    batching rules 500
    centralized payment landscape 52
    configuration 498
    decentralized banking landscape 51
    functionalities 51
    release procedures 504
    release steps 503
    workflow activation 502
SAP BTP, Cloud Foundry environment 354
SAP Business Integrity Screening 341
    configuration 343
SAP Digital Currency Hub 599, 613
    tokens 616
SAP digital payments add-on 608, 610
    credit card payments 611
SAP Document Management service 179
SAP In-House Cash 43, 260, 289
    cash pooling 82
    COBO 76
    intercompany payment 69
    POBO 73
    ROBO 79
SAP Multi-Bank Connectivity 34, 218, 349
    authorizations 415
    backend authorizations 415
    BAdIs 413
    certificate exchange 393
    configuration 387
    connection monitor 367
    connector 352
    connector to tenant 399
    EBIC 2.5 configuration 407
    EBICS 3.0 configuration 404
    ERP-to ERP routing 398
    frontend authorizations 416
    inbound/outbound processing 402, 403
    landscape 351
    local routing 397

SAP Multi-Bank Connectivity (Cont.)
    number ranges 392
    onboarding 385
    pickup files 379
    preboarding 386
    pull messages 377
    push messages 378
    routing 397
    scoping 385
    sender/receiver ID mapping 402
    sensitive data 382
    SSF application parameters 395
    SSF profile data 396
SAP S/4HANA Cloud Private Edition 609
SAP S/4HANA Cloud Public Edition 533, 609
    payment factory sidecar 618
SAP S/4HANA Finance 27, 261
    configuration 387
SAP Watch List Screening 343
    configuration 346
Sarbanes-Oxley (SOX) 639
Schedule Jobs for Flow Builder app 448
Search strings 579
Secure File Transfer Protocol (SFTP) 361
Segment SLA 247
Selection variants 390
SEPA B2B Direct Debit 128
SEPA Core Direct Debit 126
SEPA Instant Credit Transfer 124
Service-Level Agreement app 246, 247
Short-Term Cash Positioning app 438, 444
Simple Object Access Protocol (SOAP) 361
Single Euro Payment Area (SEPA) 641
Single European Payment Area (SEPA) 130
    IBAN only 131
    request to pay 132
    verification of payee 135
Source application accounting 477
Subledger clearing 543
Submit Bank Account Applications app 154
SWIFT 95, 365
    Alliance Gateway 96
    Alliance Lite2 for Business Applications 98
    BIC 104
    cloud connectivity 99
    configuration 410
    connectivity options 96
    contracting and membership 99
    FileAct 103
    Financial Network (FIN) 102
    forwarding 120
    GPI 106

SWIFT (Cont.)
    InterAct 102
    key transmission services 101
    MA-CUG 100
    message types 144, 512, 608, 625
    MT940 .... 53, 542
    onboarding 104
    private infrastructure 96
    qualification criteria 101
    SCORE 99
    shared infrastructure 97
    Treasury Counterparty (TRCO) 101
SWIFTRef 107, 383
    configuration 411

T

TARGET2 139
Tax calculations 48
TemSe 323
TIPS 140
Transaction
    /BSNAGT/FILE_SEND 378
    /BSNAGT/MONITOR 367, 374, 380
    /PF1/EH 255
    /PF1/FH_IMPORT_DIR 318
    /PF1/FH_IPM_EXPERT 219
    /PF1/FH_SHOW_DB 219
    /PF1/PO_EXPERT 233
    /PF1/RN 250
    /PF1/SLA 246
    ALRTCATDEF 507
    BIC2 384
    BNK_APP 485
    BNK_MERGE_RESET 495
    BNK_MONI 372, 486
    BNK_MONIP 493
    DRFOUT 183
    F110 53
    F111 53
    F8BT 457
    F9K1 265
    FBPM1 .... 492, 495
    FCLM_FLOW_BUILDER_2 .... 447, 471
    FEB_FILE_HANDLING 556
    FEB_IMP_FILEPATH 566
    FEB_IMP_SOURCE 567

Transaction (Cont.)
    FF.5 .... 515, 551, 554
    FI01 163
    FIBLFFP 230
    FILE 318, 381
    FJEPV_IHBASGLOBJ 337
    FJEPVC_IHB_ACC_DET 338
    FJEPVC_IHB_DERIV_CO 336
    FQM_DELETE 451
    FTE_BSM 545
    IBS_JP 558
    IBSX_JP 558
    OBAX 571
    OBPM1 388
    OBPM4 499
    OT83 574
    S_EBJ_98000208 497
    SNRO 300
    STRUST 393
Transaction banking 634
Transaction types 577
Transactional costs 601
Transfer media 318, 324
Treasury payments 83
    external 86
    internal 84
Two-factor authentication 487
Two-person verification 184

U

Upload In-House Bank Accounts app 274

V

Vendor management 24
Virtual accounts 261
Virtual bank accounts (VBAs) 30
Virtual banks 261

X

XML tag names 629

Z

Zengin 140, 542, 557

- Set up financial accounting and controlling processes in SAP S/4HANA

- Configure your system with step-by-step instructions

- Prepare for testing, go-live, and production support

Stoil Jotev

Configuring SAP S/4HANA Finance

Starting a new SAP S/4HANA Finance implementation? Get it right the first time! From setting up an organizational structure to defining master data, this comprehensive guide to configuring SAP S/4HANA Finance walks you through each key task. Follow step-by-step instructions organized by functional area: general ledger, accounts payable and receivable, margin analysis, group reporting, and more. Customize SAP S/4HANA to meet your FI/CO needs!

744 pages, 3rd edition, pub. 12/2024
E-Book: $84.99 | **Print:** $89.95 | **Bundle:** $99.99

www.sap-press.com/5920

- Configure TRM functionality for debt and investments, foreign exchange, securities, and derivatives in SAP S/4HANA
- Master exposure management, cash flow hedging, and balance sheet hedging
- Run your core treasury transactions with step-by-step instructions

Carlson, Carlson, Lasecki

Treasury and Risk Management with SAP S/4HANA

The Comprehensive Guide

Decipher the complex world of TRM! With this all-in-one guide to SAP S/4HANA Finance for treasury and risk management, you'll get the detailed, expert help you need to run your operations smoothly. Set up your core configuration elements for debts and investments, foreign exchange, derivatives, and securities. Then expand your scope with exposure management, cash flow hedging, and balance sheet hedging. With information on correspondence forms, analyzers, reports, and integration, this book is your one-stop shop for TRM!

820 pages, pub. 01/2025
E-Book: $114.99 | **Print:** $119.95 | **Bundle:** $129.99

www.sap-press.com/5907

Interested in reading more?

Please visit our website for all new book
and e-book releases from SAP PRESS.

www.sap-press.com